the Elements

Y0-AHO-269

			III	IV	V	VI	VII	O
								2 **2 He** 4.0026
			2 3 **5 B** 10.811	2 4 **6 C** 12.01115	2 5 **7 N** 14.0067	2 6 **8 O** 15.9994	2 7 **9 F** 18.9984	2 8 **10 Ne** 20.183
			2 8 3 **13 Al** 26.9815	2 8 4 **14 Si** 28.086	2 8 5 **15 P** 30.9738	2 8 6 **16 S** 32.064	2 8 7 **17 Cl** 35.453	2 8 8 **18 Ar** 39.948
2 8 16 2 **28 Ni** 58.71	2 8 18 1 **29 Cu** 63.546	2 8 18 2 **30 Zn** 65.37	2 8 18 3 **31 Ga** 69.72	2 8 18 4 **32 Ge** 72.59	2 8 18 5 **33 As** 74.9216	2 8 18 6 **34 Se** 78.96	2 8 18 7 **35 Br** 79.904	2 8 18 8 **36 Kr** 83.80
2 8 18 18 **46 Pd** 106.4	2 8 18 18 1 **47 Ag** 107.868	2 8 18 18 2 **48 Cd** 112.40	2 8 18 18 3 **49 In** 114.82	2 8 18 18 4 **50 Sn** 118.69	2 8 18 18 5 **51 Sb** 121.75	2 8 18 18 6 **52 Te** 127.60	2 8 18 18 7 **53 I** 126.9044	2 8 18 18 8 **54 Xe** 131.30
2 8 18 32 17 1 **78 Pt** 195.09	2 8 18 32 18 1 **79 Au** 196.967	2 8 18 32 18 2 **80 Hg** 200.59	2 8 18 32 18 3 **81 Tl** 204.37	2 8 18 32 18 4 **82 Pb** 207.19	2 8 18 32 18 5 **83 Bi** 208.980	2 8 18 32 18 6 **84 Po** (210)	2 8 18 32 18 7 **85 At** (210)	2 8 18 32 18 8 **86 Rn** (222)

2 8 18 25 8 2 **63 Eu** 151.96	2 8 18 25 9 2 **64 Gd** 157.25	2 8 18 27 8 2 **65 Tb** 158.924	2 8 18 28 8 2 **66 Dy** 162.50	2 8 18 29 8 2 **67 Ho** 164.930	2 8 18 30 8 2 **68 Er** 167.26	2 8 18 31 8 2 **69 Tm** 168.934	2 8 18 32 8 2 **70 Yb** 173.04	2 8 18 32 9 2 **71 Lu** 174.97
2 8 18 32 24 9 2 **95 Am** (243)	2 8 18 32 25 9 2 **96 Cm** (247)	2 8 18 32 26 9 2 **97 Bk** (247)	2 8 18 32 27 9 2 **98 Cf** (249)	2 8 18 32 28 9 2 **99 Es** (254)	2 8 18 32 29 9 2 **100 Fm** (253)	2 8 18 32 30 9 2 **101 Md** (256)	2 8 18 32 31 9 2 **102 No** (254?)	2 8 18 32 32 9 2 **103 Lw†** (257)

Atomic weights are based on carbon-12;
values in parentheses are for the most stable or the most familiar isotope.
† Symbol is unofficial.

12.85

PRINCIPLES OF PHYSICAL, ORGANIC, AND BIOLOGICAL CHEMISTRY

PRINCIPLES OF Physical, Organic, and Biological Chemistry

AN INTRODUCTION TO THE MOLECULAR BASIS OF LIFE

JOHN R. HOLUM, PH.D.
Professor of Chemistry
Augsburg College

JOHN WILEY & SONS, INC. New York · London · Sydney · Toronto

10 9 8 7 6 5 4

Library of Congress Catalog Card Number: 68-9249
SBN 471 40852 2
Printed in the United States of America

Mary

Elizabeth

Ann

Kathryn

PREFACE

The theme of this book is the molecular basis of life. It is intended for use in a two- or three-term terminal college course in chemistry for students in the humanities, the social sciences, and the paramedical sciences, including nursing, home economics, physical therapy, many areas of biology, and many programs in the agricultural sciences. A knowledge of high school chemistry is not required.

Because the world is filled with such a fascinating variety of living things, and because our knowledge of the life sciences has grown so tremendously in recent years, particularly at the molecular level, a terminal college course in chemistry should today devote considerably more attention to organic chemistry and biochemistry than has been customary. We no longer have any excuse for denying the student in a terminal science course the opportunity of studying in some depth a highly useful and intellectually satisfying way of viewing life, both in himself and in his environment, while he is at the same time learning the most fundamental of the physical and chemical principles and laws. There are, to be sure, other views of life; students become well aware of them as they study literature, philosophy, and the religions of the world. But in the physical sciences, which traditionally deal with the world as though it were not populated by living things, an idea of such power and beauty has been developed that humanists, philosophers, social scientists, and any who would claim to be educated are the poorer to the extent that they ignore or dismiss it. The idea, so simply stated that the struggle to bring it forth is not remembered, is this. Life does have a molecular basis. This concept makes physics and chemistry an important part of the world of living things and people, and it transforms the kind of study offered here into a gratifying opportunity for anyone who would know as much as possible about life in all its complexity. John W. Gardner wrote "Anyone interested in leading a creative life will have the deepest respect and concern for the marvelously intricate organism that he is."[1] David Hawkins put it another way. "Part of coming to terms, of being at home, is in the sheer familiarity of the environment and in the sureness with which one lives and walks there."[2]

[1] John W. Gardner, *Self-Renewal,* Harper and Row, New York, 1964, page 16.
[2] David Hawkins, "The Informed Vision: An Essay on Science Education," *Daedalus,* Summer 1965, page 538.

The text opens with an introduction to several physical principles, particularly energy. To develop an understanding of energy, a central question in physics is first raised. What are the simplest elements in terms of which motion may be analyzed? Three emerge—mass, distance (length), and time. It is of secondary importance that in considering these elements we must discuss weight, volume, and density. Energy is our primary concern. To help the student understand the concept of energy, it is first presented as a particular function of the elements of motion with a profoundly important property: energy gives every appearance of remaining changeless in the midst of change. The recognition or identification of anything that endures in the midst of change has always been a significant development in the history of ideas.

After introducing an analysis of motion and energy, the reductionist question is again posed about matter. What are the simplest elements in terms of which matter may be analyzed? In the ensuing discussion of chemical elements, chemical bonds, and the states of matter, the concept of enthalpy is introduced. Other traditional topics of physical chemistry are presented at a very elementary level.

Next on the list of fundamental features of our world is spontaneity. What are the simplest, most useful elements in terms of which spontaneity may be analyzed? Enthalpy and entropy emerge and with them the Gibbs free energy. The discussion is deliberately confined to systems and changes at constant temperature and pressure, partly to simplify matters but mostly because such conditions are closely approximated in living organisms. The calculus, obviously, cannot be used. I know how severely its omission limits the development of the concepts of enthalpy, entropy, and free energy, but to take the stand of the purist on this matter is to deny to large numbers of students in terminal science courses any contact with some of the greatest principles in all science. A qualitative and intuitive discussion of entropy, enthalpy, and free energy as the major elements in an analysis of spontaneity is far better than no discussion at all.

Attention is next directed to the last great aspect of physical existence to be scrutinized, permanence or equilibrium. What are the simplest elements in terms of which permanence may be analyzed? The first answer given uses the Gibbs free energy concept. Then reaction rates are introduced, and the more common view of equilibrium in terms of equal but opposing rates is presented. Activation energy and the influence of temperature on rates are studied in terms of simple collision theory. The first section of the text ends with a discussion of acid-base equilibria and solubilities.

Material for the chapters on organic chemistry was selected to provide background for the biochemistry and molecular biology to follow. Synthetic methods of organic chemistry having no counterpart in living systems are seldom mentioned. Many properties of aromatic compounds are not discussed, although the theory of resonance is introduced. Nomenclature is developed rather extensively within the text in the hope that much less time than usual need be spent on it during class. (For many situations in which this text is used I see no reason to discuss or study IUPAC nomenclature at all.)

To become, be, and stay alive an organism needs information, materials, and energy. Chapters on carbohydrates, lipids, and proteins describe the principal materials. The information needed and the forms in which it is provided are topics in the chemistry of heredity, enzymes, and hormones. Chapter 21, "Energy for Living," is one of the most important in the text. It, together with Chapter 25, "Chemistry of Heredity," dramatizes our deep and abiding kinship with all of nature.

I have reversed the traditional way of presenting the catabolism of carbohydrates. Usually the discussion starts with glucose, which is then carried through glycolysis, the citric acid cycles, and (sometimes) the respiratory chain. Living cells, however, call forth this catabolism the other way round, for the temporary absence of ATP, rather than the presence of glucose, triggers this chain of events. It therefore seems better to start with the respiratory chain and to present the citric acid cycle as the chief supplier of the elements of the element hydrogen, the fuel for the respiratory chain. Sequences that provide fuel and intermediates for the citric acid cycle are described in separate chapters on the metabolism of carbohydrates, lipids, and proteins.

Not all metabolic pathways are discussed in detail, but for those that are I have tried to show how each step is like a reaction studied earlier among simpler organic molecules. The Gibbs free energy function is applied in Chapter 21, "Energy for Living," to help explain how a free-energy-producing reaction can power a free-energy-consuming change coupled to it and thus convert low-energy molecules to higher-energy ones. The conversion of the chemical energy in ATP into muscular work is discussed in terms of the R. E. Davies theory.

I am grateful to the editor of *Nature* and to J. D. Watson and F. H. C. Crick for permission to quote extensively from two papers in *Nature* which stated the now-famous Watson-Crick theory. I am also grateful to the Philosophical Library for permission to quote from their edition of Dalton's *A New System of Chemical Philosophy*.

Many people, some unknown to me by name, helped me considerably at various stages in the preparation of the manuscript. I want especially to thank Dr. Earl Alton, my friend and colleague, for the many pleasant lunch times we shared discussing the content and the teaching of chemical principles and for his critical reading of the first eight chapters. Dr. Arne Langsjoen (Gustavus Adolphus College) and Dr. A. H. Blatt (Queens College) read the entire manuscript and taught me much that I needed to known both about chemistry and about writing. The mistakes that remain are mine. Please write when you find them and when you have any other comments and suggestions.

THE ACCOMPANYING LABORATORY WORK

I should like to recommend that you consider a manual written by Dr. Arne Langsjoen. Prepared for a course very similar to the one for which this text is designed, it is titled *Exercises in General, Organic and Biological Chemistry* (second edition, 1968). The manual can be obtained from the publisher, Burgess Publishing Company, 426 South Sixth Street, Minneapolis, Minnesota 55415.

TEACHER'S SUPPLEMENT FOR THIS TEXT

A teacher's supplement which includes the traditional features of such publications is available without charge to any teacher using or thinking about using this text. A letter requesting this supplement should be written on chemistry or science department stationery and sent to John Wiley and Sons, 605 Third Avenue, New York, New York 10016.

John R. Holum

Augsburg College
Minneapolis, Minnesota 55404
January 1969

CONTENTS

PRINCIPLES OF PHYSICAL, ORGANIC, AND BIOLOGICAL CHEMISTRY

CHAPTER ONE

Energy

Dynamos, fullbacks, waterfalls, bees, hurricanes, and small boys indisputably have one characteristic in common, energy. Even though it is not an object like matter, occupying space, energy is so commonplace and apparently so well understood that the word is learned and used at an early age. We do not see energy, taste, smell, hear, or feel it, but we experience its effects. Man virtually had to invent the term so that he could better describe features common to a variety of his experiences with matter.

Energy is something like a talent. A person with a talent has an ability to do something and do it more or less well. But a talent is not a thing. It is a characteristic of certain things, people. Just as a talent is related to an ability, so too is energy. A person or other object that we say has energy has the ability or the potential to do something—to do work. In fact, the word "energy" comes from the Greek *en* plus *ergon,* "in work." But the word "work" is not a synonym for energy. Work is usually thought of as the doing of something in a purposeful way. Energy is needed, but not all expenditures of energy actually work. It is useful to think of energy as available in several forms: mechanical, thermal (heat), electric, electromagnetic (light), sound, chemical, and potential. The presence of chemical energy on this list makes this chapter necessary. Chemical energy can be converted into all the other forms, and it happens that analyzing mechanical energy is one of the better ways to begin a study of chemical energy.

THE ANALYSIS OF MOTION

Elements of Motion. If energy is used to move an object from one point to another, we say that "mechanical work" has been done. Any object in motion is said to possess mechanical energy. In fact, the whole modern notion of energy developed from attempts to analyze the motion that nature exhibits in almost endless variety. We see movements and motions everywhere, as we look out the window, ride along in a car or bus, or watch a football game. Although the list of motions in the universe seems infinite, a meaningful analysis is still possible. Certain features are apparently common to all kinds of motion.

For centuries the analysis of motion has occupied the attention of scientists in one branch of physics, those studying mechanics. It is often true in science, as elsewhere, that asking the right questions is an essential part of progress. The right questions in mechanics read something like this. What are the simplest elements or the simplest aspects in terms of which motion can be described? What are the simplest elements to which all forms of motion can be reduced? Two of the most helpful "elements" have to do with position—position of the object in space and position of the object in time. A description of a moving body has to state where it is and when. How an object moves can be described by showing how changes in these two positions relate to each other. The best description is an equation by means of which we may predict where the object will be at some future time or where it was at some previous time. The movements of stars, planets, and moons have been successfully fitted into such equations. The remarkable accuracy with which men and equipment are hurled into orbits and recovered is a dramatic illustration of how successful the analysis of motion has been.

References. The notion of position is meaningless without some set, agreed-upon reference position. When we describe the position of an object, we really specify the path we would take to reach it from some starting point. We may say "The car is over there," pointing. The reference is our position. Or we may say "It's along the north side of the building"; by implication one of our references is the North Star, or the north pole, and any agreed-upon system of latitude, longitude, city map, etc., that describes where the building is. For more exact descriptions the distance from the reference point has to be stated together with directions. Units for describing distance or length are obviously necessary.

The standard length universally used in scientific work, and in virtually all countries, is the *meter*. Until late in 1960 the world's standard meter was universally agreed to be the distance between two thin scratches on a long bar of platinum-iridium alloy stored in a locked subbasement at Sèvres, near Paris, France. When first proposed, it was meant to be one ten-millionth of the distance along the surface of the earth from the North Pole to the equator. In October 1960 the World Conference on Weights and Measures decided to replace the standard meter with a reference that would not be subject to corrosion, to thermal expansion, or to loss by any possible means, and one that was more portable. This new standard meter is defined as 1,650,763.73 wavelengths of the orange-red light given off by the electrically excited isotope krypton-86, a rare gas found in trace amounts in the atmosphere all over the world and available to all scientists (see Figure 1.1).

The meter is but one of the standards defined in the metric system of weights and measures. The great convenience of this system is the decimal relations of its units, relations always implied by adding a suitable prefix to the name of the unit itself. Table 1.1 contains a summary of the important prefixes in the metric system.

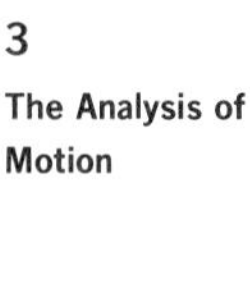

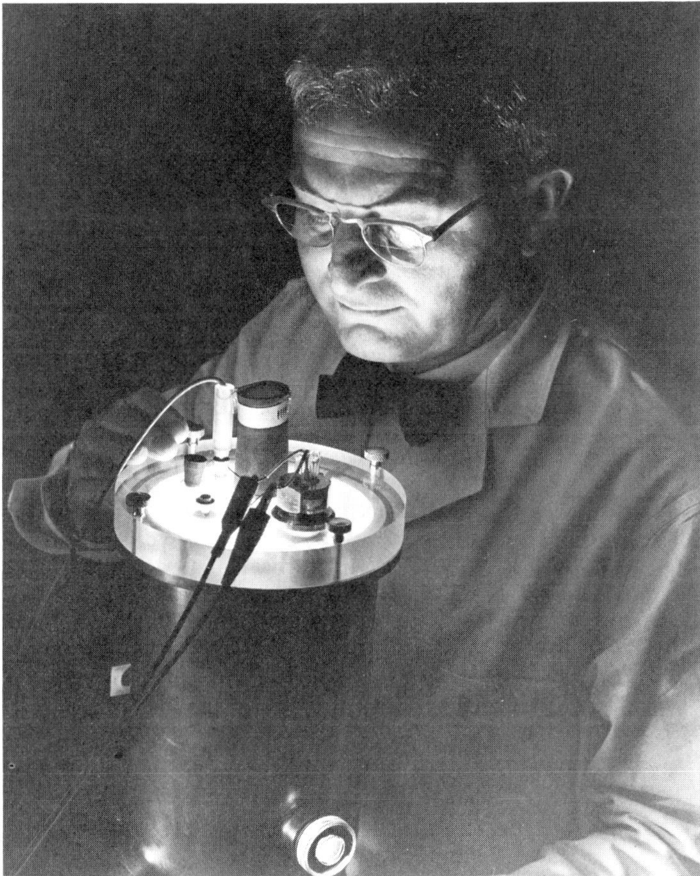

Figure 1.1 A National Bureau of Standards scientist adjusts a krypton-86 lamp in its liquid nitrogen bath. The wavelength of the orange-red light emitted by the lamp has been adopted as the international standard of length. (Courtesy of the National Bureau of Standards, Washington, D.C.)

To specify position in time, a unit of time is needed, and this unit is the *second.* The standard for the second has undergone change, the most recent being in 1964 For virtually all the affairs of most people, including navigators, the second may be taken as the *mean solar second,* which is 1/86,400 of the mean solar day—the average time it takes the earth to make one complete turn on its axis. If a perfect clock had been set on this basis in the year 1900, it would now be off about 30 seconds. This amount does not seem like much, but for physicists, astronomers, and space scientists such accuracy is not good enough. Therefore in 1964 another General Conference on Weights and Measures redefined the second in terms of the transition between two specific energy levels in cesium-133. Greater detail than this is unnecessary for our purposes.

Table 1.1 Prefixes in the Metric System

The names of multiples and submultiples of the units are formed with the following prefixes.*

Multiples			Submultiples		
Factor by Which Unit is Multiplied	Prefix	Symbol	Factor by Which Unit is Multiplied	Prefix	Symbol
10	deka	da	10^{-1}	deci	d
10^2	hecto	h	10^{-2}	**centi**	c
10^3	**kilo**	k	10^{-3}	**milli**	m
10^6	mega	M	10^{-6}	**micro**	μ
10^9	giga	G	10^{-9}	nano	n
10^{12}	tera	T	10^{-12}	pico	p
			10^{-15}	femto	f
			10^{-18}	atto	a

* The most common are indicated by boldface.

Kinematics. Thus far we have found two elements of motion, position and time. We have ignored what moves and, surprisingly, we can go on ignoring it for a while. Our analysis is simpler, up to a point, if we treat "what moves" as a geometric point and then proceed to think about how a point might move in both space and time. Our discussion, then, is of the geometry of motion, or kinematics. With just two elements of motion, kinematics gives us two extremely useful combinations of them, speed (and velocity) plus acceleration.

Suppose that a point moves in a straight line and that it passes some reference point, a "mile zero" so to speak. Let us start a stopwatch at "time zero." The following observations illustrate what would be observed if the motion were uniform.

The point was at	when the stopwatch read
0 cm	0 sec
4 cm	2 sec
7 cm	3.5 sec
10 cm	5 sec
14 cm	7 sec

A quick calculation at each reading shows that in the midst of this change in both location and time something remains unchanged—the ratio of distance traveled to time elapsed. Man has been fascinated by both change and permanency. He has found by experience that anything that remains unchanged in the midst of change bears looking into. Often he invents a name for it. This ratio of distance traveled to time elapsed is called the *speed* of the particle:

$$\text{Speed} = \frac{\text{distance traveled}}{\text{time elapsed}}$$

$$s = \frac{L}{T}$$

$$s = LT^{-1} \qquad (1.1)$$

Here we use L to represent distance or length and T to stand for time. (Later, in other contexts, T will stand for temperature, but the context will clearly indicate how the symbol is used.)

Thus just two elements of motion in one of their possible combinations give us a very useful property of a moving object, its speed. We can improve on this particular combination by specifying its direction, which is as important as its speed. When a hurricane warning or a tornado alert is sounded, we want to know both the speed and direction of the storm. When both speed and direction are specified, the quantity is called *velocity*.[1] By definition velocity has both magnitude and direction.

Actually, very few motions of real objects correspond to uniform motion in a straight line, for velocity is not usually constant throughout the movement. Either the speed or the direcion, or both, change. In one important situation the velocity changes smoothly with time, and we shall look into this, since it is obviously easier to examine than a velocity that changes erratically with time. To illustrate what is meant by smooth uniform change in velocity, the following values represent typical data:

The velocity of the point was	when the stopwatch read
0 cm/sec	0 sec
4 cm/sec	2 sec
6 cm/sec	3 sec
10 cm/sec	5 sec
18 cm/sec	9 sec

It is assumed that the motion is in some fixed direction. The velocity is clearly changing with time, but the ratio of velocity to elapsed time at any moment does not and is the same for all measurements, $\frac{2 \text{ cm/sec}}{\text{sec}}$ (2 cm per sec per sec). We say that the point is undergoing an *acceleration,* and we may define acceleration in terms of a ratio:

$$a = \text{acceleration} = \frac{\text{velocity}}{\text{time}}$$

$$a = \frac{L/T}{T}$$

$$a = LT^{-2} \tag{1.2}$$

Here, then, is another useful combination of but two of the elements of motion, a phenomenon well known to drag racers and astronauts and to anyone who has been pressed back into his seat when the car started or the plane took off.

Other combinations of L and T, although possible on paper, are not useful. For a more extensive analysis of motion we must change from a moving point to a real object, introducing a decidedly nongeometric property.

Inertia. Objects that look alike in shape, size, color, and volume can differ greatly in how easy it is to make them move or, if they are moving, to make them

[1] Velocity is therefore a *vector* quantity, speed a *scalar* quantity. We have no need for these distinctions in this book, but many readers will be familiar with vectors and scalars.

veer to one side or to stop. A cubic meter block of wood sheathed in thin lead foil looks exactly like a cubic meter block of pure lead, but imagine trying to move them. Suppose that they are suspended by some stout wire to reduce frictional problems. Making the block of solid lead swing would be much harder than making the other block perform the same motion. The block of solid lead possesses something that is profoundly important in any analysis of its possible movements. Whatever that something is, it is independent of the other elements of motion (at least when velocities are small compared with the velocity of light). That something is a new element of motion which we must consider if we are to move real objects or analyze their motions. It is called *inertia,* and it is an idea so important that the word is part of ordinary usage. Any person, any society, any object that resists change is said to have inertia. The solid block of lead has more inertia than the lead-coated wood. To continue our analysis of motion we must have a measure for inertia or differences in inertia.

Mass. As the science of motion developed, the quantitative measure of inertia came to be called *mass*. An object with a large mass, such as the block of pure lead, has a large inertia. The reference unit of mass in the metric system is the kilogram, the mass of a particular cylinder of platinum-iridium alloy, called the International Prototype Kilogram, which is preserved in the Sèvres (p. 2) vault by the International Bureau of Weights and Measures (Figure 1.2). In chemical work the gram (one-thousandth of the kilogram) and the milligram (one-thousandth of a gram) are the most common mass units.

With these three elements of motion—mass, distance, and time (M, L, and T), together with reference standards—a number of different motions of real objects have been successfully analyzed. Success here means finding an equation that includes nothing more than these three variables to describe the motion. With the right equations programmed into high-speed calculators, Jet Propulsion Laboratory scientists routinely put to the machines such questions as "If the lunar landing module was at point A in space at time T, traveling so many meters per second, where will it be at time T plus one hour?"

FORCE

Let us turn our attention to the effect of a push or a pull on an object. Human power is not the only source of pushes and pulls. Blowing wind pushes on the slanted vanes of a windmill, flowing water can be made to push on the paddles of a paddle wheel, and a harnessed horse pushes on its harness, thereby pulling a plow or a wagon. But a large block of lead with its huge mass would take a powerful push to overcome its inertia. If the block of lead is at rest, and if we push on it hard enough, it will go from a velocity of zero to some definite velocity. In other words, applying the push changes its velocity, which is another way of stating the first of the three laws of motion expressed by Isaac Newton:

> Newton's First Law of Motion. Every body continues in its state of rest, or in a state of uniform motion in a straight line, except insofar as it is compelled by forces to change that state.

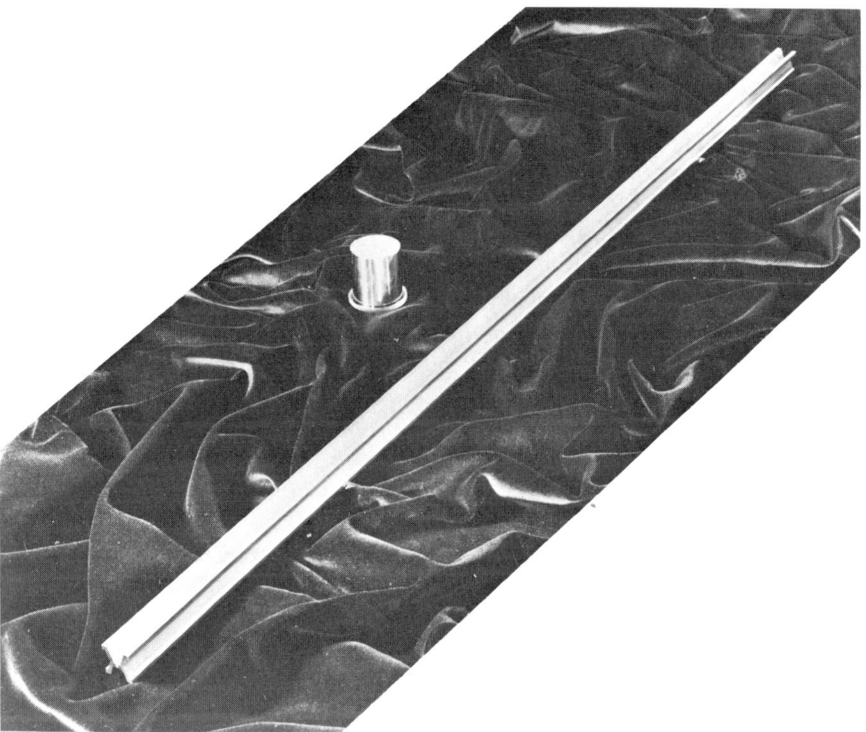

Figure 1.2 The United States standard of mass, Prototype Kilogram Number 20, the cylinder shown above, is a duplicate of the international standard kept in France. Also shown is Prototype Meter Number 27, a distance marked off on the platinum-iridium bar, which served the United States as its standard of length until the new standard, based on the krypton-86 lamp, was adopted. The bar will remain as a secondary standard because of the ease with which it can be used for certain types of measurements. (Courtesy of the National Bureau of Standards, Washington, D.C.)

This law gives us a qualitative definition of force. A force is what will cause a body to change its motion. (Force is a better term than push or pull because these two words imply living creatures at work; force is more general.)

We learned earlier that a change in velocity is called an acceleration. The greater the applied force, the greater the change in velocity—that is to say, the greater the acceleration. But if the mass of the object receiving this force is large, the change in velocity will not be great. Switching to simpler terms, mathematical terms, for expressing these ideas, we have

(1.3) $$a \propto F$$ The symbol $\propto$ means "is proportional to." Here "acceleration is proportional to the force" for a given mass.

(1.4) $$a \propto \frac{1}{M}$$ "Acceleration is inversely proportional to the mass," for a given force.

These may be combined:

(1.5) $$a \propto \frac{F}{M}$$ "Acceleration is proportional to the ratio of the applied force to the mass."

The first statement, equation 1.3, gives Newton's second law of motion:

> Newton's Second Law of Motion. Change in motion is proportional to the force, and the change occurs in the direction of the straight line along which the force acts.

Newton meant "change in motion" to mean "change in quantity of motion," and to him this meant momentum, a combination of M, L, and T to be described later.

Sir Isaac Newton (1642–1727). Born the year in which Galileo died, Newton brought to fruition the scientific revolution begun by Copernicus and continued by Kepler and Galileo. Newton is not just one of the greatest scientists who ever lived; he is one of the greatest figures in all history. The success that Newtonian physics had in explaining such an enormous variety of experiences and events in nature was so overwhelming that it has dominated thought not only in physics but also in the social sciences up until the present century. Even today Newtonian physics is an essential part of our understanding of our world and continues to dominate an important area of experimental psychology. Newton's monumental book, *Principia* (full title, *Philosophiae Naturalis Principia Mathematica*), published in 1687, still stands as the greatest scientific work ever written. Newton did not discover gravity, but it was his genius to recognize that gravity is universal, that what is said to act on an apple to impel it toward the earth extends not just to the top of the tree but all the way to the moon and beyond. On this assumption Newton derived equations for expressing the motion of the moon and other bodies, and the equations checked with observations. He never tried to speculate on the cause of gravity; he believed it enough to know that gravity does really exist. In addition to his contributions to mechanics, he made great discoveries in optics, including the ability of a prism to disperse white light into its colors.

The third proportionality, equation 1.5, can be converted to an equation by inserting a proportionality constant:

$$a \propto \frac{F}{M}$$

becomes

$$a = k \cdot \frac{F}{M}$$

where k is some constant.[2] Or we could ignore this proportionality constant by *defining* force with the simple equation

[2] The dot in the equation is one symbol we sometimes use to signify multiplication; the other is the more common times sign, $\times$.

(1.6) $$a = \frac{F}{M}$$

or by rearranging terms,

(1.7) $$F = M \cdot a$$

In metric terms the unit of force is the *newton,* defined as the force that will give a mass of one kilogram an acceleration of one meter per second squared. Equation 1.7 is really a useful combination of all three elements of motion because in terms of them acceleration a has the dimensions of LT^{-2} (equation 1.2). Substituting these into equation 1.7, we find that the dimensions of force are

(1.8) $$F = MLT^{-2}$$

(The newton has the specific units kilogram-meter per second squared.)

To minimize possible confusion, let us review a few of the terms that have been used. The fundamental *elements of motion* are mass, distance, and time, and they have concrete meaning only in terms of arbitrary reference standards and the multiples or subunits of them. The *dimension* of a physical quantity (e.g., velocity, acceleration, force, etc.) is the description of it in terms of the fundamental elements of motion rather than in other terms. Thus, although force may be defined as the product of mass and acceleration, the dimensions of force are obtained by multiplying the dimension of mass (M) by the dimensions of acceleration (LT^{-2}). The result is MLT^{-2}. The *units* of a physical quantity depend on the units of mass, distance, and time selected for the measurements or for the calculations. In the meter-kilogram-second system of units (mks system) the units of force are kilogram-meter per second squared. In many instances, especially in the older literature of physics, a cgs system or centimeter-gram-second system was the basis for definitions. Thus the dyne is a unit of force that imparts to a one-gram mass an acceleration of one centimeter per second squared. The *units* of force may vary, but the *dimensions* of force do not. There are many choices of measuring units; there is now only one definition of force, and the definition fixes the dimensions.

Force, then, is simply one particular combination of the three elements of motion. Others are equally useful, but before going on to them, we shall digress again to inquire into one particular origin of a force and how we capitalize on it.

Gravitational Force. Weight. When a stone is dropped, it falls. There is nothing very profound in this; or is there? Why, after all, should it fall? Why not move up? We could say, as almost everybody does, that it falls because of gravity, but what have we gained in understanding? "Gravity" is just a new label for the mystery, for what is gravity? Is it what makes stones fall? The argument is circular. We could say that the stone falls because it is in the nature of stones to fall. The trouble with this explanation is that people, some people at least, tend to proceed with the questioning. Why is it the nature of stones to fall? These questions "Why?" really help us not at all. They may be interesting, but they tell us nothing new about falling stones.

We may try easier questions. How does a stone fall to the ground? Not why but how? Does it fall, for example, with a constant velocity? Whatever we may later decide makes the stone fall, we may ask whether it does so with constant velocity so that we can at least make measurements. Since the stone starts with zero velocity, being motionless just before it is dropped, and since this velocity begins to change when it is released, we would guess that the velocity of the stone is not constant. At least at the very beginning of its fall, its velocity cannot possibly be constant. The velocity has to change or the stone would not move at all. It could be, of course, that very soon after the stone has been released it reaches a constant velocity which holds for the rest of the fall. But casual experiences suggest and actual measurements prove that the velocity of a falling stone changes, it increases all the while the stone is in free fall until it thumps to the ground. (To be strictly true, the last sentence should have contained a qualifying clause about the effect of air resistance, but we ignore that problem here.)

What could possibly act on a stone, of a given mass, to make it undergo this acceleration, this change in velocity with time? We just learned a name for it, force. By definition, an unbalanced force acting on an object changes its velocity. Consistent, logical application of this definition to the stone falling toward the earth leads us to a most remarkable conclusion. The earth exerts a force on the stone, a force that is not balanced by anything else. Plain, ordinary planet earth exerts a force on anything reasonably near it. At least so we say. We have not explained the force, but we have described it in a language that is useful. And we must go a step further, said Newton. He asserted that if we are to be able to derive the properties of motion in any mechanical system, we must assume the truth of a third proposition, his third law of motion:

> Newton's Third Law of Motion. The force that one body exerts on another body must always be equal in magnitude and opposite in direction to the force that the second body exerts on the first.

In other words, if it is useful to say that the earth exerts a force on a stone, or on a moon, that stone or that moon must exert an equal but oppositely directed force on the earth. For example, the force that the moon exerts on the earth causes the tides. It cannot be of equal effect everywhere around the earth. With the rotation of the earth, the effect of this force at any point rises and falls, and so do the tides.

Without explaining the cause of the force of attraction between two objects, we can at least name it. It is called *gravity*. The space in which such a force is experienced also has a special name, *gravitational field*. We speak, for example, of the earth's gravitational field or the sun's gravitational field. If two objects of mass M_1 and M_2 are separated by a distance L, then according to *Newton's universal law of gravitation*, the gravitational force acting between them is given by the equation

$$F = G \cdot \frac{M_1 M_2}{L^2} \tag{1.9}$$

where G is a constant having the value of 6.67×10^{-11} newton-meter²/kilogram². This equation states that the force of attraction F between two bodies is proportional to the product of their masses M_1 and M_2 and inversely proportional to the square of the distance L separating them. We include this equation only to indicate that, for this gravitational force to be of any consequence, the distance L separating the two objects cannot be enormously large. Furthermore, the masses of the objects must be large for the force to be of any size. One mass at least has to be very large, such as the mass of the earth, if the other mass is to be small, such as a stone's. Two pebbles presumably exert a gravitational force on each other, but only the most sensitive of instruments could detect it.

Objects in free fall, in the absence of wind and air resistance, experience a steady change in velocity. Although the velocity is changing, the value of the acceleration, because of gravity, is not. It is a constant. If a force, in general, is defined by equation 1.7 ($F = Ma$), a gravitational force is given by equation 1.10, with the symbol g for gravitational constant replacing a for acceleration:

$$F = M \cdot g \tag{1.10}$$

Thus, for a particular mass, since M is a constant so is $M \cdot g$. In other words, the gravitational force exerted on an object of fixed mass is a constant. Here again we have something that can be constant in the midst of change, and we give the name *weight* to this constant gravitational force exerted on an object. The gravitational constant g is known to vary slightly from place to place on the earth's surface; it is less at the equator (978 cm/sec^2) than at the poles (983 cm/sec^2). On the surface of the moon, which has a much smaller total mass than the earth, the gravitational constant is about one-sixth that on earth. Thus an astronaut on the moon, although he will have lost none of his mass in getting there, will weigh about one-sixth what he weighs on earth. At some stages of his flight he will have no weight at all. Mass and weight are different. We take the word mass to mean the amount of matter in a body as related to some standard reference mass. Its weight at a particular location is the gravitational force exerted on it because of its mass and the mass of the body nearby, usually the earth.

The Measurement of Mass. The chemist takes a much greater interest in the concept of mass than he does in that of weight because mass gives a measure of "amount of matter" (as related to the reference). For purposes of measurement, however, masses are compared by comparing their weights. The mass of some object A, in relation to the internationally accepted standard of mass, is determined by comparing the weight of A with the weight of the standard. One means of doing this involves an equal-arm balance (Figure 1.3).

Initially, the empty pans are balanced. That is, the gravitational forces acting on the pans are equal and the pointer points to zero. If the pans are set in motion, the pointer will swing equal units on either side of the zero mark. (Frictional losses change this slightly.) If an object is now placed on one pan, the gravitational force on that side will be greater, and the change in force of gravity is given by

$$\begin{aligned} M_A g &= \text{unbalanced force acting on the object } A \text{ of mass } M_A \\ &= \text{the weight of } A \end{aligned}$$

To rebalance the system, standard masses, all related to the International Prototype Mass, either as accurately known fractions or as multiples of it, are placed on the other pan until the pointer again points to zero or swings equally to either side of the zero mark. The force acting on the standards (the "weights") is given by

$$M_s g = \text{force acting on standards of total mass } M_s$$
$$= \text{weight of standard masses}$$

At the point of balance these two forces must be equal:

$$\text{Gravitational force on } A = \text{gravitational force on } B$$

or

$$M_A g = M_s g$$

Canceling the g's, we have

$$M_A = M_s$$

Thus by comparing weights (i.e., gravitational forces) we can compare masses. The mass values of the standard masses used in the operation are summed, the unit is specified, and the result is recorded; and here common usage has us do a peculiar thing. The result is called a *weight* in nearly all situations. The verb is "to weigh" rather than "to mass", and the procedure is called "weighing" rather than "massing." As long as we understand and know what we are doing, we may go along with this usage. Thus when the word "weight" is used in this text, "mass" is meant.

Are Newton's Laws of Motion Truly Experimental or Are They Postulates? There is some difference of opinion among scientists on the correct reply to this

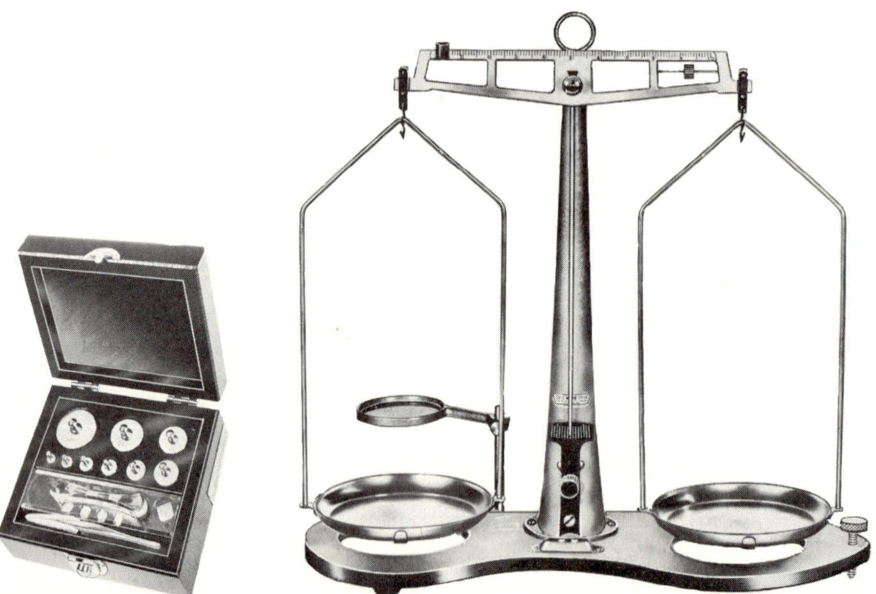

Figure 1.3 An equal-arm balance. (Balance, courtesy of Welch Scientific Company; weights, courtesy of Voland and Sons.)

question. Many great philosopher-scientists, men such as Poincaré, Mach, and Hertz, believed that these laws must be taken as postulates. They are, to be sure, compatible with experience and are certainly suggested and supported by numerous experiments. A full, unambiguous test of the laws, however, is out of the question. In the final analysis they are taken as assumptions, something like the axioms of plane geometry. They may not be quite as self-evident as these axioms, but apparently they were almost so to Newton. Taking them as assumptions, Newton made deductions from them. Many other physicists did the same, and the deductions when checked against experiments were so satisfactory that the original ideas became that much more credible. The laws of motion are not derived from simpler laws—they are the simple laws. Confidence in their truth has been built by repeated successes in using them. Two of these victories are the subject for the continuation of our elementary discussion of mechanics.

MOMENTUM

Mass multiplied by acceleration (i.e., $M \cdot LT^{-2}$) gives us force. Mass multiplied by velocity (i.e., $M \cdot LT^{-1}$) gives us still another useful combination of the three elements of motion. Called *momentum,* this technical concept, like energy and inertia, is mentioned frequently in everyday speech. ("The election campaign is gaining momentum.") What interests us most about momentum is that it is the subject of one of the conservation laws, the law of conservation of momentum.

Suppose that we have two masses M_1 and M_2 which exert a gravitational force on each other, and suppose that other forces are absent. The law of conservation of momentum says that the sum of the momenta of the two masses will remain a constant. Thus if the momentum of the first mass P_1 is $M_1 v_1$ and that of the second mass P_2 is $M_2 v_2$, then

$$P_1 + P_2 = \text{a constant}$$

If the two particles collide, they will change courses. Having new velocities, they will have new momenta P_1' and P_2'. But the sum of these will still be the same. Thus

$$P_1 + P_2 = P_1' + P_2'$$

or

$$M_1 v_1 + M_2 v_2 = M_1 v_1' + M_2 v_2'$$

where v_1 and v_2 are velocities before collision and v_1' and v_2' are velocities after collision. The object here is not to acquire skill in handling calculations illustrating this law but to see that this law illustrates something significant about the nature of motion. It is part of the givenness of our natural world, that is to say, it is in the nature of things, that out of one particular combination of the three elements of motion, $M \cdot v$ or MLT^{-1}, we find something constant in the midst of change—namely, the sum of the momenta of the objects in motion. To give an object a particular momentum, a force must be made to act on it until its velocity becomes the required value. The momentum of an object is therefore the cumulative effect of a force acting on it for a certain period of time. We next inquire what the effect of an unbalanced force acting on an object throughout a certain distance would be, quite without regard for how long it takes the force to move the object through the distance.

ENERGY

The multiplication of force and time gave use momentum:

$$\begin{aligned}\text{dimensions of force} &= MLT^{-2}\\ \text{(dimensions of force)(time)} &= (MLT^{-2})(T)\\ &= MLT^{-1}\\ &= M(LT^{-1})\\ &= \text{(mass)(dimensions of velocity)}\\ &= \text{momentum dimensions}\end{aligned}$$

The multiplication of force and distance results in the following:

$$\begin{aligned}\text{(dimensions of force)(distance)} &= (MLT^{-2})(L)\\ &= ML^2T^{-2}\\ &= M(L/T)^2 \qquad (1.11)\\ &= \text{(mass)(dimensions of velocity)}^2\end{aligned}$$

These shifts to alternate groupings of symbols would be no more than an idle exercise were it not for the fact that the grouping in equation 1.11 represents an extremely important combination of the elements of motion. This combination defines and describes something else that is constant in the midst of change, namely energy. If the mass and the velocity of a moving object are known, the product of the mass and the square of the velocity defines the energy it has by virtue of its motion. It took almost two centuries for this idea to emerge, and in its final form the energy of a moving object is called its *kinetic energy.* It is *defined* as

$$\text{Kinetic energy} = \tfrac{1}{2}MV^2 \qquad (1.12)$$

(The reason why MV^2 is multiplied by one-half lies in the calculus of mechanics and will not be discussed.) Let us examine again the two masses M_1 and M_2 used to illustrate the conservation of momentum. If no external forces act on them, the sum of their kinetic energies before a collision is the same as the sum after collision. Energy is neither created nor destroyed—and this is one way of stating the law of conservation of energy. One of the metric units of energy is the *joule,* which is the energy given to a body when a force of one newton is exerted on it for a distance of one meter in the direction of the force.

The purpose again is not to acquire skills in computing kinetic energies but to identify kinetic energy as a particular combination of the three elements of motion. Equation 1.12 is a definition of kinetic energy. It has not been proved here that the sum of kinetic energies of colliding objects is the same before and after collision. In fact, as stated, this supposition would be difficult to prove because we have specified that no external forces be present. The gravitational field is present, however, and we must therefore broaden the concept of energy to take it into account.

Potential Energy. Imagine a steel ball held a meter from the ground. It is motionless and has no kinetic energy. As soon as it starts to fall, it picks up more and more kinetic energy as it accelerates in the earth's gravitational field. While gaining kinetic energy, however, the ball is losing something—its ability or potential for falling that initial distance. Just before it starts to drop, we must credit it with having something of interest—a potential for delivering the energy of motion. It possesses this potential because it has both mass and a certain position with respect to the ground. When it reaches the ground, this potential is entirely gone. As it loses potential energy it gains kinetic energy. During free fall it has something of both at any given instant. At any moment the sum of its kinetic energy and potential energy is a constant which is called the *total mechanical energy* of the object:

$$\text{Kinetic energy} + \text{potential energy} = \text{total mechanical energy} \quad \text{(a constant)}$$

When the steel ball hits the ground, assuming it does not bounce, its energy of motion is delivered into other forms—heat, mechanical agitation of ground particles, and sound. Since we cannot add terms unless their dimensions are the same, potential energy must have the same dimensions as kinetic energy. This is true of all forms of energy. For convenience we may from time to time define other quantities to represent energy, such as the calorie for thermal energy, but these new quantities must ultimately reduce to the dimensions of energy, ML^2T^{-2}.

An object may have energy, therefore, by virtue of its motion (kinetic energy) or by virtue of its position (potential energy). We can give kinetic energy to an object by subjecting it to an unbalanced force acting throughout a distance, and we can give potential energy by placing the object at any particular height in relation to the earth. We next ask in what ways besides applying the mechanical forces just described we can give energy to a system. These ways will include applying heat, chemical, electric, and light energy.

HEAT

The Motive Power of Heat. In a steam engine heat is converted into work, which is but one example of the fact that it is possible to obtain mechanical energy, as work, from heat. But what is heat? In one sense the word "heat" is our answer to the question "What flows or passes from a hot object to a cooler one touching it until the temperature of the two objects has evened out?"

For the moment we shall define temperature simply as what is measured by a thermometer; Figure 1.4 describes the two most familiar temperature scales, centigrade and Fahrenheit. Temperature is an intensive property; it does not depend on how much matter is present. Heat, however, whatever it really is, is an extensive property. How much heat is present depends directly on how much matter is present. If a fifty-gram cube of red-hot steel is dropped into a beaker of water, the rise in water temperature will be greater than when a one-gram cube of red-hot steel is used.

The *calorie* measures the quantity of heat. One calorie is the amount of heat required to raise the temperature of one gram of water from 14.5 to 15.5°C, that

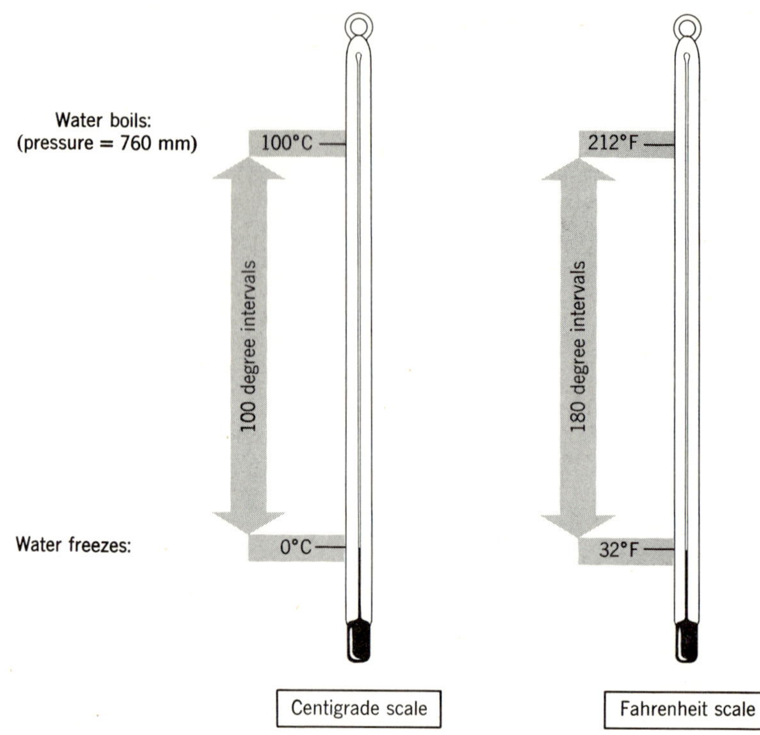
Water boils:
(pressure = 760 mm)
100°C
212°F
100 degree intervals
180 degree intervals
Water freezes:
0°C
32°F
Centigrade scale
Fahrenheit scale

Figure 1.4 Relation between Fahrenheit and centigrade scales of temperature. If it is necessary to convert a temperature on one scale to its equivalent on the other, the following equations may be used:

$$°C \text{ to } °F: \quad °F = °C(\tfrac{9}{5}) + 32°$$
$$°F \text{ to } °C: \quad °C = (°F - 32°)\tfrac{5}{9}$$

Average room temperature is 25°C (77°F); a hot day is 38°C (100°F); normal body temperature is 36.6°C (98.6°F); anesthetic ether boils at 35°C (94°F).

is, one degree at this particular point on the scale. Both one gram and one degree are very small quantities, and so is the calorie on this scale. A larger unit, the kilocalorie (kcal) is the amount of heat that will raise the temperature of one kilogram (one liter) of water one degree on the centigrade scale (specifically the degree between 14.5 and 15.5°C).

At this stage, all we venture to say is that heat is a form of energy. It used to be regarded as a substance, an imponderable fluid that was virtually weightless but could flow in a more literal sense than we now mean when we say that heat flows. It is part of the givenness of nature that heat always moves from a warmer body to a colder one. Heat has never been reliably reported to flow spontaneously in the reverse direction, nor has anyone ever reported, for example, a bar of iron of uniform temperature spontaneously becoming warmer at one end and colder at the other. These observations, possible and impossible, are about the *direction* of natural changes or of what we call spontaneous events. From such events man has been able to obtain energy to supplement his own. In this chapter we have been analyzing motion, and soon we shall analyze matter. Eventually we must come to grips with an analysis of spontaneity, for which a study of heat and its workings is a good introduction.

As long as we are dealing just with motion, three basic instruments for measuring things are sufficient: a meter stick, an equal-arm balance, and a stopwatch. Fancier instruments exist, of course, instruments that not only are more convenient but also are capable of great accuracy. They add nothing fundamentally new to our knowledge, however. When we shift from a study of motion to a study of heat, we add a new instrument, the thermometer, and encounter a new field of science, thermodynamics. The laws of Newton formed the basic assumptions for dynamics. For thermodynamics we retain Newton's laws and add some more, the laws of thermodynamics. Only as we delve into these laws, which we shall do step by step in several of the next chapters, will we gain better insights into what heat is. One reason why heat energy is of central importance is that all the other forms of energy can be converted quantitatively into it, providing a means for measuring these other forms. It is known, for example, that 4.184 joules of mechanical energy are equivalent to 1 calorie of heat (a good indication of how small a unit the joule is).

CHEMICAL ENERGY

A substance can have energy by virtue of its ability to undergo chemical reactions. A pile of newspapers, the surrounding air, and a match can in one minute be at room temperature and in the next minute be at a much higher temperature. To all outward appearances energy is constantly being created when a pile of paper burns. But this would be a violation of the law of conservation of energy. The law is worth saving, however, for its usefulness, and it is saved by assigning to the combination of papers, air, and match the possession of a particular form of potential energy, chemical energy. This energy is potential, but not through any particular position with respect to the earth. It is potential because of the particular properties of the chemicals involved. Just *how* chemicals can have this potential energy is a matter for study in succeeding chapters, but the fact that they do have it forever weds chemistry and biology to physics and mathematics.

ELECTRIC ENERGY AND ELECTRIC FORCES

The ancients noted that if a piece of amber is rubbed against a cloth, tiny pieces of lint, straw, or chaff jump to the amber and stick there for a while. Since the rubbed amber has the ability to change the velocity of the lint or straw (from zero velocity), we say that it exerts a force (force is what can change velocity), one that is an attraction but yet is unlike gravity. This mystery puzzled people for centuries. How could a rubbed piece of amber act as if invisible hands reached out and drew lint to itself? Actually, the action is no more mysterious than the "attraction" of a stone (or a person) by the earth. It is just much less common and therefore not familiar and not taken for granted. A rubbed piece of glass will also exert the same force on dust and lint.

The Greek word for amber is *electron,* and from this term come all our common words with "electr-" in them. The rubbed piece of amber is said to be electrified. Eventually it was discovered that the electrification condition of amber is not quite the same as that of rubbed glass. Two small pieces of electrified amber seemingly repel each other, and so too do two small electrified pieces of glass. Rubbed pieces of glass and amber, however, behave as if they attract each other. The two substances are said to be electrically charged, but they have charges of opposite character as implied by the phenomena of attraction and repulsion. To designate kinds of charge, plus and minus signs are used, and from a long history of study, two simple laws of nature have emerged:

—Like-charged particles repel each other.
—Unlike-charged particles attract each other.

Because an electrically charged particle can exert forces on certain kinds of objects in its vicinity, this vicinity is called an electric field. It is somewhat analogous to a gravitational field, and an electrically charged object in an electric field can have energy through both its motion in that field and its position.

We now know that an electric current is a flow of electrons, which are particles bearing negative electric charge (see Chapter 2). It is in connection with a flow of electrons that we speak of electric energy, because electricity can be made to turn motors, heat homes, and do a host of other things for us.

Magnetism. It seems likely that some ancient shepherd of the Magnetes tribe in the land called Magnesia in Thessaly discovered a stone (now called magnetite) that can attract pieces of iron to it. Here, again, is a force acting at a distance to produce a change in velocity. The magnetite can impart energy to the iron pieces, and if it can *give* energy, it must have some. We call this *magnetic energy.* If a piece of iron is rubbed carefully against the magnetite, the iron is magnetized too. The familiar concept of oppositely behaving magnetic poles is the counterpart of oppositely charged objects in electricity. A *magnetic field* is simply a region in space, usually near a magnet, in which magnetic forces can be detected.

ELECTROMAGNETIC ENERGY. LIGHT

One peculiarity of an electric current is that when it is first turned on, it will affect a magnetized compass needle nearby. During the time that an electric field is building up near the current in the wire, this field is not steady in the way that a gravitational field is reasonably steady. The effect on the magnet suggests that associated with a changing electric field is a changing magnetic field. By rapidly reversing the direction of the flow of the electric current, the electric field can be made to oscillate, and accompanying this oscillating electric field will be an oscillating magnetic field. If the oscillations in field strength occur with the right frequency, the disturbance is sensed as light when it reaches the eye. Frequency is defined here as the number of times the field goes from some maximum intensity through a minimum value and back to a maximum—in other words, through one cycle—in the period of one second. More briefly,

$$\text{Frequency} = \text{number of cycles per second}$$

Defined in this way, the higher the frequency, the more energy associated with the disturbance. As the disturbance moves away from its source, the sites of its maximum values will be spaced regularly apart along the line in which it moves at any given instant. The distance separating these maxima at that instant is called the wave length (Figure 1.5):

$$\text{Wavelength} = \text{centimeters per cycle} \qquad \text{(or some other distance per cycle)}$$

If we multiply frequency by wavelength,

$$\frac{\text{centimeters}}{\text{cycle}} \cdot \frac{\text{cycles}}{\text{second}} = \frac{\text{centimeters}}{\text{second}}$$

we obtain the velocity with which the disturbance moves from the source to the detector (e.g., an eye) expressed in the dimensions of velocity. It is part of the givenness of nature that this velocity is a constant in a particular medium—in air the velocity of light is 3×10^{10} cm/sec.

When a detector receives the disturbance, that is, the oscillating electric and magnetic fields, it may become warmer or do other things. The point is that this electromagnetic disturbance transmits energy, a form of energy that is particularly important in chemistry because chemicals can be made to emit or absorb light of particular and accurately measurable energies. By knowing what values of energy a chemical can absorb or can be made to emit, we can deduce much about the inner workings and structure of that chemical.

Electromagnetic Spectrum. Because electromagnetic disturbances oscillate regularly as they move through space from their source, we speak of them as waves. In principle they can have any frequency from very, very few cycles per second to very high values. Because different ranges of frequencies affect different kinds of instruments in different ways, these ranges have been classified in what is called the *electromagnetic spectrum*. Figure 1.6 shows the chief classes. At the low-frequency end of the spectrum are radio and television waves, at the high-frequency end the X-rays and gamma rays. Perhaps you have noticed that when driving near high-power lines your car radio has considerable static. The transmission of alternating

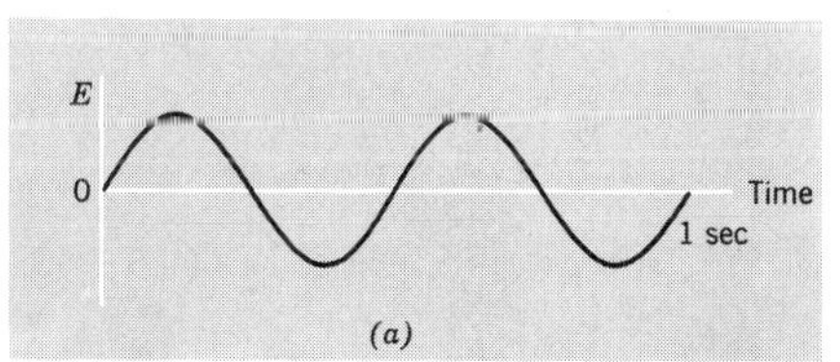
E
0
Time
1 sec
(a)

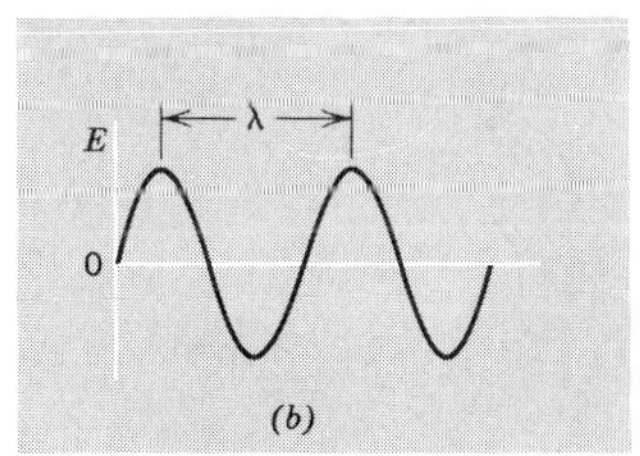
E
λ
0
(b)

Figure 1.5 The meaning of frequency and wavelength in electromagnetic radiations. (a) This curve indicates how the electric field E at any *fixed point* along the line of propagation will vary with time. As drawn, this curve illustrates radiation of 2 cycles/sec.

(b) This curve indicates how the electric field E at any *fixed time* or instant varies along the direction of propagation. The wavelength λ corresponds to the distance between maxima.

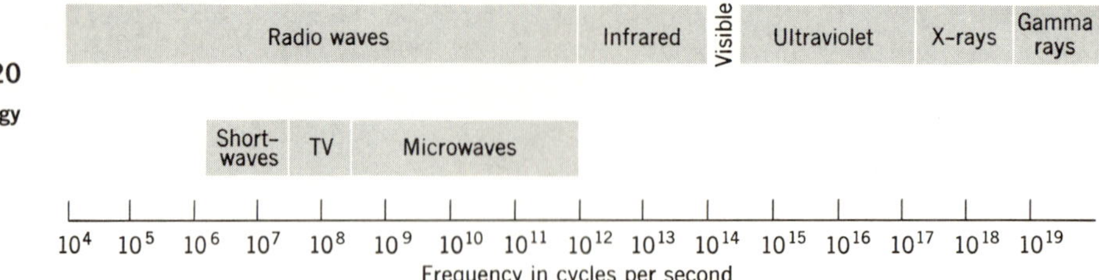
Radio waves
Infrared
Visible
Ultraviolet
X-rays
Gamma rays
Short-waves
TV
Microwaves
10^4 10^5 10^6 10^7 10^8 10^9 10^{10} 10^{11} 10^{12} 10^{13} 10^{14} 10^{15} 10^{16} 10^{17} 10^{18} 10^{19}
Frequency in cycles per second

Figure 1.6 The electromagnetic spectrum.

electric current inevitably and unavoidably sets up an electromagnetic field radiating away from the wires; this field interferes with the regular radio waves transmitted by the station.

The very narrow range from about 4.3×10^{14} to 7×10^{14} cycles/sec constitutes the visible range of the spectrum and is detected by the eye. White light is a mixture of all frequencies in this range, but when it is passed through a prism, it is dispersed into its components (Figure 1.7). Each color corresponds to a particular narrow band of frequencies in the visible light range.

Sunlight includes all frequencies of the visible range plus frequencies in both the ultraviolet and the infrared ranges. Not all glowing objects, however, behave this way. Glowing sodium metal, for example, emits only a few frequencies, and when light from a sodium vapor lamp is passed through a prism (or other dispersing device), these relatively few frequencies become separated. If they are made to impinge on a photographic plate, lines corresponding to the frequencies emitted by the "excited" sodium will appear when the plate is developed. Such lines constitute a line spectrum and characterize many substances (Figure 1.8). We shall have occasion to use such information in connection with theories about atomic structure.

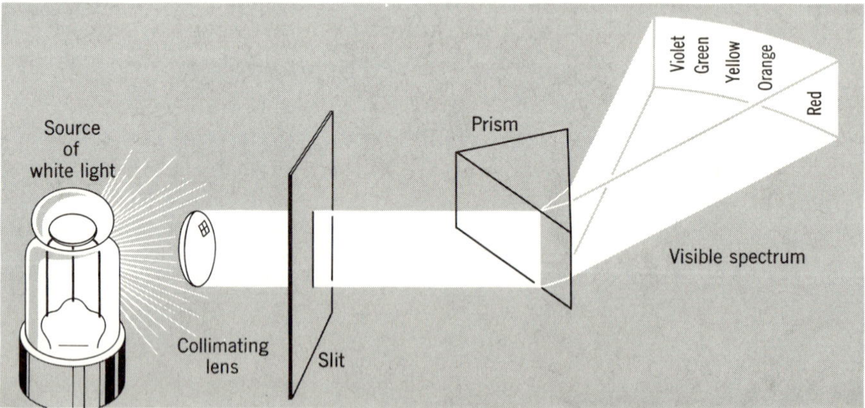
Violet
Green
Yellow
Orange
Red
Prism
Source of white light
Visible spectrum
Collimating lens
Slit

Figure 1.7 A prism will disperse the frequencies of white light and display the visible spectrum of colors. For white light, with all frequencies present, the spectrum is said to be continuous.

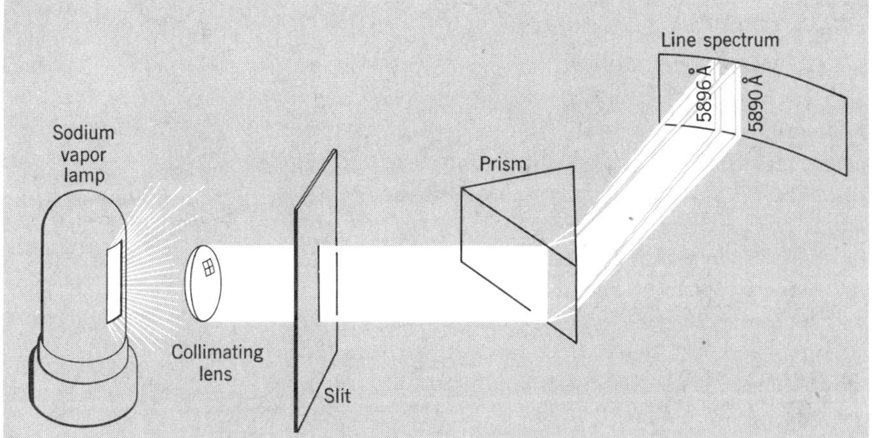
Line spectrum
5896 Å
5890 Å
Sodium vapor lamp
Prism
Collimating lens
Slit

Figure 1.8 When certain substances such as sodium (or mercury and other elements) are excited by heat or electricity, the visible light they emit does not give a continuous spectrum. Only certain frequencies are present and these show up on a photographic plate as separated line images of the slit. The result is called a line spectrum. In this illustration, only the two most prominent lines of sodium light are indicated. They are closely spaced in the yellow region of visible light. Many cities use sodium vapor lamps for lighting streets, and they give off the characteristic yellow light of these frequencies. Other cities use mercury vapor lamps whose emitted frequencies are in the blue-violet region.

The point here is that certain kinds of matter can be made to emit electromagnetic energy, and this energy can be separated or dispersed into the frequencies present. Both frequencies and intensities can be measured. From such data we can construct theories in answer to the question "What must the structure of this piece of matter be in order that it behave in this measurable way?"

The Dual Nature of Light. In some experiments light behaves as if it were an electromagnetic oscillation in space and has a wavelike motion. In other experiments light behaves as if it were made up of tiny packets of energy or light quanta. What light really is has not been fully resolved. It is both wave and particle, depending on the nature of the experiment. The equation that comes closest to unifying wave and particle is one identifying the amount of energy in a radiated quantum as being proportional to the frequency of the light detected as if it were a wave:

$$E = h\nu$$

where E = energy
ν = frequency
$h = 6.6256 \times 10^{-27}$ erg-sec

This h is Planck's constant, a universal constant in nature named after Max Planck (1858–1947), a German physicist who ushered in the quantum theory, one of the greatest developments in the whole history of science. Planck received the 1918 Nobel prize for physics.

A BRIEF SUMMARY

The emphasis in this chapter has not been on learning skills in doing calculations. You have undoubtedly heard of all these technical terms before—force, momentum, inertia, mass, energy, etc. The concepts of mass and energy are those most important for future uses, but the idea of forces, particularly electric forces, will also be applied. What you should carry from this chapter to the rest of the book are a few ideas. Thus our concepts of force, weight, momentum, and energy grew out of efforts to reduce motion to its simplest elements. Force came to be defined as what will change a velocity. Weight is the name for the specific force, gravity, that we say acts on the object in question. We compare masses by comparing their weights, and we understand the need for a reference standard. (And we noted that the words mass and weight are used interchangeably.)

Energy is the name that came to be associated with the cumulative result of an unbalanced force acting on an object throughout a certain distance. Energy, in another view, is simply one particular combination of the values of the three elements of motion (M, L, and T) that a moving body happens at any moment to have. Kinetic energy specifically was identified as $\frac{1}{2}MV^2$, and the dimensions for this were ML^2T^{-2}. We generalized by saying that *we shall use the term energy with any data that must be given these dimensions.* We discussed several forms of energy, but heat is the central one because all other forms can be converted into heat quantitatively. Heat cannot be *quantitatively* converted into the other forms, however.

We saw that an object can have energy in a gravitational field by virtue of its motion and its position. It can also have energy in an electric field if it is electrically charged—again by virtue of motion or position. A magnetic field and a magnetized object in the field are still another way in which energy may be possessed. The electric and magnetic fields combine in an oscillating form in electromagnetic radiation. Finally, a particle is said to have chemical energy if it has a potential for reacting chemically and thereby producing other forms of energy.

Before closing this chapter, we must for future need define two other combinations of the elements of motion, those of volume and density.

VOLUME

In 1901 the International Committee of Weights and Measures defined the unit of volume to be the volume occupied by one kilogram of water under a pressure of one atmosphere at the temperature of its maximum density, about 4°C. This volume was called the *liter,* and it was originally intended that it be identical in value to the cubic decimeter (a cube 10 cm on each side). There was a discrepancy of 0.0028%, however. In 1963 the International Committee recommended that the old definition be abandoned and that the liter be recognized simply as a special name for one cubic decimeter (or for 1000 cubic centimeters). By this definition one cubic centimeter (1 cc) is exactly the same as one milliliter (1 ml), and we shall use them interchangeably. Table 1.2 contains a summary of the more commonly used metric units and equivalents in the English system.

Table 1.2 Some Common Metric Units and English Equivalents

Unit	Abbreviation	Definition	English System Equivalents
		Length	
Meter	m	Standard meter (Figures 1.1 and 1.2)	39.37 in. (slightly over a yard)
Centimeter	cm	$\frac{1}{100}$ meter	0.394 in. (2.54 cm per inch)
		Mass	
Kilogram	kg	Standard kilogram (Figure 1.2)	2.2 lb.
Gram	g	$\frac{1}{1000}$ kg (0.001 kg)	0.035 oz (1 oz per 28.35 g) (453.6 g per pound)
Milligram	mg	$\frac{1}{1000}$ g (0.001 g)	There are 64.8 mg per grain
		Volume	
Liter	l	1 cubic decimeter (old standard: volume of 1 kg water at 4°C and 760 mm pressure)	1.057 liquid qt; 33.8 oz (approx. 1 qt)
Milliliter	ml	$\frac{1}{1000}$ liter	29.6 ml per fluid ounce (16.231 minims) (volume of 1 g water at 4°C and 760 mm pressure)
Cubic centimeter	cc	identical with the milliliter in the new standards	

Miscellaneous Units of Measure

Commercial Dry Measure

1 bushel = 4 pecks = 32 dry quarts (qt) = 64 pints (pt) = 35.24 liters

Commercial Weight—Avoirdupois

1 pound (lb) = 16 ounces (oz av) = 256 drams (dr av) = 7000 grains (gr) = 453.6 g

Apothecaries' Fluid Measure

1 fluid ounce (fl oz, or f℥) = 8 fluid drams (fl dr, or fʒ) = 480 minims (min, or ♏) = 29.6 ml

Apothecaries' Weight

1 ounce (oz ap, or ℥) = 8 drams (dr ap, or ʒ) = 24 scruples (s ap, or ℈)
= 480 grains (gr) = 31.1 g

DENSITY

Density is defined as the amount of mass contained in a unit of volume, the traditional units being grams per cubic centimeter:

$$\text{Density} = \frac{\text{mass}}{\text{volume}} \tag{1.13}$$

The density of lead is 11.2 g/cc, of aluminum 2.70 g/cc. (In such expressions the slash line / is read as "per.") Thus it would be wrong to say that lead is heavier than aluminum. Rather, it is denser.

Nearly all substances expand in volume when they are heated and contract when they are cooled, but their masses do not change. Density is therefore different at different temperatures, although the change is usually not much for solids and liquids. At 0°C, mercury has a density of 13.5955 g/cc, and at 100°C it is 13.351 g/cc. The problem is handled by recording the temperature at which the measurements were made. Thus for water $d_4 = 1.000$ g/cc, meaning that the density of water at 4°C is 1.000 g/cc.

Specific gravity is the ratio of the mass of any object to the mass of an equal volume of water at the same temperature (unless otherwise specified). This definition automatically makes the specific gravity of water equal to one. (Why?) Knowledge of the specific gravity of another substance informs us at once whether it is more or less dense than water.

REFERENCES AND ANNOTATED READING LIST

At the end of each chapter an annotated list of books and articles will be found. Some are references of such importance to the writing of this text that they must be acknowledged, although their level may be too high for use except by the instructor. Others are suggested for supplementary reading. Most of the books are paperbacks, and most of the articles have appeared in *Scientific American,* the *Journal of Chemical Education,* and *Chemistry.* Omission of books and articles from these lists does not imply adverse judgment of their worth. They may simply have been missed.

George Gamow. *Biography of Physics.* Harper Torchbook TB567, Harper and Row, New York, 1961. The author manages to convey an idea of what physics is and what kinds of people physicists are. The book is by no means an advanced treatise. Written for the young reader, it includes principles of physics as well as highly interesting historical material. (Paperback.)

E. N. da C. Andrade. *Sir Isaac Newton.* Science Study Series S1, Doubleday Anchor Books, Garden City, N.Y., 1958. In this short book (150 pages) Newton's personality and his scientific genius engage the reader, and the principles of his discoveries are simply explained. (Paperback.)

Alexander Vavoulis and A. Wayne Colver. *Science and Society: Selected Essays.* Holden-Day, San Francisco, 1966. The humanities and social science student will be especially stimulated by these essays written by philosophers, educators, and scientists. The focus is on the basic principles of science and its methods and impact on other areas.

Floyd W. Matson. *The Broken Image. Man, Science and Society.* George Braziller, New York, 1964. Newtonian physics was so triumphantly successful that social scientists after his day investigated the same program used by physicists: What are the most useful elements in terms of which man and society may be understood and described? The answer for most was Newton's laws, and they came to view man as a machine. Matson traces this development and argues for a reconstruction of Whole Man to replace this Machine Man, the man of the "broken image." Students of the social sciences and humanities may not always agree with Matson, but they will surely find him stimulating.

Thomas S. Kuhn. *The Structure of Scientific Revolutions.* University of Chicago Press, Chicago, 1962. For the student interested not just in the nature of nature but also in the nature of science as a human activity, this book explores the meaning, causes, and consequences of revolutions in scientific concepts.

PROBLEMS AND EXERCISES

1. What do the following words mean?
   (*a*) kinematics
   (*b*) meter
   (*c*) velocity
   (*d*) acceleration
   (*e*) inertia
   (*f*) force
   (*g*) mass
   (*h*) weight
   (*i*) kilogram
   (*j*) density
   (*k*) specific gravity
   (*l*) calorie
   (*m*) extensive property
   (*n*) momentum
   (*o*) kinetic energy
   (*p*) potential energy
   (*q*) total mechanical energy
   (*r*) frequency (of light)
   (*s*) wavelength
   (*t*) line spectrum
   (*u*) quantum
2. What are the ways that an object can have energy in a gravitational field?
3. Using an equal-arm balance, we find that an object has a mass of 0.5 kg. If the same balance were used to measure this object on the moon, what would the mass of the object be? Explain.
4. One common kind of balance is a spring balance, as illustrated in Figure 1.9. Suppose that the scale is calibrated on earth against the standard kilogram and its fractional units

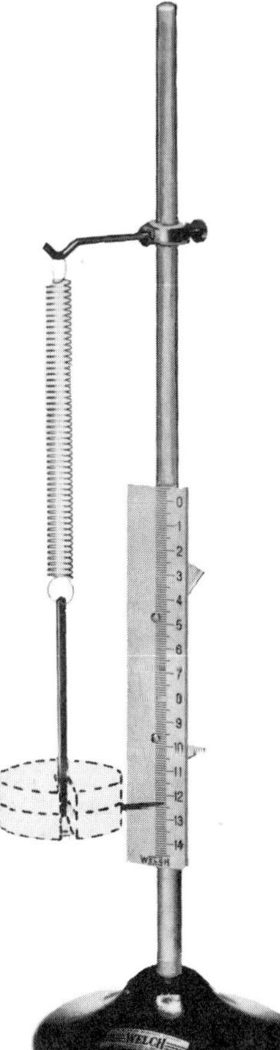

Figure 1.9 A spring balance. When no object occupies the holder attached to the spring, the pointer reading is marked zero. By placing known weights on the pan, new pointer marks can be made. (Courtesy of the Welch Scientific Company.)

so that marks placed on the scale correspond to these masses. Suppose that, using such a scale on earth at the same place where it was calibrated, we find an object has a mass of 6 kg. What would the reading be if the same scale were used to determine the mass of the same object on the moon? Explain.

5. The following example illustrates a calculation involving momenta. A rifle of mass 4 kg fires a 10-g bullet at a muzzle velocity of 800 m/sec. If the rifle is free to recoil (is not held firmly against the shoulder), at what velocity will it recoil and hit the shoulder? The law of conservation of momentum is assumed to hold, and it applies to this problem as follows:

$$M_{\text{rifle}}V_{\text{recoil}} + M_{\text{bullet}}V_{\text{bullet}} = M_{\text{rifle}}V_{\text{before firing}} + M_{\text{bullet}}V_{\text{before firing}}$$

$$= 0$$ (since velocities of both rifle and bullet before firing are zero)

or $4000\ \text{g} \times V_{\text{recoil}} + 10\ \text{g} \times 800\ \text{m/sec} = 0$

$$V_{\text{recoil}} = \frac{-(10)(800)}{4000}\ \text{m/sec}$$

$$= -2\ \text{m/sec}$$ The minus sign indicates that the rifle recoils in a direction opposite that of the bullet.

The recoil velocity is 2 m/sec which is roughly 4½ miles/hr.

Problem. A 4-kg piece of wood is resting on a horizontal surface so smooth that it is frictionless. A 10-g bullet is fired horizontally into it at a velocity of 800 m/sec and is embedded in the wood. With what velocity do block and bullet move from the block's resting position?

Ans. 1.99 m/sec.

6. Convert −40°C into degrees Fahrenheit.
7. Calculate the number of kilocalories required to raise the temperature of 4 liters of water 10 degrees on the centigrade scale.
8. How many grams of lead can be contained in a volume of 1000 cc? How much aluminum?
9. A German nurse reported that a patient's temperature was 39°C. Was the patient sick or healthy?

CHAPTER TWO

Atoms and Elements

The atomic theory, born in the speculations of ancient philosophers, has in this century become one of the great accomplishments in the history of human thought. In the previous chapter we analyzed briefly the bewildering varieties of motion and we saw that motions of real objects can in principle be resolved into three elements, mass, distance, and time.

If matter is taken as anything that occupies space and has mass, then with matter we are faced with a variety surpassing even the variety of motions. Let us ask the same type of question about matter that we asked about motion: What are the simplest elements in terms of which matter can be analyzed?

The history of the atomic theory makes fascinating reading, and much has been written about it both by scientists and by historians of science. The reading list at the close of this chapter has several such histories, most in inexpensive paperback form. We leave to the discretion of teacher and student assignments in this area. What follows in this chapter is not a history, although some of the key experiments will be described. We start with the assumption that matter in any of its forms which are detectable by the senses (or such extensions of the senses as microscopes, weighing balances, photographic plates, spectroscopes, etc.) is constructed from invisible particles called atoms. Our major goal for the rest of this book will be to acquire some understanding of macroscopic forms of matter in terms of the structure and organizations of these building blocks. We shall see that we can build a hierarchy of organizations from just a few "elements" of matter, a hierarchy that will start, not with the atom but with subatomic particles. Quite by coincidence, just as there are three fundamental elements of motion, at the simplest level there are three "elements" of matter, the subatomic particles called the electron, the proton, and the neutron. Taking these three in varying proportions and numbers, we move to the next level of organization, the atom. Then we move to combinations of atoms which are called molecules or are designated in some other way we need not worry about now. The next level is that of the detectable, usually visible and weighable, sample of matter. Living things constitute the highest, most complex levels of organization, cells, tissues, and organisms.

ATOMS

Greek Speculation. In the fourth and fifth centuries B.C., two of the many questions Greek philosophers speculated about were whether permanence and stability could exist in the midst of change, and whether change itself was the only "permanent" feature of the universe. Many believed that there had to be something stable, something enduring. Some thought that this something really had to be matter and so they questioned how matter *should be* (not *is*) constituted if there is to be permanence. One line of thought, first introduced by Democritus (460–370 B.C.), considered matter to be made up of very small, invisible particles called atoms. The atoms may be moved around, as in the imperceptible wearing away of a marble staircase, but they are considered indestructable. If we take a piece of gold, for example, how many times can it be cut (at least mentally) with the new, smaller pieces still being gold. Can it be cut indefinitely? In other words, is matter continuously divisible or are there ultimate, discrete particles? Democritus and others of like mind asked questions such as this, and they believed that there is a limit, that there must be a particle of gold which cannot be cut at all. The Greek word for "not cut" is *atomas,* and from it we have the word "atom" for this uncuttable particle, this smallest sample of gold possible. Aristotle (384–322 B.C.) disagreed. He rejected this atomic theory and insisted that matter was indefinitely divisible. A student of Plato and the teacher of Alexander the Great, Aristotle achieved such a position of eminence among learned men that theories contrary to his own in any area of thought met considerable opposition. The idea of atoms went into a long eclipse, not to be seriously revived in the history of human thought until the sixteenth and seventeenth centuries.

The Idea of a Model in Science. The atomic notion of matter survived after its revival because an increasing quantity and variety of observations seemed to make better sense in terms of an atomic model of matter than in terms of a continuous model. *Model,* as used here, means a mental construct, often pictured on paper, that we devise to explain observations. Observations concerning air and other "airs" (gases), for example, were difficult to understand in any terms other than atoms, and not merely atoms but atoms in motion. Christina Georgina Rossetti composed these lines:

> Who has seen the wind?
> Neither you nor I:
> But when the trees bow down their heads,
> The wind is passing by.

She uses an observation, trees bowing, to support the idea of something that cannot be seen, the wind, but she begs the question "What is the wind?" We might answer in terms of a model and say "The wind is like . . . ," with the rest of the sentence constituting our model for wind. A good model must fit most of the visible data. At the same time it must offer both helpful explanations for other observations not seemingly connected with winds and fruitful predictions of hitherto unobserved information. The better the model performs, the more correct it will seem. What models, then, are useful in understanding atoms? How would we complete the sentence "An atom is like . . ."?

COMPOSITION OF ATOMS

Subatomic Particles. Whatever atoms are, they must be incredibly small. In terms of our best model, one cubic centimeter of gold contains about 6×10^{22} atoms of gold. In spite of their smallness, they can be "cut," something that no one could have discovered until machines capable of delivering the high energy needed were developed. When matter is subjected to high enough energies, streams consisting of particles can be detected moving away from it. Different kinds of matter can all be made to give off the *same* streams; it is as though the different kinds of matter were made from common pieces, and it is these that are called subatomic particles.

Three subatomic particles are important in the study of chemistry, the electron, the proton, and the neutron. Many other particles are known, but at this stage in the history of chemistry it is possible to construct a very satisfactory theory of chemical behavior by ignoring all but the three listed.

The most useful statements we can make about electrons, protons, and neutrons describe their behavior under various conditions. All three behave as if they have mass. When they stream away from the site of formation, their mass and motion together give them both momentum and energy, quantities that can be detected, quantitatively measured, and used to calculate other properties. Two of the particles, the electron and the proton, behave as though they are electrically charged. The charge on the proton is defined as one unit of positive charge. The electron has the other kind of charge, but it is of the same intensity. Hence we say the electron has one unit of negative charge. Since charged particles exert forces that can be measured, the detection and measurement of electrons and protons are relatively easy. The third subatomic particle, the neutron, has no charge, is much more difficult to detect, and not surprisingly was the last of the three to be discovered.

The masses of these particles as measured by indirect methods are summarized in Table 2.1. In grams they are so small that for convenience a new unit is defined, the atomic mass unit (amu). One atomic mass unit is equivalent to 1.66043×10^{-24} g, which is exactly one-twelfth the mass of one atom of carbon-12, the reference standard. Defined this way, the masses of the proton and the neutron, rounded off, are each 1 amu. On this scale the mass of the electron is $\frac{1}{1823}$ amu.

ATOMIC STRUCTURE

The Thomson Model. In 1904, before the discovery of either the proton or the neutron, J. J. Thomson proposed what others nicknamed the "raisin pudding" model for atomic structure. He reasoned that an atom must have enough positive

Sir Joseph John Thomson (1856–1940). At the young age of 28 this English physicist was honored by the Royal Society's electing him a Fellow. In the same year he became Cavendish Professor of Experimental Physics at Cambridge. In 1897 he announced that he had identified cathode rays as streams of negatively charged particles which he called corpuscles but which were later named electrons. For this and other discoveries he received the 1906 Nobel prize in physics.

charge somewhere in it to neutralize the negative charges of the electrons present. He thought of a sphere of positive electricity in which were embedded "corpuscles" of negative electricity (as he initially called electrons). Hence the analogy with a raisin pudding. Thomson even speculated that the electrons might be in motion in circular orbits about the center of the sphere, but that if their motions were too energetic they would fly out and the atom would disintegrate. This was one problem with the model, but there were others.

The Rutherford Model. Many features of Thomson's model came under scrutiny and criticism. It logically led, for example, to certain predictions about the behavior of very thin metal foils toward atomic-sized "bullets," such as alpha particles. Streams of alpha particles come from uranium or radium in rays initially called simply alpha rays because no one knew their nature, yet some name was needed. Ernest Rutherford identified alpha rays[1] as consisting of atomic-sized particles with a mass of 4 amu and a charge of 2^+.

The thinnest preparable metal foil is several atoms thick, and Figure 2.1 shows how a foil of Thomson atoms would appear in a cross-sectional, atomic-level view. Because electrons are very small and the positive charge is postulated not to be concentrated in any one particular portion of the sphere, the Thomson atom should present no formidable target to an alpha particle. A stream of alpha particles should go straight through, and some in the stream might be deflected through very small angles. Such are the predictions that would have to be made if the

Table 2.1 Three Subatomic Particles

Name	Mass* in grams	Mass* in atomic mass units (amu)	Electrical Charge	Common Symbols
Electron	9.1091×10^{-28}	$\frac{1}{1823}$	-1	${}_{-1}^{0}\beta$, β^-, e^-
Proton	1.67252×10^{-24}	1.007277	$+1$	p, p^+, H^+, or ${}_{1}^{1}H$
Neutron	1.67482×10^{-24}	1.008665	0	n, or ${}_{0}^{1}n$

* Values are for rest masses of the particles. In experiments designed to detect these particles, they are traveling at very high speeds approaching that of light. Under these conditions their masses are higher in accordance with relativity theory. In most uses the masses of the proton and the neutron are rounded off to 1 amu each. In symbols such as ${}_{-1}^{0}\beta$ or ${}_{1}^{1}H$ the top number is the mass in atomic mass units (rounded off), and the bottom number is the atomic number.

[1] They are now recognized as the nuclei of helium atoms.

Ernest Rutherford (1871–1937), the "father of nuclear science" and often called the founder of modern atomic physics, was born in New Zealand. He came to England in 1895 with a scholarship to work in the famous Cavendish Laboratory at Cambridge under J. J. Thomson, the Director. It was the year Wilhelm Röntgen at Würzburg, Germany, discovered X-rays. The next year was to see the discovery of radioactivity by Henri Becquerel. In 1898, at the age of 27, Rutherford accepted the post of Professor of Physics at McGill University, Montreal. Working on the nature of radioactivity, he characterized the alpha particle as having a charge of 2^+ and a mass of 4 amu. He also pioneered in unraveling radioactive disintegration series. In addition, he worked at using radioactive decay rates for calculating reasonable values for the age of the earth. The Nobel prize in chemistry went to him in 1908 for his work with alpha particles. Thus this highest honor came even before one of his most distinguished accomplishments, the theory of a nuclear atom (1911). In 1919 he succeeded J. J. Thomson to the Cavendish Chair of Physics at Cambridge, a post he held until his death. In 1925 he received the Order of Merit, England's highest distinction. In 1930 he was elevated to the peerage as Baron Rutherford of Nelson, which gave him the title Lord Rutherford.

Thomson model were correct. But predictions did not match observations. Hans Geiger and his student Ernest Marsden working in Rutherford's laboratory in Manchester found that when streams of alpha particles hit thin metal foils, most went straight on through, but a significant number were deflected at very large angles. Some were even deflected backward. When Geiger told Rutherford about the backward deflections of alpha rays, Rutherford was incredulous, remarking that it was almost as unbelievable as if one had fired a 15-inch artillery shell at a piece of tissue paper and it came back and hit the gunner. Rutherford reasoned that only

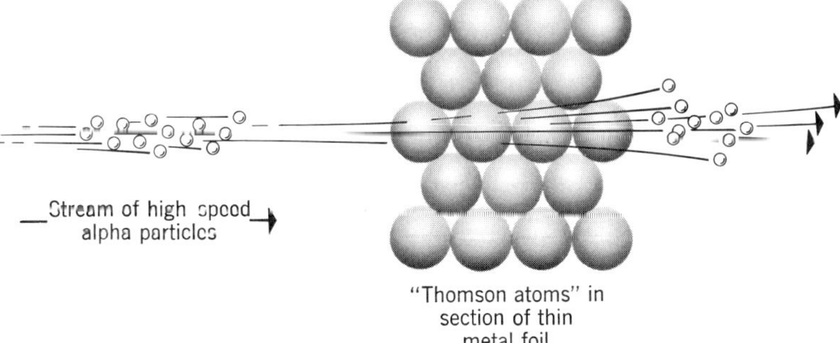
Stream of high speed alpha particles
"Thomson atoms" in section of thin metal foil

Figure 2.1 Predicted behavior of alpha particles bombarding a thin metal foil consisting of "Thomson atoms." Thomson envisioned the atom as a sphere of positive electricity in which electrons were embedded. Although a thin foil is several atoms thick, Thomson atoms would present no localized concentration of anything massive enough to turn back a high-speed alpha particle. At best, only a small fraction of alpha particles would be deflected and then only through small angles, a prediction not borne out by experiment.

something extremely massive could stop an alpha particle and turn it back. The fact that *most alpha particles went through* without deflection indicated that whatever an atom is, it must be mostly empty space. The fact that positively charged alpha particles could be *deflected* indicated that the deflector was like-charged. (Otherwise a near miss would cause no change in path direction.) By 1911 Rutherford was ready to propose a nuclear model of the atom, and with it the nuclear age was born.

Rutherford proposed that all the mass of an atom and all its positive charge were concentrated into one small, dense packet at the atom's center, the nucleus. Surrounding this nucleus at some distance from it were the electrons that made the whole particle electrically neutral. Just how they could be out there and not be attracted into the nucleus was left unanswered.

Electrons might possibly keep from collapsing into the nucleus by being in such rapid circular motion that the force attracting them in is balanced by the force throwing them out. But the electron is electrically charged. Classical theories of electricity and magnetism and experiences with electric currents had shown that whenever electricity is circulating, energy is radiating away from it in the form of electromagnetic radiation. If electrons were in motion about the nucleus in circular paths, they would constitute a miniature circulating electric current. Atoms should therefore constantly radiate energy, which is contrary to observation, and, contrary to experience, this energy should change smoothly. *Line* spectra, not continuous spectra, are observed, however, and then only under special circumstances. Moreover, as the electron loses energy by such radiation, it should spiral into the nucleus—again, contrary to observations. Atoms are more stable than that.

The Bohr Model. In 1913 a twenty-seven-year-old Danish physicist, Niels Bohr (working in Rutherford's laboratory at Manchester), proposed a tradition-shattering

---

Niels Henrik David Bohr (1885–1962), Danish physicist whose bold adventures in the realm of theory earned for him the Nobel prize in physics (1922) and the Atoms for Peace award (1957). His doctorate was taken at the University of Copenhagen (1911), after which he continued his studies at the Cavendish Laboratory at Cambridge, England, under J. J. Thomson and Ernest Rutherford. Einstein, in commenting on Bohr and his work, said that he was without question one of the greatest discoverers of our scientific age.

---

solution. Classical theories of energy and electricity were already under attack. Max Planck had introduced the idea of energy "quanta" or corpuscles of energy which he called "energy elements," and Albert Einstein had used this idea to explain for the first time a phenomenon (the photoelectric effect) that defied efforts to fit it into the molds of classical theory. With Planck's work the quantum age was born. Some of the grand old men of physics could not follow Planck and Einstein. Bohr was young enough that his thinking was not so rooted and grounded in classical traditions that he could not also think of alternatives to classical theories. Bohr started with two assumptions: first, that Rutherford was right, that all the positive charge and all the mass of an atom were in a nucleus and all the electrons were somehow, somewhere outside this; and second, that Planck and Einstein were right, that energy could be emitted or absorbed by matter in "packets" or quanta. Bohr also had to start with two indisputable facts: first, that the atom is reasonably stable, and second, that the atom does not continuously radiate energy under normal circumstances.

Bohr's first postulate was that an atomic system of a proton and an electron—to begin with the simplest possible atomic system—possesses several *states* or arrangements in which no electromagnetic energy is radiated or absorbed even if the electron is in motion relative to the proton. As long as the electron stays in a particular state, the atom neither picks up nor releases energy. In energy terms, it is in a *stationary state*. This postulate amounted to the rejection of classical electrodynamic theory for electrons in atoms. It said, in effect, that classical theory may serve well enough for large systems, but in a system as small as an atom it is not correct.

If and when an atom makes a transition from one stationary state to another, it means that an electron has changed from one energy state or "level" to another, and that the atom has either emitted electromagnetic energy or absorbed it. (One way to make matter emit electromagnetic energy is to heat it. Iron, for example, when hot enough will "glow." As it glows its atoms, excited to higher energy states by the heat, emit energy and return to more stable, lower energy states.) The frequency of the emitted radiation is proportional to the difference in energy of the two states:

$$\nu \propto E_{\text{state 1}} - E_{\text{state 2}} \qquad \text{where } \nu = \text{frequency}$$

or

$$\nu \propto \Delta E \qquad \text{where } \Delta \text{ is a symbol commonly used to represent a difference between two absolute values}$$

The proportionality constant that will convert this relation to an equation is Planck's constant (cf. p. 21). Thus

$$\text{(2.1)} \quad \Delta E = h\nu \qquad \text{where } h = \text{Planck's constant}$$

One of the ways of detecting such radiated energy is by its effect on a photographic plate, each frequency being brought to bear on a different part of the plate by diffraction, as described on page 20 and in the next section.

The Spectrum of the Hydrogen Atom. The most striking success of the Bohr model of the hydrogen atom was the almost exact correlation of observed values with values calculated for the differences in energy between the postulated energy levels. We shall digress briefly here to study these energy differences, how they are measured, and how they support Bohr's theory.

If a sample of hydrogen is energized in the absence of air by heating it or by passing a spark through it, it emits electromagnetic radiations. This emitted light is focused through a slit onto a prism or something that substitutes for one (e.g., a diffraction grating). It is then allowed to fall on a photographic plate. When the plate is developed, it is found that the light does not affect it all over. Only certain sharply defined line images of the slit appear. Each line corresponds to a very narrow range of frequencies, so narrow in fact that the line is said to represent light of just one frequency or of one wavelength. In other words energized hydrogen, like sodium (p. 20), does not emit every color of the rainbow, only a few. The group of lines obtained in this manner is called the line spectrum of hydrogen.

One series of lines from excited hydrogen lies in the visible region (Figure 2.2). A Swiss mathematical physicist, Johann Balmer (1825–1898), had noted a regularity in the spacing of these lines. Their frequencies neatly fitted the equation

$$\text{(2.2)} \qquad \text{Balmer series:} \quad \nu = \frac{1}{\lambda} = R\left(\frac{1}{2^2} - \frac{1}{n^2}\right)$$

where ν = frequency
λ = wavelength
R = 109,677.7 (now called the Rydberg constant)
n = 3, 4, 5, 6, . . .

Until Bohr proposed his theory, the constant R was just a number that made the observed frequencies fit the equation in this form. When Bohr made calculations based on his postulates, the results were such that he could *predict* that one group of lines in the spectrum of hydrogen should fit the equation

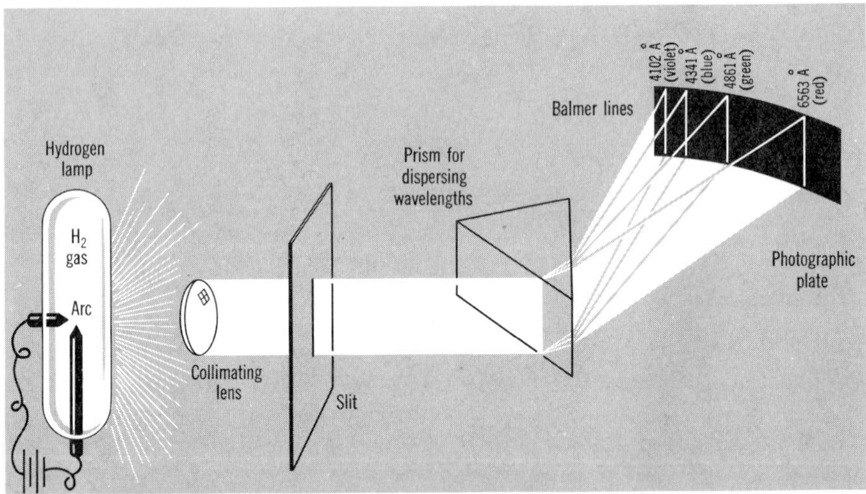
4102 Å (violet)
4341 Å (blue)
4861 Å (green)
6563 Å (red)
Balmer lines
Hydrogen lamp
H_2 gas
Arc
Prism for dispersing wavelengths
Photographic plate
Collimating lens
Slit

Figure 2.2 Colored line images of the slit appear on the photographic plate when the light from a hydrogen lamp is focused through it and onto a prism which disperses or separates the light into whatever frequencies are present. The Balmer series of lines, which occurs in the visible region of the electromagnetic spectrum, is shown here. Special prisms or diffraction gratings and specially sensitized photographic plates or other devices locate many other lines, some in the ultraviolet region and some in the infrared region, that belong to other series.

(2.3) Bohr's equation: $$\nu = \frac{1}{\lambda} = \frac{2\pi^2 me^4}{h^3}\left(\frac{1}{2^2} - \frac{1}{n^2}\right)$$

where m = mass of electron
e = charge on electron
h = Planck's constant
n = 3, 4, 5, . . .

The similarity between equation 2.2 (experimental) and equation 2.3 (predicted) is quite striking. The first is merely a device for storing the values for the observed frequencies of the Balmer series, but the second is a theoretically derived equation for predicting expected frequencies—expected if the assumptions are right. The two equations would be identical if only

(2.4) $$R = \frac{2\pi^2 me^4}{h^3} = \text{the Rydberg constant}$$

Using the best values for the mass m and the charge e of an electron plus Planck's h, all determined in independent experiments, Bohr found that the right-hand group of quantities in 2.4 calculated to a value of 109,750—less than 0.1% away from the Rydberg constant, 109,677.7! This result astonished the scientific world.

Besides the Balmer series, there are many other groups of lines in the spectrum of hydrogen, some in the ultraviolet and some in the infrared (Figure 2.3). The Balmer series corresponds to quantum "jumps" from energy levels higher than the *second* level down to that level. That is the significance of the 2 that is squared in the $1/2^2$ term of the Balmer equation 2.2. The values of n in the $1/n^2$ term in equation 2.2 correspond to higher energy levels. A general equation for all the series of spectral lines from hydrogen is given by the following equation:

(2.5) $$\nu = \frac{1}{\lambda} = R\left(\frac{1}{k^2} - \frac{1}{n^2}\right) \qquad n = (k+1), (k+2), (k+3), \cdots$$

when k = 1 Lyman series of lines
= 2 Balmer series
= 3 Paschen series
= 4 Brackett series

These series are also indicated on Figure 2.3. In this figure the curved sections are of circles (and do not overlook the smallest circle). Bohr postulated that his energy states, or energy levels, correspond to a circular path along which an electron can move if it is at the particular level. These paths he called orbits because of their obvious resemblance to orbits of planets about the sun (Figure 2.4). Bohr's model of the atom thus came to be called the planetary model, and it is unfortunate that so much emphasis has been placed on this feature of his theory. It does not really matter what kind of path or motion an electron may make. What is important is that only so-called allowed values of energy, not just any value, may be possessed by an electron. Energies can be measured, and only with measurements do we have science.

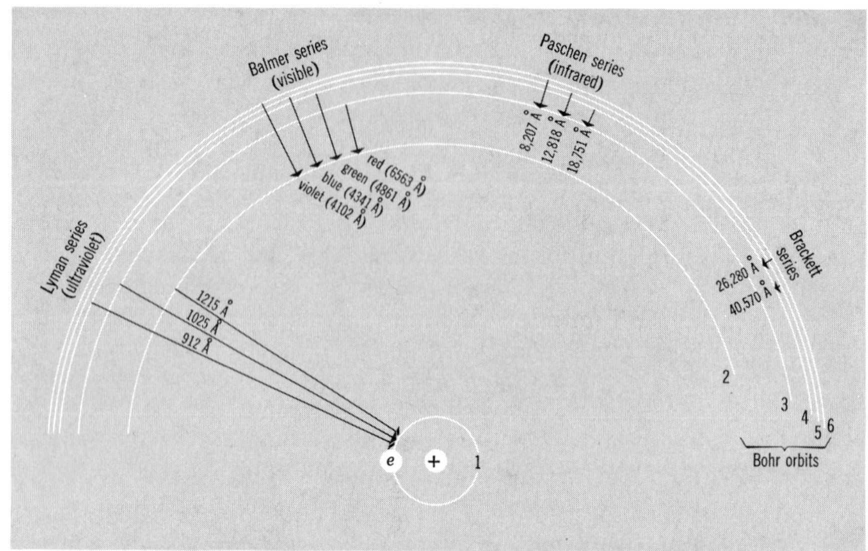
Balmer series
(visible)
Paschen series
(infrared)
red (6563 Å)
green (4861 Å)
blue (4341 Å)
violet (4102 Å)
8,207 Å
12,818 Å
18,751 Å
Lyman series
(ultraviolet)
1215 Å
1025 Å
912 Å
Brackett
series
26,280 Å
40,570 Å
2
3
4
5
6
Bohr orbits
e
+
1

Figure 2.3 Line spectra of hydrogen as interpreted in terms of the Bohr model of the atom. When hydrogen atoms are thermally or electrically excited, their electrons move to higher energy levels. In the huge population of any actual sample of hydrogen, all the excited states will be represented. As electrons in these excited states return to lower states, the "jumps" may be one, two, or more states at a time. Those that terminate at the lowest energy level produce the Lyman series of lines in the spectrum, those that stop at the second level give the Balmer series, etc. (A fifth series, the Pfund, produced by electrons terminating at the fifth Bohr level, is not shown.)

The series of lines represented by equation 2.5 as well as in Figure 2.3 can all be calculated (and predicted) from equations derived from Bohr's postulates. The accuracy of these predictions galvanized the attention of scientists. Not only did Bohr have to be taken seriously when he proposed that the atomic world requires departures from classical physics, but Planck also had to be on the right track with his energy quanta. (It is interesting that Planck did not receive his Nobel prize until 1918, *after* the introduction of the Bohr theory and some eighteen years after Planck first proposed his theory.) Bohr's successful postulates, which assumed

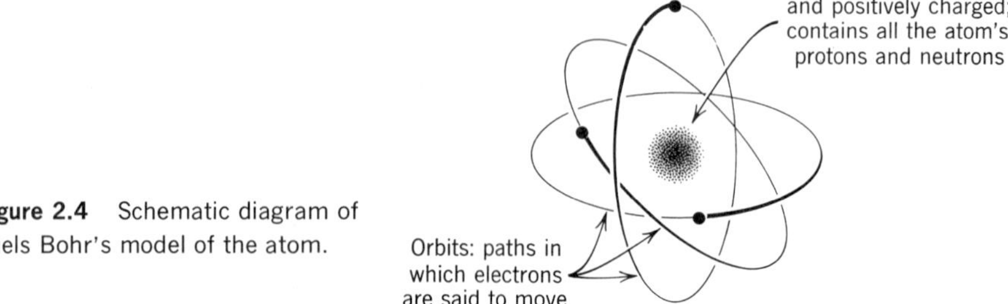
Nucleus: very dense
and positively charged;
contains all the atom's
protons and neutrons
Orbits: paths in
which electrons
are said to move

Figure 2.4 Schematic diagram of Niels Bohr's model of the atom.

Planck was right and which included Planck's constant, virtually demolished all opposition. It compelled belief in the basic quantum postulate. The year 1913, the year of Bohr's theory, was one of the most important in the history of scientific thought. It was a climax to several decades of revolution in science. Just to list a few developments dramatizes the point:

1887	Hertz discovers the photoelectric effect.
1895	Röntgen discovers X-rays.
1896	Becquerel discovers radioactivity.
1897	Thomson discovers the electron.
1900	Planck proposes the idea of an energy quantum.
1905	Einstein explains the photoelectric effect in terms of Planck's quantum hypothesis, working all the while on the theory of relativity.
1907	Thomson proposes his model of the atom.
1909	Geiger, Marsden, and Rutherford discover scattering of alpha particles.
1911	Rutherford introduces the nuclear atom.
1913	Bohr proposes his model of the atom.

In Bohr's planetary or solar system model of the hydrogen atom, the orbits were assumed to be circular. This model did not work out too well for atoms more complicated than hydrogen, although Arnold Sommerfeld, a German physicist, rescued the theory for a time by showing that electron orbits could be elliptical as well as circular. But the scientific world was in turmoil, and half measures would not do. Hard on the heels of the Bohr theory came new ideas. The basic postulate of Bohr, that electrons could "reside" in an atom only in allowed energy states, was retained, but the solar system model was eclipsed.

The Orbital Model. During the 1920s scientists relinquished the idea of orbits. The conception of an electron as a very hard, very small particle also lost some of its strength. In the search for an equation or a set of equations that would permit the calculation of energy states, there emerged from the work of de Broglie, Schrödinger, Heisenberg, and many others a method for calculating, not how electrons moved

---

Louis Victor de Broglie (1892–) extended the dual character of light (wave and corpuscular) to matter. In 1924 he proposed that an electron in motion (as in a Bohr orbit) had a wave associated with it. C. J. Davisson, L. H. Germer, and G. P. Thomson found experimental evidence for this wave nature of particles in 1927. De Broglie received the 1929 Nobel prize in physics for his work.

Werner Heisenberg (1901–), German physicist, received the 1932 Nobel prize in physics for the creation of quantum mechanics and the mathematical apparatus that is a part of it. He is best known for his formulation of the uncertainty principle in 1927, which states that it is impossible to determine with full accuracy both the position and the momentum of a particle *simultaneously*.

Erwin Schrödinger (1887–1961), Austrian mathematical physicist, extended the ideas of de Broglie by supposing that the matter waves of the particles in an atom could be superimposed on one another. Although Bohr had introduced quantum numbers arbitrarily, Schrödinger found that to have physically real solutions to wave equations such numbers were essential. The Schrödinger equation, one of the most famous in all of science, is the best model we have for the hydrogen atom. He shared the 1933 Nobel prize in physics with Paul Dirac (1902–), another pioneer in the field of quantum mechanics, who did much to integrate the work of Heisenberg and Schrödinger.

---

but where they most probably would be when they had the allowed values of energy. Equations were found for calculating the probability that an electron would be at a particular location in relation to the nucleus. For each allowed energy state there was one such equation or one set of equations. For a particular state, at certain distances and angles from the nucleus, the probability would be high; at other distances and angles it would be very low. Regions with, for example, at least a 90% probability of having an electron of a particular energy were enclosed by an imaginary envelope, and these regions came to be called *atomic orbitals.*

Atomic Orbitals. An atomic orbital is a particularly shaped volume of the space near an atomic nucleus in which an electron having a particular allowed energy might most likely be. Several aspects of the idea of an atomic orbital must be emphasized. First, it is a particular portion of space. It does *not* have to have an electron in it. An orbital is a certain region in all the space enclosing an atomic nucleus. We think of it as having a definite shape. Shortly we shall look at these possible shapes, but each shape arises from wrapping an imaginary envelope about just that much of the space whose chances are arbitrarily high (e.g., at least 80 to 90%) of having an electron of a particular energy within it—if an electron is to have such energy at all and be associated with that nucleus. These shapes have been calculated with great precision for only one very simple system, namely one proton and one electron, the hydrogen atom. *Exact* calculations for a system of one proton and two electrons have not been possible, and there appears to be no hope that they can ever be made. (Approximate solutions, as contrasted with exact solutions, have been worked out.) Unfortunately, few interesting kinds of matter are as simple as hydrogen. In order to apply what has been found to be so exact for the hydrogen atom to more complicated systems, we must *assume* that the orbitals available for electrons in other systems are similar to the hydrogen orbitals. Over the years great confidence has been built in this assumption because it works so well. Thus we use hydrogen-like atomic orbitals for more complicated atoms.

Calculations for the system of one proton and one electron show that there are many allowed states. The electron can be at several energy levels without leaving the proton or collapsing into it. These energy levels, sometimes called shells, are given the *principal quantum numbers,* 1, 2, 3, 4, etc. Earlier, when they were called shells, they were lettered K, L, M, N, . . . , with K corresponding to 1, L to 2, etc. At the first level, number 1 or K-shell, there is only one orbital, and its "probability envelope" encloses a sphere whose center is the nucleus (Figure 2.5). The

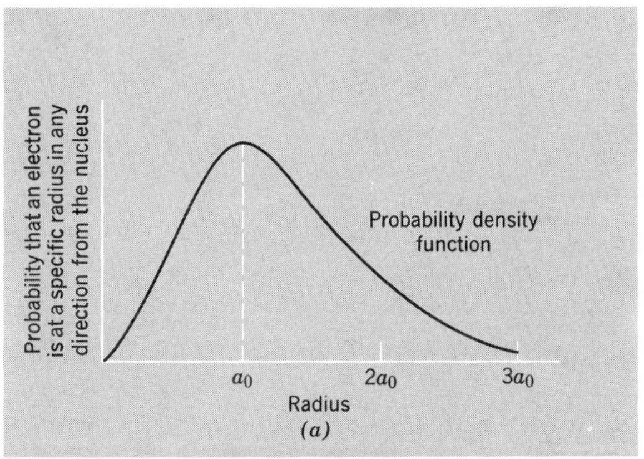
Probability that an electron is at a specific radius in any direction from the nucleus
Probability density function
a_0
$2a_0$
$3a_0$
Radius
(a)

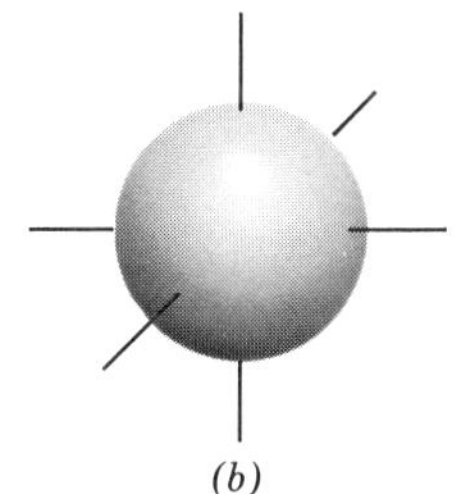
(b)

Figure 2.5 The 1*s* orbital. (*a*) Imagine the space surrounding a proton to be made up of layer on layer of infinitesimally thin, concentric, spherical shells. What is the probability that an electron will be located in each of these? This curve is the answer. At radius equal to a_0 (0.529 Å) the probability is a maximum. This result of calculations in wave mechanics corresponds exactly with the radius of the first orbit in the Bohr model.

(*b*) One of the microspheres just described will enclose a space within which the total probability of finding an electron is 90%. The space in this "probability envelope" is the 1*s* orbital. If the electron of a hydrogen atom is to be in its lowest energy level, it will occupy a space such as this.

envelope is spherically symmetrical. Whatever the probability is of finding an electron at a specific spot within the envelope, that is, at a specific distance *and direction* from the nucleus, the probability of finding an electron at another spot at the same distance but in any other direction will be exactly the same. Any orbital of this symmetry is called an *s orbital,*[2] specifically the 1*s* orbital. The fact that there is only one orbital corresponding to the first energy level means that one and only one equation for calculating the lowest energy of a system of one electron and one proton is discoverable from the postulates of Schrödinger. It is *as though* (and now we give a very crude model of limited usefulness) in a gymnasium *one* large basket is placed low on one wall (a basket, not an open net). Higher up is a horizontal row of baskets. A basketball higher up on the wall will have more energy, be at a higher level. If the basketball is to be somewhere on the wall and yet have the lowest possible energy, there is only one place it can be. There is only one "orbital" for it (temporary end of analogy—we shall pick it up again later).

At the next energy level, number 2 or the *L*-shell, there are four orbitals. An electron can have the next highest allowed energy in four ways. Four equations exist from which the same value of energy can be calculated and from which four orbitals can be drawn. It is *as if* (to return to our crude analogy) next up on the gymnasium

[2] We shall encounter the letters *s, p, d,* and *f* as names for certain kinds of orbitals. These symbols stand for the words "sharp," "principal," "diffuse," and "fine"—words that have particular meaning to spectroscopists. What that meaning is will not be explained, for it requires greater knowledge of the subject than we need for our purposes.

wall from the first basket there is a horizontal row of four baskets. A ball can be at this energy level in any one of four ways. And now the analogy has really become weak. The four hydrogenic orbitals at energy level 2 are not of identical shape. One of them does have the spherical symmetry we described earlier for the lone orbital at the first level. Therefore it is called an *s* orbital, but at energy level 2 it is the 2*s* orbital (Figure 2.6). The other three orbitals at level 2 have the shapes shown in Figure 2.6. These "dumbbell" shapes are identical, but their main axes are mutually perpendicular. An orbital with this dumbbell shape having symmetry around only one axis is called a *p orbital;* to give the three 2*p* orbitals separate names we call them $2p_x$, $2p_y$, and $2p_z$ orbitals (x, y, and z corresponding, of course, to the three mutually perpendicular axes).

At the third energy level there are nine orbitals, nine regions where an electron can have the energy of this level, and nine equations all leading to the calculation of the same energy. Each equation, however, leads to the calculation of a particular orbital shape. One orbital, the 3*s*, is spherically symmetrical; three have the symmetry and projections into space of the *p* orbitals, $3p_x$, $3p_y$, and $3p_z$. (They have the *symmetry,* but the actual shapes are somewhat different.) The remaining five, designated as *d* orbitals, have the shapes and orientations in space shown in Figure 2.7.

At the fourth energy level there are sixteen orbitals, 4*s*, $4p_x$, $4p_y$, and $4p_z$, plus five orbitals of the *d* type, plus seven orbitals designated as *f orbitals*. The orbitals

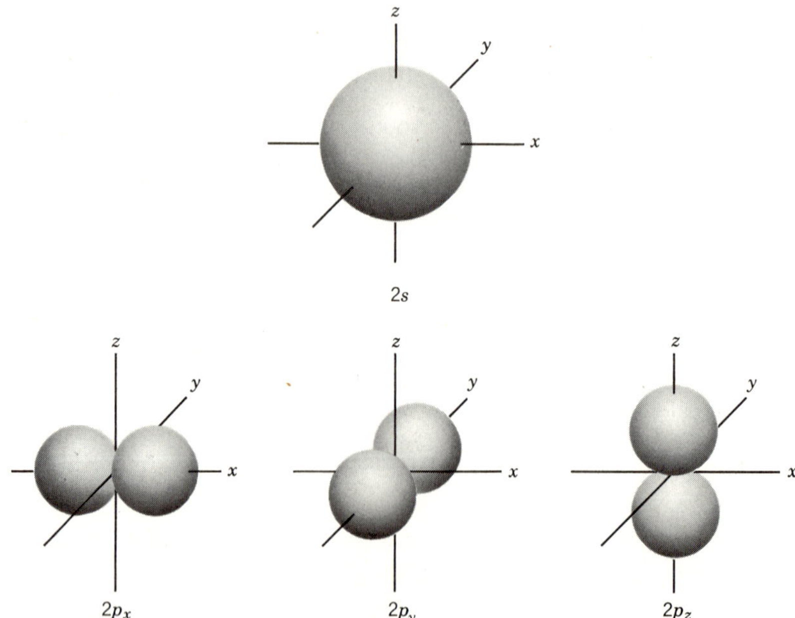
z
y
x
2s
z
y
x
2px
z
y
x
2py
z
y
x
2pz

Figure 2.6 Orbitals at principal energy level 2 (the *L*-shell). For the hydrogen atom the electron may be at the second level in any one of four ways, in the 2*s* orbital or in one of the mutually perpendicular 2*p* orbitals.

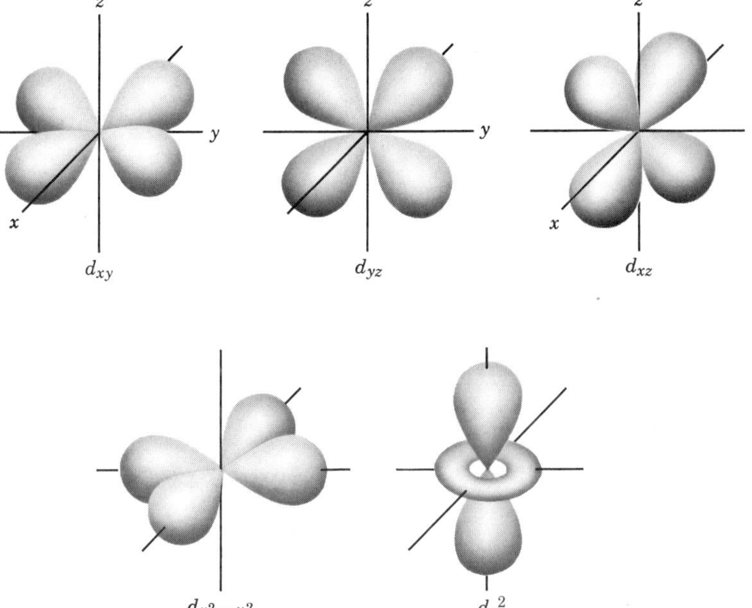
z
y
x
d_{xy}
z
y
d_{yz}
z
x
d_{xz}
$d_{x^2-y^2}$
d_{z^2}

Figure 2.7 Shapes of the $3d$ orbitals and their orientations in relation to each other and the same set of axis.

of the d and f types will scarcely be encountered in our future study. These orbitals are important to the chemistry of a large number of metallic elements, but in a survey of basic principles much can be learned without a detailed understanding of how f and d orbital electrons affect the chemistry of metals.

SPLITTING OF PRINCIPAL ENERGY LEVELS IN ATOMS OTHER THAN HYDROGEN

Calculations for the hydrogen atom, which has only one proton and one electron, result in the orbitals depicted in Figures 2.5, 2.6, and 2.7. We shall now introduce a necessary modification of this theory so that we can use these orbitals in connection with atoms having nuclei of higher charge and more electrons. For hydrogen and hydrogen alone, the four orbitals at the second energy level are of *equal* energy (and then only in the absence of a magnetic field). For other atoms the $2s$ orbital is of slightly less energy than any one of the $2p$ orbitals. It is *as if* a small splitting of the second energy level occurs so that in atoms more complicated than hydrogen there are two sublevels, the s and the p. At the third main energy level there are three sublevels, one orbital at the s, three at the p, and five at the d. The fourth energy level is split into four sublevels and, interestingly, the $4s$ level is of lower energy than the $3d$ for the first twenty elements. Figure 2.8 illustrates energy relations among the sublevels. (For hydrogen this splitting occurs only if the hydrogen atoms are subjected to an external magnetic field.) The reader may note that we have not explained how these splittings occur but have simply stated that

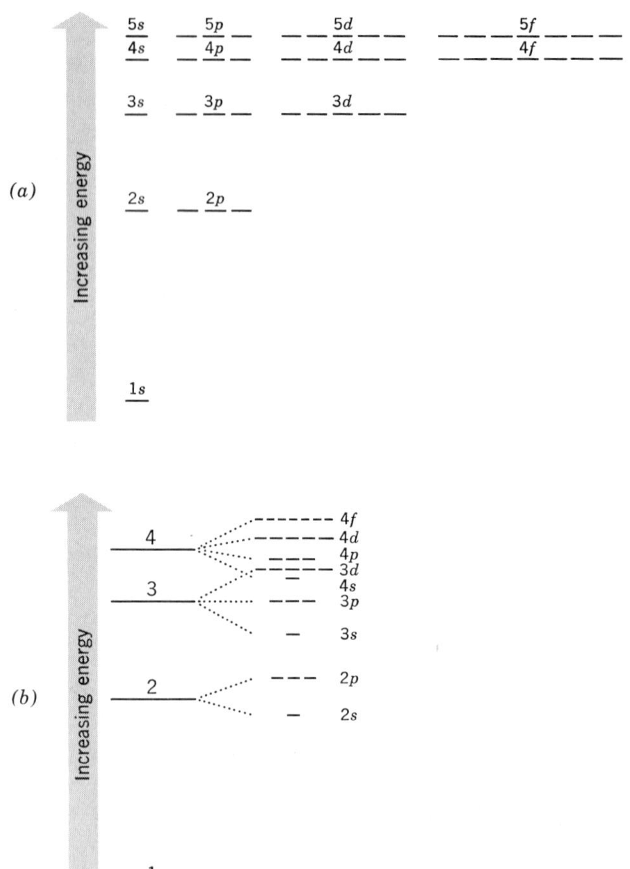
(a)
Increasing energy
5s
4s
3s
2s
1s
5p
4p
3p
2p
5d
4d
3d
5f
4f
(b)
Increasing energy
4
3
2
1
4f
4d
4p
3d
4s
3p
3s
2p
2s
1s

Figure 2.8 Energy levels. (*a*) The relative energies of the atomic orbitals of the hydrogen atom. Note that the sublevels are not "split," and the levels are therefore said to be degenerate.

(*b*) Relative energies of atomic orbitals when splitting of the principal energy levels occurs. The shapes of the orbitals (not shown here) remain basically as they are for hydrogen, but their relative energies are changed. (The splitting of the fifth level is not shown. No inferences should be made about the relative separations of the sublevels, for the figure implies no quantitative information.)

they do. Generally speaking, energy levels split because of repulsions between electrons when there are many in a system.

In our model for studying how electrons are distributed about a nucleus in a particular atom, we have a nucleus containing all (or virtually all) the mass of an atom and all the positive charge. Surrounding this core we have particularly shaped volumes of space or orbitals in which electrons are allowed to be. Only certain allowed values of energy are permitted for the electrons (Bohr's basic idea still holds). We have not given actual values to these energies (e.g., stating them in calories), but we have given them index numbers, the principal quantum numbers,[3] and letters s, p, d, and f. (If and when we need actual values we shall introduce them.) Our next step is to work from the simplest kind of atom, hydrogen, to the progressively more complex and examine just how electrons are distributed in them. To do this systematically, we conceive of a step-by-step *aufbau* (German, "building up").

Aufbau Principles. The most stable form of the hydrogen atom is one in which the electron is in the lowest allowed energy level, the $1s$ state. If it were in the $2s$ state, it could and eventually would jump to the $1s$ state and deliver a quantum of electromagnetic energy to the surroundings. Considering how we define stability and energy, spontaneous events in nature tend to go in the direction of increasing stability, and increasing stability is associated with minimum energy. If, in our hypothetical proton-electron combination, the electron jumped from the $2s$ to the $1s$ state, the combination would now have less energy than before (the difference being radiated) and it would be more stable. Thus the *first aufbau rule* for working out the distributions of electrons about a nucleus is that electrons are "fed" one at a time, first into the allowed orbitals of lowest energy and then successively up through higher levels as required. The order of filling, specified in Figure 2.9, follows the order of relative energies of the sublevels.

The *second aufbau rule* limits to two the number of electrons a particular orbital can have. It was first recognized by Wolfgang Pauli (1900–1958; Nobel prize in physics, 1945), an Austrian-born physicist, and is now known as the Pauli exclusion principle: an orbital can hold at most only two electrons and these only if they are spinning in opposite directions. Electrons behave as if they spin about an axis; to designate the two possible ways of spinning, clockwise or counterclockwise, arrows pointing in opposite directions are used, ↑ and ↓.

Another aufbau principle will be introduced soon, when we discuss the subject matter to which it is most relevant. The Pauli principle immediately places limits on the maximum number of electrons that can be part of a given principle energy level or any particular sublevel. These data are summarized in Table 2.2.

Table 2.2 Principal Energy Levels

Principal quantum number (energy level number)	1	2	3	4	5	6	7
Letter designation	*K*	*L*	*M*	*N*	*O*	*P*	*Q*
Maximum number of electrons (found in nature)	2	8	18	32	32	10	2

Maximum Population of Sublevels

s	2	(1 orbital)
p	6	(3 orbitals)
d	10	(5 orbitals)
f	14	(7 orbitals)

[3] Students who took particularly good physics and chemistry courses in high school or elsewhere will know that there are three other kinds of quantum numbers, the orbital, the azimuthal, and the spin quantum numbers. Our future needs for studying some aspects of the molecular basis of life do not depend on our understanding them. Still, they are implicit in the theory under discussion.

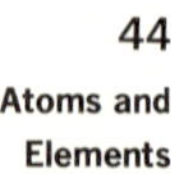

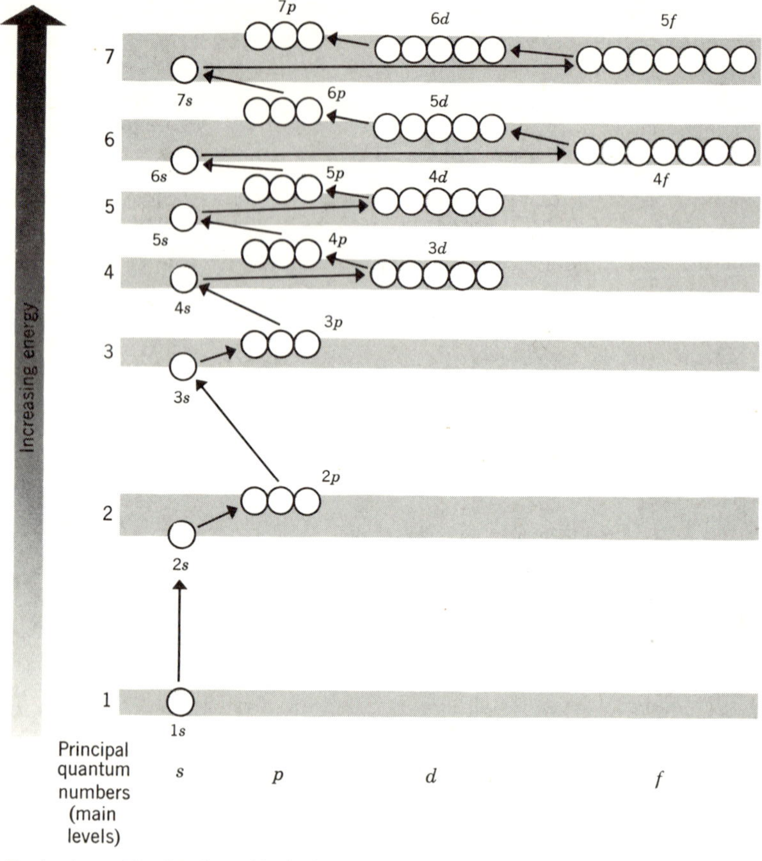
7p
6d
5f
7
7s
6p
5d
6
6s
5p
4d
4f
5
5s
4p
3d
4
4s
3p
3
3s
2p
2
2s
1
1s
Increasing energy
Principal quantum numbers (main levels)
s
p
d
f

Figure 2.9 Beginning with the $1s$ orbital, the general order of filling successive orbitals is shown by the paths of the arrows.

Electronic Configurations. Elements Defined. The most probable and the most stable (the lowest-energy) distribution of electrons among the available orbitals of an atom is called its electronic configuration. Only 103 fundamentally different kinds are now known, and of these only 89 occur naturally. The remainder have been prepared in the laboratory by means of costly electronic devices. When all the atoms present in a sample of matter have identical charges on their nuclei and identical electronic configurations, we call that substance an *element.* The atomic nuclei of different elements have different nuclear charges, and such elements have different electronic configurations. Each element may be identified by a number called its *atomic number* which is identical to the positive charge on the nucleus of one of its atoms. An atom is defined today to be an *electrically neutral* particle of one and only one nucleus together with enough electrons to make the whole particle neutral.

Part of the mass of an atom is made up of the protons present. An infinitesmal amount is contributed by the electrons, but this is ignored because it would take about 1823 electrons to provide the mass equivalent to one proton. No atom has more than 103 electrons. In every element except hydrogen the mass of the atom is greater than the sum of the proton masses; the rest of the mass is said to be made up of neutrons in the atom's nucleus. Since it is the size of the positive charge on the nucleus that determines the number of electrons in an atom, we shall study electronic configurations without regard to the mass of the atom, that is without regard to the number of neutrons present. Later we shall discuss the significance of the neutrons.

We shall use boxes to represent orbitals, and the higher up on the page a box is, the higher its energy. The electronic configuration of the hydrogen atom (atomic number 1) is represented simply as follows:

Hydrogen

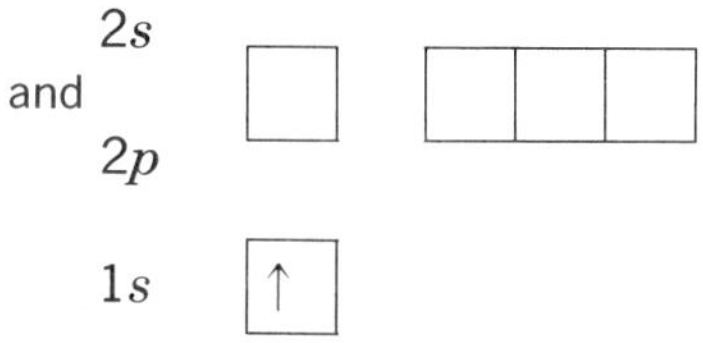

$2s$ and $2p$	[ ]	[ ][ ][ ]	In hydrogen atoms the $2s$ and $2p$ "sublevels" correspond to the same energy.
$1s$	[↑]		H, condensed symbol: $1s$

The condensed symbol for hydrogen is the customary way of writing its electronic configuration. The pattern will become clearer as additional examples are developed. (Some authors express $1s$ as $1s^1$, a practice we shall not follow.)

The element of atomic number 2 is helium. Having a nuclear charge of 2^+, its atoms must each have two electrons. According to the aufbau rules, its electronic configuration is

Helium

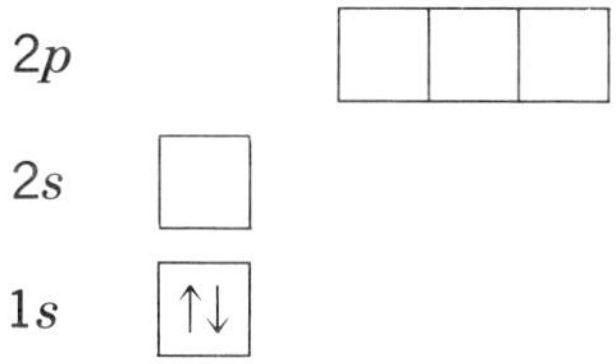

$2p$		[ ][ ][ ]	
$2s$	[ ]		
$1s$	[↑↓]		He, condensed symbol: $1s^2$

The condensed symbol says that there are two electrons (the superscript) in the $1s$ state.

Because the $1s$ state is now filled, the next electron will have to go into the next highest orbital, the $2s$. The electronic configuration of lithium, atomic number 3, is represented as

Lithium

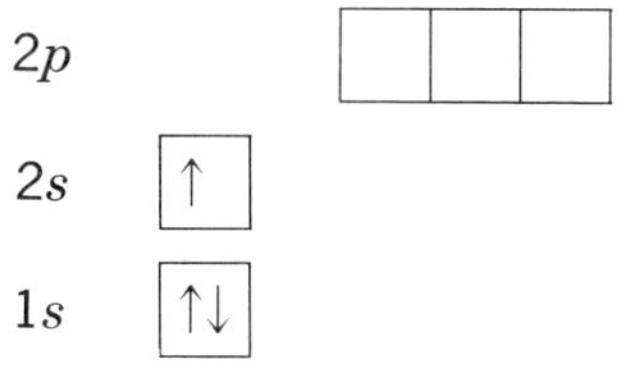

$2p$		[ ][ ][ ]	
$2s$	[↑]		
$1s$	[↑↓]		Li, condensed symbol: $1s^22s$

Carrying this development to atoms of higher and higher atomic number, in accordance with the aufbau rules and the order of filling orbitals (Figure 2.9), the following electronic configurations show how the second energy level is filled.

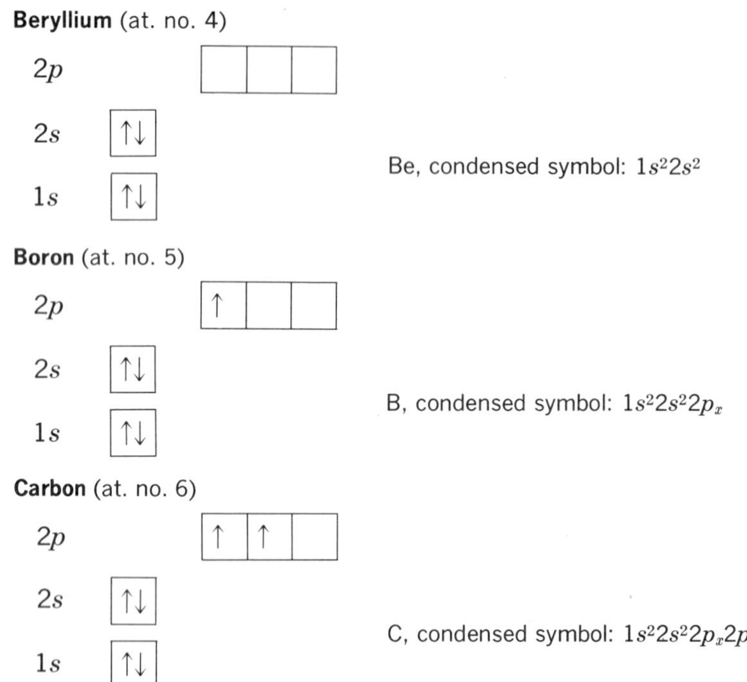
Beryllium (at. no. 4)
2p
2s ↑↓
1s ↑↓
Be, condensed symbol: $1s^2 2s^2$
Boron (at. no. 5)
2p ↑
2s ↑↓
1s ↑↓
B, condensed symbol: $1s^2 2s^2 2p_x$
Carbon (at. no. 6)
2p ↑ ↑
2s ↑↓
1s ↑↓
C, condensed symbol: $1s^2 2s^2 2p_x 2p_y$

With carbon another aufbau rule, known as Hund's rule, is applied. It says that if empty orbitals of the *same* energy are available, electrons will be distributed as evenly among them as possible. Carbon is therefore not shown as $1s^2 2s^2 2p_x^2$. The sixth and last electron is not put into the same p orbital as the fifth but in a different $2p$ orbital. In this arrangement it can be farther away from the fifth electron, repulsions between the two will be less, and the system is more stable. The Pauli exclusion principle *permits* two electrons (and no more) in the same space orbital, but it does not suggest that two paired electrons in such a circumstance have some special, extra stability. They do not, but the system is more stable that way than if the second electron of the pair were in an orbital of higher energy. Hund's rule states that the second electron will go to a different orbital of the *same* energy if one is vacant. We can now proceed to atoms of higher atomic number.

Nitrogen (at. no. 7)

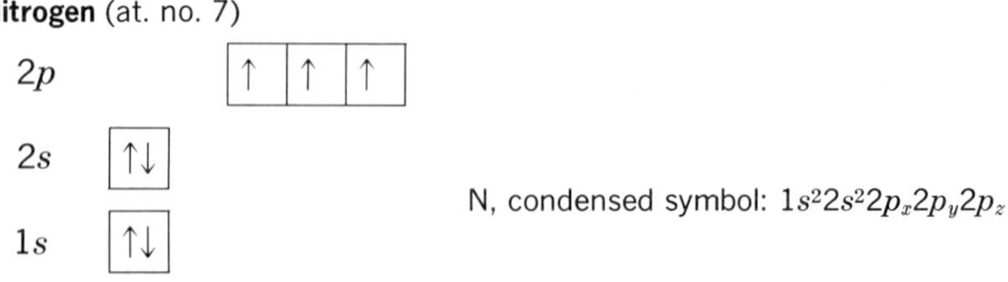
2p ↑ ↑ ↑
2s ↑↓
1s ↑↓
N, condensed symbol: $1s^2 2s^2 2p_x 2p_y 2p_z$

Oxygen (at. no. 8)

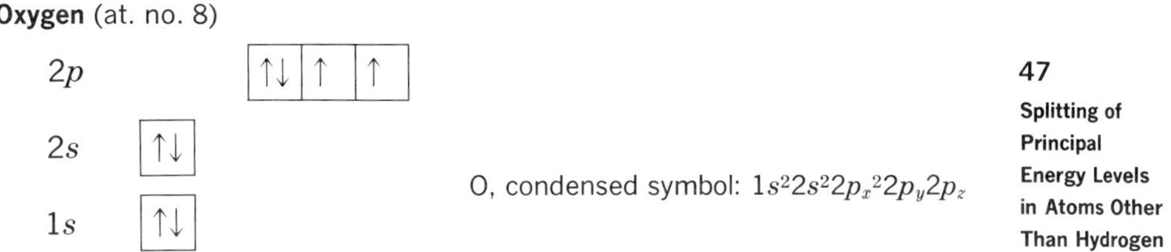
2p
2s
1s
O, condensed symbol: $1s^22s^22p_x{}^22p_y2p_z$

Note that the eighth electron is put into a $2p$ orbital with a spin opposite that of the electron already there. Although this arrangement means some crowding of like-charged particles, it is more stable than if the eighth electron were placed in the empty but higher-energy $3s$ orbital.

Fluorine (at. no. 9)

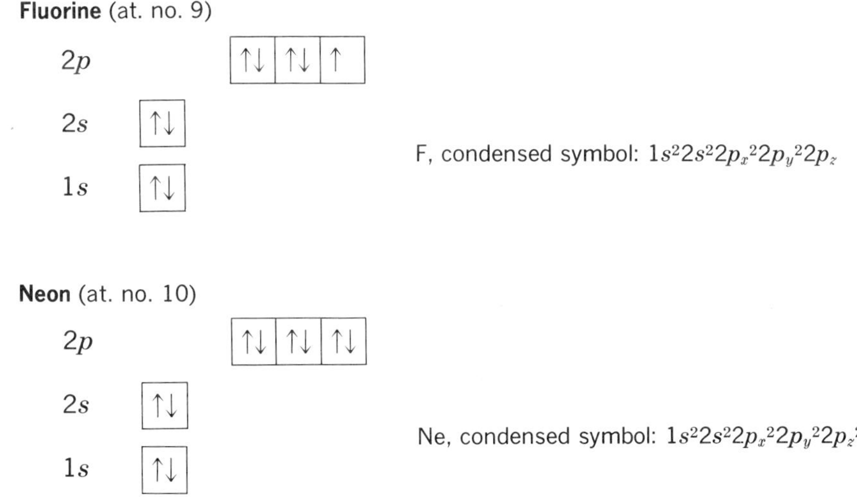
2p
2s
1s
F, condensed symbol: $1s^22s^22p_x{}^22p_y{}^22p_z$
2p
2s
1s
Ne, condensed symbol: $1s^22s^22p_x{}^22p_y{}^22p_z{}^2$

Neon (at. no. 10)

With this atom the orbitals in the second main level are filled.

For the element of atomic number 11, sodium, we shall have to start using orbitals of the third principal level.

Sodium (at. no. 11)

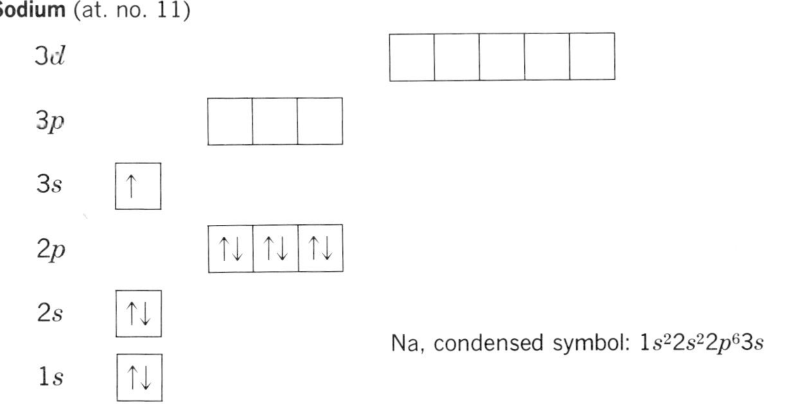
3d
3p
3s
2p
2s
1s
Na, condensed symbol: $1s^22s^22p^63s$

For convenience, we switch now to tabular organizations of the data (see Table 2.3). When we come to the element of atomic number 19, we must remember that the $4s$ orbital provides a lower-energy location for an electron than the $3d$ orbitals. Hence electron 19, and electron 20 when we get to the element of atomic number 20, will go into the $4s$ state. Only in elements of atomic number 21 and above are the $3d$ states filled. Table 2.4 summarizes the electronic configurations of all the 103 elements. It may be noted that a few exceptions to the aufbau rules occur; sometimes the orbitals do not fill up in the regular way predicted by these rules. These exceptions will be discussed later. By and large, we shall be interested primarily in elements numbered 1 to 20 because, except for a few trace minerals, it is among these that we find virtually all the elements important in the substances of the body.

Isotopes. The table of atomic weights on the inside of the back cover shows that only rarely are the masses of atoms stated in simple whole numbers when they are given in atomic mass units. But the masses should be whole numbers if the mass of an atom is simply the sum of the number of neutrons and protons, each with mass of 1 amu. Although more than one factor causes this departure from whole number values for atomic masses, we shall discuss only the major cause here, the existence of isotopes.[4]

An atom of a particular element must have only one electronic configuration for its lowest energy state and only one value for the size of the positive charge on the nucleus, but an element may consist of atoms having varying numbers of neutrons in their nuclei. Two (or more) atoms having identical nuclear charges but different numbers of neutrons are said to be isotopes. An element may therefore consist of atoms of slightly different atomic masses.

The element chlorine (atomic number 17), for example, consists principally of two isotopes designated as chlorine-35 and chlorine-37, or as $^{35}_{17}Cl$ and $^{37}_{17}Cl$, with the subscript the atomic number and the superscript the atomic mass. Although both have atomic number 17, the mass of the first is 35 amu, that of the second 37 amu.

Table 2.3 Electronic Configurations of Elements, Sodium to Potassium

Atomic Number	Element	$1s$	$2s$	$2p_x$	$2p_y$	$2p_z$	$3s$	$3p_x$	$3p_y$	$3p_z$	$3d$	$4s$
11	sodium	2	2	2	2	2	1					
12	magnesium	2	2	2	2	2	2					
13	aluminum	2	2	2	2	2	2	1				
14	silicon	2	2	2	2	2	2	1	1			
15	phosphorus	2	2	2	2	2	2	1	1	1		
16	sulfur	2	2	2	2	2	2	2	1	1		
17	chlorine	2	2	2	2	2	2	2	2	1		
18	argon	2	2	2	2	2	2	2	2	2		
19	potassium	2	2	2	2	2	2	2	2	2		1

[4] Another factor is the slight loss of mass that accompanies the packing of protons and neutrons into the small space of an atomic nucleus. Some of their mass is converted into a proportional amount of energy as given by Einstein's equation, $E = mc^2$ where m is mass and c is the velocity of light.

Table 2.4 Electronic Configurations of the Elements

Element	1s	2s	2p	3s	3p	3d	4s	4p	4d	4f	5s	5p	5d	5f	5g
1. H	1														
2. He	2														
3. Li	2	1													
4. Be	2	2													
5. B	2	2	1												
6. C	2	2	2												
7. N	2	2	3												
8. O	2	2	4												
9. F	2	2	5												
10. Ne	2	2	6												
11. Na	2	2	6	1											
12. Mg	2	2	6	2											
13. Al	2	2	6	2	1										
14. Si	2	2	6	2	2										
15. P	2	2	6	2	3										
16. S	2	2	6	2	4										
17. Cl	2	2	6	2	5										
18. Ar	2	2	6	2	6										
19. K	2	2	6	2	6		1								
20. Ca	2	2	6	2	6		2								
21. Sc	2	2	6	2	6	1	2								
22. Ti	2	2	6	2	6	2	2								
23. V	2	2	6	2	6	3	2								
24. Cr	2	2	6	2	6	5	1								
25. Mn	2	2	6	2	6	5	2								
26. Fe	2	2	6	2	6	6	2								
27. Co	2	2	6	2	6	7	2								
28. Ni	2	2	6	2	6	8	2								
29. Cu	2	2	6	2	6	10	1								
30. Zn	2	2	6	2	6	10	2								
31. Ga	2	2	6	2	6	10	2	1							
32. Ge	2	2	6	2	6	10	2	2							
33. As	2	2	6	2	6	10	2	3							
34. Se	2	2	6	2	6	10	2	4							
35. Br	2	2	6	2	6	10	2	5							
36. Kr	2	2	6	2	6	10	2	6							
37. Rb	2	2	6	2	6	10	2	6			1				
38. Sr	2	2	6	2	6	10	2	6			2				
39. Y	2	2	6	2	6	10	2	6	1		2				
40. Zr	2	2	6	2	6	10	2	6	2		2				
41. Nb	2	2	6	2	6	10	2	6	4		1				
42. Mo	2	2	6	2	6	10	2	6	5		1				
43. Tc	2	2	6	2	6	10	2	6	6		1				
44. Ru	2	2	6	2	6	10	2	6	7		1				
45. Rh	2	2	6	2	6	10	2	6	8		1				
46. Pd	2	2	6	2	6	10	2	6	10						
47. Ag	2	2	6	2	6	10	2	6	10		1				
48. Cd	2	2	6	2	6	10	2	6	10		2				
49. In	2	2	6	2	6	10	2	6	10		2	1			
50. Sn	2	2	6	2	6	10	2	6	10		2	2			
51. Sb	2	2	6	2	6	10	2	6	10		2	3			
52. Te	2	2	6	2	6	10	2	6	10		2	4			
53. I	2	2	6	2	6	10	2	6	10		2	5			
54. Xe	2	2	6	2	6	10	2	6	10		2	6			

Table 2.4 **Electronic Configurations of the Elements** *(continued)*

Element	K	L	M	4s	4p	4d	4f	5s	5p	5d	5f	5g	6s	6p	6d	6f	6g	6h	7
55. Cs	2	8	18	2	6	10		2	6				1						
56. Ba	2	8	18	2	6	10		2	6				2						
57. La	2	8	18	2	6	10		2	6	1			2						
58. Ce	2	8	18	2	6	10	2	2	6				2						
59. Pr	2	8	18	2	6	10	3	2	6				2						
60. Nd	2	8	18	2	6	10	4	2	6				2						
61. Pm	2	8	18	2	6	10	5	2	6				2						
62. Sm	2	8	18	2	6	10	6	2	6				2						
63. Eu	2	8	18	2	6	10	7	2	6				2						
64. Gd	2	8	18	2	6	10	7	2	6	1			2						
65. Tb	2	8	18	2	6	10	9	2	6				2						
66. Dy	2	8	18	2	6	10	10	2	6				2						
67. Ho	2	8	18	2	6	10	11	2	6				2						
68. Er	2	8	18	2	6	10	12	2	6				2						
69. Tm	2	8	18	2	6	10	13	2	6				2						
70. Yb	2	8	18	2	6	10	14	2	6				2						
71. Lu	2	8	18	2	6	10	14	2	6	1			2						
72. Hf	2	8	18	2	6	10	14	2	6	2			2						
73. Ta	2	8	18	2	6	10	14	2	6	3			2						
74. W	2	8	18	2	6	10	14	2	6	4			2						
75. Re	2	8	18	2	6	10	14	2	6	5			2						
76. Os	2	8	18	2	6	10	14	2	6	6			2						
77. Ir	2	8	18	2	6	10	14	2	6	7			2						
78. Pt	2	8	18	2	6	10	14	2	6	9			1						
79. Au	2	8	18	2	6	10	14	2	6	10			1						
80. Hg	2	8	18	2	6	10	14	2	6	10			2						
81. Tl	2	8	18	2	6	10	14	2	6	10			2	1					
82. Pb	2	8	18	2	6	10	14	2	6	10			2	2					
83. Bi	2	8	18	2	6	10	14	2	6	10			2	3					
84. Po	2	8	18	2	6	10	14	2	6	10			2	4					
85. At	2	8	18	2	6	10	14	2	6	10			2	5					
86. Rn	2	8	18	2	6	10	14	2	6	10			2	6					
87. Fr	2	8	18	2	6	10	14	2	6	10			2	6					1
88. Ra	2	8	18	2	6	10	14	2	6	10			2	6					2
89. Ac	2	8	18	2	6	10	14	2	6	10			2	6	1				2
90. Th	2	8	18	2	6	10	14	2	6	10			2	6	2				2
91. Pa	2	8	18	2	6	10	14	2	6	10	2		2	6	1				2
92. U	2	8	18	2	6	10	14	2	6	10	3		2	6	1				2
93. Np	2	8	18	2	6	10	14	2	6	10	5		2	6					2
94. Pu	2	8	18	2	6	10	14	2	6	10	6		2	6					2
95. Am	2	8	18	2	6	10	14	2	6	10	7		2	6					2
96. Cm	2	8	18	2	6	10	14	2	6	10	7		2	6	1				2
97. Bk	2	8	18	2	6	10	14	2	6	10	8		2	6	1				2
98. Cf	2	8	18	2	6	10	14	2	6	10	10		2	6					2
99. Es	2	8	18	2	6	10	14	2	6	10	11		2	6					2
100. Fm	2	8	18	2	6	10	14	2	6	10	12		2	6					2
101. Md	2	8	18	2	6	10	14	2	6	10	13		2	6					2
102. No	2	8	18	2	6	10	14	2	6	10	14		2	6					2
103. Lw	2	8	18	2	6	10	14	2	6	10	14		2	6	1				2

The lighter has seventeen protons and eighteen neutrons, the heavier seventeen protons and twenty neutrons. In nature these isotopes occur in a ratio of about 4:1. The average mass may then be estimated by the following simple calculation.

4 of chlorine-35	$4 \times 35 =$	140 amu
1 of chlorine-37	$1 \times 37 =$	37 amu
Total mass, five atoms		177 amu
Average mass, five atoms, $\frac{177}{5}$		35.4 amu

The measured value for chlorine is 35.453. Part of the small discrepancy is that the ratio is not exactly 4:1. The natural isotopic compositions of a few elements are shown in Table 2.5. These proportions are remarkably constant whatever the terrestrial source of the element sample. Isotopes are very common, for about 300 are known to occur naturally. In massive nuclear reactors physicists and chemists have succeeded in making about 900 more, some of which are highly useful in medicine.

Isotopes of the same element have virtually identical chemical properties because, as we shall see later, the chemical properties of elements are determined by their electronic configurations, not by their masses. Isotopes of the same element having identical atomic numbers must of course have identical electronic configura-

Table 2.5 Isotopes of Some Common Elements

Atomic Number	Element	Isotope	Percent Natural Abundance	Atomic Mass (in amu)*	Atomic Weight of Element
1	Hydrogen	$^{1}_{1}H$	99.985	1.007825	1.00797
	(deuterium)	$^{2}_{1}H$	0.015	2.01410	
	(tritium)	$^{3}_{1}H$			
6	Carbon	$^{12}_{6}C$	98.89	12.00000	12.01115
		$^{13}_{6}C$	1.11	13.00335	
7	Nitrogen	$^{14}_{7}N$	99.63	14.00307	14.0067
		$^{15}_{7}N$	0.37	15.00011	
8	Oxygen	$^{16}_{8}O$	99.759	15.99491	15.9994
		$^{17}_{8}O$	0.037	16.99914	
		$^{18}_{8}O$	0.204	17.99916	
29	Copper	$^{63}_{29}Cu$	69.09	62.9289	63.54
		$^{65}_{29}Cu$	30.91	64.9278	
34	Selenium	$^{74}_{34}Se$	0.87	73.9225	78.96
		$^{76}_{34}Se$	9.02	75.9192	
		$^{77}_{34}Se$	7.58	76.9199	
		$^{78}_{34}Se$	23.52	77.9173	
		$^{80}_{34}Se$	49.82	79.9165	
		$^{82}_{34}Se$	9.19	81.9167	
36	Krypton	$^{78}_{36}Kr$	0.35	77.9204	83.80
		$^{80}_{36}Kr$	2.27	79.9164	
		$^{82}_{36}Kr$	11.56	81.9135	
		$^{83}_{36}Kr$	11.55	82.9141	
		$^{84}_{36}Kr$	56.90	83.9115	
		$^{86}_{36}Kr$	17.37	85.9106	

* Based on 1961 standard, C = 12.00000 amu

tions. In Table 2.5 the last three isotopes of selenium (78, 80, 82) have essentially the same masses as the first three of krypton, illustrating again that it is atomic number rather than atomic mass that is important in determining chemical properties.

ELEMENTS

KINDS OF ELEMENTS. SYMBOLS

Many familiar substances are elements. Metals such as gold, silver, copper, platinum, iron, aluminum, chromium, nickel, zinc, mercury, tin, and lead are all elemental in nature. That is, all the nuclei present in samples of any one of them have identical charges. Other familiar substances such as the gases oxygen, nitrogen, neon, helium, hydrogen, and chlorine are also elements. Only two elements are liquids at room temperature, bromine and mercury; eleven are gases and the rest are solids.

The names of elements have been abbreviated to symbols whose usefulness will become clear when we study substances, compounds, having nuclei from two or more elements. Table 2.6 lists several of the more common elements, together with their symbols, and these must be learned.

FAMILIES OF ELEMENTS. THE PERIODIC LAW

Superficially it appears that the work nature gave the investigators of motion is easier than that given the investigators of matter. Motion reduces itself to only three observable "elements", mass, distance, and time. With matter, however, there are over thirty-three times as many, the chemical elements. If each of these were uniquely different from all the others, the undertaking for both the original investi-

Table 2.6 Symbols for Common Elements

B	Carbon	Be	Beryllium	Cl	Chlorine	Ag	Silver (*argentum*)
F	Fluorine	Al	Aluminum	Mg	Magnesium	Cu	Copper (*cuprum*)
H	Hydrogen	Si	Silicon	Mn	Manganese	Fe	Iron (*ferrum*)
I	Iodine	Ca	Calcium	As	Arsenic	Pb	Lead (*plumbum*)
N	Nitrogen	Li	Lithium	Cr	Chromium	Hg	Mercury (*hydrargyrum*)
O	Oxygen	Br	Bromine	Sr	Strontium	K	Potassium (*kalium*)
P	Phosphorus	Ba	Barium	Zn	Zinc	Na	Sodium (*natrium*)
S	Sulfur	Ni	Nickel	Pt	Platinum	Au	Gold (*aurum*)
		Ra	Radium			Sn	Tin (*stannum*)

Notes. The elements of the first column have the simplest symbols, the capitalized first letter of the name. The elements of the second column have symbols consisting of the first two letters. Whenever two letters comprise a symbol, the first is always capitalized and the second is always lowercase. (Beginning students often make mistakes on this point. Consider, however, the two symbols Co and CO. The first, Co, is the symbol for the element cobalt. The second, CO, is the symbol for carbon monoxide.) Some elements have names beginning with the same first two letters, for example, magnesium and manganese. The third column lists some examples for which the first letter and some letter other than the second are used. A few important elements were named when Latin was the universal language of educated men. Common examples of these elements whose symbols are derived from their Latin names are listed in the fourth column (with Latin names in parentheses).

gator of matter and the student would be massive. One of the great simplifying developments in chemistry, however, was the discovery that elements can be grouped together into a few families. Chemical elements that belong to the same family have quite similar chemical behavior. Since it is just such behavior that we are about to examine, a brief study of how these families are organized will help us take advantage of certain unifying principles.

The organization of the elements into families and the discovery of the periodic law spanned a century. It started with Johann Dobereiner of Germany in 1817 and reached its essentially final form with H. G. J. Moseley of England in 1913. Much of the work was done in the middle of the nineteenth century by John Newlands (England), Lothar Meyer (Germany), and Dimitri Mendeleev (Russia), who is usually credited with the greatest contribution.

Periodic Law. *Among the chemical elements, characteristic properties are periodic functions of their atomic numbers.* As the elements are arranged one after the other by increasing atomic number, every now and then in a fairly regular or periodic way properties very similar to those of lower-numbered elements reappear among those with higher numbers. This periodicity is apparent in both chemical and physical properties. Instead of smooth increases (or decreases) in the value of a particular property with increasing atomic number, wide fluctuations are apparent.

Physical and Chemical Properties and Changes: Preliminary Definitions. Some preliminary definitions are in order. The distinguishing characteristics of a substance are referred to as its *properties,* generally specified as two types, chemical and physical. *Physical properties* are those that can be observed and measured without inducing a chemical change. A *chemical change* will come to mean for us any reasonably enduring rearrangement or redistribution of electrons in relation to atomic nuclei. An observed event will be classified as a *physical change* if we have reason to believe that no deep-seated changes have occurred in the electronic configurations of the substances involved. It will be classified as a chemical change or a *chemical reaction* if we believe electrons have been rearranged or redistributed in relation to nuclei. Chemical reactions, by definition, result in the formation of new substances. It follows, therefore, that at least one physical property must have changed simultaneously, for otherwise we would not recognize that anything had happened. This physical change can be reversed only by a chemical process. Some of the common physical properties include color, luster, physical state (gas, liquid, solid), density, specific gravity, electrical conductivity, melting point, boiling point, ductility (can the substance be drawn into a wire), malleability (can it be hammered into a sheet), and transparency.

Periodicity and Physical Properties. To return to the changes in physical properties with increasing atomic number, Figure 2.10 is a graph of the boiling points of the first twenty elements versus their atomic numbers, and Figure 2.11 is a plot of ionization potentials of these same elements versus atomic numbers. The ionization potential is the energy required to remove an electron from an atom when it is in the gaseous state. How this is done is another story; the data are what interest us here. These values might also be arranged as follows:

			H				He
			314				567
Li	Be	B	C	N	O	F	Ne
124	215	191	260	336	314	402	497
Na	Mg	Al	Si	P	S	Cl	A
119	176	138	188	254	239	300	363
K	Ca						
100	141						

(The numbers beneath the atomic symbols are the ionization potentials in kilocalories per 6×10^{23} atoms, a number about which we shall say more later.) Where

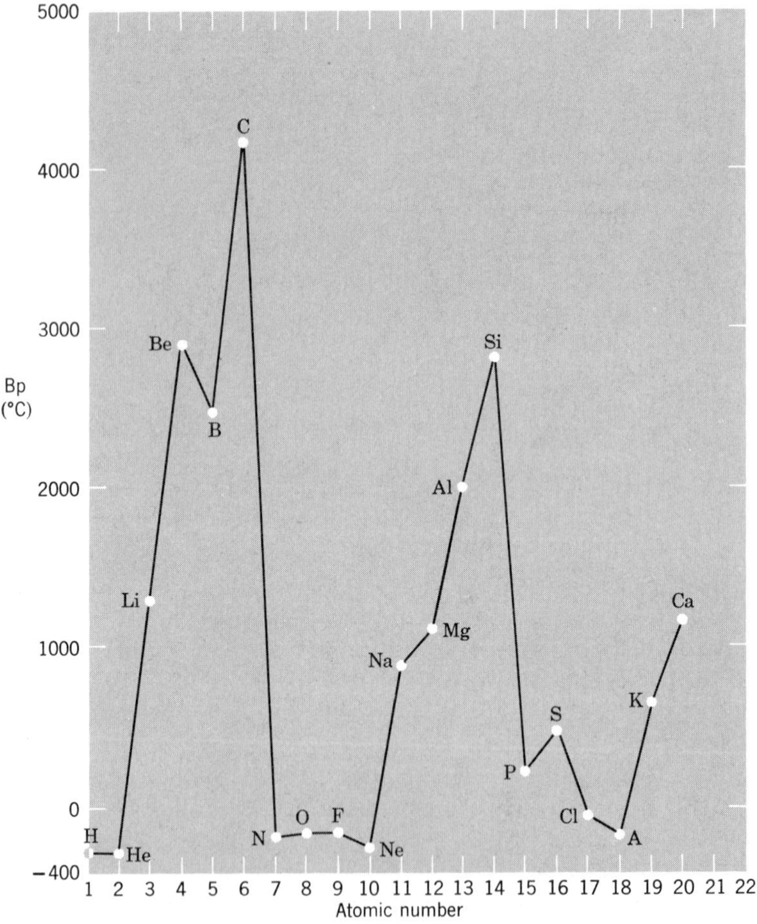
5000
4000
3000
2000
1000
0
−400
Bp
(°C)
C
Be
B
Li
H
He
N
O
F
Ne
Na
Mg
Al
Si
P
S
Cl
A
K
Ca
1 2 3 4 5 6 7 8 9 10 11 12 13 14 15 16 17 18 19 20 21 22
Atomic number

Figure 2.10 The boiling points of the elements (1 to 20) do not change smoothly, either up or down, with increasing atomic number. They fluctuate widely in a roughly periodic manner.

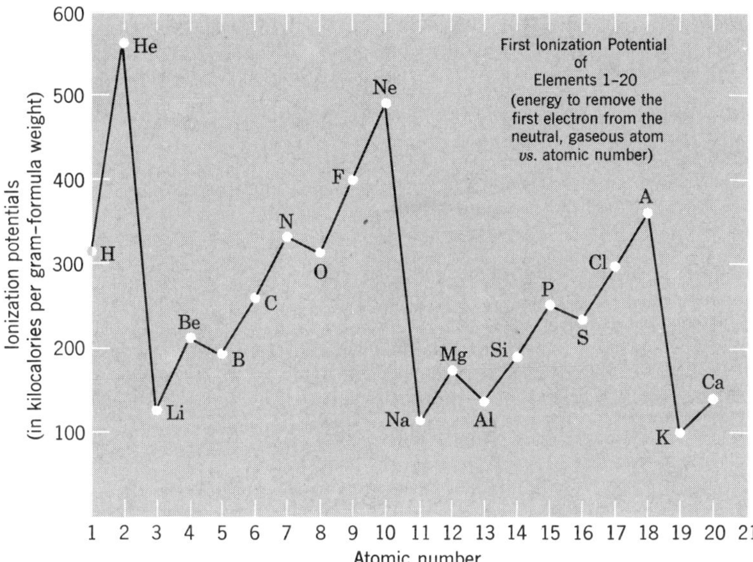
First Ionization Potential of Elements 1-20 (energy to remove the first electron from the neutral, gaseous atom *vs.* atomic number)
Ionization potentials (in kilocalories per gram-formula weight)
Atomic number
600
500
400
300
200
100
1 2 3 4 5 6 7 8 9 10 11 12 13 14 15 16 17 18 19 20 21
H
He
Li
Be
B
C
N
O
F
Ne
Na
Mg
Al
Si
P
S
Cl
A
K
Ca

Figure 2.11 The energy needed to remove one electron from an atom does not change smoothly, either up or down, with increasing atomic number. It fluctuates widely in a periodic manner. Plotted vertically is the energy to remove one electron, not just from one atom but from each of about 6×10^{23} atoms (one gram-formula weight).

the plot in Figure 2.11 has its biggest break, a new row of elements is started to make elements of similar physical properties appear in vertical columns.

More data besides these two physical properties would be necessary to make a strong case for this arrangement and the periodic law it implies. Not only would we need more physical and chemical information, but the work should be extended to include all the elements. The information has of course been collected and the work done. One of the modern forms of the periodic arrangement, a *periodic chart,* is displayed on the inside front cover.

Periodic Chart. Each box in the chart contains the following information:

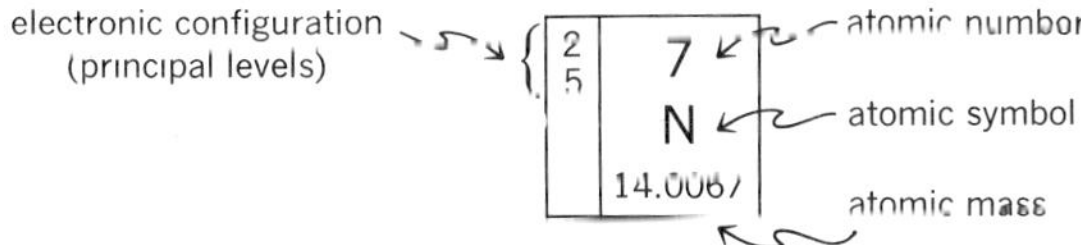
electronic configuration (principal levels)
2
5
7
N
14.0067
atomic number
atomic symbol
atomic mass

The numbers arranged vertically along the left-hand side of each box give from top to bottom the numbers of electrons in the principal energy levels (shells) from the lowest to the highest level (roughly, from the shell closest to the nucleus and out). Populations of sublevels are not shown. For this information, when needed, refer to Table 2.4.

As the elements are tabulated from left to right in order of increasing atomic number, rows are interrupted, new rows started, or both to make it possible for elements of similar properties to be placed beneath each other. *Thus each vertical column consists of one family of elements.* They may be called families or *groups,* with each identified by number and some by a descriptive name. Four of these groups are shown in Table 2.7. Portions of this table are shaded to call attention to the most important single feature of the groups. *Within* each family the shaded numbers, the numbers of electrons in the outside shells, are identical. (Exception: In group 0, the noble gases, the outside shell, the K-shell, of helium cannot hold more than two electrons.) *Between* families these numbers are different. We shall be able to trace similarities in chemical properties within a family to the fact that its members have identical numbers of electrons in their highest occupied principal energy levels.

Transition Elements. The horizontal rows in the periodic chart are called *periods.* The first period is very short, and hydrogen is not placed with any particular family. The second and third periods have eight elements each, but the rest of the periods are longer. The shaded portion of the chart in Figure 2.12 contains elements classified as the transition elements. Why it is reasonable for the periods to lengthen may be seen by referring to the electronic configurations of some of these elements. Those for elements 20 to 31 are given in Table 2.8. (Elements 20 and 31 are included for reference; they are not properly transition elements themselves.) Because the 4s state is of slightly lower energy than the 3d state, the configuration

Table 2.7 Electronic Configurations of Atoms in Several Common Families

Family	Element	Atomic Number	Electronic Configuration (Principal Levels)						
			K	L	M	N	O	P	Q
Group 0: Noble gases	Helium	2	2						
	Neon	10	2	8					
	Argon	18	2	8	8				
	Krypton	36	2	8	18	8			
	Xenon	54	2	8	18	18	8		
	Radon	86	2	8	18	32	18	8	
Group I: Alkali metals	Lithium	3	2	1					
	Sodium	11	2	8	1				
	Potassium	19	2	8	8	1			
	Rubidium	37	2	8	18	8	1		
	Cesium	55	2	8	18	18	8	1	
	Francium	87	2	8	18	32	18	8	1
Group II: Alkaline earth metals	Beryllium	4	2	2					
	Magnesium	12	2	8	2				
	Calcium	20	2	8	8	2			
	Strontium	38	2	8	18	8	2		
	Barium	56	2	8	18	18	8	2	
	Radium	88	2	8	18	32	18	8	2
Group VII: Halogens	Fluorine	9	2	7					
	Chlorine	17	2	8	7				
	Bromine	35	2	8	18	7			
	Iodine	53	2	8	18	18	7		
	Astatine	85	2	8	18	32	18	7	

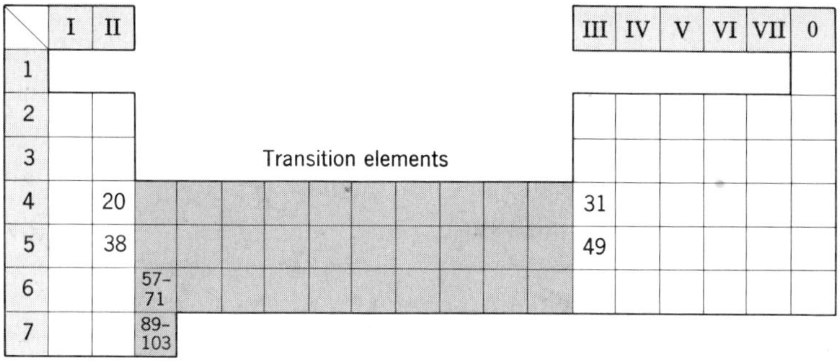

	I	II											III	IV	V	VI	VII	0
1																		
2																		
3			Transition elements															
4		20											31					
5		38											49					
6			57–71															
7			89–103															

Figure 2.12 Location of the transition elements in the periodic system.

of calcium shows an empty $3d$ sublevel and two electrons in its $4s$ sublevel. Now this sublevel is full. Because the $4p$ level is of higher energy than the $3d$, after calcium electrons go into the $3d$ sublevel, one by one, from scandium (21) to zinc (30). Apparently having a d sublevel exactly half filled, d ↑ ↑ ↑ ↑ ↑, represents a special condition of stability, for chromium (24) has at its outermost sublevels a distribution of $3d^5 4s$ instead of $3d^4 4s^2$, a configuration that would make the series

Table 2.8 Transition Elements, Fourth Period

	Element		Electronic Configuration									
			1	2		3			4			
At. No.	Name	Symbol	s	s	p	s	p	d	s	p	d	f
20	Calcium	Ca	2	2	6	2	6		2			
21	Scandium	Sc	2	2	6	2	6	1	2			
22	Titanium	Ti	2	2	6	2	6	2	2			
23	Vanadium	V	2	2	6	2	6	3	2			
24	Chromium	Cr	2	2	6	2	6	5	1			
25	Manganese	Mn	2	2	6	2	6	5	2			
26	Iron	Fe	2	2	6	2	6	6	2			
27	Cobalt	Co	2	2	6	2	6	7	2			
28	Nickel	Ni	2	2	6	2	6	8	2			
29	Copper	Cu	2	2	6	2	6	10	1			
30	Zinc	Zn	2	2	6	2	6	10	2			
31	Gallium	Ga	2	2	6	2	6	10	2	1		

from 21 to 30 look more uniform. Similarly, in copper (29) an entirely filled 3-shell rather than a filled 4*s* sublevel is the more stable arrangement. Copper has, at its outer fringes, a distribution of $3d^{10}4s$ instead of $3d^{9}4s^{2}$. In summary, for the transition metals in the fourth period of the periodic chart the 3*d* sublevel fills while the 4*s* level has one or two electrons.

Electronic configurations for the transition elements of the fifth period are given in Table 2.9. Again we have an inner sublevel filling, the 4*d* this time, while the outer sublevel, the 5*s*, has one, two, or no electrons. The 4*f* sublevel remains empty until the 5*p* and 6*s* sublevels fill up (elements 49 through 56). Then the 4*f* level starts filling again, and the elements with these configurations comprise the *lanthanide series,* named after the first member lanthanum (57). We note here that the order of filling of sublevels corresponds quite closely with the order of increasing energy as summarized in Figure 2.9. The lanthanide series is usually placed outside the main body of the periodic chart because it will not conveniently fit into the pattern that vertical columns list families. In a sense it is a separate family, sometimes called the *rare earths.* Another similar series is the *actinide series,* or the "transuranium elements," numbers 89 to 103. With these, as shown in Table 2.4, the following sublevels near the fringes are filled: the 5*s*, 5*p*, and 5*d*, the 6*s* and 6*p*, and the 7*s*. At element number 88 (radium) the 5*f* and the 6*d* and 6*f*

Table 2.9 Transition Elements, Fifth Period

	Element		Electronic Configuration													
			1	2		3			4				5			
At. No.	Name	Symbol	*s*	*s*	*p*	*s*	*p*	*d*	*s*	*p*	*d*	*f*	*s*	*p*	*d*	*f*
38	Strontium	Sr	2	2	6	2	6	10	2	6			2			
39	Yttrium	Y	Krypton configuration for all inner sublevels of these elements								1		2			
40	Zirconium	Zr									2		2			
41	Niobium	Nb									4		1			
42	Molybdenum	Mo									5		1			
43	Technitium	Tc									6		1			
44	Ruthenium	Ru									7		1			
45	Rhodium	Rh									8		1			
46	Palladium	Pd									10					
47	Silver	Ag									10		1			
48	Cadmium	Cd									10		2			
49	Indium	In									10		2	1		

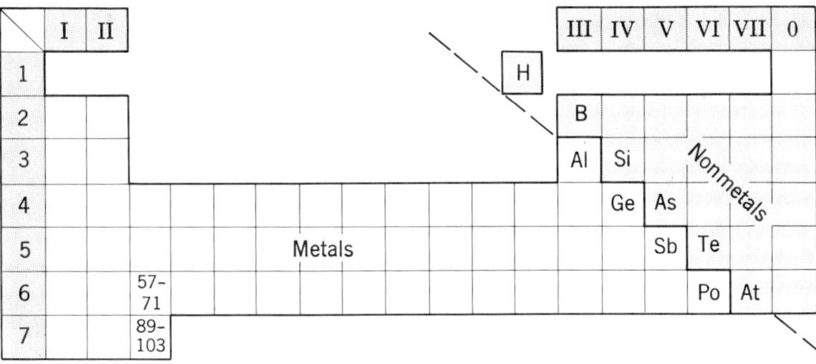
I
II
III
IV
V
VI
VII
0
1
2
3
4
5
6
7
H
B
Al
Si
Ge
As
Sb
Te
Po
At
Nonmetals
Metals
57–71
89–103

Figure 2.13 Locations of the metals and nonmetals in the periodic system of the elements.

are empty. In the actinide series the $5f$ sublevel fills while the $6d$ sublevel holds one, two, or no electrons.

In general, the transition elements all have one, two, or no electrons in the s orbitals while the d or f orbitals at lower levels are filling.

Metals and Nonmetals. If the periodic chart is divided in two by the stepped line shown in Figure 2.13, the portion to the left of the line includes all the metals, that to the right the nonmetals. Actually, the border line is not this sharp, for the elements along it have some metallic properties and some nonmetallic properties. Thus carbon is very brittle but it conducts electricity. Elements on the border line are sometimes called *semimetals,* sometimes metalloids. All the gaseous elements are nonmetals. If we consider only the principal energy levels and ignore for the moment the sublevels, the elements that have but one, two, or three electrons in their highest principal levels are metals. Those with four to eight electrons are generally nonmetals. Perhaps the most characteristic property of metals is that they are good conductors of electricity. Moreover, this conductivity *decreases* slowly with increasing temperature. The nonmetals as a group are very poor conductors of electricity, and their conductivities, usually very small, *increase* with rising temperature if they are measurable.

ATOMIC WEIGHTS AND FORMULA WEIGHTS. THE MOLE CONCEPT

Atoms, being so small, obviously cannot be counted one by one; yet in our study of chemistry we shall often need to accomplish an effect similar to counting them. Chemists count particles by weighing chemicals, which is basically what the mole concept is all about. The mole concept allows us to count invisible particles by weighing large collections of them, but we must begin with an understanding of atomic weight.[5]

Atomic Weights. The atomic weight of an element is a number that provides a way to compare the weight of an atom of that element with the weight of some

[5] From here on we lapse into the usage common to chemists and speak of weights when masses are meant. If we use standard masses as counterbalances for one pan of an equal-arm balance, the result is the measurement of mass. The reader is referred to the discussion of this point on page 11.

reference standard. An atomic weight is a *relative* weight. It does not refer to the actual weight of an individual atom. The reference standard since 1961 has been the carbon-12 isotope which is *arbitrarily* assigned the relative weight of 12.00000. On this scale the relative weight of the lightest atom, hydrogen, is 1, the smallest whole number. This means that one atom of hydrogen is one-twelfth the weight of one atom of carbon-12. Similarly, atoms of oxygen, atomic weight 16, are sixteen-twelfths the weight of one carbon-12 atom. Thus the atomic weight value provides a numerator for an "understood" denominator of 12. If carbon had been assigned a relative weight of 6.00000, then hydrogen, one-twelfth as heavy, would have had a relative weight of 0.5. There would be nothing fundamentally wrong in using this value and this scale, but it is not quite so neat and tidy as the system that has been adopted.

The *actual* weight of a carbon-12 atom is about 2×10^{-23} g by indirect measurements. The atomic weight of sulfur is 32; therefore one atom of sulfur is thirty-two twelfths the weight of one atom of carbon-12, or

$$\tfrac{32}{12} \times (2 \times 10^{-23}\ \text{g}) = 5.3 \times 10^{-23}\ \text{g}$$

These calculations dramatize the point that the numbers listed in a column of atomic weights (see the inside back cover) are not actual weights of anything; they are relative weights. If one atom of carbon weighed 12 tons, one atom of hydrogen would weigh 1 ton, because *in relation to carbon* hydrogen is one-twelfth as heavy.

Instead of one ton let us inquire about one gram. "Gram" is a unit of choice for scientists; "one" corresponds to something familiar about hydrogen, its relative weight, that is, its atomic weight. For carbon, we would inquire about 12 grams. Our basic question is this: How many atoms will give us 1 gram of hydrogen? If we find that number, the same number of carbon atoms will give 12 grams of carbon. (Carbon atoms are twelve times as heavy.) That number has been measured and its value is

$$6.0238 \times 10^{23} \qquad \text{(Avogadro's number)}$$

Named after Amedeo Avogadro, an Italian physicist who first suggested ways of measuring equal numbers of atoms or molecules, *Avogadro's number* refers to 6.0238×10^{23} in the same way that the word *dozen* refers to 12, or the word *million* stands for 1×10^{6}. Avogadro's number is therefore as much a pure number as others to which we have given special names:

12	dozen
20	score
1×10^{6}	million
1×10^{9}	billion
1×10^{15}	quadrillion
1×10^{21}	sextillion
6.0238×10^{23}	Avogadro's number

The number need not have any physical or chemical significance. We could speak of Avogadro's number of stars, for example. But when we speak of Avogadro's number of atoms of a particular element, we are talking about a collection of its atoms that is large enough to be weighed and is exactly large enough that the collection weighs *in grams* something familiar about the element, its atomic weight.

The Mole. The mole is the scientist's fundamental unit for "amount of substance." In 1965 the Commission on Symbols, Terminology and Units of the International Union of Pure and Applied Chemistry (IUPAC) recommended that the IUPAC endorse the following definition:

> "A mole is an amount of a substance, of specified chemical formula, containing the same number of formula units (atoms, molecules, ions, electrons, quanta, or other entities) as there are atoms in 12 grams (exactly) of the pure nuclide ^{12}C."

In other words, one mole contains Avogadro's number of formula units. One mole of sodium, formula Na, contains the same number of sodium atoms (Avogadro's number) as there are carbon atoms in one mole of carbon, formula C. In general, *equal numbers of moles contain equal numbers of formula units.*

To prepare for situations in which formula units are not atoms, we shall adopt the general policy of using the expression *formula weight* instead of atomic weight or molecular weight. The formula weight of carbon is 12, that of argon 40 (rounded off). A sample of argon weighing 40 grams will have the same number of atoms of argon as there are atoms of carbon in 12 grams of carbon-12. The formula weight refers to the relative weight of whatever kind of formula unit is involved—be it atoms, molecules, ion groups, etc.

In general, when a sample of a chemical of any size is weighed, the number of moles present will be found by the equation

$$\text{Number of moles present in sample} = \frac{\text{number of grams present}}{\text{gram-formula weight of substance}}$$

where "gram-formula weight" is simply the formula weight given in grams. Stated more briefly,

$$\text{Number of moles} = \frac{\text{grams}}{\text{gram-formula weight}}$$

RADIOACTIVITY

Nuclei of the transuranium elements are unstable, making these substances radioactive. Some of the important evidence for atomic structure came from studying radioactivity.

THE BASIS FOR ATOMIC RADIATIONS

The spontaneous processes of nature are sources of useful energy. Certain of them, such as waterfalls, are produced by universal gravitation. Others, for example, combustion of fossil fuels such as gasoline, stem from rearrangements of outer-shell planetary electrons into more stable configurations about atomic nuclei. (Virtually all the rest of the book will be devoted to this topic.) Chemical energy is possible because certain configurations of planetary electrons can achieve more stable arrangments. *Atomic energy is possible because certain configurations of particles in atomic nuclei can achieve more stable arrangements.* Atoms that possess such nuclei are said to be radioactive because, in the spontaneous process they undergo to achieve nuclear stability, they emit streams of radiation. The term *radiation,* ordinarily used in connection with light, is appropriate because, like light, these streams can affect photographic film. In fact, that is how radioactivity was discovered.

In 1896 the French physicist A. H. Becquerel chanced to store some photographic plates in a drawer containing samples of uranium ores. When the plates were developed it was as though they had been used to photograph fog. Becquerel might have blamed the accident on faulty plates or careless handling had it not been for the fact that X-rays, discovered the year before by Wilhelm Röntgen, were known to be very potent film foggers. Becquerel found that a natural source of radiation resembling X-rays had fogged his plates.

Further research by Becquerel revealed that these radiations were emitted by any compound of uranium but, gram for gram, most intensely by uranium metal itself. They were not produced by a chemical reaction, for the usual conditions that influence the speed of such a reaction—temperature, pressure, and state of chemical combination—had absolutely no detectable effect on the intensity of the radiation.

After several years of work two British physicists, Rutherford and Soddy, were able to explain radioactivity in terms of events happening in certain kinds of atomic nuclei that are unstable. In acquiring stability, these atomic nuclei undergo small disintegrations, and tiny nuclear bits or fragments, such as protons, electrons, and helium nuclei, are thrown out into space. An X-ray-like radiation sometimes appears also.

As a result of most radioactive disintegrations, the size of positive charges on nuclei change. The resulting atoms are obviously different elements. Radioactive decay is therefore usually accompanied by what is called *transmutation* of an element.

RADIATIONS AND NUCLEAR REACTIONS. NATURAL RADIOACTIVITY

The three kinds of radiations from natural sources are alpha rays, beta rays, and gamma rays; these names were given before the rays had been more specifically identified.

Alpha Rays. Alpha rays consist of particles with a mass of 4 amu and a charge of 2^+, which means that each particle possesses two protons and two neutrons. In other words, the alpha particle is the nucleus of a helium atom. When initially emitted alpha particles are moving at very high speeds, upward of 10^9 cm/sec. (The speed of light is 3×10^{10} cm/sec.) After traveling in air for only a few centimeters, they are stopped by collisions with air molecules. They cannot penetrate cardboard. Should alpha particles strike *outside* the body, they are relatively harmless for they cannot get through the outer layer of dead cells on the skin. If they are produced inside the body through the agency of some accidentally ingested radioactive material, they can raise havoc among the body's chemicals with which they collide.

Beta Rays. These rays also consist of particles, in this instance electrons that have been produced within and then thrown out of nuclei. Since electrons are several thousand times smaller than alpha particles, they can more effectively penetrate matter before chance collisions stop them. Beta rays will penetrate living cells of the skin, for example, and the skin will appear "burned." They cannot get as far as internal organs *from the outside,* however. But if beta-emitting radioactive dust reaches the lungs, lung cells will obviously be damaged. Similarly, absorption of beta emitters in the diet may place these rays in many parts of the body.

Gamma Rays. Gamma rays do not consist of particles. They are identical with powerful X-rays, except that X-rays are artificially produced and gamma rays come from natural sources. When radioactive elements emit either alpha or beta rays, they usually send out gamma rays as well. Like powerful X-rays, they are extremely penetrating and can interact with living matter in lethal ways if the dosage is high enough.

Units of Radiation Measurement. There are two common ways of describing radiation measures, physical and biological. Physical measurement is related to the number of particles per second emitted by a source. The *curie,* named after Marie and Pierre Curie, is one standard, it is defined as a unit of activity equal to 3.7×10^{10} disintegrations per second, which is the number of disintegrations per second occurring in a one-gram sample of radium. In describing radioactive fallout, smaller units are normally used: the *millicurie* (10^{-3} curie), the *microcurie* (10^{-6} curie), and the *micromicrocurie* or more correctly the picocurie (10^{-12} curie).

With biological measurements we attempt to describe the amount of energy absorbed by a living substance. Tissue is damaged principally by the removal of electrons from the chemicals of cells or the addition of electrons to them. As we shall see in Chapter 3, it is possible for atoms and groups of atoms to gain or lose electrons and become negatively or positively charged. These new charged particles are called ions. This preliminary definition is given here because one of the special units of radiation dose is defined in terms of the number of pairs of ions produced.

The *roentgen* (symbol: r) produces approximately 2×10^9 ion pairs per cubic centimeter of air. The unit applies to X-rays or gamma rays only. If one roentgen is absorbed by living tissue instead of air, it generates about 1.8×10^{12} ion pairs per gram of tissue. Apparently, the number of ions produced by one roentgen varies in different types of substances.

For types of radiation other than X-rays or gamma rays other units are usually employed. The *rad* (radiation absorbed dose) corresponds to the absorption of 100 ergs[6] of energy by 1 gram of tissue. It is roughly the same as the roentgen. The *millirad* (mrad) is 10^{-3} rad. A 500-rad dose to the body is usually lethal. Even though this represents an infinitesimal amount of energy, it wreaks havoc not as heat but by causing the random formation of unstable ions among key body chemicals.

[6] The erg is an extremely small unit of energy. It takes over 40 billion ergs to make up 1 kilocalorie. Specifically, the erg is the energy resulting from the action of a force of one dyne over a distance of one centimeter, with one dyne being defined as the force needed to impart an acceleration of one centimeter per second per second to a mass of one gram.

A unit called the *rem* (roentgen equivalent for man) is sometimes used in describing radiation energy absorbed specifically by a human being. One rem is the quantity of ionizing radiation that generates, when absorbed by man, an effect equivalent to the absorption of one roentgen. One millirem (mrem) is 10^{-3} rem.

Different kinds of radiations vary in their *relative biological effectiveness* (RBE). Large, doubly charged alpha particles are particularly dangerous, for on a scale giving the RBE of beta rays a value of 1, alpha rays have an RBE of 20.

Nuclear Reactions. The most common isotope of uranium is uranium-238 or $^{238}_{92}U$. Being an alpha emitter, when one atom of uranium decays (emits an alpha particle), it loses two units of plus charge and four units of mass. The remaining particle will therefore have an atomic number of 90 and a mass of 234. It is now the 234-isotope of thorium. Such a change is called a *nuclear reaction.* (*Chemical* reactions, by definition, do not involve such changes in nuclear composition. They have to do only with electronic configurations.)

Nuclear reactions are best described in nuclear equations. Like chemical equations, a nuclear equation separates symbols for the initial substances from the final substances by an arrow pointing from the former to the latter. Unlike chemical equations, the symbols in a nuclear equation always carry numbers specifying atomic numbers and atomic masses. For example, the decay of uranium-238 is symbolized as

$$^{238}_{92}\text{U} \longrightarrow {}^{234}_{90}\text{Th} + \underset{\text{Alpha particle}}{^{4}_{2}\text{He}}$$

A nuclear equation is said to be balanced if the sum of the atomic numbers on the left equals that on the right ($92 = 90 + 2$) and the sum of the mass numbers on the left equals that on the right ($238 = 234 + 4$).

When a beta particle, an electron, is emitted, the net effect in the nucleus is to *increase* the plus charge by one unit. It is *as though* a neutron is made up of a proton and an electron; when the electron leaves, the proton is left. Departure of an electron amounts to increasing the number of protons by one. A neutron very likely is not simply a proton glued to an electron, but the net effect of beta emission is what has been described. The thorium-234 isotope produced in the previous nuclear reaction is a beta emitter:

$$^{234}_{90}\text{Th} \longrightarrow {}^{234}_{91}\text{Pa} + {}^{0}_{-1}\beta + \text{gamma radiation}$$

Note that the equation is balanced: $90 = 91 + (-1)$; $234 = 234 + 0$.

Comparing Nuclear Instabilities. Half-lives. Radioactive decay occurs when a nucleus is inherently unstable. New emerging theories that explain how some isotopes have more stable nuclei than others are currently being tested. In order to describe relative stabilities, whatever the theory that explains them, physicists have invented a measurement known as the *half-life, the time it takes for one-half of an initial quantity of radioactive material to decay.* The half-life of uranium-235, for example, is about 5 billion years. If, for example, we start with 100 g of ^{238}U, after 5 billion years there will be 50 g of ^{238}U left. The rest will have decayed and transmuted to other elements. Obviously, half-lives are not measured by waiting for one entire half-life period. From measurements of *rates* of decay over a brief period, it is possible to calculate the half-life. An isotope of radon, ^{222}Rn, has a relatively short half-life period, about 4 days. Of an initial 100 g, only 50 g will remain as ^{222}Rn after 4 days. After the next 4 days half of 50 g or 25 g of ^{222}Rn will be left, and in another 4 days only 12.5 g. Thus, after three half-life periods of 4 days each, only 12.5% of the original ^{222}Rn remains.

The enormous variations in nuclear stabilities are seen in the data of Table 2.10, which lists some common, naturally occurring isotopes and their half-lives. The half-lives of some rare isotopes are fractions of seconds.

Radioactive Disintegration Series. The decay of an atom of ^{238}U to ^{234}Th does not answer the requirements of its nucleus for stability. The nucleus of ^{234}Th is not stable either. It decays, and still a stable nucleus is not formed. In fact, starting with ^{238}U, a whole succession of radioactive disintegrations takes place until a stable isotope of lead, ^{206}Pb, forms. This succession, called a *radioactive disintegration series,* is shown in Figure 2.14.

Estimating the Age of Rock Formations. When scientists recognized that the decay of naturally occurring isotopes eventually stops with some lower-numbered, nonradioactive isotope, they realized that they had a means for determining the ages of rock formations in which these materials occurred. For example, in common lead the abundance of ^{206}Pb is about 26%. If a very old crystal of uraninite (a mineral of approximate composition UO_2) is studied, it will have a higher percentage of ^{206}Pb than ordinary lead would have. We can estimate how old the crystal is on the basis of how long it would have taken ^{238}U to produce this extra ^{206}Pb by decay.

Our presentation of the method for dating rock samples is oversimplified, for we have discussed only the general outline of the principle involved. Many other isotope pairs exist which, by an analysis of their relative abundance, yield ages for the minerals containing them. Thus it is often possible to check the age of the rock with more than one isotope pair. Scientists using these methods are aware that certain factors must be considered. Account must be taken, for example, of the amount of final, stable isotope that might have been present initially. There must be evidence that no loss or gain of either the starting isotope or the terminal isotope has occurred throughout the lifetime of the rock sample. It is therefore important that several analyses of a number of rock samples from many sources be made for a

Table 2.10 Half lives and Radiations of Some Naturally Occurring Isotopes

Element	Isotope	Half-life ($t_{1/2}$)	Radiations
Potassium	$^{40}_{19}K$	200 million years	beta, gamma
Radon	$^{222}_{86}Rn$	3.82 days	alpha
Radium	$^{226}_{88}Ra$	1590 years	alpha, gamma
Thorium	$^{234}_{90}Th$	25 days	beta, gamma
	$^{230}_{90}Th$	80,000 years	alpha, gamma
Uranium	$^{235}_{92}U$	800 million years	alpha, gamma
	$^{238}_{92}U$	4.7 billion years	alpha

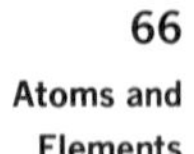

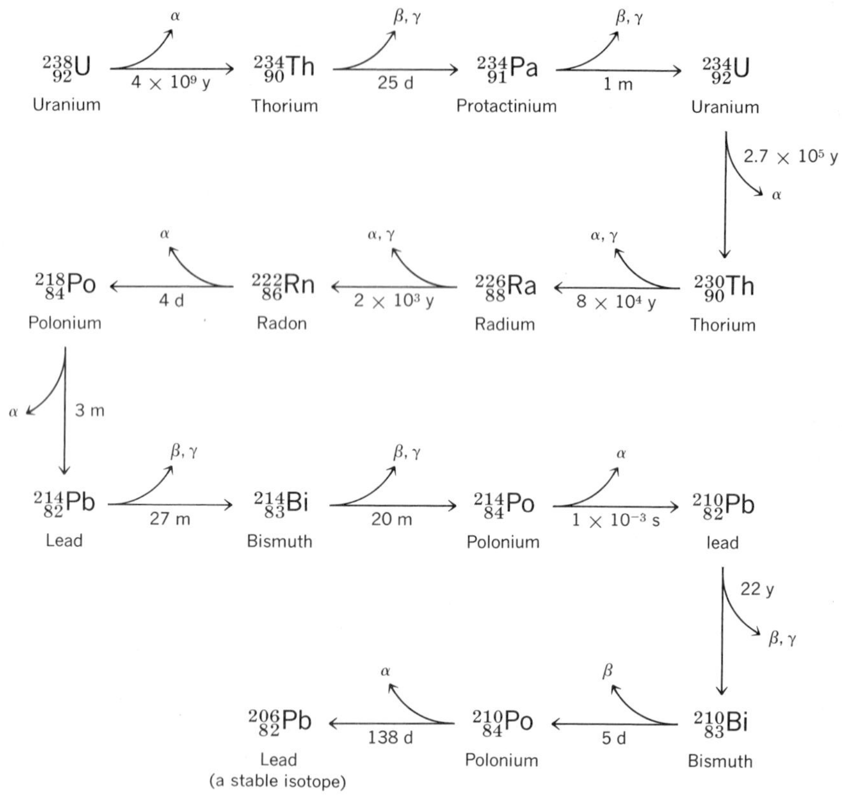
$^{238}_{92}U$
Uranium
α
4 × 10⁹ y
$^{234}_{90}Th$
Thorium
β, γ
25 d
$^{234}_{91}Pa$
Protactinium
β, γ
1 m
$^{234}_{92}U$
Uranium
2.7 × 10⁵ y
α
$^{230}_{90}Th$
Thorium
α, γ
8 × 10⁴ y
$^{226}_{88}Ra$
Radium
α, γ
2 × 10³ y
$^{222}_{86}Rn$
Radon
α
4 d
$^{218}_{84}Po$
Polonium
α
3 m
$^{214}_{82}Pb$
Lead
β, γ
27 m
$^{214}_{83}Bi$
Bismuth
β, γ
20 m
$^{214}_{84}Po$
Polonium
α
1 × 10⁻³ s
$^{210}_{82}Pb$
lead
22 y
β, γ
$^{210}_{83}Bi$
Bismuth
β
5 d
$^{210}_{84}Po$
Polonium
α
138 d
$^{206}_{82}Pb$
Lead
(a stable isotope)

Figure 2.14 Uranium-238 radioactive disintegration series. The number beneath each arrow signifies the half-life of the preceding isotope: y = years, d = days, m = minutes, s = seconds. The small arrow that curves away from each main arrow indicates the kind (or kinds) of radiation emitted by the preceding isotope.

variety of isotope pairs. The references at the close of this chapter give additional details.

INTERACTIONS OF RADIATIONS WITH MATTER

Induced Transmutations. As radiations penetrate matter, some of the particles in the rays may be on collision course with nuclei in the atoms of the matter. If the angles and energies are just right, the decay particle may bury itself within the target nucleus and, in so doing, expel a small fragment. Lord Rutherford was the first to observe such an event. In 1919, while studying the effects of alpha rays on nitrogen gas, he noted the production of high-speed particles later identified as protons. Still later it was found that an isotope of oxygen had also been produced. This was the first artificial transmutation of an element, and the nuclear equation for it is

$$^{14}_{7}N + \underset{\text{High-speed }\alpha\text{-particle}}{^{4}_{2}He} \longrightarrow \ ^{17}_{8}O + \underset{\text{High-speed proton}}{^{1}_{1}H}$$

Not only does this reaction transmute nitrogen into oxygen, but it also makes available another high-energy ray, a stream of protons. A stream of neutrons may be generated by letting alpha rays bombard beryllium atoms:

$$^{9}_{4}\text{Be} + {}^{4}_{2}\text{He} \longrightarrow {}^{12}_{6}\text{C} + {}^{1}_{0}n$$

High-speed α-particles; Stable isotope of carbon; High-speed neutrons

Artificial Radioactivity. With high-speed neutrons, protons, and other nuclear "bullets"as well as the natural sources of alpha and beta particles, we have a roster of high-energy rays available for studying the effects of bombarding matter. The rays consisting of charged particles can be given even higher energy by accelerating the particles to higher velocities. Linear accelerators, cyclotrons, bevatrons, synchro-cyclotrons, etc., have been designed and built for this purpose.

Since Rutherford's first discovery in 1919, hundreds of artificial nuclear reactions have been achieved. Most of them have produced radioactive isotopes, in contrast to the nonradioactive oxygen-18 isotope made by Rutherford. The decay of man-made isotopes is called artificial radioactivity, but since it does not differ in principle from natural radioactivity, the term is of limited usefulness.

Detection of Radiation. The Geiger-Müller Counter. The Geiger-Müller tube and counter is one of several ways by which radiations may be detected. As shown in Figure 2.15, sealed into the tube is a thin wire running down the middle. Surrounding the tube's inner wall is a thin metallic cylinder. The cylinder and the wire are connected on the outside to a powerful battery or its equivalent. In operation, the voltage of the battery is adjusted to be not quite powerful enough to force a spark to jump the gap between the two metals. When a product of radioactive decay plows into the tube, it leaves in its wake a track of gaseous, electrically charged particles, ions, which conduct a current. Thus each radioactive particle, gamma ray, X-ray, or cosmic ray that penetrates the tube makes possible a brief surge of electricity between the two metals. In other words, the circuit is momentarily closed when radiation enters the tube. With clever engineering the brief flow of current can be converted into a light flick, a click, or the next larger number on a counter dial. Highly accurate, reproducible counts of radiation are possible.

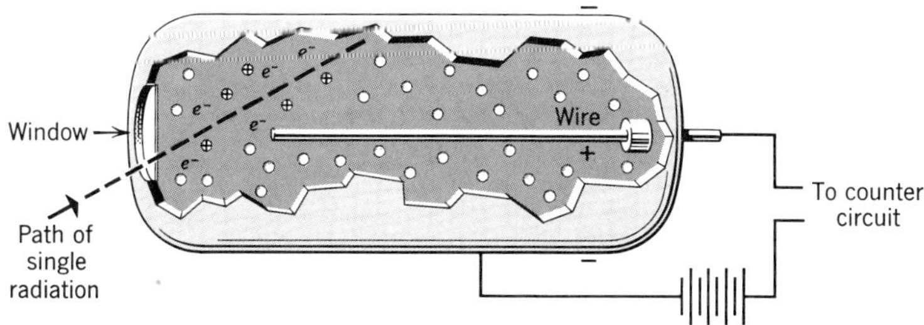
Window
Path of single radiation
Wire
To counter circuit

Figure 2.15 Operating principle of a Geiger-Müller tube.

REFERENCES AND ANNOTATED READING LIST

BOOKS

A. G. van Melsen. *From Atomos to Atom.* Harper Torchbook TB517, Harper and Row, New York, 1960. This historical survey of the concept atom was written for those seeking both historical and philosophical presentations. It begins with the earliest Greek speculations and the questions that drove them to concepts of matter, and it courses through modern atomic science in language the average college student can understand. (Paperback.)

J. J. Lagowski. *The Structure of Atoms.* Houghton Mifflin Company, Boston, 1964. Evidence for the existence of atoms and the subatomic particles is presented largely through excerpts from the original writings. The author has provided a careful selection and a skillful commentary to help beginners understand difficult points. (Paperback.)

A. Romer. *The Restless Atom.* Science Study Series S12, Doubleday Anchor Books, Garden City, N.Y., 1960. The author presents a history of the scientific progress in the twenty years before World War I that marked the birth of the nuclear age. (Paperback.)

D. L. Anderson. *The Discovery of the Electron.* Momentum Book 3, Van Nostrand, Princeton, N.J., 1964. How we came to think that electricity is atomic in nature, rather than continuous, is traced through the history of the discovery of the electron. (Paperback.)

E. N. da C. Andrade. *Rutherford and the Nature of the Atom.* Science Study Series S35, Doubleday Anchor Books, Garden City, N.Y., 1964. This biography about the father of nuclear physics by another distinguished physicist recapitulates the excitement of those years before and after World War I when Lord Rutherford was one of the dominant figures in science. (Paperback.)

D. J. Hughes. *The Neutron Story.* Science Study Series S1, Doubleday Anchor Books, Garden City, N.Y., 1959. The discovery of the neutron, its properties, and its importance in nuclear fission, medicine, and industry are related by a physicist of the Brookhaven National Laboratories. (Paperback.)

H. H. Sisler. *Electronic Structure, Properties, and the Periodic Law.* Reinhold Publishing Corporation, New York, 1963. How some of the more fundamental properties of atoms vary with atomic structure is described and correlated with their locations in the periodic system. (Paperback.)

R. T. Sanderson. *Chemical Periodicity.* Reinhold Publishing Corporation, New York, 1960. A large number of figures, graphs, and tables illustrate about sixty applications of chemical periodicity.

W. F. Kieffer. *The Mole Concept in Chemistry.* Reinhold Publishing Corporation, New York, 1962. This short book ties together the scattered but very frequent applications of the concept of a mole of substance, one of the most significant ideas in all chemistry. Until this concept is fully grasped, the student is not "over the hump" in the study of chemistry. This book could be listed as a reference for the next several chapters. (Paperback.)

P. M. Hurley. *How Old Is the Earth.* Science Study Series S5, Doubleday Anchor Books, Garden City, N.Y., 1959. The author discusses at a commonsense level the clues that have been used to estimate the age of the earth, those from geology and from analyses of the moon and the planets, earthquakes, volcanoes, and the most promising of all, natural radioactivity. The carbon-14 method of dating organic remains is also described. (Paperback.)

ARTICLES

E. N. da. C. Andrade. "The Birth of the Nuclear Atom." *Scientific American,* November 1956, page 93. The experiments that led Rutherford to propose the existence of the nucleus are described.

F. L. Lambert. "Atomic Orbitals from Wave Patterns." *Chemistry,* February 1968, page 10, and March 1968, page 8. By an easy, step-by-step discussion of simple vibrations and wave motions, such as those resulting from a plucked string, a qualitative picture of the various kinds of atomic orbitals emerges.

John E. Frey. "Discovery of the Noble Gases and Foundations of the Theory of Atomic Structure." *Journal of Chemical Education,* Vol. 43, 1966, page 371. The noble gases are the least reactive of all elements. Their discovery depended on some sleuthing of remarkable quality, and this article is a brief recount of this adventure.

E. G. Mazurs. "Ups and Downs of the Periodic Table." *Chemistry,* July 1966, page 7. There is no such thing as *the* periodic chart. Mazurs describes the many varieties, including one of his own making.

C. L. Dunham. "Isotopes—Explorers in Medical Science." *Chemistry,* January 1965, page 12. The use of radioactively labeled or "tagged" molecules in medicine is described.

Editors of *Chemistry.* "The Elements—Materials for a Growing World of Knowledge." *Chemistry,* November 1956, page 12. This table of the elements summarizes in one place information about their dates of discovery, the discoverers, and the origin of their names, as well as listing atomic mass and atomic numbers.

PROBLEMS AND EXERCISES

1. A number of terms have been introduced in this chapter, and you should write out definitions in your own words. Where appropriate, illustrations should be given.
   (*a*) electron (*b*) proton (*c*) neutron
   (*d*) stationary state (*e*) atomic orbital (*f*) principal quantum number
   (*g*) Pauli exclusion principle (*h*) Hund's rule (*i*) isotopes
   (*j*) atomic number (*k*) physical property (*l*) chemical reaction
   (*m*) atomic weight (*n*) Avogadro's number (*o*) α-particle
   (*p*) β-particle (*q*) γ-ray (*r*) radioactivity
   (*s*) transmutation (*t*) half-life (*u*) curie
   (*v*) roentgen (*w*) RBE
2. Several important scientists were named in this chapter. Briefly describe the major contribution of each.
   (*a*) Thomson (*b*) Rutherford (*c*) Geiger
   (*d*) Bohr (*e*) Balmer (*f*) Röntgen
   (*g*) Becquerel (*h*) Schrödinger (*i*) Heisenberg
   (*j*) Mendeleev
3. How did the Thomson model of the atom differ from Rutherford's?
4. In general terms, describe the strong and weak points of the Rutherford model.
5. What key experiment lead to the discarding of the Thomson model?
6. As far as classical theory was concerned, what was the principal objection to the Bohr model? How did Bohr resolve this problem?
7. Describe in general terms how orbitals for atoms other than hydrogen are like those of hydrogen and how they are different.

8. How many protons must the nucleus of an atom having the electronic configuration $1s^22s^22p^63s^23p^63d^54s$ contain?
9. What is its atomic number?
10. Still referring to the configuration in question 8, is this element more likely be a transition element or a member of another group? Explain. (You should be able to handle such a question without looking at a periodic chart or table of elements.)
11. Without referring to tables or charts, you should be able to construct electronic configurations for any element from 1 to 20.

    (*a*) Using condensed symbolism (cf. question 8), write electronic configurations for elements of atomic number 3, 7, 5, 19, 14, 10, 20, 9, 5, 17, and 13.

    (*b*) Without referring to tables or charts, predict on the basis of electronic configurations just written for part *a* whether the element in question is more likely to be a metal or a nonmetal. Explain.
12. Write the electronic configuration, and this time include the composition of the nucleus, for a reasonable isotope of an element of atomic number 12 and atomic mass 25.
13. The accompanying diagram shows a section of a periodic chart. Hypothetical atomic symbols have been used. You should be able to answer the following questions without having to refer to an actual periodic chart. (The numbers are atomic numbers.)

		7	8	9	10
		15 W	16 X	17 Y	18 Z
		33	34	35	36

    (*a*) Write the atomic number of an element in the same family as Y.

    (*b*) Write the atomic number of an element in the same period as W.

    (*c*) Are the elements shown more likely to be metals or nonmetals?

    (*d*) If there are six electrons in the outside shell (highest occupied principal energy level) of X, how many would the outside shells of elements 8, 15, and 36 hold? (You should be able to deduce the answers without constructing electronic configurations.)
14. What would be the weight of a sample of sodium having Avogadro's number of sodium atoms? (State the unit.)
15. What would be the weight of a sample of krypton, having 3.01×10^{23} atoms of krypton?
16. What are the weights of the following samples (include correct unit of weight)?[7]

    (*a*) 1 mole Li (*b*) 0.34 mole S

    (*c*) 25.6 moles Na (*d*) 25.6 moles Li

    (*e*) 25.6 moles H (*f*) 0.002 mole Ar

    (*g*) 3.4×10^{-6} mole U (express in micrograms)

    (*h*) 1.56×10^{-3} mole Hg (express in milligrams)

    (*i*) 9.78×10^{2} moles Ca (express in kilograms) (*j*) 6.02×10^{22} atoms of C
17. What volume is occupied by 1 mole Na (d_{20} 0.97 g/cc)?
18. What volume is occupied by 1 mole He at 0°C and 1 atmosphere pressure if its density under these conditions is 0.1785 g/liter? Note the units of density in this case. Helium is a gas.[8]

    Ans. 22.42 liters.
19. What is the volume occupied by 1 mole of each of the following gases at 0°C and 1 atmosphere pressure if under these conditions they have the densities given?

    (*a*) Ne, $d = 0.9002$ g/liter

    (*b*) Ar, $d = 1.7824$ g/liter

    (*c*) Kr, $d = 3.708$ g/liter

    Ans. (*a*) 22.420 liters/mole Ne.

[7] In calculations involving atomic weight data taken from the table on the inside book cover, round atomic weight values to the nearest whole numbers, except for chlorine use 35.5.

[8] In problems 18 and 19 use exact values of atomic weights.

20. Reviewing the results of your calculations in questions 18 and 19, what appears to be substantially constant among the properties of these different gases? (Be sure to specify correct units.)
21. Assuming that this constant pertains for other gases, such as nitrogen, oxygen, chlorine, and fluorine, what is the formula weight of nitrogen if its measured density is 1.2506 g/liter? How does this answer compare with the atomic weight of nitrogen?
22. How many moles would there be in each of the following samples?
(*a*) 23 g Na (*b*) 6 g C (*c*) 40 g Ne
(*d*) 500 g K (*e*) 1 kg Fe (*f*) 500 mg Al
(*g*) 1 lb Li (*h*) 28.6 g S *Ans.* (*g*) 65.4 mole Li.
23. Supply symbols for the bombardment particles used in the following transmutations.
(*a*) $^{27}_{13}\text{Al} + \_\_\_\_ \longrightarrow {}^{30}_{15}\text{P} + {}^{1}_{0}n$
(*b*) $^{32}_{16}\text{S} + \_\_\_\_ \longrightarrow {}^{32}_{15}\text{P} + {}^{1}_{1}\text{H}$
24. When nitrogen-14 is bombarded with α-particles to produce oxygen-17 and protons, it is believed that an intermediate atom forms with a "compound nucleus" that then decays to the observed products:

$$^{14}_{7}\text{N} + {}^{4}_{2}\text{He} \longrightarrow \_\_\_\_ \longrightarrow {}^{17}_{8}\text{O} + {}^{1}_{1}\text{H}$$

Write the atomic symbol, including both atomic number and atomic mass, for this intermediate.
25. When boron-10 is bombarded with α-particles, nitrogen-13 forms:

$$^{10}_{5}\text{B} + {}^{4}_{2}\text{He} \longrightarrow {}^{13}_{7}\text{N} + {}^{1}_{0}n$$

Nitrogen-13 is also radioactive ($t_{1/2}$ = 10 minutes). It decays to yield carbon-13 plus a new particle called a positron:

$$^{13}_{7}\text{N} \longrightarrow {}^{13}_{6}\text{C} + \text{positron}$$

Figuring from the equation given, what is the electrical charge on a positron? Its mass? Compare it with an electron.
26. Suppose you have a 100-g sample of a radioactive element whose half-life is 10 days. How much of this element will be left after five half-life periods?

CHAPTER THREE

Substances and Change

SIMPLE VERSUS COMPOUND SUBSTANCES

Laws of Chemical Combination. In the previous chapter an element was defined largely in terms of a concept, not in terms of a set of directions or operations for finding out whether a sample of matter is or is not an element. A *conceptual definition* rather than an *operational definition* was used. The concept had to do with atoms and with the notion that all the atoms in a sample of an element must have the same atomic number and identical electronic configurations. Implied in this understanding of an element was the notion of simplicity. By definition, the term element in chemistry is reserved for "simple substances," although the nuclear physicist, of course, would be aghast at this use of the word "simple." If either physical or chemical means[1] have not been found and presumably do not exist for breaking a given sample of matter into simpler substances, the sample is classified as an element. This gives us an operational definition of an element. An element is defined in terms of the operations we perform on the substance; if they are unsuccessful in the hands of enough skilled people, the scientific community with confidence labels the substance an element.

The application of heat is one of the operations. It is seldom alone successful, but one example of its use that is very simple serves to illustrate several laws of chemical combination. When a substance called platinum(IV) oxide is heated at about 600°C, it decomposes. A metal of high density remains, identified by the name platinum. A gas that has escaped as well can be identified by mildly sophisticated techniques as oxygen. (The gas, for example, will support combustion.) According to the definitions of physical and chemical changes given on page 53,

[1] By definition, we exclude nuclear reactions (decay, transmutation) from the meaning of the terms physical and chemical means.

this event is clearly a chemical reaction. An old substance with its characteristic physical properties has disappeared to be replaced by new substances, each with its own unique set of properties.

Without knowing anything more about platinum, oxygen, and platinum(IV) oxide than their physical properties by which we identify them, plus the fact that heat decomposes the latter, we are inclined to say that platinum and oxygen are both simpler than the substance liberating them. It would take more effort to show that neither platinum nor oxygen can be further decomposed into even simpler substances that are stable enough to be stored, but such work has been done. These two are elements.

In the history of thought it took a long time for students of nature to realize the significance of another aspect of a reaction such as this, the relative changes in weights of the substances involved.

Law of Conservation of Mass. The following data, reconstructed for purposes of illustration, indicate what a study of the weight relations in chemical changes may lead to. It is supposed that weighed samples of platinum(IV) oxide from different sources are heated at the decomposition temperature and that the platinum and the oxygen are individually collected and weighed.

Weight of Sample of Platinum(IV) Oxide Used	Weights of Recovered Substances: Platinum	Weights of Recovered Substances: Oxygen	Sum of Weights of Products	Ratio of Weights (grams platinum to grams oxygen)
10.0 g	8.59 g	1.41 g	10.0 g	6.09:1
12.5	10.7	1.76	12.5	6.08:1
15.0	12.9	2.12	15.0	6.08:1
			Average ratio	6.08:1

Two features stand out. The sum of the weights of the products in each case equals the weight of the starting material. The French chemist, Antoine Lavoisier, is credited with producing the first concrete evidence for what is now called the law of conservation of mass, that in chemical changes the total mass is conserved. Matter is neither created nor destroyed in chemical changes; only its form changes.

Law of Definite Proportions. Joseph Louis Proust, another French chemist, is credited with discovering and proving another feature of pure substances, their constant proportions. In its present form the law of definite proportions states that in a given pure substance the elements are always combined in the same proportions by weight, regardless of where or how the compound was obtained. In platinum(IV) oxide the ratio of platinum to oxygen is evidently 6.08:1 by weight.

Pure Substances versus Mixtures. The idea of a pure substance has just been introduced in an explicit way. Before continuing our study of the laws of chemical combination, we need a provisional definition of this term, at least enough of a definition to help us distinguish pure substances from mixtures and solutions. At this stage an operational definition is best. A substance is usually judged to be pure if it cannot by *physical means* producing physical changes be resolved or separated into other substances. In other words, if all efforts to purify a substance finally yield a material which, on further application of these or other methods, undergoes no change in its physical properties, the substance is said to be pure. The actual purity of a substance is therefore limited by the methods of purification and the skill of the purifier. Our knowledge of how pure a sample is is further limited by the accuracy of the instruments used to measure physical properties. Thus in practical work we accept different degrees of purity as the situation varies. "Pure" tap water, if used repeatedly, will ruin most steam irons.

Antoine Laurent Lavoisier (1742–1794). Although trained as a lawyer, Lavoisier's interests in the sciences, particularly chemistry, dominated his life. He was born into a world in which the best theory of chemical elements described just four—the age-old elements of the ancients, earth, air, fire, and water. Furthermore, all substances were believed to be transmutable, gases were immaterial and spiritous instead of material, the phlogiston theory of combustion dominated thinking in this area, and the chemical compositions of both water and air were not even suspected. In twenty years of careful study Lavoisier changed all this. He placed chemistry on a firm quantitative basis. He pioneered in the reform of chemical nomenclature and symbolism, replacing such irrational terms as vitriol of Venus and butter of antimony by the systematic names copper sulfate and antimony chloride. His book, *Traité élémentaire de Chimie, présenté dans un ordre nouveau et d'après les découvertes modernes* (Elements of Chemistry, in a new systematic order, containing all the modern discoveries), was published in 1789. It was to the development of chemistry almost what Newton's *Principia* was to physics, and Lavoisier is deservedly called the father of modern chemistry. He lived at the time of the French revolution. On a trumped-up charge he and several colleagues were arrested and convicted, without trial, of plotting with enemies of France. They went to the guillotine on May 8, 1794.

Joseph Louis Proust (1754–1826). A French chemist who did much of his research at the Royal Laboratory in Madrid, Proust carried on a long controversy with Claude Louis Berthollet (1748–1822) who disputed the notion that pure substances are characterized by definite compositions. Although Proust apparently did not formulate the law of definite proportions, his data were the essential experimental evidence that Dalton needed.

It is not enough to say that a pure substance is homogeneous, for a solution of salt in water is that. A drop of this solution will have the same properties and analyses as any other drop. Yet we know that water and salt may be separated from this solution by simple physical means, for example, by evaporation or distillation. We also know that we can prepare salt water in a wide variety of proportions. The solution is in a special class of matter called mixtures, which are distinguished from chemically pure substances by the fact that they can be prepared in a variety of proportions and their components can be separated by physical methods. A pure substance will form in only a definite proportion of the starting substances.

Law of Multiple Proportions. To return to platinum and oxygen, a compound other than platinum(IV) oxide can be made to yield these elements through the action of heat. This substance, called platinum(II) oxide, has weight relations illustrated by the following data. We note that the law of conservation of mass is obeyed, and

Weight of Sample of Platinum(II) Oxide Used	Weights of Recovered Substances		Sum of Weights of Products	Ratio of Weights (grams platinum to grams oxygen)
	Platinum	Oxygen		
10.0 g	9.24 g	0.76 g	10.0 g	12.1:1
15.0	13.9	1.14	15.0	12.2:1
18.0	16.6	1.36	18.0	12.2:1
			Average ratio	12.2:1

the law of definite proportion holds. John Dalton was the first to see the significance of a third feature of chemicals when two elements can be made to form two (or more) compounds. In platinum(IV) oxide the weight ratio of platinum to oxygen is 6.08:1, in platinum(II) oxide 12.1:1. The ratio 6.08 to 12.1 is 1:2, a ratio composed of small whole numbers. Tin and oxygen also form two "compound substances." The analysis of one, as reported by John Dalton, was 88.1% tin and 11.9% oxygen for a ratio of 7.4:1. The other analyzed for 78.7% tin and 21.3% oxygen for a ratio of 3.7:1. The ratio of 7.4:3.7 is the same as 2:1, again a ratio of small whole numbers. Results such as these are generalized by the law of multiple proportions: If two elements form more than one compound, the different weights of one that can combine with the same weight of the second are in the ratio of small whole numbers.

John Dalton (1766–1844). This English chemist and physicist is known best for his atomic theory. In connection with some studies of the properties of gases, Dalton had to make several analyses of compounds. He came to realize the significance of weight relations, for they confirmed in his mind the conviction that the properties of matter he knew about could be explained only by an atomic theory.

DALTON'S THEORY

Dalton went one very important step further. He speculated about the most satisfactory model for understanding these empirical laws of chemical combination.[2] He asked, in effect, what must matter be like to obtain the kinds of weight relations we do. The idea of atoms, of course, had been part of the history of thought for more than two millenniums. Dalton was by no means the first to speculate about an atomic theory. In the empirical laws just described, however, he thought he saw evi-

[2] An empirical law is a generalization about data or experiences. The truth it expresses is independent of any theory proposed to explain it in a larger context.

dence that *compelled* belief in atoms, even though they could not be seen. More particularly, these atoms had to have certain characteristics, and a list of these made up much of his theory. Before attending to this list, we should mention that Dalton's "atoms" actually included some particles of matter we now call molecules. To Dalton an atom was simply a fundamental particle. His postulates are given in condensed form in the left column, quotations from his book *A New System of Chemical Philosophy*[3] in the right.

1. Matter consists of atoms.

"These observations have tacity led to the conclusion which seems universally adopted, that all bodies of sensible magnitude, whether liquid or solid, are constituted of a vast number of extremely small particles, or atoms of matter bound together by a force of attraction, which is more or less powerful according to circumstances. . . ."

2. Atoms are indestructible, and chemical reactions are nothing more than rearrangements of atoms.

"Chemical analysis and synthesis go no farther than to the separation of particles from one another, and to their reunion. No new creation or destruction of matter is within reach of chemical agency."

3. All atoms of one element are identical with respect to weight and other properties.

"Therefore we may conclude that the ultimate particles of all homogeneous bodies are perfectly alike in weight, figure, etc. In other words, every particle of water is like every other particle of water; every particle of hydrogen is like every other particle of hydrogen, etc."

4. Different elements consist of different kinds of atoms, the most significant difference being their weights.

5. In the formation of a compound from its elements, a definite but small number of atoms of each element join to form compound particles.

"It is one great advantage of this work to show the importance and the advantage of ascertaining the relative weights of the ultimate particles, both of simple and compound bodies, the number of simple elementary particles which constitute one compound particle, and the number of less compound particles which enter into the formation of one more compound particle."

[3] This book was published in 1808 and has been reprinted by Philosophical Library (New York, 1964) with an introduction by Alexander Joseph. The quotations from Dalton are taken from this reprint by permission.

The appearance of Dalton's theory intensified several activities in science: the search for new elements, the determination of the relative combining weights of atoms, and speculation about the nature of the "compound particles" and the forces that hold atoms together within them. Before we study these topics, we must first have methods for symbolizing the compound particles and the chemical changes they undergo.

CHEMICAL FORMULAS AND EQUATIONS. SYMBOLS

In terms of what we have already studied about formula weights and moles, the weight data for the two oxides of platinum may be reinterpreted as mole data. A weight ratio of 6.08:1 means a mole ratio of

$$\frac{6.08 \text{ g Pt}}{195 \text{ g Pt/mole Pt}} : \frac{1 \text{ g O}}{16 \text{ g O/mole O}}$$

or

$$0.0312 \text{ mole Pt} : 0.0625 \text{ mole O}$$

or

$$1 \text{ mole Pt} : 2 \text{ moles O}$$

Since equal numbers of moles contain equal numbers of atoms, the weight ratio translates into a ratio of one atom of platinum to two atoms of oxygen. To express the formula for platinum(IV) oxide we therefore write the symbol

$$PtO_2$$

The number 1 is "understood." — This subscript refers to O only.

By convention, the metallic element's symbol is usually placed first. One compound particle of platinum(IV) oxide consists of one platinum atom and two oxygen atoms. We have said nothing about the forces holding these atoms together, for we are concerned only with symbolizing pure compounds.

The data for platinum(II) oxide translate from a weight ratio of 12.1:1 in grams to a ratio of 1:1 in moles. The correct symbol is therefore PtO.

The symbols PtO and PtO_2 may each be used for purposes other than representing the *ratio* of constituent atoms. They may represent one *formula unit* of the substance in question. One formula unit of PtO would consist of one platinum nucleus, one oxygen nucleus, and enough electrons to make the unit electrically neutral. In frequently encountered contexts these formulas also represent one *mole* of the substance in question. One mole of platinum(II) oxide, symbol PtO, would consist of Avogadro's number of platinum nuclei, Avogadro's number of oxygen nuclei, and enough electrons to make the whole sample electrically neutral. Again, our concern here is not with the electronic configuration of this collection of particles but rather with the way to symbolize it, whatever the arrangement.

Chemical Equations. The conventional shorthand symbolism for the decomposition of platinum(II) oxide is as follows:

$$PtO \longrightarrow Pt + O$$

Except for one fact, this equation correctly states the event with particular attention to the proportions of the elements. The fact not accounted for is the formula unit for the element oxygen, which is O_2 rather than O. Since chemical equations must show correct symbols for formula units, the decomposition of platinum(II) oxide should be represented as

$$PtO \longrightarrow Pt + O_2$$

In this form, however, the equation is *unbalanced;* two oxygens on the right are more than could be obtained from what is shown on the left. Since matter is neither created nor destroyed in a chemical change, atomic nuclei are conserved (as are electrons). To balance the equation, we place coefficients, whole numbers, before the formulas requiring them for final balance:

$$2PtO \xrightarrow[\text{heat}]{} 2Pt + O_2$$

The equation is now balanced and, as is sometimes done, the special condition needed to accomplish the event, here heat, is indicated beneath (or above) the arrow.

Balanced equations should be written only if they represent actual chemical changes, or at least changes that are overwhelmingly probable. To balance an equation, the formulas for all reactants and products must be known. In practice, we may need to perform chemical analyses or search through the reference literature of chemistry; or perhaps our memory will serve correctly. Unless all can be written, and written correctly, no meaningful equation can be formulated and balanced. (Nonsense equations are always possible.) Once the correct formulas are set down, none of the *subscripts* may be changed. If subscripts are changed, the formula becomes that of another compound or of one that cannot or does not exist. *Balancing is the last step.* It is performed after the correct formulas have been set down and is accomplished by a trial and error juggling of *coefficients.* For the time being, the important skill to acquire is not the ability to balance an equation but the ability to recognize one and to interpret it in terms of the mole ratios of reactants and products.

Formula Weights of Compounds. Because of our very considerable confidence in the law of conservation of mass, and for additional experimental reasons, the formula weight of a compound is accepted as the sum of the atomic weights of the elements appearing in the symbol for the formula unit.

Examples CO_2: Formula weight $= 44 = \underset{C}{12} + \underset{O}{16} + \underset{O}{16}$

PtO_2: Formula weight $= 227 = \underset{Pt}{195} + \underset{O}{16} \times 2$

We shall encounter some formulas containing parentheses; thus

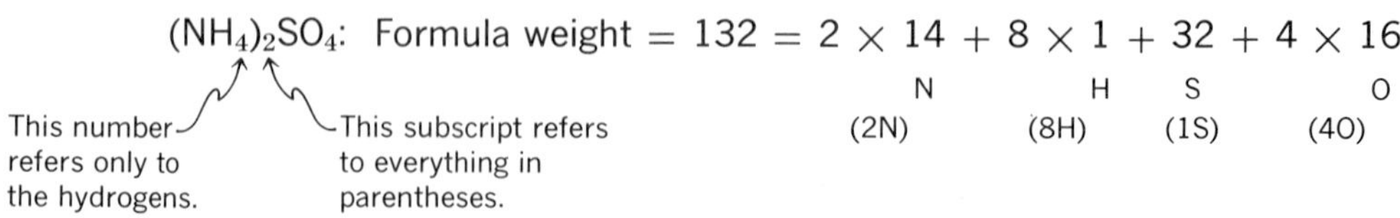
$(NH_4)_2SO_4$: Formula weight $= 132 = 2 \times 14 + 8 \times 1 + 32 + 4 \times 16$
N (2N)
H (8H)
S (1S)
O (4O)
This number refers only to the hydrogens.
This subscript refers to everything in parentheses.

A formula weight for CO_2 of 44 simply means that one formula unit of CO_2 weighs forty-four twelfths as much as one formula unit (one atom) of carbon-12. Its gram-formula weight is 44 g. One mole of CO_2 weighs 44 g and contains Avogadro's number of formula units.

Exercise 3.1 Determine the total number of atoms of all kinds in one formula unit of each of the following: (*a*) $(NH_4)_2CO_3$, (*b*) $Mg_3(PO_4)_2$, (*c*) $Al_2(HPO_4)_3$. *Ans.* (*a*) 14.

Exercise 3.2 Using atomic weight data from the table inside the back cover, and rounding off to the nearest whole number, calculate the formula weights of each of the compounds in exercise 3.1. *Ans.* (*a*) 96.

Exercise 3.3 Calculate the number of moles present in the following weights of compounds: (*a*) 100 g of compound *a*, exercise 3.1; (*b*) 5 g of compound *b*, exercise 3.1; (*c*) 2 lb of compound *c*, exercise 3.1. *Ans.* (*a*) 1.04 moles.

Exercise 3.4 Which would have the greater number of moles, 100 g NaCl or 100 g NaBr? Show calculations.

Exercise 3.5 Consider the following balanced equation for the decomposition of water into its elements:

$$\underset{\text{water}}{2H_2O} \xrightarrow[\text{electrolysis}]{} \underset{\text{hydrogen}}{2H_2} + \underset{\text{oxygen}}{O_2}$$

(*a*) Two formula units of water will produce how many formula units of hydrogen? Of oxygen? *Ans.* One O_2.

(*b*) If the object is to make one mole of oxygen, how many moles of water are necessary? How many moles of hydrogen will also necessarily be obtained? *Ans.* 2 moles H_2.

(*c*) If the object is to make 64 g of oxygen, how many grams of water are necessary? (*Hint.* Whenever the problem specifies weight units, *always convert to moles,* for the coefficients of the equation reveal *mole ratios,* not weight ratios. Therefore temporarily rephrase the question to how many moles of water are necessary. When this has been correctly calculated, convert the moles into the requested weight unit.)

Follow this procedure. First, 64 g of oxygen is the same as 2 moles:

$$\text{moles } O_2 = \frac{\text{weight in grams}}{\text{gram-formula weight}} = \frac{64}{32} = 2$$

In the balanced equation the ratio of water to oxygen is

$$2 \text{ moles } H_2O : 1 \text{ mole oxygen}$$

Hence, 4 moles H_2O are needed to make 2 moles oxygen.

$$4 \text{ moles } H_2O = \frac{\text{weight water}}{18 \text{ g } H_2O/\text{mole}}$$

$$\text{weight water} = 4 \text{ moles} \times 18 \text{ g/mole} = 72 \text{ g}$$ *Ans.* 72 g water.

If you can learn the trick of carrying along all the correct units and of canceling units, you will have a built-in check on whether you are using mathematical equations correctly.)

(*d*) If the object is to make 6 g of hydrogen, how many grams of water will be needed? How many grams of oxygen will be produced? *Ans.* (*d*) 54 g water, 48 g oxygen.

STATES OF MATTER

The idea that matter consists of atoms or of "compound particles," to use Dalton's phrase, is consistent with many observations already noted. Among the many other experiences with matter with which this theory should be consistent is the existence of the three states of matter, solids, liquids, and gases. Historically, in fact, experiments with gases were important in the development of an atomic theory.

We can change a substance from one state to another in two ways, by heating or cooling it or by changing the pressure applied to it (or some combination of these). The temperature of a system[4] is usually, but not always, changed by externally adding heat or removing it. A change in pressure is usually, but not always, brought about with a mechanical pump or similar device. Heat and pressure are two important agencies for initiating or preventing chemical reactions and for controlling their rates. How these agencies affect the states of matter is therefore important to an understanding of matter itself.

THE GASEOUS STATE

Kinetic Theory. We account for several properties of gases with the kinetic theory of gases. Basically, it adds to the idea of the existence of atoms or compound particles one more assumption, that these particles are in a state of constant motion and agitation. The theory starts with a model of what the particles of a gas should be like to behave as they do. It postulates an *ideal gas*. After a brief description of the model, the properties of *real gases* will be discussed and compared with that model.

The Model for an Ideal Gas—Kinetic Theory of Gases

1. A gas consists of an extremely large number of minute particles in a state of constant, chaotic, utterly random motion.
2. These particles are very hard and perfectly elastic (meaning that when they collide there are no frictional losses whatsoever).
3. Between collisions these particles move in straight lines. (In other words, there are no forces of attraction between the particles.)
4. The motions of individual particles are in accordance with all the laws of classical mechanics.

It is not within the intended scope of this book to discuss the derivations and calculations that follow from this model. The behavior of real gases will be described and compared with the predicted behavior of an ideal gas. (The derivations may be found in many freshman college chemistry text books and in virtually all physical

[4] The word *system* has special meanings in science. In chemistry it is taken to apply to that part of the physical universe being investigated. Its *boundaries* may be physically real, or they may be imaginary, and they separate the system from all the rest of the universe, the *surroundings*. Some systems are said to be *isolated*, that is, the boundary prevents interaction between the system and its surroundings. Other systems are said to be *open*, that is, matter moves across the boundaries. In a *closed system* this would not happen.

chemistry textbooks.) The actual behavior of gases is summarized in terms of the following gas laws.

Boyle's Law. If the pressure for a given weight of gas kept at constant temperature is increased, the volume will decrease. Robert Boyle first observed this important property of gases. The relation, simply put, is that the volume is inversely proportional to the pressure (weight of gas and temperature, constant):

$$\text{(3.1)} \qquad \text{Boyle's law:} \qquad V \propto \frac{1}{P} \qquad (t,\ w\text{: fixed})$$

where V = volume
P = pressure
t = temperature
w = weight of gas

Put another way, for a fixed amount of gas at a fixed temperature, the product of the pressure and the volume of the gas is a constant:

$$\text{(3.2)} \qquad PV = \text{a constant}$$

By trapping a gas in a sealed tube above a column of mercury (Figure 3.1), we have a means of changing the pressure of the confined gas and thus altering and measuring its volume.

If a variety of readings of volume and pressure are taken for real gases, the product of these two is substantially a constant. For an ideal gas the product will be an exact constant. Under the conditions of Boyle's law real gases behave more like ideal gases at the lower pressure ranges.

The kinetic theory of gases explains pressure in terms of the cumulative effect of the innumerable collisions per second that moving gas molecules make per unit area on the walls of the container (Figure 3.2). If this area is made to shrink, that is, if the volume of the container is reduced, the frequency of collisions with the remaining wall will increase. The pressure on the wall, in other words, will be higher.

Robert Boyle (1627–1691). Born during the lifetime of Galileo Galilei (1564–1642), the great Italian astronomer, physicist, and mathematician, and dying just four years after the publication of *Principia* in 1687 by Sir Isaac Newton (1642–1727), Robert Boyle's life and work stood at the threshold of modern science. He was both a physicist and chemist. His work on the relation between the pressure and the volume of a gas culminated in what we now call Boyle's law. In his book *The Sceptical Chemist* (1661) he theorized that matter is fundamentally composed of "corpuscles" in a variety of sizes and sorts which can clump together into groups, each such group constituting a chemical substance. Clearly Dalton's debt to Boyle was great, and Boyle is regarded as one of the founders of modern chemistry.

Joseph Louis Gay-Lussac (1778–1850). This man's research ranged from terrestrial magnetism through meteorology, properties of gases, the search for new elements (codiscoverer of boron), a theory of acids, industrial chemistry, and

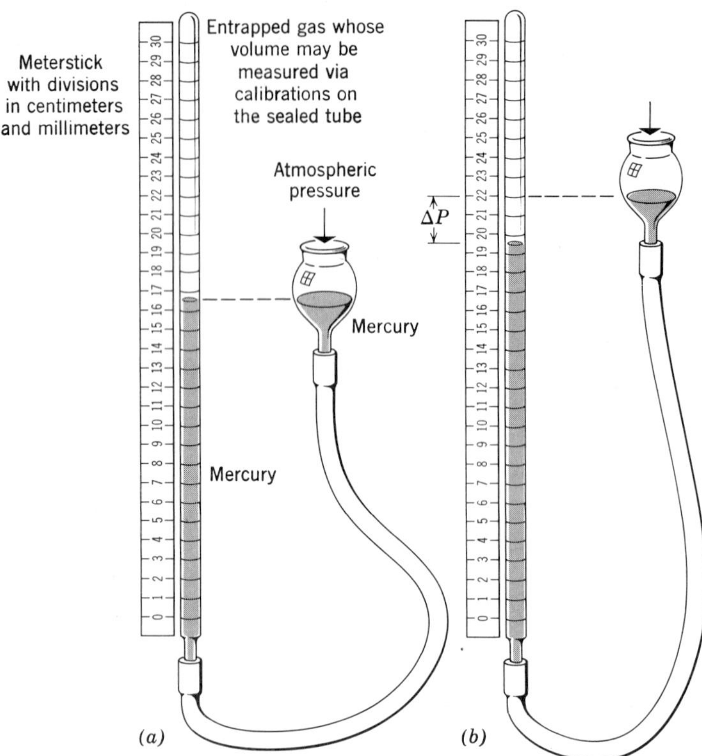
Meterstick with divisions in centimeters and millimeters
Entrapped gas whose volume may be measured via calibrations on the sealed tube
Atmospheric pressure
Mercury
Mercury
ΔP
(a)
(b)

Figure 3.1 Trapping a given weight of gas in a sealed tube over a column of mercury. By raising or lowering the bulb, the pressure exerted on the gas may be varied. (*a*) When the level of mercury in the bulb is the same as in the tube, the pressure inside the tube is the same as atmospheric pressure. If the bulb and tube were clamped into fixed positions, and if the atmospheric pressure on the mercury in the bulb increased, the mercury level in the tube would change. By measuring this change, following suitable calibration of the scale markings, the changes in atmospheric pressure can be observed.

(*b*) In Boyle's experiments the bulb is raised (or lowered). The difference between the two mercury levels is the change in pressure ΔP. The new volume may be measured by means of the calibrations for volume on the tube. In this way pressure-volume data at a given temperature for a given amount of gas can be obtained.

fermentation. His discovery that when gases combine with one another they do so in simple, whole-number volume ratios is known as Gay-Lussac's law.

Jacques Alexandre César Charles (1746–1823). Although he was more a mathematician than a chemist, Charles made some important experimental contributions to the physical sciences. He was the first to use hydrogen for inflating balloons.

Charles–Gay-Lussac Law. For a given weight of gas kept at constant pressure, the volume increases with temperature. Gases expand when heated and contract

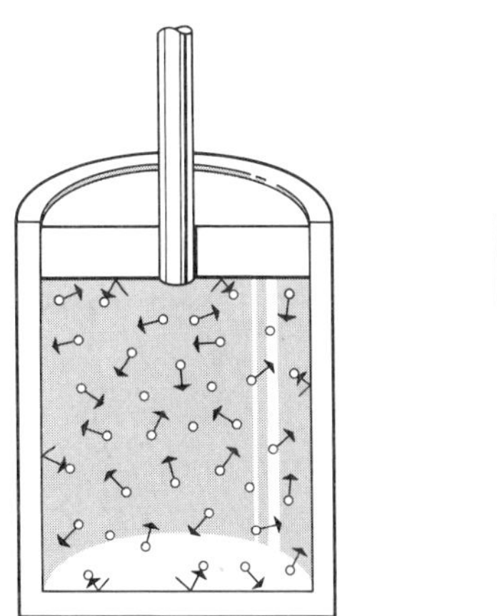

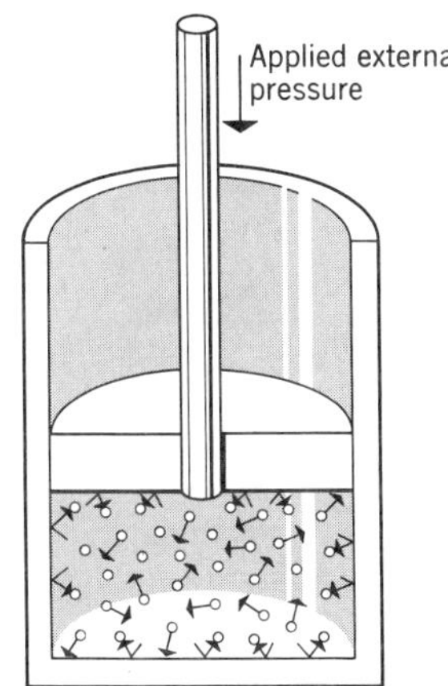
Applied external pressure

Figure 3.2 At a given temperature and for a given amount of gas, the pressure exerted by a gas on the walls of its container is determined by the frequency of collisions at the walls. This frequency (hence the internal pressure) will increase if the surface area is reduced by pushing the piston in and reducing the volume. Qualitatively, this is the picture provided by the kinetic theory to explain Boyle's law.

when cooled. Figure 3.3 shows how quantitative measurements can be made. The relation between volume and temperature, at constant P and w, is not one of *direct* proportionality if the centigrade scale for temperature is used. Rather, the change fits the equation

$$V = V_0(1 + \alpha t) \tag{3.3}$$

where V_0 = volume of gas at 0°C (for the given weight of gas and at the fixed pressure)
V = volume of gas at some other temperature t
t = temperature in degrees centigrade
α = a constant, corresponding to the relative change in volume per degree change in temperature

The French scientist Gay-Lussac was the first to find that the volume of a fixed weight of gas at constant pressure changes *linearly* with temperature, in substantial agreement with this equation. Figure 3.4 illustrates this equation with a plot of data for volume versus temperature. Another French chemist, Charles, was apparently the first to recognize the relation in a qualitative way and realize that the constant α is substantially the same for all gases. The equation is therefore the mathematical form of what is now known as the law of Charles and Gay-Lussac.

The value of α is 1/273.15 when the centigrade scale is used; therefore the unit for α must be degrees centigrade. The equation may be rearranged as follows:

$$V = V_0\left(1 + \frac{1}{273.15}t\right) = V_0\left(\frac{273.15 + t}{273.15}\right)$$

$$= (\text{constant})(273.15 + t) \qquad \text{where the constant is simply } \frac{V_0}{273.15}$$

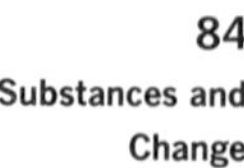

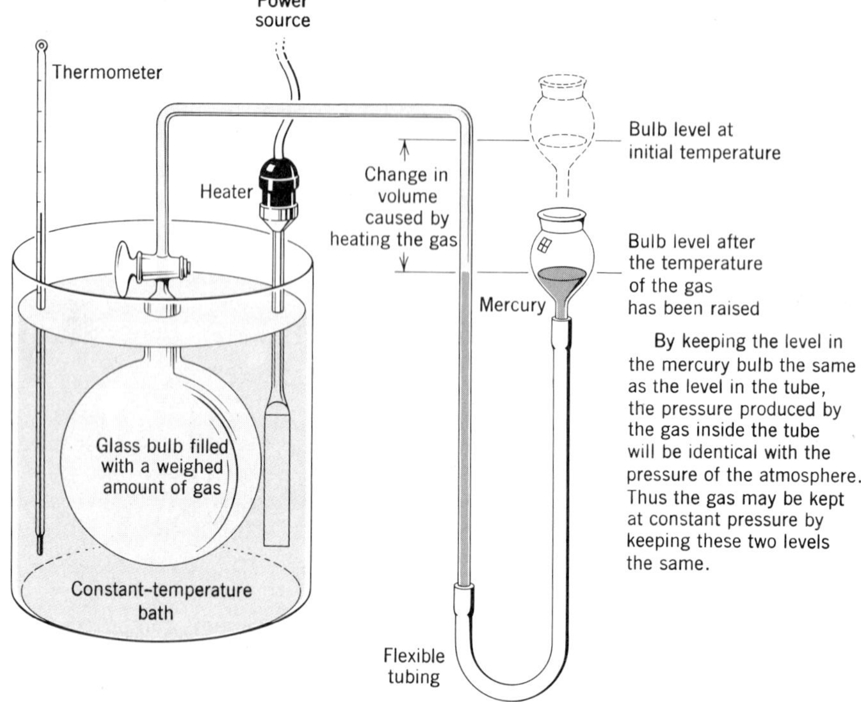
Power source
Thermometer
Heater
Change in volume caused by heating the gas
Bulb level at initial temperature
Bulb level after the temperature of the gas has been raised
Mercury
By keeping the level in the mercury bulb the same as the level in the tube, the pressure produced by the gas inside the tube will be identical with the pressure of the atmosphere. Thus the gas may be kept at constant pressure by keeping these two levels the same.
Glass bulb filled with a weighed amount of gas
Constant-temperature bath
Flexible tubing

Figure 3.3 Measuring how the volume of a fixed amount of gas changes with temperature while the pressure is kept constant. By changing the temperature of the bath, and then holding it constant at the new value until the gas in the bulb reaches the same new temperature, the change in volume of the gas is measured by the change in the height of the mercury column.

In other words, the volume of a gas (of fixed weight and pressure) is proportional to something that has the units of temperature. It could be made directly proportional to temperature if we redefine the temperature scale, letting

$$T = 273.15 + t \tag{3.4}$$

where T is the temperature in a new unit, *degrees Kelvin,* abbreviated °K. (Tempera-

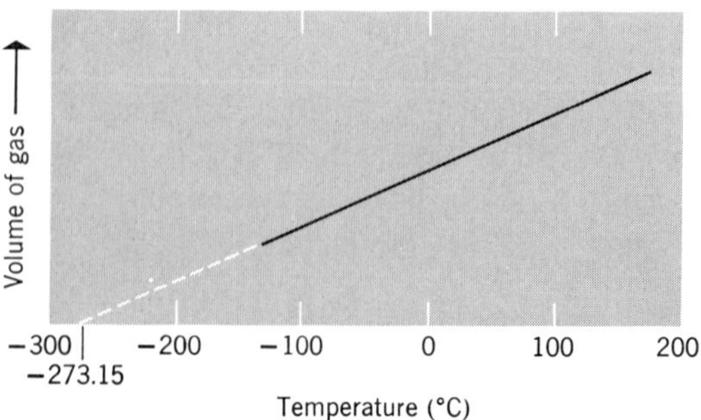
Volume of gas
−300
−273.15
−200
−100
0
100
200
Temperature (°C)

Figure 3.4 Law of Charles and Gay-Lussac. The solid line shows how an ideal gas would behave if a fixed weight were heated (or cooled) at a fixed pressure. The significance of the dotted extension of the line is discussed in the text.

tures on the Kelvin scale are sometimes called absolute temperatures, and in some texts and references the degrees are called *degrees absolute,* still abbreviated °K but sometimes seen as °A.)

Figure 3.4 depicts another way of understanding the absolute scale. If the solid line were extended backward, that is, along the dotted line, it would intersect the temperature axis at −273.15°C. If the line were to be extended any farther back, it would dip below the temperature axis into the domain of negative volumes. Matter cannot have a negative volume. Therefore efforts to achieve a temperature lower than −273.15°C should prove fruitless. One of the difficulties in verifying this supposition experimentally is that real gases condense to liquids before this point is reached. Nevertheless, on the basis of other more sophisticated arguments and experiments, it does seem certain that −273.15°C is the absolute minimum temperature. It is the lowest temperature attainable, and it is therefore designated as zero degrees Kelvin (0°K). Efforts to reach 0°K have come within a thousandth of a degree of it.

William Thomson, Baron Kelvin of Largs (1824–1907). Better known now by the title Lord Kelvin, Thomson was one of the greatest scientists of the nineteenth century. In 1848 he proposed the absolute scale of temperature, and in succeeding years he made major contributions to the formulation of the laws of thermodynamics, his major scientific accomplishments. In this and other fields the advances made by him were so important that he was elevated to the peerage in 1892, elected to the presidency of the Royal Society in 1902, and given the Order of Merit in 1902. He virtually revolutionized submarine telegraphy, invented instruments for determining all the known electrical quantities and made major improvements in the mariner's compass and in gauges for measuring, analyzing, and predicting tides. A professor of natural philosophy at the University of Glasgow, Scotland, for fifty-three years, he was honored in death by burial in Westminster Abbey.

With the definition of a new temperature scale, the Charles–Gay-Lussac relation becomes

$$V = \text{constant} \times T$$

or for a given gas

$$\frac{V}{T} = \text{constant} \qquad (P,\ w\text{: fixed}) \tag{3.5}$$

Ideal Gas Law. Let us summarize the gas laws thus far discussed.

Boyle's law: $PV = C$ (C depends on the values of the *fixed* T and *fixed* w of the gas)

Charles–Gay-Lussac law: $\frac{V}{T} = C'$ (C' depends on the values of the *fixed* P and *fixed* w of the gas)

Experimental data show that a more general law exists for gases:

$$\frac{PV}{T} = C''$$ (C'' depends on the value of the *fixed* w of the gas)

This new constant C'', varies directly with the number of moles of the gas. If we switch from using weight for measuring the amount of gas to using moles, we have

$$C'' \propto n \qquad \text{where } n = \text{number of moles of gas}$$

To convert the proportionality to an equation, we insert the proportionality constant R:

$$C'' = nR$$ where R is the *universal gas constant*

Thus the more general gas law is expressed as

Ideal gas law: $$\frac{PV}{T} = nR$$

or

(3.6) $$PV = nRT$$

By definition a gas with a behavior fitting this equation exactly is an ideal gas. Some real gases "fit" it better than others; that is, some real gases approach ideal behavior better than others.

Molar Volumes and the Universal Gas Constant. The actual value of R depends on the units chosen for expressing the other quantities in the ideal gas equation. Chemists have found it useful to work with certain units and, furthermore, to have a set of reference conditions called *standard pressure* and *standard temperature,* symbolized as STP. The standard pressure is one atmosphere, defined as 760 mm Hg. Standard temperature is 273.15°K (usually rounded in most calculations to 273°K). If we next arbitrarily select n to be one mole, all we have left in the ideal gas law equation, besides the constant R, is a value for the volume V. In other words, we must know what volume one mole of a gas will occupy at STP, a quantity called the *molar volume* of the gas. Different real gases have slightly different molar volumes. Those that chemists have reason to believe should be most like an ideal gas at STP have molar volumes that round off to 22.4 liters/mole. For example, in liters per mole, molar volumes for some typical gases are as follows: nitrogen, 22.403; hydrogen, 22.431; oxygen, 22.392. (See also problems 18–21, Chapter 2.) On other grounds there is reason to take the value of 22.414 liters/mole for the molar volume of an ideal gas. What has been implied here is something of importance in itself. The data show that *equal volumes of gases contain equal numbers of formula units,* a statement known as *Avogadro's law,* named after the Italian physicist who first found evidence for it.

Taking, then, the data

$$P = 1 \text{ atmosphere}$$
$$V = 22.414 \text{ liters}$$
$$n = 1 \text{ mole}$$
$$T = 273°\text{K}$$

we calculate R:

$$R = \frac{PV}{nT} = \frac{(1 \text{ atm})(22.4 \text{ liters})}{(1 \text{ mole})(273°\text{K})}$$

(3.7) $R = 0.082$ (liter)(atm)/(°K)(mole) (units must be included)[5]

By taking other units, the gas constant will have other values; later on we shall have need for one of them, to be introduced at that time.

The Kinetic Theory and an Interpretation of Temperature and Heat Energy. The kinetic theory was at least qualitatively useful in explaining Boyle's law. What about the law of Charles and Gay-Lussac? From an "atom's-eye view" what is the effect of heat on a gas? Stated very briefly, a gas absorbs heat by undergoing an increase in the average kinetic energy of its atoms. This is not a postulate of the kinetic theory of gases, but it is derived from that theory as a conclusion and an interpretation. To demonstrate this statement fully would require doing what is not within the earlier stated scope of this chapter, for we would have to present the complete physics and mathematics of the kinetic theory of gases. As a compromise between rigor and brevity we shall simply state without derivation or proof one of the conclusions of the kinetic theory. This conclusion is an equation expressing the pressure of a gas, not in terms of the ideal gas equation but in terms of the three elements of motion that the ideal gas particles would have. By comparing this equation from the field of mechanics, which says nothing about temperature, with the ideal gas equation, which includes temperature, we shall arrive at an interpretation of temperature and heat energy.

Based on the model of an ideal gas, the kinetic theory *predicts* (successfully) that the pressure of a collection of N gas particles, each having a mass m and all confined in a volume V, will be

(3.8) $$P = \frac{1}{3}\frac{N}{V} m\overline{v^2}$$

Thus pressure may be expressed solely as a function of the three elements of motion, mass, length, and time. (Volume has the dimensions of length cubed; velocity, the dimensions of length divided by time.) The equation includes velocity v in a strange way, however. The symbol $\overline{v^2}$ stands for the *mean squared velocity* of all the particles, which requires some explanation. In giving it we encounter some concepts that will be very useful in our later study of rates of chemical reactions.

[5] As we shall discover on page 94, the product of volume and pressure, (liter) (atm), has the dimensions ML^2T^{-2}, which are those of energy. Thus the gas constant has dimensions of energy per degree per mole.

Intuition suggests that in a box of innumerable, randomly moving particles there is no *one* velocity with which *all* are moving. Some will be moving slowly, some very rapidly, some few not at all at any given instant. Figure 3.5 is a graph showing how the fraction of gaseous particles having a particular speed varies with those speeds. (Speed, which has no direction, is of more interest here than velocity, which is directional. See page 4.) A vanishingly small fraction at any given instant will have zero speed. Similarly a very small fraction will at any instant have very high speeds. Thus the curve in Figure 3.5 starts at a low value for the fraction, rises to a maximum, and then drops back down toward zero.

If the speeds of individual molecules are squared and if the squared values are then added and averaged, the result will be the "mean squared speed" of the particles, $\overline{c^2}$. If we use velocity rather than speed, we have at the end "mean squared velocity" $\overline{v^2}$ instead.

The quantity in parentheses in equation 3.8 looks suspiciously like the combination of the elements of motion that was defined on page 14 as kinetic energy. In fact, if equation 3.8 were doctored a bit without changing the equality (multiply both numerator and denominator by 2), the expression for pressure would be

$$P = \frac{2}{3}\frac{N}{V}\left(\frac{1}{2}\, m\overline{v^2}\right) \tag{3.9}$$

But for a single particle of mass m, and velocity v, the kinetic energy is defined as

$$\text{K.E.} = \tfrac{1}{2}mv^2$$

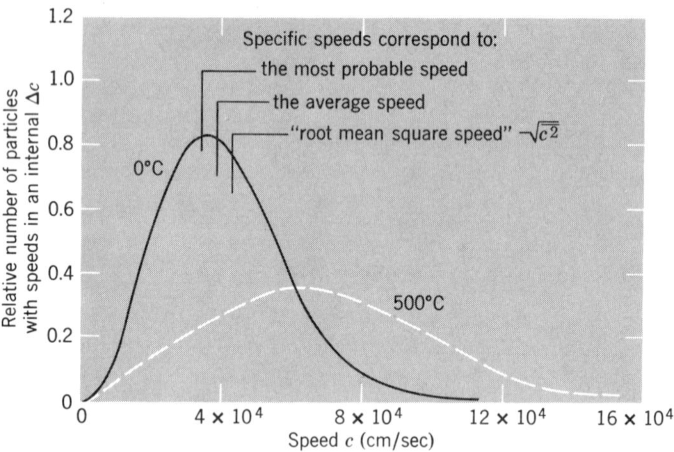
1.2
1.0
0.8
0.6
0.4
0.2
0
Relative number of particles with speeds in an internal Δc
Specific speeds correspond to:
the most probable speed
the average speed
"root mean square speed" $\sqrt{\overline{c^2}}$
0°C
500°C
0
4 x 10⁴
8 x 10⁴
12 x 10⁴
16 x 10⁴
Speed c (cm/sec)

Figure 3.5 Distribution of speeds of gas particles. The vertical scale, which is relative, is for the fraction of gas particles that have speeds between each value of c (on the horizontal scale) and $c + \Delta c$, where Δc is an increment of speed that can be made as small as desired. Mean square *speed* is plotted instead of mean square *velocity* to avoid the problem of the direction of motion. The solid line gives an idea of the distribution of speeds of oxygen molecules at 0°C. The dotted line indicates how the shape of the curve changes as the temperature is raised, approaching the shape shown at a temperature of about 500°C. (For purposes of comparison, a speed of 4×10^4 cm/sec is equivalent to 1312 ft/sec or 894 miles/hr.)

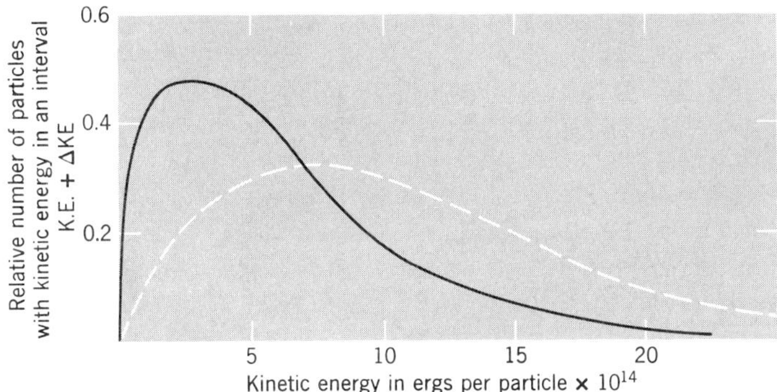
Relative number of particles with kinetic energy in an interval K.E. + ΔKE
0.6
0.4
0.2
5
10
15
20
Kinetic energy in ergs per particle × 10^14

Figure 3.6 Distribution of kinetic energy among gas particles. The solid curve shows how the relative number of particles of a particular kinetic energy varies with values of kinetic energy at a temperature of 300°K, which is, roughly, room temperature. The dotted line indicates how the shape of the curve changes as the temperature is raised by showing what it would be like at a higher temperature.

The quantity in parenthesis in equation 3.9 must have the same units and must be an energy term. It is called the *mean kinetic energy* of the collection of particles in the gas. Figure 3.6 shows how the fraction of particles having a particular kinetic energy varies with values of that energy. Some, being at one instant virtually motionless, have almost zero kinetic energies; others, having very high speeds, have very high kinetic energies.

From the kinetic theory we therefore obtain for an *ideal gas a picture of pressure being proportional to the mean kinetic energy of the gas particles.* The kinetic theory, however, being based on mechanics, says nothing about temperature. It is from the behavior of *real* gases and the gas laws that we obtain a notion of pressure being related to temperature. Thus by rearranging the terms of the ideal gas law, equation 3.6, we have

$$P = \frac{nR}{V} T \tag{3.10}$$

Since the same pressure is referred to in both approaches, it must be true that, comparing equations 3.9 and 3.10,

$$\frac{2}{3} \frac{N}{V} \left(\frac{1}{2} m\overline{v^2}\right) = \frac{nR}{V} T \tag{3.11}$$

The volumes V may be canceled; for convenience, the total number of gaseous particles N may be taken as Avogadro's number or one mole, n; both the fraction one-third and the gas constant R are constants, of course, The conclusion is therefore quite inescapable. If the kinetic theory is correct,

$$\tfrac{1}{2}m\overline{v^2} \propto T \tag{3.12}$$

The mean kinetic energy of the gaseous particles is proportional to the absolute temperature.

We must remind ourselves that the model is based on all the particles of the gas having the same atomic-sized, perfectly round shape, each of almost zero volume. The theory leads us to the conclusion that a collection of such particles can have energy only by virtue of motions from one point to another—called *translational motion.* If such a gas is heated, it can absorb energy only by experiencing an increase in the mean kinetic energy of translation. Because the masses of the particles are fixed, the mean kinetic energy of translation for such a gas can increase only if the mean speeds of the particles increase. By striking the walls of the container with increasing energy, the particles can force a movable wall (as in a piston-cylinder device) to recede and make more room. Thus we understand at an atom's-eye view what happens when gases behave according to the law of Charles and Gay-Lussac.

Among the real gases, only those belonging to the noble gas family are believed to have perfectly rounded particles, as postulated in the kinetic theory. Other gases have at least two nuclei per molecule. In many respects the nuclei behave *as though* (enter here another model) they are two masses attached to the ends of a coiled spring (Figure 3.7). A particle with this design can have energy *while remaining in one spot.* The particle does not of course remain in one spot, but it can have energy in more ways than just through translational motion. The two masses, for example, can vibrate, giving the particle as a whole *vibrational energy.* The particle can also rotate about its center of gravity, even when that center does not move—somewhat like a dumbbell or a baton being rotated about its center of gravity—giving the particle as a whole *rotational energy.* This description is leading us to a qualitative picture (model) of *internal energy.*

How a System Can Have Energy. Suppose we release a balloon filled with hydrogen gas, a gas less dense than air, inside a large field house. It rises and thumps against the ceiling, illustrating one way in which a system can have energy. The entire system has a net translational motion in one direction. Just before we released the balloon, it had none of this "whole system" kinetic energy, but obviously it did have the potential for delivering such energy. It had, in other words, potential energy, which is another way a system as a whole can have energy by virtue of its position. Now let us go inside the system. Let us inquire about the ways it can have internal energy, energy quite without regard to the location of the system and its gross movement.

Within the balloon the applicability of the kinetic theory almost compels us to believe that the individual molecules of hydrogen are moving about randomly and at high speeds between collisions. The system can therefore have an internal translational energy E_{trans}. The most useful model for the hydrogen molecule is that of two masses (two protons, actually) held together by a spring and vibrating back and forth along the axis of the spring. The molecule therefore has vibrational energy, E_{vib}. Hydrogen molecules can also rotate much as a baton can be twirled, thus possessing rotational energy E_{rot}. It is fairly common knowledge that hydrogen can react with oxygen to form water with the release of considerable energy. This event is accompanied by a redistribution of electrons with respect to the nuclei available. In other words, electronic configurations change. Hydrogen therefore apparently possesses

Translational motion

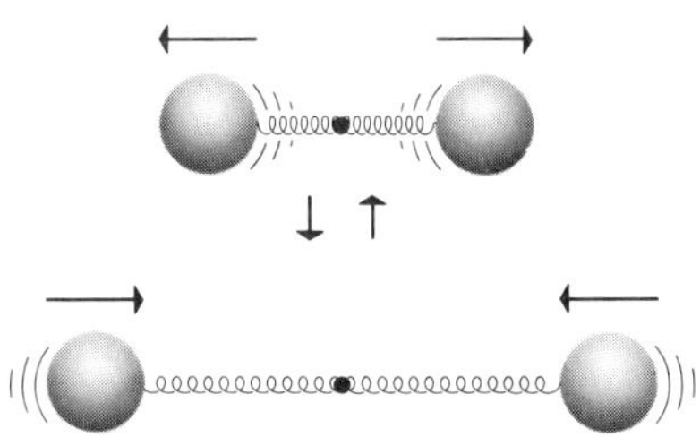

Vibrational motion

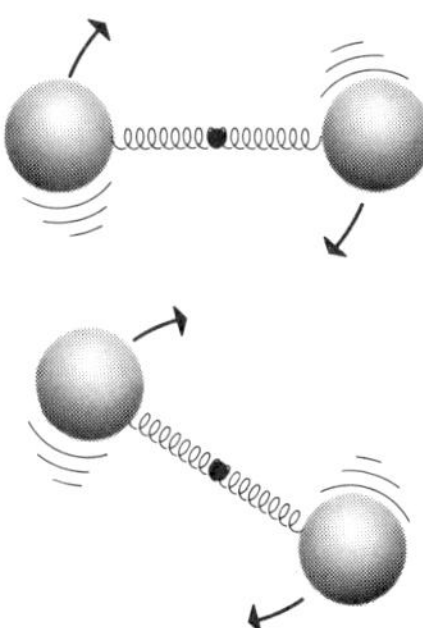

Rotational motion

Figure 3.7 A very good model of a gas particle having more than one nucleus is that of two (or more) balls held together by a spring (or springs). The particle may now have energy by more than just translational motion, which changes the position of its center of gravity in space. It may also have energy through vibrational and rotational motions. All these motions plus other factors (electronic and nuclear) contribute to the total internal energy of a gas.

energy by virtue of its electronic configuration, E_{elect}. This energy may also be inserted into hydrogen molecules by forcing electrons to transfer to higher electronic energy levels. Finally, it is fairly common knowledge that under very special conditions hydrogen nuclei may be made to fuse to form helium. This process consists of more than one step, but it is possible. Part of the enormous energy produced by the sun each day comes from nuclear fusions. For our purposes we need note only that a system may possess energy by virtue of its nuclei, E_{nuc}.

Our model, then, is of a system that can have energy in a number of ways simultaneously, and the total energy E_{total} is the sum of energy terms: a combination of energies:

$$E_{\text{total}} = E_{\text{whole system}} + E_{\text{internal}} \tag{3.13}$$

In chemical changes we are most interested in the internal energy:

$$E_{\text{internal}} = E_{\text{nuc}} + E_{\text{elect}} + E_{\text{rot}} + E_{\text{vib}} + E_{\text{trans}} \tag{3.14}$$

Moreover, in a chemical *change* it is the way in which the energy changes that is of greatest interest. In fact, we need not know the absolute values of the systems we study, just the differences between them. Since in chemical changes E_{nuc} remains the same, energy changes must take place among the other terms.

We therefore represent the change in internal energy accompanying a chemical reaction as

$$\Delta E = E_{\text{int},1} - E_{\text{int},2} \tag{3.15}$$

where ΔE is the increment, plus or minus, in internal energy accompanying the reaction. For a particular change ΔE depends only on the initial and final state, however few or many intermediate steps are needed or used to effect the change.

How a System Can Gain (or Lose) Energy. For a system to undergo a change in its energy, it *must* interact in some way with its surrounding. If a cat (a system) somehow sits down on a hot stove lid (surroundings), energy will move across the boundary. Whatever energy is delivered to the cat is lost by the surroundings. Even here we believe that the law of conservation of energy applies. This example illustrates just one way *energy* transfer can take place between a system and its surroundings, by the transfer of *heat.* If the cat and the stove lid had both had the same initial temperature, no heat would have been transferred.

A system can also gain, or lose, energy through the agency of mechanical work. Imagine that the system is a gas confined in a cylinder-piston device, with the piston connected to a machine in some useful way. If the gas pressure is already great enough, it can push on the piston, producing *work* on the surroundings and an equivalent loss of energy by the system. The converse is also possible. By doing work *on* the system, the work energy is converted into an equivalent amount of system energy. Finally, a system can give or receive energy through electrical processes, but we shall ignore them because of their complexity and because we have no immediate need to understand them.

THERMODYNAMICS

"Thermo-" refers to heat, "-dynamics" to motion or movement. Thermodynamics is therefore a branch of physics and chemistry in which the transfer of heat between a system and its surroundings is studied. From these studies have emerged four statements now called the laws of thermodynamics; we shall study two of them now.

These laws are not derived from more fundamental principles, nor have they been proved in one experiment for all things and for all time. They are basic axioms which, because they are so successful, are regarded as having been proved. In fact, a host of common experiences make them, in retrospect, seem almost intuitively true. Historically, two of them emerged from repeated failures to make various kinds of perpetual motion machines. One such failure by one individual might be regarded as a failure of the person. Repeated failures by many people made scientists wonder eventually whether nature itself rather than the erstwhile engineer or inventor was at fault. Perhaps something in the givenness of nature forever rules perpetual motion machines out of the question. So it overwhelmingly seems.

Zeroth Law of Thermodynamics. A natural law was assigned this strange number because the other laws had been extensively studied and accepted before the importance and the most logical position of this one were realized. Another name for it is the "law of thermal equilibrium." Qualitatively it is quite simple. Two systems in thermal equilibrium with a third are in thermal equilibrium with each other. How do we recognize thermal equilibrium? When the temperatures of two systems have reached the same constant value, they are in thermal equilibrium. If system A is placed in contact with a thermometer, the mercury level will soon reach a constant value. If system B is found to have the same temperature, then A and B are in thermal equilibrium. No net thermal energy will pass spontaneously from one to the other.

First Law of Thermodynamics. In the discussion on page 92 of the general ways in which a system can gain or lose energy, the possibility of producing work from heat was raised. We use the symbol q for heat and w for work. Suppose that a piston-cylinder device containing a gas is our system. If its surroundings are at a higher temperature, heat will flow across the boundary into the system. The heat raises the internal energy of the system and also causes the piston to move out, which produces work elsewhere in the surroundings. The heat q has brought about two changes, a change in the internal energy ΔE and the production of work w. The first law of thermodynamics summarizes the relations between the three quantities q, ΔE, and w as

$$q = \Delta E + w$$

or, as it is commonly seen,

$$\Delta E = q - w \tag{3.16}$$

In words, equation 3.16 says that the change in internal energy of a system is the difference between the heat *absorbed* from the surroundings and the work *done* on the surroundings. We now need sign conventions. Systems, after all, might both give off heat and have work done on them.

q	positive when the system absorbs heat (heat is added: +) negative when the system loses heat (heat leaves: −)
w	positive when the system does work negative when the system has work done on it[6]

The sign of ΔE will obviously depend on the algebraic sum of q and w.

[6] These are conventions, not laws of nature. Unfortunately, there is no international or even local agreement about the most appropriate sign for w. Some references will say that if the system does work, that is, "gives *out*" work, the sign should be the same (minus) as when it gives out heat. Thus the usage for this chapter is arbitrary and conforms with the author's habits as well as the conventions in a large number of references.

Equation 3.16 is a general statement of the law of conservation of energy. It allows for the conversion of energy from one form to another, but it says that the total quantity of energy of all forms is the same before and after the change.

Enthalpy. In the piston-cylinder device we have been using to illustrate principles, the moving of the piston (and thereby the doing of work out in the surroundings) is brought about by some increase in pressure that is relieved by the enlargement of the volume. A change in pressure and volume, in this situation, produces work which chemists usually call *pressure-volume work* (to distinguish it from electrical work, for example). If there were no piston—if the gas were confined in a space with unyielding walls—no work could be done. With $w = 0$, all the heat added to the system would go to increasing its internal energy:

$$q = \Delta E$$

But this situation is so rare that we have no further need for it. In chemistry the most common systems are open to the atmosphere—open beakers, flasks and the like. In living cells the pressure throughout the system is substantially atmospheric. For our future work we shall need a term that is more inclusive than ΔE. Suppose, for example, that a reaction occurs in a flask and a gas is evolved or the level of the matter (e.g., a liquid) in the flask rises. The atmosphere above the flask may not be a piston, but it gets pushed just the same. Work is done on it, pressure-volume work. The heat put into the system changes its internal energy and does some pressure-volume work. The sum of these two forms of energy is defined as the *enthalpy* of the system.

By definition,

$$\text{enthalpy} = \text{internal energy} + \text{energy of a pressure-volume}^{7}\text{ kind}$$

$$H = E + PV \qquad (3.17)$$

By studying changes in enthalpy rather than just changes in internal energy, we can move (figuratively) among the systems that have the most relevance for our future needs—those systems substantially open to the atmosphere, at least to being at the same pressure as the atmosphere.

[7] That the product of P and V has the dimensions of energy is verified by the following analysis:

pressure = force per unit area

force: has the dimensions of MLT^{-2} (see p. 9)

area: has the dimensions of L^2 (length × length = area)

force per unit area: $\dfrac{MLT^{-2}}{L^2} = ML^{-1}T^{-2}$

volume: has the dimensions of L^3

energy: has the dimensions of ML^2T^{-2} (see p. 14)

Thus the answer to the question whether the product of pressure and volume has dimensions of energy is yes:

$$\begin{matrix}(ML^{-1}T^{-2}) & \times & L^3 & = & ML^2T^{-2} \\ P & \times & V & = & \text{energy}\end{matrix}$$

The defining equation for enthalpy, equation 3.17, expresses absolute values for the quantities present, but it is only *changes* in these that are measurable and of interest to us. We are concerned not about absolute values for H but about values of ΔH,

$$\Delta H = H_{\text{final state}} - H_{\text{initial state}}$$

Like ΔE, ΔH depends only on the initial and final states.

The enthalpy change ΔH amounts to nothing more than the thermal energy absorbed by a system (or released) when it is kept at constant pressure, quite without regard to how that heat is partitioned between changing the systems' internal energy and participating in pressure-volume work:

$$\Delta H = q_p \qquad (3.18)$$

refers to constant pressure

When the system loses enthalpy, ΔH is negative and the process is said to be *exothermic*. When the system gains enthalpy, ΔH is positive and the process is *endothermic*.

The difference between ΔH and ΔE is very small except when gases are involved. Chemical reactions carried out, for example, in solutions open to the atmosphere produce essentially miniscule changes, if any, in the volume of the solution and the pressure. At low pressures (e.g., one atmosphere) in liquid and solid systems, the pressure-volume work is virtually zero, and if no other kind of work (e.g., electrical) is done, then, as a good approximation,

$$\Delta E \simeq \Delta H \qquad \text{(for reactions involving liquids and solids only)} \qquad (3.19)$$

Spontaneity in Nature. With this introduction to the first two laws of thermodynamics and to the concepts of internal energy and enthalpy, we can consider some important truths. We look to nature to help us carry some of our burdens, and we tap her resources for useful work. Generally speaking, it is among the spontaneous processes of nature, many of them chemical, that we find energy for doing our work. The relation between work and energy is of such fundamental importance in civilization that wars have been fought over the control of sources of energy. Eventually engineers and physicists and chemists began to analyze the spontaneous processes of nature, asking what features, if any, are common to all of them. In a sense they were asking *what the simplest elements are in terms of which spontaneity can be analyzed.* The existence of *motion* is one of the major aspects of the givenness of nature. Its analysis is carried on in the field of mechanics, and we find three fundamental elements, mass, distance, and time. The existence of *matter* is another major aspect of nature. Its analysis is carried on in the fields of chemistry and nuclear physics, and at one level we find two elements, nuclei and electrons. A third major feature of nature, inseparable from motion and matter, is *spontaneity*. What factors govern the direction of an event? What is it about falling stones, burning gasoline, flowing water, and growing (and dying) plants and animals that these phenomena should be spontaneous in the directions we know? As for chemical changes, it appears that a great percentage of the spontaneous reactions are exothermic and that they are accompanied by the loss of enthalpy from the chemical system. On the surface it seems that enthalpy might be at least one of the "elements of spontaneity," but there is another. We should know more chemistry before attacking it, however, and for that reason we shall delay our study of the next law of thermodynamics. We began a study of thermodynamics with the gas laws and the states of matter, and we now return to these topics.

DEPARTURES FROM IDEAL GAS BEHAVIOR

Van der Waals' Forces. A real gas deviates from ideal gas behavior for two important reasons. The model of the ideal gas assumes that the particles of a gas occupy no volume, but of course they do. This net volume is so small that the error introduced by ignoring it is slight, except at very high pressures where the actual volume of the particles hinders further compression.

Another assumption is that the gaseous particles exert no forces on each other, but for most real gases this is not true. Gas particles are not perfectly elastic, and they are not totally incapable of exerting electric forces on each other. In most situations these are forces of attraction. However slight, they are real. We must leave to Chapter 4 an explanation of their origin, but we can at least acknowledge their special name, van der Waals' forces. The Dutch scientist J. D. van der Waals in 1873 introduced a modification of the ideal gas equation to take into account both the forces of attraction and the effect of particle volume.

We have no need to know the van der Waals equation, only that forces of attraction can exist between otherwise neutral particles. When they are strong enough, the gas no longer remains a gas and coalesces to a liquid or a solid.

THE LIQUID STATE

In contrast to gases, liquids do not expand to fill whatever container holds them. So much is fact. The theory is that slower-moving molecules with stronger forces of attraction between them characterize liquids. The molecules stick together strongly enough to have neighbors in contact virtually all the time, but not so strongly that these neighbors cannot change constantly.

Vapor Pressure. Molecules that happen to be moving upward at the surface of a liquid may escape. If they are not knocked back by collisions with air molecules or other vapor molecules, the escape is permanent. Thus we have the phenomenon of evaporation. Heat, which makes the molecules of a liquid move more rapidly, increases the rate of evaporation. When some liquid (e.g., water) covers the surface of the skin, thermal energy in the layer of skin helps molecules of the liquid overcome attractive forces between them. Those closest to the surface moving most rapidly outward escape in greatest numbers, leaving those with lower energy behind. Lower energy here means lower temperature, and thus evaporation has a cooling effect.

The escape of molecules of a solid or a liquid into the surrounding space pushes on the atmosphere. The "escaping tendencies" of surface molecules vary widely from substance to substance, and these tendencies may be increased or

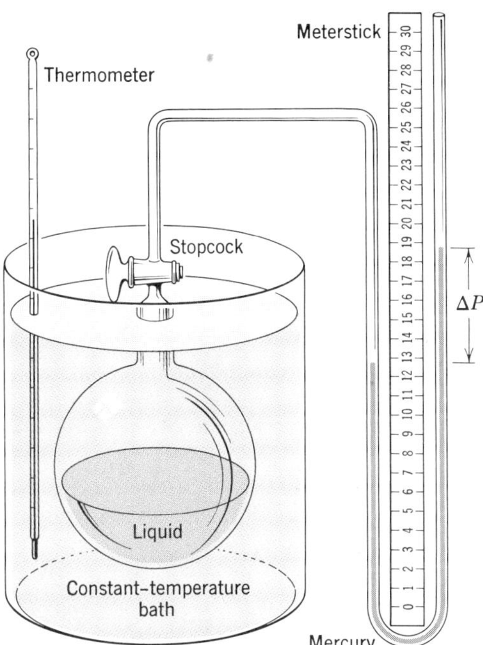
Meterstick
Thermometer
Stopcock
ΔP
Liquid
Constant-temperature
bath
Mercury

Figure 3.8 Measuring the vapor pressure of a liquid at various temperatures. The experimenter can make the bath temperature whatever he wishes by regulating a heater (not shown). Allowing time for thermal equilibrium to be reached within and without the flask, he then opens the stopcock. The change in pressure caused by partial evaporation of the liquid eventually becomes a steady value. Equilibrium between the vapor and the liquid is reached. (The vapor pressure of mercury is so low that, except for the most careful work or at greatly elevated temperatures, it may be ignored. At 100°C, for example, the vapor pressure of mercury is only 0.27 mm.)

decreased by simply changing the temperature of the substance. To measure escaping tendency, the concept of *vapor pressure* has been devised (Figure 3.8). Figure 3.9 shows how vapor pressures of some substances vary with temperature. The dotted line on the graph cuts the vertical axis at 760 mm Hg pressure, the standard atmospheric pressure. When the vapor pressure of a liquid equals atmospheric pressure, whatever that pressure happens to be on the particular day, the liquid "escapes" into the vapor state not just at the surface but virtually everywhere throughout the whole body of the liquid. The bubbles of new vapor forming beneath the surface move upward in a mechanical turbulance called boiling. A liquid boils when its vapor pressure equals the atmospheric pressure. By definition, the *normal boiling point* is the boiling point at exactly one atmosphere (760 mm Hg) pressure.

The vapor pressures plotted in Figure 3.9 are the *equilibrium vapor pressures*. They may be measured in a device very much like that shown in Figure 3.3, but since the change in pressure, rather than the change in volume, is now to be measured, Figure 3.8 better illustrates the method. We can understand what is

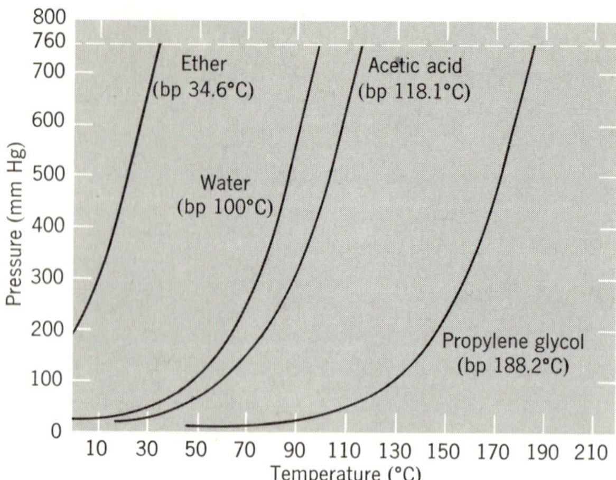
800
760
700
600
500
400
300
200
100
0
Pressure (mm Hg)
Ether
(bp 34.6°C)
Water
(bp 100°C)
Acetic acid
(bp 118.1°C)
Propylene glycol
(bp 188.2°C)
10 30 50 70 90 110 130 150 170 190 210
Temperature (°C)

Figure 3.9 Variation of equilibrium vapor pressures with temperature for some liquids. The boiling points are the *normal* boiling points—the temperatures at which the vapor pressure equals 760 mm Hg. Room temperature is about 20 to 25°C. The relatively high vapor pressure of ether in this range means that it is much more volatile than water. By comparison, proplylene glycol is almost nonvolatile. (Ether is an anesthetic. Acetic acid is the "sour principle" in vinegar. Propylene glycol is a constituent in some brands of permanent antifreeze.)

meant by "equilibrium" by referring to Figure 3.10. Before evaporation starts, no molecules of the liquid have pushed their way in and among the air molecules (the atmosphere) above the surface. As evaporation proceeds, the total number of molecules (those from the liquid plus air molecules) trying to occupy the available space increases, thus raising the total pressure. The change in pressure is due entirely to the appearance of molecules from the liquid.[8] As more and more appear, the chances increase that some will return to the liquid state. The change from vapor to liquid is called *condensation*. Eventually, the number of molecules going from liquid to vapor is exactly balanced by the number returning to the liquid. At this

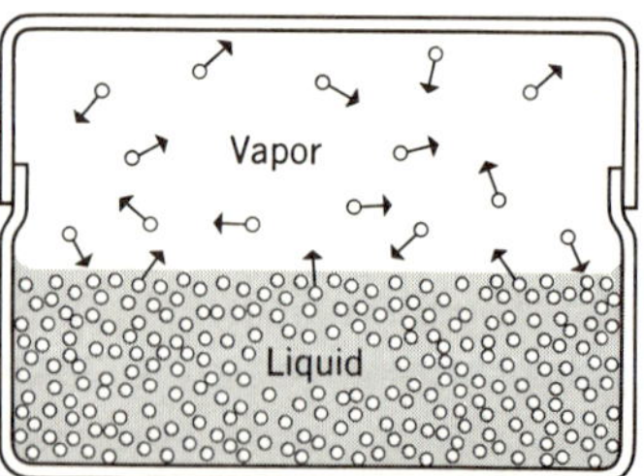
Vapor
Liquid

Figure 3.10 Liquid-vapor equilibrium. When the number of molecules entering the vapor state per unit time equals the number returning in the same time, a condition of dynamic equilibrium exists.

[8] Another of the contributions of John Dalton is his law of partial pressures: The total pressure of a mixture of two or more gases equals the sum of pressures that the individual gases would exert if each alone occupied the given volume. These individual pressures are called partial pressures, symbolized as P_1, P_2, etc. Thus

$$P_{\text{total}} = P_1 + P_2 + P_3 + P_4 + \cdots$$

point a *dynamic equilibrium* exists. There is much coming and going (hence "dynamic"), but no *net* change (hence "equilibrium"). This vapor pressure, the equilibrium vapor pressure, can be measured. More importantly, our study of vapor pressure has provided a simple introduction to one of the most important topics in all chemistry, that of equilibrium phenomena. We are not interested merely in features that change, for the permanent aspects of nature are of equal importance. An old question might be rephrased: What are the simplest elements in terms of which permanence may best be analyzed? The concept of dynamic equilibrium is one of them, and we shall encounter it repeatedly.

THE SOLID STATE

Solids hold their shapes, in contrast with liquids and gases. This much is fact. The theory is that the particles in the solid have fixed rather than changing positions (Figure 3.11). To be sure, they may vibrate about those fixed sites, but they never leave them as long as the substance remains a solid. The forces of attraction between the particles are now strong enough to hold them in place. If the solid is heated, its molecules vibrate more and more violently until eventually their thermal energies are great enough to overcome the forces of attraction. Neighbors start to slip by each other. The solid melts. Particles of the solid at the surfaces may escape into the vapor state directly, a phenomenon called *sublimation*. Crystals of the element iodine are known to sublime readily.

ENERGY RELATIONS IN CHANGES OF STATE

Such changes as solid to liquid or liquid to gas (or vice versa) are called changes of state. Figure 3.12 shows how the temperature will change with *time* if heat is added slowly but steadily to a solid to take it through the liquid and vapor states. The temperature rises steadily with time except for two important points at which the system is apparently able to absorb heat without experiencing a rise in temperature. The first is the melting point of the solid, when it is changing from solid to liquid. The second is the boiling point, when it is changing from liquid to vapor. The heat absorbed during melting or boiling does not change the mean kinetic energy of the particles, because the temperature is constant during these changes of state. Rather the heat energy is required to overcome forces of attrac-

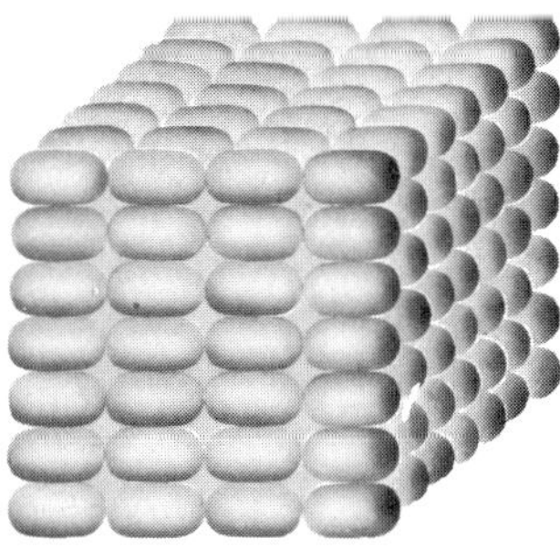

Figure 3.11 The solid state is characterized by an orderly aggregation of particles. Forces of attraction between them are strong enough to hold them in fixed positions, although they vibrate about these points.

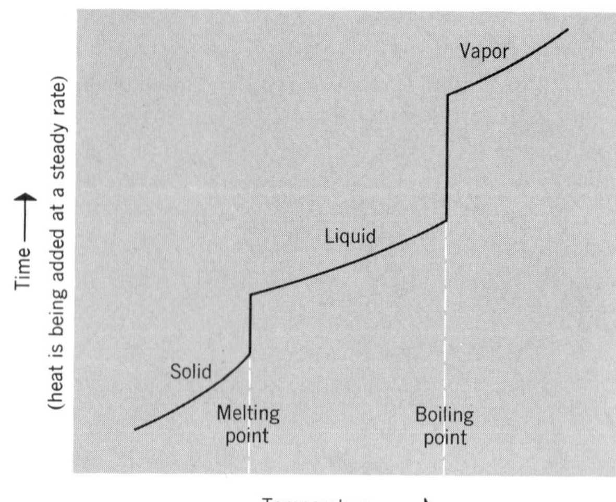
Vapor
Liquid
Solid
Melting point
Boiling point
Time
(heat is being added at a steady rate)
Temperature

Figure 3.12 Heating diagram for changes of state. It is assumed that the rate at which heat is added to the system is constant. Beginning with the solid, the addition of heat changes its temperature until the melting point is reached. Assuming that heat is added slowly enough, no change in temperature will be detected while the solid is in virtual (but not exact) equilibrium with the liquid. Similarly, at the boiling point, the temperature will not change with time while the liquid is in virtual equilibrium with the vapor.

tion between the particles and to cause the vapor to push out against the atmosphere.

The amount of heat required to change a solid to a liquid at the melting point is called its *heat of fusion.* It may be stated in calories per gram, in which case it is called the *specific heat of fusion,* or in calories per mole, the *molar heat of fusion.* Whatever the units, since this is strictly an enthalpy change, the symbol is ΔH_{fus}. For water $\Delta H_{fus} = +79.7$ cal/g, one of the highest specific heats of fusion known. For ethyl alcohol $\Delta H_{fus} = +25$ cal/g, for carbon tetrachloride, $+4$ cal/g.

The *heat of vaporization* of a liquid is the heat needed to convert it to a vapor at its boiling point. Symbolized as ΔH_{vap}, it may be expressed as either the specific or the molar heat of vaporization, in either calories per gram or calories per mole. For water $\Delta H_{vap} = +540$ cal/g, for ethyl alcohol, $+204$ cal/g, for carbon tetrachloride, $+46$ cal/g. Again water has an unusually high ΔH_{vap}, but for all substances the heats of vaporization are much larger than the heats of fusion.

Thus far we have viewed the changes of state as going in one direction. Vapors may condense to liquids, of course, and liquids may freeze to solids. When such changes of state occur, the heats of vaporization and heats of fusion are given off by the system to the surroundings.

REFERENCES AND ANNOTATED READING LIST

BOOKS

A. G. van Melsen. *From Atomos to Atom.* Harper Torchbook TB517, Harper and Row, New York, 1960. Section II traces the history of the concept of the atom from shortly before Dalton. The significance of Dalton's work and the importance of attending to weight relations in chemical changes are placed in proper historical and philosophical perspective.

W. F. Kieffer. *The Mole Concept in Chemistry.* Reinhold Publishing Corporation, New York, 1962. (See page 68.)

L. E. Strong and W. J. Stratton. *Chemical Energy.* Reinhold Publishing Corporation, New York, 1965. Requiring only a knowledge of college algebra, this short book gives freshmen college students an excellent introduction to the concepts of energy and their application in understanding chemical systems. This reference can serve as a supplement to this and several succeeding chapters. (Paperback.)

Harrison Brown. *The Challenge of Man's Future.* Viking Press, New York, 1954. "If we choose to remain ignorant of the facts in this great book, we choose a fool's paradise." So said H. J. Müller, Nobel laureate and geneticist. Brown's concern is with our future needs for energy and food and things without which human life simply cannot progress beyond the ragged edge of existence.

H. A. Bent. *The Second Law.* Oxford University Press, New York, 1965. Without demanding prior knowledge of calculus, the author takes a sometimes whimsical, usually informal, but always rigorous journey through the laws of thermodynamics. This hardcover book contains many problems with worked examples and answers.

MATHEMATICS SELF-HELP BOOKS

A. Vavoulis. *Chemistry Calculations with a Focus on Algebraic Principles.* Holden-Day, San Francisco, 1966. Intended for any beginning chemistry student who has difficulty in applying algebra to chemistry problems, this book discusses all the topics of algebra that are applicable to general chemistry. Many worked examples with applications in chemistry are provided. (Paperback.)

R. J. Flexer and A. S. Flexer. *Programmed Reviews of Mathematics.* Harper and Row, New York, 1967. This series includes the following short volumes: 1. "Fractions." 4. "Exponents and Square Roots." 5. "Logarithms." They are designed to help the user regain quickly and easily simple mathematical tools he may not have used since high school. (Paperbacks.)

ARTICLES

F. J. Dyson. "What is Heat?" *Scientific American,* September 1954, page 58. Heat is discussed as disordered energy in contrast to the ordered energy of work.

M. Wilson. "Count Rumford." *Scientific American,* October 1960, page 158. The overthrow of the caloric theory that heat is a substance is described in this article. Count Rumford's supervision of the boring of cannon in a Bavarian arsenal led him to the notion that heat could not be a substance. Heat seemed to be produced in unlimited quantities as the drill bit into the iron, and the only apparently unlimited aspect of the operation was the motion of the horses powering the drill. Heat somehow had to be related to motion, not substance.

S. W. Angrist. "Perpetual Motion Machines." *Scientific American,* January 1968, page 115. The fruitless efforts to make perpetual motion machines, described in this article, contributed to the discoveries of the laws of thermodynamics.

G. H. Wannier. "The Nature of Solids." *Scientific American,* December 1952, page 39. Evidence for the orderliness in crystals is discussed. The illustrations are excellent.

B. J. Alder and T. E. Wainwright. "Molecular Motions." *Scientific American,* October 1959, page 113. Beautifully illustrated, the article is a discussion of the idea of molecules in motion.

J. W. Westwater. "The Boiling of Liquids." *Scientific American,* June 1954, page. 64. The phenomenon of boiling is shown to be not as simple as it might appear.

PROBLEMS AND EXERCISES

1. Review the meanings of the following terms by writing definitions for them in your own words.

(*a*) chemically pure substance	(*b*) formula weight	(*c*) gram-formula weight
(*d*) mole	(*e*) system	(*f*) surroundings
(*g*) isolated system	(*h*) ideal gas law	(*i*) universal gas constant
(*j*) STP	(*k*) molar volume	(*l*) internal energy
(*m*) enthalpy	(*n*) exothermic	(*o*) endothermic
(*p*) van der Waals' forces	(*q*) equilibrium vapor pressure	(*r*) specific heat of fusion

2. Enumerate the ways in which the postulates of Dalton's atomic theory are now believed to be wrong.
3. In order to solve practical weight relations problems with ease, it is good to be able to handle certain routine operations with a minimum of trouble. One of these operations is the calculation of formula weights from formulas and the information in a table of atomic weights. For practice, calculate formula weights for each of the following compounds. Use atomic weights to the nearest whole numbers, but for chlorine use 35.5.

	Set 1	Set 2	Set 3
(*a*)	S_8	Br_2	$NaOH$
(*b*)	$Al(NO_3)_3$	H_3PO_4	$Ca(NO_3)_2$
(*c*)	$(NH_4)_2CO_3$	$Mg(H_2PO_4)_2$	$Ba(HSO_4)_2$
(*d*)	$C_9H_8O_4$ (aspirin)	$C_{27}H_{46}O$ (cholesterol)	C_8H_{18} (octane)
(*e*)	$Al_2(SO_4)_3$	$Ca_3(PO_4)_2$	$Sn_3[Fe(CN)_6]_2$

Ans. Set 1: (*a*) 256, (*b*) 213, (*c*) 96, (*d*) 180, (*e*) 342.

4. Another important operation is calculating the number of moles of a substance present in any particular sample size. For practice, calculate the number of moles that would be present in each situation. (Formulas given here appeared in exercise 3.)

	Set 1	Set 2	Set 3
(*a*)	1000 g $Al(NO_3)_3$	10 g $Mg(H_2PO_4)_2$	25 g $Ca(NO_3)_2$
(*b*)	100 g NaOH	9.8 g H_3PO_4	10 g $Ba(HSO_4)_2$
(*c*)	1.6 g Br_2	2 g NaOH	10^{-3} g $Ca_3(PO_4)_2$
(*d*)	9.6 g $(NH_4)_2CO_3$	0.387 g cholesterol	193 mg cholesterol
(*e*)	4.5 g aspirin ($C_9H_8O_4$)	1 lb $(NH_4)_2CO_3$	1 kg Br_2

Ans. Set 1: (*a*) 4.7 moles, (*b*) 2.5 moles, (*c*) 0.01 mole, (*d*) 0.1 mole, (*e*) 0.025 mole.

5. Instruments for weighing amounts less than 1×10^{-6} g are very specialized and rare. How many formula units (molecules) of water (H_2O) would be present in 1×10^{-6} g,

assuming that Avogadro's number is 6.02×10^{23}? (If you could count one per second, how many centuries would it take to count the molecules of water in such a small sample?)

6. If one atom of carbon-12 weighs 2×10^{-23} g, how much does one molecule of water weigh? *Ans.* 3×10^{-23} g.
7. Another frequently encountered operation in chemistry is calculation of the number of grams present in any given number of moles of a compound. For practice, calculate the number of grams present in each of the following. (The formulas all appeared in exercise 3.)

	Set 1	Set 2	Set 3
(*a*)	0.1 mole aspirin	0.002 mole $Ca(NO_3)_2$	0.5 mole NaOH
(*b*)	0.3 mole C_8H_{18}	1.69 moles $Ca_3(PO_4)_2$	1.17 moles Br_2
(*c*)	0.02 mole H_3PO_4	25 moles cholesterol	10^{-6} mole cholesterol
(*d*)	2.63 moles Br_2	0.125 mole $(NH_4)_2CO_3$	0.25 mole Br_2
(*e*)	5 moles NaOH	1.11 moles H_3PO_4	0.33 mole H_3PO_4

8. Determine how many grams of each of the following substances would make 0.1 mole: (*a*) HF, (*b*) HCl, (*c*) HBr, (*d*) HI.
9. How many moles would there be in 100 g of each of the compounds in exercise 8?
10. Calculate the percentage compositions of the elements in the following substances. A worked example is given.

    (*a*) H_2O (water) (*b*) CH_4 (methane) (*c*) NH_3 (ammonia)
    (*d*) C_2H_6O (ether) (*e*) $CHCl_3$ (chloroform) (*f*) $C_{12}H_{22}O_{11}$ (table sugar)

    Ans. (*a*) 11.1% H, 88.9% O.

    Example. C_2H_6O.

    Step 1. Calculate the formula weight. (The constituents are some percentage of precisely this number.) For C_2H_6O the formula weight is 46.
    Step 2. Write down the fractions of this weight contributed by each element. These fractions must total one.

    For two carbons (at. wt. 12): $\frac{24}{46}$
    For six hydrogens (at. wt. 1): $\frac{6}{46}$
    For one oxygen (at. wt. 16): $\frac{16}{46}$
    (*Note.* Sum of fractions $= \frac{46}{46} = 1$)

    Step 3. Convert the fractions to percentages which must total 100%.

    For carbon: $\frac{24}{46} \times 100 = 52.2\%$
    For hydrogen: $\frac{6}{46} \times 100 = 13.0$
    For oxygen: $\frac{16}{46} \times 100 = 34.8$
    100.0%

11. Using the ideal gas equation, prove that the formula weight of a gas may be calculated from the following equation:

$$\text{F. wt.} = w\frac{RT}{PV} \qquad \text{where } w = \text{weight, in grams, of the sample of gas}$$

12. A sample of an "unknown" gas weighing 0.065 g occupies a volume of 100 cc at a temperature of 27°C and a pressure of 760 mm Hg. Calculate its formula weight. *Note.* To use the ideal gas equation, or the form of it given in exercise 11, you must put all data into the correct units for a value of $R = 0.082$ (liter)(atm)/(°K)(mole)

13. The gas of exercise 12 was analyzed and found to contain its elements in the following percentages: carbon, 75.0%; hydrogen, 25.0%. Compute its molecular formula.

Example. Compute the atom ratio of the elements in a sample containing 11.1% hydrogen, 88.9% oxygen.

Step 1. Treat percentage data as though they told you how many *grams* of the element are present in 100 g of the substance, which is really what they do say.

Hydrogen: 11.1 g H per 100 g substance
Oxygen: 88.9 g O per 100 g substance

Step 2. Convert the grams to moles:

$$11.1 \text{ g H at } 1 \text{ g/mole} = 11.1 \text{ moles}$$

$$88.9 \text{ g O at } 16 \text{ g/mole} = \frac{88.9 \text{ g}}{16 \text{ g/mole}} = 5.55 \text{ moles}$$

Step 3. Reduce the mole ratio to the set of smallest *whole* numbers that still express that ratio:

$$11.1{:}5.55 \quad \text{as} \quad \frac{11.1}{5.55}{:}\frac{5.55}{5.55} \quad \text{as} \quad 2{:}1$$

The atom ratio is 2H to 1O, the most likely formula H_2O.

14. Compute the percentage compositions of the elements in formulas HO and H_2O_2. The first is hypothetical as written, and the second is the actual formula of hydrogen peroxide.

Note. This problem introduces a major exception to the rule (p. 77) that the smallest whole-number ratio is to be used for representing a substance. The smallest whole-number ratio represents the *empirical formula;* the *molecular formula* may be some simple multiple of the empirical formula, and it will always show the actual composition of a formula unit. Benzene (C_6H_6) and acetylene (C_2H_2) have different molecular formulas but the same empirical formula CH.

(*a*) How do the percentage compositions of HO and H_2O_2 compare?

(*b*) What further experimental quantitative data would be necessary in addition to the percentage compositions to know which of the two had been analyzed?

(*c*) Compute the percentage compositions of benzene and acetylene.

15. An unknown organic compound, a gas, was analyzed and the following data were obtained: carbon, 85.7%; H, 14.3%. (Could any other elements be present? How can you tell?) A sample of this gas weighing 0.171 g occupied a volume of 100 cc at a temperature of 27°C and a pressure of 1 atmosphere. (*a*) Calculate its empirical formula. (*b*) Calculate its molecular formula.

16. Write balanced equations to symbolize the following statements.

(*a*) Hydrogen (H_2) reacts with chlorine (Cl_2) to form hydrogen chloride (HCl).

(*b*) Sodium combines with chlorine (Cl_2) to form sodium chloride (NaCl).

(*c*) Methane (CH_4) reacts with oxygen (O_2) to form carbon dioxide (CO_2) and water (H_2O).

17. If a fixed weight of a gas is kept at a constant temperature, and if its volume at 760 mm Hg is 100 cc, calculate its volume at each of the following pressures: (*a*) 800 mm Hg, (*b*) 380 mm Hg, (*c*) 2 atmospheres. *Ans.* (*a*) 95 cc.

18. If a fixed weight of a gas is kept at a constant pressure, and if its volume at 25°C is 100 cc, calculate its volume at each of the following temperatures: (*a*) 0°C, (*b*) 100°C, (*c*) 200°C. Remember to convert to the Kelvin scale. *Ans.* (*a*) 91.6 cc.

19. How many calories are required to convert 1 kg of ice to water at 0°C?

20. The same amount of heat that is the answer to exercise 19 will convert how many grams of ethyl alcohol from its solid state to its liquid state at its melting temperature?
21. What is the enthalpy change for the conversion of 100 g of water to steam at 100°C, the boiling point of water?
22. Using the mechanical model of atoms and molecules in motion and certain features of the states of matter ascertained through this model, offer explanations for each of the following common observations. Explain them, in other words, from a molecule's-eye view.
    (*a*) Moisture evaporates faster in a breeze than in still air.
    (*b*) Ice melts faster if it is crushed than if it is left in one block.
    (*c*) Warm water evaporates faster than cool water.
    (*d*) Wet clothes, hung out to dry in freezing weather, will eventually become completely dry even though they "freeze" first.
    (*e*) Sandwich bread will dry out in a matter of minutes in Death Valley when the temperature stands at 90°C, but at exactly the same temperature in Atlantic City it will take longer. (Assume no differences in wind conditions.)

CHAPTER FOUR

Substances and Structure. Chemical Bonds

TYPES OF SUBSTANCES

When the properties of hundreds and thousands of substances are surveyed, it becomes more and more apparent that these substances should be grouped into classes. Figure 4.1 names many of these classes and shows, in general, how they are related.

Some of the border lines of our categories are fuzzy, which is nearly always a problem with classification schemes. In Chapter 2, for example, we found an indistinct border line between metals and nonmetals. In this chapter, in fact, throughout the rest of the book, the distinction between covalent substances and ionic substances will sometimes be a problem. But it is easier to live with apparent exceptions and substances that defy classification than to have no schemes at all. And in the final analysis we need not worry whether a classification scheme is exact or inexact. We ask only that it be useful.

In Chapters 2 and 3 we looked principally at elements, but compounds were introduced. In daily life we see only a few elements, for we live and move among compounds. Our coins have silver, nickel, and copper in them. Jewelry may contain gold. Aluminum and iron products are common. We live bathed in an invisible atmosphere of oxygen and nitrogen, although we cannot see them. From Figure 4.1 we learn that there are two main families, covalent compounds and ionic compounds. The membership of a particular compound in either of these two families is *determined* by the characteristics tabulated in Figure 4.2, and it is *understood* in terms of the kinds of bonds that are said to exist in the compound. With this statement we make our first major move into the domain of one of the grand unifying principles of chemistry. *Whatever properties a compound has are determined mainly by its structure, that is, the arrangement of electrons and nuclei.* To understand structures we must understand what holds these particles together. Whatever this turns out to be, we shall call it a *bond.*

Chemical Bonds. The idea of a chemical bond has been developed in answer to a basic question. Unless the question is understood, the answers will not mean much regardless of the level of the discussion. To begin with, we know that in compound particles two or more nuclei coexist together with enough electrons to render the particles both neutral and reasonably stable. The basic question is this. How is it possible for like-charged particles such as atomic nuclei to keep from flying apart? We may also ask why the electrons do not repel each other to infinity. How can these particles "coalesce" into substances of enough durability that we say *bonds* exist?

The natural behavior for nuclei is to repel each other. Electrons repel each other too. Yet somehow in chemical compounds positively charged nuclei from two or more elements become fixed in relation to each other. They neither touch each other nor adopt perfectly rigid positions. Even though there are internal vibrations and rotations, it is still possible to say that nuclei in chemical compounds are stationed at some very small average distance from each other and that bonds exist between them. Thus the central problem in understanding a chemical bond is in discovering what prevents nuclei in compounds from flying apart, and what prevents the electrons from flying apart?

Stating the problem this way offers the first faint glimmer of an answer. It is

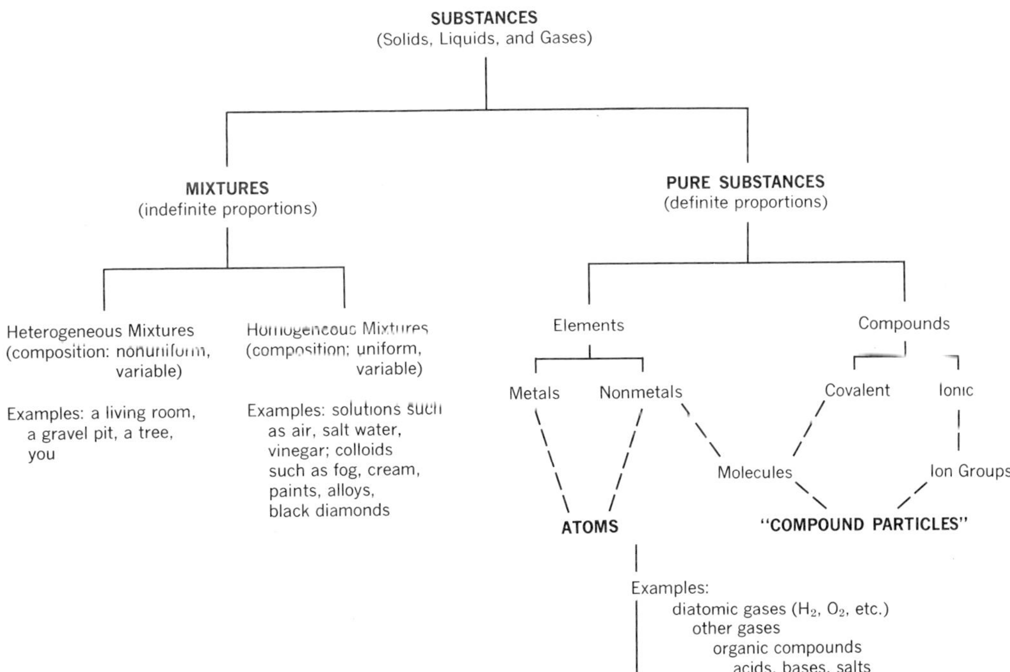
SUBSTANCES
(Solids, Liquids, and Gases)
MIXTURES
(indefinite proportions)
PURE SUBSTANCES
(definite proportions)
Heterogeneous Mixtures
(composition: nonuniform, variable)
Examples: a living room, a gravel pit, a tree, you
Homogeneous Mixtures
(composition: uniform, variable)
Examples: solutions such as air, salt water, vinegar; colloids such as fog, cream, paints, alloys, black diamonds
Elements
Metals
Nonmetals
Compounds
Covalent
Ionic
Molecules
Ion Groups
ATOMS
"COMPOUND PARTICLES"
Examples:
diatomic gases (H_2, O_2, etc.)
other gases
organic compounds
acids, bases, salts

Figure 4.1 A scheme for classifying matter.

IONIC SUBSTANCES	COVALENT SUBSTANCES
A Common Characteristic	
In the pure state at room temperature only very rarely will substances of either category conduct electricity. (Metals are omitted from both families; carbon, a nonmetal, does conduct electricity, but it would be classified as covalent.)	
Most Significant Differences	
—If soluble in water, the solution will conduct electricity.	—If soluble in water, the solution will not conduct electricity (with only a few exceptions).
—If insoluble in water but fusible, the molten mass will conduct electricity.	—If insoluble in water but fusible, the melt will not conduct electricity.
Other Common Differences	
—Percentage of members of this class that are somewhat soluble in water is relatively high.	—Small percentage is soluble in water.
—Virtually no members dissolve in solvents such as carbon tetrachloride, gasoline, benzene, ether, alcohol, etc. (organic solvents).	—Members are generally much more soluble in the organic solvents.
—Rarely does the element carbon furnish nuclei for these compounds (exceptions: bicarbonates, carbonates, cyanides).	—It is very common to find carbon nuclei in these compounds. (This class includes virtually all organic compounds.)
—At room temperature all are solids.	—This class includes all the gases, all the liquids, and innumerable solids (room temperature being assumed).
—Virtually none will burn.	—Nearly all will burn.
—Melting points are almost always above 350° C and are usually much higher.	—Melting points are usually well below 350°C, but some char and decompose before they melt.
Examples	
—Salts (e.g., sodium chloride, barium sulfate) —Metal oxides (sodium oxide, iron oxides) —Carbonates and bicarbonates	—Water, alcohols, sugars, fats and oils, lacquers, perfumes, most drugs and dyes

Figure 4.2 Ionic and covalent substances, general descriptions.

common knowledge that two north magnetic poles on two magnets behave as though they repel each other. South magnetic poles act in the same way. Yet we have no difficulty in making two magnetized bars of iron stick together. In fact, they could not be made to cling at all if they were not magnetized. It is just that they will cling *only in a particular way.* Proper *configuration* is the answer. In one arrangement of the two magnets repulsive forces are minimized and attractive forces are maximized, resulting in a net force of attraction.

The situation in a chemical compound is not identical, but it is analogous. Basically, to understand how a collection of atoms can become organized into a compound particle, we develop models that take advantage of the behavior of electrified particles of opposite charge.[1] The reorganization of the electrons and the atomic nuclei that occurs when compound particles form (or undergo reactions with each other) is an event we call a chemical reaction. The two classes of substances, ionic and covalent compounds, are understood as the outcomes of two generally different courses this reorganization can take. To explain the two classes we think in terms of two kinds of bonds, ionic bonds and covalent bonds.

IONIC COMPOUNDS

The time-honored example for illustrating the formation and nature of an ionic bond is the reaction of the elements sodium and chlorine to form table salt, sodium chloride. The balanced equation for this reaction, which, incidentally is violent, is

$$2\text{Na(solid)} + \text{Cl}_2\text{(gas)} \longrightarrow 2\text{NaCl(solid)}$$

Ignoring for the moment the fact that the element chlorine has the formula Cl_2 rather than simply Cl, we look first at the reaction as a rearrangement of electrons in relation to nuclei. Thus one effect of the reaction is the transfer of an electron from the sodium atom to the chlorine atom; a simple Bohr picture will help us visualize this:

11+ 12n | 2 8 1 | K L M + 17+ 18n | 2 8 7 | K L M ⟶ [11+ 12n | 2 8 | K L]$^+$ + [17+ 18n | 2 8 8 | K L M]$^-$

Sodium atom | Chlorine atom | Sodium ion | Chloride ion

$$\text{Na} + \text{Cl} \longrightarrow \text{Na}^+ + \text{Cl}^-$$

Sodium, by itself, is very stable[2] and so is chlorine. In the presence of each other, however, they have a more stable way of arranging a total of twenty-eight electrons between their two nuclei. The two new particles are still atomic in size, and they carry net electric charges. (Verify this by comparing in each case the sums of minus

[1] Before pursuing this subject in detail, we must explain common usage for certain words and phrases in chemistry. Chemists say, for example, that elements or atoms are "present" in compounds, or they speak of the compounds as "consisting of atoms." These are manners of speaking. Compounds are not collections of neutral atoms in the way that gravel is a collection of stones and pebbles. The *parts* of atoms make up compounds. Only in a *chemical* sense do we say that table salt, for example, *contains* sodium and chlorine. It contains the *nuclei* of these elements, but not the atoms themselves. (Enough electrons, of course, are present to render the whole sample electrically neutral.) Thus, when phrases such as "atoms in a molecule" or "atoms in a compound" are used, regard these as convenient manners of speaking. The pieces of atoms are present but not the intact atoms.

[2] As used here, stability means the nonexistence of an alternative arrangement of electrons and nuclei of less internal energy (to be qualified in Chapter 7).

and plus charges.) Such particles are called *ions.* Because they are of opposite electric charge, they must inevitably attract each other as soon as they are made. In an actual reaction hundreds of billions of each form. To minimize repulsions between like-charged sodium ions and between like-charged chloride ions and to maximize attractions between sodium and chloride ions, these aggregate in an orderly way (Figures 4.3 and 4.4). Each positive ion is surrounded by negative ions as closest neighbors, and each negative ion is surrounded by positive ions. When enough ions have aggregated, a visible crystal of an *ionic compound* appears. *Ionic compounds, in general, consist of orderly aggregations of oppositely charged ions assembled in an arrangement of maximum stability and in a definite ratio to assure overall electrical neutrality.* Crystals have regular shapes because the components have assembled in a regular way.

Redox Processes. Terminology. Of the types of reactions that chemists recognize, the reaction of sodium with chlorine is classified as an *oxidation-reduction* or a *redox* reaction. By definition, the loss of an electron (or more) by a particle is called *oxidation.* In the example given sodium atoms are *oxidized* by chlorine, the *oxidizing agent.* The gain of an electron (or more) is called *reduction.* The chlorine atom is *reduced* by sodium, the *reducing agent.*

Electron transfer processes are the basis for all electric batteries. The battery is the engineering solution to the problem of making the transfer of electrons occur through an *external* circuit where the flow of electrons can be forced to do work. It illustrates how man has successfully tapped spontaneous processes in nature for energy.

Thermochemistry of Reactions. Hess's Law. Besides being highly exothermic, the reaction of sodium and chlorine is also spontaneous. In Chapter 3 it was suggested that one of the elements of spontaneity might be the enthalpy change in a reaction. Therefore the *direction* of this reaction might be understood in terms of the enthalpy changes that accompany it. The ΔH for the reaction is a fairly large *negative* number, -98.23 kcal/mole. (This energy may be measured directly by means of a calorimeter, a device that in essence makes possible the translation of

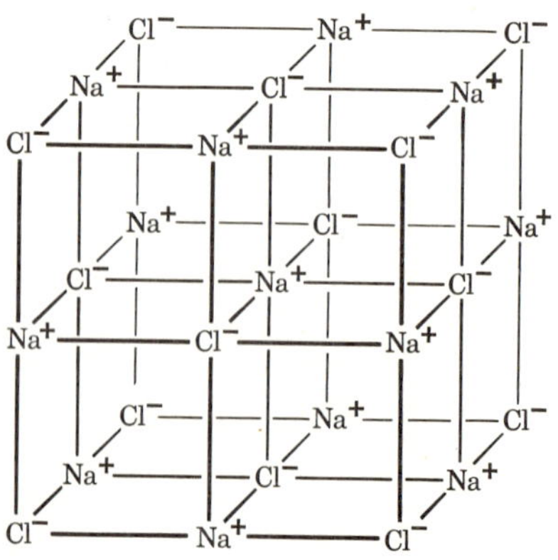

Figure 4.3 The relation of ions in sodium chloride (table salt).

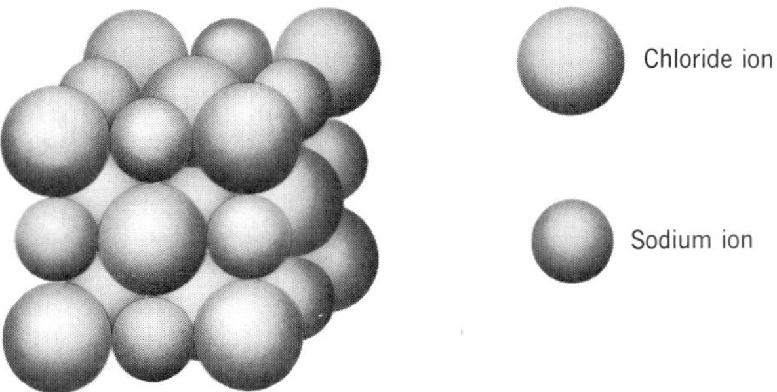
Chloride ion
Sodium ion

Figure 4.4 Orderly aggregation of ions in crystalline sodium chloride.

the temperature change accompanying a reaction into the heat that must have been evolved, or absorbed, to cause the temperature change.)

We can look upon the reaction of sodium with chlorine as occurring in steps, each step having its own independently measured enthalpy change. The algebraic sum of the enthalpy changes for the steps is the net enthalpy change for the overall reaction. This is true regardless of the particular series of steps or paths used in the analysis as long as the end result is the same. Whatever the series of steps between the initial and final state of the system, the "resting" internal energies of the chemicals chosen initially and of the products finally formed must be independent of the route from reactants to products. The condition or the *state* of chemicals in separate bottles at any given temperature must be a function only of that temperature, the pressure, and the nature of the chemicals themselves. Sodium chloride, whether it is made by a combination of the elements, by the evaporation of seawater, or by any of several other alternative paths, is still sodium chloride. Therefore the net *heat of formation* of sodium chloride must be independent of the path used to make it, provided all the enthalpy measurements are related to some common standard. Our intuitive approach is leading us to a rule about the thermodynamics of a chemical change or the *thermochemistry* of the change. Called *Hess's law of constant heat summation,* the rule states that the heat evolved or absorbed at constant pressure for any chemical reaction is the same no matter what path is selected to bring about the change. The choice of path may affect the *rate* of the change, how fast or how slow the reaction takes place, but it cannot affect the net energy change.

We can envision for the formation of sodium chloride the following steps, at least on paper.

1. Sodium metal at 25°C is vaporized to free its atoms from the solid state. *Energy of sublimation* is required:

$$\mathrm{Na}(s) \xrightarrow[\text{energy}]{\text{sublimation}} \mathrm{Na}(g)$$

where s and g designate solid and gas (l would designate liquid)

2. Then we imagine the insertion of enough energy, the *ionization energy,* to pull an electron out of the gaseous sodium atom:

$$\text{Na}(g) \xrightarrow[\text{ionization energy}]{} \text{Na}^+(g) + e^-$$

3. Now it is chlorine's turn. First we imagine the use of enough energy, *dissociation energy,* to convert Cl_2 at 25°C into chlorine atoms:

$$\tfrac{1}{2}\text{Cl}_2(g) \xrightarrow[\text{dissociation energy}]{} \text{Cl}(g)$$

4. Then the chlorine atom is given the electron "produced" in step 2. A chlorine atom has an affinity for an electron. This is actually an energy-releasing step:

$$\text{Cl}(g) + e^- \xrightarrow[\text{electron affinity}]{} \text{Cl}^-(g)$$

5. Then we imagine the gaseous ions combining to form a solid crystalline lattice at a final temperature of 25°C:

$$\text{Na}^+(g) + \text{Cl}^-(g) \xrightarrow[\text{lattice energy}]{} \text{NaCl}(s)$$

Each step will either require or release energy. The algebraic sum must be reasonably close to the enthalpy of formation of sodium chloride directly from the elements.

	Reaction	
1.	$\text{Na}(s) \xrightarrow[\text{sublimation energy}]{} \text{Na}(g)$	$\Delta H = 26.0$ kcal/mole Na
2.	$\text{Na}(g) \xrightarrow[\text{ionization energy}]{} \text{Na}^+(g) + e^-$	$\Delta H = 118.0$ kcal/mole Na
3.	$\tfrac{1}{2}\text{Cl}_2(g) \xrightarrow[\text{dissociation energy}]{} \text{Cl}(g)$	$\Delta H = 28.9$ kcal/mole Cl
4.	$e^- + \text{Cl}(g) \xrightarrow[\text{electron affinity}]{} \text{Cl}^-(g)$	$\Delta H = -87.3$ kcal/mole Cl
5.	$\text{Na}^+(g) + \text{Cl}^-(g) \xrightarrow[\text{lattice energy}]{} \text{NaCl}(s)$	$\Delta H = -183.8$ kcal/mole NaCl
Algebraic sum of steps and ΔH values	$\text{Na}(s) + \tfrac{1}{2}\text{Cl}_2(g) \xrightarrow[\text{heat of formation}]{} \text{NaCl}(s)$	From the steps: $\Delta H_f^\circ = -98.2$ kcal/mole NaCl From the direct conversion: $\Delta H_f^\circ = -98.23$ kcal/mole NaCl

Many features are noteworthy in this summing up. First, various descriptive names are introduced to identify the energies associated with certain processes. Second, we may *add chemical steps algebraically* provided cancellation of particles on opposite sides of the arrows is restricted to those *in the same state.* We may cancel Cl(g) with Cl(g) but not with $\tfrac{1}{2}Cl_2(g)$ or with Cl(s) or Cl(l) if these states were possible and represented. Third, the chloride ion Cl^- is apparently much more stable than the chlorine atom (ΔH is large and negative for step 4). For this reason the energy term in step 4 is called *electron affinity;* the chlorine atom as well as the atoms of all the group VII elements, the halogens, have this affinity for an electron. Energy is given off when halogen atoms become ions.

Finally, the biggest contributor to stability is the formation of the crystalline lattice, the orderly aggregation of Na^+ and Cl^- ions. Sodium ions are *less* stable than sodium atoms (assuming that no other chemical is about). It takes considerable energy, 118.0 kcal/mole, to force an electron away from a sodium atom. The problem created by this energy demand is erased by the apparently huge stabilizing influence of the crystalline environment in which sodium ions are surrounded by oppositely charged chloride ions. The *lattice energy is our measure of this influence.* The energy relations in a change such as this are diagramed in a Born-Haber cycle (Figure 4.5).

The formation of sodium chloride from the elements is accompanied by a large negative enthalpy change, the heat of formation at constant pressure. So much is fact. That the reaction is spontaneous is another fact. We shall therefore keep active our theory that a negative enthalpy change is an important element of spontaneity. It may be a major element.

Standard States. Throughout our discussion of the heat of formation of sodium chloride as calculated by Hess's law, the starting temperature was speci-

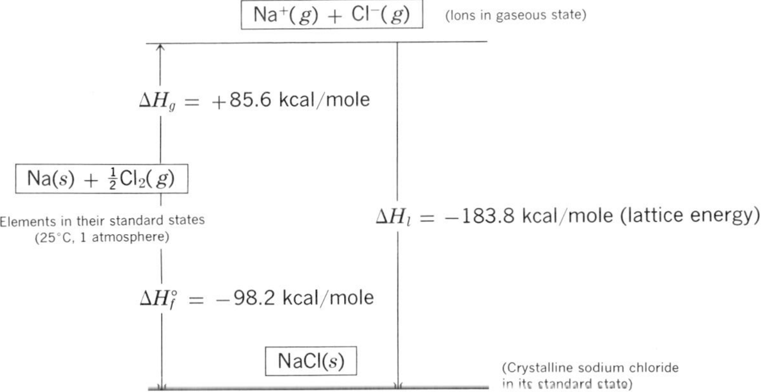
$Na^+(g) + Cl^-(g)$
(Ions in gaseous state)
$\Delta H_g = +85.6$ kcal/mole
$Na(s) + \frac{1}{2}Cl_2(g)$
Elements in their standard states
(25°C, 1 atmosphere)
$\Delta H_l = -183.8$ kcal/mole (lattice energy)
$\Delta H_f^\circ = -98.2$ kcal/mole
$NaCl(s)$
(Crystalline sodium chloride
in its standard state)

Figure 4.5 The Born-Haber cycle for sodium chloride. The energy-consuming enthalpy change for the formation of the separated gaseous ions, designated ΔH_g, is actually the algebraic sum of steps 1 to 4, page 112. The enthalpy change for the formation of the lattice of orderly arranged ions, the lattice energy ΔH_l is of such a high negative value that the net enthalpy change ΔH_F° is also negative. Thus sodium nuclei and chlorine nuclei plus all their electrons can be in each other's presence only by arranging electrons in such a manner that there are no longer sodium and chlorine atoms but sodium and chloride ions. Cycles such as these have been worked out for several reactions. They are named in honor of Max Born (1882–) and Fritz Haber (1868–1934) who made contributions to the theory of solid states and in other areas. Both received Nobel prizes, Born the physics prize in 1954 (with Walther Bethe) for statistical studies on wave functions, and Haber the chemistry prize in 1918 for the first direct synthesis of ammonia from its elements.

fied as 25°C. At the end we brought the product NaCl down to that temperature. Obviously, some temperature must be selected for reference. Less obvious, but equally important, is the need to select a *standard state,* a reference state for an element. When a substance is in its most stable form at 1 atmosphere pressure and at a temperature of 25°C (298°K), it is defined as being in its standard state. All enthalpy *changes* are therefore measured from this reference. It is as though the absolute enthalpy of an element were zero under these standard conditions and enthalpy changes were measured from this "ground zero of energy." Thus enthalpies of formation of sodium chloride by either the direct or step route are for the following overall change: *The elements sodium and chlorine under standard conditions are converted into crystalline sodium chloride at 25°C and 1 atmosphere pressure.* The heat evolved has been dissipated and measured and the product is at standard conditions before we compare the heats of formation of different routes. When determined under these standard conditions, the result is called the *standard heat of formation* (more correctly, the standard *enthalpy* of formation). The symbol is ΔH_f° with the degree superscript commonly meaning "standard."

Why $NaCl_2$ Cannot Form. Referring to step 2 in the step-by-step formation of sodium chloride, or referring to the Born-Haber cycle (Figure 4.5), why cannot enough energy be inserted into sodium to force it to release two electrons and generate Na^{2+} ions? These would attract Cl^- ions even more strongly and thereby make an even more stable crystal. This does not happen, but why not? The difficulty is that it would take 1088 kcal/mole more than the 118 kcal/mole already invested to move the second electron out of the sodium atom:

$$Na(g) \xrightarrow[+118\text{ kcal/mole}]{\nearrow\text{ first } e^-} Na^+(g) \xrightarrow[+1088\text{ kcal/mole}]{\nearrow\text{ second } e^-} Na^{2+}(g)$$

Even though the lattice energy (step 5) would be more favorable (larger *negative* value), it would apparently not be enough to compensate for the huge energy cost of moving the second electron out. Therefore NaCl forms, not $NaCl_2$. This simple example is a very striking illustration of how a study of energy relations can contribute to an understanding of chemical phenomena.

Noble Gas Configurations. The group 0 elements, He, Ne, Ar, Kr, Xe, and Rn, are exceptionally stable. They form very, very few compounds. For review, their electronic configurations are repeated in Table 4.1 together with those of other families and their stable ions. Whatever the cause, these electronic configurations of the noble gases are extremely stable. We infer this from the poverty of their chemistry, and we have hard evidence for it from another source, their ionization energies, the energy needed to remove an electron from an atom of the element. How these gases compare with the group I elements as well as the rest of the periodic system may be seen in Figure 4.6. The noble gas elements appear at the peaks, the alkali metals in the valleys. The alkali metals have the lowest ionization energies of all, but when they are forced to yield a second electron, the energies needed for *second*-electron ionization are very high, as we have seen for sodium. It is almost as though the common *ions* of the group I elements were the *atoms* of the noble gas elements. They are very stable.

The electronic configurations of these singly charged, positive *ions* of the alkali metals, shown in Table 4.1, are the same as those of the nearest noble gases. The sacrifice of electrical neutrality seems to purchase stability (at least in a crystalline lattice). When we move to group II, the alkaline earths, the cost of removing the second electron after the first has gone is much less than in group I. For example,

$$\begin{array}{llll} Mg(g) \longrightarrow Mg^{+}(g) + e^{-} & & \Delta H = 176 \text{ kcal/mole} \\ Mg^{+}(g) \longrightarrow Mg^{2+}(g) + e^{-} & & \Delta H = 347 \text{ kcal/mole} \\ \hline Mg(g) \longrightarrow Mg^{2+}(g) + 2e^{-} & \text{Total} & \Delta H = 523 \text{ kcal/mole} \end{array}$$

The lattice energy made available (about -600 kcal/mole) when *doubly* charged magnesium ions aggregate with chloride ions more than repays the cost of making them. This lattice energy is much more than would be available if only singly charged magnesium ions "condensed" with chloride ions to form MgCl. The overall enthalpy advantage lies with taking two electrons from magnesium atoms so that $MgCl_2$ can form rather than with the initially cheaper route of taking only one electron and forming MgCl.

Table 4.1 Configurations of Ions and of Atoms of the Noble Gases

Group	Common Element	Nuclear Charge (at. no.)	Electronic Configurations *K*	*L*	*M*	*N*	*O*	*P*	Ion	Electronic Configurations *K*	*L*	*M*	*N*	*O*	*P*	Comparable Noble Gas
I:	Li	3+	2	1					Li^+	2						Helium
Alkali	Na	11+	2	8	1				Na^+	2	8					Neon
metals	K	19+	2	8	8	1			K^+	2	8	8				Argon
II:	Mg	12+	2	8	2				Mg^{2+}	2	8					Neon
Alkaline	Ca	20+	2	8	8	2			Ca^{2+}	2	8	8				Argon
earth metals	Ba	56+	2	8	18	18	8	2	Ba^{2+}	2	8	18	18	8		Xenon
VI:	O	8+	2	6					O^{2-}	2	8					Neon
	S	16+	2	8	6				S^{2-}	2	8	8				Argon
VII:	F	9+	2	7					F^-	2	8					Neon
Halogens	Cl	17+	2	8	7				Cl^-	2	8	8				Argon
	Br	35+	2	8	18	7			Br^-	2	8	18	8			Krypton
	I	53+	2	8	18	18	7		I^-	2	8	18	18	8		Xenon
O:	He	2+	2													
Noble	Ne	10+	2	8												
gases	Ar	18+	2	8	8				The noble gases do not form stable ions.							
	Kr	36+	2	8	18	8										
	Xe	54+	2	8	18	18	8									

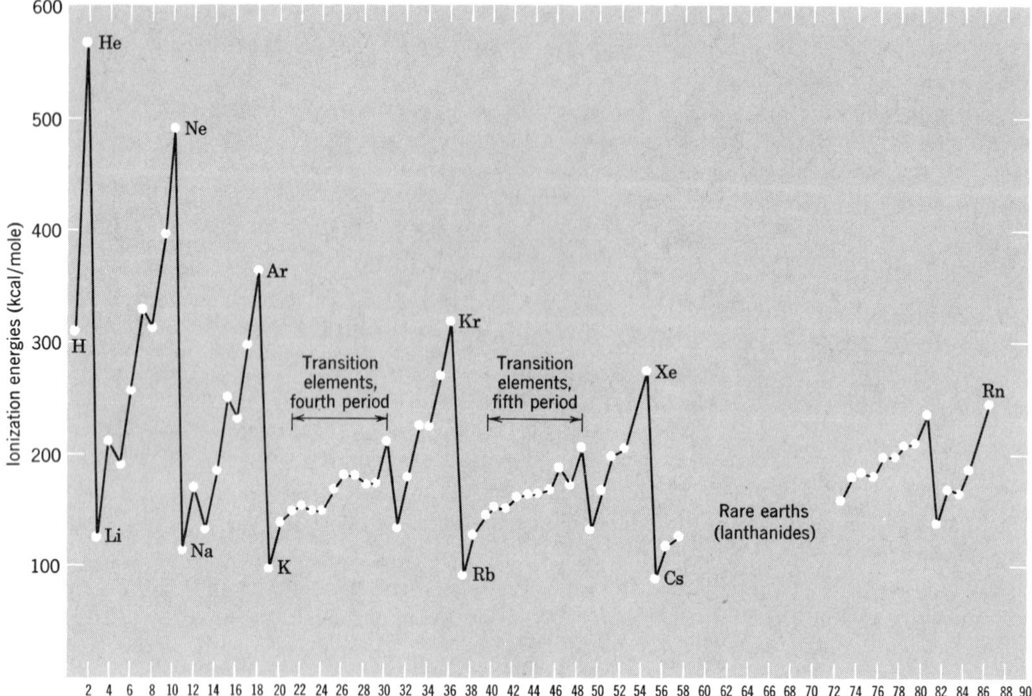

Ionization energies (kcal/mole)
600
500
400
300
200
100
He
Ne
Ar
Kr
Xe
Rn
H
Li
Na
K
Rb
Cs
Transition elements, fourth period
Transition elements, fifth period
Rare earths (lanthanides)
2 4 6 8 10 12 14 16 18 20 22 24 26 28 30 32 34 36 38 40 42 44 46 48 50 52 54 56 58 60 62 64 66 68 70 72 74 76 78 80 82 84 86 88 90
Atomic number

Figure 4.6 Ionization energies in kilocalories per mole for the loss of the first electron from the neutral gaseous atom.

Let us look at this event diagrammatically, using a simple Bohr model:

12+ 12n 2 8 1 1 K L M; 17+ 18n 2 8 7 K L M; 17+ 18n 2 8 7 K L M ⟶ [12+ 12n 2 8 K L]$^{2+}$ [17+ 18n 2 8 8 K L M]$^{-}$ [17+ 18n 2 8 8 K L M]$^{-}$

Magnesium atom — Chlorine atoms — Magnesium ion — Chloride ions

$$Mg + 2Cl \longrightarrow Mg^{2+} + 2Cl^{-}$$

Again, the resultant electronic configurations are those of noble gases. The rest of the group II elements behave like magnesium in this respect: their most stable ions have double positive charges. The rest of the halogens (group VII) behave as chlorine: the most stable ions have unit negative charges. In fact, all the halogens have electron affinities; their atoms all release thermal energy by taking on an electron and acquiring a unit negative charge, as the data in Table 4.2 show.

Table 4.2 Electron Affinities of Gaseous Atoms for One Electron

Element	Electron Affinity* (kcal/mole)
Group VII	
Halogens	
Fluorine	−79.6
Chlorine	−83.3
Bromine	−77.5
Iodine	−70.6
Group VI	
Oxygen	−34
Sulfur	−48

* ΔH for the reaction:

$$\text{atom} + e^- \longrightarrow \text{ion}^-$$

The usual practice in listing electron affinities is to omit the minus sign, its presence being understood. (Electron affinities in kilocalories per mole were obtained by converting into these units data from F. A. Cotton and G. Wilkinson, *Advanced Inorganic Chemistry,* second edition, Interscience, New York, 1966, page 33.)

The group III elements, when they form ions, go as far as the 3+ stage. Each atom in this family has three electrons in its outside shell. By losing all three, it acquires a noble gas configuration. Group IV and V elements do not generally form ions of sufficient stability that they can, like the ions of groups I and II elements, exist independently as discrete particles. The ionization energies needed to pull four electrons are far higher than can be repaid by lattice stability. Group VI and VII elements can form ions, and they do so *by taking on electrons* until noble gas structures are reached, as indicated by Table 4.1.

A very useful rule of thumb has emerged, the *octet rule: The attainment of a noble gas configuration is the correct guide for predicting which ion of an element will be the most stable for the first twenty elements.* As corollaries to this rule, elements in groups I to III give up electrons to form positive ions; elements in groups VI and VII take on electrons to form negative ions; elements in groups IV, V, and 0 generally do not form ions. By giving up one, two, or three electrons (rather than picking up seven, six, or five), elements in groups I, II, and III acquire outer shells of eight electrons, the "outer octets" possessed by the noble gases.[3] On the other side of the chart, elements in groups VI and VII, by picking up two or one electrons rather than by losing six or seven, acquire outer octets and become negatively charged.

Among the transition elements and those in other families, near the bottom of the periodic chart, exceptions to the octet rule will be found. It is still a useful rule, and we shall refer to it frequently. Its success lies in this usefulness, which is explained by the energy considerations given for the two examples discussed.

In summary, ionic compounds result from the transfer of electrons; atoms that can give up electrons interact with those that can accept them, and oppositely charged ions emerge. Crystal lattices form with enough release of enthalpy to pay whatever enthalpy costs are incurred in creating the ions. The ions form in a definite ratio symbolized by the formula of the compound that is made. (They cannot aggregate in any ratio other than the one that ensures overall electrical neutrality.)

Ionic Bond. When the electrons in sodium atoms have become redistributed among chlorine atoms so that sodium ions and chloride ions have formed, these new oppositely charged particles attract each other. The nuclei are shielded and act to prevent the electrons from flying apart. The attraction between the oppositely charged *ions* keeps the like-charged nuclei from flying apart. We call this electrostatic attraction between ions of opposite charge an *ionic bond.* Evidently it is very strong. All ionic compounds are solids at room temperature and take their place in the families of compounds called *salts.* Table salt, NaCl, is but one of hundreds of examples.

Names and Symbols for Ionic Compounds. Table 4.3 contains the names and symbols for several of the more important ions derived from single atoms. Many ions are derived from several atoms and contain two or more nuclei. They have special names and symbols that are best studied when we see how they are held together. The positively charged ions in Table 4.3 are named after the parent element. Names of all negatively charged ions in the list add the suffix "-ide" to a base word taken from the name of the parent element. So much for the names. The symbols are those of the original elements except that the ionic charge for each is indicated in the superscript position. *It cannot be emphasized too strongly that ions have properties profoundly different from those of their corresponding neutral atoms,* even though the symbols are very similar. We must therefore exercise great care in making statements about ions and atoms. For example, except when sodium is the first word in a two-part name of a compound, it refers to the metal, an element, or to an *atom* of that element. It *never* refers to the sodium ion. If we mean "sodium ion," then "sodium" simply will not do, and vice versa.

Ionic compounds are named after their constituent ions (with the word *ion* omitted). The name of the positively charged ion, by convention, appears first. Table 4.4 shows several names of ionic compounds. Shorthand symbols for ionic

[3] If the outer shell is the K-shell, the first principal energy level, the rule is amended to focus not on an outer octet but rather on a filled K-shell as the *outside* shell.

Table 4.3 Some Important Ions*

Group		Element	Symbol for Neutral Atom	Symbol for Its Most Stable Ion	Name of Ion
Alkali Metals	I	Lithium	Li	Li^+	lithium ion
		Sodium	Na	Na^+	sodium ion
		Potassium	K	K^+	potassium ion
Alkaline Earth Metals	II	Magnesium	Mg	Mg^{2+}	magnesium ion
		Calcium	Ca	Ca^{2+}	calcium ion
		Strontium	Sr	Sr^{2+}	strontium ion
		Barium	Ba	Ba^{2+}	barium ion
		Radium	Ra	Ra^{2+}	radium ion
	III	Aluminum	Al	Al^{3+}	aluminum ion
	VI	Oxygen	O	O^{2-}	oxide ion
		Sulfur	S	S^{2-}	sulfide ion
Halogens	VII	Fluorine	F	F^-	fluoride ion
		Chlorine	Cl	Cl^-	chloride ion
		Bromine	Br	Br^-	bromide ion
		Iodine	I	I^-	iodide ion
Transition Elements		Chromium	Cr	Cr^{3+}	chromium ion
		Iron	Fe	Fe^{2+}	iron(II) ion (ferrous ion)†
				Fe^{3+}	iron(III) ion (ferric ion)
		Nickel	Ni	Ni^{2+}	nickel ion
		Copper	Cu	Cu^{1+}	copper (I) ion (cuprous ion)
				Cu^{2+}	copper(II) ion (cupric ion)
		Zinc	Zn	Zn^{2+}	zinc ion
		Silver	Ag	Ag^{1+}	silver ion
		Mercury	Hg	Hg^{1+}	mercury(I) ion (mercurous ion)
				Hg^{2+}	mercury(II) ion (mercuric ion)

* Other common ions, derived from more than one element, are given in Table 4.4.
† Names in parentheses represent older practice.

compounds are *empirical formulas* (cf. problem 3.14) and show what ions are present and in what ratio. The smallest whole numbers that will express the ratio are used, and the particular collection of ions represented in the formula is the *formula unit,* the ion pair or the ion group used in calculating formula weights and moles.

Table 4.4 Writing Names and Formulas for Ionic Compounds

Constituent Ions	Ratio of Aggregation (to insure neutrality)	Compound Formula	Name of Compound
Ca^{2+} and F^-	1:2	CaF_2	Calcium fluoride
Li^+ and O^{2-}	2:1	Li_2O	Lithium oxide
Mg^{2+} and Br^-	1:2	$MgBr_2$	Magnesium bromide
Fe^{3+} and O^{2-}	2:3	Fe_2O_3	Iron(III) oxide ("ferric oxide")

Exercise 4.1 Using the information given in Table 4.3, write correct empirical formulas and names for ionic compounds that could be expected to form from the aggregation of the following ions. If no compound is to be expected, write "None."

(*a*) sodium ion and bromide ion.
(*b*) iodide ion and calcium ion
(*c*) oxide ion and sodium ion
(*d*) potassium ion and sulfide ion
(*e*) zinc ion and barium ion
(*f*) lithium ion and bromide ion
(*g*) copper(II) ion and sulfide ion
(*h*) nickel ion and iodide ion
(*i*) silver ion and chloride ion
(*j*) mercury(I) ion and sulfide ion
(*k*) iron(II) ion and oxide ion
(*l*) aluminum ion and oxide ion
(*m*) chromium ion and barium ion

COVALENT SUBSTANCES

In the previous section we noted but did not explain the fact that the formula of chlorine is Cl_2, not Cl. The smallest stable sample of the element chlorine consists of two chlorine nuclei and enough electrons to make this compound particle stable and electrically neutral. Such a particle is called a *molecule*. We use this word, in general, for any electrically neutral, nonionic particle consisting of two or more nuclei (plus enough electrons to ensure neutrality) that is the smallest stable sample of the substance. Other diatomic molecules are found among the gaseous elements: H_2, O_2, F_2, and N_2. The ionic bond is an inadequate concept for dealing with these. We would have to postulate such ions as Cl^+ and Cl^- existing in pairs rather than in crystalline lattices. Not only have we seen the importance of the lattice energy in stabilizing ions,[4] but we have noted the apparent importance of outer octets of electrons. The electronic configuration of Cl^+—$1s^22s^22p^63s^23p_x^2 3p_y3p_z$—shows a sextet rather than an octet at the outside (or third) level.

For ionic bonds to form, electrons must be transferred between dissimilar atoms—atoms of elements generally on opposite sides of the chart, such as metals and nonmetals. Ionic bonds do not usually form between metals and metals or between nonmetals and nonmetals. Between the former we have the metallic bond, between the latter the covalent bond.

Covalent Bond. The octet theory was advanced by Walther Kossel of Germany and G. N. Lewis and Irving Langmuir of the United States between 1916 and 1920.

[4] This phrase is obviously a manner of speaking. Lattice energy is stabilizing because it is energy that *the lattice does not have*, for it has been released to the surroundings during the assemblage of the oppositely charged ions. It is in this sense that we speak of the importance of the lattice energy in stabilizing ions in crystals.

Because major contributions to the concept of covalent bonds were initially developed by Lewis, it is his name that we generally associate with the first modern theory of chemical bonding. Lewis' paper appeared only a short while after Bohr's solar system model of the atom was published. Lewis objected to the model because it provided no reasonable way to account for chemical bonds. Although Bohr talked about energy levels, he also talked about electrons moving in orbits, and the two are not necessarily the same thing. Lewis insisted that nothing in the facts requires that the electrons be in motion, and he proposed a static model in which electrons are stationary.

Lewis subdivided an atom into two domains, illustrated by an iodine atom:

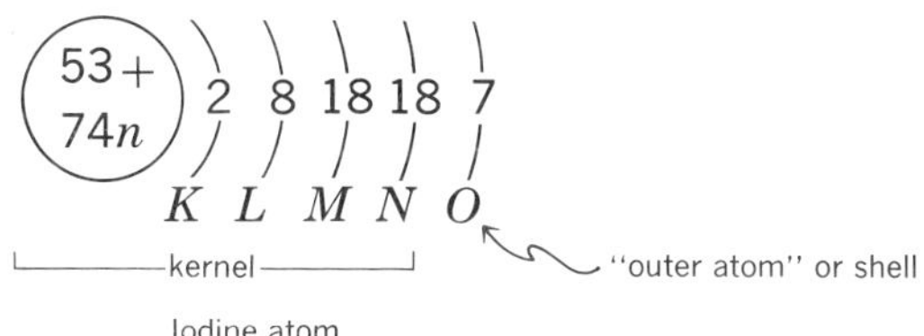
53+
74n
2 8 18 18 7
K L M N O
kernel
"outer atom" or shell

Iodine atom

The inner he called the essential *kernel,* the nucleus and the inner-shell electrons which remain unaltered in all ordinary chemical changes. The other domain is an "outer atom" or *shell,* the valence shell. Both Lewis and Kossel realized the apparent importance of eight electrons in this shell. It was Lewis who postulated that *the shells of different atoms are interpenetrable.* An outer-shell electron may become part of the shell of two different atoms and henceforth cannot be said to belong exclusively to either.

Lewis proposed that atomic kernels be represented by the usual atomic symbols and that the outer-shell electrons be shown as dots grouped as much as possible in pairs about the kernel. Thus two fluorine atoms would be depicted as

$$:\ddot{\underset{\cdot\cdot}{\mathrm{F}}}\cdot \qquad \cdot\ddot{\underset{\cdot\cdot}{\mathrm{F}}}:$$

The fluorine molecule, F—F, would be written as

$$:\ddot{\underset{\cdot\cdot}{\mathrm{F}}}:\ddot{\underset{\cdot\cdot}{\mathrm{F}}}:$$

This suggestion became one of the most successful diagrammatic ways of showing that the noble gas configurations are or are not achieved in a particular structure, and we shall use it often in this text.

Scientists had for decades used a straight line joining atomic symbols to represent a bond, for example, H—H, Cl—Cl, F—F, Br—Br. Lewis identified this straight line as a shared pair of electrons. Langmuir proposed that the bond resulting from this sharing of electrons be called a covalent bond. It was he who first used the term "octet theory" when writing on this subject, and he extended the work of Lewis (and Kossel) to the entire domain of the periodic chart.

Multiple Bonds. Before Lewis' time many substances containing atoms with unused bond-forming capacities were known. Hydrogen, which could form only one bond to anything else, was said to have a bond-forming capacity or *valence* of one. Carbon, as in methane, CH_4, was assigned a valence of four. In the graphical symbols used before Lewis, and still used, methane is represented roughly as

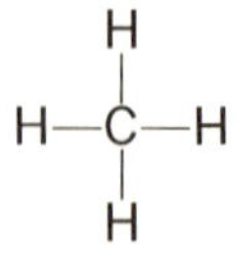

```
    H
    |
H—C—H
    |
    H
```

Methane

Each valence of carbon, in other words, is represented by a line going away from it; each valence of hydrogen (only one) is also represented by the line. Since all the valences of carbon in methane are used to bind the maximum number of neighbors, it is said to be *saturated.* In a covalent substance such as ethylene, however, each carbon is bound to only three other groups. The graphical structure in common use both now and before Lewis' time is

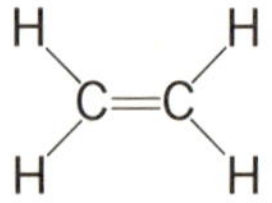

```
H        H
 \      /
  C═C
 /      \
H        H
```

Ethylene

Such a substance is said to be unsaturated because not all the valences of carbon are taken up in bonds to four other groups. To preserve the idea of a valence of four for carbon, two lines are drawn between the carbons, and a *double bond* is said to exist. When Lewis identified the lines in the structural formulas as shared pairs of electrons, he brought the unsaturated compounds under the umbrella of his theory by saying that two pairs (double bond) and three pairs (triple bond) may be shared.

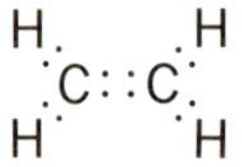

```
H.     .H
 .C::C.            :N:::N:
H.     .H

 Ethylene          Nitrogen

   :O:
   ::
   .C.             H:C:::C:H
 H.   .H

Formaldehyde       Acetylene
```

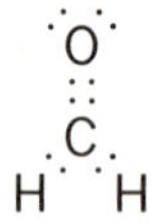

Gilbert Newton Lewis (1875–1946). For most of his adult life a professor of physical chemistry at the University of California at Berkeley, Lewis' fame rests largely on three contributions: his theory of chemical valency, his theory of acids and bases, and his coauthorship (with Merle Randall) of one of the most widely used and influential treatises on chemical thermodynamics, *Thermodynamics and the Free Energy of Chemical Substances* (1923).

Irving Langmuir (1881–1957). The 1932 Nobel laureate in chemistry for his work on surface chemistry, Langmuir from 1909 to 1950 was a scientist with the General Electric research laboratories. He is seldom remembered for his contribution to chemical valency, which at the time was substantial. His name is much more closely associated with the development of vacuum tubes, later used in television sets, atomic hydrogen welding, cloud seeding, and molecular films.

Walther Kossel (1888–). Although Lewis proposed a "cubical atom," Kossel, a physicist in Danzig at the time, placed the electrons around a ring and was principally concerned with ionic substances.

Molecular Orbitals. Since the advent of quantum mechanics and wave mechanics, the covalent bond has been interpreted as the overlapping of atomic orbitals. We theorize that the hydrogen molecule is produced when the two $1s$ orbitals of the separate atoms partially merge as illustrated in Figure 4.7. The merger of two

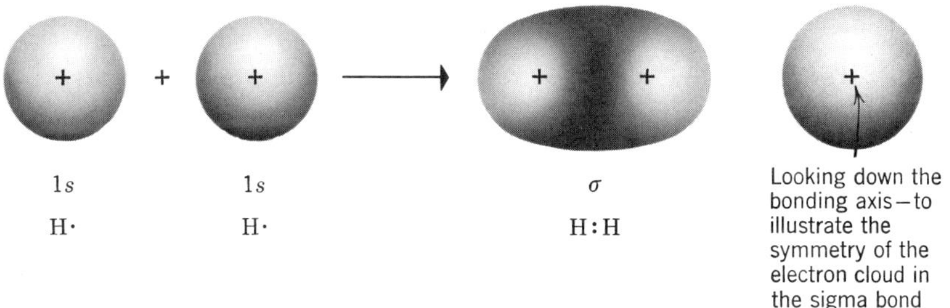
+
+
+
+
+
+
1s
H·
1s
H·
σ
H:H
Looking down the bonding axis—to illustrate the symmetry of the electron cloud in the sigma bond

Figure 4.7 The formation of the covalent bond in H_2 is shown as the overlap or the interpenetration of the $1s$ atomic orbitals of the hydrogen atoms. The relative darkness of the shading in the structure of the hydrogen molecule gives a qualitative idea of where the electron cloud is densest, between the two nuclei. Only the bonding orbital of H_2 is described.

atomic orbitals produces two new molecular orbitals.[5] Each can accommodate a pair of electrons provided the spins are not parallel. When the pair of electrons is in one of these new orbitals, the *bonding orbital*, the system has *less* energy than it did when the electrons were in separate $1s$ orbitals. The other molecular orbital corresponds to *more* energy and is called an *antibonding orbital.* The aufbau rules are substantially the same for molecular orbitals as for atomic orbitals. The two $1s$ electrons from the hydrogens naturally go into the lower-energy molecular orbital. The system is therefore more stable this way than with the atoms separated. The electrons, in a sense, move with respect to and are attracted to two nuclei instead of one, which helps explain why the molecule has lower energy and greater stability than the separate atoms. The formation of the covalent bond between the hydrogens is

[5] It is as if there were a "law of conservation of orbitals."

an energy-lowering process that occurs spontaneously; to break this bond requires 103 kcal/mole.

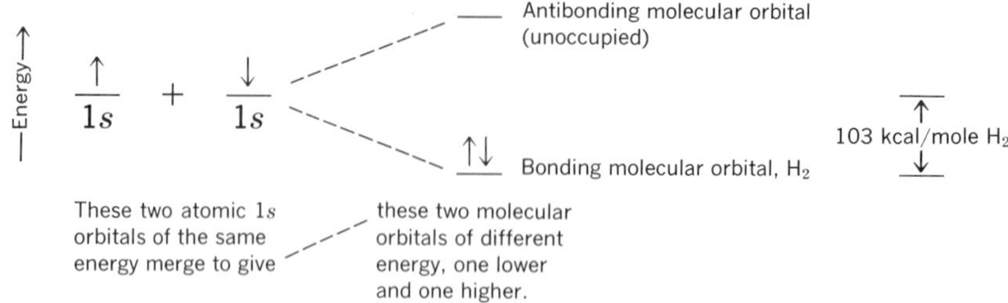
Energy
1s
+
1s
Antibonding molecular orbital (unoccupied)
Bonding molecular orbital, H_2
103 kcal/mole H_2
These two atomic 1s orbitals of the same energy merge to give
these two molecular orbitals of different energy, one lower and one higher.

Although two molecular orbitals are available in the hydrogen molecule, only one is populated with the two electrons. It is the shape of this one that is usually described in discussions of a single covalent bond. The other, the higher-energy orbital, although of great importance when ways in which the molecule can absorb energy, particulary visible and ultraviolet light, are discussed, is ignored when the topic is the chemistry of the substance.

The electron distribution in a hydrogen molecule is often symbolized as an "electron cloud." (This is one model.) Embedded in the cloud are the two nuclei. They cannot fly apart because the attractive forces between them and the electrons keep them within the electron cloud. The density of this cloud, greatest between the two nuclei, effectively shields them from repelling each other, and the attractive forces between the electrons and the nuclei keep the electrons within the system.

The bonding orbital created by the overlap or interpenetration of the two $1s$ orbitals is symmetrical about the line joining the two nuclei, the *bonding axis*. Covalent bonds that have this symmetry are called *sigma bonds* or σ-bonds.

Among higher atoms interpenetration of other types of orbitals is equally possible; s to p and p to p are two examples. The F_2 molecule has the latter type. The electronic configuration of a fluorine atom (atomic number 9) is $1s^22s^22p_x{}^22p_y{}^22p_z$. The bonding between two fluorine atoms is envisioned as the overlap of the two p orbitals at the second level (Figure 4.8). The shape of the new bonding molecular orbital (and only this one is described) is different from the shape that results from the overlap of two s orbitals. The *symmetry,* however is the same, a symmetry along and around the bonding axis. Therefore a p–p orbital overlap of the type shown in Figure 4.8 is also labeled a sigma bond.

By far the largest number of covalent substances consist of molecules in which unlike atoms have become bonded together. An example of this is H—F (Figure 4.9).

POLAR MOLECULES

As long as there are only two nuclei per molecule, and these are identical, the sharing of the electron pair has to be exactly 50:50. In H—F, however, the bonding pair is, in effect, more under the control of the higher-charged nucleus than of the nucleus of the hydrogen. The sharing is not 50:50. Suppose, now, that in H—F the "sharing" is 0:100. That is, suppose that the hydrogen has lost complete charge of

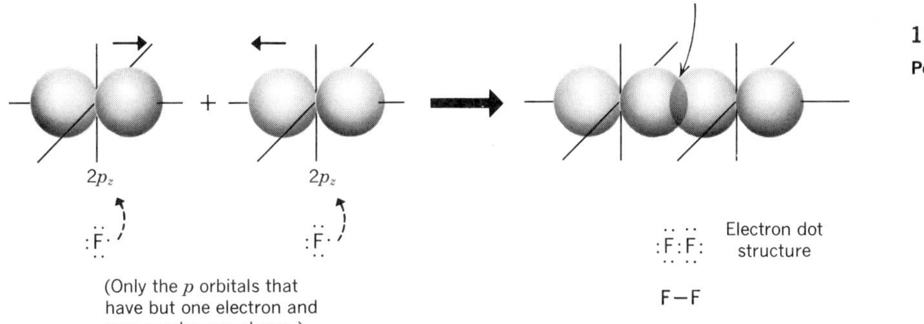
Overlap region: p_z—p_z (σ-bond)
$2p_z$
$2p_z$
:F·
:F·
:F:F:
Electron dot structure
F—F
(Only the p orbitals that have but one electron and may overlap are shown.)

Figure 4.8 The overlap of the two p_z orbitals of two fluorine atoms gives a bond that is symmetrical about the bonding axis. Thus p–p overlap of this kind also gives rise to a sigma bond. The shape of the resultant molecular orbital is oversimplified. A better model would look more like the following:

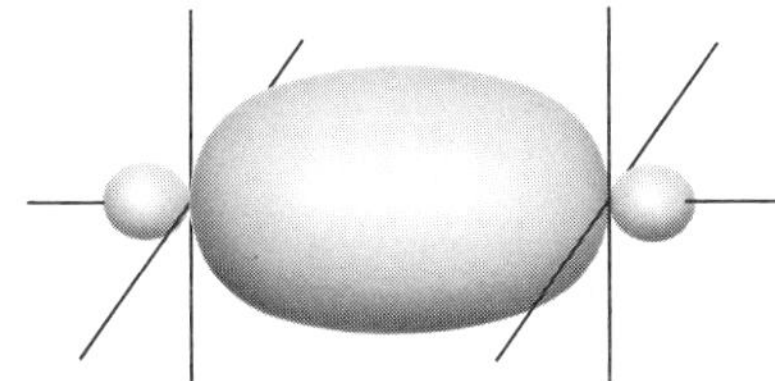

In other words, the electron density is more concentrated between the nuclei than is indicated in the simplified picture.

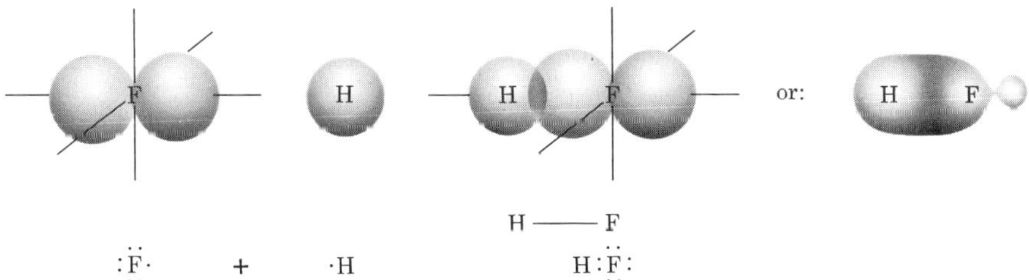
F
H
H
F
or:
H
F
H —— F
:F· + ·H
H:F:

Figure 4.9 A bond that is symmetrical about the bonding axis may also form from overlap between a p orbital and an s orbital, as illustrated here for hydrogen fluoride, H—F. Note that the maximum electron density does not fall halfway between the two nuclei. This maximum is shifted toward the fluorine end of the bond, and the electron density in the immediate vicinity of the nucleus of hydrogen is *less* than it was in the hydrogen atom. Similarily, the total electron density in the vicinity of the nucleus of fluorine is greater than in the fluorine atom. The H—F molecule is therefore polar as described in the text.

its one electron and that the fluorine has assumed full control. This is the same thing as a full electron transfer, and instead of writing hydrogen fluoride as a molecule, H—F, we should have to depict it as an orderly aggregation of oppositely charged ions, H^+ and F^-. If the ratio were reversed, if instead of 0:100 it were 100:0, we would again have ions, but now they would be H^- and F^+. The extremes may be represented as follows:

$$(\mathrm{H}\!:^{-} + \ddot{\underset{..}{\mathrm{F}}}\!:^{+}) \qquad \mathrm{H}\!:\!\ddot{\underset{..}{\mathrm{F}}}\!: \qquad (\mathrm{H}^{+} + :\!\ddot{\underset{..}{\mathrm{F}}}\!:^{-})$$

Hydrogen fluoride is a gas and in this state a nonconductor. There is simply no evidence to assign to it an ionic structure. Yet we do have evidence that the center of density of positive charge does not coincide with that of the negative charge in its molecules. The molecule is *polar,* and although the bond is best described in terms of a sharing rather than a transferring of electrons, the sharing is unequal. Instead of the extremes just shown, the sharing may be something like 30:70. Sometimes instead of calling a bond covalent we describe it as *polar covalent,* and to represent it partial charges may be written thus:

$$\overset{\delta+}{\mathrm{H}}\text{—}\overset{\delta-}{\mathrm{F}}$$

The lowercase Greek letter *delta,* δ, is a common symbol for partial or fractional. The algebraic sum of the partial charges is zero, making the molecule as a whole neutral.

The fact that the molecules of most covalent substances are more or less polar is indispensible to our understanding how electrically neutral molecules can stick together (Figure 4.10). If there were no possibility for forces of attraction between

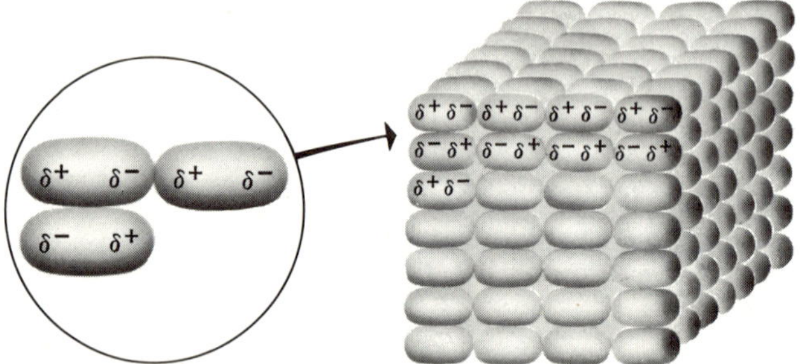
δ+ δ−
δ+ δ−
δ− δ+
δ+ δ− δ+ δ− δ+ δ− δ+ δ−
δ− δ+ δ− δ+ δ− δ+ δ− δ+
δ+ δ−

Figure 4.10 Electrically neutral molecules may stick to each other if they are polar, as illustrated here. Only an orderly aggregation would minimize the forces of repulsion between sites of like partial charge and maximize the forces of attraction between the unlike partial charges. If the net force of attraction is strong enough, the substance will be a solid at room temperature. If it is relatively weaker, the substance will be a liquid; if virtually zero, a gas. The shape of the polar molecule will be important in determining the magnitude of the net force of attraction. Molecules with highly irregular shapes may not be able to fit close enough together to take advantage of possible forces of attraction. Irregularly shaped molecules are often encountered among organic substances.

H						
2.20						
Li	Be	B	C	N	O	F
0.97	1.47	2.01	2.50	3.07	3.50	4.10
Na	Mg	Al	Si	P	S	Cl
1.01	1.23	1.47	1.74	2.06	2.44	2.83
					Se	Br
					2.48	2.74
					Te	I
					2.01	2.21

Figure 4.11 Electronegativities for some elements. (Data from A. L. Allred and E. G. Rochow, *Journal of Inorganic and Nuclear Chemistry,* Vol. 5, 1958, page 264.)

molecules, most covalent substances (most organic compounds), including water, would be gases at room temperature. That the *molecules* of most are polar is a fact, which we explain by theorizing that individual *bonds* are polar. The polarity of a molecule is the resultant of the individual bond polarities.[6] Generally speaking, bonds between dissimilar atoms, if one is hydrogen, will have partial minus signs at the *heavier* atom end, but better approaches to predicting locations of partial charges are available.

Electronegativity. Several scientists have worked out schemes representing the relative tendencies of atoms to pull electrons toward themselves when they are covalently bound. The greater the ability to attract electrons along a covalent bond, the greater the *electronegativity* of that atom. The scale of relative electronegativities for a number of elements is given in Figure 4.11. It should be noted that, in general, electronegativities *within the same family* are highest at the top of the periodic chart. *Within the same period* (horizontal rows) electronegativities increase from left to right, which is consistent with our knowledge that the families on the left have lower ionization potentials than those on the right. In fact, the atoms of the group VII elements (halogens) have electron affinities of negative values, meaning that when they gain an electron, the energy of the system is lowered (p. 117). The following specific relations for the elements most frequently encountered in covalent compounds should be learned. In terms of relative electronegativities,

$$F > Cl > Br > I$$
$$F > O > N > C > H$$

The Directionality of Covalent Bonds. The water molecule is quite polar, but if its three nuclei were all on a line, it could not be. Even if, on the basis of electronegativities, we write partial charges as

$$\overset{\delta+}{H}—\overset{\delta-}{O}—\overset{\delta+}{H}$$

[6] Those familiar with vectors will realize that this resultant is simply the vector sum.

the whole molecule could not be polar. On the basis of the symmetry of this structure, the center of density of negative charge would be in the center of the oxygen, the exact spot for the center of density of positive charge. When these two centers coincide, the molecule cannot be polar. Yet the water molecule is highly polar, which suggests that something is wrong with the structural representation given. X-ray diffraction studies have shown that the angle between the two bonds from oxygen is not 180° but 104.5°. Water should therefore be represented as (partial charges are again shown)

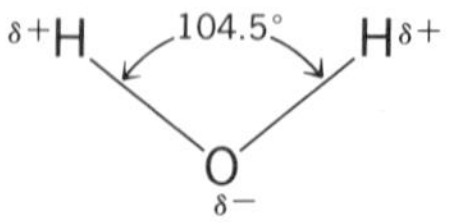

$$\overset{\delta+}{H}\quad \angle 104.5^\circ \quad \overset{\delta+}{H}\quad \text{bonded to}\quad \underset{\delta-}{O}$$

The centers of density of opposite charges cannot now coincide and the molecule as a whole is polar.

The Lewis theory could not satisfactorily account for such directionality of covalent bonds. Modern theories have been more successful. We shall study one model now.

The electronic configuration of the oxygen atom (atomic number 8) is

$2p$		↑↓ | ↑ | ↑
$2s$	↑↓	
$1s$	↑↓	

or: $1s^22s^22p_x{}^22p_y2p_z$

There are two p orbitals with single electrons apiece in them. Drawing only these, and bringing up two hydrogens ($1s$ orbitals), we have the bonding shown in Figure 4.12. Overlap can occur only where orbitals are available. To achieve maximum overlap, that is, to produce the strongest bonds when oxygen combines with two hydrogens, the resultant bond angle should be 90°, the angle between the axis of the two p orbitals. By bringing the two protons in this close, however, mutual repulsions between them increase and tend to spread them apart. This, then, is one explanation for the bond angle in the water molecule. The "predicted" angle is 90° because that is the angle of the orbitals originally available in the oxygen atom. Once p-to-s overlaps occur between the oxygen and the two hydrogens, the nuclei of the latter repel each other to spread the predicted angle to the observed value. Consistent with this model is the fact that the bond angle in H_2S is 92°, much closer to the predicted 90°. Sulfur, being in the same family as oxygen, has the same outside-shell configuration. Its full configuration is $1s^22s^22p^63s^22p_x{}^22p_y2p_z$. As with oxygen, two p orbitals are available for overlap with the $1s$ orbitals of the hydrogens in H_2S. Sulfur, however, has a larger radius than oxygen. Because the hydrogens do not have to be as close to each other to overlap with the sulfur, the spreading of the bond angle is not so great.

Ionic bonds are not directional. The force of attraction (or repulsion) is "radi-

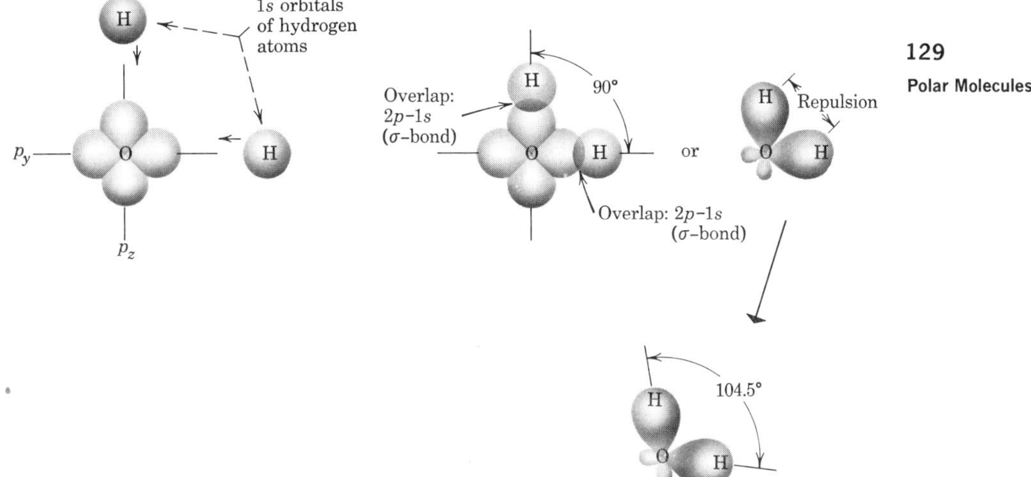
1s orbitals
of hydrogen
atoms
H
H
O
p_y
p_z
Overlap:
2p–1s
(σ–bond)
90°
H
O
H
Overlap: 2p–1s
(σ–bond)
or
H
Repulsion
O
H
104.5°
H
O
H

Figure 4.12 Molecular orbital picture of the water molecule. According to one model, the bonds are created from the overlap of p orbitals in oxygen with s orbitals provided by the hydrogens. Then repulsions between the hydrogen nuclei and repulsions between the electron clouds of the new molecular orbitals spread the bond angle from the "predicted" value of 90° to the actual value of 104.5°.

ated" equally in all directions from any given ion. In the molecules of covalent substances, resultant forces of attraction are localized in particular directions.

Coordinate Covalent Bonds. In many structures the shared pair of electrons does not include one electron from each of the bonded atoms. Both electrons of the shared pair may originate from one atom, from one donor. Lewis, for example, was the first to explain how an ammonia molecule can have a residual binding capacity. In his electron dot symbolism ammonia originates in the following way:

$$\mathrm{H\cdot} \;\rightarrow\; \cdot\ddot{\mathrm{N}}: \;+\; 2\,\mathrm{H\cdot} \longrightarrow \mathrm{H}:\underset{\textstyle \mathrm{H}}{\overset{\textstyle \mathrm{H}}{\ddot{\underset{\cdot\cdot}{\mathrm{N}}}}}:$$

Ammonia

It had long been known that a neutral ammonia molecule can bind still another proton (remember that a proton is the nucleus of the hydrogen atom), producing an electrically charged particle, an ion. But how could this fourth proton be held? Lewis suggested that the two unshared electrons on the neutral ammonia molecule were both "donated" to form the bond, the shared pair, to the fourth hydrogen:

$$\mathrm{H}:\underset{\textstyle \mathrm{H}}{\overset{\textstyle \mathrm{H}}{\ddot{\underset{\cdot\cdot}{\mathrm{N}}}}}: + \mathrm{H^+} \longrightarrow \left[\mathrm{H}:\underset{\textstyle \mathrm{H}}{\overset{\textstyle \mathrm{H}}{\ddot{\underset{\cdot\cdot}{\mathrm{N}}}}}:\mathrm{H}\right]^+$$

Ammonium ion

Once the bond is formed, it is indistinguishable from a covalent bond, but in some instances it is useful to have a different name, the *coordinate covalent bond* (sometimes dative bond). In the chemical vernacular, the nitrogen in the ammonia molecule is said to *coordinate with the proton;* or the molecule is said to have been *protonated.* Nitrogen retains its outer octet, and each hydrogen atom, in effect, experiences an outer filled K-shell.

Complex Ions. Radicals. The ammonium ion is but one of many ions which, although they consist of groups of nuclei held together by covalent bonds, are still electrically charged. These ions are stable enough to be able to survive many chemical changes intact, and to stable things it is natural to assign names. As a group they are called *ionic radicals,* or simply radicals. Names, formulas, and charges for several that should be learned are listed in Table 4.5.

VALENCES

As a matter of convenience, it is useful to learn the relative combining abilities of elements and radicals in terms of simple numbers. If the element is to participate in the formation of an ion, we speak of its electrovalence. Values of electrovalences coincide exactly with the kind and amount of charge on the stable ions. Thus the electrovalence of sodium is 1+, of magnesium 2+, of oxygen 2−, of chlorine 1−, etc. The signs and numbers designating electrical charges in the next to the last column of Table 4.3 are the electrovalences of these elements. Table 4.4 indicates electrovalences of radicals.

For covalent substances we use closely related covalence numbers. These coincide exactly with the number of pairs of electrons that a given atom shares with its neighbors in a particular structure. These numbers are not given plus or minus signs because charged ions are not involved. The covalence number can also be de-

Table 4.5 Common Ionic Radicals

Name of Ion	Formula	Electrovalence of Radical
Hydroxide ion	OH^-	1−
Carbonate ion	CO_3^{2-}	2−
Bicarbonate ion	HCO_3^-	1−
Sulfate ion	SO_4^{2-}	2−
Bisulfate ion (or hydrogen sulfate ion)	HSO_4^-	1−
Nitrate ion	NO_3^-	1−
Nitrite ion	NO_2^-	1−
Phosphate ion	PO_4^{3-}	3−
Monohydrogen phosphate ion	HPO_4^{2-}	2−
Dihydrogen phosphate ion	$H_2PO_4^-$	1−
Cyanide ion	CN^-	1−
Permanganate ion	MnO_4^-	1−
Chromate ion	CrO_4^{2-}	2−
Dichromate ion	$Cr_2O_7^{2-}$	2−
Sulfite ion	SO_3^{2-}	2−
Bisulfite ion (or hydrogen sulfite ion)	HSO_3^-	1−
Ammonium ion	NH_4^+	1+

Table 4.6 Covalences of Common Elements

Element	Atomic Number	Electronic Configuration					Covalence*
		K	L	M	N	O	
Hydrogen	1	1					1
Carbon	6	2	4				4
Nitrogen	7	2	5				3
Oxygen	8	2	6				2
Sulfur	16	2	8	6			2
Fluorine	9	2	7				1
Chlorine	17	2	8	7			1
Bromine	35	2	8	18	7		1
Iodine	53	2	8	18	18	7	1

* Only the most common covalence numbers are listed. Many elements have multiple covalences.

fined as the number of electrons an atom needs to acquire, by a sharing process, to have a stable outside-shell configuration (filled K-shell, if that is the outside shell; otherwise an outer octet). As a working definition, covalence numbers coincide with the number of straight lines (each representing a shared pair) that may extend from an atom in a structural formula. Table 4.6 has a list of covalences of common elements.

THE METALLIC BOND

Metallic elements have two common features important to our understanding of their properties. They all have relatively low ionization energies, less than 220 kcal/mole, except for mercury which has 241 kcal/mole. Most have from one to three electrons in the highest principal energy levels, which means that at these levels many orbitals are unoccupied.

In studying the covalent bond we learned of the energy-lowering effect of placing bonding electrons in a molecular orbital so that they can move (if they do move) with respect to two nuclei instead of one. In the solid state the valence orbitals of metals overlap in three directions to produce large multicentered "molecular" orbitals, in which the valence electrons move more or less freely. This mobility accounts in a qualitative way for the fact that metals are good conductors of electricity. It is as though a metal consists of a collection of metallic ions immersed in a sea of valence shell electrons.

REFERENCES AND ANNOTATED READING LIST

BOOKS

J. W. Linnett. *The Electronic Structure of Molecules. A New Approach.* John Wiley and Sons, New York, 1964. Linnett proposes a significant modification of the Lewis-Langmuir theory. Although this change has not been incorporated into this chapter, Linnett's Chapter 2, "Chemical Binding," is a good discussion of the use and abuse of the octet rule.

C. Bunn. *Crystals. Their Role in Nature and in Science.* Academic Press, New York, 1964.

Crystals are interesting for both scientific and aesthetic reasons. Writing for the scientific layman, Bunn describes the growth and the forms of crystals, including jewels, and their interaction with light and X-rays. (Paperback.)

ARTICLES

R. L. Fullman. "The Growth of Crystals." *Scientific American,* March 1955, page 74. Crystals do not grow layer by layer, as in the building of a brick wall, but in a spiral fashion in a continuous process.

Sir Lawrence Bragg. "X-ray Crystallography." *Scientific American,* July 1968, page 58. The author shared with his father, W. H. Bragg, the 1915 Nobel prize in physics for work on analyzing structures of crystals by X-rays. He tells how it is done in this article.

H. Selig, J. G. Malm, and H. H. Claassen. "The Chemistry of the Noble Gases." *Scientific American,* May 1964, page 66. In 1962 these three authors together discovered that the supposedly inert noble gas xenon could be made to form xenon tetrafluoride with surprising ease.

F. H. Verhoek. "What Is a Metal?" *Chemistry,* November 1964, page 6. This article discusses the properties of metals in relation to the "electron sea" model.

K. Lonsdale. "Disorder in Solids." *Chemistry,* December 1965, page 14. Crystalline solids are not the perfect arrays of highly ordered aggregations of constituents. The existence of disorder and its importance are discussed.

PROBLEMS AND EXERCISES

1. For any of the first twenty elements, you should be able to write the correct electronic configuration, given only the atomic number, and answer the following questions.
   (*a*) In what family is the element? Give the roman numeral and the family name if there is one, for example, halogens, noble gases, alkali metals, alkaline earth metals.
   (*b*) Can the element occur in the form of electrically charged particles (ions)? If so, what is the electronic configuration of the *ion*? What electrical charge does it bear? What is its correct symbol? What is its electrovalence?
   (*c*) Can the element participate in electron-sharing processes, that is, in covalent or coordinate covalent bonds? If so, what is its covalence?
   Exercise. Without referring to charts or tables, but by constructing and studying electronic configurations, answer these questions for elements of each of the following atomic numbers: 20, 9, 5, 17, 2, 6, 13, 18, 19, 8, 16
2. Explain why it would be unreasonable for fluorine to have an electrovalence of 7+.
3. Sodium chloride may be prepared from the elements by the following two-step path (standard conditions are assumed):

$$\tfrac{1}{2}\mathrm{H_2}(g) + \tfrac{1}{2}\mathrm{Cl_2}(g) \longrightarrow \mathrm{H{-}Cl}(g) \qquad \Delta H = -22.06 \text{ kcal/mole}$$
$$\mathrm{Na}(s) + \mathrm{H{-}Cl}(g) \longrightarrow \mathrm{NaCl}(s) + \tfrac{1}{2}\mathrm{H_2}(g) \qquad \Delta H = -76.17 \text{ kcal/mole}$$

   (*a*) Write the algebraic sum of these two reactions.
   (*b*) What is the net enthalpy of formation for the reaction you wrote in part *a*?
   (*c*) How does its value compare with ΔH_f° for sodium chloride?
4. Given the following information for lithium fluoride, LiF, compute its lattice energy.

$$\mathrm{Li}(s) \longrightarrow \mathrm{Li}(g) \qquad \Delta H = +\ 37 \text{ kcal/mole Li}$$
$$\mathrm{Li}(g) \longrightarrow \mathrm{Li^+} + e^- \qquad \Delta H = +124 \text{ kcal/mole Li}$$

$$\tfrac{1}{2}F_2(g) \longrightarrow F(g) \qquad \Delta H = +18 \text{ kcal/mole F}$$
$$F(g) + e^- \longrightarrow F^-(g) \qquad \Delta H = -80 \text{ kcal/mole F}$$
$$Li(s) + \tfrac{1}{2}F_2(g) \longrightarrow LiF(s) \qquad \Delta H_f^\circ = -146 \text{ kcal/mole LiF}$$

5. Construct a Born-Haber cycle for the formation of LiF.
6. Define atom, ion, and molecule in such a way that you clearly explain how they are alike and how they are different.
7. What is the difference between a molecule and a mole?
8. How are a molecule and an ionic radical alike and how are they different?
9. Within the rules of covalence, what is a likely structure for HNO_2?
10. Given the electron dot structure for the sulfate ion:

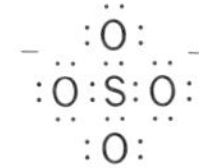

(*a*) What is the total number of electrons, including the inner-shell electrons not represented by dots but still there?
(*b*) What is the total plus charge, summing over all nuclei? Do these two answers agree with the net charge shown?
(*c*) Where would two H^+ ions most likely become attached to this radical?
(*d*) Assuming that when they do become attached to the radical, covalent bonds form, write the structure of H_2SO_4 using a straight line to represent each shared pair of electrons.
(*e*) Examine this structure and check to see whether the rules of covalence are all "obeyed." Note particularly whether the number of lines going away from each atom equals the covalence number.
(*f*) There are two coordinate covalent bonds in H_2SO_4. Which ones would they be in your structure?
(*g*) What conclusion do these questions lead to about situations in which the covalence numbers of Table 4.5 may not apply?

11. Given that no coordinate covalent bonds are present in the molecular formulas listed, follow the rules of covalence and write reasonable structures for each. First use straight lines to represent bonds. Then use Lewis' electron dot symbolism which displays octets (or filled outer K-shells).

(*a*) H_2O_2	(*b*) H_2S	(*c*) C_2H_6
(*d*) HCN	(*e*) NH_3O	(*f*) CH_4O
(*g*) C_2F_6	(*h*) CH_5N	(*i*) PCl_3

12. Write *empirical formulas* for each of the following.

(*a*) sodium dihydrogen phosphate	(*b*) calcium hydroxide
(*c*) aluminum phosphate	(*d*) strontium nitrate
(*e*) potassium dihydrogen phosphate	(*f*) aluminum sulfide
(*g*) iron(III) oxide	(*h*) copper(II) sulfate
(*i*) potassium dichromate	(*j*) barium carbonate
(*k*) lithium phosphate	(*l*) sodium chromate
(*m*) magnesium chloride	(*n*) calcium carbonate
(*o*) strontium chloride	(*p*) sodium cyanide
(*q*) potassium hydrogen sulfite	(*r*) sodium sulfite
(*s*) ammonium phosphate	(*t*) ammonium carbonate

13. Write complete electronic configurations, including compositions of nuclei, for each of the following particles. Refer to tables for atomic numbers and atomic weights. The configurations you write must reflect the fact that electrically charged particles have unequal numbers of protons and electrons.
(a) H^- *(b)* H *(c)* H^+
(d) Ca^{2+} *(e)* O^{2-} *(f)* K^+
(g) Al^{3+} *(h)* F^- *(i)* HO^-

14. Pairs of elements given hypothetical symbols of X and Y have the listed atomic numbers. Predict the empirical formula of the ionic compound that each pair is expected to form. If none is predicted, write "None." You should not need to refer to the actual elements having these atomic numbers.

Example. Y (at. no. 4), X (at. no. 9) Formula YX_2
X (at. no. 4), Y (at. no. 8) Formula XY

Part	Atomic Number of X	Atomic Number of Y
(a)	9	12
(b)	18	14
(c)	3	9
(d)	17	13
(e)	20	8
(f)	4	11

15. Boiling point and melting point information for the chlorides of the second and third-period elements are listed in Table 4.7.
(a) Which are probably ionic compounds?
(b) Which are probably covalent compounds?
(c) What geometry would BCl_3 have if it were nonpolar?

Table 4.7 Chlorides of the Second- and Third-Period Elements: Physical Constants

Compound		Mp (°C)	Bp (°C)
Lithium chloride	LiCl	614	1325–1360
Beryllium chloride	$BeCl_2$	405	520
Boron trichloride	BCl_3	−107	13
Carbon tetrachloride	CCl_4	−23	77
Nitrogen trichloride	NCl_3	−40	71 (explodes at 95°C)
Oxygen dichloride	OCl_2	−20	3.8 (explodes)
Fluorine chloride	FCl	−155	−101
Sodium chloride	NaCl	801	1413
Magnesium chloride	$MgCl_2$	708	1412
Aluminum chloride	$AlCl_3$ (or Al_2Cl_6)	190*	183
Silicon tetrachloride	$SiCl_4$	−70	58
Phosphorus trichloride	PCl_3	−112	76
Sulfur dichloride	SCl_2	−78	decomposes at 59°C
Chlorine	Cl_2	−105	−35

* At 2.5 atmospheres pressure.

16. (*a*) Write a "Lewis structure" for ammonia, NH_3.
(*b*) Write the electronic configuration of nitrogen, $1s^22s^22p$, etc., distinguishing the $2p$ sublevels by $2p_x$, $2p_y$, and $2p_z$. Which orbitals are available for overlapping with $1s$ orbitals from hydrogen atoms?
(*c*) How many hydrogen *atoms* could combine with nitrogen?
(*d*) Draw a picture, giving it a three-dimensional perspective, for the orbital structure of ammonia, assuming the orbitals of the hydrogen atoms overlap where atomic orbitals of nitrogen are available.
(*e*) What are the "predicted" bond angles (H–N–H angles)?
(*f*) The H–N–H bond angles in ammonia are all 106°47′. How might this departure from the "predicted" be explained?
(*g*) The following bond angles in the nitrogen family have been measured:

NH_3	106°47′
PH_3	93°30′
AsH_3	92°0′
SbH_3	91°30′

Explain this trend.
(*h*) Offer an explanation for the relatively large change in bond angle in going from NH_3 to PH_3 and the much smaller changes in the remainder of the series.
17. Outline the essential features of the Lewis theory.
18. Define the following terms.

(*a*) kernel	(*b*) polar covalent bond	(*c*) overlap
(*d*) ionic bond	(*e*) covalent bond	(*f*) sublimation energy
(*g*) electron affinity	(*h*) lattice energy	(*i*) standard state
(*j*) standard heat of formation	(*k*) unsaturated	(*l*) molecular orbital
(*m*) electronegativity	(*n*) coordinate covalent bond	(*o*) electrovalence
(*p*) radical	(*q*) metallic bond	

CHAPTER FIVE

Solutions and Colloidal Dispersions

Solutions and colloidal dispersions are an integral part of all living things and many other systems as well. The bloodstream is mostly water, and in it are dissolved ions and clumps of ions, molecules and clumps of molecules, ion-molecule complexes, molecular species so huge that they are called macromolecules and are said to be colloidally dispersed, and, finally, several cellular bodies. The digestive juices are mostly water, but they contain indispensible chemicals in solution. Fluids inside cells are solutions and colloidal dispersions. The body itself lives and moves in another solution—oxygen, nitrogen, carbon dioxide, and traces of other substances all mixed up. We call it air. Other interesting and important examples of solutions are carbonated drinks, steel, antifreezes, and honey. Whipped cream, marshmallows, mayonnaise, milk, butter, jellies, and black diamonds are colloidal dispersions. Before going into greater detail, we shall search for our bearings in distinguishing between these two special classes of mixtures, solutions and colloidal dispersions. The difference lies primarily in particle size.

PARTICLE SIZES: TRUE SOLUTIONS, THE COLLOIDAL STATE, AND SUSPENSIONS

The diameters of ordinary ions and molecules are on the order of 0.5 to 2.5 Å, where 1 Å = 1 angstrom unit = 1×10^{-8} cm. Particles of this size cannot be seen directly or indirectly by optical or electron microscopes. A solution is a uniform, intimate, and stable mixture of the smallest particles—atoms, ions, molecules—from two or more substances. Since there are three states of matter, there are nine kinds of solutions. These are listed together with examples in Table 5.1. In this chapter we direct our attention to solutions in which water is the solvent or the dispersing medium, that is, to *aqueous solutions.* Substances dissolved in a solvent

Table 5.1 Kinds of Solutions

Kinds	Common Examples
Gas in a liquid	Carbonated beverages (carbon dioxide in water)
Liquid in a liquid	Vinegar (acetic acid in water)
Solid in a liquid	Sugar in water
Gas in a gas	Air
Liquid in a gas	Humid air
Solid in a gas	Certain kinds of smokes
Gas in a solid	Alloy of palladium and hydrogen
Liquid in a solid	Benzene in rubber (e.g., rubber cement)
Solid in a solid	Carbon in iron (steel)

are called *solutes*. Solute particles of the sizes just described form *true solutions,* which means that they are transparent (although sometimes colored), nonfilterable, homogeneous, and stable. The force of gravity causes no settling out.

At the next higher level of particle size are clusters of several hundred to a few thousand particles, with average diameters ranging from 10 to 1000 Å. Although such particles are still seldom observable with any optical microscope, they may be photographed with the aid of an electron microscope. Matter consisting of particles having average diameters of 10 to 1000 Å is said to be in the *colloidal state.*[1] ("Colloidal" comes from the Greek word for "gluelike.") When particles of this size are scattered in water, the product is not a true solution but a *colloidal dispersion.* It is nonfilterable and is usually not transparent, although the more dilute the dispersion, the more likely that it will appear transparent. Examples of colloidal systems are given in Table 5.2. Colloidal particles are large enough to reflect and scatter light, and colloidal dispersions in water exhibit a phenomenon known as the *Tyndall effect* (Figure 5.1). True solutions do not show the Tyndall effect.

[1] Although *average* diameters have been used in these descriptions, a particle of colloidal size need not be spherical. It may have any shape and still be classified in this way provided *one* dimension is 10 to 1000 Å. The soap film in a soap bubble is only a few molecules thick, yet soap in this state is classified as colloidal.

Table 5.2 Colloidal Systems

Type	Dispersed Phase*	Dispersing Medium†	Common Examples
Foam	Gas	Liquid	Suds, whipped cream
Solid foam	Gas	Solid	Pumice, marshmallow
Liquid aerosol	Liquid	Gas	Mist, fog, clouds
Emulsion	Liquid	Liquid	Cream, mayonnaise, milk
Solid emulsion	Liquid	Solid	Butter, cheese
Smoke	Solid	Gas	Dust in air
Sol	Solid	Liquid	Starch in water, jellies,‡ paints
Solid sol	Solid	Solid	Black diamonds, pearls, opals, alloys

* The colloidal particles constitute the dispersed phase.
† The continuous matter into which the colloidal particles are scattered is called the *dispersing medium.*
‡ Sols that adopt a semisolid, semirigid form (e.g., gelatin desserts, fruit jellies) are called *gels.*

Figure 5.1 Tyndall effect. The bottle nearer the light source holds a concentrated solution of copper(II) sulfate. The second bottle holds a colloidal dispersion prepared by adding one drop of India ink to water. The light beam is visible in the latter, but it passes through the first container unscattered. (Used by permission from Charles W. Keenan and Jesse H. Wood, *General College Chemistry,* third edition, Harper and Row, New York, 1966, page 278.)

If colloidal dispersions of a solid in a gas (smoke) or a liquid are viewed with the aid of a microscope, scintillations which are reflections of light from randomly moving particles may be seen. This behavior was first reported in 1827 by an English botanist, Robert Brown, who observed with a microscope the random, chaotic movements of pollen grains in water. Randomly unequal buffeting of colloidal particles by motions of the molecules of the dispersing phase (e.g., air molecules) creates this phenomenon, now called the *Brownian movement.* The buffeting action partly accounts for the stability of colloidal dispersions, for it overcomes the tendency of larger colloidal particles to settle out.

In the most stable colloidal dispersions the dispersed particles all bear the same kind of electrical charge. Being like-charged, they repel each other and cannot coalesce into particles large enough to settle out. Many proteins in the blood are in electrically charged states, for example. Some common colloidal systems achieve their stability through the mediation of a third substance, a *stabilizing agent* or a *protective colloid.* Oil, for example, can be colloidally dispersed in soapy water, for soap molecules form a negatively charged "skin" about each tiny oil droplet and prevents them from coalescing. Mayonnaise is a colloidal system, an emulsion of olive oil or corn oil in water. It is stabilized by egg yolk, whose protein molecules form a skin around the oil droplets and keeps them separated. Milk is an emulsion of butterfat in water with the protein casein acting as the stabilizing substance. Agents that stabilize emulsions are called emulsifying agents.

If particles are much larger than 1000 Å in average diameter, the force of gravity becomes effective and they settle. Clay shaken in water, for example, forms a *suspension* which is filterable, unstable, and nontransparent.

Table 5.3 Types of Dispersions: Solutions, Colloidal Dispersions, Suspensions

Property	Solution	Colloidal Dispersion	Suspension
Average diameter of dispersed particles	0.5–2.5 Å	10–1000 Å	Larger than 1000 Å
Behavior toward gravity	Does not separate (kinetic motions)	Does not separate (Brownian movement)	Separates (directed movement under influence of gravity)
Behavior toward light	Transparent	Usually translucent or opaque (Tyndall effect)	Translucent or opaque
Filterability	Nonfilterable	Nonfilterable	Filterable
Homogeneity	Homogeneous	Border line	Heterogeneous
Number of phases* present	One	Two	Two
Example	Sugar or salt in water	See Table 5.2	Clay in water

* A *phase* is defined as any part of the system that is separated from other parts of the system by physically distinct boundaries. Ice in water, for example, is a system consisting of two phases, one solid, the other liquid. Oil on water is also a two-phase system. In this instance both phases are liquid.

The distinguishing features of solutions, colloidal dispersions, and suspensions are compared in Table 5.3.

AQUEOUS SOLUTIONS

PROPERTIES OF WATER

Polarity. Hydrogen Bonds. The structure of the water molecule is represented in two ways in Figure 5.2, together with some of water's important physical properties. One of the keys to understanding the behavior of water is the great polarity of its molecules. They are attracted to each other so strongly, in fact, that chemists find it convenient to think in terms of an actual bond existing *between* separate molecules of water. The bond has a special name, *hydrogen bond*. As illustrated in Figure 5.3, where it is represented by a dotted line, it extends from a hydrogen in one polar water molecule to the oxygen of a different molecule. The hydrogen bond, in other words, is a force of attraction between opposite partial charges, one of them, $\delta+$, being on a hydrogen. The hydrogen atom is like a bridge from one electronegative atom, O, to another in a separate molecule. It is *not* a covalent bond within a molecule, although its name is admittedly misleading on this point. For one thing, the hydrogen bond is much weaker than a covalent bond. From 90 to 100 kcal of energy are required to break a mole of covalent bonds, but only about 5 kcal will rupture the same number of hydrogen bonds.

The relative weakness of a hydrogen bond does not diminish its importance; it magnifies it. The most important link in a chain is the weakest one. By no means

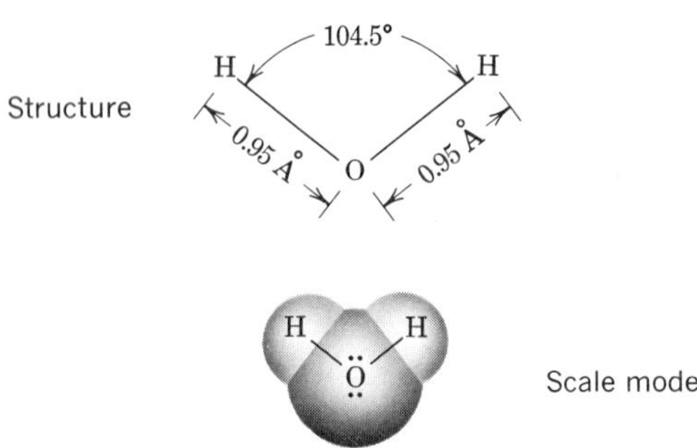
Structure
104.5°
H
H
O
0.95 Å
0.95 Å
H
H
O
Scale model

Formula weight	18.015
Melting point	0°C
Boiling point	100°C (at 760 mm Hg pressure)
Density	1.000000 g/cc at 4.08°C (temperature of maximum density) 0.99998 g/cc at 20°C
ΔH_{fusion}	1.44 kcal/mole; 79.7 cal/g
$\Delta H_{\text{vaporization}}$	9.72 kcal/mole; 539.6 cal/g
Dielectric constant	78.5 (at 25°C)

Figure 5.2 Physical constants for water.

restricted to water molecules, hydrogen bonds form weak links in the structures of key organic compounds of the body—genes, enzymes, proteins, for example—thereby determining much of their stability.

Dielectric Constants as Measures of Polarity. The structure and polarity of the water molecule have another important effect. They make water as a dissolving medium capable of reducing forces of attraction between oppositely charged particles (or of reducing forces of repulsion between like-charged particles). As a measure of this reduction, the property of the *dielectric constant* has been defined:

$$\text{Dielectric constant} = \frac{\text{force between two charges in a vacuum}}{\text{force between same charges in the medium}}$$

Because the units of both numerator and denominator are the same (units of force), they cancel in the ratio and the dielectric constant has no units. We may assume that the distance separating the two charges is the same when the measurement is taken under vacuum and when the medium is placed between them. In a series of measurements the numerator will therefore be a constant, and any changes in dielectric constant will have to reflect some change in the medium. Forces between charges are at their maximum in a vacuum. The dielectric constant can be greater than one only if the medium *reduces* the force between the charges, as compared to the vacuum. Water is very good at this, for its dielectric constant is 78.5. That for petroleum oil is only 2.2. Other values are listed in Table 5.4. Water has one of the highest dielectric constants of all pure substances. A model illustrating the general principle of one way of measuring and comparing dielectric constants is given

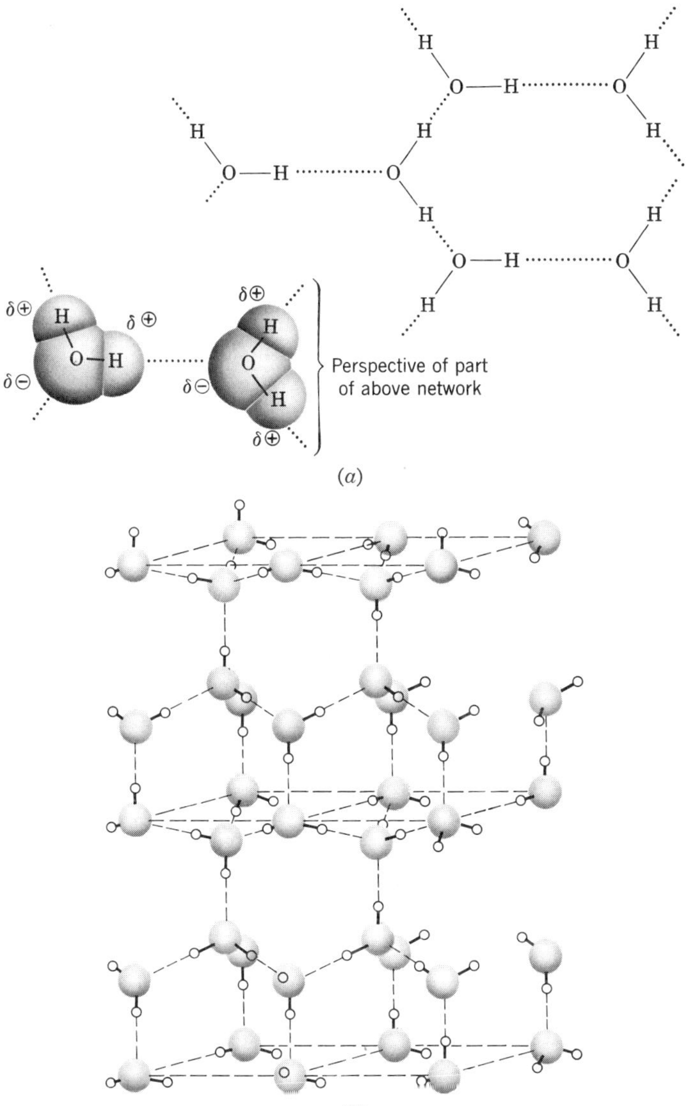

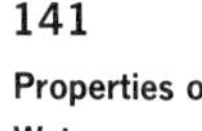

Perspective of part of above network

(a)

(b)

Figure 5.3 Forces of attraction between polar water molecules are strong enough that bonds, hydrogen bonds—represented in part *a* by dotted lines—are said to exist between them. The high degree of order implied in part *a* does not apply to liquid water. When water freezes to ordinary ice, however, the molecules form an intricate network of "cages" with hydrogen bonds furnishing the forces of attraction to hold the crystalline network in part *b* together. To build this lattice network, more volume is required than for the liquid state, and ice is less dense than water. (Crystal structure of ice from L. Pauling, *The Nature of the Chemical Bond,* third edition, © 1960 by Cornell University Press, Ithaca, N.Y. Used by permission of Cornell University Press.)

Table 5.4 Dielectric Constants of Some Common Substances

Substance	Dielectric Constant (at specified temperature, °C)
Water	78.5 (25°)
Hydrogen fluoride (liquefied)	84 (0°)
Hydrogen chloride (liquefied)	4.6 (28°)
Ammonia (liquefied)	25 (−78°)
Organic Solvents	
Those that mix uniformly with water in all proportions and are considered to be polar:	
Methyl alcohol (wood alcohol)	32.6 (25°)
Ethyl alcohol (grain alcohol)	24.3 (25°)
Propyl alcohol (1-propanol)	20 (25°)
Acetone	20.7 (25°)
Those that are insoluble in water and are considered to be relatively nonpolar; these substances are the so-called "hydrocarbon solvents," or grease solvents:	
Ether (ethyl ether)	4.3 (20°)
Benzene	2.3 (20°)
Carbon tetrachloride	2.2 (20°)
Gasoline	2.2 (20°)

in Figure 5.4. This ability of water to reduce forces between charged particles is a major factor in its excellent solvent action on many ionic compounds.

Surface Tension. The high polarity of a water molecule helps explain the phenomenon of surface tension and, indirectly, answers questions such as why are detergents needed for cleansing action in water and why should people whose systems do not produce bile have low-fat diets.

The surface of water, whether in contact with air or with oils and greases, behaves as though it consists of an elastic membrane. This is why fine powders, or wee insects, or even small, clean steel needles can float on water even though they may be denser than water. Beneath the surface, water molecules experience electric forces of attraction toward neighbors equally in all directions. At the surface, however, there are no water molecules outward, nor are there molecules of comparable polarity if air or oils and greases are beyond. A water molecule at a surface cannot therefore experience forces of attraction equally in all directions. There is a net downward pull (Figure 5.5). Surface molecules, of course, cannot really be pulled very far, so they jam together more tightly at the surface than elsewhere in the liquid. This jamming produces the effect of a thin, invisible, elastic membrane at the surface, a phenomenon called surface tension.

This surface "membrane" accounts in part for the poor ability of water alone to break up particles of grease or oil. These substances act as "glues" to bind dirt to fabrics and skin. They also coat food particles in the digestive tract. Since grease molecules are only slightly polar, at the interface between water and grease virtually no force originates on the surface of the grease to break up the "membrane" and permit water to penetrate the grease. All the chemical reactions in the digestion of

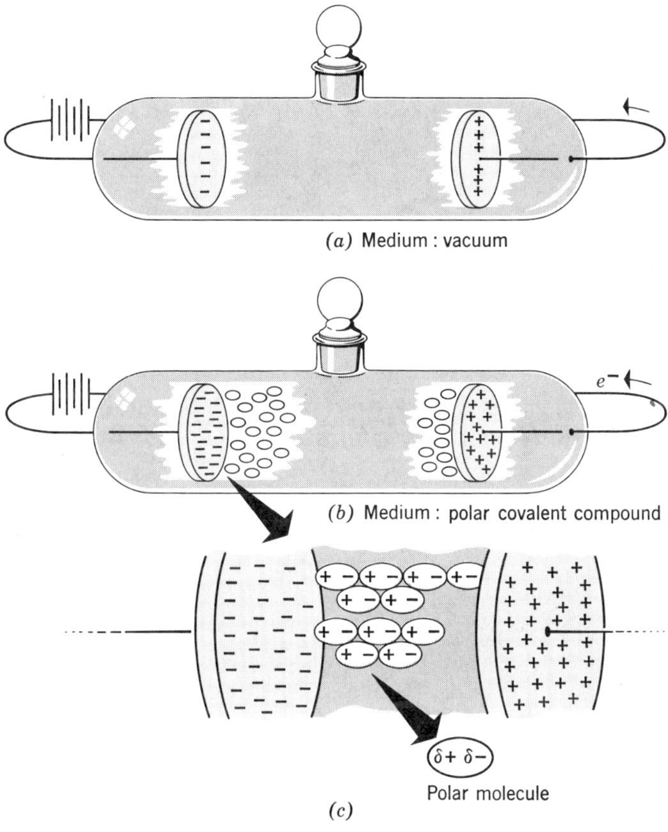

(a) Medium : vacuum
e^-
(b) Medium : polar covalent compound
δ+ δ−
Polar molecule
(c)

Figure 5.4 Measuring dielectric constants of liquids. (*a*) Two metallic plates are connected externally by a wire to a battery. When the switch is closed, the battery forces electrons into one plate and pulls them from another. Then the surge of current stops because the circuit is not complete. This arrangement of two plates and a source of current is called a condenser, and its capacity for electricity can be measured.

(*b*) The capacity is increased greatly if a polar liquid is placed between the two plates. As seen in inset *c*, the polar molecules align themselves. The positive ends of their dipoles help to retain electrons on one plate; at the other end the negative ends of dipoles help to stabilize positive charges (absences of electrons). The increase in the capacity of the condenser can be measured, and the data can be translated into a dielectric constant for the polar liquid.

foods involve water. Therefore, oil-coated food particles are poorly digested. Fortunately, bile, one of the juices secreted into the digestive tract, contains a "soap" (bilt salt) that markedly reduces the surface tension of water. Soaps, in general, are *surface-active agents.*

The beading of water on a waxy surface and the meniscus effect of water in glassware are also understood in terms of water's polarity. The wax, like a grease or an oil, consists of relatively nonpolar molecules. When water is splattered on

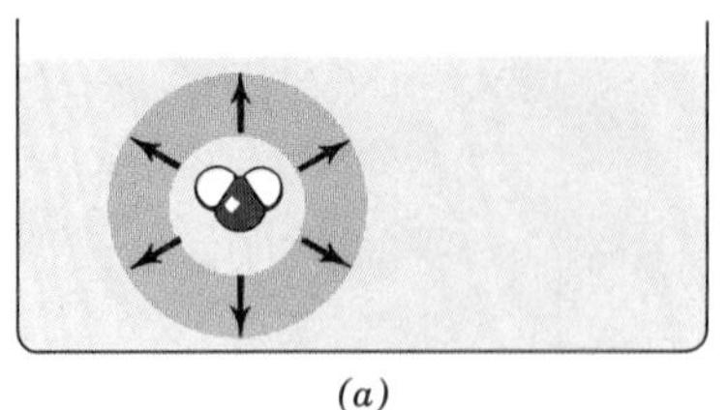

(a)

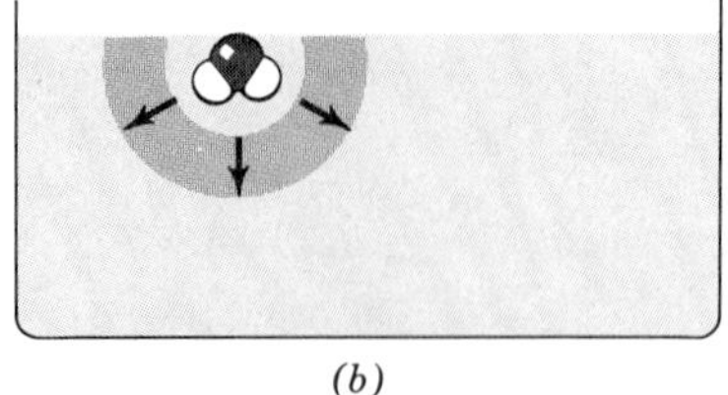

(b)

Figure 5.5 The origin of the surface tension of water. (*a*) In the body of the liquid, molecules of water are attracted by neighbors somewhat equally in all directions. (*b*) Water molecules at the surface experience a net downward attraction, for they now have no neighboring molecules of any appreciable polarity above them.

such a surface, it beads as illustrated and explained in Figure 5.6. The surface of a liquid in contact with glass curves because water molecules are attracted by the polar "molecules" in the glass itself, as illustrated and further explained in Figure 5.7.

Water as a Chemical. Water is usually thought of as a coolant or as a solvent. Its role as a chemical is equally important, but its chemical properties are more conveniently studied in connection with the substances with which it reacts. One property, somewhat on the border line between chemical and physical, is participation in the formation of hydrates.

Hydrates. Many chemically pure substances contain water molecules held more or less firmly in definite proportions. The formulas of a few are listed in Table 5.5. Called hydrates as a group, they are classified as compounds because they obey the law of definite proportions. Their formulas, however, are written to indicate that intact molecules of water are present. Since the action of heat will usually bring about the expulsion of this *water of hydration* and convert these compounds into their *anhydrous forms,* it would be misleading to write the formula of copper sulfate pentahydrate, for example, as $CuH_{10}SO_9$. Instead it is written as $CuSO_4 \cdot 5H_2O$.

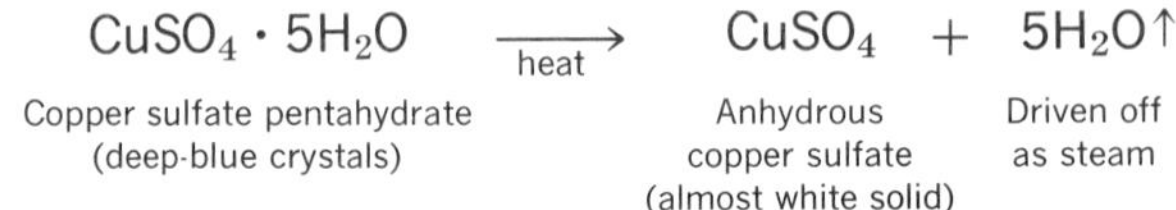

$$CuSO_4 \cdot 5H_2O \xrightarrow[\text{heat}]{} CuSO_4 + 5H_2O\uparrow$$

Copper sulfate pentahydrate (deep-blue crystals) — Anhydrous copper sulfate (almost white solid) — Driven off as steam

Some anhydrous compounds readily take up water and re-form their hydrates. Plaster of paris, for example, although not completely anhydrous, contains relatively

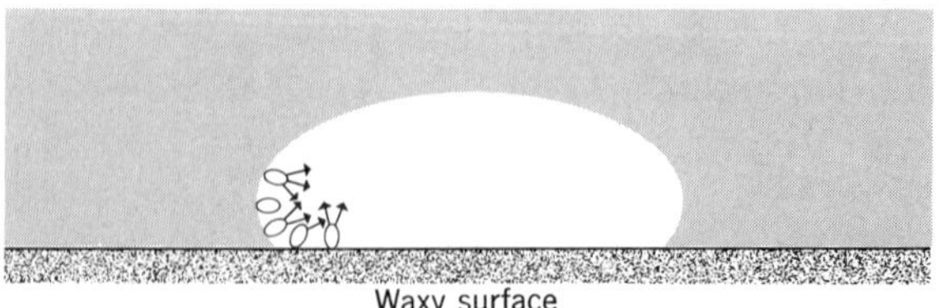
Waxy surface

Figure 5.6 Water beads on a waxed or greased surface because the unbalanced forces acting on water molecules at the surface pull them inward as indicated.

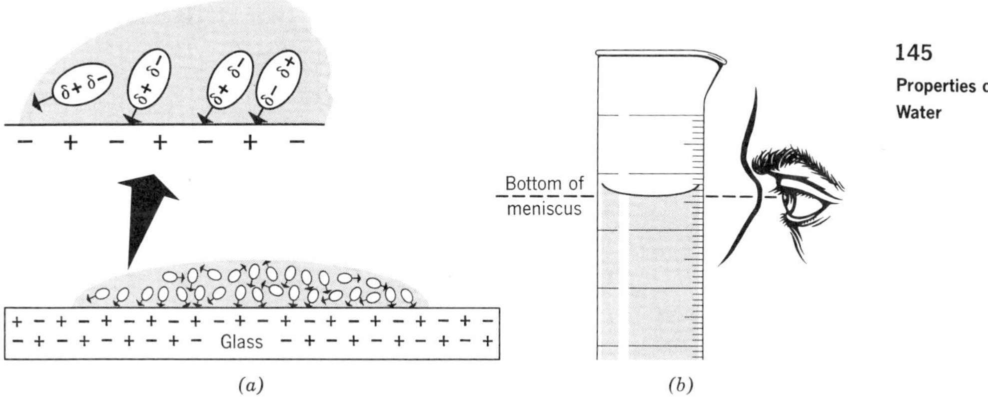
δ+ δ−
Bottom of
meniscus
Glass
(a)
(b)

Figure 5.7 The spreading of water on a polar surface such as glass. (*a*) Forces of attraction between water and glass are slightly greater than between water molecules within the droplet. The droplet therefore tends to spread. (Charges shown in the glass are partial charges.)

(*b*) When the glass surface is vertical, the initially unbalanced force of attraction between water and glass causes the water to creep up the glass wall until the force becomes balanced by the extra height of the water. The resulting curved surface is called a convex meniscus; in taking measurements with burets and graduated cylinders, standard practice is to read the bottom of the meniscus.

Table 5.5 Some Common Hydrates

Formulas	Names	Decomposition Modes* and Temperatures (°C)	Uses
$(CaSO_4)_2 \cdot H_2O$	Calcium sulfate hemihydrate (plaster of paris)	$-\frac{1}{2}H_2O$ (163)	Casts, molds
$CaSO_4 \cdot 2H_2O$	Calcium sulfate dihydrate (gypsum)	$-2H_2O$ (163)	Casts, molds, wallboard
$CuSO_4 \cdot 5H_2O$	Copper(II) sulfate pentahydrate (blue vitriol)	$5H_2O$ (150)	Insecticide
$MgSO_4 \cdot 7H_2O$	Magnesium sulfate heptahydrate (epsom salt)	$-6H_2O$ (150) $-7H_2O$ (200)	Cathartic in medicine Used in dyeing and tanning
$Na_2B_4O_7 \cdot 10H_2O$	Sodium tetraborate decahydrate (borax)	$-8H_2O$ (60) $-10H_2O$ (320)	Laundry
$Na_2CO_3 \cdot 10H_2O$	Sodium carbonate decahydrate (washing soda)	$-H_2O$ (33.5)	Water softener
$Na_2SO_4 \cdot 10H_2O$	Sodium sulfate decahydrate (Glauber's salt)	$-10H_2O$ (100)	Cathartic
$Na_2S_2O_3 \cdot 5H_2O$	Sodium thiosulfate pentahydrate (photographer's hypo)	$-5H_2O$ (100)	Photographic developing

* Loss of water is indicated by the minus sign before the symbol.

less water than gypsum. When it is mixed with water, it soon sets into a hard crystalline mass, according to the reaction

$$\underset{\text{Plaster of paris}}{(CaSO_4)_2 \cdot H_2O} + 3H_2O \longrightarrow \underset{\text{Gypsum}}{2CaSO_4 \cdot 2H_2O}$$

Anhydrous calcium chloride, $CaCl_2$, a common dehumidifier for damp basements and humid rooms, forms its hydrate by drawing water molecules from the atmosphere. Any substance that can perform this service is called a *desiccant.* If it acts by forming a hydrate, it is said to be *hygroscopic.* Anhydrous sodium sulfate, Na_2SO_4, and magnesium sulfate, $MgSO_4$, are common examples, forming the decahydrate and the heptahydrate, respectively. Some anhydrous substances are able to draw so much water from moist air that a liquid solution forms, and they are said to be *deliquescent.*

DISSOLVING ACTION OF WATER

Hydration. Sodium chloride does not melt until a temperature of 801°C is reached, a temperature normally beyond the heating capacity of an ordinary bunsen burner. It does not boil until the temperature reaches 1413°C. When sodium chloride melts, its ions become mobile. When it boils, they enter into the vapor state, not so much as isolated ions but as ion pairs and small ion clusters. The point is that the separation of sodium ions from chloride ions requires considerable energy. Water manages to bring about this separation of the oppositely charged ions at room temperature and below. Two approaches are helpful in understanding this remarkable fact.

The dielectric constant of water is unusually high. It is therefore a medium that greatly reduces the forces between charged particles. How it does this may be understood by examining Figure 5.8. On the left in this figure is a molecule's-eye view of the dynamic aspects of dissolving action. When randomly moving water molecules bombard the surface of a crystal of sodium chloride, the most probable collisions are those in which opposite charges make contact. Thus the end of the polar water molecule with a partial negative charge will strike sodium ions. Chloride ions in the lattice will be hit by parts of molecules of water that have partial positive charges. This bombardment therefore means that small attractive forces are at work to pull ions away. As soon as this happens, in fact, while it happens, several water molecules surround each of the ions, as shown on the right in Figure 5.8. Such behavior is natural, for unlike charges attract each other. The attraction is not between two *full* opposite charges, however, but between a fully charged ion and one end of the *dipole* of the solvent molecule—an *ion-dipole interaction.* When water is the solvent, the solute particles are said to be hydrated and the phenomenon is called *hydration.* For solvents in general it is called *solvation.* This phenomenon is the major factor in understanding the solubility of ionic compounds in water, and an action very much like it makes polar covalent compounds soluble. The more polar covalent compounds dissolve in water through *dipole-dipole interactions* (Figure 5.9). Hydrogen bonding is a special kind of dipole-dipole interaction.

Ammonia, NH_3, an important example of a soluble polar covalent compound, is a gas at room temperature but dissolves in water to a far greater extent than most gases. Nitrogen is more electronegative than hydrogen, and fractional charges may be assumed to be at the ends of the nitrogen-hydrogen bonds in ammonia molecules. Dipole-dipole interactions (Figure 5.10), here hydrogen bonding, may therefore occur between water and ammonia molecules.

Gases normally are only very slightly soluble in water at room temperature. Those that dissolve do so either by hydrogen bond formation (e.g., ammonia) or by an actual reaction with water to form solvated ions, a topic for a later chapter.

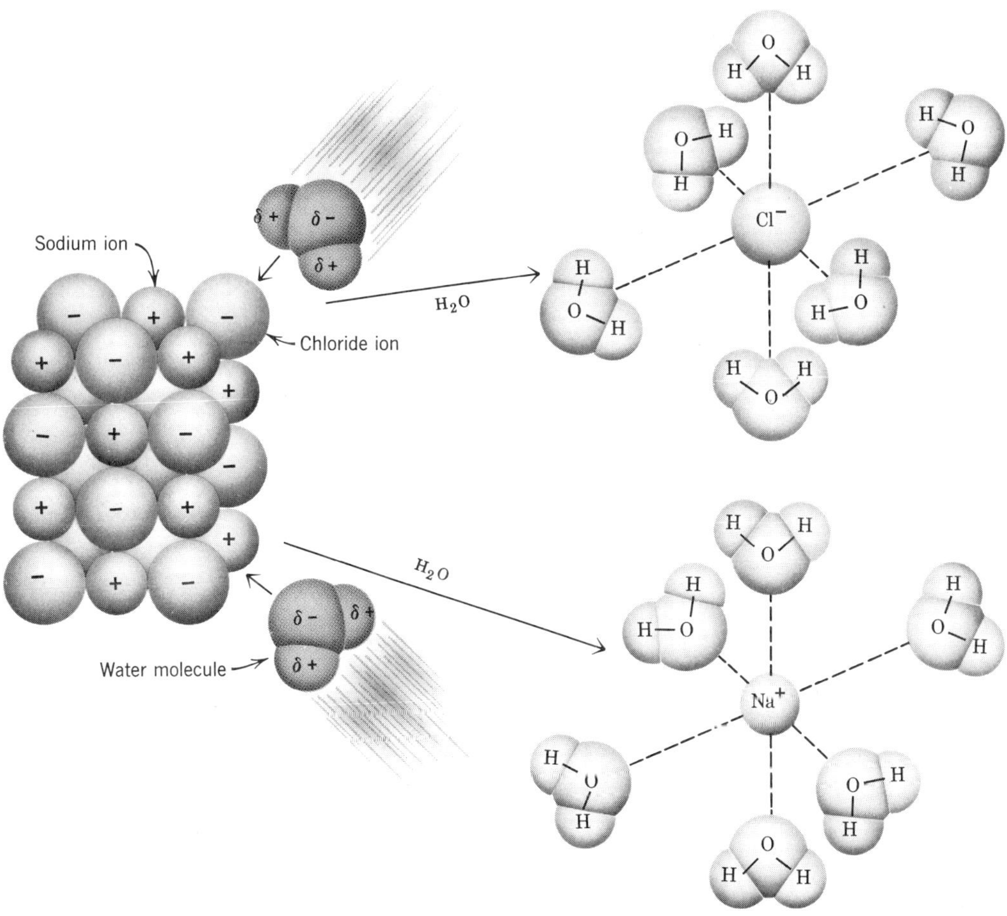
Sodium ion
Chloride ion
Water molecule
δ+
δ−
H_2O
Cl^-
Na^+

Figure 5.8 Factors making possible the dissolution of sodium chloride in water. On the left, polar water molecules are seen bombarding the surface of the crystal. A combination of mechanical bumping and electric forces of attraction dislodges ions on the surface and the edges of the crystal, and they go into solution. As they do so, on the right, they become surrounded by water molecules. Hydrated forms of the sodium and chloride ions are shown.

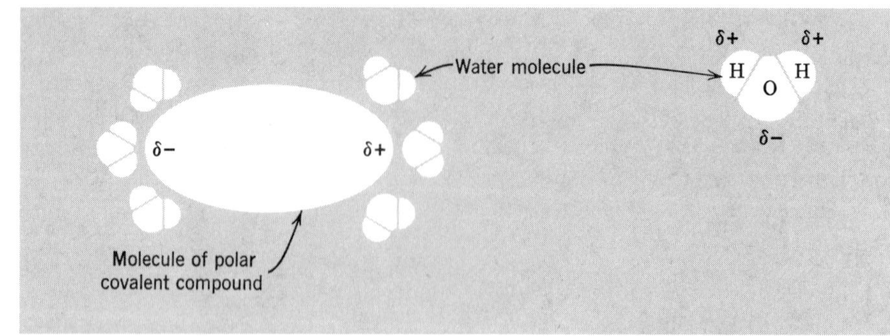
Water molecule
δ+
δ+
H
H
O
δ−
δ−
δ+
Molecule of polar covalent compound

Figure 5.9 Molecules of covalent compounds can be dissolved by water if they are polar, particularly if they have structural features that make hydrogen bonds possible.

METHODS OF EXPRESSING CONCENTRATIONS

Qualitative Methods. There is usually a limit to how much solute a given volume of solvent can hold at a given temperature. This limit varies with the solute, the solvent, and the temperature as the data in Table 5.6 show. When a solution contains all it can of a solute at a given temperature, it is said to be saturated. An *unsaturated solution* contains any smaller concentration of solute. Other qualitative expressions for concentration are *dilute* and *concentrated.* A concentrated solution is very nearly saturated, but the expression is usually reserved for solutions that can hold considerable amounts of solute. A dilute solution is one that contains a very small amount of solute per unit volume.

Dynamic Equilibria and Saturated Solutions. As more and more solid solute is added to a solution and as more and more of it dissolves, the rate at which dissolved solute particles *return* to the crystalline form increases. Eventually, the number that return will exactly equal the number that leave it elsewhere. At this point the rate of crystal regrowth equals the rate of crystal dissolution. In spite of much "coming and going," there is no *net* change either in the weight of the undissolved solute or in the amount of solute out in solution. Dynamic equilibrium has been reached. No more will dissolve at that temperature. A saturated solution is therefore defined as one in which a state of dynamic equilibrium exists between the undissolved solute and its dissolved state. It may be described by the symbolism of a chemical reaction:

$$\text{Undissolved solute} \rightleftharpoons \text{dissolved solute}$$

Two oppositely pointing arrows are the chemist's way of stating that a condition of dynamic equilibrium exists (Figure 5.11).

$$^{\delta+}H-{}^{\delta-}O-\overset{\delta+}{H}\cdots\overset{\delta-}{N}(H^{\delta+})H_2$$

Hydrogen bond

Figure 5.10 The solubility of ammonia in water is made possible by the existence of opportunities for hydrogen bonding. Ammonia molecules can slip into the hydrogen bond network of water.

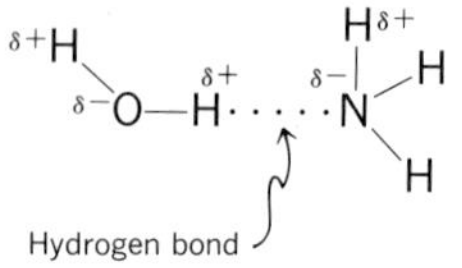

Table 5.6 Solubilities of Some Substances in Water

Solute	Solubilities (in grams per 100 g water)			
	0°C	20°C	50°C	100°C
Sodium chloride, NaCl	35.7	36.0	37.0	39.8
Sodium hydroxide, NaOH	42	109	145	347
Barium sulfate, $BaSO_4$	0.000115	0.00024	0.00034	0.00041
Calcium hydroxide, $Ca(OH)_2$	0.185	0.165	0.128	0.077
Oxygen, O_2	0.0068	0.0043	0.0027	0
Carbon dioxide, CO_2	0.335	0.169	0.076	0
Nitrogen, N_2	0.0029	0.0019	0.0012	0

If a saturated solution of a solid, with undissolved solute present, is heated, molecular motions of all particles increase. Solute particles hold to fixed positions in the crystal with increasing difficulty, and more go into solution. It is a general rule that the solubility of *solids* in water increases with temperature. An exception, calcium hydroxide, is given in Table 5.6.

If a saturated solution of a solid whose solubility is less at lower temperatures is cooled, the rate of crystal regrowth normally exceeds the rate of crystal dissolution. Dissolved solute *precipitates* until the rates of the opposing events are once again equal. If no crystalline surface is available for such regrowth, a saturated solution may cool without any precipitation. Such a cooled solution is said to be *supersaturated.* The addition of a "seed" crystal is usually all that is needed to bring about swift separation of the excess solute, for it provides a crystalline nucleus for further crystal building. Often crystallization can be induced by scratching the inner wall of the container with a glass rod. Honey, jams, and jellies are frequently supersaturated solutions of sugar. On long standing the sugar may crystallize.

Gases, in contrast with solids, become less and less soluble as the temperature rises, as the data in Table 5.6 show. Heat increases molecular motions for all kinds of solute particles, but gas molecules can escape from solution. One way to prevent this escape is to increase the pressure of the gas above the solution.

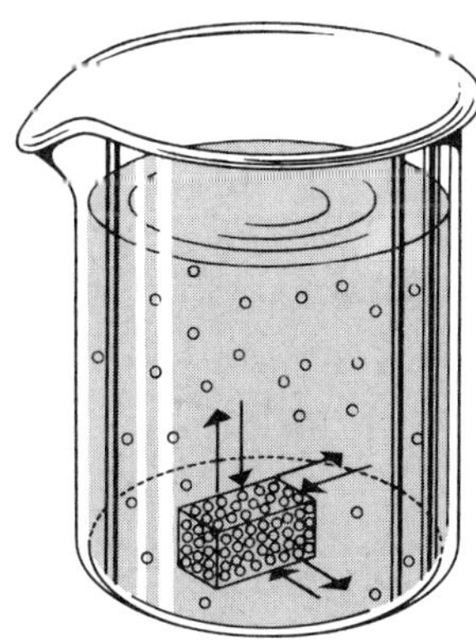

Figure 5.11 In a saturated solution, with undissolved solid solute in contact with the solution, solute particles return to the crystal as fast as they leave elsewhere. Dynamic equilibrium exists.

Quantitative Methods of Expressing Concentrations. For further insight into the nature and properties of solutions, we need some quantitative ways for expressing concentrations. One of the chief values of a solvent in facilitating chemical reactions between substances is its ability to bring into intimate contact two or more potential reactants. Ionic substances, in fact, rarely react with each other unless their ions are free to intermingle and collide with each other. They have this freedom in solution. We have learned that reactants interact on a particle-to-particle basis, and that the mole concept permits us to measure substances in ratios of *particles* by weighing them. When chemicals are in the dissolved state, we can determine moles by measuring *volumes* of the solution. Volumetric measurements are especially important when solutions are used in quantitative analysis.

Chemists usually express concentration in terms of *molarity,* the number of moles of solute per liter of solution. Concentration expressed in any units is basically a ratio, the amount of solute to some amount of solvent or of final solution. Molarity expresses concentration specifically in moles per liter, per liter of final solution, that is. A solution is said to be one molar (abbreviated $1M$) if it contains a ratio of one mole solute to one liter of solution.[2] Figure 5.12 illustrates how a liter of solution of this concentration is made. In a $0.5M$ solution the ratio is 0.5 mole solute to 1 liter of solution. Some worked examples and problems will illustrate applications of the concept.

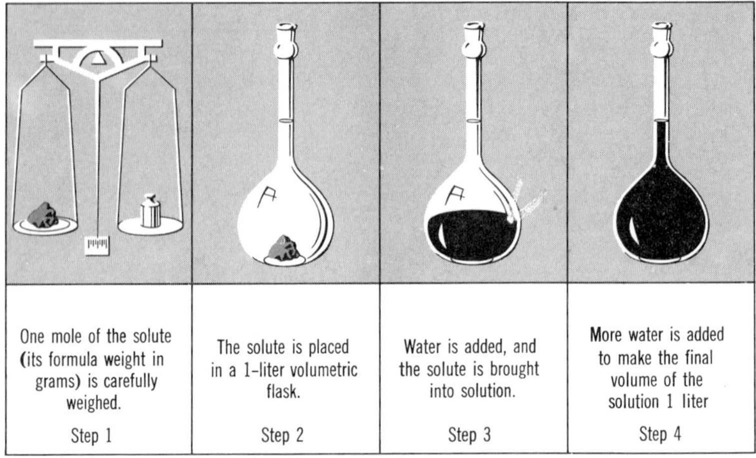

One mole of the solute (its formula weight in grams) is carefully weighed.	The solute is placed in a 1-liter volumetric flask.	Water is added, and the solute is brought into solution.	More water is added to make the final volume of the solution 1 liter
Step 1	Step 2	Step 3	Step 4

Figure 5.12 The preparation of 1 liter of a $1M$ solution.

Example 1 A bottle contains a sodium hydroxide solution labeled $1M$ NaOH. A reaction requiring 0.5 mole dissolved sodium hydroxide is to be performed. How much of the solution should be measured to obtain this amount of solute?

[2] Some references use the letter F from gram-*f*ormula weight instead of M from *m*olar, and this is very acceptable. A $3F$ solution and a $3M$ solution have identical concentrations. The word *formal* is sometimes used as a substitute for *molar*.

Solution. If you are a novice, and until you gain experience and familiarity, always translate the very brief description of concentration into its full meaning: $1M$ NaOH on the label means that the solution contains one mole sodium hydroxide in every liter. But, the next thing to be noted, the problem requires half a mole. If there is 1 mole in 1 liter, there will be half a mole in half a liter. *Ans.* $\frac{1}{2}$ liter or 500 cc.

Example 2 A sugar solution labeled $0.1M$ sucrose is available. A chemist needs 0.4 mole sucrose dissolved in water. How much of the stock solution should be taken?

Solution. Make the translation: $0.1M$ sucrose means 0.1 mole sucrose in every liter of solution. To obtain four times as much sucrose, 0.4 mole, the chemist must take 4 liters of the stock solution. *Ans.* 4 liters.

These very simple examples illustrate basic common sense for solving such problems. If the numbers are not so easily multiplied or divided, this common sense is best translated to the terms of the following equation:

$$\begin{array}{c}\text{Number of moles solute}\\ \text{per liter of solution}\end{array} \times \begin{array}{c}\text{number of liters of}\\ \text{solution taken}\end{array} = \begin{array}{c}\text{number of moles}\\ \text{solute obtained}\end{array}$$

or

$$\text{Molarity} \times \text{volume (in liters)} = \text{number of moles solute}$$

An analysis of the units gives

$$\frac{\text{Moles solute}}{\text{Liter solution}} \times \text{liter solution} = \text{moles solute}$$

Exercise 5.1 The chemical stockroom is supplied with the following stock solutions: $0.1M$ HCl, $0.2M$ NaOH, $0.01M$ HNO_3. What volume of stock solution is needed to obtain the amount of solute specified in each of the following.

(*a*) 0.2 mole HCl (*b*) 0.2 mole NaOH (*c*) 0.2 mole HNO_3
(*d*) 0.01 mole HCl (*e*) 4 moles NaOH (*f*) 0.005 mole HNO_3

Ans. (*a*) 2 liters.

Exercise 5.2 Assume the availability of pure substances NaCl, NaOH, and glucose ($C_6H_{12}O_6$) and volumetric flasks of sizes 100 cc, 250 cc, 500 cc, and 1 liter. Determine the grams of solute you must weigh out to prepare solutions, using the flasks specified and the final concentrations given for the labels.

(*a*) 1 liter flask, label: $0.5M$ NaOH
(*b*) 500 cc flask, label: $1M$ NaOH
(*c*) 250 cc flask, label: $1M$ NaOH
(*d*) 100 cc flask, label: $0.1M$ NaCl
(*e*) 100 cc flask, label: $0.3M$ glucose
(*f*) 250 cc flask, label: $0.05M$ NaOH
(*g*) 500 cc flask, label: $0.03M$ NaCl
(*h*) 250 cc flask, label: $0.25M$ glucose

Ans. (*b*) 20 g NaOH, (*c*) 10 g NaOH.

Another method of expressing concentration, used when exact molarities are not needed but some measure of concentration is, is *weight-volume percentage.* This is simply a weight of solute per unit volume of final solution; because it is expressed as a percent, it amounts to the *number of grams solute in 100 cc of solution.* Thus a 10% glucose solution will have 10 g glucose in 100 cc of solution. There may be a gallon of this solution or a few drops, but the *ratio* of solute to final volume of solution is 10 g/100 cc. A 0.9% salt solution will have 0.9 g salt in 100 cc of solution.

Example 1 How is 200 cc of a 4% glucose solution prepared?

Solution. Translate the label into its full meaning: 4% glucose means 4 g glucose in 100 cc of solution. For 200 cc, twice this amount, you will need twice 4 g or 8 g glucose dissolved in water and made up to a final volume of 200 cc.

Example 2 How much salt is present in 50 cc of a 3% solution?

Solution. Translate 3% into 3 g salt per 100 cc of solution. For 50 cc (half of 100) you will need half of 3 g or 1.5 g.

Exercise 5.3 Describe how you would prepare each of the following solutions, assuming the availability of suitable volumetric flasks and the necessary pure chemicals.

(*a*) 100 cc 2% potassium permanganate, $KMnO_4$

(*b*) 25 cc 1% NaOH

(*c*) 1 liter 5% KOH

(*d*) 500 cc 0.9% NaCl

(*e*) 250 cc 0.001% $AgNO_3$ *Ans.* (*e*) 2.5 mg $AgNO_3$.

Exercise 5.4 Various solutions bear the labels 10% NaCl, 10% NaBr, and 10% NaI. Translate these into labels with concentrations expressed as moles per liter, or molarity.

For very dilute solutions it is common practice to describe concentrations in *parts per million* or ppm. This abbreviated phrase means *parts solute per million parts of solvent* when it refers to solutions. The volume of very dilute solutions is for all practical purposes identical with the volume of the solvent used to make it. The word "parts" may refer to any weight unit the operator chooses, because in the *ratio* the units cancel. Usually the unit selected for preparing the solution is grams. If a solution, for example, contains 50 ppm sodium ion (plus, of course, some other negative ion), it contains 50 g sodium ion in 1 million g of water. By converting to different units, this becomes approximately 50 mg/liter which is a useful way of translating ppm: *ppm means milligrams per liter* (if the solvent is water for which 1 cc = 1 g).

Still another method for expressing the concentration of a dilute solution is as *milligram-percent,* or the number of milligrams solute per 100 cc of solution. This as well as the ppm system are often encountered in clinical work. Other methods of expressing concentrations (e.g., molality and normality) will be introduced where they will be most helpful.

COLLIGATIVE PROPERTIES OF SOLUTIONS

The chemical properties of solutions depend on the nature of the solute and the solvent and, to a certain extent, on their concentrations. The solubility of a compound in a solvent also depends on the nature of solute and solvent. Some properties of solutions, however, depend not so much on the nature of solute and solvent as on the concentration of solute particles, whatever they may be. Such properties are called *colligative,* and there are four: lowering the vapor pressure, raising the boiling point, lowering the freezing point, and creation of an osmotic pressure.

Colligative Properties of Dilute Solutions Containing Nonvolatile Covalent Solutes

1. **Lowering the vapor pressure.** If the equilibrium vapor pressure of a pure solvent is P_0 and the *mole fraction* of the solvent is X, the equilibrium vapor pressure of the solution, P, is given by the equation form of *Raoult's law*:

$$P = P_0X \tag{5.1}$$

The mole fraction of solvent, X, is simply the ratio of the moles of solvent to the total moles (solute plus solvent) present in the solution:

$$X = \frac{\text{moles solvent}}{\text{moles solute} + \text{moles solvent}} \tag{5.2}$$

Most solutions "obey" the relation expressed by Raoult's law only when they are dilute. Equation 5.1, in fact, is a way of defining an *ideal solution;* any solution whose vapor pressure fits the equation is ideal. In the molecule's-eye view equilibrium vapor pressure is lowered because the presence of nonvolatile solute particles reduces the population of solvent molecules at the surface (Figure 5.13). With fewer opportunities for the escape of solvent molecules, fewer are in the gas phase at any one time. Hence the vapor pressure will have to be less than for the pure

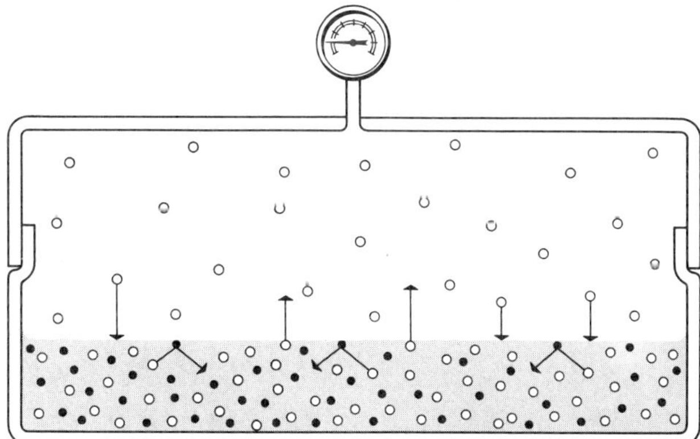

Figure 5.13 Escape of molecules of the liquid into the vapor state is hindered by the presence of particles of the nonvolatile solute (dark figures). Thus the equilibrium vapor pressure is lowered, by Raoult's law.

solvent. Figure 5.14 shows graphically how the vapor pressure is changed by the presence of the solute, which leads us directly to the second colligative property.

Francois Marie Raoult (1830–1901). French chemist who discovered how the presence of a nonvolatile solute affects the vapor pressure and the freezing point of a liquid.

---

2. **Raising the boiling point.** As Figure 5.14 shows, a higher temperature is needed to bring the vapor pressure of a solution up to atmospheric pressure. In other words, the boiling point of the solution is higher than that of the solvent alone.[3] The elevation of the boiling point Δt is directly proportional to the concentration of the solute, regardless of the chemical nature of the covalent solute, provided the concentration is expressed in a new way, *molal* units. A 1 molal solution has 1 mole solute per 1000 g solvent. It amounts to expressing concentration in terms of the number of particles of solute per fixed number of particles of solvent. Since colligative properties are determined by *particle* ratios without regard to the chemical nature of the particles, the molal unit is particularly convenient when dealing with them. (Remember, in this section we are limiting the discussion to *nonvolatile covalent* substances.) The equation is

[3] The temperature being measured is that of the *liquid* which is boiling, not that of the vapor above the liquid. If a thermometer is placed in the vapor and if vapor condenses on it so that at the thermometer bulb there is equilibrium between condensed liquid and vapor, the thermometer reading will be the boiling point of the pure liquid, because no solute is present in the vapor or its condensate.

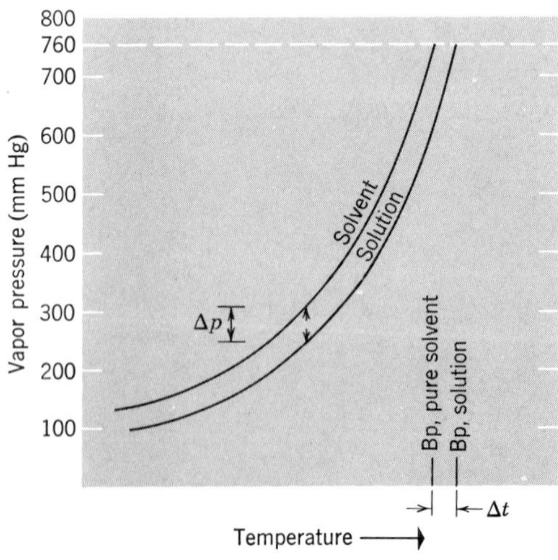
800
760
700
600
500
400
300
200
100
Vapor pressure (mm Hg)
Solvent
Solution
Δp
Bp, pure solvent
Bp, solution
Δt
Temperature

Figure 5.14 The addition of a nonvolatile solute to a solvent causes a drop in the solvent's vapor pressure Δp at any given temperature. It therefore takes a higher temperature to reach a vapor pressure that equals the atmospheric pressure. The presence of the solute, in other words, elevates the boiling point by a value of Δt.

$$\Delta t = K_b m \tag{5.3}$$

where $\Delta t = \text{bp}_{\text{solution}} - \text{bp}_{\text{solvent}}$
m = molal concentration of solute
K_b = molal boiling point elevation constant

Table 5.7 lists K_b values for various common solvents. With these constants known, with formula weights for these solvents also known, and with the obvious possibility of measuring Δt, the elevation of the boiling point provides a means of determining formula weights for solutes.

Example A solution prepared by dissolving 10.00 g of an unknown nonvolatile covalent substance in 100 g water was found to boil at 100.45°C at 1 atmosphere. What was the formula weight of the solute?

From Table 5.7, for water $K_b = 0.51$

From the data $\Delta t = 0.45°$

From equation 5.3 $m = \dfrac{\Delta t}{K_b}$

$= \dfrac{0.45}{0.51} = 0.88$ = number of moles solute per 1000 g water (by the definition of m)

From the data, 10.00 g solute per 100 g water is the same ratio as 100 g solute per 1000 g water. But 100 g solute must correspond to 0.88 mole. Therefore

$$\frac{100 \text{ g}}{0.88 \text{ mole}} = 114 \text{ g/mole} = \text{formula weight}$$

Exercise 5.5 A solution prepared by dissolving 11 g of a nonvolatile covalent solute in 100 g chloroform (bp 61.20°C) was found to boil at 64.30°C. What was the formula weight of the solute? *Ans.* 129.

---

3. **Depression of the freezing point.** One of the ways of converting seawater into water fit for drinking is to freeze it, collect the ice, and thaw it. This method works, in principle, because the frozen solvent is pure and does not contain the

Table 5.7 K_b and K_f Values for Various Solvents

Solvent	Bp (°C)	K_b	Mp (°C)	K_f
Water	100	0.51	0	1.86
Acetic acid	118.3	3.07	16.6	3.57
Benzene	80.2	2.53	5.45	5.07
Chloroform	61.2	3.63	—	—
Camphor	—	—	178.4	37.7

solute. (In practice, the energy cost is financially formidable.) From the use of antifreezes and the sprinkling of salt on icy streets to melt the ice, it is fairly common knowledge that solutions have lower freezing points than pure solvents. In the molecule's-eye view (Figure 5.15) these facts are related. The presence of solute particles in the solution interferes with the return of solvent particles to the solid phase. But since the frozen solvent does not contain solute particles, its molecules are not hindered from going out into the solution. To have equilibrium between the frozen state and the liquid solution, the temperature must be lower than the usual melting temperature. At the lower temperature the solvent molecules in solution slow down a bit and their "capture" by the solid phase is therefore more likely.

On a quantitative basis, the depression of the freezing point (fp) is proportional to the molality of the solution:

$$\Delta t = K_f m \tag{5.4}$$

where $\Delta t = \text{fp}_{\text{solvent}} - \text{fp}_{\text{solution}}$
m = moles solute/1000 g solvent
K_f = molal freezing point constant

Again, as with all colligative properties, we need no information about the nature of the solute. Its particle concentration is what is important, not the chemical nature of the particles. (Again the reminder, we are concerned here only with nonvolatile covalent solutes.) Values for K_f are listed in Table 5.7.

This phenomenon, as expressed quantitatively in equation 5.4, provides another way of measuring formula weights. In many respects it is superior to the elevation of the boiling point method because freezing points are not detectably sensitive to atmospheric pressure within the ranges observed in the open atmosphere.

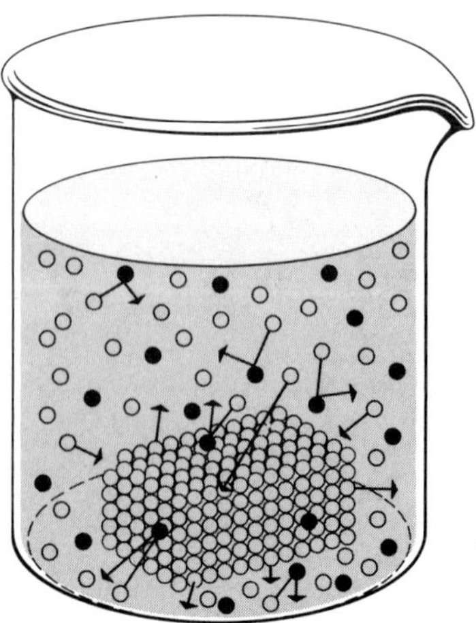

Figure 5.15 Depression of the freezing point of a liquid (open circles) by the presence of a solute (solid circles.) The solute does not enter the crystalline lattice of the solvent, but it interferes with the return to this lattice of solvent molecules that are out in the liquid. These solute particles do not interfere in the same way with the departure of molecules from the crystal and into the liquid state. In order, therefore, to lower the rate of crystal disappearance by this mechanism to the point that it equals the lower rate of crystal regrowth, the temperature must be lowered. Thus the equilibrium freezing temperature is lower than in the absence of the solute.

Exercise 5.6 A solution prepared by dissolving 1.5 g of an unknown compound in 100 g benzene was found to freeze at 5.12°C. What was its formula weight? *Ans.* 230.

Exercise 5.7 (*a*) Ethylene glycol, $C_2H_6O_2$, is the basic constituent of several commercial permanent antifreezes. For water-cooled automobile engines in the northern tier of American states as well as in Canada, it is common practice to seek protection for −40°F. How many moles of ethylene glycol are needed per 1000 g water to insure this protection? In other words, what would be the molality of a solution of ethylene glycol in water if its freezing point were −40°F? *Ans.* 21.5 molal.

(*b*) The density of ethylene glycol is 1.11 g/cc. How many cubic centimeters of ethylene glycol per 1000 g water does this amount to.

(*c*) By looking up conversion units, translate this into quarts of ethylene glycol per quart of water.

4. **Osmotic pressure and osmosis.** Among living things, boundaries between the internal systems and the surroundings have to serve many functions. These boundaries, called membranes, must first of all keep living units, cells, intact while at the same time allowing for the exchange of materials back and forth between the cell's interior and its surrounding fluids and the neighboring cells and tissues. A highly refined selection process must be at work so that only the correct materials come and go. Selection is accomplished in two ways. First, by so-called *active transport* energy-consuming reactions in effect pass certain selected ions and molecules from one side of the boundary to the other. Second, the membranes themselves are so constructed that only certain molecules can go through and the rest cannot. Not being permeable to all things, the membranes are called *semipermeable* (Latin *permeare,* to go through). Such membranes vary considerably in their selectivity. Some will permit the passage of molecules of solvent but not of dissolved solutes of any kind. If such a membrane is used to separate a dilute solution from a concentrated one, there will be a net flow of solvent into the more concentrated side. The system seemingly "moves" to equalize the concentrations. As water solvent leaves the more dilute side, the remaining solution becomes more concentrated. As the solvent moves into the more concentrated side, this solution becomes more dilute. When and if the concentrations become the same, there will be no net flow of solvent in either direction through the membrane. This net flow of solvent from a dilute solution (or pure solvent) into a more concentrated solution through a semipermeable membrane is called *osmosis.*

In the molecule's-eye view this phenomenon can be explained by Figure 5.16. On the basis of this model, semipermeability is attributed to the extreme smallness of the pores in the membrane. Small solvent molecules can move through them but larger hydrated ions or larger polar molecules cannot. These larger molecules of solute get more in the way of solvent molecules on one side of the membrane than they do on the other. Solvent movement out of the side having the less-concentrated solution is impeded less than solvent movement in the opposite direction. There is a net flow into the more concentrated side.

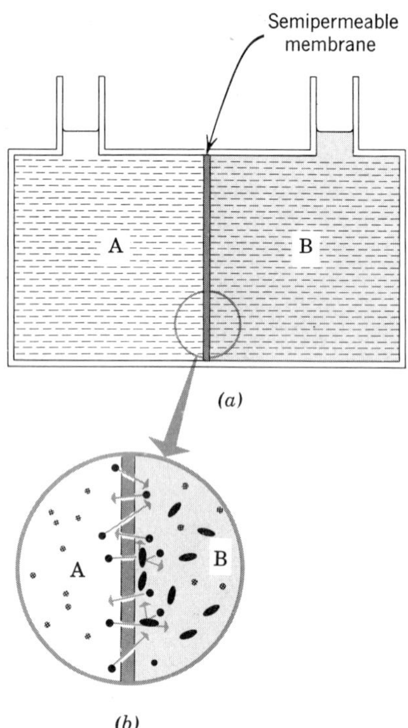
Semipermeable membrane
A
B
(a)
A
B
(b)

Figure 5.16 Osmosis. (*a*) Enlarged view of two fluid compartments separated by a semipermeable membrane. (*b*) Close-up of a small section of the membrane and its immediate environment. Water molecules are represented by dots, solute molecules in compartment B, by shaded ovals. In the "sieve" theory of osmosis, the membrane is said to have pores large enough to permit passage of water molecules but small enough to stop solute molecules (or ions). As drawn, of every five molecules of water that go from A to B, only three return. Two others are shown colliding with solute molecules. The result is a net flow of water from A to B, and level B rises.

Another model of osmosis postulates that solvent molecules, but not the solute, "dissolve" into the membrane and come out on the other side. From one point of view, it appears as though in the phenomenon of osmosis an unbalanced pressure is created within the solution that is the more dilute, a pressure that acts selectively to squeeze solvent out. This is a strange kind of pressure, or rather it seems a strange way to use the word. The pressure does not, for example, originate from the application of any outside force to any piston. It does not even exist until solutions of unequal concentration are separated by a semipermeable membrane, and then it is the relatively higher concentration on one side that causes the *effect* of the pressure from the *other* side. The resulting osmosis can be prevented only by applying an external pressure to the rising liquid on the more concentrated side, which provides us with a means for measuring *osmotic pressure* experimentally. By definition, the exact pressure needed to prevent osmosis from occurring when a solution is separated from pure solvent by a semipermeable membrane is called the osmotic pressure of that solution.

Osmotic pressure P_{os} is directly proportional to the *particle* concentration of the solution. When we speak of the osmotic pressure of a solution, we are referring to the pressure that would be generated if we separated this solution from pure solvent by the membrane, at least the pressure we would have to exert on the solution side to prevent osmosis. One way of talking about particle concentration is in terms of moles per liter, or n/V. Dealing only with covalent solutes, we find the osmotic pressure is directly proportional to this term:

$$P_{os} \propto \frac{n}{V}$$

or

$$P_{os} = \frac{n}{V} \times \text{constant} \tag{5.5}$$

By a slight rearrangement, equation 5,5 takes on the appearance of the Boyle's law equation, equation 3.2:

$$P_{os}V = n \times \text{constant} = \text{constant}' \quad (\text{if } n \text{ is fixed}) \tag{5.6}$$

It is also known experimentally that the osmotic pressure varies directly with the absolute temperature:

$$P_{os} = T \times \text{constant}'' \tag{5.7}$$

When combined, the relation between osmotic pressure, concentration, and temperature has the form

$$P_{os} = \frac{nT}{V} \times \text{constant} \tag{5.8}$$

Equation 5.8 resembles very much the ideal gas law equation, $PV = nRT$. What could not have been anticipated and is a surprising fact, the constant in 5.8 is the same as the gas constant R. Thus for osmotic pressure we have equation 5.9 for the ideal solution separated from the pure solvent by an ideal semipermeable membrane:

$$P_{os}V = nRT \tag{5.9}$$

The equation does not include information about the chemical nature of the solute, and osmosis with its attendant osmotic pressure is a colligative property. It depends only on the particle concentration without regard for the identity of the particles. The calculation of n obviously requires knowledge of the formula weight, which of course does depend on information about the solute—but only to give us particle numbers or ratios of these numbers.

The ideal osmotic membrane to use when water is the solvent is difficult to make and does not occur naturally. The best one is made by precipitating copper(II) ferrocyanide inside unglazed porcelain cups. The membranes in living systems are all less selective. They permit not just the passage of water but those of other chemicals as well. This is how it must be. Water is obviously important to life, but what it carries in the way of nutrient molecules in one direction and waste chemicals in another is equally important.

When solvent plus certain kinds of dissolved substaces experience a net flow through a semipermeable membrane, the phenomenon is called *dialysis* and the membrane is called a dialyzing membrane. Cellophane tubing is an example. It will permit the passage of water and of most particles small enough to be in true solution, but larger particles of a colloidal size will not get through. Thus the net *direction* of dialysis is determined by the imbalance in concentrations of these larger particles.

One practical effect of dialysis is the separation of substances in true solution from colloidal particles, a procedure often used in purifying colloidal substances. For this purpose cellophane, collodian, or an animal bladder may be chosen as the dialyzing membrane. Thus if a mixture of colloidally dispersed protein (perhaps an important enzyme) known to be contaminated by salts is placed in a dialyzing "bag," and water is circulated around it (Figure 5.17), the ions from the salts will soon appear in the water outside the bag where they can be washed away. The protein remains inside, becoming gradually purer. Its molcules are too large to get out.

The membranes in the body act as dialyzing membranes. Blood vessels, for example, are not ordinarily permeable to large protein molecules (e.g., serum albumins and globulins), but are permeable to water, carbon dioxide, oxygen, ions of salts, and small organic molecules fed into the system from the digestive tract.

Colligative Properties of Ionic Substances as Compared with Covalent Substances. Earlier we learned that when sodium chloride, NaCl, dissolves in water its ions separate. If each formula unit counts as one hypothetical compound particle in the crystalline lattice, it produces twice as many real particles (ions) when it dissolves. What is true of sodium chloride is believed to be true of salts in general. When crystalline lattices of salts dissolve in water, they do so by a separation of the component ions. Thus magnesium chloride, $MgCl_2$, should release three ions per formula unit, as should sodium sulfate, Na_2SO_4. Since colligative properties depend only on particle concentrations, ionic substances cannot be treated in terms of their formula units but rather in terms of the number of ions present in one formula unit. A very interesting and significant feature of solutions of ionic substances is revealed by the following example.

A 1 molal solution of sodium chloride is really 2 molal in terms of the concentration of particles (ions) present. Instead of 1 mole of NaCl molecules per 1000 g

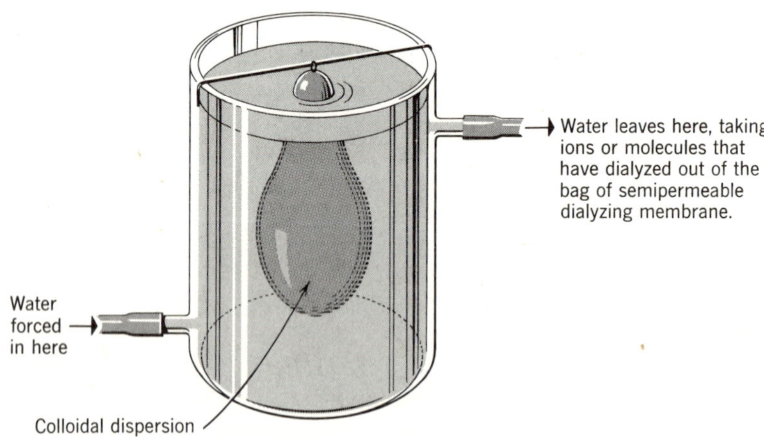
Water leaves here, taking ions or molecules that have dialyzed out of the bag of semipermeable dialyzing membrane.
Water forced in here
Colloidal dispersion

Figure 5.17 An apparatus for dialysis.

solvent, there are 2 moles of *ions* ($Na^+ + Cl^-$) per 1000 g solvent. Since the K_f value for water is 1.86(°C)(kg)/mole this solution will not freeze at -1.86°C but at $-1.86 \times 2 = -3.72$°C. The solution actually freezes at -3.37°C. Instead of a ratio of $3.72/1.86 = 2 =$ number of particles per formula unit, theoretically, we have a ratio of $3.37/1.86 = 1.81$.

This ratio is called the *mole number*[4] for sodium chloride at this particular concentration, and in general the mole number i is defined by the equation

$$\text{(5.10)} \quad \text{Mole number} = i = \frac{\text{value of colligative property for ionic compound}}{\text{value of colligative property for ideal covalent compound}}$$

Apparently, a 1 molal concentration of sodium chloride does not give the effect of a 2 molal concentration of particles, perhaps because of incomplete separation of the ions. Ion pairs, for example, may exist in solution. The ions are hydrated, and this interaction with the solvent is a departure from ideal behavior. Supporting these ideas is the fact that the mole number depends on the concentration of the solution. The more dilute it is, the more nearly does the mole number approach the theoretical whole-number value. The explanation is that the more dilute the solution, the less likely there will be any interactions between ions that would make ion pairs or ion clusters, and if these do not form the solute is, in effect, 100% ionized.

Because of this departure from ideal behavior, because the effective concentration of an ion in solution is usually less than that based on the molar amounts taken of its parent compound, chemists have invented the concept of the *activity* of an ion. Called the *activity coefficient,* it is some fraction of the expected or apparent molar or molal concentration:

$$\text{(5.11)} \qquad a = M\gamma$$

where a = activity
M = apparent molar (or molal) concentration
γ = activity coefficient (a unitless fraction)

The more dilute a solution is, the closer to 1.00 is the activity coefficient. In dilute solutions only a slight error is introduced by using the value of M for a in equation 5.11.

The topics of activity and activity coefficients have been introduced, but we shall not be scrupulously rigorous about their application. We shall, in fact, adopt the simplifying policy of using molarities or molalities instead of activities. Circumstances force this policy to a certain extent, for the values for activity coefficents are often unknown. Still, in dealing with solutions of ions, we should at least be aware of the existence of this problem.

CONDUCTIVITY

An aqueous solution in which ions are dispersed is a good conductor of electricity. In general, the greater the ionic concentration, the better the solution con-

[4] Also called the van't Hoff factor after Dutch chemist Jacobus Hendricus van't Hoff (1852–1911), the first Nobel laureate in chemistry.

ducts electricity. Solutions of covalent substances fail to conduct, assuming they have not reacted with water to produce ions. If an unknown solid is so insoluble in water that no test can be made on an aqueous solution, it is in principle possible to make the test on the pure compound in the molten state. Molten ionic compounds are conductors and are therefore called *electrolytes*. A very simplified explanation of conductivity is given in Figure 5.18.

HARD WATER

One of the most abundant minerals in the earth's crust is limestone, a form of calcium carbonate, $CaCO_3$. It is not very soluble in water. Rainwater that contains dissolved carbon dioxide has a leaching action on limestone, bringing calcium ions into the groundwater. The following reaction occurs:

$$\underset{\substack{\text{Calcium carbonate} \\ \text{(limestone)}}}{CaCO_3} + CO_2 + H_2O \longrightarrow \underset{\text{Calcium bicarbonate}}{Ca(HCO_3)_2}$$

The product is more soluble in water than calcium carbonate. Another metallic ion often present in groundwater is Mg^{2+}. In certain areas ions from iron, Fe^{2+} or Fe^{3+}

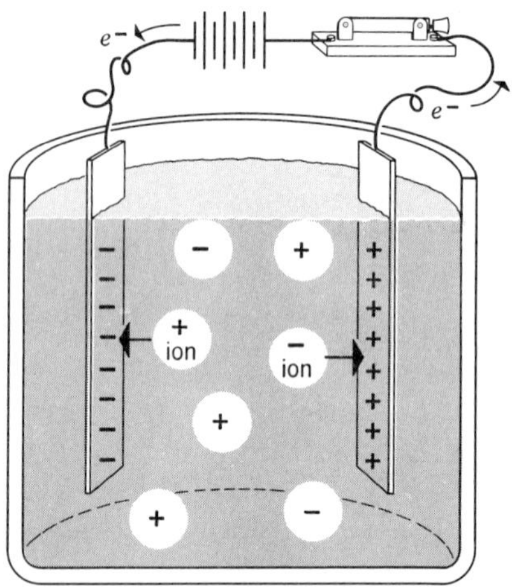
e−
e−
−
+
+
ion
−
ion
+
+
−

Figure 5.18 Electrolysis, simplified view. The source of electric current (e.g., a battery) forces electrons toward one metal strip or electrode and withdraws electrons from the other electrode. Unless electrons can be removed from the one plate and given to the other, there can be no closed circuit and no continuous flow. The result is a condenser as depicted in Figure 5.4. But when ions are present, electrons can be removed from one plate and placed on the other. Positive ions in the solution take electrons. Negative ions give other electrons on the other side. The electrode toward which positive ions are attracted, the negative electrode, is called the cathode, and positive ions are called cations. The other electrode is the anode, and negative ions are called anions.

or both, are also present. (The common negative ions besides bicarbonate are chloride and sulfate.) Water that contains any of or all the metallic ions mentioned —Ca^{2+}, Mg^{2+}, Fe^{2+}, and Fe^{3+}—interacts with soap to form an insoluble curd which is very difficult to wash out of fabrics once they pick it up. Water containing these metallic ions is therefore called *hard water*. *Soft water* lacks these specific ions, although it may have others (e.g., sodium or potassium ions). The softening of hard water in one way or another effectively removes the offending metallic ions.

In the home a very common water-softening agent is an excess of soap. Some of the soap is wasted as it precipitates with the metallic ions, and the remainder of the soap does the cleansing work. Even more common agents are the synthetic detergents ("syndets") which do not precipitate in hard water. Washing soda and household ammonia, other softening agents, provide an ion with which the metallic ions of hard water will precipitate. The following equations illustrate how these softeners work. The sulfate ion has arbitrarily been chosen for the negative ion, but it could just as well have been the chloride ion.

$$CaSO_4 + Na_2CO_3 \longrightarrow CaCO_3\downarrow + Na_2SO_4$$

$CaSO_4$	Na_2CO_3	$CaCO_3$	Na_2SO_4
Present in hard water	Washing soda in solution	Calcium carbonate (insoluble)	Sodium sulfate (soluble)

$$MgSO_4 + 2NH_4OH \longrightarrow Ca(OH)_2\downarrow + (NH_4)_2SO_4$$

$MgSO_4$	$2NH_4OH$	$Ca(OH)_2$	$(NH_4)_2SO_4$
Present in hard water	Ammonium hydroxide[5]	Calcium hydroxide (insoluble)	Ammonium sulfate (soluble)

More modern techniques of water softening take advantage of naturally occurring porous substances (zeolites) or of synthetic ion exchange resins. As hard water trickles through columns of these materials, the troublesome metallic ions are exchanged for sodium ions or others that do not interfere with soaps. These techniques do not normally remove bacteria.

A commonly encountered hard water contains bicarbonate ions which decompose when heated in water according to the following equation:

$$2HCO_3^- \xrightarrow[\text{heat}]{} CO_3^{2-} + CO_2\uparrow + H_2O$$

$2HCO_3^-$	CO_3^{2-}
Bicarbonate ion	Carbonate ion

As soon as carbonate ions form, they combine with magnesium or calcium ions also present and precipitate as magnesium or calcium carbonate. When hard water containing bicarbonate ions is heated in boilers and circulated as hot water, the carbonates precipitate as a hard scale on the inside of the pipes or the boiler. The resulting water is of course soft, but the cost of softening this way is high, for the scale not only tends to plug pipes (Figure 5.19) but also interferes with the transmission of heat from the source into the boiler. Higher and higher operating temperatures are required to overcome the effect of the boiler scale, and eventually

[5] Household ammonia consists of a dilute solution of ammonia in water. Just how it provides NH_4OH will be discussed on page 177.

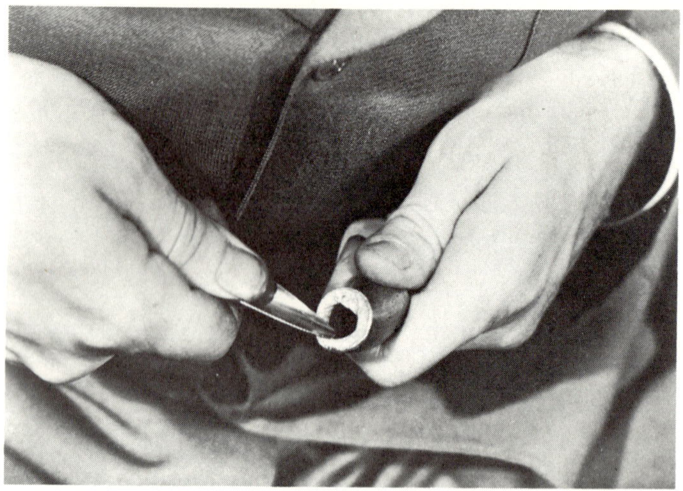

Figure 5.19 Carbonate scale in a hot-water pipe. The inside diameter of this pipe has been reduced about half by the accumulation of this scale. (Courtesy of the Permutit Company, a division of Ritter Pfaudler Corporation.)

breakdowns occur. Water that is hard only until it is boiled is called *temporary hard water,* and the predominate negative ion in it must be *bicarbonate.*

Water containing iron ions leaves rust deposits on porcelain and enamelware because the exposure of such water to the oxygen of the air causes the precipitation of the iron ions as their insoluble oxides, rust.

REFERENCES AND ANNOTATED READING LIST

BOOKS

C. J. Nyman and G. B. King. *Problems for General Chemistry and Qualitative Analysis.* John Wiley and Sons, New Vork, 1965. Chapter 9 ("Expressing Concentrations of Solutions") and Chapter 10 ("Colligative Properties of Solutions") cover the topics of this chapter with discussion, worked examples, and many exercises. (Paperback.)

K. S. Davis and J. A. Day. *Water. The Mirror of Science.* Science Study Series S18, Doubleday Anchor Books, Garden City, N.Y., 1961. While providing a good discussion of the unusual physical, chemical, and biological properties of water, the authors manage to illustrate a great deal about the nature of science as well.

ARTICLES

R. D. Jackson and C. H. M. van Bavel. "Solar Distillation of Water from Soil and Plant Materials: A Simple Desert Survival Technique." *Science,* Vol. 149, September 17, 1965, page 1377. A yield of 1.5 liters per day of drinking water can be obtained from a "survival still" made from a large piece of clear plastic and a collecting cup. Readers who venture into the desert may someday owe their lives to these authors.

A. M. Buswell and W. H. Rodebush. "Water." *Scientific American,* April 1956, page 77. Properties of water are related to its composition and structure in a relatively nontechnical way.

A. E. Snyder. "Desalting Water by Freezing." *Scientific American,* December 1962, page 41. Several ways of removing salt and other solutes from seawater by freezing are described.

H. F. Walton. "Ion Exchange." *Scientific American,* November 1950, page 48. The author describes how some of the ion exchange resins work to remove solutes from water.

P. F. Kerr. "Quick Clay." *Scientific American,* November 1963, page 132. This is a fascinating account of the properties and the dangerous implications of a certain kind of soil called "quick clay." Actually a type of colloid called a gel, quick clay can suddenly and without warning liquify and flow, carrying with it houses, barns, cattle, and people. A 1957 slide in Göta, Sweden, killed three and destroyed most of a pulp mill.

G. R. Choppin. "Water. H_2O or $H_{180}O_{90}$?" *Chemistry,* March 1965, page 7. The molecular structure of liquid water is discussed.

PROBLEMS AND EXERCISES

1. How do solutions, colloidal dispersions, and suspensions differ? What experimental evidence might we look for to identify these differences?
2. Define or identify each of the following terms
   (*a*) colloidal state (*b*) colloidal dispersion (*c*) Tyndall effect
   (*d*) Brownian movement (*e*) emulsifying agent (*f*) emulsion
   (*g*) sol (*h*) gel (*i*) hydrogen bond
   (*j*) dielectric constant (*k*) surface tension (*l*) surface-active agent
   (*m*) dessicant (*n*) hydration (*o*) ion-dipole interaction
   (*p*) solvation (*q*) dipole-dipole interaction (*r*) unsaturated solution
   (*s*) dynamic equilibrium (*t*) molarity (*u*) colligative property
   (*v*) mole fraction (*w*) ideal solution (*x*) Raoult's law
   (*y*) molality (*z*) osmotic pressure
   (*aa*) mole number
   (*bb*) activity
   (*cc*) activity coefficient
   (*dd*) electrolyte
   (*ee*) hard water
3. Why are hydrates classified as compounds rather than as wet mixtures?
4. How is the surface tension of water affected by raising its temperature?
5. Carbon tetrachloride is a nonpolar solvent. What does this mean?
6. If the molecules of carbon tetrachloride do not have as strong forces of attraction between them as are present between water molecules in water, why does carbon tetrachloride not dissolve in water?
7. The speed with which a solid can be dissolved in a solvent increases if (*a*) the solid is crushed to a powder, (*b*) the mixture is stirred, and (*c*) the mixture is heated. Using the kinetic theory, explain these facts. (This problem has nothing to do with how much can be put into solution, even though heat affects the amount. It has only to do with how fast a solution can be made.)
8. The dissolving of an ionic solid in water may be represented by the following equation in

which a symbol such as $Na^+(aq)$ refers to the hydrated form of the ion in solution (aq = aqueous):

$$NaCl(s) \xrightarrow[\text{water}]{} Na^+(aq) + Cl^-(aq)$$

The enthalpy for this particular change is +930 cal/mole at 25°C.

(*a*) Is the process of dissolving sodium chloride in water exothermic or endothermic?

(*b*) Referring to page 112, and given the enthalpy of the solution of sodium chloride in water, calculate the enthalpy change for the following reaction:

$$Na(s) + \tfrac{1}{2}Cl_2(g) \xrightarrow[\text{water}]{} Na^+(aq) + Cl^-(aq)$$

9. Ammonia, NH_3, is soluble in water; methane, CH_4, is insoluble in water.
   (*a*) What does this suggest about the ability of methane to form hydrogen bonds?
   (*b*) How do relative electronegativity differences in the C—H system compare with the N—H system? (See Figure 4.11.)
   (*c*) How is this related to the fact that CH_4 is insoluble in water?
10. Temporary soft water is softened by boiling. Can it be softened by ammonia water or by washing soda? Explain.
11. Explain why liquid water will not admit randomly dispersed molecules of a nonpolar covalent compound.
12. A solution of sodium chloride at 50°C was found to contain 36.5 g sodium chloride per 100 g water. Was it saturated, unsaturated, or supersaturated?
13. A solution of barium sulfate at 20°C was found to contain 0.00024 g barium sulfate per 100 g water. Would this solution be described as saturated, unsaturated, or supersaturated? Would it also be described as dilute or concentrated?
14. Describe in detail the "recipe" you would follow if you were to go to the laboratory and make up each of the following solutions.
   (*a*) 100 cc 1*M* glucose
   (*b*) 100 g 1 molal glucose
   (*c*) 100 cc 1% glucose
   (*d*) 100 cc of solution containing 100 ppm glucose
   (*e*) 100 cc of solution labeled 25 mg % glucose
   (Describe not only the weights and volumes but also the kinds of containers you would select for mixing the solutions.)
15. Explain in detail why it is much more convenient to use molality than molarity in dealing with colligative properties.
16. Suppose you do not know the concentrations of saturated solutions of some salt at various temperatures, but you still want to prepare and store a saturated solution. Without making any elaborate analyses of concentrations, how can you make a solution of a salt that at room temperature will be saturated, beyond any doubt?
17. Explain briefly how having a high dielectric constant makes a liquid a better solvent for ionic compounds and at the same time makes it a *poorer* solvent for nonpolar compounds.
18. For each of the following pairs of solutions, which member of the pair will have the higher osmotic pressure? Ignore activity coefficients by assuming them to be unity. If the first solution in each pair is placed in compartment A, Figure 5.16, and the second in B, which level will rise? If the pressures will be the same, state so.
   (*a*) 0.1*M* glucose and 0.2*M* glucose
   (*b*) 1% sodium chloride and 1*M* sodium chloride
   (*c*) 5% potassium chloride and 5% potassium iodide

19. A solution is prepared by dissolving 18 g glucose (f. wt., 180) in 1000 g water.
    (*a*) What is its molality?
    (*b*) What is the mole fraction of glucose?
    (*c*) What is the mole fraction of water?
    (*d*) What is the sum of your answers in parts *b* and *c*? What should it be?
20. The vapor pressure of water at 25°C is 23.8 mm Hg. Assuming that the solution of glucose in water described in problem 19 could be ideal, what would be its vapor pressure?
21. What would be the freezing point of a two molal solution of a covalent compound in water? What would be its boiling point?
22. For the measurement of formula weights of organic compounds by the method of the freezing point depression, benzene is more often used than water. More organic compounds are soluble in benzene at temperatures just above its freezing point than are soluble in water, but this is not the only factor. Even if solubility were no problem, what might be the most significant reason for choosing benzene over water? (*Hint.* The reason has to do with relative K_f values and the practical problem of measuring small changes of temperature.)
23. Glycerol ($C_3H_8O_3$, f. wt., 92) is a liquid that for all practical purposes is nonvolatile. It is very soluble in water. If 46 g are dissolved in 250 g of water:
    (*a*) What is the boiling point of the solution at 760 mm Hg?
    (*b*) What is its freezing point?
    (*c*) What is the vapor pressure at 25°C? (Vapor pressure of pure water is 23.8 mm Hg at this temperature.) *Ans.* (*a*) 101.86°C.
24. A solution of 12.00 g of an unknown dissolved in 200 g benzene froze at exactly 3.45°C. What was the formula weight of the solute?
25. A 0.5 molal solution of a compound of formula weight 100 froze at 12.00°C. The pure solvent has a freezing point of 14°C. What is the K_f value for the solvent?

CHAPTER SIX

Important Ionic Substances

Our present study of the principal ion-producing substances and some of their properties will be largely descriptive and qualitative. Quantitative aspects of ionic equilibria will be studied in Chapter 8. We must emphasize that in this chapter we study *ions,* not atoms, although we shall encounter the latter as ion-producers. Both chemically and physically, ions are vastly different from their corresponding neutral atoms. The symbols are very similar, which can cause confusion, but except in empirical formulas of ionic compounds the symbol of an ion *always* shows the kind and amount of charge it bears. The difference between Na(s) and $Na^+(aq)$ is the difference between something violently explosive in the presence of water (sodium atoms) and something so stable that it has virtually no chemical reactions (hydrated sodium ions); the difference between something we store away from both air and moisture and something without which we could not live.

Aqueous solutions of ions are for the most part made in two ways, by dissolving an ionic solid in water or by permitting certain covalent compounds or certain elements to react with water or with aqueous solutions of other compounds. The first process, the separation, solvation, and diffusion of preexisting ions, is called *dissociation.* The formation or synthesis of ions by chemical reactions is sometimes called *ionization.* The distinction between these terms is not important, and they are often used interchangeably.

PRINCIPAL ION-PRODUCING SUBSTANCES

The principal ion-producing substances are classified as acids, bases, and salts, and there are three common ways of defining them. The Arrhenius theory is the oldest and the least comprehensive, the Brønsted-Lowry theory is broader than the first, and the Lewis concept is the most general, subsuming the first two. There are advantages in knowing all three.

ARRHENIUS THEORY OF ACIDS AND BASES

Acids. Substances that liberate hydrogen ions in water are acids, according to the Arrhenius theory. The properties of acids are largely the properties of the hydrogen ion. As we now know, the hydrogen ion, H^+ (bare proton with no electrons), does not exist in aqueous solution, or anywhere else except in certain atom-smashing experiments. We shall frequently use the symbol H^+ and the phrase "hydrogen ion" as though the ion existed, but we permit ourselves this convenience only after learning a better approximation of the truth.

Svante August Arrhenius (1859–1927). Swedish scientist whose theory of how solutions of certain compounds can conduct electricity was found just barely acceptable as a doctoral dissertation in 1884 but who in 1903 received for it the Nobel prize in chemistry.

The truth is that hydrogen ions (protons) are attached to water molecules in aqueous solution in the form of *hydronium ions,* H_3O^+. For example, when gaseous hydrogen chloride is bubbled into water, an exothermic reaction occurs in which virtually all the hydrogen chloride molecules react with water (Figure 6.1). In the newly formed hydronium ions the extra proton is attached to oxygen by means of a covalent bond, but once formed all three oxygen-to-hydrogen bonds are equivalent. There are two reasons why it is acceptable to use H^+ for H_3O^+. First, none of

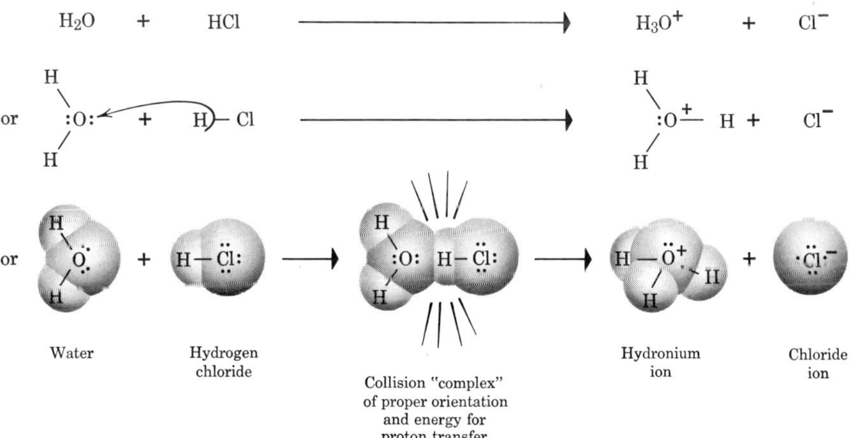
H2O + HCl → H3O+ + Cl−
or
or
Water
Hydrogen chloride
Collision "complex" of proper orientation and energy for proton transfer
Hydronium ion
Chloride ion

Figure 6.1 When molecules of gaseous hydrogen chloride dissolve in water, they react with water molecules to form hydronium ions and chloride ions. The proton initially bound to chlorine is now more strongly bound to oxygen, and the new ions are stabilized by hydration (p. 146).

the O—H bonds in H_3O^+ is very strong, and the proton is easily transferred to some acceptor. Furthermore, the hydronium ion is hydrated. We should write it as $H_3O^+ \cdot (H_2O)_n$ with n some small number probably six or smaller. In fact, we could almost call the hydronium ion a hydrated proton and write it as $H^+ \cdot H_2O$. In common usage we therefore lapse into writing H^+ and hydrogen ion, meaning of course H_3O^+ and hydronium ion if water is the solvent.

Some of the most common Arrenhius acids are listed in Table 6.1. *Strong acids* are those that ionize to a high percentage in water, even in fairly concentrated solutions. *Weak acids* ionize to only a small percentage in water. At room temperature only about 1% of the acid molecules are ionized in $0.1M$ acetic acid, a typical organic acid. (Of the hydrogens in acetic acid, $CH_3—\overset{O}{\overset{\|}{C}}—OH$, only the one attached to oxygen is ionizable.)

In Chapter 8 we shall put this theory on a quantitative basis, but a qualitative idea of the dynamic equilibrium present in an ionized acid is helpful at this point. When molecules of an acid intermingle with water molecules, collisions of all sorts occur. Figure 6.2 illustrates the kind of collision that can lead to the formation of a hydronium ion in an aqueous solution of a typical weak acid, acetic acid. If the acid is strong, virtually every collision of proper orientation will give a hydronium ion. But acetic acid is weak, and only the more violent of the on-target collisions result in ionization. Because only a small fraction of all the collisions occuring at any moment are violent enough, at any moment only a small fraction of the possible hydronium ions are generated. At the same time, collisions between the newly formed hydronium ions and the negative ion (acetate in our example) need not be violent for a transfer of hydrogen back again. Therefore most of these collisions are

Table 6.1 Common Arrhenius Acids and Their Percent Ionizations*

Name	Formula	Percent Ionization
Strong acids		
Perchloric acid	$HClO_4$	~100
Hydrochloric acid	HCl	~100
Hydrobromic acid	HBr	~100
Hydriodic acid	HI	~100
Nitric acid	HNO_3	~100
Sulfuric acid†	H_2SO_4	60 (in $0.05M$ solution)
Moderate acid		
Phosphoric acid	H_3PO_4	27
Weak acids		
Acetic acid	$CH_3\overset{O}{\overset{\|}{C}}O—H$	1.3
Carbonic acid	H_2CO_3	0.2
Boric acid	H_3BO_3	0.01

* Data are for $0.1M$ solutions of the acids in water at room temperature.

† Concentrated sulfuric acid is a particularly dangerous chemical, not only because it is a strong acid but also because it is a powerful dehydrating agent. This action generates considerable heat at the reaction site, and at higher temperatures sulfuric acid becomes even more dangerous. Moreover, it is a thick, viscous liquid that does not easily wash away from skin or fabric.

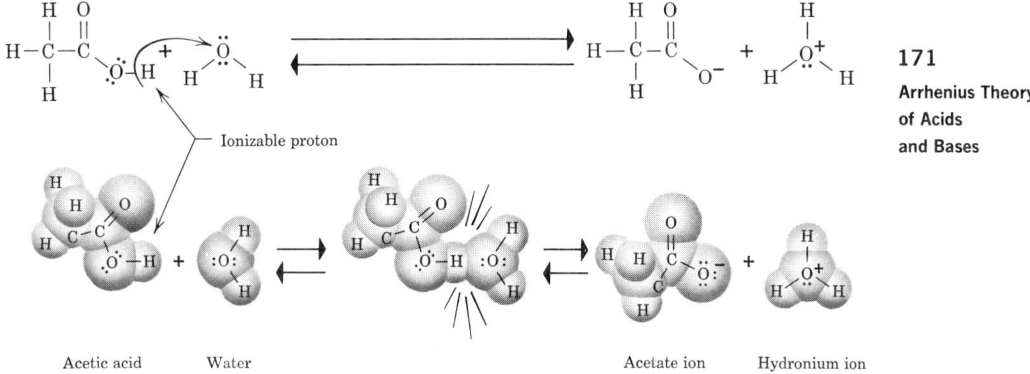
Ionizable proton
Acetic acid
Water
Acetate ion
Hydronium ion

Figure 6.2 When acetic acid molecules dissolve in water, a small percentage reacts with water molecules at any given instant to form a low concentration of hydronium ions and acetate ions. Only the more violent, on-target collisions lead to proton transfer. The fraction of all the collisions that are violent enough is much lower for acetic acid than for hydrogen chloride because the proton to be transferred is rather weakly bound to chlorine in H—Cl but is more strongly bound to oxygen in acetic acid.

successful in retransferring hydrogen, but the total number of collisions is small because few hydronium ions have been generated. On balance, dynamic equilibrium is eventually established at some particular concentration of each species, and for a poorly ionized species, or a weak acid, this balance is represented qualitatively by arrows of unequal length:

$$CH_3{-}\overset{\displaystyle O}{\overset{\|}{C}}{-}O{-}H + H_2O \rightleftarrows CH_3{-}\overset{\displaystyle O}{\overset{\|}{C}}{-}O^- + H_3O^+$$

In contrast, the interaction of a potentially strong acid such as hydrogen chloride is represented as

$$H{-}Cl + H_2O \rightleftarrows Cl^- + H_3O^+$$

Solutions of Arrhenius acids have the following characteristic properties, all of which are really properties of the hydronium ion in water.

1. To our taste they are sour. Organic acids in vinegar, lemon juice, rhubarb, and grapefruit are responsible for their tartness.

2. They react with active metals to generate salts and hydrogen gas. For example, zinc metal reacts with hydrochloric acid to form zinc chloride (a typical salt) and hydrogen. This reaction may be represented in any one of the following ways.

(*a*) $Zn + 2HCl \longrightarrow ZnCl_2 + H_2\uparrow$ This way of writing the reaction shows the formulas of the actual chemicals involved and the proportions in which they interact.

(*b*) $Zn + 2(H^+ + Cl^-) \longrightarrow (Zn^{2+} + 2Cl^-) + H_2\uparrow$ This expression emphasizes the fact that the reaction involves atomic zinc and ionized hydrogen chloride, and that zinc ions and molecular hydrogen form. Chloride ions do nothing; they are merely passive "observers."

(*c*) $Zn + 2H^+ \longrightarrow Zn^{2+} + H_2\uparrow$ This equation lays bare the essence of the reaction, the only chemical event that occurs. Note that it is a redox process.

Equation *c* is an example of an *ionic equation,* a type often used to isolate and emphasize the chemical event and to put into the background other particles (ions or molecules) that are just part of the environment. A complete ionic equation requires two kinds of balance, material and electrical. Both the number and kinds of nuclei and the *net* number and kind of electric charge on both sides of the equation must balance.

Metals differ widely in their tendencies to react with hydrogen ions. When they do, the *atoms* of the metal are oxidized; they lose electrons and become metal *ions.* The electrons are transferred to hydrogen ions, which are reduced, made electrically neutral, and emerge as molecules of hydrogen gas. Elements in the alkali metal family (group I), such as sodium and potassium, include the most reactive metals of all. Not only do they react with hydrogen ions, but they also displace hydrogen from water. The following reaction of sodium is extremely violent:

$$\underset{\text{Sodium metal}}{2Na} + \underset{\text{Water}}{2H_2O} \longrightarrow \underset{\text{Sodium hydroxide}}{2NaOH} + \underset{\text{Hydrogen}}{H_2\uparrow}$$

Platinum and gold, on the other hand, are stable not only toward water but also toward hydrogen ions. This variation in reactivity of metals suggests that it should be possible to arrange metals by their tendency to become ionic. Tendency is not a very good scientific term, and an activity series of metals could hardly be made accurate and therefore useful without some quantitative means for measuring their activities. Such methods are available in the field of electrochemistry, a field we are for the most part omitting; Table 6.2 summarizes the results. Atoms of any metal above hydrogen in this activity series will transfer electrons to hydrogen ions to form hydrogen gas, leaving the metal ions behind. Tin and lead, however, react

Table 6.2 Activity Series of the Metals

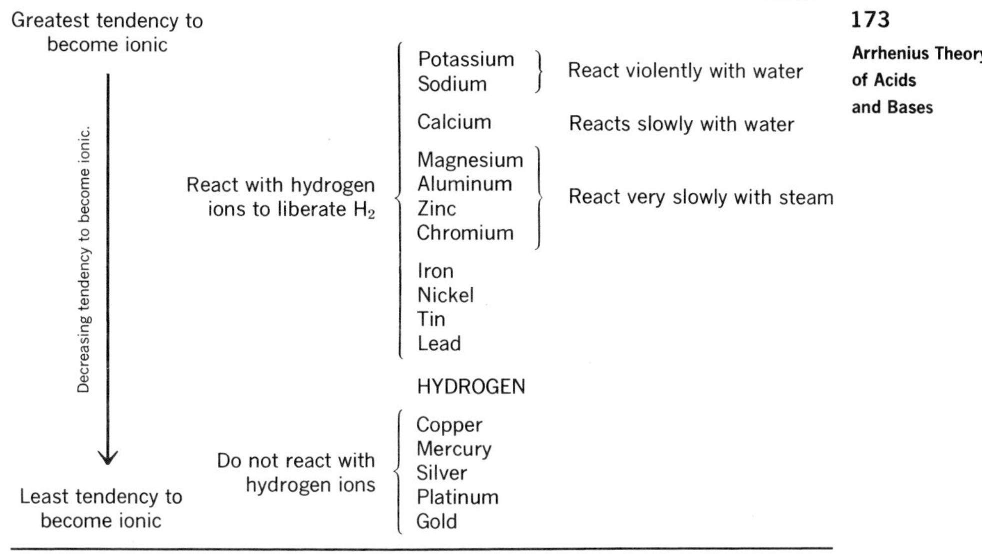

Tendency to become ionic	Reaction with hydrogen ions	Metal	Reaction with water
Greatest tendency to become ionic	React with hydrogen ions to liberate H_2	Potassium	React violently with water
		Sodium	
↓ Decreasing tendency to become ionic.		Calcium	Reacts slowly with water
		Magnesium	React very slowly with steam
		Aluminum	
		Zinc	
		Chromium	
		Iron	
		Nickel	
		Tin	
		Lead	
		HYDROGEN	
	Do not react with hydrogen ions	Copper	
		Mercury	
		Silver	
		Platinum	
Least tendency to become ionic		Gold	

only very, very slowly. The metals below hydrogen on the scale cannot be converted into their ions by the action of hydrogen ions. Better electron acceptors are required and are available.

3. Acids react with metal hydroxides to form salts and water. For example, sodium hydroxide and nitric acid react as follows:

$$\underset{\text{Sodium hydroxide}}{NaOH} + \underset{\text{Nitric acid}}{HNO_3} \xrightarrow[\text{water}]{} \underset{\text{Sodium nitrate}}{NaNO_3} + H_2O$$

The only chemical event, however, is the following:

$$OH^- + H^+ \longrightarrow H_2O$$

If the solution resulting from this reaction were evaporated to dryness, crystals of sodium nitrate, $NaNO_3$, a typical salt, would remain.

4. Acids react with *metal oxides* to form salts and water. This reaction is confined to attack by the acid not on a solution of the oxide but on its solid form. Those oxides that do dissolve in water react with it to form metal hydroxides, for example,

$$\underset{\text{Sodium oxide}}{Na_2O} + H_2O \longrightarrow \underset{\text{Sodium hydroxide}}{2NaOH}$$

Oxide ions, O^{2-}, do not exist in water. They react with it as follows, in the ionic equation for the previous reaction:

$$O^{2-} + H{-}OH \longrightarrow {}^{-}OH + {}^{-}OH \quad (\text{or } 2OH^{-})$$

Oxide ion in the crystal of the metal oxide

The reaction of a hydronium ion with an oxide ion in a crystalline material is very similar, with the important difference that the hydronium ion is a much better proton releasor than a water molecule:

$$O^{2-} + H{-}\overset{+}{O}(H)_2 \longrightarrow {}^{-}OH + H_2O$$

Oxide ion in the crystal of the metal oxide

Since the hydroxide ion is produced in the vicinity of unchanged hydronium ions, it will compete with oxide ions to form water:

$$^{-}OH + H_3O^{+} \longrightarrow 2H_2O$$

The reaction of acid with iron rust (approximate formula, Fe_2O_3) illustrates the overall effect:

$$Fe_2O_3 + 6HCl \longrightarrow 2FeCl_3 + 3H_2O$$

Iron(III) oxide (ferric oxide) — Iron(III) chloride (ferric chloride)

5. Acids react with *metal bicarbonates* to form salts, water, and carbon dioxide. Sodium bicarbonate, for example, reacts with hydrochloric acid as follows:

$$NaHCO_3 + HCl \longrightarrow NaCl + H_2O + CO_2\uparrow$$

Sodium bicarbonate — Hydrochloric acid

To understand this reaction better, we must know the structure of the bicarbonate ion and realize that carbonic acid is unstable, decomposing into carbon dioxide and water.

$$H{:}\ddot{O}{:}\overset{\ddot{O}}{C}{:}\ddot{O}{:}H \qquad H{:}\ddot{O}{:}\overset{\ddot{O}}{C}{:}\ddot{O}{:}^{-} \qquad ^{-}{:}\ddot{O}{:}\overset{\ddot{O}}{C}{:}\ddot{O}{:}^{-}$$

or

$$H{-}O{-}\overset{\overset{O}{\|}}{C}{-}O{-}H \qquad H{-}O{-}\overset{\overset{O}{\|}}{C}{-}O^{-} \qquad ^{-}O{-}\overset{\overset{O}{\|}}{C}{-}O^{-}$$

H_2CO_3	HCO_3^{-}	CO_3^{2-}
Carbonic acid	Bicarbonate ion	Carbonate ion

Exercise 6.1 Verify that the electric charges shown for H_2CO_3, HCO_3^-, and CO_3^{2-} are correct.

The reaction of bicarbonate ion with hydronium ion may be represented as

$$H{-}O{-}\overset{\displaystyle O}{\overset{\|}{C}}{-}O^- + H{-}\overset{+}{O}H_2 \longrightarrow \left[H{-}O{-}\overset{\displaystyle O}{\overset{\|}{C}}{-}O{-}H\right] + H_2O$$

$$\left[H_2CO_3\right] \xrightarrow{\text{decomposes}} O{=}C{=}O + H_2O$$

Carbon dioxide CO_2

Actually, an equilibrium exists between carbon dioxide and water,

$$O{=}C{=}O + H{-}O{-}H \rightleftarrows O{=}C(OH)(OH)$$

Carbonic acid (elements of water circled)

but the *position of equilibrium* favors free carbon dioxide (and water) rather than carbonic acid. (Carbonic acid can exist in aqueous solution in small concentrations, a fact of great importance in the chemical processes whereby living organisms remove waste carbon dioxide made in their cells.) When metal bicarbonates are mixed with acids, fizzing occurs as carbon dioxide forms and bubbles away.

6. Acids react with *metal carbonates* in essentially the same way as with

$$CaCO_3 + 2HCl \xrightarrow[\text{water}]{} CaCl_2 + H_2O + CO_2$$

$$^-O{-}\overset{\displaystyle O}{\overset{\|}{C}}{-}O^- + H{-}\overset{+}{O}H_2 + H{-}\overset{+}{O}H_2 \longrightarrow \left[H{-}O{-}\overset{\displaystyle O}{\overset{\|}{C}}{-}O{-}H\right] + 2H_2O$$

$$\left[H_2CO_3\right] \longrightarrow CO_2\uparrow + H_2O$$

bicarbonates. Carbonates occur widely in the earth's crust, and they may be distinguished from other minerals by the fact that when treated with acid the odorless, colorless gas carbon dioxide evolves. In the calcite group of minerals are such carbonates as calcite itself ($CaCO_3$), magnesite ($MgCO_3$), siderite ($FeCO_3$), and smithsonite ($ZnCO_3$). Marble and limestone are largely carbonates of calcium and magnesium.

7. Acids turn blue litmus paper red. This is one of the most common techniques for detecting acids, although many substances besides litmus will indicate their presence. Called *indicators,* they will be discussed in detail in Chapter 8.

Exercise 6.2 Complete and balance each of the following equations by writing the structures of the expected products and then adjusting coefficients to obtain the correct balance. Water is assumed to be the solvent for the acid and for all other chemicals given except metal oxides and metals.

(*a*) $KOH + HCl \longrightarrow$
(*b*) $K_2O + HCl \longrightarrow$
(*c*) $KHCO_3 + HCl \longrightarrow$
(*d*) $K_2CO_3 + HCl \longrightarrow$
(*e*) $H_3O^+ + OH^- \longrightarrow$
(*f*) $Zn + HNO_3 \longrightarrow$
(*g*) $Na_2CO_3 + HBr \longrightarrow$
(*h*) $NiO + H_2SO_4 \longrightarrow$
(*i*) $BaO + HNO_3 \longrightarrow$

Bases in the Arrhenius Theory. The hydroxide ion is the one basic species in the Arrhenius theory. The principal providers of such ions are metal hydroxides (e.g., NaOH, KOH, LiOH) and a few metal oxides (Na_2O, etc.) which react with water as we have seen. These solutions have the following properties.

1. They are bitter to the taste and slippery or soapy to the touch.
2. They turn red litmus blue.
3. As described earlier, they react with hydrogen ions to form water:

$$OH^- + H^+ \longrightarrow H_2O$$

or more accurately,

$$OH^- + H_3O^+ \longrightarrow 2H_2O$$

Called acid-base neutralization, much of Chapter 8 is devoted to this reaction.

A *strong base* is one for which the percentage of dissociation in water is high; for *weak bases* this percentage is low. Table 6.3 lists several common bases classified as strong or weak. The terms *strong* and *weak* when applied to acids and bases do not necessarily describe their behavior toward tissue or cloth. Damage by alkalies or acids to certain fabrics or to the skin occurs quickly only when fairly concentrated solutions of hydroxide ions or hydrogen ions are involved. Acetic acid dissolves in water in all proportions, and high concentrations of the molecules of acetic acid are possible. The percentage of these that are at any instant ionized is very low, however, and acetic acid is a weak acid. It is also "weak" in its attacks on at least certain fabrics or skin. In contrast, calcium hydroxide and magnesium hydroxide are both classified as strong bases chemically because their percentages of ionization *in solution* are very high. But very little can be put into solution; their solubilities

Table 6.3 Common Arrhenius Bases

Name	Formula	Solubility in Water*	Percent Dissociation
Strong bases			
Sodium hydroxide	$NaOH$	109	91 (in 0.1M solution)
Potassium hydroxide	KOH	112	91 (in 0.1M solution)
Calcium hydroxide (an aqueous solution is called "lime-water")	$Ca(OH)_2$	0.165	100 (saturated solution)
Magnesium hydroxide (a slurry in water is called "milk of magnesia")	$Mg(OH)_2$	0.0009	100 (saturated solution)
Weak base			
Ammonium hydroxide	$NH_3(aq)$ ("NH_4OH")	89.9 g NH_3 at 0°C	1.3 (at 18°C)

* Solubilities are in grams of solute per 100 g water at 20°C except where noted otherwise.

in water are so low that, in spite of 100% ionization, the actual molar concentration of hydroxide ions is very low. Hence limewater, $Ca(OH)_2$ solution, and milk of magnesia, $Mg(OH)_2$ solution, are not to be feared physiologically in the way that lye, NaOH, is. Ammonia, on the other hand, is very soluble in water, but because the percentage of hydroxide ion present is very low, it is a weak base. Thus the terms strong and weak have qualitative significance only and are of limited usefulness because of the confusion between their chemical sense and common, everyday physiological sense. These terms are used in the same way to describe the ability of certain solutions to conduct electricity. Since solutions of acids, bases, and salts conduct an electric current, these ion producers are called *electrolytes*. A *strong electrolyte* gives a high percentage of ions, a *weak electrolyte* a low percentage, and a *nonelectrolyte* no ions at all.

Ammonium hydroxide, as it is commonly called, is really a solution of ammonia gas in water. As random collisions occur, some of them as indicated below may be violent enough to effect the transfer of a proton from water to the ammonia molecule:

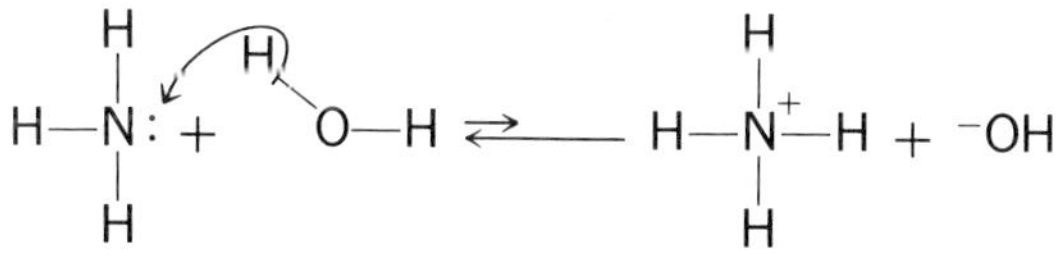

$$H_3N: + H{-}O{-}H \rightleftarrows H{-}\overset{+}{N}H_3 + {}^{-}OH$$

or

$$NH_3 + H_2O \rightleftarrows \underset{\text{Ammonium ion}}{NH_4^+} + OH^-$$

The pure substance of formula NH_4OH is unknown, but ammonium ions are very common. As the unequal lengths of the arrows indicate, the dynamic equilibrium favors the species on the left. Only a small percentage of hydroxide ions exist at any instant.

BRØNSTED-LOWRY CONCEPT OF ACIDS AND BASES

In the Arrhenius theory the only acid is the hydrogen ion. The theory applies only to water as a solvent, and hydrogen ions must not just be potentially available but must actually be there. The only base is the hydroxide ion, and again the solvent is restricted to water. Brønsted and Lowry extended and broadened the concept of acids and bases by classifying as an acid *any substance that is capable of donating a proton to any other substance.* The hydronium ion clearly qualifies, but so do the bicarbonate ion and a host of other *proton donors.* The bicarbonate ion, for example, can react with the hydroxide ion as follows:

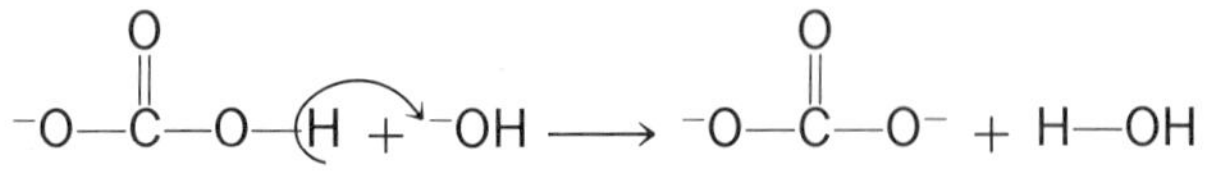

$$^{-}O-\overset{\overset{O}{\|}}{C}-O-H + {}^{-}OH \longrightarrow {}^{-}O-\overset{\overset{O}{\|}}{C}-O^{-} + H-OH$$

Johannes Nicolaus Brønsted (1879–1947). Danish physical chemist who stressed the role of the hydrogen ion in systems of both acids and bases. With his theory the problem of how buffers work became clearer.

Thomas Martin Lowry (1847–1936). English physical chemist who, independently of Brønsted, proposed essentially the same theory.

A base is defined as any substance that can accept a proton from any other compound. Bases are proton acceptors. Thus the bicarbonate ion is a base too, for it can accept a proton from the hydronium ion:

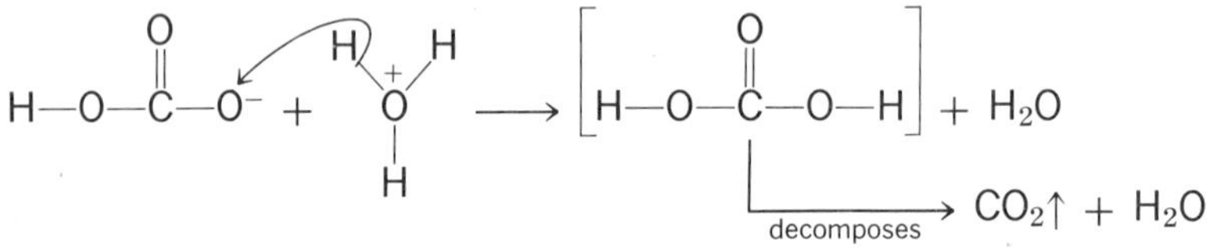

$$H-O-\overset{\overset{O}{\|}}{C}-O^{-} + H_3O^{+} \longrightarrow \left[H-O-\overset{\overset{O}{\|}}{C}-O-H\right] + H_2O$$

$$\xrightarrow{\text{decomposes}} CO_2\uparrow + H_2O$$

The hydroxide ion, of course, is just one of many possible bases. The "neutral" ammonia molecule is also a base, a Brønsted base, for short:

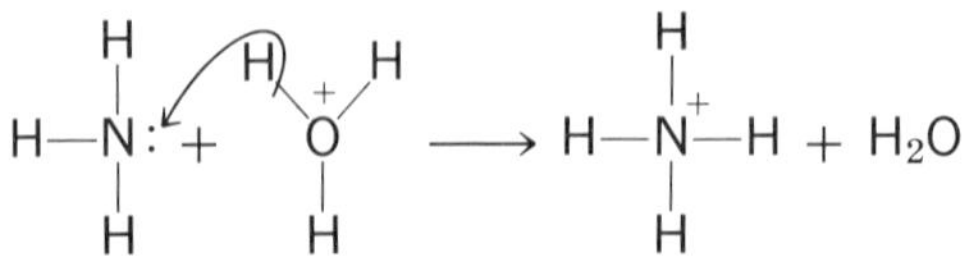

$$H-\underset{H}{\overset{H}{N}}: + H_3O^{+} \longrightarrow H-\underset{H}{\overset{H}{N^{+}}}-H + H_2O$$

In fact, in so-called ammonium hydroxide ("ammonia water," household ammonia) the principal acid-neutralizing substance is the ammonia molecule, not the hydroxide ion.

In the Brønsted-Lowry concept many ionic species fall into one or the other of the acid-base categories, and it is useful to think in terms of acid-base pairs. For example, when gaseous hydrogen chloride is bubbled into water, the proton donor (the acid) is the hydrogen chloride molecule, the proton acceptor (the base) the water molecule. The latter is converted into what in other situations would be a proton donor, the hydronium ion. If a water molecule has acted as a base in this reaction, the hydronium ion is called the *conjugate acid*. Similarly, the chloride ion is called the *conjugate base* of hydrogen chloride:

$$\text{H—Cl} + H_2O \rightleftarrows Cl^- + H_3O^+$$

H—Cl	H₂O	Cl⁻	H₃O⁺
Hydrogen chloride, the stronger acid	Water, the stronger base	Chloride ion, the weaker base (Cl^- is the conjugate base of H—Cl)	Hydronium ion, the weaker acid (H_3O^+ is the conjugate acid of H_2O)

In the Brønsted-Lowry concept a strong acid is a good proton donor; it is a particle having a relatively weak hold on the proton it can donate. A strong base is a good proton acceptor and a strong proton binder. A weak acid, on the other hand, is a poor proton donor and has a relatively strong hold on the potential proton. A weak base is a poor proton binder and has a tendency to release its proton once it receives it. Although these are qualitative ways of expressing concepts of strength and weakness, they are highly useful. As noted in the reaction just given, acid-base equilibrium will favor the substances that are the relatively weaker base and acid. A molecule of hydrogen chloride is a strong proton donor; the chlorine end has a weak hold on the proton *in relation to the hold a water molecule can have.* The water molecule is a stronger base than a chloride ion; it can bind a proton (as H_3O^+) much better than can a chloride ion (as H—Cl). From these considerations there emerge three very useful generalizations:

—If an acid is strong, its conjugate base is weak.
—If a base is strong, its conjugate acid is weak.
—Acid-base neutralizations always tend to produce the weaker of the possible acids and bases.

To illustrate these generalizations, a typical organic acid, acetic acid, may be described. We have learned that it is a weak acid, which means it is ionized only to a low percentage in water. To the extent that it ionizes at all, it has reacted with water molecules. Acetic acid acts as an acid, water as a base. But are these the weaker acid-base set or the stronger?

$$CH_3\text{—}\overset{\overset{\large O}{\|}}{C}\text{—O—H} + \text{H—OH} \rightleftarrows CH_3\text{—}\overset{\overset{\large O}{\|}}{C}\text{—}O^- + H_3O^+$$

Acetic acid — Water — Acetate ion — Hydronium ion

The unequal arrows pointing the way that they do express, in a rough way, an experimental fact. Sorting out conjugate pairs, we find that acetic acid is a potential proton donor and that the acetate ion must be its conjugate base. The water molecule is a proton acceptor, the hydronium ion its conjugate acid. With the arrows pointing the way that they do, the strongest proton donor must be the hydronium ion, and the strongest base or proton acceptor must be the acetate ion. We should therefore make the following labels:

$$\underset{\text{Weaker acid}}{CH_3{-}\overset{\overset{\displaystyle O}{\|}}{C}{-}O{-}H} + \underset{\text{Weaker base}}{H{-}OH} \rightleftharpoons \underset{\text{Stronger base}}{CH_3{-}\overset{\overset{\displaystyle O}{\|}}{C}{-}O^-} + \underset{\text{Stronger acid}}{H_3O^+}$$

Suppose that we present to the acetic acid molecule not a weak base such as water but a strong base such as the hydroxide ion. Now the *relative* positions have changed:

$$\underset{\text{Stronger acid}}{CH_3{-}\overset{\overset{\displaystyle O}{\|}}{C}{-}O{-}H} + \underset{\text{Stronger base}}{OH^-} \rightleftharpoons \underset{\text{Weaker base}}{CH_3{-}\overset{\overset{\displaystyle O}{\|}}{C}{-}O^-} + \underset{\text{Weaker acid}}{H_2O}$$

These two examples show the relative nature of the designations strong and weak. To say that an acid is *strong* does not mean much unless we know *in relation to what*. The direction of proton transfer and its extent depend on these relative proton-donating and proton-binding abilities of the potential acids and bases involved. Part of the factual information of chemistry that should be assimilated is a short list of which acids and bases are strong in aqueous media. If the strong acids and bases in Table 6.2 and 6.3 are memorized, and if we assume that *all* others are weak, the memory task is lightened and we have in mind sufficient information for a variety of situations. (Acetic acid is representative of a whole family of organic acids, all of which are weak in water.)

Exercise 6.3 The lengths of the arrows and their directions indicate the side favored for the following reactants and products in equilibrium.

1. Identify conjugate acid-base pairs.
2. Rewrite the equations and beneath each formula write the suitable designation, for example, weaker acid, weaker base, stronger acid, stronger base. The example indicates what is wanted.

Example. $H_3PO_4 + H_2O \rightleftharpoons H_2PO_4^- + H_3O^+$

Ans. Conjugate acid: H_3PO_4, conjugate base: $H_2PO_4^-$
conjugate acid: H_3O^+, conjugate base: H_2O

$$\underset{\text{Weaker acid}}{H_3PO_4} + \underset{\text{Weaker base}}{H_2O} \rightleftharpoons \underset{\text{Stronger base}}{H_2PO_4^-} + \underset{\text{Stronger acid}}{H_3O^+}$$

(*a*) $HClO_4 + H_2O \rightleftharpoons ClO_4^- + H_3O^+$
(*b*) $H_2SO_4 + H_2O \rightleftharpoons HSO_4^- + H_3O^+$
(*c*) $H_2S + H_2O \rightleftharpoons HS^- + H_3O^+$
(*d*) $HSO_3^- + H_2O \rightleftharpoons SO_3^{2-} + H_3O^+$
(*e*) $HSO_4^- + H_2O \rightleftharpoons SO_4^{2-} + H_3O^+$
(*f*) $H_2 + H_2O \longleftarrow H{:}^- + H_3O^+$
(*g*) $H_2O + H_2O \rightleftharpoons HO^- + H_3O^+$

Exercise 6.4 For the following list of potential proton donors, write the formula of the conjugate base.

Example. $H_2PO_4^-$

Ans. Conjugate base: HPO_4^{2-} (In other words, remove a proton and leave behind the rest plus one more minus charge.)

(*a*) NH_3 (*b*) $CH_3—O—H$ (Hydrogens attached to carbon are not as potentially available as those attached to oxygen.)
(*c*) HBr
(*d*) HPO_4^{2-} (*e*) CH_4 (*f*) $H—C\equiv N$
(*g*) HO^- (*h*) HS^-

Exercise 6.5 For the following list of potential proton acceptors, write the formula of the conjugate acid.

Example. $H_2PO_4^-$.

Ans. Conjugate acid: H_3PO_4 (In other words, add a proton and change the charge by 1^+.)

(*a*) $H:^-$ (*b*) CO_3^{2-} (*c*) NO_3^-
(*d*) I^- (*e*) PO_4^{3-} (*f*) H_2O
(*g*) NH_3 (*h*) HCO_3^- (*i*) $CH_3—O^-$
(*j*) HS^- (*k*) HO^-

Relative strengths of several acids and bases in the Brønsted-Lowry concept are indicated in Table 6.4. Above the hydronium ion all the acids are of almost the same strength when in aqueous solution; all the bases above water are also of comparable (and very weak) strength in aqueous solution. The reasons for these comparable strengths and weaknesses are as follows. When these acids are dissolved in water, they react with it at once, and the only acidic species present is actually the hydronium ion. This ion is, in fact, the strongest proton donor that can exist in any significant concentration in water. Any better potential proton donor would give up its proton to water as soon as contact was made. Below the hydroxide ion, the conjugate base of water, any base is capable of reacting rapidly with water to take one of its protons, leaving behind hydroxide ion. For example,

$$NH_2^- + H—OH \longrightarrow NH_3 + OH^-$$

Amide ion		Ammonia molecule	
Stronger base	Stronger acid	Weaker acid	Weaker base

Consequently, the hydroxide ion is the strongest base that can be present to any appreciable extent in water. Any stronger base would take a proton from water and make a hydroxide ion. Special methods are used to make the bases that fall below the hydroxide ion in Table 6.4. The amide ion is made by allowing sodium or potassium metals to react with liquid ammonia. These metals react with liquid ammonia very much as they do with water:

Table 6.4 Relative Strengths of Brønsted Acids and Their Conjugate Bases

Reciprocal relations:
—If acid is strong, conjugate base is weak.
—If acid is weak, conjugate base is strong.
—If base is strong, conjugate acid is weak.
—If base is weak, conjugate acid is strong.

Conjugate Acid		Conjugate Base	
Name	Formula	Formula	Name
Perchloric acid	$HClO_4$	ClO_4^-	Perchlorate ion
Hydrogen iodide	HI	I^-	Iodide ion
Hydrogen bromide	HBr	Br^-	Bromide ion
Sulfuric acid	H_2SO_4	HSO_4^-	Hydrogen sulfate ion
Hydrogen chloride	HCl	Cl^-	Chloride ion
Nitric acid	HNO_3	NO_3^-	Nitrate ion
Hydronium ion	H_3O^+	H_2O	Water
Hydrogen sulfate ion	HSO_4^-	SO_4^{2-}	Sulfate ion
Phosphoric acid	H_3PO_4	$H_2PO_4^-$	Dihydrogen phosphate ion
Acetic acid	$CH_3\overset{O}{\overset{\|\|}{C}}OH$	$CH_3\overset{O}{\overset{\|\|}{C}}—O^-$	Acetate ion
Carbonic acid	H_2CO_3	HCO_3^-	Bicarbonate ion
Hydrogen sulfide	H_2S	HS^-	Hydrogen sulfide ion
Ammonium ion	NH_4^+	NH_3	Ammonia
Hydrogen cyanide	HCN	CN^-	Cyanide ion
Bicarbonate ion	HCO_3^-	CO_3^{2-}	Carbonate ion
Water	H_2O	OH^-	Hydroxide ion
Methyl alcohol	$CH_3—O—H$	$CH_3—O^-$	Methoxide ion
Ammonia	NH_3	NH_2^-	Amide ion
Hydrogen	H_2	$H:^-$	Hydride ion
Methane	CH_4	$CH_3:^-$	Methyl carbanion

↑ Increasing acid strength (upward) — Increasing base strength (downward) ↓

$$2Na + 2NH_3(l) \longrightarrow \underset{\text{Sodium amide}}{2NaNH_2} + H_2\uparrow$$

The methoxide ion, CH_3O^-, is made in a similar manner by allowing these active metals to react with anhydrous methyl alcohol:

$$2Na + \underset{\text{Methyl alcohol}}{2CH_3—O—H} \longrightarrow \underset{\text{Sodium methoxide}}{2CH_3—O^-Na^+} + H_2\uparrow$$

The hydride ion occurs in certain metal hydrides such as lithium hydride, LiH.

Exercise 6.6 On the basis of qualitative data in Table 6.4, correctly place the arrowheads on the lines:

(*a*) $CH_3^- + H_2O$ —— $CH_4 + OH^-$ (*b*) $CH_3O^- + HCN$ —— $CH_3OH + {}^-CN$
(*c*) $ClO_4^- + NH_4^+$ —— $HClO_4 + NH_3$ (*d*) $OH^- + NH_4^+$ —— $NH_3 + H_2O$

In the Arrhenius concept acid-base neutralization was confined solely to a combination of hydrogen ions with hydroxide ions to produce water. In the enlarged concept of Brønsted and Lowry, any reaction of a stronger acid with a stronger base to produce the conjugate weaker acid and weaker base is called neutralization. Our study, however, will for the most part be confined to aqueous solutions. The neutralizing that will be important in our work is proton binding.

THE LEWIS CONCEPT OF ACIDS AND BASES

The Arrhenius concept was severely limited, admitting only one acid species and one basic species. The Brønsted-Lowry concept enlarged this to proton donors and acceptors in general. Proton binding, of course, is just a special example of forming a covalent bond by an electron-sharing process. The Brønsted bases have a pair of electrons to share with a proton. The Brønsted acids can donate something capable of accepting a share of this pair, but this something is only one species, a proton. Thus the Brønsted-Lowry concept is also rather restricted. G. N. Lewis, known for his octet theory, removed this restriction by proposing a much more general, all-inclusive concept which focuses attention on the ability of a species to donate or to accept a share in a pair of electrons. According to the Lewis concept,

—An acid is any species that can accept a share in a pair of electrons.
—A base is any species that can donate a share in a pair of electrons.

The term neutralization may still be used, but now it means the process of sharing the electron pair. Since this process results inevitably in the formation of a covalent bond, usually a coordinate covalent bond, it is better to use the word *coordination* instead of neutralization. The distinction is not important, and we use both on various occasions. As a typical example, the interaction of ammonia with boron trifluoride may be cited:

```
   F         H              F H
   |         |              | |
F—B  +  :N—H  ——→  F—B:N—H
   |         |              | |
   F         H              F H
```

Boron trifluoride   Ammonia   Boron trifluoride–ammonia complex or adduct

Boron trifluoride has a sextet of valence shell electrons, not an octet. By coordinating with ammonia, it completes its octet without sacrificing that of ammonia.

Bases in the Lewis concept are essentially the same as those in the Brønsted-Lowry system. In the Brønsted-Lowry system they are restricted to coordinating with a proton, but in the Lewis system they may coordinate with any species that can accept a share of a pair of electrons. The proton, H^+, is just one such species, and boron trifluoride is another. The list may be extended to *any species that has at least one unfilled orbital in the outer shell or can make one available*. Any particle with an incomplete octet (e.g., BF_3, BCl_3)[1] will fit this definition.

[1] Any metallic ion could also be classified as a Lewis acid.

SALTS

It is not easy to generalize about salts. They do not possess an ion common to all. In the Arrhenius definition, a salt is an ionic compound resulting from the aggregation of metallic and nonmetallic ions. *Neutral salts* make available in aqueous media only these two oppositely charged ions. Examples are NaCl, KBr, $Mg(NO_3)_2$, and Na_2SO_4. Their solubility in water varies widely, but some useful rules of thumb have been developed.

—All sodium, potassium, and ammonium salts are soluble in water.
—All nitrates and acetates are soluble.
—All chlorides, except those of lead, silver, and mercury(I), are soluble.
—Salts not in these categories are generally insoluble, or, at best, only slightly soluble.

What we mean by "soluble" also varies, but usually if the substance will dissolve to the extent of at least 3 g in 100 cc of water at room temperature, it may be called soluble. There are several exceptions to these rules of thumb, but we shall not very often be wrong in applying them.

When salts dissolve in water, to whatever extent, the degree of dissociation is high, approaching 100% in dilute solutions. In concentrated solutions the *activity* of each ion may be significantly less than its expected molarity, as shown by freezing point depression experiments and the like (p. 160.)

All Arrhenius salts are prepared by the action of an Arrhenius acid on any one of the substances whose reactions were studied earlier—active metals, oxides, hydroxides, bicarbonates, and carbonates. Evaporation of the resulting solutions to dryness yields the crystalline salt. Arrhenius salts may also be prepared by a *double decomposition* or a "change of partners" reaction. If oppositely charged ions that ordinarily constitute an insoluble salt are mixed together, these ions will aggregate and precipitate. This is one of many instances in which the solubility rules are helpful. Consider, for example, an experiment in which a solution of sodium sulfate is added to a solution of barium nitrate. Since both are soluble in water, the ions Ba^{2+}, NO_3^-, Na^+, and SO_4^{2-} are initially present in the dissolved (and hydrated) state. To predict what might happen, all possible combinations of oppositely charged ions must be examined.

$Ba^{2+} + 2NO_3^- \xrightarrow{?} Ba(NO_3)_2$ This reaction is obviously not possible, for barium ions and nitrate ions do not precipitate together from water. ("All nitrates are soluble.")

$2Na^+ + SO_4^{2-} \xrightarrow{?} Na_2SO_4$ This reaction is also not possible. ("All sodium salts are soluble.")

$Na^+ + NO_3^- \xrightarrow{?} NaNO_3$ Again, "All sodium salts are soluble."

$Ba^{2+} + SO_4^{2-} \xrightarrow{?} BaSO_4\downarrow$ Barium sulfate is not in any of the categories of water-soluble compounds. We therefore conclude that it is insoluble.

The prediction is therefore that barium ions and sulfate ions will aggregate and precipitate from the solution. The equation may be written as

$$Ba(NO_3)_2 + Na_2SO_4 \longrightarrow 2NaNO_3 + BaSO_4\downarrow$$

which shows the "change of partners" feature of the reaction. The only event of importance, however, is

$$Ba^{2+} + SO_4^{2-} \longrightarrow BaSO_4\downarrow$$

This new salt, precipitated barium sulfate, can be collected by filtration. The clear filtrate, if evaporated to dryness, will yield the other new salt, sodium nitrate.

If we mix solutions of salts and no interaction between ions occurs, the new solution will be nothing more than a mixture. The mixing of the solutions to make it will not be described as producing a chemical change. Thus mixing together solutions of potassium bromide and sodium chloride would give only a new solution containing the hydrated ions Na^+, K^+, Cl^-, and Br^-.

The formation of barium sulfate illustrated one situation in which ions will interact—when there is the possibility of precipitating an insoluble salt. There are others, and it is important that we know them. Although examples have already been indicated in reactions cited earlier, the following list generalizes them and provides rules of thumb for predicting a very large number of reactions.

In general, reactions between ionic substances (or those that can produce ions on demand) will occur if there is the possibility of forming

1. A precipitate. Example: Formation of $BaSO_4$.
2. An un-ionized, although soluble, species. Example: Formation of water when H^+ and OH^- react.
3. A gas. Example: Action of acids on carbonates and bicarbonates.

COORDINATION COMPOUNDS

The Arrhenius neutralization of an acid by a base produces the kinds of ionic salts described in the last section. In the Brønsted and Lewis systems, especially the Lewis system, coordination rather than neutralization has seemed to be the better term. We could speak of Lewis salts (boron trifluoride–ammonia adduct is an example), but we more often call these coordination compounds. Of particular interest are those resulting from the interaction of simple metallic ions with Lewis bases.

We have already encountered examples of such coordination compounds, the hydrated forms of the ions. True covalent bonds have not formed in such complexes, but it is nevertheless still convenient to classify these hydrated ions as coordinated complexes. Obviously, a sodium ion or a potassium ion does not hold the water molecules strongly. These ions are very weak Lewis acids. Metallic ions of larger positive charge are usually stronger Lewis acids. The aluminum ion, Al^{3+}, for example, coordinates strongly with water to form the *hexaaquo* ion, $Al(H_2O)_6^{3+}$.

When the Lewis acid is a metallic ion, we call the Lewis base the *ligand* (*ligare,* to bind or tie). Two of the most important ligands are water and ammonia.

But in the field of organic chemistry are a large number of water-like or ammonia-like compounds (alcohols and amines) all having unshared pairs of electrons either on oxygen or on nitrogen. All, therefore, are potential Lewis bases or ligands. These water-like or ammonia-like compounds occur throughout the vast domain of molecular biology, and complexes between them and metallic ions are some of the most important in nature. Examples of these compounds are chlorophyll, heme (in hemoglobin), and many enzymes (biological catalysts without which life as we know it would be impossible), including the all-important cytochromes which transport electrons from organic intermediates to molecular oxygen for the formation of water. Although most ligands are organic compounds, several inorganic species as well as water and ammonia deserve mention.

The cyanide ion, CN^-, is a powerful ligand. The ferricyanide ion, for example, $Fe(CN)_6^{3-}$, has six of these ions about a central iron(III) ion. The strength of the cyanide ion as a ligand helps to explain how it acts as a potent poison. The Fe^{3+} and Fe^{2+} ions as well as Cu^{2+} occur as central ions in the cytochromes, mentioned as essential to the use of oxygen in the body. When the cyanide ion coordinates with these metallic ions more strongly than can the enzyme, the ability of the enzyme to do its job is virtually destroyed. Oxygen cannot be used, and death occurs rapidly. Carbon monoxide, an electrically neutral substance, is also a strong ligand, stronger than oxygen. It acts as a respiratory poison by coordinating with Fe^{2+} ions in heme (or hemoglobin). When carbon monoxide molecules have become complexed with the iron ions, oxygen transport fails and death is swift.

A huge section of the detergent industry depends on the ability of certain phosphates to coordinate with those metallic ions responsible for hard water (Ca^{2+}, Mg^{2+}). The diphosphate and the triphosphate (or tripolyphosphate ion), shown in their fully ionized forms, are called *sequestering agents* because of their ability to

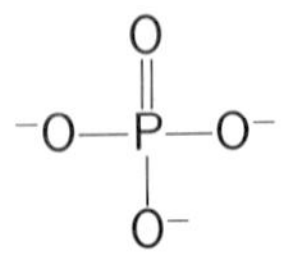

Phosphate ion (not a sequestering agent)

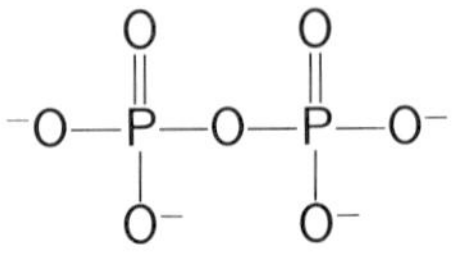

Pyrophosphate ion

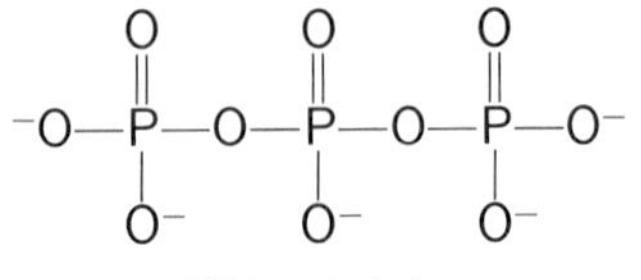

Triphosphate ion (or tripolyphosphate ion)

sequester or bind up such metallic ions *without forming a precipitate.* A more expensive, and therefore much less used, sequestering agent is the disodium salt

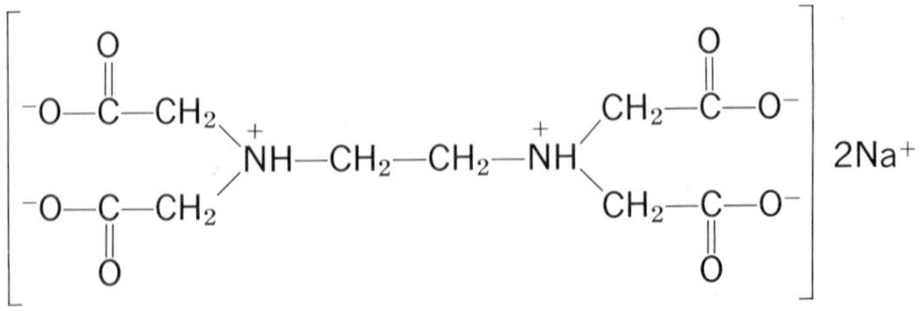

Ethylenediaminetetraacetic acid disodium salt (EDTA)

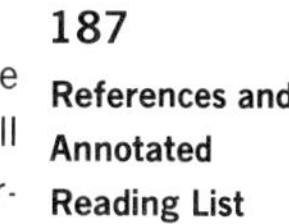

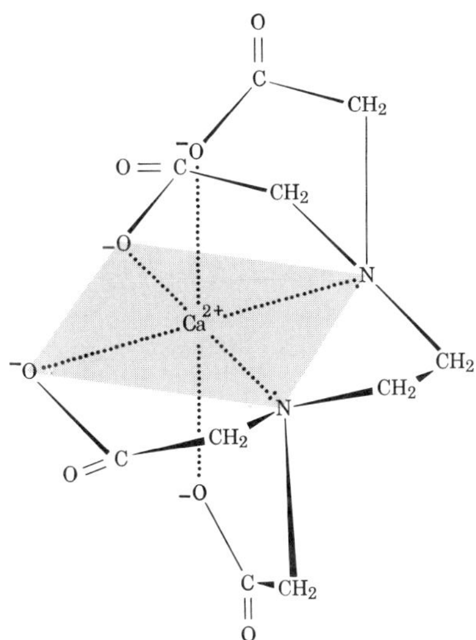
O
C
CH2
O
O = C
CH2
O
N
Ca2+
O
CH2
CH2
N
CH2
O = C
O
C
CH2
O

Figure 6.3 EDTA complex with Ca^{2+}. The calcium ion is surrounded by sites of full negative charge (on the oxygens) and partial negative charge (on the nitrogens). The dotted lines are drawn between the sites experiencing mutual forces of attraction. The net charge on the entire complex is 2^-. Not shown are two sodium ions which would also be present if the disodium salt of EDTA were used. This material is sometimes called a *chelating agent* (Latin *chele,* claw) because the central ion is kept in a clawlike hold. Both chlorophyll and heme are "chelate compounds," the former with magnesium ion and the latter with iron(II) ion. Many other examples occur in living cells.

of ethylenediaminetetraacetic acid (EDTA).[2] Called variously EDTA, Trilon B, Versene, or Sequestrene, it is one of the most powerful sequestering agents known and has found many uses in the quantitative analysis of metallic ions. Figure 6.3 shows how EDTA forms a 1:1 complex with calcium ion. It will form complexes like this with most of the dipositive metallic ions. The entire complex has a net charge of 2^- and can as a whole be solvated by water molecules. Thus the sequestered (or chelated —cf. Figure 6.3) metallic ion is kept in solution.

REFERENCES AND ANNOTATED READING LIST

BOOKS

C. A. Vanderwerf. *Acids, Bases, and the Chemistry of the Covalent Bond.* Reinhold Publishing Corporation, New York, 1961. This reference is an excellent supplement to the material of this chapter and Chapter 8. (Paperback)

ARTICLES

M. R. Bloch. "The Social Influence of Salt." *Scientific American.* July 1963, page 89. This interesting, relaxing article delves into the role of salt in history.

G. G. Schlessinger. "A Summer Short Course in Coordination Chemistry. I. Bonding Theories." *Chemistry,* June 1966, page 8. "II. Nomenclature and Physical Properties." *Chemistry,* July 1966, page 13. "III. Applications." *Chemistry,* August 1966, page 15. For more information and theory about coordination compounds, this series of short articles is an outstanding starting point. It is written at a level virtually all college freshmen should be able to understand.

J. Schubert. "Chelation in Medicine." *Scientific American,* May 1966, page 40. A good discussion of coordination compounds in clinical applications.

[2] Eth'yl·ene-di'a·mine-tet'ra-a·cet'ic acid.

PROBLEMS AND EXERCISES

1. Provide brief definitions and/or illustrations of each of the following:
   (*a*) dissociation (*b*) ionization
   (*c*) Arrhenius acid (*d*) Arrhenius base
   (*e*) strong acid (*f*) strong base
   (*g*) weak electrolyte (*h*) balanced ionic equation
   (*i*) relative activity of a metal (*j*) acid-base neutralization (Arrhenius)
   (*k*) indicator (*l*) Brønsted base
   (*m*) Brønsted acid (*n*) ammonia water
   (*o*) conjugate acid-base pair (*p*) Lewis acid
   (*q*) Lewis base (*r*) coordination compound
   (*s*) ligand (*t*) sequestering agent
2. Write two types of balanced equations, complete equations and ionic equations, for the reaction of hydrochloric acid with each of the following substances. If salts form, write their names beneath their formulas. If no reaction is to be expected, write "None."

	Set 1	Set 2	Set 3
(*a*)	$NaHCO_3$	ZnO	Mg
(*b*)	NH_4Cl	Cu	$NaBr$
(*c*)	BaO	K_2CO_3	$SrCO_3$
(*d*)	Li_2CO_3	$Mg(OH)_2$	$Ba(OH)_2$
(*e*)	Al	$KHCO_3$	$Mg(HCO_3)_2$
(*f*)	NH_3	$LiNO_3$	CaO

3. Assume that you have separate solutions of each compound in each of the following pairs. Predict what will happen, chemically, if the two solutions in each pair are mixed. Write an ionic equation for the reaction. If no reaction occurs, write "None." The way to approach each problem is to make a list of the ions that will initially be mixed together, try all combinations of oppositely charged ions, and ask yourself whether each possible combination will give (*a*) an insoluble salt, (*b*) an un-ionized but still soluble molecule, or (*c*) a gas. In this exercise you will apply not only the solubility rules for salts but also your knowledge of strong and weak acids and bases, especially in relation to the Arrhenius concept.

	Set 1	Set 2	Set 3
(*a*)	K_2SO_4 and $CaCl_2$	$CaCl_2$ and $AgNO_3$	NH_4Cl and $NaOH$
(*b*)	$Mg(HCO_3)_2$ and HI	$NaHCO_3$ and H_2SO_4	NH_4Br and $AgNO_3$
(*c*)	KOH and HBr	$Ca(OH)_2$ and HCl	$KHCO_3$ and HNO_3
(*d*)	$LiCl$ and $AgNO_3$	K_2CO_3 and $MgCl_2$	Na_3PO_4 and HCl
(*e*)	Na_2CO_3 and $Ca(NO_3)_2$	$LiBr$ and $MgCl_2$	$LiOH$ and H_2SO_4

4. Explain how the concepts of acids and bases become successively broader in the Arrhenius, Brønsted-Lowry, and Lewis concepts.
5. Boron trifluoride reacts with hydrofluoric acid as follows:

$$BF_3 + H{-}F \longrightarrow BF_4^- + H^+$$

Identify the Lewis acid and the Lewis base in this reaction.

CHAPTER SEVEN

The Direction of Chemical Changes

THE ANALYSIS OF SPONTANEITY

Nature is a rich provider of energy, energy which from our point of view is free for the taking. The taking is not always easy, however, but throughout man's history human cleverness has continually found more fruitful ways. Ever more sophisticated toolmaking is certainly evidence of growing resourcefulness. Equally important is recognizing which of nature's events can most profitably be bent to our laborsaving advantage. It is among the spontaneous changes in nature, the changes that occur naturally and of themselves, that we seek supplies of such energy. This much is almost obvious, but which will serve us best? Which are most dependable? Which can be controlled? Many such questions arise, and they strongly indicate that in the history of Western civilization, indeed of all modern industrialized societies, the analysis of spontaneity has been one of the most significant but least-heralded human activities.

Although by spontaneous we roughly mean "happening without human intervention," the starting of an event may involve something a person does. The burning of coal, although spontaneous, may have started because someone ignited it. But lightening could also have ignited the coal. Once the burning has started, as long as both coal and air (or oxygen) are available in suitable proportions, the following reaction continues in the direction shown:

$$C + O_2 \longrightarrow CO_2 + \text{heat}$$

Under the circumstances of its formation, the newly made carbon dioxide does not spontaneously revert to carbon and oxygen. The reverse of this reaction, going from right to left, is not a natural one. Spontaneous events in nature thus have a direction to them. Any successful analysis of spontaneity must uncover some factor or factors that corresponds to this directionality.

While making this analysis, we should not forget the law about the conservation of energy, the first law of thermodynamics. We must go beyond it, however, for it is not enough. Nothing in the first law, for example, would prevent the building of a shed in the manner of James Frankfort's cartoon. This law gives us a relation between q, w, and ΔE, but it tells us nothing about the natural direction of any interconversions among them. According to the law, if the materials were all together and the bomb were big enough to furnish all the needed energy, it should be possible to build a shed by tossing a bomb into a pile of lumber and nails. The net enthalpy change could be made as large and negative as we please simply through the design of the bomb.

(Reprinted with permission from James Frankfort and the *Saturday Evening Post.*

It is very important to reexamine the first law because with it we developed the concept of enthalpy. And in a surprising number of events there appears to be a correlation between the algebraic *sign* of the enthalpy change and the natural *direction* of the change. Many spontaneous events happening to systems deliver energy to their surroundings. Forest fires, avalanches, windstorms, volcanoes, tides, waterfalls, and reactions between acids and bases are a few examples. Spontaneous changes are often exothermic, and for these ΔH is negative. But are spontaneous events always exothermic? Is a *negative* ΔH always associated with spontaneity? Does a positive (or zero) ΔH inevitably mean that the event as written will not occur without the intervention of some external agency that can supply the energy needed? If we could unequivocally answer "yes" to these questions, we should have to seek no further for additional elements of spontaneity. We would have the only necessary one in ΔH. But knowing the sign of ΔH of an event is not enough. For some events ΔH is positive (or zero) and the change still happens. For others ΔH is negative but the change does not take place spontaneously.

If ammonium chloride is added to a beaker of water that has a thermometer in it, the spontaneous event is simultaneously the dissolving of the salt and a *drop* in the temperature. (Both the water and the salt were initially at the same temperature, e.g., room temperature.) To keep the temperature constant we would have to *add* heat *to* the system. A large number of salts are like this one. When they dissolve, the event is endothermic. They have *positive* enthalpies of solution. But they still dissolve spontaneously.

In some spontaneous processes the enthalpy change is zero. If two ideal gases (approximated very closely by helium and neon) are separated as illustrated in Figure 7.1 and then the stopcock is opened, they will eventually become randomly

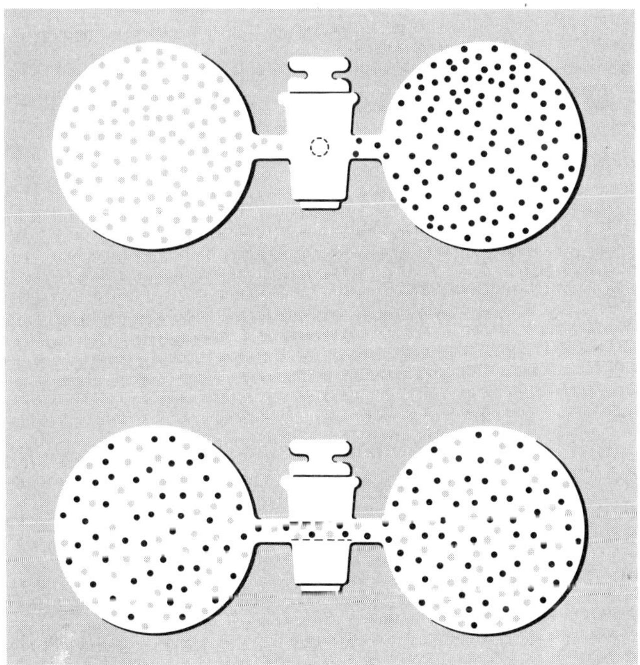

Figure 7.1 Order to chaos with two ideal gases. (*a*) The two gases, one represented by light dots, the other by dark dots, are separated by a closed stopcock. (*b*) The stopcock has been open for some time and the two gases have diffused into each other's original spaces. This arrangement is much more random and much more probable than the arrangement in part *a*, yet there has been no energy gained or lost during this spontaneous process. For an analogy, imagine that flies are in one bulb and mosquitos in the other. If the stopcock is opened, what do you expect will happen?

and uniformly mixed. This diffusion of the two gases into each other occurs without any net change in the internal energy. No temperature or pressure changes take place. For this event ΔH is zero, but the spontaneous diffusion most assuredly happens. Two ideal solutions would also mix in similar circumstances and, assuming no chemical change, the mixing would occur with zero ΔH. The *direction* of the changes in these examples is from a relatively ordered situation to a relatively disordered one, from arrangement to "mixed-upness," from order to chaos. Atoms or molecules are *relatively* more ordered when one kind is on one side of the flask and the other kind is on the other side—even though within these two regions the individual particles are moving about as described by the kinetic theory.

We shall continue to consider enthalpy a possible element of spontaneity, but clearly it cannot be the only one.

ENTROPY

The mixed-up distribution of the gaseous particles in Figure 7.1 certainly has no less internal energy than the more ordered distribution in which all atoms of helium are on one side and all atoms of neon on the other. The net internal energy in either situation will be the same, but the more mixed-up, random distribution is the more *probable* of the two. And *if the way is open* for a system to put itself into a more probable situation, it will *inevitably* do so, even though the event may not constitute, so to speak, "falling down an internal energy hill." (The way is open in our example because the atoms do move about.) The chances that two mixed-up gases will ever *spontaneously* become separated are even more infinitely remote. The number of atoms of a gas in even a cubic centimeter, roughly 10^{20}, ensures the validity of these statistics. Left to itself the system of Figure 7.1 will invariably become mixed up and remain so. All our experience supports this conviction. The random mixing is the spontaneous process in this example, and it occurs even though there is no driving force for it *other than* the tendency to maximize chaos as the particles move about.

Ludwig Boltzmann (1844–1906), Austrian theoretical physicist, attacked the problem of explaining the second law of thermodynamics in terms of the mathematical theory of probability and the atomic theory of matter. His equation relating entropy to probability is engraved on the monument at his grave in Vienna. Although attacked by other eminent scientists who at the time (around the 1870s) did not believe in the existence of atoms, his work has survived and he is credited with being the father of statistical mechanics.

This effect, this drift toward "mixed-upness," is called the *entropy effect. Entropy is the thermodynamic quantity that is used as a measure of the mixed-upness or chaos of a system.* (This is an interpretation of entropy; and we shall say more about interpreting thermodynamics later.) Entropy is related to the disorderliness of a system. The greater the amount of chaos, the greater the entropy. This does not mean that entropy and chaos are synonyms, but they are related; Ludwig Boltzmann was the first to undertake a theoretical study of just how they might be connected. He arrived at the following equation (stated here without his arguments):

$$S = 2.303k \log W + C \tag{7.1}$$

where S = entropy
W = probability
k = a constant, the *Boltzmann constant,* equal to the ratio of the gas constant divided by Avogadro's number, R/N
C = another constant

In other words entropy S is specifically a *logarithmic* function of probability W. (If you do not know what logarithms are, refer to Appendix II. Equation 7.1, incidentally, will be used only once more.) Thus if the probability W is high that a particular system will have a certain distribution or arrangement, its entropy S will also be relatively large. Since logarithms do not have units, the units of the Boltzmann constant k determine the units of entropy as defined in equation 7.1. Since

$$k = \frac{R}{N}$$

and since R is expressed in units of cal/(°K)(mole) and N in units of molecules/mole

$$k = \frac{R}{N} = \frac{\dfrac{\text{cal}}{(°\text{K})(\text{mole})}}{\dfrac{\text{molecules}}{\text{mole}}} = \frac{\text{cal}}{(°\text{K})(\text{molecule})}$$

Therefore entropy S has the units cal/(°K)(molecule). Because the entropy of just one molecule is not an experimental quantity, chemists usually speak of a mole or some related quantity. Hence the units of experimental significance for entropy are cal/(°K)(mole). The other constant in equation 7.1, namely C, was shown to be zero by Max Planck. The relation between entropy and probability is therefore very simply expressed by

$$S = 2.303k \log W \tag{7.2}$$

Nothing has been said here about how we can calculate probabilities W. Equation 7.2 says only that if W can be calculated, then S can be also. We are deliberately remaining at a very qualitative level in this discussion. (To better understand the rest of this treatment of entropy, the reader is advised to review the definitions of *system, surroundings, boundary,* and the other related terms found on page 80).

To return to the mixing of two ideal gases or of two ideal solutions, as the random motions of the atoms produce ever more random distributions, they bring into existence ever more probable situations. (To be randomly mixed is more probable than to be separated.) This means that the entropy of the system is increasing during the mixing. It also means that the *change in entropy,* which we symbolize as

ΔS, *is positive.* As the mixing continues and the entropy gradually increases, *it will eventually reach a maximum.* Sooner or later the mixing must be complete, and from that time on no further observable change will happen spontaneously. From that time on the *change* in entropy ΔS will be zero. Of course, from that time on we have a condition of equilibrium. In this example at least, the *sign* of the change in entropy is positive and only positive during the spontaneous change itself. Only when equilibrium is reached does ΔS become zero. At no time is ΔS negative. *In fact* ΔS *can never be negative* in this example. The only other event that could conceivably happen would be for the two gases to become unmixed, but this is not the way things occur in nature. Such an event just does not happen in systems having so many particles in random motion, but *if* it did ΔS would have the only algebraic sign remaining, minus.

In this example of the mixing of the two ideal gases (Figure 7.1), we are dealing with both a closed system and an isolated system. Neither matter nor energy transfers into or out of the container. The temperature remains constant, and the mixing is accompanied by no change whatsoever in the internal energy or enthalpy of the system. (See page 90 for the factors that make up the energy; volume is not one of them.) Under these circumstances—closed and isolated system and zero ΔH —the only event that can happen spontaneously is one accompanied by an increase in entropy, an increase in probability. Entropy is clearly an important element of spontaneity.

If the system under study is in fact not isolated and closed with respect to its surroundings, a more common situation, any change occurring in the system may produce some change in the surroundings. A full evaluation of the entropy changes taking place must therefore include both system and surroundings.

The Second Law of Thermodynamics. When both system and surroundings are included in the analysis of a spontaneous event in nature, all the experiments that have been reported support the truth of the following statement about the world in which we live:

The total amount of entropy in the universe is increasing.

This is one of the ways of stating the *second law of thermodynamics.* It is not a law that we or anyone else has mathematically derived from more fundamental principles but is itself a fundamental principle. We do not know it to be true without exception, for the simple reason that all the conceivable tests have not been made and cannot be made. The universe is a big place. The tests that have been made, and they are many, all strongly suggest this law; it has stood the test of time so well that for all practical purposes it may be considered proved. With entropy related to probability, and that to randomness and chaos, the second law says that with the passage of time the universe is becoming more chaotic. It says that eventually this randomness will reach a maximum and after that no further observable change will occur. At least these are interpretations placed on the second law. Their verifiability is obviously impossible because no one has even seriously predicted *when* they will have come to pass. It has been said only that we are tending in this direction.

Does the Entropy Concept Stand or Fall on Our Theories about the Structure of Matter? Boltzmann, whose statistical approach started with the assumption that atoms exist, worked during a period in history when not all the best minds in science believed in the atomic hypothesis, and at no time in the nineteenth century did a satisfactory model for the structure of an atom exist. Boltzmann also accepted the concept of entropy. He did not invent the concept, but he sought an interpretation of it. This concept and *all* the concepts of thermodynamics were developed with no reference whatsoever to an atomic or molecular hypothesis. In fact, important advances even took place when heat was believed to be some form of matter. We have not developed the concept of entropy as it emerged in classical theory but have started at once with the interpretation of entropy that depends on belief in the existence of atoms. At this stage in the development of science we seem to be on extraordinarily safe ground, but the truth of the matter is that all the functions in thermodynamics—heat, work, energy, enthalpy, temperature, and entropy (there are other functions besides these)—deal with certain properties of matter in bulk quite without regard to any notions about atoms and chemical bonds.[1] None of these functions stands or falls on the question of atoms, and this is one of the great strengths of thermodynamics.

Does the Second Law Force Us Always to Work with the Entire Universe? The statement of the second law given earlier says that the entropy of the universe is always increasing, and that this increase is an important driving force in nature. But suppose we want to deal with just the tiniest sections of the universe, a system that might be a person, a plant, a cell, or the relatively simple supposition of an ideal gas in a tube. Can the entropy concept and the second law be made useful in dealing with just a *system?* Must we always work with the rest of the universe as the surroundings in all our calculations or reasonings?

Scientists have found a way of avoiding this problem and being as exclusive as they please in focusing attention on a particular system. With this focus, however, an event may be spontaneous even when the entropy of the *system* decreases. The *total* entropy change is the algebraic sum of the individual entropy changes for system and surroundings:

$$\Delta S_{\text{system}} + \Delta S_{\text{surroundings}} = \Delta S_{\text{universe}}$$

Thus in a natural process $\Delta S_{\text{universe}}$ can still be what the second law says it must be, namely positive, even if ΔS_{system} is negative, *provided* $\Delta S_{\text{surroundings}}$ is positive enough.

If, for example, we take as our system a plant plus the air space around it, plus the soil penetrated by its roots, the growth of the plant is really a change from disorder to order, in apparent violation of the second law. Randomly moving molecules of carbon dioxide and water and certain minerals are converted into the more

[1] Entropy, for example, was originally defined in terms of the heat q involved in a certain kind of process and the temperature of the system. This definition will not be given here because it involves calculus notation, but neither q nor T requires a conception of atomic and molecular structure. The original definition, incidentally, is still the most commonly used because it deals in quantities that can be measured experimentally. The probability term W in equation 7.2 is much more difficult to evaluate.

highly ordered molecules of the plant and its cellular structures. The plant cannot grow without sunlight, however. Accompanying the production of solar energy on the sun, there is believed to be a tremendous increase in the sun's entropy. Thus, for the universe as a whole, the unfavorable entropy change in the plant is compensated by a highly favorable entropy change on the sun, and the net change in entropy (sun plus plant) is positive. As the plant grows, the entropy of the universe is presumably increasing in accordance with the second law.

Our example of the plant shows us that if we wish to focus attention on a particular system, the direction of any spontaneous change it undergoes must be determined by an interplay of the two elements of spontaneity, energy and entropy. Something can happen spontaneously to a *system* even when its entropy is decreasing, when order is being formed out of chaos. Thus, for the system alone, although entropy is surely an element of spontaneity, it cannot serve as the exclusive criterion. We must take into account the energy (or enthalpy) change too.

In our analysis of spontaneous processes we have asked what the simplest and most useful elements are in terms of which spontaneity can be analyzed. Two such elements have been found, energy and entropy. Although our progress is gratifying, we are still faced with the annoying fact that an event may happen without human intervention if ΔH is plus, minus, or zero; or if ΔS is plus, minus, or zero, as far as the system alone is concerned. We do not have one element that alone is invariably associated with the *direction* of the change in a system. We could hope for just one term that represents the net effect of the energy and the entropy factors for a particular system, one composite term that has only one algebraic sign for a spontaneous change. For certain situations, events in systems at constant T and V or in systems at constant T and P, such terms do exist. More natural events, including those in living systems, come closer to being at constant T and P than at constant T and V. We shall therefore seek only the composite term applicable in these circumstances.

GIBBS FREE ENERGY

In this section we assume that any system mentioned is at constant temperature and pressure, which means that if a chemical change produces heat, it is passed out to the surroundings. If the change requires heat, the surroundings provide it. The temperature somehow is kept constant. By constant pressure is usually meant a system open to the atmosphere. If a reaction in a solution causes some small change in the volume of that solution, some small pressure-volume work will be done. By using enthalpy rather than internal energy, we automatically take this work into account, as discussed on page 94.

For a change occurring at constant temperature and constant pressure, the following equation defines a new thermodynamic function, the *Gibbs free energy change,* symbolized by ΔG:

(7.3) $\Delta G = \Delta H - T\Delta S$ for constant, absolute T and P conditions

This equation gives us the one term we seek, ΔG, the term with just one algebraic sign, a minus sign, to associate with any spontaneous change in a system at constant T and P. This equation is set forth without derivation because none really exists. Equation 7.3 is a *defining equation.* By it ΔG is defined, thus begging the question why we use this particular combination of ΔH, ΔS, and T. In the final analysis this combination is the one that works and gives us the composite function for changes at constant T and P that we sought in the preceding section. The presence of T ensures that the term $T\,\Delta S$ will have the same units as ΔH, units of energy:

$$T\,\Delta S = (^\circ\text{K})\left[\frac{\text{cal}}{(^\circ\text{K})(\text{mole})}\right] = \frac{\text{cal}}{\text{mole}}$$

Moreover, temperature is sometimes a factor in determining the direction in which a change will occur. At $T > 373^\circ$K (1 atm) it is natural for water to boil, not condense. At $T < 373^\circ$K (1 atm) it is natural for water to condense, not boil.

Equation 7.3 summarizes the net effects of enthalpy and entropy. It gives us the net balance between the tendency in natural events for the system to release internal energy and thus reach a lower energy level and the tendency for a system to become more mixed up and chaotic and therefore *more probable. The criterion for a spontaneous change at constant T and P is that*

(7.4) $\qquad \Delta G < 0 \qquad$ That is, ΔG is less than zero or negative.

It is part of the givenness of nature that for constant T and P events the spontaneous change a system can undergo is the one that lowers its Gibbs free energy. For any other event to happen, an external source of energy (and information) must be supplied. In other words, a nonspontaneous event for a system can be made to occur provided energy is furnished in sufficient quantities and in the appropriate form. If the system has no changes open to it whereby its free energy may be lowered, it is in a state of equilibrium, and nothing observable happens. For example, suppose a system consists of the chemicals A, B, X, and Y in some solvent. Suppose, further, that X and Y are formed from A and B, but that they can also reconvert to A and B. The situation may be represented by the following equation:

$$\text{A} + \text{B} \rightleftharpoons \text{X} + \text{Y}$$

The left-to-right reaction will have some ΔG associated with it; let us symbolize it by $\overrightarrow{\Delta G}$. The right to left reaction has its Gibbs free energy change $\overleftarrow{\Delta G}$. The net ΔG will be the algebraic sum of these two, and if this sum is zero, equilibrium exists. *The criterion for equilibrium at constant T and P is that*

(7.5) $$\Delta G = 0$$

To help us use the Gibbs free energy function to predict the direction of a reaction, Table 7.1 summarizes the net effects of enthalpy and entropy, the final balance between the two when only the system is considered (at constant T and P). We shall study each of the four possibilities in turn.

1. ΔH is negative and ΔS is positive. If the system loses enthalpy $(-\Delta H)$ and gains entropy $(+\Delta S)$, the change in the Gibbs free energy must be negative:

Table 7.1 Net Effects of Enthalpy and Entropy Changes (constant T and P)

Sign of		Net Effect	
ΔH	ΔS	ΔG	Remarks
$-$	$+$	$-$	Both elements of spontaneity favor the change.
$+$	$-$	$+$	Neither element of spontaneity favors the change.
$-$	$-$	?	The net effect will depend on relative values of ΔH and ΔS and the temperature; usually ΔH is negative enough to offset an unfavorable (minus) entropy effect.
$+$	$+$	?	The net effect will depend on relative values of ΔH, ΔS, and T, but the favorable entropy effect here does not often compensate for the unfavorable enthalpy change.

$$-\Delta H - (+T)(+\Delta S) = -\Delta G$$

This means that if a system, by undergoing some particular change, can simultaneously acquire a lower enthalpy and a higher disorder, *both* elements of spontaneity favor the change, and ΔG is appropriately negative.

2. ΔH is positive and ΔS is negative. Gain of enthalpy ($+\Delta H$) coupled with loss in entropy ($-\Delta S$) could never give a net negative ΔG:

$$+\Delta H - (+T)(-\Delta S) = +\Delta G$$

In other words, if a possible change for a system is analyzed to be endothermic and also to produce order from chaos, the postulated change cannot be spontaneous. Neither element of spontaneity is favorable.

It must be emphasized that changes for which ΔH is positive and ΔS is negative can be made to occur. It is just that they will not occur without intervention by some aspect of the surroundings (e.g., human labor). A shed will never be built by leaving a pile of lumber and nails lying about, even if they are "energized" by an explosion. But a carpenter, by eating regularly, consumes foodstuffs rich in chemical energy which eventually provide him with the strength for moving the lumber about (to provide positive ΔH to the lumber and nails) and for organizing the pieces into the more ordered form of a shed (to provide, in other words, the negative ΔS to the system of lumber and nails). This homely example provides an analogy for events occurring constantly in living organisms. Chemical reactions that are producers of free energy are able to power changes that are consumers.

3. ΔH is negative and ΔS is negative. A minus ΔH favors spontaneity, but a minus ΔS does not. The balance will depend on the relative sizes of ΔH, ΔS, and the temperature. If the ΔH is negative enough (i.e., favorable enough), it might compensate for the entropy effect:

$$-\Delta H \quad - \quad T(-\Delta S) \quad = \quad -\Delta G$$

$-\Delta H$	$T(-\Delta S)$	$-\Delta G$
If this is a large enough negative enthalpy change,	the resultant $+T\,\Delta S$ will not be larger numerically, and the algebraic sum	will be negative.

This situation is in fact quite common. Enthalpy changes are usually considerably greater than entropy effects.[2] The formation of water from its elements illustrates the point. If the reaction is conducted so that the temperature is maintained at 25°C (298°K), ΔH, ΔS, and ΔG have the values shown, with the superscripts and subscripts of these symbols making them into *standard* enthalpy, standard entropy, and standard Gibbs free energy of *formation*. (See page 113 to review the concept of the standard state and how it is used.)

$$H_2(g) + \tfrac{1}{2}O_2(g) \longrightarrow H_2O(l) \qquad \begin{aligned} \Delta H_f^\circ &= -68{,}320 \text{ cal/mole} \\ \Delta S_f^\circ &= -39 \text{ cal/(°K)(mole)} \\ \Delta G_f^\circ &= -56{,}690 \text{ cal/mole} \end{aligned}$$

The enthalpy change is large, favorable, and negative. The entropy change is small, unfavorable, and also negative. The balance of these two in terms of the Gibbs free energy change is struck as follows:

$$\begin{aligned} \Delta G_f^\circ &= \Delta H_f^\circ - T\,\Delta S_f^\circ \\ \Delta G_f^\circ &= -68{,}320 \text{ cal/mole} - (+298\text{°K})(-39 \text{ cal})/(\text{°K})(\text{mole}) \\ &= -68{,}320 \text{ cal/mole} + 11{,}630 \text{ cal/mole} \\ &= -56{,}690 \text{ cal/mole} \end{aligned}$$

Thus the large, negative, and favorable enthalpy change easily swamps the small, negative, and unfavorable entropy change, and ΔG is negative. The reaction is spontaneous.

4. ΔH is positive and ΔS is positive. In the previous example we saw how a favorable enough ΔH overcame an unfavorable ΔS. We now consider how a favorable ΔS may outweigh an unfavorable ΔH.

$$+\Delta H \quad - \quad T(+\Delta S) \quad = \quad -\Delta G$$

If this is a small enough plus enthalpy change,	and if ΔS is large enough in a favorable (plus) sense, making $T\,\Delta S$ a large quantity,	then ΔG can still be negative, and the change will be spontaneous.

As an example, when ammonium chloride is made to dissolve in water, heat must be *added* if a constant temperature is to be maintained. The dissolving action does not liberate heat but consumes it. Although the process is endothermic and ΔH is positive, the fact remains that this crystalline solid dissolves. The crystal, of course, is highly ordered, and the oppositely charged ions are aggregated in an organized fashion. In solution these ions are randomly dispersed; there is much more chaos in the dissolved state than in the crystal. (To be sure, the ions are solvated; some of the water molecules are more or less tied to the ions, which actually represents *more* order for some of the water. But the high disorder represented by the dissolved ions is not counterbalanced by the smaller additional order in the solvent.) The entropy change for dissolving ammonium chloride is so favorable (i.e., highly positive) that the $T\,\Delta S$ term is large enough to make ΔG negative.

[2] We use $T\,\Delta S$ as a measure of the entropy *effect* and ΔS alone as the entropy change.

It is not particularly common for spontaneous chemical reactions to be endothermic (have a $+\Delta H$). Usually they are exothermic enough to overcome the entropy effect if it is unfavorable. For this reason we may often as a rule of thumb, rely on just the one element of spontaneity, the enthalpy change, as our single criterion. If there is any basis for predicting that a reaction will be exothermic, then in all likelihood it will be spontaneous.

ΔG as a Measure of the Useful Work to Be Obtained from a Change. In addition to predicting the direction of a change in a system at constant T and P, the Gibbs free energy function is also a measure of the possible usefulness of the change as a source of energy to save our own labors. The very qualitative nature of our discussion in this chapter forces us once again to examine conclusions of scientists without the benefit of the reasoning leading to them. (Reading the references at the end of the chapter and of Chapter 3 will make up this omission.) It is stated without proof that *the Gibbs free energy change ΔG represents the maximum amount of energy that a change can deliver in the form of usable work.* The maximum is almost never actually obtained, but the Gibbs function is still the best single measure for comparing changes for their possible usefulness as energy producers. Whenever an event occurs spontaneously, the capacity of the system for delivering useful energy is *decreasing.* When water drops spontaneously from top to bottom of a waterfall, the capacity of a particular amount of water to turn a paddlewheel is much less when it reaches the bottom than when it was still at the top. Consider another example. Two chemicals that can react and deliver free energy have a greater capacity for doing so at the moment of mixing than later when equilibrium is reached.

Thus if ΔG can be calculated or estimated for a change (at constant T and P), whether the change will occur in the direction initially expected is known from the *sign* of ΔG, and its possibility of producing (or consuming) useful work is revealed in the *size* of ΔG. The Gibbs free energy is clearly a very important thermodynamic function.

Josiah Willard Gibbs (1839–1903) was a professor at Yale whose studies of energy-entropy relations were so significant that he is regarded as one of the greatest scientists ever produced by the United States. Scientists have designated the free energy function, discussed here as the Gibbs free energy function, to honor him. In many sources, however, including the majority of references in biochemistry that are only a few years old, ΔF is used instead of ΔG. The symbol ΔG is endorsed by the International Union of Pure and Applied Chemistry.

Dependence of Free Energy on Concentration. Suppose that we placed an ideal solution in contact with its solvent. After a while the random motions of solvent and solute molecules would produce a completely uniform distribution with one concentration. In the natural event, that is, the spontaneous change, the solute molecules "spread out," become more randomly dispersed. For convenience, let us suppose that the initial solution contains one mole of solute. The difference between a mole of solute in the initial solution and the same mole in a more dilute solution is really just a difference in the degree of randomness. The change in entropy ΔS is simply the difference

$$\Delta S = S_{\text{final}} - S_{\text{initial}}$$

where S_{final} and S_{initial} are the absolute entropies per mole of solute in the concentrated and dilute solutions. The total amount (one mole) of solute has not changed in this process, nor has its total energy (or enthalpy). But the total entropy of that mole has changed. By this somewhat intuitive argument it is reasonable to suppose that *entropy has an important dependence on concentration.* For the ideal solute ΔH has no dependence on concentration. Neither total internal energy nor enthalpy for an ideal solute in solution varies with its state of dilution, but entropy does. We must therefore assume that free energy changes also depend on concentration, for ΔG includes an entropy term.

The change in free energy ΔG is really the difference between two terms. We may imagine the existence of absolute values of Gibbs free energy, even though they cannot be measured absolutely:

$$\Delta G = G_{\text{final}} - G_{\text{initial}}$$

where G_{final} = absolute value of the Gibbs function for the final product
G_{initial} = absolute value of the Gibbs function for the starting material

To continue the logic, since the absolute Gibbs free energy G must be made up of enthalpy and entropy terms,

(7.6) $G = H - TS$ (the defining equation for the Gibbs free energy)

and since entropy S does depend on concentration, *G must be some function of concentration* too. Understanding just what that function is constitutes our next goal. We cannot derive it from fundamental principles. If we dealt exclusively with ideal gases, one form could be so derived, but from then on with more ordinary substances we should have to resort to making an assumption and then seeing how it works. This, in fact, is what has been done, with the following results.

We define, first, a *standard free energy $G°$ as the Gibbs free energy of one mole of a substance at standard conditions,* one atmosphere and 25°C. (It is strongly emphasized that this is the Gibbs free energy *per mole.*) The actual free energy per mole of a substance is related to its standard free energy per mole and its concentration and temperature by the following equation:

$$G = G° + 2.303RT \log C \qquad (7.7)$$

where G = free energy per mole of the substance at temperature T and concentration C, in moles per liter
R = gas constant per mole
C = concentration in moles per liter
$G°$ = standard free energy per mole

At this point we pause to see where we have been as well as to look ahead. We have discussed spontaneity by asking what the simplest, most useful elements are in terms of which spontaneity can be analyzed. Two elements, enthalpy and entropy, were found. We concentrated almost exclusively on constant T and P processes. To avoid an analysis of the entire universe, we looked at how ΔH and ΔS terms together might be related to the direction of a change for a *system*. The Gibbs free energy was the result. In terms of the sign of ΔG we found a single criterion both for spontaneity and for the direction of some change within a system. We have just now examined how the Gibbs free energy per mole varies with concentration.

This dependence of G on concentration will, as we shall see, enable us to analyze *permanence*, that is, equilibrium or the lack of observable change. We have analyzed motion, matter, and spontaneity. We next look at equilibrium, and for that equation 7.7 will be very useful.

EQUILIBRIUM AND THE EQUILIBRIUM CONSTANT

First let us review the phenomenon of substances A and B interacting to produce X and Y (cf. p. 197). The reaction is reversible; X and Y, once formed, can interact to make A and B. A specific example is the ionization of a weak acid, acetic acid:

$$\underset{\substack{\text{Acetic acid}\\ \text{(a weak acid)}}}{CH_3\overset{\overset{\displaystyle O}{\|}}{C}OH} + H_2O \rightleftharpoons H_3O^+ + \underset{\text{Acetate ion}}{CH_3\overset{\overset{\displaystyle O}{\|}}{C}O^-}$$

$$A \quad + \quad B \rightleftharpoons X \quad + \quad Y$$

Each of the substances will have its own particular standard free energy per mole: G_A°, G_B°, G_X°, G_Y°. At any given instant each will have its own particular concentration: C_A, C_B, C_X, C_Y. (We should use activities instead of concentrations as described on page 161, but because activity coefficients are difficult to know, concentration is the best and easiest approximation.) Applying equation 7.6, we find that each substance will have its own particular free energy per mole at any particular instant, G_A, G_B, G_X, G_Y, given by the following expressions:

$$G_A = G_A^\circ + 2.303RT \log C_A \tag{7.8}$$

$$G_B = G_B^\circ + 2.303RT \log C_B \tag{7.9}$$

$$G_X = G_X^\circ + 2.303RT \log C_X \tag{7.10}$$

$$G_Y = G_Y^\circ + 2.303RT \log C_Y \tag{7.11}$$

The free energy change for the forward reaction is the difference between the total free energy of the products and the total free energy of the reactants. (Taking the difference this way rather than the other way around is arbitrary but standard practice.)

(7.12) $$\Delta G = \underbrace{(G_X + G_Y)}_{\text{Total free energy of products}} - \underbrace{(G_A + G_B)}_{\text{Total free energy of reactants}}$$

Since equations 7.8 through 7.11 give us free energies in terms of standard free energies and concentrations, we substitute these into 7.12 and obtain

(7.13) $$\Delta G = \underbrace{[(G_X^\circ + 2.303RT \log C_X) + (G_Y^\circ + 2.303RT \log C_Y)]}_{\text{Total free energy of products}}$$

$$- \underbrace{[(G_A^\circ + 2.303RT \log C_A) + (G_B^\circ + 2.303RT \log C_B)]}_{\text{Total free energy of reactants}}$$

By rearranging terms, we can simplify equation 7.13 to

(7.14) $$\Delta G = \underbrace{(G_X^\circ + G_Y^\circ) - (G_A^\circ + G_B^\circ)}_{\equiv \Delta G^\circ} + 2.303RT(\log C_X + \log C_Y) - (\log C_A + \log C_B)$$

or

(7.15) $$\Delta G = \Delta G^\circ + 2.303RT \log \frac{C_X C_Y}{C_A C_B}$$

With equation 7.15 to help us, we next ask what the situation at equilibrium is.

When a condition of equilibrium is reached, there are no longer any net changes in concentrations. As far as taking advantage of the system to "extract" any further work, it is used up; there is no further change in the Gibbs free energy. At equilibrium $\Delta G = 0$ (equation 7.5). Therefore at equilibrium we have from 7.15

(7.16) $$\Delta G^\circ = -2.303RT \log \frac{C_X C_Y}{C_A C_B}$$

where the values of C_X, C_Y, C_A, and C_B are all equilibrium concentrations. Since the concentrations are all now constant, the ratio in equation 7.16 that involves these terms is a constant, designated as K. Thus

(7.17) $$\Delta G^\circ = -2.303RT \log K$$

We call this K the *equilibrium constant:*

(7.18) Equilibrium constant: $$K = \frac{C_X C_Y}{C_A C_B}$$

where all concentrations are in moles per liter (as mentioned earlier, activities should be used for accuracy).

To recapitulate, when A and B are mixed and they start to react, a certain amount of free energy is made available and may be tapped for useful work. The maximum available is ΔG. Eventually a state of equilibrium is reached. The system has "run down," so to speak. No further free energy is available, and ΔG is henceforth zero. Under this circumstance there is a definite and extremely important relation between the equilibrium concentrations, a relation expressed by equation 7.18.[3]

Since chemical analysis can be used to determine concentrations, we are able to calculate the equilibrium constant K. Then by using equation 7.17, we have a way of determining ΔG°. Alternatively, if there is a way to measure or at least to make a good estimate of ΔG°, equation 7.17 can be used to calculate an equilibrium constant. The equilibrium constant will be considered in greater detail in Chapter 8. As mentioned earlier, its value can be calculated from data for equilibrium concentrations, and such concentrations can usually be measured. This constant is also obviously a means for predicting the direction a reaction may take. If K is large, for example, the numerator in 7.18 must be large in relation to the denominator. In other words, by the time equilibrium is reached, the reactants will have substantially disappeared and the products will have formed as the major species. Of course, from equation 7.17 a large value of K means a large negative value of ΔG°, corresponding to precisely the direction of the reaction. In short, *if a reaction goes for the most part to completion, K will be large and ΔG° will have a large negative value.*

We must note here that we have been asking what direction a change will take if there is one. Nothing, however, has been said or implied about *how fast* the predicted change will occur. We have been busy with the problem of direction and with the situation at equilibrium. We must now consider the question how fast. It is entirely possible for a change to have associated with it an enormous negative free energy change—in other words, to be highly favorable—and at the same time to move as slowly as a glacier. If a very rapid reaction is needed, and many changes in living cells demand such speed, it does no good to have a highly favorable free energy situation but a sluggish reaction. We shall therefore analyze rates of chemical reactions and ask what the simplest and most useful elements are in terms of which such rates can be analyzed.

[3] In the general case for an equilibrium,

$$a\mathrm{A} + b\mathrm{B} + c\mathrm{C} + \cdots \rightleftharpoons m\mathrm{M} + n\mathrm{N} + p\mathrm{P} \cdots$$

$$K = \frac{C_\mathrm{M}{}^m C_\mathrm{N}{}^n C_\mathrm{P}{}^p \cdots}{C_\mathrm{A}{}^a C_\mathrm{B}{}^b C_\mathrm{C}{}^c \cdots}$$

(Equation 7.15 would be changed accordingly.) As an example, suppose that the reaction for the synthesis of hydrogen iodide from the elements is the following:

$$\mathrm{H_2 + I_2 \rightleftharpoons 2HI}$$

$$K = \frac{[\mathrm{HI}]^2}{[\mathrm{H_2}][\mathrm{I_2}]}$$

Another example is the equation for the formation of ammonia from the elements:

$$\mathrm{N_2 + 3H_2 \rightleftharpoons 2NH_3}$$

$$K = \frac{[\mathrm{NH_3}]^2}{[\mathrm{N_2}][\mathrm{H_2}]^3}$$

Squared brackets designate the concentration in moles per liter of the substance whose formula is enclosed.

CHEMICAL KINETICS

The question how fast is in the domain of *kinetics* (Greek *kinetikos,* to move), a branch of chemistry wherein efforts are made to relate the rates of reactions to properties of the reactants and such aspects as the volume, the temperature, and the pressure.

We shall assume that atoms and molecules exist and that the kinetic theory of gases, liquids, solids, and solutions is substantially correct. Our approach will be similar to that used for entropy. We shall discuss rates of chemical reactions in terms of what atoms or molecules are believed capable of doing and are believed actually to do. Thus we start with an interpretation of data about rates rather than with methods for obtaining the data and for expressing rates in useful forms. This approach is dictated largely by the initial decision not to assume that the student has a working knowledge of the calculus.

Suppose that we have a chemical change whereby A reacts with B—C to produce A—B and C:

$$\text{A} + \text{B—C} \longrightarrow \text{A—B} + \text{C} \tag{7.19}$$

We start by mixing together samples of A and of B—C; in any actual experiment (given the size of Avogadro's number) hundreds and thousands of billions of molecules of both types will be involved. The molecules are not all moving in the same direction since the distribution of the directions of motion is random, nor are they all moving at the same speeds since the distribution of speeds is also random. Some are moving slowly, some moderately, and some extremely rapidly. Some may be momentarily motionless. Random collisions take place at varying angles. At least we assume this description of behavior to be true. The molecules cannot actually be seen to act this way, but the evidence collected from observed events makes the best sense if such movement at the molecular level is assumed. (Thus what we see and what we think about what we see are two different things.)

To continue with this theory, it follows that some collisions between molecules of A and B—C may be mere nudges, some moderate whacks, and others violent crashes. It all depends on how rapidly both particles are moving just before impact and whether they collide head on. Each collision will have a particular *collision energy,* and these values will vary from very low to very high. Few collisions involve the very low values of energy, and few are at the other extreme. Figure 7.2 shows graphically how the fraction of collisions having a particular energy varies with that energy.

Energy of Activation. For the reaction of equation 7.19 to take place, the bond between B and C in B—C must be broken, requiring energy. A new bond between A and B must form, releasing energy. The trouble is that the *timing* of the release and that of the demand are seldom so perfectly synchronized that A has only to tap the B end in B—C to make it all happen. The collision must usually be of some violence. The molecule A has to penetrate somewhat into the electron cloud of B—C before the old bond (in B—C) will let go and the new bond (in A—B) will start to form. In other words, we expect that below some definite value of collision energy no

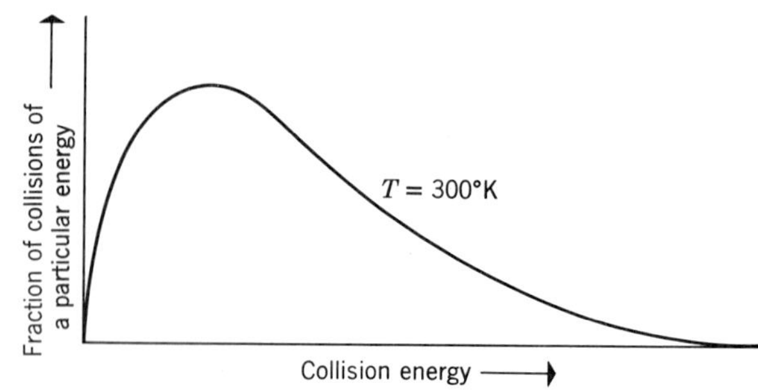
Fraction of collisions of a particular energy
T = 300°K
Collision energy

Figure 7.2 Fraction of collisions of a particular energy versus the collision energy.

chemical change occurs and no reallocation of bonding electrons takes place. Above this value the collision will be successful, that is, lead to products. This minimum value of collision energy is called the *energy of activation.* If it is high, only the most violent of collisions between reactants can be successful. These, of course, occur relatively infrequently, and the rate of the reaction is slow. The situation is further explained in Figure 7.3. If the energy of activation is relatively low, the frequency of collisions above this value will be high. A high rate of successful collisions is the same thing as a high rate of reaction.

The actual value of the energy of activation depends on the particular reaction. It is but one factor that affects the rate, and once a particular reaction has been picked, the energy of activation has been picked, too. The variable cannot be changed at will by the experimentalist, with one extremely important exception, the phenomenon of catalysis which will be taken up shortly. Other factors affecting reaction rates the scientist can rather easily manipulate, and we shall turn our attention to them next. They are best discussed in terms of one way of analyzing the rate of a reaction.

Experimentally, a reaction rate can be measured as the rate of appearance of some product(s) or the rate of disappearance of some reactant(s). At measured intervals of time a portion of the reacting mixture is analyzed for relative concentrations of it participants, and a mathematical equation that fits the data is then sought.

Conceptually, a reaction rate may be discussed in terms of some theory about molecules in collision and may be defined as

(7.20)
Rate = number of successful collisions per second per cubic centimeter

The number of successful collisions will be some fraction of the total number of collisions, and our task now is to identify the factors that give us the appropriate fraction(s):

$$(7.21)\quad \text{Rate} = \frac{\text{number of successful collisions}}{\text{(sec)(cc)}}$$
$$= \frac{\text{total number of collisions}}{\text{(sec)(cc)}} \times \begin{matrix}\text{energy}\\ \text{factor}\end{matrix} \times \begin{matrix}\text{probability}\\ \text{factor}\end{matrix}$$

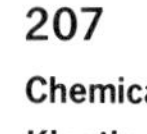

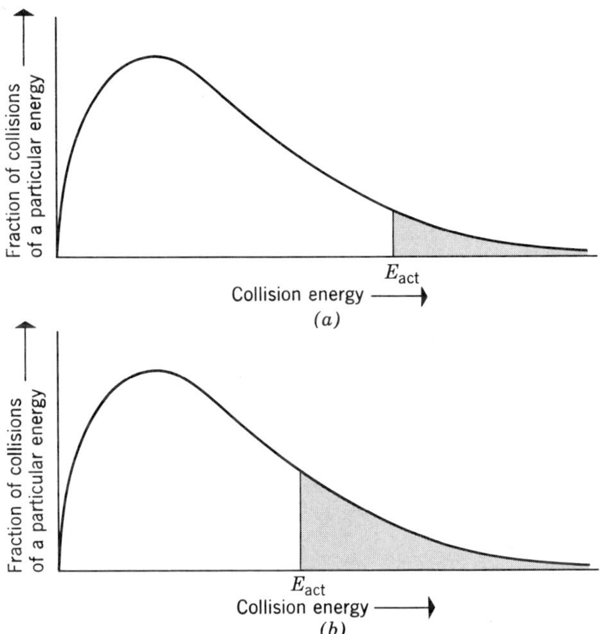
Fraction of collisions of a particular energy
E_{act}
Collision energy
(a)
Fraction of collisions of a particular energy
E_{act}
Collision energy
(b)

Figure 7.3 Energy of activation versus fraction of *successful* collisions. The total number of collisions per cubic centimeter per second is represented by all the area under the curve in either part *a* or part *b*. Only the shaded area in each represents the total number of *successful* collisions, those leading to products. The *fraction* of successful collisions is the ratio of the shaded area to the total area. When the energy of activation E_{act} is relatively high, as in part *a*, this fraction is relatively small and the rate of reaction is low. But when E_{act} is low as in part *b*, the fraction of successful collisions is larger and the rate is larger.

From the total number of collisions of all degrees of violence and of all possible orientations, only a fraction will have enough energy (energy of activation), and this fraction is the energy factor in equation 7.21. Only a fraction of the collisions (regardless of their energy) will be properly oriented. In reaction 7.19, A must strike B—C at the B end not the C end, and it must not be just a glancing blow. Ideally it should be head on along the B—C bond axis. Only a fraction of the collisions are of this orientation, giving us the probability factor. Not much can be done experimentally to increase or decrease this fraction once a particular reaction has been selected. If one reactant is a large molecule that has to be hit at one and only one site to undergo reaction, this fraction will be smaller than it would be if the reactant(s) were little molecules or atoms. Thus of the three terms to the right of the equal sign in equation 7.21, the last one (the probability factor) is difficult to change by manipulating the conditions of the reaction.

The first term (total number of collisions per second per cubic centimeter) is easily changed. By increasing the *concentrations* of reactants, for example, the chances of collisions of all orientations and energies are increased. By increasing the *temperature*, the reactant molecules are made to move more rapidly, and collisions of all orientations and energies become more frequent. Thus concentration and temperature are important factors in determining rates of reactions.

Effect of Temperature on Rate. Increasing the temperature increases not only the total frequency of collisions of all types but also the fractions of the more violent types disproportionately. Figure 7.4 shows how the curve of Figure 7.2 shifts when the temperature is increased. This shift increases the fraction of collisions of enough energy. Thus temperature increases not only the total frequency of collisions but also the second term to the right of equation 7.21, the energy factor.

This factor is exponential in nature, having the form $e^{-E_{act}/RT}$, and the student is not expected to be able to evaluate it. Table 7.2 shows the results of sample computations of this term for varying values of E_{act} (energy of activation) at different temperatures. These results support the following qualitative statements concerning the influences of various factors on rates of reactions.

1. Relatively small changes in E_{act} can produce very large changes in the energy factor. It depends on the starting value of E_{act}, but for situations commonly encountered a small percentage drop in E_{act} produces a large percentage increase in rate.

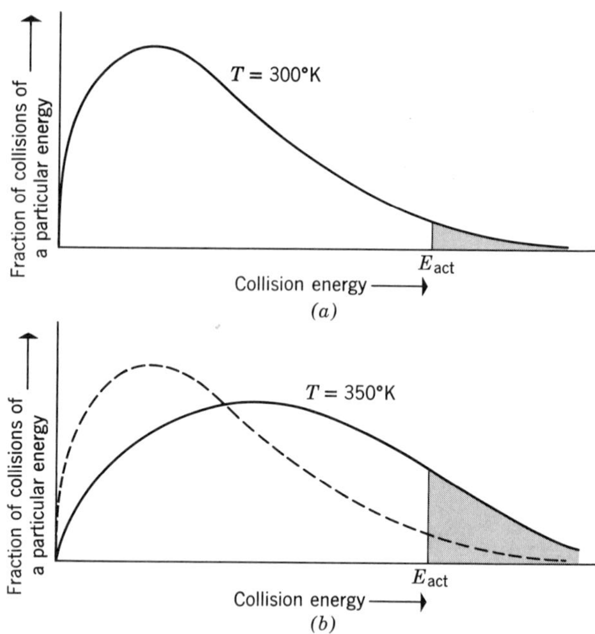
Fraction of collisions of a particular energy
T = 300°K
E_{act}
Collision energy
(a)
Fraction of collisions of a particular energy
T = 350°K
E_{act}
Collision energy
(b)

Figure 7.4 Effect of temperature on the energy factor. The principal factor causing the difference between curves a and b is the temperature. The curve tends to flatten, and its maximum shifts to the right as the temperature is raised. (The change has been exaggerated for purposes of illustration.) Assuming that the energy of activation is unchanged by this rise in temperature, the shaded area is considerably larger in part b than in part a. Thus the fraction of successful collisions at the higher temperature is much larger, and the rate of the reaction is much faster.

Table 7.2 Evaluations of the Energy Factor, $e^{-E_{act}/RT}$

	E_{act} (cal/mole)	Temperature (°K)	Energy factor, $e^{-E_{act}/RT}$	Comments*
Very low E_{act} values (rare)	600	300	0.367	Over a third of all the collisions will have proper energy, a very large fraction.
	10% drop in E_{act}			
	540	300	0.407	The change is slight, compared with the first.
Larger E_{act} values (more common)	6000	300	0.000045	Of every 22,200 collisions only one will have enough energy for reaction.
	10% drop in E_{act}			
	5400	300	0.000123	Now one collision in every 8100 will have enough energy. In other words, the energy factor and therefore the rate are increased about two and a half times by a mere 10% drop in the E_{act} at this level.
Effect of temperature	6000	300	0.000045	A 10° rise in temperature, in this range of E_{act} and temperature, causes the energy factor and therefore the rate to increase about one and a half times.
		10° rise in temperature		
	6000	310	0.000062	

* The gas constant has a value of 1.987 cal/(°K)(mole); in the calculations for this table it was rounded to 2.

2. Relatively small changes in the temperature of the reaction can produce very large changes in the energy factor. Again, it depends on the E_{act} as well as the initial value of the temperature. For commonly encountered situations a small percentage rise in temperature can produce a large increase in rate. (One rule of thumb says that a 10°C rise in temperature will double and often triple a rate.)

The effect of temperature on rate is very important to living creatures. For higher animals to remain healthy, they must have built-in mechanisms for maintaining quite constant internal temperatures, even while the temperature of the environment varies between wide extremes. Just a small rise in body temperature (e.g., from 98.6 to 103°F) brings about serious increases in rates of reactions in the body. Some of the reactions are just those needed to handle the problem, but a sustained high body temperature is fatal.

Temperature has its greatest effect on rate by changing the energy factor. Its effect on the total collision frequency, although important, is not dramatic.

Effect of Concentration on Rate. Manipulation of concentration can affect only the total collision frequency. For some reactions the rates are directly proportional to the concentration of *each* reactant. In the general reaction,

$$\mathrm{A + B \longrightarrow X + Y}$$

if the concentration of A is doubled, the rate is doubled. If the concentration of B is halved, the rate is halved. Such facts may be expressed mathematically as follows:

$$\text{rate} \propto [\mathrm{A}][\mathrm{B}] \tag{7.22}$$

where [A] and [B] represent concentrations (ideally, activities) in moles per liter. By inserting a proportionality constant, we have

$$\text{rate} = k[\mathrm{A}][\mathrm{B}] \tag{7.23}$$

where k is the *rate constant.* If the coefficients in the reaction are not equal to one, that is, if for example

$$\mathrm{A + 2B \longrightarrow AB_2}$$

the rate equation may take the form

$$\text{rate} = k[\mathrm{A}][\mathrm{B}]^2 \tag{7.24}$$

This is not *necessarily* the case. In a real reaction, the rate equation must show how the rate actually does depend on concentrations. But if the rate data do fit equation 7.23, the reaction is said to be *first order* in A and *first order* in B, and it is overall a *second order* reaction. If the data, however, fit equation 7.24, the reaction is said to be first order in A, second order in B, and overall third order. The order of a reaction with respect to a particular reactant is therefore the exponent to which its concentration is raised in the rate equation. The overall order is the sum of these exponents. Many reactions have fractional orders, and many more reactions have rates that cannot be made to fit simple equations such as 7.23 or 7.24. We shall not venture into these territories.

The Equilibrium Constant, Another Approach. If we deal with a reversible reaction such as we have done earlier (p. 202),

$$\mathrm{A + B \rightleftharpoons X + Y}$$

(for example, $\mathrm{H{-}A + H_2O \rightleftharpoons A^- + H_3O^+}$, where H—A is a weak acid), we can imagine that each reaction, the forward and the reverse, has its own rate. In the simple case each rate may be expressed as follows:

$$\overrightarrow{\text{rate}} = \vec{k}[\mathrm{A}][\mathrm{B}]$$
$$\overleftarrow{\text{rate}} = \overleftarrow{k}[\mathrm{X}][\mathrm{Y}]$$

At equilibrium the two "opposing" rates must be equal. In no other way could there be no *net* changes in concentrations; for every A that disappears in the forward reaction, a new A is formed by the reverse. But if the two rates are equal, then

$$\overrightarrow{k}[A][B] = \overleftarrow{k}[X][Y]$$

By rearranging these terms, we have

$$\frac{\overrightarrow{k}}{\overleftarrow{k}} = \frac{[X][Y]}{[A][B]}$$

But the left side of this equation is a ratio of constants, and it must therefore be a constant. In fact, it is the equilibrium constant we derived earlier from a consideration of free energy changes:

$$\frac{\overrightarrow{k}}{\overleftarrow{k}} = K = \frac{[X][Y]}{[A][B]} \tag{7.25}$$

Progress of Reaction Diagrams. In discussing many kinds of reactions, we find it convenient to use a diagram that looks like a graph but has imprecisely defined coordinates. The "Y-axis" may represent the Gibbs free energy G (if the process is for constant T and P), or it may represent internal energy E. No numbers will be assigned to intervals along this axis. It will be purely qualitative. A high position will mean high internal energy (or Gibbs function), and a low position will mean less. The "X-axis," even more vague, stands for "progress of reaction," which means that from left to right along the axis the reaction takes place as indicated. The example given in Figure 7.5, has internal energy for the vertical axis.

In Figure 7.5 the sum of the absolute values of the internal energies for the two reactants, A + B—C, is suggested by the line indicated. We imagine then that a molecule of A collides with B—C. If a head-on collision develops, a collision that is going to be successful, the total internal energy of the particles involved in the collision rises along the curve as the kinetic energies become converted to potential or internal energy. If the collision develops the minimum energy needed for reaction,

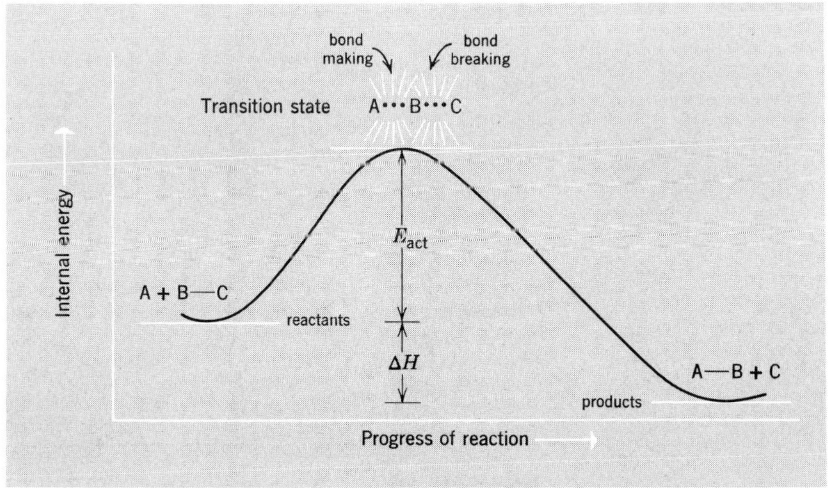
bond making
bond breaking
Transition state
A•••B•••C
Internal energy
E_{act}
A + B—C
reactants
ΔH
A—B + C
products
Progress of reaction

Figure 7.5 Progress of reaction versus internal energy for an exothermic reaction.

this curve peaks. Momentarily, at this point of minimum internal energy needed for successful collision (but a maximum point on the curve), an unstable particle consisting of both A and B—C exists. The bond between A and B is beginning to form; the old bond in B—C is breaking. The electron cloud of A has made a significant penetration of the electron cloud at the B end of B—C. From this point on, C starts to break away and A—B remains. Together, these two may have very high *kinetic* energies, much higher than those of A + B—C. Heat is released. As the reaction progresses, the products will therefore be at a lower level of total internal energy than were the reactants. The energy released in the overall process is the difference between the initial and final levels. The energy of activation for the forward reaction is the separation of the initial level and the *transition state;* the transition state consists of the unstable aggregate of reactants, the collection of nuclei and electrons with minimum internal energy that is still high enough to be able to break down to give products. The higher the energy barrier, that is, the higher the energy of activation, the slower the rate of the reaction. Diagrams such as Figure 7.5 rapidly convey a qualitative picture of the energetics of an exothermic reaction. Figure 7.6 shows roughly what an endothermic reaction would look like on such a diagram.

Catalysis. Potassium chlorate, $KClO_3$, melts at 368°C. Slightly above this temperature, at about 400 to 420°C, it decomposes, releasing oxygen and potassium chloride:

$$2KClO_3 \xrightarrow[\text{heat to about 400°C}]{} 2KCl + 3O_2$$

If a trace amount of manganese dioxide, MnO_2, is added before heating, however, potassium chlorate decomposes at a much lower temperature (about 270°C). If in the presence of manganese dioxide the temperature is brought up to about 400°C, oxygen will form at a vastly greater rate than in its absence.[4] A substance such

[4] The reaction described is potentially quite dangerous. Potassium chlorate is a vigorous oxidizing agent. If traces of organic substances, even bits of filter paper, cork dust, or rubber, come in contact with hot potassium chlorate, an explosion may occur.

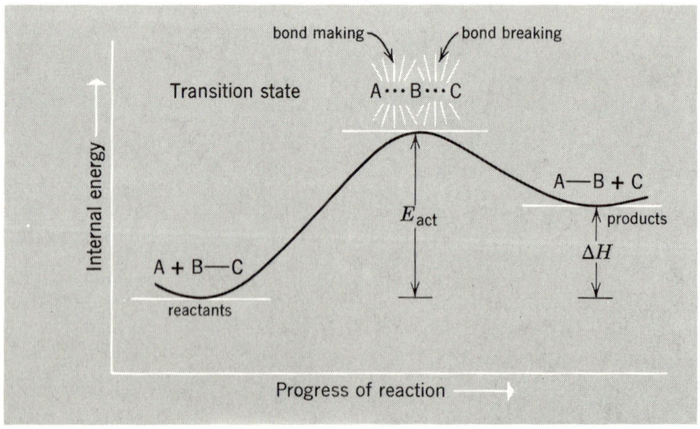
bond making
bond breaking
Transition state
A···B···C
Internal energy
A—B + C
E_{act}
products
ΔH
A + B—C
reactants
Progress of reaction

Progress of reaction

Figure 7.6 Progress of reaction versus internal energy for an endothermic process.

as manganese dioxide that will in relatively trace amounts accelerate a reaction without itself undergoing permanent change is called a *catalyst*. The general phenomenon involving such substances and their behavior is called *catalysis*. In general terms, a catalyst functions by making available to the reactants a path to the products that requires a lower energy of activation. It may perform this function in several ways but, generally speaking, the catalyst participates temporarily in the reaction by combining with one or more of the reactants, and the intermediate complex then reacts further to yield products and regenerate the catalyst. If a catalyst is found, the cost in the coinage of energy to make a reaction go and go rapidly may be greatly reduced. In fact, without the catalyst, many reactions would not be feasible at all. Catalysts are necessary in the manufacture of plastics, drugs, dyes, margarine, synthetic gasoline—the list is almost endless. Entire industries depend on the existence of catalysts. Life itself could not be imagined without the participation of hundreds of catalysts in the reactions of living beings. In the body the catalysts are called enzymes, and the properties of enzymes, their formation, control, and misbehavior, are some of the most important topics in all molecular biology. Certain vitamins are known to provide molecular pieces for making enzymes in the body. Certain hormones are known to exert their dramatic influences by regulating enzymes.

REFERENCES AND ANNOTATED READING LIST

BOOK

L. E. Strong and W. J. Stratton. *Chemical Energy*. Reinhold Publishing Corporation, New York, 1965. Although in a few places a knowledge of very elementary calculus is assumed, for the most part this reference is the best single choice for coverage of the subject of chemical energy at a relatively easy level. (Paperback.)

ARTICLE

G. Porter. "The Laws of Disorder." *Chemistry*, May 1968, page 23. Based on the booklet written to accompany a British Broadcasting Corporation television series on the laws of disorder, this article by a Nobel laureate is Part 1 in a series in *Chemistry*.

For other references, see Chapter 3.

PROBLEMS AND EXERCISES

1. Write out definitions of the following terms. Some of them appeared first in earlier chapters and may be reviewed as necessary.
   (*a*) natural event (*b*) enthalpy
   (*c*) entropy (*d*) system
   (*e*) surroundings (*f*) boundary
   (*g*) first law of thermodynamics (*h*) second law of thermodynamics
   (*i*) the Gibbs free energy change (*j*) endothermic reactions

(*k*) exothermic reactions (*l*) standard state
(*m*) standard free energy of formation (*n*) kinetics
(*o*) energy of activation (*p*) catalyst
(*q*) equilibrium

2. What does a minus sign signify when it is associated with each of the quantities ΔH, ΔS, and ΔG?
3. What does a plus sign signify when it is associated with each of the quantities in problem 2?
4. What does a value of zero signify when it is associated with each of the quantities in problem 2?
5. How can an endothermic process for a system be spontaneous?
6. How can a process for a system take place even if the entropy change is negative for the system?
7. When are we justified in using ΔH in lieu of ΔG as a criterion of spontaneity?
8. What is meant when we say that a "reaction has gone to equilibrium?"
9. Suppose that we have a reaction such as $A + B \rightleftharpoons AB$.
   (*a*) Write the expression for the equilibrium constant for this reaction.
   (*b*) Suppose that the equilibrium constant is 10^2 at 25°C. What is $\Delta G°$?
   (*c*) Suppose that the equilibrium constant is 10^{-2} at 25°C. What is $\Delta G°$?
   (*d*) How does the sign of $\Delta G°$ that you calculated for parts *b* and *c* correlate with the position of equilibrium?
10. Ammonia can be made to form from the elements according to the following equation:

$$N_2(g) + 3H_2(g) \longrightarrow 2NH_3(g)$$

The following thermodynamic data are known for NH_3, and they refer to the synthesis of *one* mole of ammonia:

$$\Delta H_f^\circ = -11{,}040 \text{ cal}$$
$$\Delta S_f^\circ = -23.7 \text{ cal/°K}$$
$$\Delta G_f^\circ = -3976 \text{ cal}$$

(*a*) Considering the equation for the formation of ammonia, explain how it is reasonable for ΔS_f° to have the algebraic sign it does.
(*b*) Explain how it is reasonable for the reaction to occur, in spite of your answer to part *a*.
(*c*) Calculate the equilibrium constant for the reaction at 25°C.
(*d*) This synthesis of ammonia is used industrially. In the Haber process nitrogen and hydrogen are heated to a temperature of 400 to 500°C under a pressure of 200 to 1000 atmospheres in the presence of a catalyst. Does the equilibrium constant, or the sign of $\Delta G°$, give any indication that such severe conditions are necessary to make the reaction take place?
(*e*) In general, how would it be possible to have a reaction that is favored by the thermodynamics (i.e., large negative ΔG) but is still very slow kinetically? (Use a diagram showing progress of reaction in your discussion.)
(*f*) Using encyclopedias and histories of science, find out how the Haber process for making ammonia affected the conduct of World War I.

CHAPTER EIGHT

Ionic Equilibria

The three concepts of acids and bases described in qualitative terms in Chapter 6 may now be discussed quantitatively. First let us review the three systems.

Arrhenius system of acids and bases (cf. p. 169):

—Acids are substances that produce hydrogen ions in water; their solutions turn litmus red and have a sour taste.

—Bases are substances that produce hydroxide ions in water; their solutions turn litmus blue, have a brackish taste, and feel soapy.

—Neutralization is the combination of these two ions:

$$H^+ + OH^- \longrightarrow H_2O$$

Brønsted-Lowry system of acids and bases (cf. p. 178):

—Acids are substances that can donate protons.

—Bases are proton acceptors.

—Neutralization is acceptance of a proton from a proton donor (B: represents a base that can bind a proton more strongly than the donor):

$$\underset{\text{Stronger base}}{B:} + \underset{\text{Stronger acid}}{H_3O^+} \xrightarrow[\text{water}]{} \underset{\text{Weaker acid}}{B{-}H^+} + \underset{\text{Weaker base}}{H_2O}$$

Lewis system of acids and bases (cf. p. 183):

—An acid can accept a share in a pair of electrons.

—A base can donate a share in a pair of electrons.

—The word neutralization is not used so much here. When a Lewis acid combines with a Lewis base, the product is often described as a coordination compound. The word ''ligand'' is frequently used for the word ''base,'' and the reaction is often referred to as coordination.

For the remainder of this chapter it will be assumed that water is the solvent. The terms "hydrogen ion" and "hydronium ion" will be used interchangeably, with the first term always implying the second. A *strong acid* is one whose percentage of ionization in water is high, which means that the proton is not being bound in the acid molecule as well as it can be bound by a water molecule. We may also say that, to continue this review of terms, the *conjugate base* of the acid is a weaker base than water. A *strong base* is meaningfully defined in percentage of ionization only with a few substances, metal hydroxides. These are ionic compounds; when they dissolve in water, they do so as hydrated ions with a very high percentage of ionization. But it is proper to define a strong base in other ways. For example, a strong base is a strong proton binder and has a weak *conjugate acid.* At this point Table 6.4 (p. 182) should be consulted for familiar examples of conjugate acids and bases and their reciprocal relations.

THE IONIZATION OF WATER

When an aqueous solution has neither acidic nor basic properties in the Arrhenius sense, it is said to be *neutral.* Hydrogen ions and hydroxide ions are not completely absent, but both are present in equal concentrations. From our study of collision theory in Chapter 7, we should expect that water molecules constantly collide with neighbors and that at any instant a fraction of these collisions will be violent and "on target." If so, the following change will occur:

$$\text{(8.1)} \qquad H_2\ddot{O}: + H\ddot{O}\text{—}H \rightleftharpoons H_2\ddot{O}^{+}\text{—}H + {}^{-}:\ddot{O}\text{—}H$$

Another view of this reaction is shown in Figure 8.1.

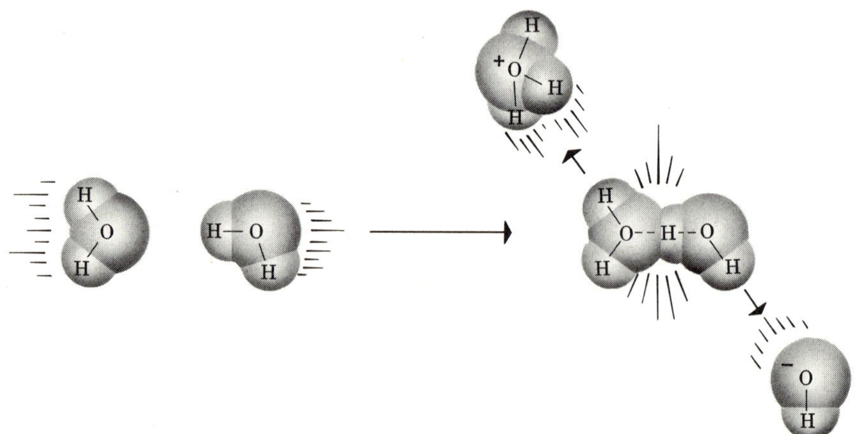
H
+O
H
H
H
O
H
H—O
H
H
O--H--O
H
H
−O
H

Figure 8.1 A collision of sufficient energy and proper orientation between two water molecules will produce a hydronium ion and a hydroxide ion. These are immediately solvated (not shown). At room temperature the fraction of these successful collisions is evidently very low, for at any instant only about 10^{-7}% of all water molecules have ionized.

For every hydronium ion that forms, a hydroxide ion *must* also form. This reaction inevitably produces them in equal numbers, and pure water is therefore neutral. The short arrow in equation 8.1 implies that at dynamic equilibrium a very small percentage of all water molecules have formed these ions, actually, just slightly more than 10^{-7} %. This amount is so small that pure water is a nonconductor of electricity, a nonelectrolyte.

The Equilibrium Constant and the Ionization Constant of Water. If we represent the ionization of water as simply

$$H{-}OH \leftrightarrows H^+ + OH^-$$

the equilibrium constant is[1]

$$(8.2) \qquad K = \frac{[H^+][OH^-]}{[H_2O]} = \frac{(10^{-7})(10^{-7})}{55.3} = 1.8 \times 10^{-16} \qquad (24°C)$$

In this calculation the concentration of water in moles per liter at 24°C—55.3—is so imperceptibly affected by its slight ionization that the latter does not reduce it. In other words, in pure water and even in dilute solutions of acids, bases, and salts, the denominator is substantially a constant. We may combine it with the constant K, already formed, and define a new constant K_w, called the *ionization constant* of water:

$$(8.3) \qquad \begin{aligned} K_w &= [H^+][OH^-] && \text{(defining equation for } K_w\text{)} \\ &= 10^{-14} && \text{(at room temperature)} \end{aligned}$$

Equation 8.3 is substantially true for water and for dilute aqueous solutions. In such solutions at room temperature (24°C) the product of the hydronium-ion concentration and the hydroxide-ion concentration is the constant value 10^{-14}, whether the solution is acidic, basic, or neutral. If $[H^+] = 10^{-1}$ mole/liter, then $[OH^-] = 10^{-13}$; $10^{-1} \times 10^{-13} = 10^{-14}$. Another way of looking at this example is in terms of equation 8.1. A relatively high concentration of hydronium ion makes much more frequent the conversion of hydroxide ion back to water, and the concentration of hydroxide ion must be much lower. In other words, to use the terminology of the field, the relatively high hydronium-ion concentration suppresses the ionization of water.

In any solution in which the hydronium-ion concentration is larger than $10^{-7}M$ (e.g., $10^{-6}M$), the hydroxide-ion concentration must be smaller than $10^{-7}M$ (e.g., $10^{-8}M$), and the solution is acidic.[2] When the situation is reversed, when the concentration of hydroxide ion exceeds $10^{-7}M$, the solution is basic.

[1] Squared brackets in this text are a symbol for concentration in moles per liter of the substance whose formula is enclosed. Throughout our study we continue to simplify matters by assuming that concentration equals activity (p. 161).

[2] Henceforth, unless otherwise specified, it is assumed that all concentration data refer to aqueous solutions at room temperature, which is taken to be about 72 to 77°F (22.2 to 25.0°C), and that $K_w = 10^{-14}$ in this range. Strictly speaking, only at 24°C does $K_w = 10^{-14}$, and 24°C = 75.2°F. See Table 8.1.

Exercise 8.1 (*a*) Calculate the hydroxide-ion concentration in a solution in which, at 25°C, $[H^+] = 5 \times 10^{-6}$. State whether the solution is acidic, basic, or neutral.

Solution.

$$(5 \times 10^{-6})[OH^-] = 1 \times 10^{-14}$$

$$[OH^-] = \frac{1 \times 10^{-14}}{5 \times 10^{-6}} = \frac{10 \times 10^{-15}}{5 \times 10^{-6}} = 2 \times 10^{-9} \text{ mole/liter}$$

The solution is acidic.

(*b*) For each of the following, calculate the hydroxide-ion concentration at 25°C and state whether the solution is acidic, basic, or neutral.

(1) $[H^+] = 4 \times 10^{-9}$ mole/liter
(2) $[H^+] = 1.1 \times 10^{-7}$ mole/liter
(3) $[H^+] = 9.9 \times 10^{-8}$ mole/liter
(4) $[H^+] = 1$ mole/liter

pH

The ionization of water at room temperature furnishes a hydrogen-ion concentration of only 1×10^{-7} mole/liter (and the same concentration of hydroxide ion). Ordinarily, trace concentrations such as this would be ignored, but chemists and physiologists have discovered that hydrogen-ion concentrations of this order of magnitude and smaller are extremely important, partly because these ions act as catalysts for many reactions. In the body they affect the catalytic activity of enzymes. Catalysts, as we have stated earlier, may exert their influence in very small concentrations.

Because of the importance of trace concentrations of hydrogen and hydroxide ions, chemists and technologists routinely make hundreds and thousands of measurements. As a matter of simple convenience, the manipulation of such awkward figures as negative exponents (e.g., 10^{-7}) or their decimal equivalents (e.g., 0.0000001) is tedious, and chemists long ago adopted a shortcut. Instead of the entire figure with its negative exponent, they use only the exponent itself. This exponent is defined as the pH of the solution. Mathematically,

$$[H^+] = 1 \times 10^{-\text{pH}} \qquad \text{(defining equation for pH)} \tag{8.4}$$

As a defining equation for pH, equation 8.4 is severely limited in its application to situations for which the 1 standing before the 10 is just that and no other number. If $[H^+] = 2 \times 10^{-6}M$, then equation 8.4 cannot be used as written; the pH of this solution is definitely not 6. To handle this common problem, equation 8.4 is recast into an alternative defining equation for pH:

$$\text{pH} = \log \frac{1}{[H^+]} = -\log [H^+] \tag{8.5}$$

Exercise 8.2 Prove that equations 8.4 and 8.5 are equivalent.

Exercise 8.3 Calculate the pH of a solution in which at 25°C the concentration of hydrogen ion in moles per liter is each of the following. State whether the solution is acidic, basic, or neutral.

Example. 4.5×10^{-7}.

Solution. $\text{pH} = -\log[\text{H}^+] = -\log 4.5 \times 10^{-7} = (-7 + \log 4.5)$

$$= (-7 + 0.65)$$

$$\text{pH} = 6.35$$

The solution is acidic.

(*a*) 4.5×10^{-8} (*b*) 4.5×10^{-6} (*c*) 6.3×10^{-13}
(*d*) 2.5×10^{-2} (*e*) 3.6×10^{-8}

Ans. (*a*) 7.35, (*b*) 5.35, (*c*) 12.2.

Exercise 8.4 Calculate the pH of a solution at room temperature in which the hydroxide-ion concentration in moles per liter is

(*a*) 2.22×10^{-7} (*b*) 2.22×10^{-9} (*c*) 1.59×10^{-2}
(*d*) 4.00×10^{-13} (*e*) 2.78×10^{-7}

Ans. (*a*) 7.35.

The ionization constant of water is affected by temperature as shown by the data in Table 8.1. As it changes, the hydrogen-ion concentration corresponding to a neutral solution also changes.[3]

Exercise 8.5 Taking normal body temperature to be 98.6°F or 37°C, calculate (*a*) $[\text{H}^+]$ (in moles per liter) for pure water at this temperature and (*b*) the pH of pure water at this temperature. (*c*) Is the water neutral at this temperature?

Ans. (*a*) 1.56×10^{-7}, (*b*) pH = 6.81, (*c*) of course, $[\text{H}^+] = [\text{OH}^-]$.

Table 8.1 Ionization Constant of Water at Various Temperatures

Temperature (°C)	Ionization Constant
10	2.92×10^{-15}
20	6.81×10^{-15}
24	1.00×10^{-14}
30	1.47×10^{-14}
37 (normal body temperature)	2.42×10^{-14}

[3] Pure water, having a pH = 7 at room temperature, is not the easiest material to prepare and keep. Carbon dioxide from the air dissolves in water and reacts to produce a weak acid, H_2CO_3, or carbonic acid:

$$H_2O + CO_2 \rightleftharpoons H_2CO_3 \rightleftharpoons H^+ + HCO_3^-$$

The ionization of this alters the pH of water. Water that is boiled and kept protected from the atmosphere has a pH nearer 7.

When $[H^+]$ exceeds 10^{-7} mole/liter, the solution is acidic, which means that pH values *less* than 7 correspond to acidic solutions. Basic solutions have pH values greater than 7 at room temperature.

DETECTION OF HYDROGEN AND HYDROXIDE IONS. INDICATORS. pH MEASUREMENTS

Indicators. Certain organic dyes show one color above a characteristic pH and another color below that pH (or pH range). Litmus is a well-known example of these dyes. In acid, below the pH range 4.5 to 8.5, it is red. In base, above that range, it is blue. Litmus is usually supplied in the form of porous strips of paper impregnated with the dye.

Any dye that will perform the service of indicating the pH of a solution is called an *indicator*. Several common ones and the colors they impart to solutions at various pH intervals are listed in Table 8.2. By using two or more of these on test portions of a solution, we can determine the pH within one to two units. Suppose, for example, that a solution is found to be colorless to phenolphthalein. Its pH must therefore be some value less than 9, and it could be any smaller value. But if this solution is also light blue to bromthymol blue, then by Table 8.2 its pH cannot be less than 8. The pH of the solution must therefore fall between 8 and 9.

Exercise 8.6 Deduce the range of values within which the pH of the solution must lie if its reaction is (*a*) red to phenolphthalein and yellow to thymolphthalein, (*b*) blue to litmus and yellow to thymol blue, (*c*) yellow to methyl orange and red to methyl red.

Table 8.2 Colors of Some Common Indicators at Various pH Values and pH Values for Some Common Substances (at 25°C)

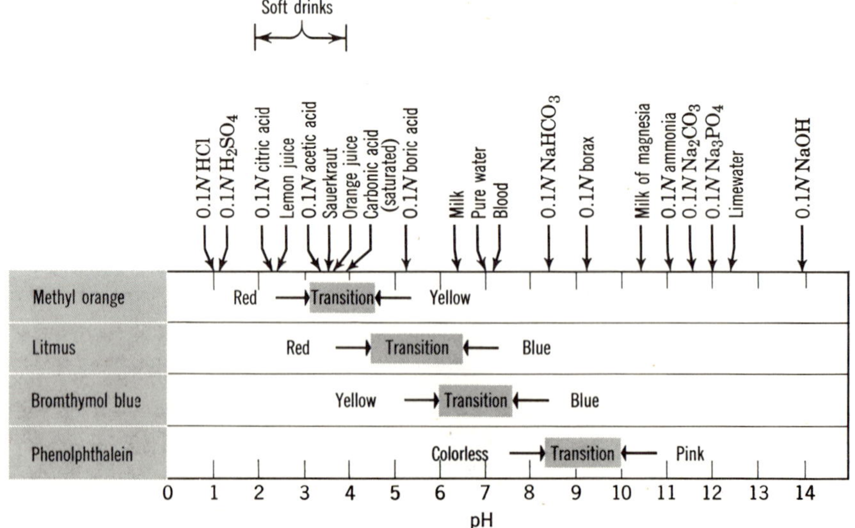
Soft drinks
0.1 N HCl
0.1 N H_2SO_4
0.1 N citric acid
Lemon juice
0.1 N acetic acid
Sauerkraut
Orange juice
Carbonic acid (saturated)
0.1 N boric acid
Milk
Pure water
Blood
0.1 N $NaHCO_3$
0.1 N borax
Milk of magnesia
0.1 N ammonia
0.1 N Na_2CO_3
0.1 N Na_3PO_4
Limewater
0.1 N NaOH
Methyl orange
Red
Transition
Yellow
Litmus
Red
Transition
Blue
Bromthymol blue
Yellow
Transition
Blue
Phenolphthalein
Colorless
Transition
Pink
0 1 2 3 4 5 6 7 8 9 10 11 12 13 14
pH

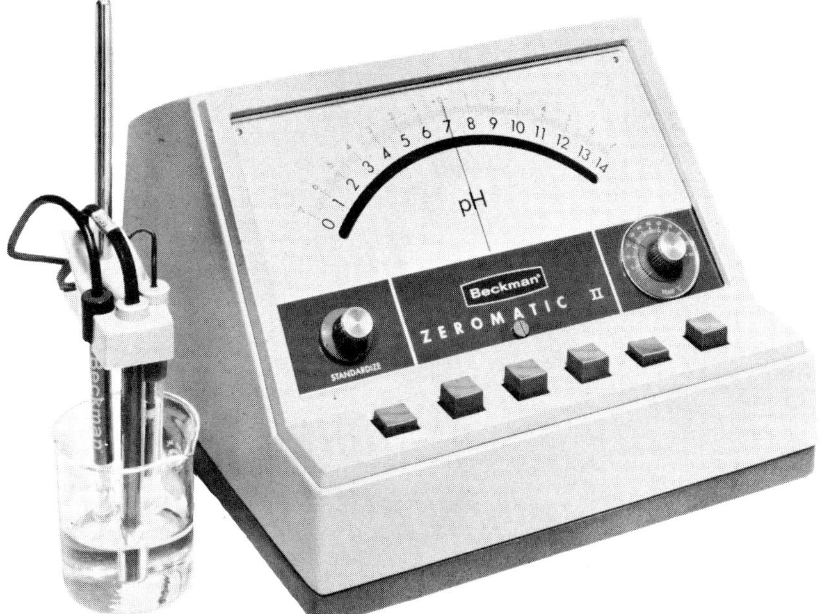
pH
Beckman
ZEROMATIC II
STANDARDIZE

Figure 8.2 pH meter. (Courtesy of Beckman Instruments.)

Specially prepared papers containing several indicator dyes are commercially available. The color they turn when dipped into the test solution is compared with a color–pH code on the dispenser. With these papers pHs to within a few tenths of a unit are easily and rapidly determined. If test solutions are highly colored, or if precise pH measurements are needed, commercial pH meters (Figure 8.2) which have electrodes specially made to dip into the test solution are used. Some are engineered for testing very small volumes. These pH meters permit rapid measurements with an accuracy to within a hundredth of a pH unit.

ACID DISSOCIATION CONSTANTS OF WEAK ACIDS. K_a

As we learned in Chapter 6, weak acids are substances that can react with water to produce hydrogen ions, but not in high percentages. If we represent any weak acid by the symbol H—A, the reaction is:

$$\text{(8.6)} \qquad H{-}A + H_2O \rightleftharpoons H_3O^+ + A^-$$

The equilibrium constant for this reaction is given by

$$\text{(8.7)} \qquad K = \frac{[H_3O^+][A^-]}{[H{-}A][H_2O]} \quad \text{(all concentrations in moles per liter)}$$

Since H—A is a weak acid and since the concentration of water is so much larger than that of H—A, reaction 8.6 leaves $[H_2O]$ essentially unchanged. Therefore in equation 8.7 $[H_2O]$ is a constant, and it may be combined with the constant K.

Table 8.3 Dissociation Constants of Acids and Bases at 25°C, K_a and K_b Values

Compound	Reaction	Dissociation Constant
Acids		K_a
Sulfuric acid	$H_2SO_4 \rightleftharpoons H^+ + HSO_4^-$	K_1 very large
	$HSO_4^- \rightleftharpoons H^+ + SO_4^{2-}$	$K_2 = 1.20 \times 10^{-2}$
Phosphoric acid	$H_3PO_4 \rightleftharpoons H^+ + H_2PO_4^-$	$K_1 = 7.52 \times 10^{-3}$
	$H_2PO_4^- \rightleftharpoons H^+ + HPO_4^{2-}$	$K_2 = 6.23 \times 10^{-8}$
	$HPO_4^- \rightleftharpoons H^+ + PO_4^{3-}$	$K_3 = 2.2 \times 10^{-13}$
Hydrofluoric acid	$HF \rightleftharpoons H^+ + F^-$	3.53×10^{-4}
Carbonic acid	$H_2CO_3 \rightleftharpoons H^+ + HCO_3^-$	$K_1 = 4.30 \times 10^{-7}$
	$HCO_3^- \rightleftharpoons H^+ + CO_3^{2-}$	$K_2 = 5.61 \times 10^{-11}$
Hydrocyanic acid	$HCN \rightleftharpoons H^+ + CN^-$	4.93×10^{-10}
Acetic acid	$CH_3\overset{O}{\overset{\|\|}{C}}—O—H \rightleftharpoons H^+ + CH_3\overset{O}{\overset{\|\|}{C}}—O^-$	1.75×10^{-5}
Formic acid	$H—\overset{O}{\overset{\|\|}{C}}—O—H \rightleftharpoons H^+ + H—\overset{O}{\overset{\|\|}{C}}—O^-$	1.8×10^{-4}
	Miscellaneous	
Oxalic acid (occurs in rhubarb)		$K_1 = 5.9 \times 10^{-2}$
Lactic acid (occurs in sour milk)		1.4×10^{-4}
Citric acid (occurs in citrus fruits)		$K_1 = 8.7 \times 10^{-4}$
Carbolic acid (or phenol)		1.3×10^{-10}
Bases		K_b
Ammonia	$NH_3 + H_2O \rightleftharpoons NH_4^+ + OH^-$	1.77×10^{-5}
Phosphate ion	$PO_4^{3-} + H_2O \rightleftharpoons HPO_4^{2-} + OH^-$	4.5×10^{-2}
Monohydrogen phosphate ion	$HPO_4^{2-} + H_2O \rightleftharpoons H_2PO_4^- + OH^-$	1.6×10^{-7}
Dihydrogen phosphate ion	$H_2PO_4^- + H_2O \rightleftharpoons H_3PO_4 + OH^-$	1.3×10^{-12}
Carbonate ion	$CO_3^{2-} + H_2O \rightleftharpoons HCO_3^- + OH^-$	1.8×10^{-4}
Bicarbonate ion	$HCO_3^- + H_2O \rightleftharpoons H_2CO_3 + OH^-$	2.3×10^{-8}

We define the new constant, the *acid dissociation constant,* abbreviated K_a.

$$K_a = \frac{[H_3O^+][A^-]}{[H—A]} \equiv \frac{[H^+][A^-]}{[H—A]} \tag{8.8}$$

Table 8.3 shows K_a values for several weak acids. The larger the value of K_a, the stronger the acid, for K_a can be large only if the terms in the numerator of equation 8.8 are large. The K_a values are temperature-dependent and in most (but not all) acids increase with temperature. Once the K_a for a weak acid is known, it can be used for calculating $[H^+]$ and pH of aqueous solutions containing the acid.

Exercise 8.7 What is the $[H^+]$ and pH of a $0.1M$ solution of acetic acid, $H—C_2H_3O_2$ ($CH_3\overset{O}{\overset{\|\|}{C}}—O—H$)? Let $x = [H^+]$; therefore $x = [C_2H_3O_2^-]$. Since for each hydrogen ion formed one acetic acid molecule must have disappeared, the concentration of acetic acid left when equilibrium is reached is $0.1 - x$:

$$H{-}C_2H_3O_2 \rightleftharpoons H^+ + C_2H_3O_2^-$$

Initial concentration:	$0.1M$	0	0
Equilibrium concentration:	$0.1 - x$	x	x

$$K_a = \frac{[H^+][C_2H_3O_2^-]}{[H{-}C_2H_3O_2]}$$

$$= \frac{(x)(x)}{0.1 - x} = 1.75 \times 10^{-5}$$

Hence $$x^2 = (0.1)(1.75 \times 10^{-5}) - (1.75 \times 10^{-5})(x)$$

All that remains is to solve for x. The last equation is a quadratic and can be solved without too much difficulty, but let us note now a simplifying assumption. Chemists have learned this tactic from experience and, although not immediately obvious to the beginner, with hindsight it appears to be. The point is that x will be a very small number in comparison with the initial concentration of weak acid, $0.1M$. (The acid is a *weak* acid.) To subtract it from 0.1 as we did will not cause much shrinkage in 0.1. The simplifying trick is to omit this subtraction, that is, to assume that $0.1 - x$ will be very close to 0.1. (The smaller the K_a value, that is, the weaker the acid, the safer this assumption is. Why?)[4] Thus, instead of

$$K_a = \frac{(x)(x)}{0.1 - x}$$

we may substitute

$$K_a \cong \frac{(x)(x)}{0.1} = 1.75 \times 10^{-5}$$

Or $$x^2 \cong 1.75 \times 10^{-6}$$

Ans. $[H^+] = x = 1.33 \times 10^{-3}$ mole/liter (or 0.00133). Note that this value is much smaller than the initial concentration of weak acid; thus our assumption is justified. To find the pH of the solution, we have

$$pH = -(-3 + \log 1.33) = -(-3 + 0.12)$$

Ans. pH = 2.88.

With this as an example, calculate the hydrogen-ion concentrations (in moles per liter) and the pH of solutions of the following concentrations.

(*a*) $0.3M$ formic acid (*b*) $0.5M$ acetic acid
(*c*) $0.6M$ acetic acid (*d*) $1M$ acetic acid
(*e*) Referring back to the example, calculate $[H^+]$ for part *d* without making the simplifying assumptions and compare the two answers.

[4] We may also neglect the negligible contribution made by the ionization of the solvent water to the total hydrogen-ion concentration. With exceptionally weak acids or with very dilute solutions of acids, however, it must be considered.

BASE DISSOCIATION CONSTANTS OF WEAK BASES. K_b

Weak bases, in water, may be handled just as weak acids are. Analogous to K_a, the *base dissociation constant* K_b can be defined in terms of the following equilibrium:

(8.9) $$B + H_2O \rightleftharpoons BH^+ + OH^-$$

Where B is the weak base and BH^+ is its conjugate acid. For example,

$$:NH_3 + H_2O \rightleftharpoons NH_4^+ + OH^-$$

(8.10) $$K_b = \frac{[BH^+][OH^-]}{[B]}$$

For ammonia,

$$K_b = \frac{[NH_4^+][OH^-]}{[NH_3]} = 1.79 \times 10^{-5} \qquad \text{(at 25°C)}$$

Table 8.3 includes K_b values for a few weak bases. It is noteworthy that the carbonate ion is a stronger base than ammonia.

Several common acids, including many in Table 8.3, are *diprotic,* meaning that each of their molecules can make available two protons, two hydrogen ions. Phosphoric acid looks as though it is tripotic—three hydrogens in H_3PO_4—but the ionization of the third hydrogen has such a low K_a value (about 10^{-13}) that H_3PO_4 is really only diprotic. The K_a values are different for each proton dissociated, and K_a for the first proton (designated K_1) is always greater than for the second proton (K_2). We have learned that the weaker an acid, the stronger its conjugate base. The K_a values for acids and the K_b values for their conjugate bases show this.

Exercise 8.8 Suggest an explanation for the fact that K_2 is less than K_1.

Exercise 8.9 Referring to Table 8.3, organize several examples that illustrate the statement that the weaker the acid, the stronger its conjugate base. For example, HPO_4^{2-} is a very weak acid, $K_a = 2.2 \times 10^{-13}$. Its conjugate base, PO_4^{3-}, is relatively strong, $K_b = 4.5 \times 10^{-2}$.

ACID-BASE TITRATIONS

The pH is a measure of the actual concentration of hydrogen ions in whatever dynamic equilibrium is present in an aqueous solution. But it does not tell us anything about what we may call the *total acidity* or the *total neutralizing capacity.* Weak acids, for example, are labeled "weak" in relation to the weak base, water. They do not give up their potential protons to water molecules to any great extent. In a $1M$ acetic acid solution, for example, only slightly more than 0.4% of all acetic acid molecules are ionized at any one time, and the pH is about 2.4. Yet the concentration of acetic acid is fairly high, 1 mole/liter.

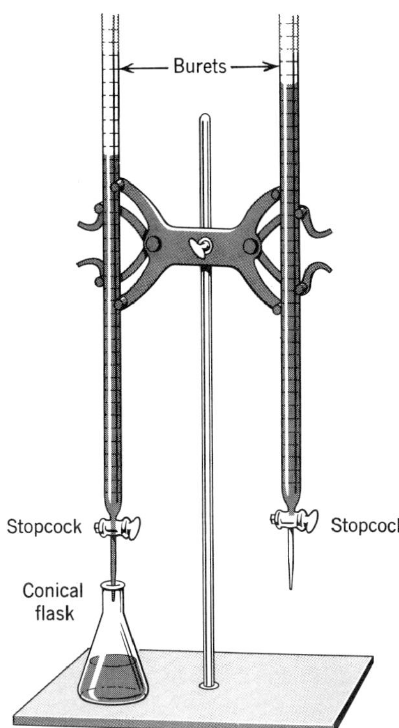
Burets
Stopcock
Stopcock
Conical
flask

Figure 8.3 A typical titration assembly.

If we make available to acetic molecules a base stronger than water, the percentage of ionization may be much larger. The hydroxide ion, for example, is a strong base—the strongest, in fact, that can exist in water.[5] It is used to measure the total neutralizing capacity of a solution containing an acid. This is done by carefully measuring the amount of base (OH^- ions) needed to react with all ionizable hydrogen ions in the solution. (We define an ionizable hydrogen ion as one that can be captured by a hydroxide ion to form water. The ionizable hydrogen ion may be in the form of a hydronium ion, or it may be held by an un-ionized molecule of the weak acid. In acetic acid, CH_3CO_2H, for example, we have noted that only the hydrogen attached to the oxygen is ionizable. Those attached to carbon are too strongly held.) We may, of course, choose instead to measure total basicity by determining the amount of acid needed to react with all available hydroxide ions in the test solution.

The most common way to conduct a measurement of this type is by *titration*. The apparatus is shown in Figure 8.3, and the general procedure is described

[5] Any stronger base, for example, NH_2^- (amide ion), will pull a proton from a water molecule, become neutralized, and leave a hydroxide ion in its place:

$$\underset{\text{Stronger base}}{NH_2^-} + \underset{\text{Stronger acid}}{H{-}OH} \longrightarrow \underset{\text{Weaker acid}}{NH_3} + \underset{\text{Weaker base}}{OH^-}$$

by explaining how the concentration of base present in an unknown solution is determined. A solution of a strong acid, for example, hydrochloric acid, is carefully prepared so that its concentration is accurately known. Whenever the concentration of a solution is accurately known, is called a *standard solution*. Whatever the operations used to know its concentration accurately, they are designated by the verb "to standardize." Two burets are now filled, one with the "unknown" solution of the base, the other with standard acid. A carefully measured volume of the unknown is drained into a conical flask, an indicator is added (more about that later), and the standard acid is slowly added and mixed until the indicator color just barely changes.

An indicator is selected so that it changes color at the pH value of the solution of the salt being formed during the titration. Solutions of the salts produced when strong acids are being titrated with strong bases have a pH of 7. Phenolphthalein comes close enough to changing color at this value, and the change is so dramatic, colorless to pink, that it is the indicator of choice for titrating strong acids and strong bases.

The precise point at which all available hydrogen ions from the acid have been neutralized is called the *equivalence point* of the titration. Usually, when this point is close at hand, the analyst adds the reagent very slowly, drop by drop (or a fraction of a drop), until the indicator undergoes its characteristic color change. When the color change occurs, we say that the *end point* has been reached because that is the end of the titration. Only if the indicator gives this color change at the pH of the equivalence point, however, will end point correspond to equivalence point. In our discussion it is assumed that such an indicator has been selected.

Normal Solutions. Gram-Equivalent Weights. "Equivalents." For purposes of acid-base titrations, chemists prefer not to deal with concentrations expressed in terms of molarity. Some acids, for example, are diprotic. A mole of sulfuric acid, H_2SO_4, makes available two moles of hydrogen ion, and it is more convenient to have a way of expressing concentration in terms of ionizable hydrogen ion rather than in terms of the particular acid. For this purpose the concept of *normality* has been invented as a substitute for molarity.

By definition, a one normal solution of an acid contains one mole of available hydrogen ion per liter. For sulfuric acid (formula weight, 98), 98 g in 1 liter of solution would be labeled $1M$ H_2SO_4. But since such a concentration makes available two moles of hydrogen ion per liter, it is also labeled $2N$ H_2SO_4, where N stands for normal, which in turn means *equivalents per liter*. ("Equivalents" will soon be defined.)

The weight in grams of an acid that makes available one mole of hydrogen ion is called its *gram-equivalent weight*. The gram-equivalent weight of sulfuric acid is $\frac{98}{2} = 49$ g H_2SO_4. By definition,

$$\text{(8.11)} \quad \text{Gram-equivalent weight of an acid} = \frac{\text{gram-formula weight of the acid}}{\text{number of ionizable } H^+ \text{ ions per formula unit}}$$

For example,

Acid	Gram-Formula Weight	Gram-Equivalent Weight
HCl	36.5 g	36.5 g
HNO_3	63 g	63 g
H_2SO_4	98 g	49 g (98 ÷ 2)

If we are dealing with bases, the gram-equivalent weight of a base is defined by the equation

$$\text{(8.12)} \quad \text{Gram-equivalent weight of a base} = \frac{\text{gram-formula weight of the base}}{\text{number of available OH}^- \text{ ions per formula unit}}$$

In a more general definition the denominator would be replaced by "number of available proton-binding sites per formula unit." For example,

Base	Gram-Formula Weight	Gram-Equivalent Weight
NaOH	40 g	40 g
KOH	56 g	56 g
$Ca(OH)_2$	74 g	37 g (74 ÷ 2)

Exercise 8.10 (*a*) Oxalic acid is a diprotic acid of molecular formula $H_2C_2O_4$. Calculate its gram-equivalent weight. *Ans.* 45 g.

(*b*) Oxalic acid is normally purchased as its dihydrate, $H_2C_2O_4 \cdot 2H_2O$. It is a solid, and if oxalic acid is weighed in this form, what weight of the *dihydrate* would furnish a mole of ionizable hydrogen ions?

In working with gram-formula weights we adopted the abbreviation *mole*. Similarly, we use the noun *equivalent* as an abbreviation for 1 gram-equivalent weight. Thus, 1 equivalent of sodium hydroxide weighs 40 g. One equivalent of sulfuric acid weighs 49 g. An equivalent is therefore a unit for amount of substance. One equivalent of an acid will provide a mole of hydrogen ions; one equivalent of a base will neutralize a mole of hydrogen ions. The defining equation is

$$\text{(8.13)} \quad \text{Number of equivalents} = \frac{\text{number of grams of substance taken}}{\text{gram-equivalent weight of substance}}$$

Exercise 8.11 How many equivalents are there present in each of the following samples.

(*a*) 20 g NaOH (*b*) 28 g KOH
(*c*) 18.5 g $Ca(OH)_2$ (*d*) 4.9 g H_2SO_4
(*e*) 15 g HNO_3 (*f*) 2 g HCl

Ans. (*a*) 0.5 equivalent, (*b*) 0.5 equivalent, (*c*) 0.5 equivalent.

The normality of a solution is the ratio of the number of equivalents of the solute per liter of solution. A one normal solution (1*N*) of an acid contains one equivalent of that acid per liter, or one mole of ionizable hydrogen ion per liter of solution. The defining equation is

$$\text{(8.14)} \qquad \text{Normality } (N) = \frac{\text{number of equivalents}}{\text{number of liters of solution}}$$

By cross-multiplying, we derive a useful variant of equation 8.14:

$$\text{(8.15)} \qquad \text{Normality } (N) \times \text{volume } (V, \text{ in liters}) = \text{number of equivalents}$$

Exercise 8.12 (*a*) What would be the normalities of each of the solutions if the amounts of substances of exercise 8.11 were dissolved in 4 liters of solution? *Ans.* (*a*) 0.125N.

(*b*) How much sodium hydroxide would you weigh in order to prepare 1 liter of a 0.100N solution? Describe what you would do to make this solution after you have weighed this amount of sodium hydroxide. *Ans.* 4 g.

(*c*) A bottle on the laboratory shelf bore the label 0.168N HCl. Would the label have to be changed if half of the solution were poured out? Explain in words exactly what that label means.

(*d*) Suppose that 500 cc of a solution labeled 0.10N HCl were diluted with pure water until the final volume was 1 liter. Would the label have to be changed? If so, how? If not, why not?

In acid-base titrations the equivalence point is reached when the equivalents of base exactly match the number of equivalents of acid. In equation form this statement becomes

$$\text{(8.16)} \qquad \text{Number of equivalents of acid} = \text{number of equivalents of base (at the equivalence point)}$$

But from equation 8.15,

$$\text{Number of equivalents of acid} = \text{normality of acid } (N_a) \times \text{volume of acid } (V_a)$$
$$\text{Number of equivalents of base} = \text{normality of base } (N_b) \times \text{volume of base } (V_b)$$

Substituting these into 8.16, at the equivalence point of a titration we have

$$\text{(8.17)} \qquad N_a \times V_a = N_b \times V_b$$

As long as the units for volume are the same for both V_a and V_b, they need not be in liters or any other particular unit; the units will cancel. Ordinarily, units of milliliters are used for volumes.

Exercise 8.13 Calculate the normality of a hydrochloric acid solution if 40 ml required 50 ml of 0.10N sodium hydroxide to neutralize it.

Solution. Collect the known facts as follows: $N_a = ?$, $V_a = 40$ ml, $N_b = 0.10$, $V_b = 50$ ml. Therefore (equation 8.17)

$$N_a \times 40 = 0.10 \times 50$$
$$N_a = 0.13N \qquad \text{(answer)}$$

(*a*) Calculate the normality of a sulfuric acid solution if 36 ml required 38 ml of $0.112N$ potassium hydroxide to neutralize it. *Ans.* $0.118N$.
(*b*) What volume of $0.855N$ nitric acid would be needed to neutralize 36 ml of a $0.862N$ sodium hydroxide solution?
(*c*) An analyst had available a supply of standard $0.101N$ hydrochloric acid. He used it to standardize a freshly prepared solution of sodium hydroxide. He found in successive trials that (1) a 36.36-ml sample of the standard acid required 35.47 ml of base; (2) a 48.55-ml sample of standard acid required 49.66 ml of base. What are the calculated normalities of the base, and what is the average?

SOLUBILITY OF SLIGHTLY SOLUBLE SALTS

In Chapter 6 we studied general solubility rules. Because solubilities of individual compounds can be measured, it is much more satisfying to have quantitative relations.

If an undissolved and slightly soluble salt is present and in contact with a saturated solution of the salt, a dynamic equilibrium exists. For example, with silver chloride,

$$AgCl(s) \rightleftharpoons Ag^+(aq) + Cl^-(aq) \qquad \text{The ions are assumed to be hydrated.}$$

As with any equilibrium, we may write the following equilibrium constant:

$$K = \frac{[Ag^+][Cl^-]}{[AgCl(s)]}$$

But the concentration of the solid silver chloride may be taken as constant, at least for all practical purposes, for this reason. As long as the solution is saturated and at least some undissolved silver chloride is present, it matters not how much is lying on the bottom of the solution. The concentrations of its ions in the dissolved state are not affected. Hence the *effect* of the solid is a constant regardless of its amount. We need not know what that effect is, quantitatively, for as long as it is constant, we may combine it with the equilibrium constant K and define a new constant, the *solubility product constant* K_{sp}. For silver chloride

$$K_{sp} = [Ag^+][Cl^-] = 1.56 \times 10^{-10} \qquad \text{(at 25°C)}$$

In general, for a generalized salt ionizing as follows,

$$M_mX_x(s) \rightleftharpoons mM^{(+)}(aq) + xX^{(-)}(aq)$$
(8.18) $$K_{sp} = [M]^m[X]^x$$

The following examples, with their solubility products, illustrate equation 8.18:

$$BaF_2: \quad K_{sp} = [Ba^{2+}][F^-]^2 = 1.7 \times 10^{-6} \qquad (18°C)$$
$$Cu_2S: \quad K_{sp} = [Cu^+]^2[S^{2-}] = 2 \times 10^{-47} \qquad (18°C)$$
$$Fe(OH)_3: \quad K_{sp} = [Fe^{3+}][OH^-]^3 = 1.1 \times 10^{-36} \qquad (18°C)$$

Table 8.4 Solubility Product for Several Salts*

Compound	Formula	K_{sp}
Barium carbonate	$BaCO_3$	8.1×10^{-9} (18°C)
Barium sulfate	$BaSO_4$	1.1×10^{-10}
Calcium fluoride	CaF_2	3.4×10^{-11} (18°C)
Copper(II) sulfide	CuS	8.5×10^{-45} (18°C)
Silver chloride	AgCl	1.56×10^{-10}
Silver bromide	AgBr	7.7×10^{-13}
Silver iodide	AgI	1.5×10^{-16}

* Unless specified otherwise, temperatures are at 25°C.

Table 8.4 lists solubility products for several common salts.

A solubility product is determined by measuring the actual solubility of the salt in question. For example, at 25°C the solubility of silver chloride is 0.00179 g/liter, and its gram-formula weight is 143.3 g/mole. The solubility therefore translates into 0.00179 g/liter $\times$ 1 mole/143.3 g $= 1.25 \times 10^{-5}$ mole/liter which is the solubility of silver chloride at 25°C. Since each formula unit of AgCl that goes into solution makes available one formula unit of Ag^+ and one of Cl^-, we have:

$$[Ag^+] = 1.25 \times 10^{-5} \text{ mole/liter}$$
$$[Cl^-] = 1.25 \times 10^{-5} \text{ mole/liter}$$
$$K_{sp} = [Ag^+][Cl^-] = (1.25 \times 10^{-5})(1.25 \times 10^{-5}) = 1.56 \times 10^{-10}$$

One of the principal uses of solubility products is in calculating how much of a slightly soluble salt can remain in solution if the concentration of but one of its ions is changed by the addition of some other salt involving that ion. More specifically, we might question how much silver ion can remain in solution if sodium chloride is present to the extent of 0.1 mole/liter. We find that the concentration of chloride ion is 0.1 mole/liter (Why?) and therefore

$$K_{sp} = 1.56 \times 10^{-10} = [Ag^+](0.1)$$

or

$$[Ag^+] = 1.56 \times 10^{-11} \text{ mole/liter}$$

In other words, whereas silver ion is soluble as silver chloride in pure water to the extent of 1.25×10^{-5} mole/liter (see above), in the presence of a large excess of one of its common ions, Cl^-, its solubility is lowered dramatically. This suppression of an ionization or of the solubility of a poorly soluble or a slightly ionized substance by one of its common ions is called the *common-ion effect.*

Exercise 8.14 At 30°C the solubility product of calcium sulfate, $CaSO_4$, is 2.38×10^{-4}. What is the solubility of calcium sulfate in water in grams per liter? *Ans.* 2.1 g/liter.

Exercise 8.15 The solubility of calcium oxalate, CaC_2O_4, in water at 25°C is 6.8×10^{-4} g per 100 cc of solution. Calculate its solubility product. (The oxalate ion is $C_2O_4^{2-}$.) *Ans.* 2.8×10^{-9}.

Exercise 8.16 The K_{sp} for silver hydroxide, AgOH, at 20°C is 1.52×10^{-8}. What is the maximum pH a solution can have without silver ions being precipitated? *Ans.* 10.09. (*Hint.* From the K_{sp} value, determine OH^- of a saturated solution of silver hydroxide. Then convert this into pH using K_w of water.)

How many drops of $1M$ NaOH added to a liter of water would make the pH change from 7 to 10? Assume that each drop has a volume of about 0.05 cc. Assume, furthermore, that the added drops produce no appreciable change in the volume of the water. (The purpose of this calculation is to show how a very small amount of base can cause a large change in the pH of water and thus, in this example of silver hydroxide, make it precipitate from an initially saturated solution.)

HYDROLYSIS OF SALTS

There are many salts which, from an inspection of their formulas, cannot in any obvious way provide either hydrogen ions or hydroxide ions in water, but which in solution are still acidic or basic. Sodium carbonate, Na_2CO_3, for example, produces a basic solution in water, basic in the traditional sense of turning litmus blue, tasting brackish, etc.; the solution must have more OH^- ions than H_3O^+ ions. Solutions of copper sulfate, $CuSO_4$, or of ammonium chloride, NH_4Cl, test acidic, and they must have more hydronium ions than hydroxide ions. The question is where the "extra" ions come from.

To answer this question we must examine all ions that can be present in these solutions and all possible behaviors of these ions toward each other. Consider a solution of sodium acetate, $NaC_2H_3O_2$. As a crystalline salt it consists of an orderly aggregation of sodium ions, Na^+, and acetate ions, $C_2H_3O_2^-$. When this salt is dissolved in water, these ions separate (and become solvated). *The solution also tests basic,* which means that $[OH^-] > [H_3O^+]$. The only source of these ions is water. The source of the *imbalance* is the salt, or at least one of the ions of the salt. If, somehow, some few of the already low concentration of hydronium ions in the original water became neutralized by something besides hydroxide ions, the OH^- ions would be in excess and the solution would have to test basic. Therefore we reexamine the ions of the salt. Sodium ions, being positively charged, could not take like-charged hydrogen ions H^+ from H_3O^+. Could acetate ions do this? To answer this question, we can use the following train of logic.

1. *If* acetate ions tied up hydrogen ions H^+ (from H_3O^+) what would the product be?
   Ans. Slightly ionized molecules of acetic acid, H $C_2H_3O_2$.
2. *Could* acetate ions do this, even to a slight extent?
   Ans. Yes, because we know qualitatively that acetic acid is a *weak* acid.
3. What does this have to do with it?
   Ans. If acetic acid is a weak acid, it is by definition capable of retaining the proton. Therefore if it can acquire a proton, it will. (In fact, in a $1M$ solution of acetic acid only about 0.4% of the acetic acid molecules are ionized.)

Thus from our knowledge of which acids are weak (all of them[6] *except* HCl, HBr,

[6] Referring only to the first ionization constant.

HI, HNO_3, H_2SO_4, $HClO_4$, and H_3PO_4), we deduce that the conjugate base of any weak acid will be a relatively strong base—capable of accepting protons from the hydronium ions formed by ionization of water and leaving OH^- in slight excess.

For sodium acetate we deduced that acetate ions would reduce $[H_3O^+]$, tending to leave OH^- ions (from water) in slight excess. But we have yet to ask whether the sodium ions tie up OH^- ions, offsetting this effect. To answer this question we employ the same logic.

1. *If* sodium ions tied up hydroxide ions, what would the product be?
   Ans. Un-ionized "molecules" of sodium hydroxide, NaOH.
2. *Could* sodium ions do this, even to a slight extent?
   Ans. No, at least not in the way acetate ions can bind hydrogen ions. We know this because sodium hydroxide is a strong base, and by definition sodium ions readily release hydroxide ions.

On balance, therefore, in a solution of sodium acetate, the acetate ions neutralize some of the hydronium ions (from the ionization of water), leaving hydroxide ions (also from the ionization of water) in slight excess. The solution tests basic. Figure 8.4 summarizes the effects in a diagrammatic way, Figure 8.5 shows how it can be inferred that a solution of ammonium chloride in water should test acidic, and Figure 8.6 deals with sodium chloride, which forms a neutral solution. Thus on the basis of qualitative knowledge concerning strong and weak acids and bases, we may deduce the effect on pH when a particular salt is added to water. If any ion from the salt interacts with water to change its pH, *hydrolysis* is said to occur (Greek: *hydro,* water; *lysis,* loosening).

In general, if a salt is derived from a weak acid and a strong base, its solution in water will be slightly basic. If a salt is derived from a weak base and a strong acid, its solution will test slightly acidic. But if the salt is made from an acid and a base

---

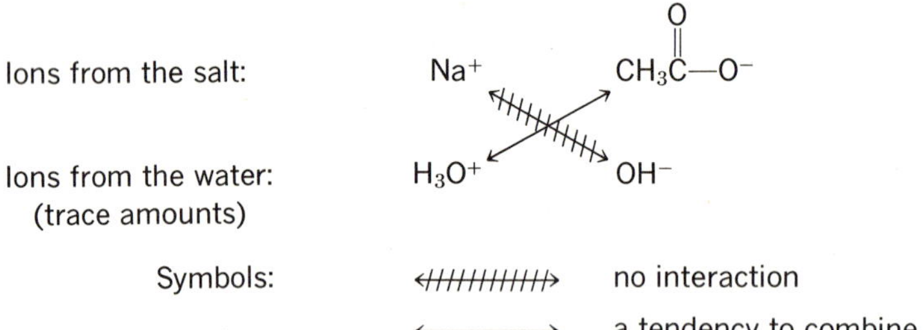
Ions from the salt:
Na+
CH3C—O−
O
Ions from the water:
(trace amounts)
H3O+
OH−
Symbols:
no interaction
a tendency to combine

Figure 8.4 The hydrolysis of sodium acetate. *Inferences.* NaOH is a strong base, with no tendency for Na^+ and OH^- to combine. Acetic acid is a *weak* acid; therefore some protons will transfer from H_3O^+ to $CH_3\overset{\overset{O}{\|}}{C}O^-$.

Conclusion. The solution is slightly basic because the transfer of a few protons from H_3O^+ ions to $CH_3—\overset{\overset{O}{\|}}{C}O^-$ ions leaves some of the OH^- ions from water in slight excess.

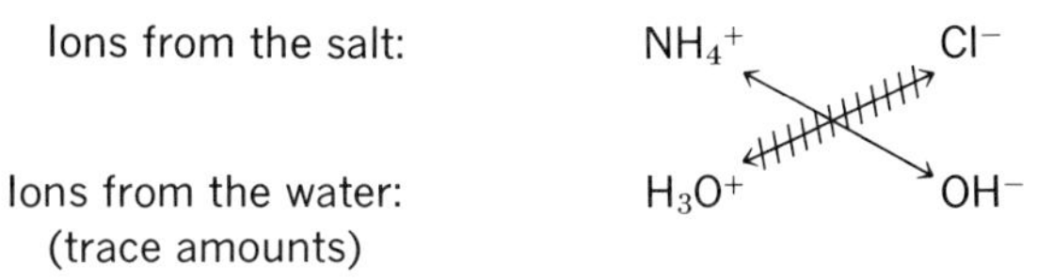

Ions from the salt: NH_4^+ Cl^-

Ions from the water: H_3O^+ OH^-
(trace amounts)

Figure 8.5 The hydrolysis of ammonium chloride. *Inferences.* HCl is a *strong* acid; there is thus no tendency for protons to be removed from H_3O^+ ions by Cl^- ions. NH_4^+ is a poorer proton binder than OH^-; there is some tendency for the following reaction to occur: $NH_4^+ + OH^- \longrightarrow NH_3 + H{-}OH$.

Conclusions. If some OH^- ions are neutralized by NH_4^+, a few H_3O^+ ions from the ionization of water will be left in excess. The solution will test acidic.

of matched strengths (both strong or both equally weak), its solution will test neutral. A more satisfactory way of judging these tendencies is in terms of the *hydrolysis constants* of salts.

Hydrolysis Constants. The reaction of acetate ion with water may be written as[7]

$$C_2H_3O_2^- + H{-}OH \rightleftharpoons H{-}C_2H_3O_2 + OH^- \tag{8.19}$$

The equilibrium constant for this reaction is expressed by

$$K_{eq} = \frac{[H{-}C_2H_3O_2][OH^-]}{[C_2H_3O_2^-][H{-}OH]} \tag{8.20}$$

But as in many previous solutions, the concentration of water is so slightly affected by the reaction that it is essentially a constant, and it may be combined with the equilibrium constant to make a new constant, the hydrolysis constant K_h:

$$K_h = \frac{[H{-}C_2H_3O_2][OH^-]}{[C_2H_3O_2^-]} = K_{eq}[H{-}OH] \tag{8.21}$$

We next perform a manipulation, discovered long ago, that reveals the hydrolysis

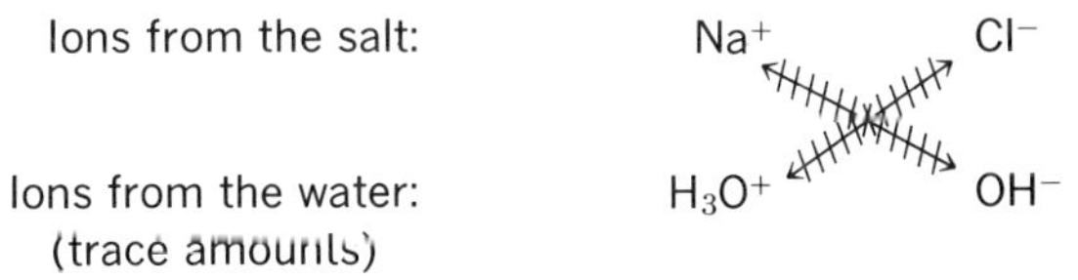

Ions from the salt: Na^+ Cl^-

Ions from the water: H_3O^+ OH^-
(trace amounts)

Figure 8.6 Sodium chloride does not hydrolyze in water. *Inferences.* NaOH is a *strong* base, with no tendency for Na^+ and OH^- ions to combine to leave H_3O^+ ions from the ionization of water in excess. HCl is a *strong* acid, with no tendency for H_3O^+ and Cl^- to react to reduce the concentration of H_3O^+ and leave OH^- ions in slight excess.

Conclusion. Since neither of the ions from the slight ionization of water is neutralized by the ions of the dissolved salt, they remain present in equal concentrations. The solution is therefore neutral.

[7] Note that this approach is slightly different. It considers the excess OH^- to develop by a reaction of acetate ion with a water molecule rather than with a hydronium ion produced by the ionization of a water molecule. The outcome is the same either way.

constant to be a ratio of two other constants we already know. The trick is to multiply both numerator and denominator in equation 8.21 by the same thing, $[H^+]$:

$$K_h = \frac{[H{-}C_2H_3O_2][OH^-][H^+]}{[C_2H_3O_2^-][H^+]} \tag{8.22}$$

But $[OH^-][H^+] = K_w$, which makes equation 8.22 read

$$K_h = \frac{[H{-}C_2H_3O_2]K_w}{[C_2H_3O_2^-][H^+]} \tag{8.23}$$

The remainder of the terms in equation 8.23 make up the K_a value for the conjugate acid of the acetate ion, namely acetic acid,

$$K_a = \frac{[C_2H_3O_2^-][H^+]}{[H{-}C_2H_3O_2]}$$

$$\frac{1}{K_a} = \frac{[H{-}C_2H_3O_2]}{[C_2H_3O_2^-][H^+]} \tag{8.24}$$

By substituting equation 8.24 into 8.23, we have the very simple relation

$$K_h = \frac{K_w}{K_a} \tag{8.25}$$

In the specific case of sodium acetate, K_a for acetic acid is 1.75×10^{-5}. Therefore for sodium acetate

$$K_h = \frac{10^{-14}}{1.75 \times 10^{-5}} = 5.72 \times 10^{-10} \tag{8.26}$$

Using the result in equation 8.26 together with the reaction represented by equation 8.19, we can now calculate the pH of a $1M$ solution of sodium acetate. To do this, we make a number of justified assumptions. For example, in the reaction for the hydrolysis, equation 8.19,

$$C_2H_3O_2^- + H{-}OH \rightleftharpoons H{-}C_2H_3O_2 + OH^-$$

we may assume that $[OH^-] \approx [H{-}C_2H_3O_2]$ for the following reason. For every molecule of un-ionized acetic acid produced, one of OH^- is also made; so much is obvious. The assumption is that we may neglect the concentration of OH^- produced from the self-ionization of water because it is supposedly very much smaller than that produced by the hydrolysis.

Next we assume that in relation to the initial concentration of acetate ion only a trace will actually disappear because of hydrolysis:

$$[C_2H_3O_2^-]_{\text{final}} \approx 1 \text{ mole/liter} \qquad \text{(the initial concentration of ion)}$$

Thus we may let $x = [OH^-] \cong [H{-}C_2H_3O_2]$ and

$$K_h = \frac{[H{-}C_2H_3O_2][OH^-]}{[C_2H_3O_2^-]} = 5.72 \times 10^{-10} \cong \frac{(x)(x)}{1}$$

$$x^2 \cong 5.72 \times 10^{-10}$$
$$x \cong 2.39 \times 10^{-5} \frac{\text{mole}}{\text{liter}} = [\text{OH}^-]$$

This[8] means that

$$[\text{H}^+] \cong \frac{10^{-14}}{2.39 \times 10^{-5}} = 4.18 \times 10^{-10} \frac{\text{mole}}{\text{liter}}$$

and

$$\text{pH} = -\log[\text{H}^+] = -(-10 + \log 4.18) = -(-10 + 0.62)$$
$$\text{pH} = 9.38$$

The pH value, we note, is on the basic side.

Exercise 8.17 Calculate the pH of a $0.1M$ solution of sodium acetate. *Ans.* 8.88.
Exercise 8.18 Calculate the pH of a $0.1M$ solution of ammonium chloride, using $K_h = K_w/K_b$ and the K_b for ammonia, 1.79×10^{-5}.

BUFFERS

The principles surrounding the hydrolysis of salts find their most important application in the phenomenon of buffer action. Aqueous solutions in the body have characteristic pHs. The pH of blood, as measured at room temperature, is normally within the range 7.3 to 7.5. Below a pH of 7.0 and above a pH of 7.9, death usually occurs. In a calculation in Exercise 8.16 we saw how a trace amount of added base can change the pH of water from 7 to 10. Life therefore hinges on the blood's ability to control a very small concentration of hydrogen ions within extremely narrow limits. That the blood can do this is all the more remarkable in view of the fact that many normal body reactions produce acids. In this section we study how the blood can perform this important task.

A solution that contains significant concentrations of both a weak acid and one if its salts is an example of a *buffer solution*. Similarly, a solution having appreciable concentrations of both a weak base and one of its salts is also a buffer solution. By absorbing and neutralizing small amounts of added acid or base, these solutes buffer such a solution against changes in pH. They do not necessarily keep the pH at 7; they act to hold the pH constant, whatever it happens to be for the solution, by neutralizing whatever extra amounts of acid or base are added.

One of the important buffer systems in blood is the pair HCO_3^-/H_2CO_3, bicarbonate ion/carbonic acid. Carbonic acid is present because of the interaction of

[8] We have confirmed our two assumptions:

1. $[\text{OH}^-]$ contributed by ionization of water ($\approx 10^{-7}$ mole/liter) is much less than that produced by hydrolysis ($\approx 10^{-5}$ mole/liter).

2. Only a trace of initial acetate ion concentration ($1M$) can disappear if only 10^{-5} mole/liter of new $[\text{OH}^-]$ is produced.

dissolved carbon dioxide with plasma water:

$$CO_2 + H_2O \rightleftharpoons H_2CO_3$$

This buffer system functions as follows.

1. One of the pair, the weak acid, is capable of handling hydroxide ions:

$$H_2CO_3 + OH^- \rightleftharpoons HCO_3^- + H_2O$$

2. The other, the negative ion or the conjugate base of the acid, can neutralize the hydrogen ion:

$$HCO_3^- + H^+ \rightleftharpoons H_2CO_3$$ (a weak acid and therefore a relatively good proton binder)

Another buffer system in blood is the pair $HPO_4^{2-}/H_2PO_4^-$, monohydrogen phosphate ion/dihydrogen phosphate ion (the "phosphate buffer"). It functions as follows.

1. The dihydrogen phosphate ion acts as a mild acid capable of neutralizing the hydroxide ion:

$$H_2PO_4^- + OH^- \rightleftharpoons HPO_4^{2-} + H_2O$$

2. The monohydrogen phosphate ion acts as a mild base capable of neutralizing the hydrogen ion:

$$HPO_4^{2-} + H^+ \rightleftharpoons H_2PO_4^-$$

Calculation of the pH That a Buffer Pair Can Maintain. Pairs of solutes acting as buffers do not necessarily keep a solution neutral, for maintaining neutrality is not the defined service of a buffer. Rather, a buffer holds the pH of a solution constant, whatever it may be, by neutralizing added acids or bases. What that pH is can be calculated from the dissociation constant of the acid (or base) and the concentrations of the acid (or base) and the salt used to make the buffer. One common combination that is especially adaptable to introducing such calculations is the acetic acid and sodium acetate buffer system. The question is this. Given initial values of $[H{-}C_2H_3O_2]$ and $[C_2H_3O_2^-]$ plus the K_a value for the acid, what will be the pH of the solution?

We have the equations

$$H{-}C_2H_3O_2 \rightleftharpoons H^+ + C_2H_3O_2^-$$

$$K_a = \frac{[H^+][C_2H_3O_2^-]}{[H{-}C_2H_3O_2]} = 1.75 \times 10^{-5}$$

By rearranging terms in the equation for K_a, we can solve for $[H^+]$:

$$[H^+] = K_a \frac{[H{-}C_2H_3O_2]}{[C_2H_3O_2^-]}$$

We now take advantage of two simplifying assumptions.[9]

[9] These assumptions are not valid if the solutions of the buffer are very dilute.

1. Assume that the concentration of $H—C_2H_3O_2$ has not been changed by (*a*) its slight ionization (it is a weak acid anyway) and (*b*) the trace amount produced by the hydrolysis of the salt.

2. Assume that the concentration of $C_2H_3O_2^-$ is identical with the initial concentration of the salt $NaC_2H_3O_2$ (thus assuming that the salt is 100% ionized and that the concentration of $C_2H_3O_2^-$ is only trivially and negligibly changed by its hydrolysis).

The equation therefore becomes

$$(8.27) \qquad [H^+] = K_a \frac{\text{initial concentration of weak acid}}{\text{initial concentration of its salt}}$$

Thus if we prepare a buffer solution that is $0.1M$ in acetic acid and $0.1M$ in sodium acetate, its hydrogen-ion concentration is calculated to be

$$[H^+] = 1.75 \times 10^{-5} \times \frac{(0.1)}{(0.1)}$$

$$[H^+] = 1.75 \times 10^{-5} \qquad \text{(answer)}$$

and

$$pH = -\log[H^+] = -(-5 + \log 1.75)$$
$$pH = 4.76$$

In other words, a solution that is $0.1M$ in acetic acid and $0.1M$ in sodium acetate will have, at room temperature, a pH of 4.76. Moreover, it will be capable of neutralizing small additional amounts of either an acid or a base and thus keeping a virtually constant pH.

For example, suppose we add 1 cc of $1M$ HCl to a liter of solution buffered as described earlier by the acetic acid/acetate pair. The number of moles of HCl added is 1 mole/liter $\times$ 10^{-3} liter = 10^{-3} mole. If this amount were added to 1 liter of pure water, the molar concentration of acid would be 10^{-3}, and since HCl is a strong acid, the pH would be about 3. Just this small amount of acid lowers the pH of pure water from 7 to 3, a change of four units. But we add the HCl to a buffered solution. Acetate ions neutralize it:

$$C_2H_3O_2^- + H^+ \rightleftarrows H—C_2H_3O_2$$

Thus we have lowered the concentration of acetate ion slightly and raised the concentration of acetic acid by the same amount. Let us assume that all the added hydrogen ions are neutralized. Then the drop in acetate-ion concentration will be 10^{-3} mole/liter, and the rise in concentration of acetic acid will be the same:

$$[C_2H_3O_2^-]_{\text{final}} = 0.1 \frac{\text{mole}}{\text{liter}} - 0.001 \frac{\text{mole}}{\text{liter}} = 0.0999 \frac{\text{mole}}{\text{liter}}$$

$$[H—C_2H_3O_2]_{\text{final}} = 0.1 \frac{\text{mole}}{\text{liter}} + 0.001 \frac{\text{mole}}{\text{liter}} = 0.101 \frac{\text{mole}}{\text{liter}}$$

Therefore, from equation 8.25,

$$[H^+] = (1.75 \times 10^{-5})\left(\frac{0.101}{0.0999}\right) = 1.77 \times 10^{-5}\ \frac{\text{mole}}{\text{liter}} \qquad \text{(answer)}$$

$$\text{pH} = 4.75 \qquad \text{(answer)}$$

The original pH was 4.76. The addition of the acid changes the pH of the buffered solution by only one-hundredth of a pH unit, whereas the same amount of acid added to pure water lowers the pH by four units. When pHs must be held constant, as in living cells and in the bloodstream, buffers are clearly capable of performing an important service.

Exercise 8.19 Calculate the pH of a solution that is $0.2M$ in acetic acid and $0.1M$ in sodium acetate. *Ans.* 4.46.

Exercise 8.20 Calculate the pH of 1 liter of the solution in exercise 8.19 after 1 cc of $1M$ HCl has been added. *Ans.* 4.45.

REFERENCES AND ANNOTATED READING LIST

BOOKS

H. N. Christensen. *pH and Dissociation.* W. B. Saunders Company, Philadelphia, 1963. This short (59 pages) programmed learning unit carries students of biological and medical sciences through the material of this chapter and considerably beyond in easy stages. (Paperback.)

K. B. Morris. *Principles of Chemical Equilibrium.* Reinhold Publishing Corporation, New York, 1965. This book treats equilibria in greater depth and breadth than was possible in this chapter. (Paperback.)

C. A. Vanderwerf. *Acids, Bases, and the Chemistry of the Covalent Bond.* Reinhold Publishing Corporation, New York, 1961.

ARTICLES

F. Szabadvary (translated by R. E. Oesper). "Development of the pH Concept—A Historical Survey." *Journal of Chemical Education,* Vol. 41, 1964, page 105. Recognition of the importance of hydrogen-ion concentration led to convenient ways of expressing it. Somewhat concurrent developments also discussed in this short article are buffers and ways of measuring pH.

Duncan A. MacInnes. "pH." *Scientific American.* January 1951, page 40. This article, by the man chiefly responsible for the development of pH meters, discusses the principles of the meter and describes the physiological importance of pH.

PROBLEMS AND EXERCISES

1. For each of the following values of concentration of hydrogen ion in moles per liter, calculate the concentration of hydroxide ion in the same units and the pH of the solution. (Unless otherwise specified, the temperature is assumed to be room temperature, 24°C.)

	Set 1	Set 2	Set 3
(*a*)	1×10^{-9}	1×10^{-6}	1×10^{-14}
(*b*)	1×10^{-3}	1×10^{-13}	1×10^{-8}
(*c*)	4×10^{-7} ($T = 20°C$)	7×10^{-7} ($T = 30°C$)	3×10^{-5} ($T = 37°C$)
(*d*)	3.67×10^{-11}	6.66×10^{-14}	7.05×10^{-9}
(*e*)	9.82×10^{-1}	0.1	10.52×10^{-3}

2. Calculate the gram-equivalent weight of each of the following:

	Set 1	Set 2	Set 3
(*a*)	HCl	$HClO_4$	HNO_3
(*b*)	$Ba(OH)_2$	H_2SO_3	$Mg(OH)_2$
(*c*)	NH_3	KOH	$NaHSO_4$

3. How many equivalents are present in each of the following samples?

	Set 1	Set 2	Set 3
(*a*)	8.1 g HBr	100 g H_2SO_4	32.7 g HNO_3
(*b*)	0.63 g HNO_3	100 g HNO_3	18.25 g HCl
(*c*)	4.9 g H_2SO_4	0.125 g $Mg(OH)_2$	8 g NaOH
(*d*)	0.128 g HI	34 g NH_3	490 g H_2SO_4

4. How many equivalents of acid are present in each of the following solutions? Note carefully the unit of concentration for each.

	Set 1	Set 2	Set 3
(*a*)	500 ml 0.5N HCl	100 ml 0.7M HCl	150 ml 0.5M H_2SO_4
(*b*)	50 ml 1.2N acetic acid	400 ml 0.1M H_2SO_4	150 ml 0.5N H_2SO_4
(*c*)	100 ml 0.1M H_2SO_4	35 ml 0.112N HNO_3	150 ml 0.5M HNO_3

5. Describe how you would prepare each of the following solutions, specifying the weight (in grams) of the solute you would use and the kind of flask you would select. Assume that all necessary solutes are available in pure form.

	Set 1	Set 2	Set 3
(*a*)	250 ml 0.25N NaOH	500 ml 0.125N KOH	100 ml 0.3N NaOH
(*b*)	100 ml 0.10N H_2SO_4	1000 ml 0.001N $Mg(OH)_2$	250 ml 1.5N H_2SO_4
(*c*)	50 ml 0.005N $Ca(OH)_2$	50 ml 0.01N H_2SO_4	50 ml 0.12N KOH

6. A 500-ml solution of sodium hydroxide is neutralized by adding 10 ml of 12N hydrochloric acid. What is the normality of the sodium hydroxide solution? How much sodium hydroxide, in grams, is dissolved in it?
7. If a 34.76-ml sample of hydrochloric acid is neutralized by 33.24 ml of 0.103N sodium hydroxide, what is the normality of the acid?
8. For each salt listed, deduce whether its aqueous solution is neutral, acidic, or basic.

	Set 1	Set 2	Set 3
(*a*)	KNO_3	NaBr	Na_2HPO_4
(*b*)	$AlBr_3$	K_2HPO_4	$NaNO_3$
(*c*)	Na_3PO_4	$FeCl_3$	NH_4Br
(*d*)	NaI	K_2CO_3	Na_3BO_3
(*e*)	$NaHCO_3$	KCl	KI

9. What is the pH at room temperature of a solution that is 0.1M sodium formate?
10. What is the pH at room temperature of a solution that is 0.2M sodium formate and 0.1M formic acid? Would this solution be buffered? Explain.

11. The solubility product of barium sulfate at room temperature is 1.1×10^{-10}. Calculate the solubility of barium sulfate in pure water, in a $0.01M$ solution of barium nitrate, and in a $0.001M$ solution of potassium sulfate.
12. Suppose you have a solution $0.1M$ in chloride ion and $0.1M$ in bromide ion. If a solution of silver nitrate is added, which of these two ions will precipitate first and why?
13. In the titration of acetic acid by sodium hydroxide, what salt will form? Will the pH of the solution at the equivalence point be 7? Explain. Which indicator, methyl orange or phenolphthalein, is the better choice for detecting the equivalence point of such a titration? Why?

CHAPTER NINE

Introduction to Organic Chemistry. Alkanes

CONCERNING THINGS ORGANIC

Living organisms possess a capacity to synthesize a great number and variety of substances. The list of organic materials extends far beyond the obvious and most important example, food, to include petroleum, dyes, fibers, drugs, and natural plastics such as gums and rubber. Crude precursory attempts at chemical technology can be seen in the efforts of early people to isolate, purify, and improve these products. Some of these efforts involved little more than using heat on natural substances. When wood is heated without burning it, wood alcohol is obtained as well as acetone and pyroligneous acid, now known to contain acetic acid.

Heat may also cause charring and decomposition. Therefore methods that could be used when heat alone was unproductive were developed. Procedures for extracting substances from barks, roots, leaves, and fruit yielded many drugs and dyes. Making tea and coffee is a simple extraction with water the solvent. Modern and comparable extractions in industry use substances such as benzene, ether, light petroleums, and carbon tetrachloride as well as water as common solvents.

The assumed or actual curative powers of natural extracts were but one reason why increasing numbers of people were drawn to the investigation of organic compounds. Dyes, perfumes, flavoring agents, and spices were lures too. Some investigations were undertaken simply because, like Mount Everest to mountain climbers, they were there to be conquered. Until the early part of the nineteenth century, very few scientists saw much hope of making the realm of the organic into a science. One of the frustrations that hounded the early workers was the difficulty of purifying organic substances, which often decompose or deteriorate during or after efforts to isolate them. It was difficult to obtain evidence that these substances even obeyed the same chemical laws that held among minerals, for example, the law of definite proportions and the law of multiple proportions. The elementary composition of organic substances did not become clear until the beginning of the nineteenth century when Lavoisier (p. 74) demonstrated that carbon and hydrogen are essential elements in organic substances. It was soon discovered that oxygen, nitrogen, phosphorus, and sulfur also occur commonly in organic compounds. Better procedures for purification as well as better quantitative techniques soon established the fact that organic substances do indeed obey the laws of chemical composition. Yet they remained mysterious. The mineral kingdom includes carbonates and water, but in the hands of the early investigators these sources of the elements carbon and hydrogen could not be used to synthesize organic compounds in the test tube.

Vitalism. Chemists came to believe that they could never make organic substances from minerals. They could burn them, transform them into other products, and analyze them, but to make them seemed in the nature of things beyond the prowess and imagination of man. The ultimate and only source of organic substances appeared to be organisms, plant and animal, and the belief grew that living systems possessed a special force acting independently of other forces—a special force that could be passed from one living thing to another in the normal processes of reproduction and growth. The force was likened to an invisible flame burning steadily within living organisms. From this source of mysterious energy—energy unlike that available from other sources—nature derived the power to put together soil nutrients, water, and carbon dioxide into sugars, proteins, fibers, drugs, dyes, perfumes, spices, and a host of other organic substances. The gulf between organic and inorganic compounds seemed to be unbridgeable. Only an accidental event could have closed the gap, and it occurred to one Friedrich Wöhler in 1828.

Wöhler's Synthesis of Urea. Wöhler had been working with cyanogen, C_2N_2, and ammonia, both of which obey the known laws of inorganic chemistry and both of which were not at the time considered to be organic. Cyanogen reacts with water to produce cyanic acid and hydrocyanic acid:

$$C_2N_2 + H_2O \longrightarrow \underset{\text{Cyanic acid}}{HCNO} + \underset{\text{Hydrocyanic acid}}{HCN}$$

When ammonia is added, ammonium ions and cyanate ions form:

$$NH_3 + HCNO \longrightarrow NH_4^+ + \underset{\text{Cyanate ion}}{CNO^-}$$

Wöhler prepared a solution containing these ions, and he hoped to obtain crystalline ammonium cyanate by heating it to drive off the water. He did obtain a crystalline, white solid, but it possessed none of the properties of the expected salt. Instead, its physical and chemical properties were those of an entirely different compound, urea. Urea had been isolated from urine half a century earlier. We now know that it is an end product of the metabolism of proteins, being made in the liver and removed from circulation at the kidneys. Although relatively simple, it is plainly of organic origin.

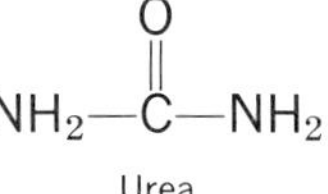

Urea

Wöhler's synthesis was checked by other scientists and rechecked by Wöhler himself. By 1830 distinguished scientists from different parts of Europe hailed his brilliant work. Still, the victory was not entirely complete. It was possible to argue that the sources of both the ammonia and the cyanogen were animal bones—dead bones, to be sure, but bones nevertheless. Thus the "vital force" might still linger in the substances used by Wöhler. This argument was taken so seriously that neither Wöhler nor his contemporaries trumpeted the overthrow of the vital force theory. It was not until 1844 when Kolbe synthesized acetic acid, CH_3CO_2H, from completely inorganic compounds that vitalism was decisively disproved. The synthesis of numerous other organic compounds without the aid of living organisms[1] soon followed.

Adolf Wilhelm Hermann Kolbe (1818–1884). A student of Wöhler's and an assistant to Bunsen, Kolbe is best remembered among organic chemists for his synthesis of salicylic acid.

Wöhler's and Kolbe's work constituted a scientific advance of immense proportions. The fields of synthetic dyes, drugs, and polymers stand as monuments to it. Because the vital force theory no longer inhibits attempts to synthesize organic substances artificially, many not known in nature are continually being made as organic chemists seek for further information about products of possible commercial or medicinal value as well as for a deeper understanding of the underlying laws that govern the behavior of all matter.

Organic chemistry has reached such a stage of development that we can study it in some depth as though living organisms, which gave the field its name and its start, did not exist. This procedure has its advantages because the organic chemicals found in living things are usually quite complicated. Since our theme is the molecular basis of life, we may not ignore them indefinitely, of course. However, by delaying a full-fledged investigation of them until we have found our bearings among simpler molecules, the final phase of our study will be much easier. The complicated molecules associated with living processes have one simplifying feature to help us—they can be understood rather well in terms of only a small number of key molecular "parts." In fact, our study of organic chemistry will evolve around the following question: What are the simplest and most useful "elements" in terms of which bio-

[1] There is irony here and George Wald has pointed it out. Wöhler's and Kolbe's experiments, as well as the work of any organic chemist, involve a very important living agency, the chemist himself. What Wöhler demonstrated was that chemists can make organic compounds externally as well as internally. There is no turning back, however. The vital force theory is defunct as far as chemists are concerned. (George Wald, *Scientific American,* August 1954, page 48.)

logical chemicals and their reactions may be understood? These "elements" are molecular parts or *functional groups* that occur in biological chemicals, and we shall study them first among simple substances.

CARBON'S UNIQUENESS

Organic chemistry is the branch of chemistry that deals with the compounds of carbon. Although other elements are present, carbon is given first rank because its atoms provide the skeletons for organic molecules. Carbon is unique among the 103 elements in that its atoms can bond to each other successively many, many times by means of strong covalent bonds. Carbon can do this at the same time that it uses its other valences in bonds to atoms of other elements. These range from such strongly electronegative elements as fluorine, chlorine, and oxygen to such a weakly electronegative element as hydrogen. Thus both ethane, **1,** and hexafluoroethane, **2,** are known. The carbon "skeleton" in these two compounds consists of only two carbons. In principle, there is no limit to the length of a chain of carbons, and the common household plastic, polyethylene, consists of molecules with carbon chains hundreds of nuclei long.

```
    H   H              F   F
    |   |              |   |
H — C — C — H      F — C — C — F
    |   |              |   |
    H   H              F   F
```

1
Ethane

2
Hexafluoroethane

STRUCTURAL ORGANIC CHEMISTRY

Geometry of Carbon Compounds. The bonds in organic compounds are almost exclusively electron pair or covalent bonds. The covalence number of carbon is four, with but very few exceptions (e.g., CO, carbon monoxide). When a particular carbon atom is bonded to four groups, its four single bonds are directed toward the corners of a regular tetrahedron (Figure 9.1). All four hydrogen atoms in methane are

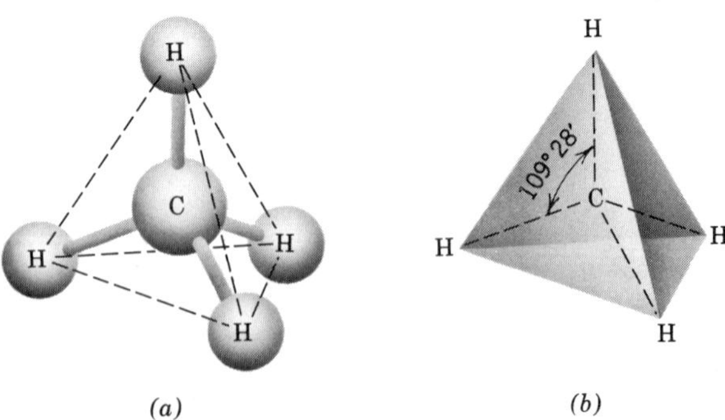
H
C
H
H
H
(a)
H
109°28′
C
H
H
H
(b)

Figure 9.1 Tetrahedral carbon atom. Part *a* shows a common ball and stick model of methane with dotted lines added to bring out the tetrahedron. A different perspective is given in part *b*.

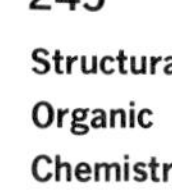

equivalent. If one were to be replaced by, say, a chlorine atom, it would not matter which was taken. Only one monochloro derivative of methane exists—methyl

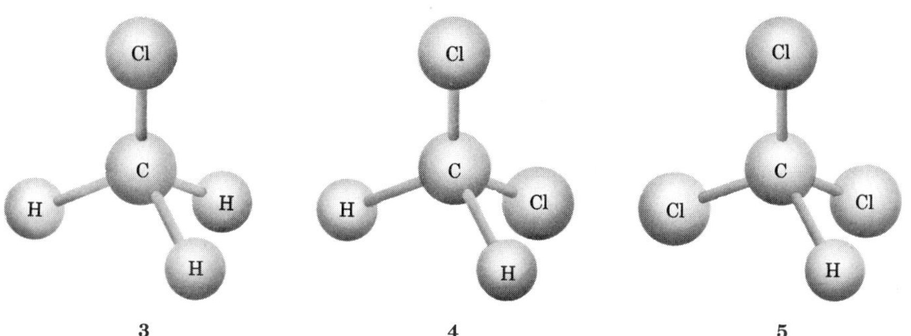
Cl
C
H
H
H
Cl
C
H
Cl
H
Cl
C
Cl
Cl
H

3 4 5

chloride, **3.** Only a symmetrical structure for methane can account for this fact. The tetrahedral orientation of the hydrogens in methane is not the only one, however, that is symmetrical. A square-planar structure, **6,** would also have four equivalent hydrogens and would also lead to only one monochloro derivative, **7.** The three hydrogens in **7,** however, are not equivalent. Two are closer to the chlorine nucleus and its surrounding large electron cloud than the other. If the square-planar structure for methyl chloride were correct, then from it we should be able to prepare two possible dichloro derivatives, **8** and **9.** But only one is known, **4,** and only one trichloro derivative, **5,** is known. These facts and many others, including results of X-ray studies, confirm that the carbon in methane and its simple derivatives has its four bonds directed not to the corners of a square but to the corners of a tetrahedron. The bond angle in methane, the angle between the lines joining any two hydrogens to the carbon, is 109°28', precisely the angle that would be calculated from the solid geometry of a regular tetrahedron.

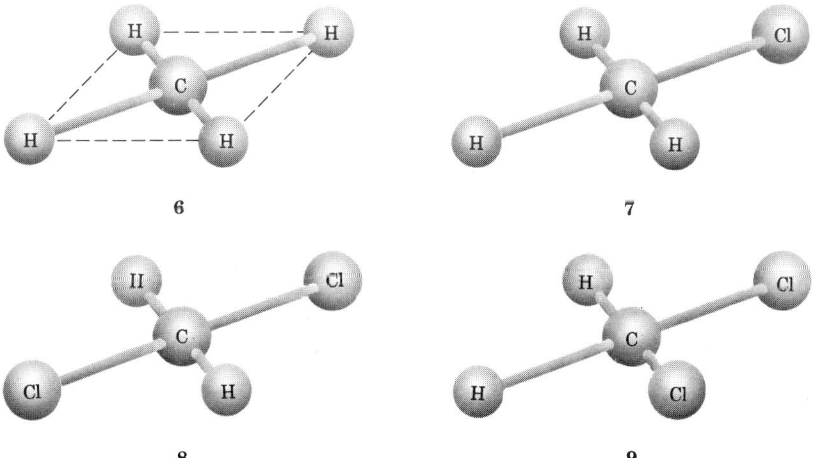
H
H
C
H
H
H
Cl
C
H
H
H
Cl
C
Cl
H
H
Cl
C
H
Cl

6 7

8 9

In virtually all compounds of carbon, wherever a carbon has four single bonds going away from it, the geometry at that point will generally be close to tetrahedral.

Moreover, groups that are joined by a single bond can rotate with respect to each other about that bond. Given the tetravalence of carbon and its tetrahedral geometry, structures **10** and **11** would seem to be two legitimate ways of orienting the nuclei in the compound known as 1-chloropropane. If **10** and **11** were each rigid structures, they would be molecules of different substances. The net polarity of **10** would not be the same as that for **11,** and **10** would therefore not have the same boiling point (or other physical properties) as **11.** The fact is, however, that only one 1-chloropropane is known. A sample of this liquid presumably contains molecules such as **10**

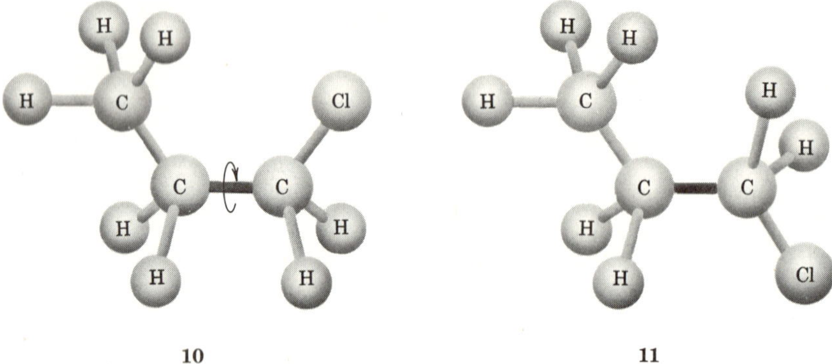
H
H
H
C
Cl
C
C
H
H
H
H
H
H
H
C
H
H
C
C
H
H
Cl

10 **11**

and **11** and all other possible forms differing only in the *relative* orientations of the large CH_3—CH_2— group and the chlorine. If we assume that these two groups are able to rotate about the bond drawn as a heavy line in **10** and **11,** the nonexistence of these two as separate compounds can be understood.

In summary, we have discussed three essential facts about carbon compounds. Carbon is tetravalent. When only single bonds are involved, it is tetrahedral. Groups attached by single bonds can rotate with respect to each other. We turn next to a theory of bonding in carbon compounds that will unify these facts.

Bonding in Carbon Compounds. The electronic configuration of carbon is $1s^2 2s^2 2p_x 2p_y$, which in terms of an energy level diagram would be represented as follows:

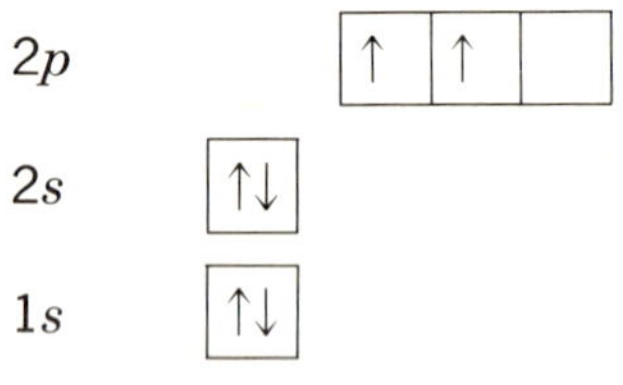
2p ↑ ↑
2s ↑↓
1s ↑↓

Atomic state of carbon

In the valence shell there are only two unpaired electrons. Only two covalent bonds rather than four would therefore be possible. The empty p_z orbital might serve in the formation of a coordinate covalent bond by accepting a share in a pair of electrons from some Lewis base. But this would give only one more bond for a total of three. Of course, the pair of $2s$ electrons might be used in donating a share to some

Lewis acid, some electron-poor species. Then the total number of bonds would be four—two ordinary covalent bonds involving the p_x and p_y orbitals, two coordinate covalent bonds involving the empty p_z orbital as an acceptor and the filled $2s$ orbital as a donor. Although this arrangement might account for the tetravalence of carbon, it would not account for the geometry of its compounds.

The most prevalent current theory accounting for both the tetravalence and the geometry of carbon in its compounds holds that, although the electronic configuration we have used for carbon may be true for its atomic state, in its compounds these valence shell orbitals are no longer present. To have four equivalent *bonds* from a carbon, it is postulated that four equivalent *orbitals* are available for overlapping with orbitals from the other groups attached to the carbon. To understand how four such equivalent orbitals can arise, the following model is used.

One of the $2s$ electrons is envisioned as promoted to the empty $2p_z$ orbital. In the bookkeeping of the energy changes, this costs energy. In eventually forming four equivalent bonds, the cost is more than repaid. The new state of carbon—it may be called a "promoted" state—will be as follows:

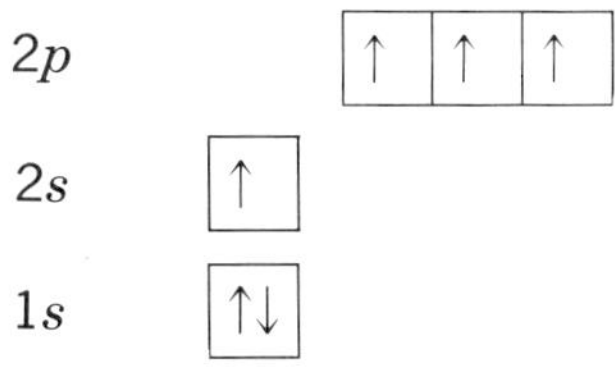
2p
2s
1s

"Promoted" state of carbon

We now have four available orbitals for forming four covalent bonds, but these would not produce the final geometry of a molecule such as methane.

Exercise 9.1 What geometry would result if the orbitals of the "promoted" state were used to form bonds to hydrogen?

From the promoted state it is postulated that the four orbitals—one s orbital and three p orbitals—become reorganized into four identical orbitals. This reorganization or "mixing" is called *hybridization*. The new orbitals are *hybrid orbitals*, hybrids of the s and the three p orbitals. Since there are ultimately several kinds of hybrid orbitals, they must be given names, and whenever they are made from one s and three p orbitals, they are called sp^3 hybrid orbitals ("s-p-three"). The new energy level diagram for carbon in this *valence state* is as follows:

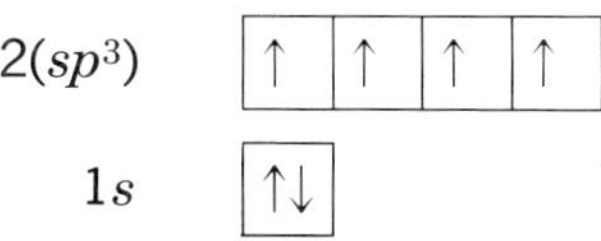
2(sp³)
1s

The shape of an sp^3 hybrid orbital is indicated in Figure 9.2. It has some s character

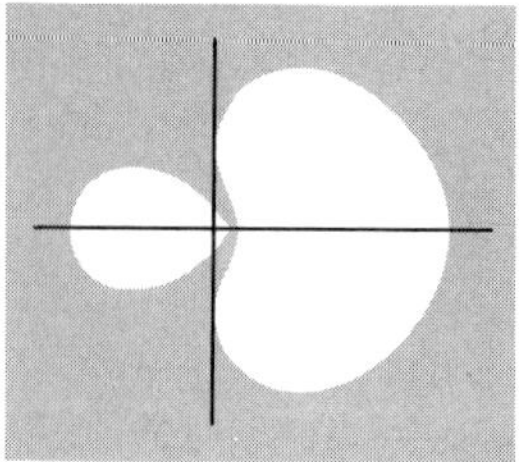

Figure 9.2 Cross section of an sp^3 hybrid orbital.

—the intersection of the axes is enclosed by the orbital—and some p character—there are two lobes. Figure 9.3 shows how these four equivalent orbitals are arranged in relation to each other. The axes of these orbitals point toward the corners of a regular tetrahedron.

This model has its basis in mathematical operations. Our models for orbitals are obtained by using particular equations and plotting points, but we have not looked at the equations or plotted the points ourselves. We merely noted in Chapter 2 that these equations exist and that from them the relative energies of orbitals may be calculated and the "probability envelopes" or orbitals determined. The process of hybridizing atomic orbitals is similarly a mathematical operation. It has been found that equations that gave us s and p atomic orbitals may be added together to form new equations from which new groups of points may be plotted, new orbital shapes may be determined, and new values of orbital energies may be calculated. As it happens, the computed (estimated) energy of a system of one carbon nucleus and four hydrogen nuclei and ten electrons (methane) is less when the orbitals of carbon are hybridized than it would be if carbon formed only two bonds to hydrogens. Less internal energy means greater stability. The calculations therefore nicely match known facts about the stability, the valence, and the geometry of carbon compounds. The question of free rotation about single bonds remains to be fitted into this theory.

Figure 9.4 illustrates how the bonds of methane are formed. Each bond from carbon to a hydrogen is the result of the overlap of an sp^3 orbital from carbon and the $1s$ orbital from hydrogen. Each bond is a sigma bond because each new molecular

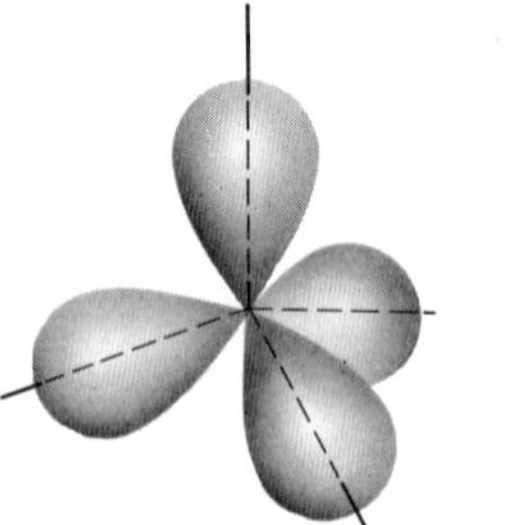

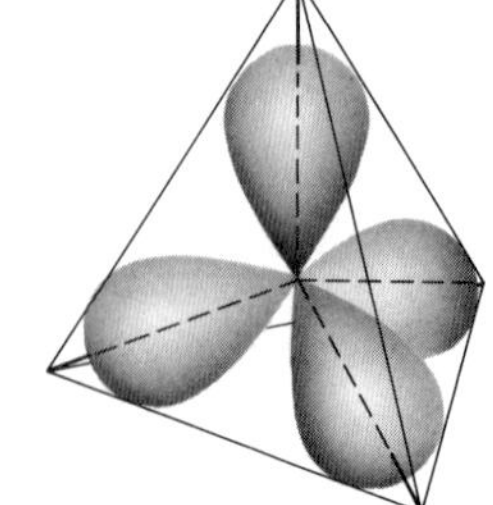

Figure 9.3 The four equivalent sp^3 hybrid orbitals of carbon are positioned so that their axes point to the corners of a regular tetrahedron. Only the large lobes of each sp^3 orbital are indicated in this figure.

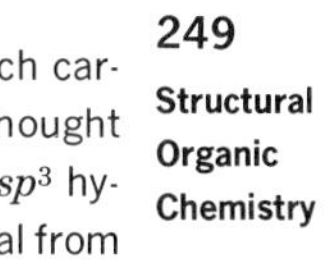

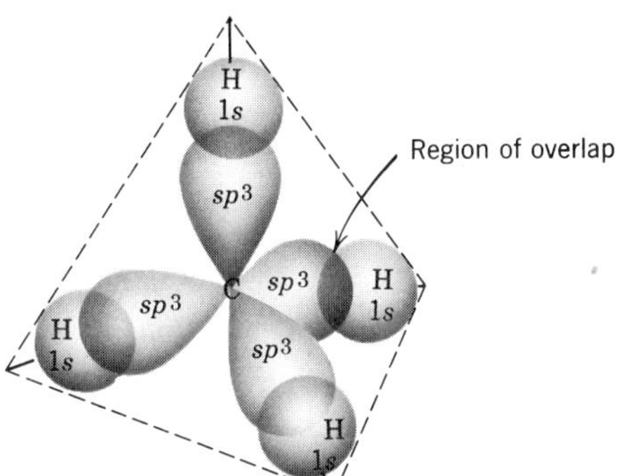
H
1s
sp3
Region of overlap
sp3
H
1s
sp3
H
1s
sp3
H
1s

Figure 9.4 Bonds in methane. Each carbon-to-hydrogen covalent bond is thought of as forming from the overlap of an sp^3 hybrid orbital from carbon and a $1s$ orbital from hydrogen. Because there is symmetry about the bonding axis, each bond is a σ-bond. The principle of maximum overlap ensures that the nucleus of each hydrogen is at the corner of the regular tetrahedron outlined by the dotted lines.

orbital has the symmetry about the bonding axis meant by that designation (cf. p. 124).

The bonding in ethane, C_2H_6, is illustrated in Figures 9.5 and 9.6, and two models of the molecule are shown in Figure 9.7. With this molecule we may return to the question how two groups attached by a single bond can rotate with respect to each other. In ethane, can the two CH_3 groups (methyl groups) rotate with respect to each other about the bond joining them? If such rotation required significant amounts of energy, the answer would be no, for the barrier to free rotation is fundamentally an energy barrier. But in ethane the overlap between the two orbitals forming the bond in question is not increased or decreased in any significant way by such rotation, and for this reason it is allowed. (That the spinning we have been describing actually happens is evident from spectral studies which we cannot discuss.) We do not imply that all possible orientations in ethane are energetically equivalent, for they are not and Figure 9.8 shows why. On the left, in the a–a''' series, the orientation is such that the six hydrogens are staggered. On the right, in the b–b''' series, the orientation has the six hydrogens mutually eclipsed; looking down the carbon-carbon bond axis, we see that a hydrogen in front eclipses one in

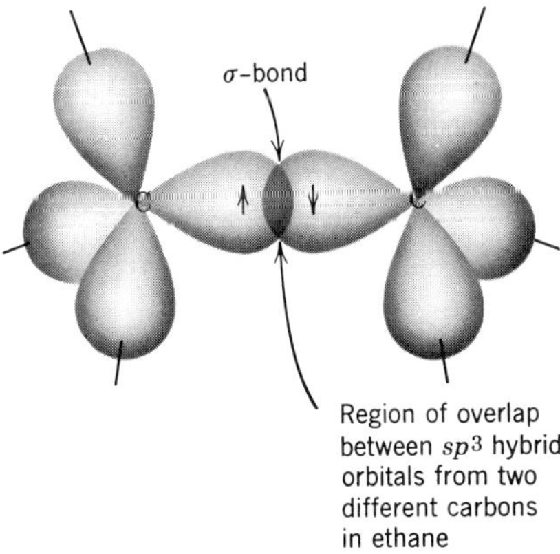
σ-bond
Region of overlap
between sp3 hybrid
orbitals from two
different carbons
in ethane

Figure 9.5 The carbon-carbon single bond in ethane. The oppositely pointing arrows symbolize that the new molecular orbital contains two electrons of opposite spin. The region of maximum electron density for these two falls between the two carbon nuclei and is the principal factor contributing to the bond between them. (The exact locations of the arrows shown in this figure are not significant.) The remaining six sp^3 hybrid orbitals are used in ethane to overlap with $1s$ orbitals from six hydrogens as shown in Figure 9.6.

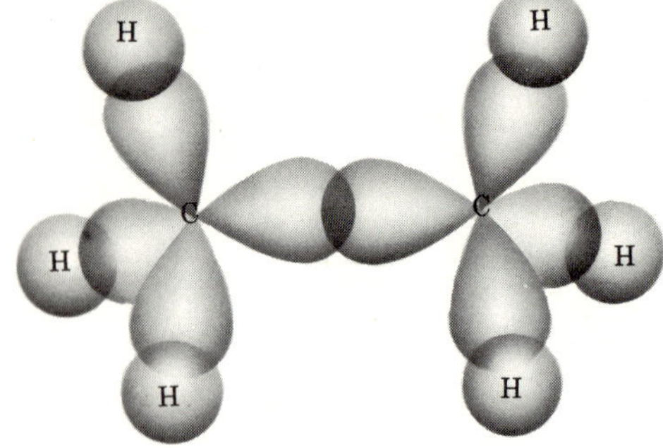
H
H
C
C
H
H
H
H

Figure 9.6 The overlap of atomic orbitals to form the single bonds in ethane, all of which are σ-bonds.

back. In this eclipsed arrangement the hydrogens on the different carbons approach each other the most closely, and the repulsion between their electron clouds as well as between their nuclei will be slightly greater than in the staggered orientation. Consequently, the staggered form will be the most stable of all possible orientations. At ordinary temperatures, however, these hydrogen-to-hydrogen repulsions are so small that, although they may inhibit free rotation, they cannot prevent it. The physical and chemical properties of ethane are therefore the result of the average effects of all possible rotational arrangements.

ISOMERISM

When Wöhler tried to make ammonium cyanate (molecular formula, CH_4N_2O; structural formula, $NH_4^+CNO^-$) and obtained, instead, urea (molecular formula, CH_4N_2O; structural formula, $NH_2-\overset{\overset{\displaystyle O}{\|}}{C}-NH_2$), he added to the small but growing list of substances that were obviously different while still having identical molecular formulas. In 1832 Jöns Jacob Berzelius proposed that such substances be called *isomeric,* from the Greek *isos,* equal, and *meros,* parts. He further suggested that the difference between compounds having identical molecular formulas must lie in a different arrangement of their parts, a prediction later confirmed.

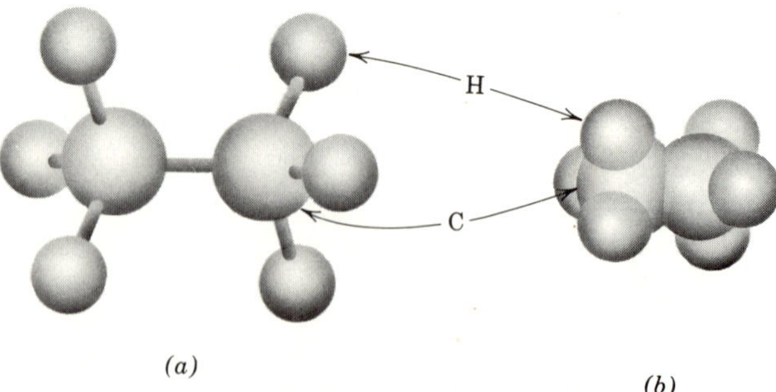
H
C
(a)
(b)

Figure 9.7 Two models for ethane. (*a*) Ball and stick model. (*b*) Scale model designed to indicate relative volumes occupied by parts of the molecule.

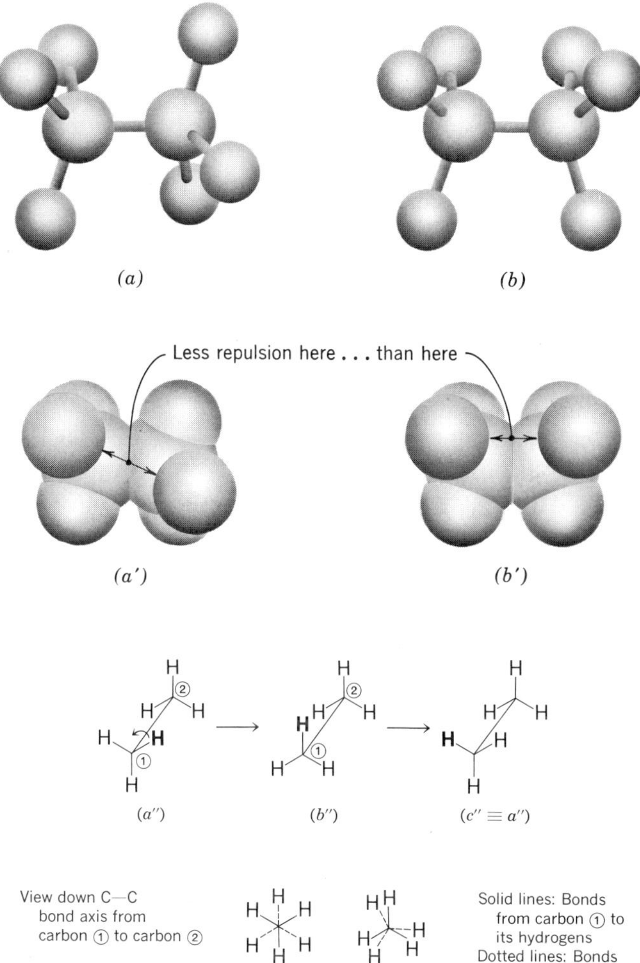
(a)
(b)
Less repulsion here . . . than here
(a′)
(b′)
(a″)
(b″)
(c″ ≡ a″)
View down C—C bond axis from carbon ① to carbon ②
(a‴)
(b‴)
Solid lines: Bonds from carbon ① to its hydrogens
Dotted lines: Bonds from carbon ② to its hydrogens

Figure 9.8 Several representations of the staggered and the eclipsed conformations of ethane. In the a to a''' series are ways of describing the staggered form; in the b to b''' series are those for the eclipsed form. In parts a''' and b''' are shown head-on views looking straight down the carbon carbon axis. The dotted lines are the bonds from the carbon *behind* the front carbon to its hydrogens. In b''' a little twist is retained to show that all hydrogens are present, but this figure indicates the origin of the word "eclipsed." The front hydrogens in b''' eclipse the rear ones in the manner of a solar eclipse.

Note that in a' hydrogen nuclei with attendant electron clouds are farther apart, all around, than in b'. Electron clouds and nuclei tend to be as far apart as possible within the limits of the bonds present, and the fact that they are closer in b' than in a' is the origin of the greater internal energy of the eclipsed form. The situation in b' is not so serious, however, that the electron clouds cannot pass by each other during a rotation about the central carbon-carbon bond. Thus at ordinary temperatures "free rotation" is said to exist about the single bonds in ethane.

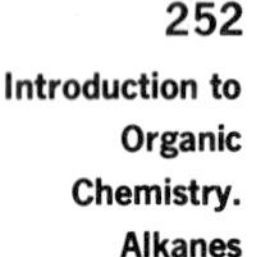

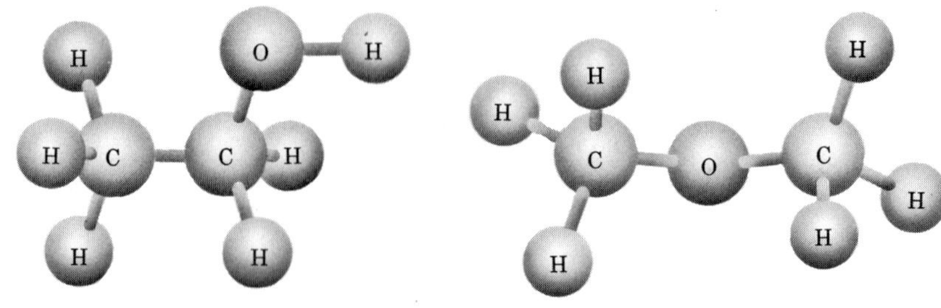
H
O
H
H
C
C
H
H
H
H
H
H
C
O
C
H
H
H
H

Ethyl alcohol

Methyl ether

Figure 9.9 Isomers of C_2H_6O, ethyl alcohol and methyl ether, in ball and stick models. Each isomer has the same molecular formula, but they differ in the sequences in which these atoms are joined together.

Consider the molecular formula C_2H_6O. *Within the rules of valence,* how many structures are possible? In how many ways within the rules of valence can the nuclei of two carbons, six hydrogens, and one oxygen be arranged? Trial and error will show just two, given as ball and stick models in Figure 9.9. Substances consisting of each of these molecules are known, ethyl alcohol and methyl ether. Table 9.1 summarizes some of their characteristic properties.

Compounds that are isomeric are said to be *isomers* of each other, and the phenomenon itself is called *isomerism. For two compounds to be related as isomers, they must have identical molecular formulas but different structures.* Experimentally, two isomers must have identical formula weights, identical analytical results for the percentages of each element, but be detectably different in at least one physical or

Table 9.1 Properties of Two Isomers: Ethyl Alcohol and Methyl Ether

Property	Ethyl Alcohol	Methyl Ether*
Molecular formula	C_2H_6O	C_2H_6O
Boiling point	78.5°C	−24°C
Melting point	−117°C	−138.5°C
Density	0.789 g/cc	2 g/liter
Solubility in water	completely soluble in all proportions	slightly soluble
Action of metallic sodium	vigorous reaction, hydrogen evolved	no reaction
Structural formula	H H H—C—C—O—H H H	H H H—C—O—C—H H H
Condensed structural formula	CH_3CH_2OH	CH_3OCH_3

* Sometimes called dimethyl ether.

chemical property. The structures must be such that simple rotations about single bonds will not convert one into the other. Although staggered and eclipsed forms of ethane that do not have the same relative locations of the nuclei or the same internal energies can be drawn, they do not correspond to different, isolable compounds. Ethane has no isomers. Free rotation assures this.

Condensed Structural Formulas. In writing the structure of ethane as CH_3CH_3 instead of as

$$\begin{array}{c} \text{H}\quad\text{H} \\ |\quad\;\; | \\ \text{H—C—C—H} \\ |\quad\;\; | \\ \text{H}\quad\text{H} \end{array}$$

considerable space and time is saved. With large structures the saving is more obvious, and we shall use such condensed structural formulas exclusively. Whenever three hydrogens are attached to the same carbon, we write the formula CH_3—. If two hydrogens are attached to the same carbon, and only two, the formula is written as —CH_2—; $-\overset{|}{\underset{|}{C}}H$ is used if the carbon has only one hydrogen. All covalent bonds are "understood" in a symbol such as CH_3CH_3, but frequently they are included, for example, CH_3—CH_3. Other conventions for writing condensed structures will be introduced as needed in subsequent chapters.

HOW ORGANIC CHEMISTRY IS ORGANIZED AS A SUBJECT

The hundreds of thousands of known organic compounds can be classified into relatively few groups on the basis of similarities in structure. Just as zoologists classify animals into families having structural likenesses, so chemists classify organic compounds according to structural similarities. And just as animals with similar structures behave in many of the same ways, so members of organic structural families exhibit many of the same chemical properties. One of the important classification schemes in organic chemistry is by structure. Our study will be organized according to this scheme. Table 9.2 shows several important classes of organic compounds. We start our study of the families of organic compounds with the alkanes, a subgroup of the hydrocarbons.

SATURATED HYDROCARBONS. ALKANES

FAMILIES OF HYDROCARBONS

A study of the hydrocarbons is the most convenient starting point for our study of other families. As the name implies, molecules in this family are derived from the two elements carbon and hydrogen. Only covalent bonds occur, and depending on whether various types of multiple bonds are present or absent, a hydrocarbon may be a member of one of several subfamilies outlined in Figure 9.10.

Table 9.2 Some Important Classes of Organic Compounds

Class	Characteristic Structural Feature
Hydrocarbons	Composed only of carbon and hydrogen, with many subclasses when single, double, or triple bonds are present Alkanes: all single bonds, e.g., CH_3CH_3 Alkenes: at least one double bond, e.g., $CH_2{=}CH_2$ Alkynes: at least one triple bond, e.g., $H{-}C{\equiv}C{-}H$ Aromatic: at least one benzenoid ring system, e.g., (benzene ring)
Alcohols	Contain the —OH group as in $CH_3CH_2{-}OH$
Ethers	Contain C—O—C system as in $CH_3{-}O{-}CH_3$
Aldehydes	Contain the $-\overset{O}{\overset{\|}{C}}-H$ group as in $CH_3-\overset{O}{\overset{\|}{C}}-H$
Ketones	Contain the $C-\overset{O}{\overset{\|}{C}}-C$ group as in $CH_3-\overset{O}{\overset{\|}{C}}-CH_3$
Carboxylic acids	Contain the $-\overset{O}{\overset{\|}{C}}-OH$ system as in $CH_3-\overset{O}{\overset{\|}{C}}-OH$
Esters	Contain the $-\overset{O}{\overset{\|}{C}}-O-C$ system as in $CH_3-\overset{O}{\overset{\|}{C}}-O-CH_2CH_3$
Amines	Contain trivalent nitrogen, with all covalent bonds single bonds as in $CH_3{-}NH_2$
Amides	Contain the grouping $-\overset{O}{\overset{\|}{C}}-\overset{\mid}{N}-$ as in $CH_3-\overset{O}{\overset{\|}{C}}-NH_2$

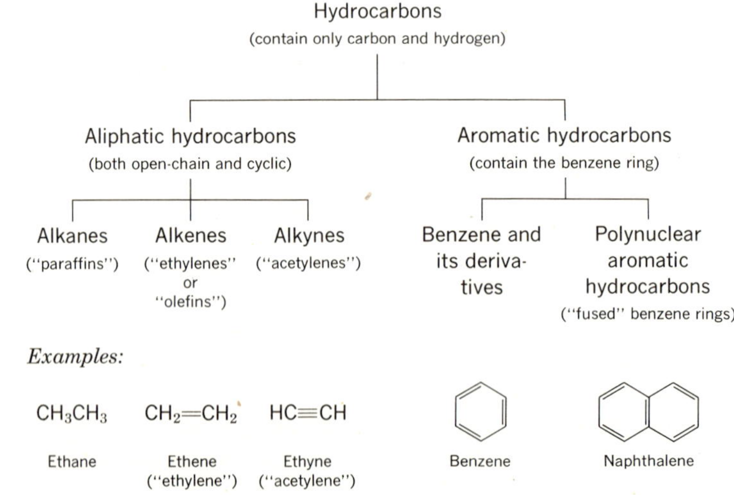

Hydrocarbons
(contain only carbon and hydrogen)
Aliphatic hydrocarbons
(both open-chain and cyclic)
Aromatic hydrocarbons
(contain the benzene ring)
Alkanes
("paraffins")
Alkenes
("ethylenes" or "olefins")
Alkynes
("acetylenes")
Benzene and its derivatives
Polynuclear aromatic hydrocarbons
("fused" benzene rings)
Examples:
CH_3CH_3
$CH_2{=}CH_2$
$HC{\equiv}CH$
Ethane
Ethene ("ethylene")
Ethyne ("acetylene")
Benzene
Naphthalene

Figure 9.10 Hydrocarbon families.

The alkanes, whose molecules consist of only carbon and hydrogen and only single bonds, are the simplest, and they furnish the basis for systems of naming members of many other families.

STRUCTURAL FEATURES OF ALKANES

Saturated Compounds. Any molecule of whatever family in which only single bonds occur is said to be *saturated*. Any molecule with one or more double or triple bonds is *unsaturated*.

Homologous Series. The structures and important physical properties of the ten smallest "straight chain" members of the alkanes are given in Table 9.3. Straight chain means only that the carbon nuclei follow one another as links in a chain, for example, C—C—C—C—C—C. In contrast, branched chain means that carbon nuclei are bonded as branches to the chain, for example,

```
C—C—C—C—C—C.
  |   |
  C   C
```

These expressions do not describe the geometry of the molecule. With free rotation about single bonds, these chains may flex, coil, and otherwise become kinky.

The series in Table 9.3 is said to be a homologous series because members differ from each other in a consistent, regular way. Here each member differs from the one just before or after it by one CH_2 unit. In chemical terminology butane, for example, is "the next higher *homolog* of propane."

Isomerism among the Alkanes. Among the alkanes, from butane and on to higher homologs, isomerism is possible. Table 9.4 lists several examples. There are two isomers of formula C_4H_{10}, three of formula C_5H_{12}, and five of formula C_6H_{14}. As the homologous series is ascended, the number of possible isomers approaches astronomical figures. There are 75 possible isomers of decane, $C_{10}H_{22}$, and an estimated 6.25×10^{13} possible isomers of $C_{40}H_{82}$. Not all possible isomers have actually been prepared in the pure state and studied, for no useful purpose would be served and time would not permit. The occurrence of isomerism again empha-

Table 9.3 Straight-Chain Alkanes

Number of Carbon Atoms	Name	Molecular Formula*	Condensed Structural Formula	Bp (°C at atmospheric pressure)	Mp (°C)	Density (in g/cc at 20°C)
1	Methane	CH_4	CH_4	−161.5		
2	Ethane	C_2H_6	CH_3CH_3	88.6		
3	Propane	C_3H_8	$CH_3CH_2CH_3$	−42.1		
4	Butane	C_4H_{10}	$CH_3CH_2CH_2CH_3$	−0.5	−138.4	
5	Pentane	C_5H_{12}	$CH_3CH_2CH_2CH_2CH_3$	36.1	−129.7	0.626
6	Hexane	C_6H_{14}	$CH_3CH_2CH_2CH_2CH_2CH_3$	68.7	−95.3	0.659
7	Heptane	C_7H_{16}	$CH_3CH_2CH_2CH_2CH_2CH_2CH_3$	98.4	−90.6	0.684
8	Octane	C_8H_{18}	$CH_3CH_2CH_2CH_2CH_2CH_2CH_2CH_3$	125.7	−56.8	0.703
9	Nonane	C_9H_{20}	$CH_3CH_2CH_2CH_2CH_2CH_2CH_2CH_2CH_3$	150.8	−53.5	0.718
10	Decane	$C_{10}H_{22}$	$CH_3CH_2CH_2CH_2CH_2CH_2CH_2CH_2CH_2CH_3$	174.1	−29.7	0.730

* The molecular formulas of the open-chain alkanes fit the general formula C_nH_{2n+2}, where n is the number of carbons in the molecule.

sizes how limited is the information in a molecular formula. Only with a structural formula can the uniqueness of a molecule be understood and correlated with its properties.

NOMENCLATURE

The earliest known organic compounds were named after their source, for example, formic acid (Latin *formica,* ants) can be made by grinding ants with water and distilling the result. Hundreds of compounds were named after their sources, but the system becomes impossibly difficult to extend to all compounds. These common names, however, are still used, and the beginning student is faced with the

Table 9.4 Properties of Isomeric Alkanes

Family	Common Name (except where noted)	Structure	Bp (°C at atmospheric pressure)	Mp (°C)	Density (in g/cc)
Butane isomers	*n*-butane	$CH_3CH_2CH_2CH_3$	−0.5	−138.4	0.622 (−20°C)
	isobutane	$CH_3CH(CH_3)CH_3$	−11.7	−159.6	0.604 (−20°C)
Pentane isomers	*n*-pentane	$CH_3CH_2CH_2CH_2CH_3$	36.1	−129.7	0.626 (20°C)
	isopentane	$CH_3CH(CH_3)CH_2CH_3$	27.9	−159.9	0.620
	neopentane	$CH_3C(CH_3)_2CH_3$	9.5	−16.6	0.591
Hexane isomers	*n*-hexane	$CH_3CH_2CH_2CH_2CH_2CH_3$	68.7	−95.3	0.659
	3-methylpentane (no common name)	$CH_3CH_2CH(CH_3)CH_2CH_3$	63.3		0.664
	isohexane	$CH_3CH(CH_3)CH_2CH_2CH_3$	60.3	−153.7	0.653
	2,3-dimethylbutane (no common name)	$CH_3CH(CH_3)CH(CH_3)CH_3$	58.0	−128.5	0.662
	neohexane	$CH_3C(CH_3)_2CH_2CH_3$	49.7	−99.9	0.649

Octane isomers, C_8H_{18}—total of 18
Decane isomers, $C_{10}H_{22}$—total of 75
Eicosane isomers, $C_{20}H_{42}$—total of 366,319
Tetracontane isomers, $C_{40}H_{82}$—estimated total of 6.25×10^{13} isomers

necessity of learning some of them. In addition, to be able to read and talk about more complicated structures without common names, the beginner must learn rules for formal or systematic nomenclature. A third system, the derived system of nomenclature, applies in a few situations.

Organic nomenclature is a rather large field of study in its own right, but it must not be confused with organic *chemistry*. Nomenclature is a necessary part of the chemistry, but it is no substitute for it, contrary to a curious opinion held by some whose only experience with organic chemistry has been with naming compounds. We shall not make a thorough study of nomenclature, but we must cover enough of the subject to be able to examine the physical and chemical properties of organic substances.

Common Names of Alkanes. The straight-chain isomers are designated the *normal* isomers, and their common names include n- for normal. Examples are n-butane, n-octane, but not n-ethane because ethane has no isomer. In Table 9.3 butane and all subsequent names would be common if n- were placed before each. As they stand, they are formal names, to be described later. Except where noted, the names in Table 9.4 are also common names. The names of all the normal alkanes through n-decane as well as the names of the isomers through the five-carbon series must be learned. The total carbon content of the alkane should be associated with the prefix portion of its name. Thus "eth-" is a word part for a two-carbon unit; "but-" signifies four carbons whether in n-butane or isobutane. The "-ane" ending is characteristic of all alkanes in any system of nomenclature. We shall turn next to the formal system, the Geneva system or International Union of Pure and Applied Chemistry (IUPAC) system, but the student is advised that common names will be used almost exclusively in this text.

Formal Names. Geneva System. IUPAC Rules for Alkanes. Representatives from chemical societies all over the world, meeting as an International Union of Pure and Applied Chemistry, with sessions held usually in Geneva, have recommended adoption of the following rules for naming alkanes.

1. The general name for saturated hydrocarbons is *alkane*.
2. The names of the straight-chain members of the alkanes are those listed in Table 9.3. (The designation n- is not included. Names going beyond the ten-carbon alkanes are, of course, available, but we shall not need them.)
3. For branched-chain alkanes, base the root of the name on the alkane that corresponds to the longest continuous (i.e., unbranched) chain of carbons in the molecule. For example, in the compound

$$\begin{array}{l} \qquad\qquad\quad CH_2{-}CH_2 \\ \qquad\qquad\quad | \qquad\quad | \\ CH_2{-}CH{-}CH_2 \quad CH_3 \\ | \qquad\quad | \\ CH_3 \quad\; CH_3 \end{array}$$

which when "straightened" is

$$\begin{array}{l} \qquad\quad CH_3 \\ \qquad\quad | \\ CH_3CH_2CHCH_2CH_2CH_2CH_3 \end{array}$$

the longest continuous chain totals seven carbons. The *last* part of the complete name for this compound will therefore be *heptane*. We next learn how to specify the location of the small CH_3— branch.

4. To locate branches, assign a number to each carbon of the longest continuous chain. Begin at whichever end of the chain will result in the smaller number or set of numbers for the carbon(s) holding branches. In our example,

$$\begin{array}{c} \quad\quad CH_3 \\ \quad\quad | \\ CH_3CH_2CHCH_2CH_2CH_2CH_3 \\ 1\quad 2\quad 3\quad 4\quad 5\quad 6\quad 7 \end{array}$$

If this chain had been numbered from right to left, the carbon holding the branch would have the number 5. Having located the branch(es), we must now be able to name it (them).

5. If a side chain or branch consists only of carbon and hydrogen linked with single bonds, it is called an *alkyl group;* "alkyl" comes from changing the "-ane" ending of "alkane" to "-yl." This change is the key to making up names for alkyl groups, and Table 9.5 lists the names and structures for the most common ones. Their structures must be learned so well that they can be recognized written backward, forward, and upside down. Each is derived (on paper) by taking an alkane and removing a hydrogen to leave an unused bond represented by a line. The site of this unused bond must be clearly known, for it is at this point that the group is attached to the chain. In learning the names and structures of the alkyl groups in

Table 9.5 Alkyl Groups—Names and Structures

Parent Alkane	Structure of Parent Alkane	Structure of Alkyl Group from Alkane	Name of Alkyl Group
Methane	CH_4	CH_3—	methyl
Ethane	CH_3CH_3	CH_3CH_2—	ethyl
Propane	$CH_3CH_2CH_3$	$CH_3CH_2CH_2$—	*n*-propyl
		$(CH_3)_2CH$—	isopropyl
n-Butane	$CH_3CH_2CH_2CH_3$	$CH_3CH_2CH_2CH_2$—	*n*-butyl
		$CH_3CH_2CH(CH_3)$—	secondary butyl (*sec*-butyl)
Isobutane	$CH_3CH(CH_3)CH_3$	$CH_3CH(CH_3)CH_2$—	isobutyl
		$CH_3C(CH_3)_2$—	tertiary butyl (*t*-butyl)

Any normal alkane: If the "free valence" extends from the *end* of the unbranched chain, change the "-ane" ending of the alkane to "-yl;" e.g., $CH_3CH_2CH_2CH_2CH_2CH_2CH_2$— is *n*-heptyl

Any alkane in general: R—H		R—	alkyl

Table 9.5, again the best advice is to associate the total number of carbons in the group with the prefix portion of the name of its "parent" alkane. The table shows, for example, four butyl groups, two related to *n*-butane and two to isobutane. All are butyl groups because each has four carbons. To distinguish them from each other, additional word parts are tacked on. The words "secondary" and "tertiary" (abbreviated *sec-* and *tert-* or simply *t-*) denote the condition of the carbon in the group having the unused bond. In the *sec*-butyl group this carbon is directly attached to *two* other carbons and is therefore classified as a *secondary* carbon. In the *t*-butyl group this carbon has direct bonds to three other carbons and is classified as a *tertiary* carbon. When a carbon is attached directly to but one other carbon, it is classified as a *primary* carbon. These classifications of carbons should be learned. In our study we shall on only two occasions use the names for the classes to make names for specific groups. Among the butyl groups, one is *sec*-butyl (a name) and one is *t*-butyl (a name). The other two are primary (a class), and the class name therefore cannot be used to name either, for it would be ambiguous to do so. But one is a straight chain with the unused bond at the *end;* it is the *n*-butyl group (or, simply, butyl). The other is called the isobutyl group.

Having learned rules for locating and naming side chains, we next consider situations in which identical groups are located on the same carbon of the main chain.

6. Whenever two identical groups are attached at the same place, numbers are supplied for each group. For example,

```
  CH3
  |
CH3CCH2CH2CH2CH3
  |
  CH3
```

Correct name: 2,2-dimethylhexane
Incorrect: 2-dimethylhexane
2,2-methylhexane

7. Whenever two or more different groups are affixed to a chain, two ways are acceptable for organizing all the name parts into the final name. The last word part is always the name of the alkane corresponding to the longest chain.

(*a*) The word parts can be ordered by increasing complexity of side chains (e.g., in order of increasing carbon content):

methyl, ethyl, propyl, isopropyl, butyl, isobutyl, *sec*-butyl, *t*-butyl, etc.

(*b*) They can be listed in simple alphabetical order. (This corresponds to the indexing system of *Chemical Abstracts,* a publication of the American Chemical Society.)

In this text we shall not be particular about the matter of order. For the following compound both names given are acceptable.

```
   CH3
   |
CH3CHCH2CHCH2CH2CH3
        |
        CHCH3
        |
        CH3
```

2-methyl-4-isopropylheptane
or 4-isopropyl-2-methylheptane

Note carefully the use of hyphens and commas in organizing the parts of names. Hypens always separate numbers from word parts, and commas always separate numbers. The intent is to make the final name one word.

8. The formal names for several nonalkyl substituents are as follows:

—F	fluoro	$—NO_2$	nitro
—Cl	chloro	$—NH_2$	amino[2]
—Br	bromo	—OH	hydroxy[2]
—I	iodo		

Several examples of compounds correctly named according to these rules are given below. In parentheses are shown some common ways in which incorrect names are often devised. As an exercise, describe how each incorrect name violates one or more of the rules.

$$\begin{array}{c} CH_3 \\ | \\ CH_3—C—CH_3 \\ | \\ CH_2 \\ | \\ CH_3 \end{array}$$

2,2-Dimethylbutane
Not 2-methyl-2-ethylpropane

$$\begin{array}{l} \quad\quad\quad\quad\quad\quad CH_3\ \ CH_3 \\ \quad\quad\quad\quad\quad\quad\quad \diagdown\ \diagup \\ CH_3—CH_2\quad CH \\ \quad\quad\quad\ \ |\quad\quad\ \ | \\ \quad\quad\quad CH_2—CH—CH_3 \end{array}$$

2,3-Dimethylhexane
Not 2-isopropylpentane

$$\begin{array}{l} \quad\quad\ \ CH_3\ \ CH_3 \\ \quad\quad\quad |\quad\quad\ | \\ CH_3—C——CH—CH_2—CH_3 \\ \quad\quad\quad | \\ \quad\quad\ \ CH_3 \end{array}$$

2,2,3-Trimethylpentane
Not 2,3-trimethylpentane
Not 2-*t*-butylbutane

$$\begin{array}{l} CH_3—CH_2—CH—Cl \\ \quad\quad\quad\quad\quad\quad\ | \\ \quad\quad\quad\quad\quad\quad Cl \end{array}$$

1,1-Dichloropropane
Not 3,3-dichloropropane
Not 3,3-chloropropane
Not 1-dichloropropane
Not 1,1-chloropropane

$$\begin{array}{l} \quad\quad\quad\ CH_3 \\ \quad\quad\quad\ \ | \\ CH_3—CH \\ \quad\quad\quad\ \ | \\ \quad\quad\quad\ CH_3 \end{array}$$

2-Methylpropane
Not 1,1-dimethylethane
Not isobutane, which is its *common* name

$$\begin{array}{l} \quad\quad\quad\quad\quad\quad\quad\quad\quad\quad\quad\quad CH_3 \\ \quad\quad\quad\quad\quad\quad\quad\quad\quad\quad\quad\quad\ | \\ CH_3CH_2CH_2CHCH_2CHCH_3 \\ \quad\quad\quad\quad\quad\quad\ \ | \\ \quad\quad\quad CH_3—C—CH_3 \\ \quad\quad\quad\quad\quad\quad\ \ | \\ \quad\quad\quad\quad\quad\quad CH_3 \end{array}$$

2-Methyl-4-*t*-butylheptane
Not 4-*t*-butyl-6-methylheptane
But 4-*t*-butyl-2-methylheptane is acceptable

[2] These two are used only in special circumstances. As we shall see in later chapters, amino compounds are usually named as amines and hydroxy compounds as alcohols, with special IUPAC rules.

Exercise 9.2 Write the structures (condensed) of each of the following.

(*a*) 1-bromo-2-nitropentane
(*b*) 2,2,3,3,4,4-hexamethyl-5-isopropyloctane
(*c*) 2,2-diiodo-3-methyl-4-isopropyl-5-*sec*-butyl-6-*t*-butylnonane
(*d*) 1-chloro-1-bromo-2-methylpropane
(*e*) 4,4-di-*sec*-butyldecane

Exercise 9.3 Write IUPAC names for each of the following.

```
(a) CH3—CH2
           \
            CH—CH3
           /
    CH2—CH2
    |
    CH3
```

```
                  CH3
                  |
              CH3—C—CH3
(b)       CH3     |
             \    |
              CH—CH—CH2—CH2—CH3
             /
      CH3—CH
          |
          CH3
```

```
         CH3        CH3          CH3
         |          |            |
(c) CH3—CH2—CH—CH—CH—CH2—CH—CH3
                |
                CH2—CH2—CH2—CH3
```

```
                  I
                  |
(d) Cl—CH2—CH—CH2—Br
```

```
                         CH3—CH—CH3
                             |
(e) CH3—CH2—CH2—CH—CH—CH2—CH2—CH3
                     |
                 CH3—C—CH3
                     |
                     CH3
```

```
                     CH3
                     |
(f) NO2—CH2—C—CH2—NO2
                     |
                     CH3
```

```
                     CH2—Cl
                     |
(g) CH3—CH2—CH—CH—CH3
                |
                CH3
```

```
         CH3                    CH2—CH3
         |                      |
(h) CH3—CH—CH2—CH2—CH—CH—CH3
                                |
                                CH3
```

```
(i) CH3
       \
        CH—CH2—CH2
       /          |
    CH3           CH3
```

```
(j) CH3—CH2—CH2        CH2—CH3
              |         |
              CH2—CH—CH—CH2CH2CH3
                  |
                  CH3
```

PHYSICAL PROPERTIES OF ALKANES

One striking feature of a homologous series in any family of organic compounds is the smooth change of a physical property as the series is ascended (cf. Table 9.3). As each additional CH_2 group is added, the boiling point, the melting point, and the density increase. These effects are attributed in part to the increased weight of the molecule and in part to small changes in forces of attraction between molecules. If molecules of higher homologs attract each other more strongly than those below them in the series, it will take more energy to break up the solid by melting. These forces will at the same time tend to make the larger molecules pack more closely together. With more weight in a given volume, the density must increase.

It is difficult to generalize about changes in these properties among the branched-chain isomers. Usually the more compact the molecules of an isomer are (as when they are highly branched), the lower its boiling point will be (cf., for example, *n*-pentane and neopentane, Table 9.4).

The data of Table 9.3 reveal one significant general physical property of alkanes. We recall that water with a formula weight of 18 boils at 100°C. In contrast, methane with nearly the same formula weight, 16, boils at −161.5°C (261.5° lower); ethane (formula weight 30) boils at −88.6°C. These experimental facts indicate that forces of attraction between molecules of alkanes must be far less than those between water molecules in liquid water. The alkanes are almost nonpolar compounds.

The alkanes are soluble in nonpolar solvents but not in water. They are less dense than water and will float on it.

What we are seeking here are general correlations between structure and property. Just as we have now learned to associate alkanes with low water solubility, we shall in the future be able to predict that *any molecule* of *any family* substantially *alkane-like* will also have this property. The alkanes are not directly involved in the chemical reactions that occur in living systems, but one of the three great classes of foods, the lipids, fats and oils, consist of molecules that are largely alkane-like. Fats and oils do not dissolve in water as the sugars do, for example, yet the great aqueous transport system of the body, the bloodstream, must somehow handle them. Before we can see how this is done, we must appreciate the root of the problem. We begin here our study of "map signs," of functional or nonfunctional groups, by which we later can "read" or predict properties of complicated systems. We have learned thus far that a molecule that is largely alkane-like will probably be insoluble in water, unless another "map sign" in the molecule, to be learned later, overrules this prediction.

We shall now examine chemical properties associated not just with the alkanes but with the *portions* of other types of molecules that are alkane-like. If we organize our study with this more general approach, the various families will not seem so isolated. Our ultimate goal is to study some examples of the molecular basis for several key processes in living systems. We are working our way toward this end by examining simpler but related systems.

CHEMICAL PROPERTIES

Alkanes are as a class chemically very unreactive. Carbon-to-carbon single bonds are extremely strong and very difficult to attack chemically; carbon-to-hydrogen bonds, *as they occur in an alkane environment,* are also exceptionally resistant to chemical attack except by fluorine, chlorine, bromine, oxygen, and hot nitric acid. Under normal circumstances, for example, at or near room temperature or below, concentrated sulfuric acid, sodium metal, and strong alkalies do not affect alkanes. It is thus understandable that alkanes were originally named paraffins (Latin *parum affinus,* little affinity). The only less reactive organic compounds are the fluorocarbons, alkanes in which all hydrogens have been replaced by fluorines. Teflon is an example.

Combustion. Oxygen is one of a few chemicals that will attack alkanes, but only at elevated temperatures. Combustion is their most useful chemical reaction. The most important product, of course, is heat. Natural gas is largely methane; propane is a common home-heating and cooking fuel known as bottle gas, or Pyrofax; heptane and octane, especially highly branched isomers of them, are found in gasoline. The twelve- to eighteen-carbon alkanes are found in kerosene, jet fuel, and tractor fuel. Diesel oil, fuel oil, and gas oil consist roughly of the twelve- to eighteen-carbon alkanes. The higher alkanes, those containing more than twenty carbons, are the principal components in refined mineral oils, lubricating oils and greases, and paraffin wax. Even asphalt and tar for roads and roofing contain large percentages of hydrocarbons.

The combustion of propane proceeds according to the following equation:

$$CH_3CH_2CH_3 + 5O_2 \longrightarrow 3CO_2 + 4H_2O \qquad \Delta H = -531 \text{ kcal/mole}$$

In the absence of enough oxygen, partial combustion may take place and produce carbon monoxide and even carbon. The former is a dangerous poison often present in exhaust fumes of cars. The latter may form deposits on motor pistons and in cylinder heads.

Chlorination and Bromination. If a mixture of an alkane and either chlorine or bromine is heated or exposed to ultraviolet radiation, a *substitution reaction* takes place—a reaction in which one atom or group already on an organic molecule is replaced (substituted) by another. In this example a hydrogen on the alkane is replaced by a chlorine atom or a bromine atom. In general:

$$\underset{\substack{(X = Cl \\ \text{or Br})}}{R{-}H + X_2} \xrightarrow[\text{ultraviolet radiation}]{\text{heat or}} R{-}X + H{-}X$$

Specific examples:

$$CH_4 + Cl_2 \xrightarrow{\text{heat}} \underset{\substack{\text{Methyl chloride} \\ (\text{bp } -24°C)}}{CH_3{-}Cl} + H{-}Cl$$

$$CH_3CH_3 + Cl_2 \xrightarrow[\text{ultraviolet radiation}]{} \underset{\substack{\text{Ethyl chloride} \\ (\text{bp } 13°C)}}{CH_3CH_2{-}Cl} + HCl$$

$$CH_3CH_2CH_3 + Cl_2 \xrightarrow{\text{light}} \underset{\substack{n\text{ Propyl chloride (45\%)} \\ (\text{bp } 47°C)}}{CH_3CH_2CH_2{-}Cl} + \underset{\substack{\text{Isopropyl chloride (55\%)} \\ (\text{bp } 36°C)}}{CH_3\underset{\substack{| \\ Cl}}{C}HCH_3}$$

Since there two different sets of hydrogens in propane that are not equivalent to each other, two monochloropropanes can and do form. This is one drawback of the reaction; if mixtures of isomers are possible, they usually form and are not always easy to separate into their pure components. If the products are industrially useful, however, time and money are invested in solving that problem.

The halogenation of alkanes presents another problem. The monohalogenated product forms in the presence of unreacted halogen and competes for it with the original, still unchanged alkane. Thus if one mole of chlorine and one mole of methane are allowed to react, the final mixture can consist of methyl chloride, dichloromethane, chloroform, and carbon tetrachloride in addition to hydrogen chloride and unchanged methane:

$$\underset{\substack{\text{Methane} \\ (\text{bp } -162^\circ\text{C})}}{CH_4} \xrightarrow{Cl_2} \underset{\substack{\text{Methyl chloride} \\ (\text{bp } -24^\circ\text{C})}}{CH_3Cl} \xrightarrow{Cl_2} \underset{\substack{\text{Dichloro-} \\ \text{methane} \\ (\text{bp } 40^\circ\text{C})}}{CH_2Cl_2} \xrightarrow{Cl_2} \underset{\substack{\text{Chloroform} \\ (\text{bp } 61^\circ\text{C})}}{CHCl_3} \xrightarrow{Cl_2} \underset{\substack{\text{Carbon} \\ \text{tetrachloride} \\ (\text{bp } 77^\circ\text{C})}}{CCl_4}$$

All products are important commercially.

PREPARATION ALKANES

In addition to natural sources such as crude oil, several laboratory methods for making alkanes have been developed. Only one of these is at all relevant to our future study of the molecular basis of life, the hydrogenation of a carbon-carbon double bond. In the laboratory a catalyst such as powdered platinum, palladium, or nickel plus high temperature and pressure bring about the combination of hydrogen with an alkene or an alkyne. For example,

$$\underset{\text{Ethylene}}{CH_2{=}CH_2} + H_2 \xrightarrow[\text{heat, pressure}]{\text{Pt or Pd}} \underset{\text{Ethane}}{CH_3CH_3}$$

$$\underset{\text{Acetylene}}{H{-}C{\equiv}C{-}H} + 2H_2 \xrightarrow[\text{heat, pressure}]{\text{Pt or Pd}} \underset{\text{Ethane}}{CH_3CH_3}$$

This reaction is discussed further in the next chapter. Reactions similar to it occur in living organisms, but enzymes are the catalysts and the hydrogen is obtained from another molecule.

CYCLOALKANES

A carbon skeleton can take the form of a closed ring as well as of an open chain. The simplest of such cyclic compounds are called the cylcloalkanes or cycloparaffins. The following examples are illustrative.

```
CH2—CH2         CH2—CH2              CH2                  CH2
  \  /          |    |             /    \              /     \
   CH2          CH2—CH2         CH2      CH2         CH2     CH2
                                  \      /            |       |
                                  CH2—CH2            CH2     CH2
                                                        \   /
                                                         CH2
```

Cyclopropane | Cyclobutane | Cyclopentane | Cyclohexane

Rings of higher carbon content (over thirty) are also known. Since the molecules in these compounds have only single bonds, they generally exhibit the same types of physical and chemical properties as the open-chain alkanes. For this reason we shall not study them further. To symbolize ring systems chemists use geometric figures as explained in Figure 9.11.

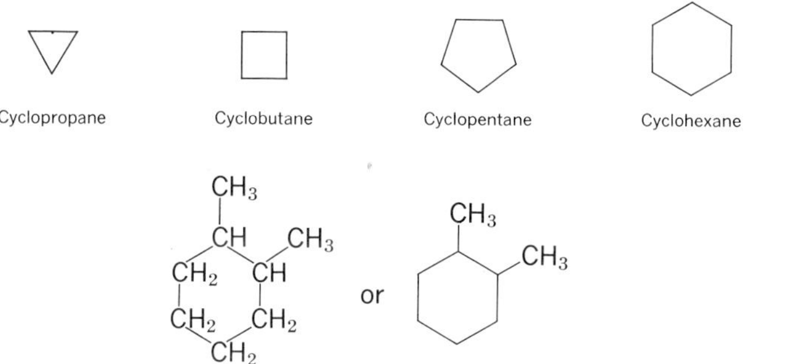
Cyclopropane
Cyclobutane
Cyclopentane
Cyclohexane
CH3
CH
CH3
CH2
CH
CH2
CH2
CH2
or
CH3
CH3
1,2-Dimethylcyclohexane

Figure 9.11 Symbolism and nomenclature in cycloalkane systems. At each corner in the geometric symbols a carbon is understood to be present. The lines from corner to corner are carbon-carbon bonds. Unless otherwise indicated, hydrogens are assumed to be present at the corner carbons in sufficient number to fill out carbon's tetravalence. When substituents occupy corners on the ring, the ring positions are numbered in a direction and from a beginning point that together yield the set of smallest numbers possible. Alkyl groups are usually given smaller numbers than halogens.

GEOMETRICAL ISOMERISM. "CIS-TRANS" ISOMERISM

Two different 1,2-dimethylcyclopropanes (Figure 9.12) are known. The lack of free rotation about single bonds in ring systems makes their formation possible.

This kind of isomerism is known as *geometric isomerism*. The difference between these isomers lies not so much *where* on a chain or skeleton substituents are located but rather in their geometric orientation. Because two different compounds cannot have the same name, chemists call the isomer in which the substituents project in generally opposite directions the *trans* isomer. In the *cis* isomer the substituents generally protrude in the same direction.

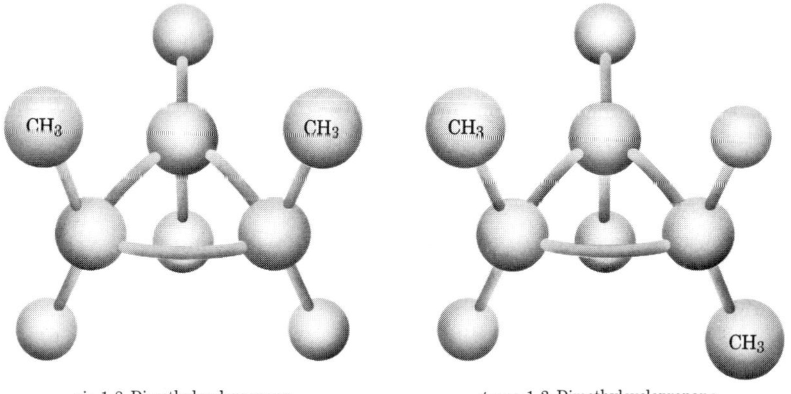
CH3
CH3
CH3
CH3
cis-1,2-Dimethylcyclopropane
trans-1,2-Dimethylcyclopropane

Figure 9.12 Geometric isomers of 1,2-dimethylcyclopropane.

Geometrical isomerism is but one type of a general kind of isomerism, *stereoisomerism* (Greek *stereos*, solid, meaning here three-dimensional shape). One other type, optical isomerism, we shall take up later. In general, stereoisomerism has profound implications for the chemical events in living organisms. An instance that is almost amusing is the difference between two geometric isomers of trimedlure, a sex attractant for the Mediterranean fruit fly. The trans isomer is a more powerful attractant than the cis isomer. Traps baited with *trans*-trimedlure and a rapid-acting insecticide have been used in Florida since 1957 to eradicate this pest, which annually used to cause severe damage to Florida's fruit crops. Although the exact chemistry of how the attractant works is not known, molecular geometry is as critical a factor as molecular structure. A likely explanation will be discussed in Chapter 19.

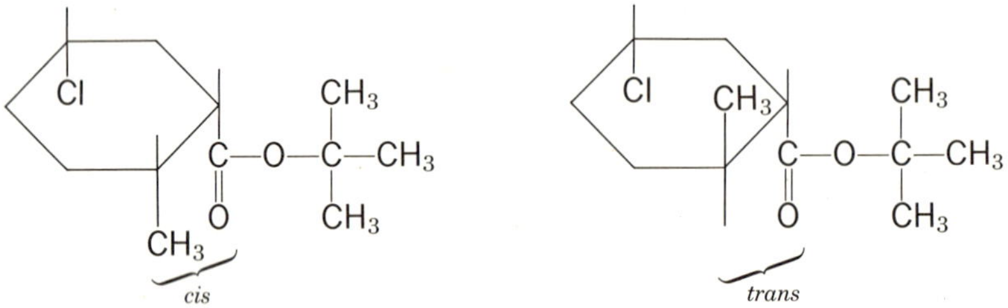

Cl CH_3 C—O—C—CH_3 O CH_3 CH_3 *cis*

Cl CH_3 CH_3 C—O—C—CH_3 O CH_3 *trans*

Trimedlure isomers

REFERENCES AND ANNOTATED READING LIST

BOOKS

O. T. Benfey. *From Vital Force to Structural Formulas.* Houghton Mifflin Company, Boston, 1964. The historical development of the structural theory in organic chemistry is presented with many interesting excerpts (in translation) from original works.

N. L. Allinger and J. Allinger. *Structures of Organic Molecules.* Prentice-Hall, Englewood Cliffs, N.J., 1965. Structures, the kinds of evidence for them, and stereoisomerism are discussed. (Paperback.)

O. Runquist. *A Programmed Review of Organic Chemistry, Nomenclature.* Burgess Publishing Company, Minneapolis, Minn., 1965. Nomenclature is easily self-taught, and this book is designed for that.

ARTICLES

C. D. Hurd. "The General Philosophy of Organic Nomenclature." *Journal of Chemical Education,* Vol. 38, 1961, page 43. Some of the reasons behind the rules of nomenclature are discussed, and a short bibliography to systems of naming compounds is given.

PROBLEMS AND EXERCISES

1. Check each of the following structures to see whether within the rules of covalence they represent possible compounds.

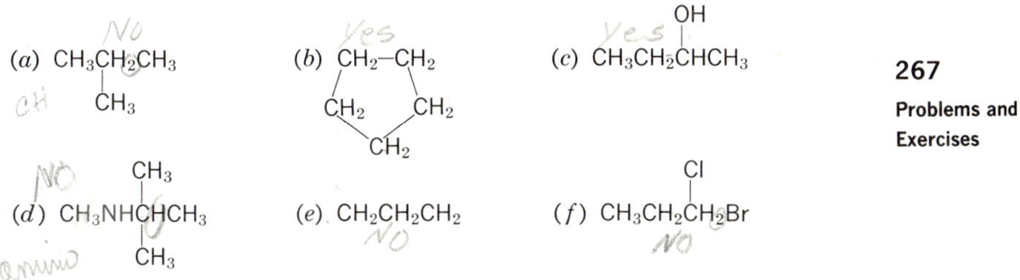
(a) CH3CH2CH3
CH3
(b) CH2—CH2
CH2 CH2
CH2
OH
(c) CH3CH2CHCH3
CH3
(d) CH3NHCHCH3
CH3
(e) CH2CH2CH2
Cl
(f) CH3CH2CH2Br

2. Convert each of the following structural formulas into a condensed structure. Straighten out the longest continuous carbon chain so that it is written on one line. (Cyclic structures, of course, are exceptions.)

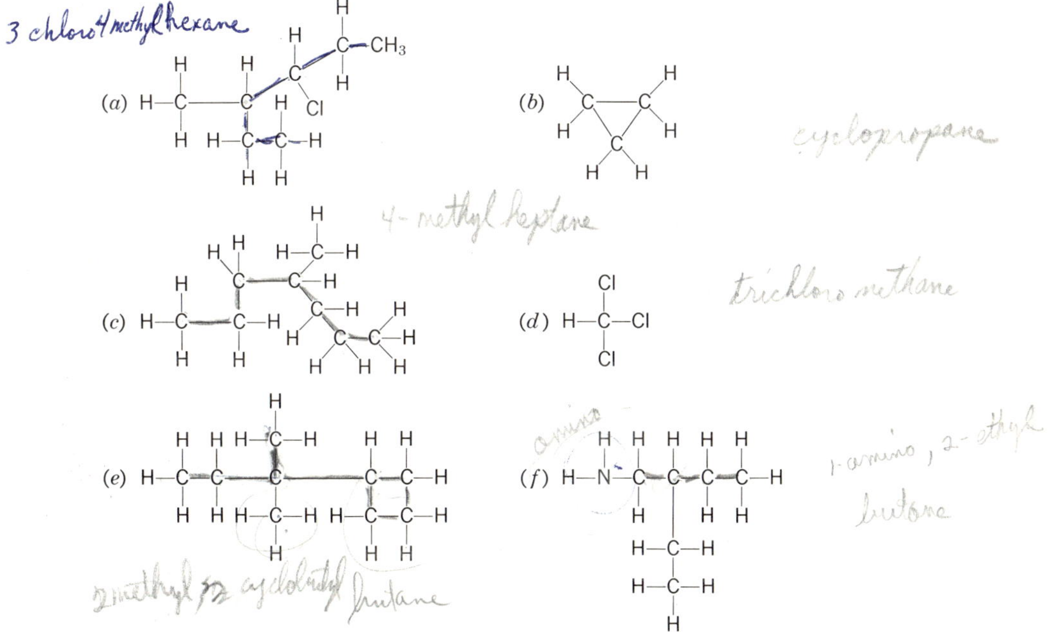

3. Write IUPAC names for structures *a* to *d* in problem 2.
4. For each of the following pairs of structures, decide whether the two are *identical, isomers,* or *neither*. With some pairs the only difference is the relative orientation of the same structure in space (i.e., on paper). Others differ only in relative rotational position within the molecule. Assume free rotation about all single bonds except those that are part of a cyclic system.

(*a*) $CH_3—CH_2(—CH_2—CH_2—CH_3)$ $CH_3—CH_2—CH_2—CH_2—CH_3$

(*b*) $CH_3—O—CH_2CH_3$ $CH_3CH_2—O—CH_3$

(*c*) 1,2-dimethylcyclohexane (CH_3, CH_3)   (CH_3, CH_3)

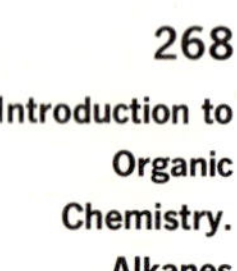

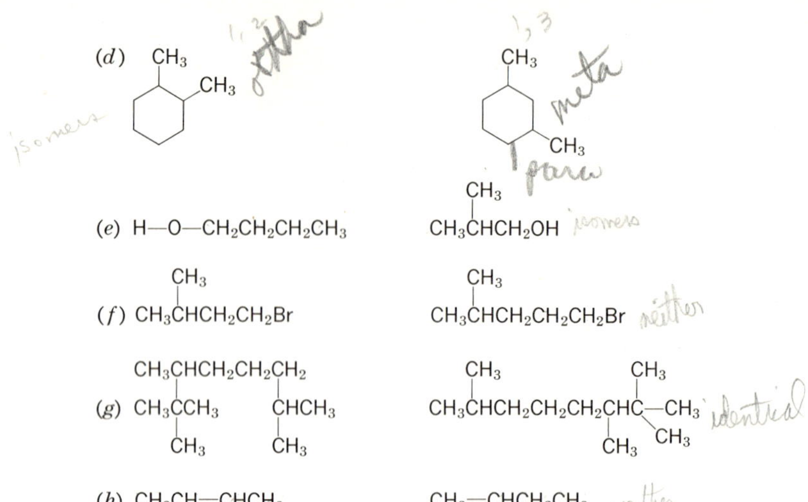

(*e*) $H—O—CH_2CH_2CH_2CH_3$ $CH_3CH(CH_3)CH_2OH$

(*f*) $CH_3CH(CH_3)CH_2CH_2Br$ $CH_3CH(CH_3)CH_2CH_2CH_2Br$

(*h*) $CH_3CH{=}CHCH_3$ $CH_2{=}CHCH_2CH_3$

5. Write condensed structures for all the isomeric monochloro derivatives of 2,2,3-trimethylpentane.
6. If the common name of chloromethane is *methyl chloride* and if the common name for chloroethane is *ethyl chloride,* what are the common names for the following?

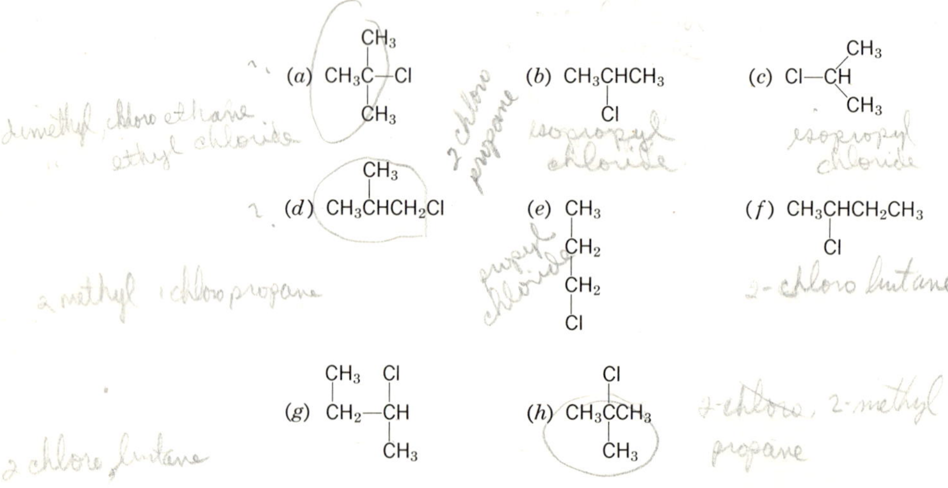

7. From the standpoint of the development of organic chemistry, what was the significance of the vital force theory?
8. In Chapter 4 we studied a model for the water molecule. From it we predicted a bond angle of 90°, but experimental fact informs us that it is 104.5°, a serious discrepancy. At that time we corrected the model by assuming that the two hydrogen nuclei repel each other enough to force the angle larger. And we took comfort in the fact that this assumption was consistent with what is known about hydrogen sulfide (see p. 128).

   An alternative model is possible; let us see if you can, with prompting, figure it out.
   (*a*) Write the electronic configuration of an oxygen atom, that is, fill in the boxes according to the aufbau rules:

$2p$ ☐☐☐

$2s$ ☐

$1s$ ☐

(*b*) Concentrate your attention on the second level, the valence shell; for the moment forget about the electrons there and consider only the orbitals. Simply note that the second level has one s orbital and three p-orbitals; the total is four orbitals. Now imagine that these are all mixed or hybridized just as they were for the carbon in methane. (The mixing of orbitals can be imagined whether or not they will all eventually contain electrons. We shall put electrons in later with due regard to the aufbau rules.) What would be the appropriate symbol for your new hybrid orbitals, sp, sp^2, or sp^3?
(*c*) Write a new set of boxes to represent all the orbitals of both principal energy levels available for the electrons of oxygen. (*Hint.* See page 247.)
(*d*) Now that you have the boxes, place a total of eight electrons in them, paying full heed to the aufbau rules.
(*e*) How many unpaired electrons do you have?
(*f*) Does your answer check with the known covalence of oxygen?
(*g*) How would these orbitals have to be arranged to minimize interelectronic repulsions (and minimize the internal energy)?
(*h*) If two hydrogen atoms, each with a $1s$ electron, were brought up to an oxygen atom in this valence state, what would the predicted bond angle be?
(*i*) How does this compare with the known angle of 104.5°? Is the discrepancy as great as we noted earlier for the other model?
A question now arises. What is the truth about the water molecule? It may annoy you at first, but the truth consists only of experimental data concerning composition, bond angles, and bond lengths. Which *model* we should use to correlate this information is not a choice the data permit us to make easily. You must realize that a model is a model, a conceptual way for organizing data—as much data for as many compounds as possible. The fact that the model we developed in this question appears to give a better prediction of the bond angle for water is in its favor. But then if we apply it to hydrogen sulfide, it is not as good as that given in Chapter 4. Part of the difficulty lies in the fact that in both models we are trying to use *hydrogen*-like orbitals for nonhydrogen nuclei.

CHAPTER TEN

Unsaturated Aliphatic Hydrocarbons

ALKENES

THE CARBON-CARBON DOUBLE BOND

The carbon-carbon double bond, which occurs in a wide variety of natural products, is a site of unsaturation in a molecule. Many of the lipids, the fats and oils, are unsaturated, as anyone who sees advertisements for "polyunsaturated cooking oils" knows. Several of the simpler alkenes are listed in Table 10.1.

The first member of the alkene family, ethylene, has the molecular formula C_2H_4. Within the rules of covalence it can have only one structure, and this must

Table 10.1 Properties of Some 1-Alkenes

Name (IUPAC)	Structure	Bp (°C)	Mp (°C)	Density (in g/cc at 10°C)
Ethene	$CH_2{=}CH_2$	−104	−169	—
Propene	$CH_2{=}CHCH_3$	−48	−185	—
1-Butene	$CH_2{=}CHCH_2CH_3$	−6	−185	—
1-Pentene	$CH_2{=}CHCH_2CH_2CH_3$	+30	−165	0.641
1-Hexene	$CH_2{=}CHCH_2CH_2CH_2CH_3$	64	−140	0.673
1-Heptene	$CH_2{=}CHCH_2CH_2CH_2CH_2CH_3$	94	−119	0.697
1-Octene	$CH_2{=}CHCH_2CH_2CH_2CH_2CH_2CH_3$	121	−102	0.715
1-Nonene	$CH_2{=}CHCH_2CH_2CH_2CH_2CH_2CH_2CH_3$	147	−81	0.729
1-Decene	$CH_2{=}CHCH_2CH_2CH_2CH_2CH_2CH_2CH_2CH_3$	171	−66	0.741
Cyclopentene		44	−135	0.722
Cyclohexene		83	−104	0.811

include a carbon-carbon double bond. Long before modern theories of chemical bonds were developed, it was known that one of the two bonds in the double bond

$$\mathrm{H_2C{=}CH_2}$$

Ethylene

is quite reactive, especially toward acids. To understand the properties of this system, we must see what modern theory has to say.

***sp*² Hybridization.** When carbon is attached to four other atoms or groups, it utilizes sp^3 hybrid orbitals. In a double bond, however, each carbon is attached to only three other groups and, in terms of the theory most commonly used by organic chemists, carbon uses different hybrid orbitals. We start with carbon in its so-called atomic state:

$2p$		↑ ↑ ( )
$2s$	↑↓	
$1s$	↑↓	

Again, as before, we envision the promotion of a $2s$ electron into the empty $2p_z$ orbital to give the "promoted state":

$2p$		↑ ↑ ↑
$2s$	↑	
$1s$	↑↓	

At this point we follow a different course than that used for sp^3 hybridization in which all four orbitals at the second level were "mixed." Instead we "mix" the $2s$ and just two of the $2p$ orbitals. Hence the designation sp^2 (s–p–two) hybridization. The third $2p$ orbital is left unhybridized. Thus in this theory the valence state for carbon when it is bonded to only three groups is, in terms of an energy level diagram,

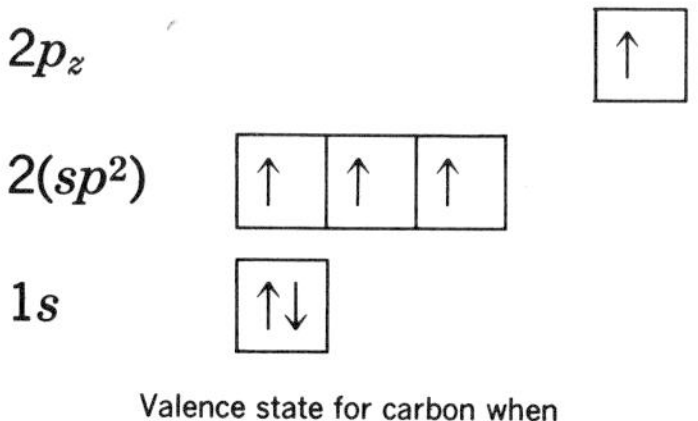
$2p_z$
↑
$2(sp^2)$
↑ ↑ ↑
$1s$
↑↓

Valence state for carbon when bonded to just three groups

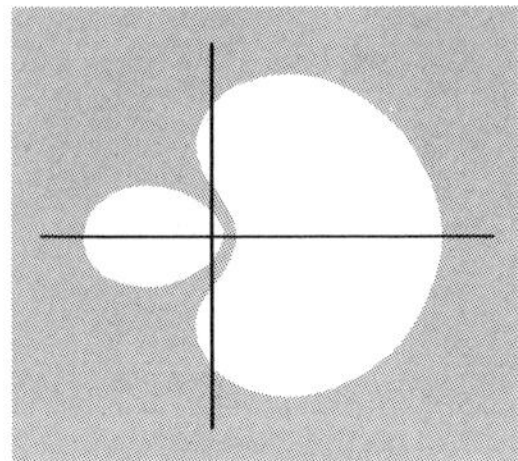

Figure 10.1 Cross section of an sp^2 hybrid orbital.

There are three new hybrid orbitals at the second level plus one unhybridized p orbital. The shapes of these new sp^2 hybrid orbitals are approximately those shown in Figure 10.1, each with both s character and p character. The arrangement of minimum energy for these orbitals, each with one electron, is shown in Figure 10.2.

The Pi Bond. Now that we have the orbital arrangement believed to exist for a carbon in a double bond, we shall assemble in Figures 10.3 and 10.4 two such carbons plus four hydrogens to show the final structure of ethylene. Of special importance is the overlap between the two unhybridized p_z orbitals (Figure 10.4). If this overlap occurs, the energy is also lowered and the two "halves" can have only one orientation in relation to each other. Any rotation about the carbon-carbon bond would break the overlap, but at room temperature the system does not have enough energy for this rupture. For all practical purposes there is no free rotation about a carbon-carbon double bond. Hence geometrical isomerism will be possible for ethylenic hydrocarbons.

One of the two bonds in a double bond is a sigma bond (or σ-bond). The overlap between an sp^2 orbital of one carbon and an sp^2 orbital of the other gives a new molecular orbital with symmetry about the bonding axis. The second bond results

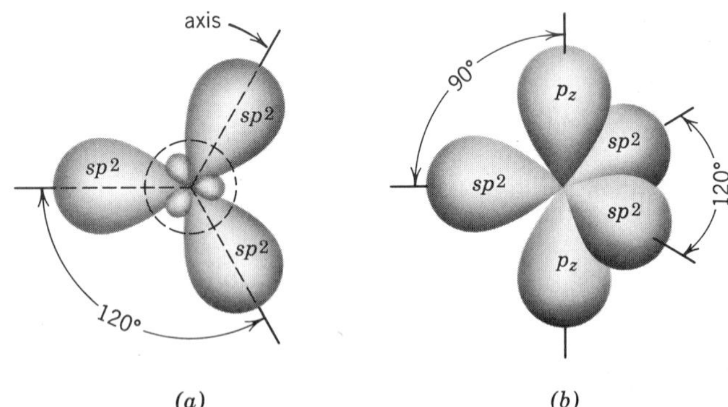
axis
sp2
sp2
sp2
120°
90°
pz
sp2
sp2
sp2
pz
120°
(a)
(b)

Figure 10.2 Bonding orbitals available from carbon when it is bonded to just three other groups. (a) Top view in which the axes of the sp^2 orbitals all lie in the plane of the paper, forming angles of 120°. The circular dashed line locates the p_z orbital that is perpendicular to the plane. (b) Perspective view, showing the larger lobes of the sp^2 orbitals and the two lobes of the unhybridized p_z orbital.

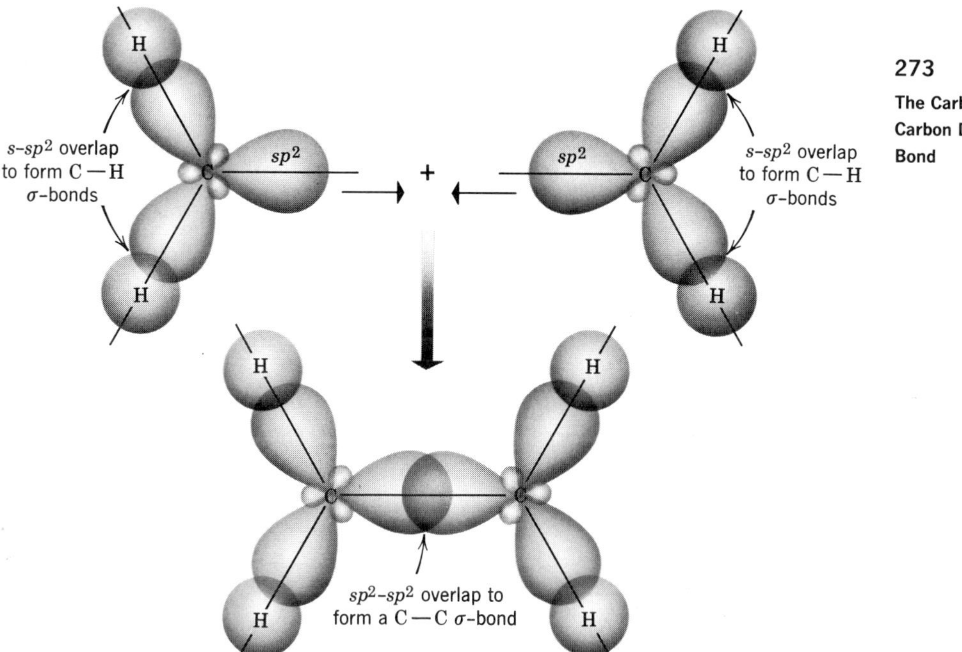
H
H
s–sp2 overlap to form C—H σ-bonds
sp2
C
+
sp2
C
s–sp2 overlap to form C—H σ-bonds
H
H
H
H
C
C
sp2–sp2 overlap to form a C—C σ-bond
H
H

Figure 10.3 The σ-bond network in ethylene consists of one carbon-carbon bond and four carbon-hydrogen bonds. The p_z orbitals are not shown, but their lobes lie above and below the plane of the page with their axes piercing the paper at each carbon.

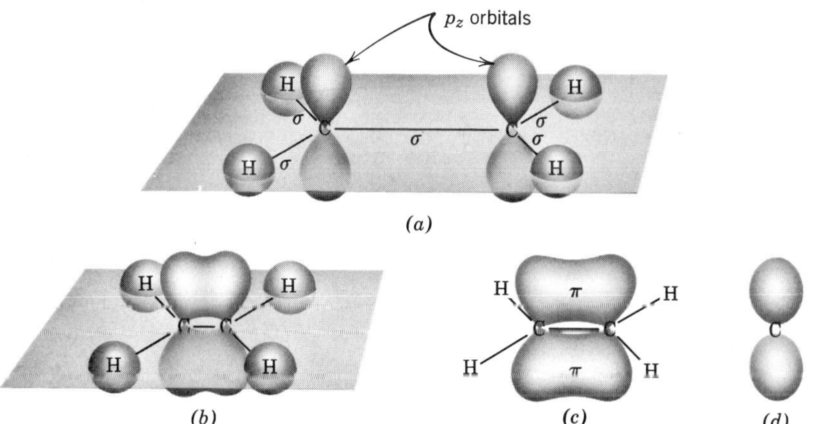
pz orbitals
H
σ
C
σ
H
σ
H
C
σ
σ
H
(a)
H
H
C
C
H
H
(b)
H
π
H
C
C
H
π
H
(c)
C
(d)

Figure 10.4 The nature of the bonding and the geometry at a carbon-carbon double bond, according to one common theory. The solid lines labeled σ are described in Figure 10.3.

(*a*) This is the hypothetical situation before the p_z orbitals overlap. (*b*) Maximum overlap of the p_z orbitals can occur only when the six nuclei shown here lie in the same plane. (*c*) The double bond in ethylene consists of one σ-bond (the heavy line) and one π-bond. The π-bond has the appearance of two sausages above and below the plane. (*d*) A cross-sectional view of the π-bond looking down the carbon-carbon axis shows its two lobes.

from the overlap of two p orbitals and does not have this symmetry. Having two lobes of identical shape, it resembles a p orbital and is called a pi bond (or π-bond). Its two sausage-shaped lobes together constitute one molecular orbital, and somewhere in this region are two electrons whose electric charges help to hold the two carbon nuclei together. The two electrons in the sigma bond plus the two in the pi bond provide a stronger attractive force for the carbon nuclei, which are closer together than in ethane. The carbon-carbon distance in ethane is 1.54 Å; in ethylene it is 1.33 Å.

The electrons in the pi bond are called the pi electrons, and they are not located as close to the skeleton as the sigma electrons are. The overlap in the pi bond is not as effective as that in the sigma bond, with the net result that the pi electrons are capable of reacting with electron-poor species (Lewis acids, including protons). As we shall see, *the key to the chemistry of the double bond is the availability of its pi electrons*.

ISOMERISM AMONG THE ALKENES

Isomers are possible in the alkene family whenever the double bond can be located differently between carbons or whenever the groups attached to the double bond permit cis-trans relations. Four isomers that have the formula C_4H_8 and are alkenes are known. Their structures are

$CH_2{=}CHCH_2CH_3$ — 1-Butene (α-butylene)

$(CH_3)(H)C{=}C(CH_3)(H)$, both CH_3 on the same side — *cis*-2-Butene (*cis*-β-butylene)

$(CH_3)(H)C{=}C(H)(CH_3)$, CH_3 groups on opposite sides — *trans*-2-Butene (*trans*-β-butylene)

$(CH_3)_2C{=}CH_2$ — 2-Methyl-1-propene (isobutylene)

In 1-butene the double bond is between the first and the second carbon. In both of the 2-butenes it is between the second and third. If there were free rotation about a double bond, 2-butene, like n-butane, could have no geometric isomers. But two

Table 10.2 Physical Properties of the Butenes and Pentenes

Name	Bp (°C)	Mp (°C)	Density (in g/cc)
Butenes			
2-Methyl-1-propene	−6.90	−140.4	0.640 (−20°C)
1-Butene	−6.26	−185	0.641
trans-2-Butene	+0.88	−105.6	0.649
cis-2-Butene	+3.72	−138.9	0.667
Pentenes			
1-Pentene	30.0	−165.2	0.641 (+20°C)
cis-2-Pentene	36.9	−151.4	0.656
trans-2-Pentene	36.4	−140.2	0.648
2-Methyl-1-butene	31.2	−137.5	0.650
3-Methyl-2-butene	20.1	−168.5	0.627
2-Methyl-2-butene	38.6	−133.8	0.662

2-butenes are known. Their physical properties are listed in Table 10.2, which is included only to demonstrate that differences in structure do mean differences in observable properties. The six isomeric pentenes are also shown in Table 10.2.

Exercise 10.1 Identify each of the following compounds by name.

(*a*) $CH_3(H)C{=}C(H)CH_3$ (CH₃ and H on upper carbon; H and CH₃ on lower carbon)

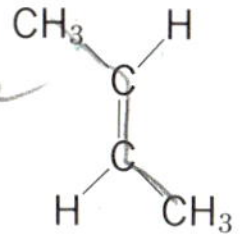

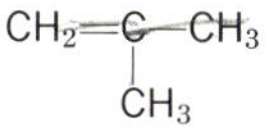

(*b*) $CH_3CH_2CH{=}CH_2$

(*c*) $CH_2{=}C(CH_3){-}CH_3$

(*d*) $(H)(CH_3)C{=}C(H)(CH_3)$ (H, H above; CH₃, CH₃ below)

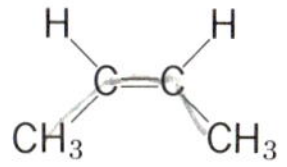

Exercise 10.2 Examine each of the following structures, write trial structures for geometric isomers, and determine whether or not cis-trans isomerism is possible.

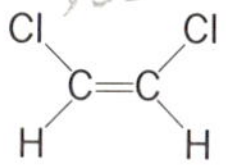

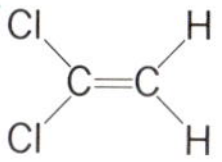

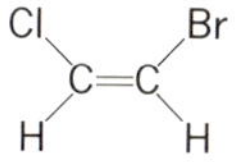

(*a*) $CH_3{-}CH_2(H)C{=}C(H)(H)$

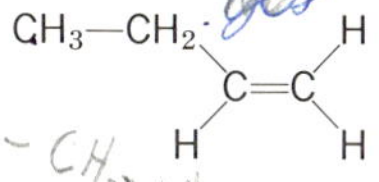

(*b*) $(Cl)(H)C{=}C(Cl)(H)$

(*c*) $(Cl)(Cl)C{=}C(H)(H)$

(*d*) $(Cl)(H)C{=}C(Br)(H)$

(*e*) $(Cl)(Br)C{=}C(H)(H)$

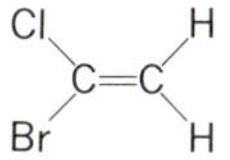

(*f*) $CH_3CH_2CH{=}CHCH_2CH_3$

(*g*) $(CH_3CH_2)(CH_3CH_2)C{=}C(H)(CH_3)$

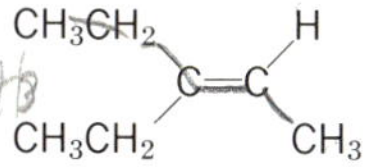

Exercise 10.3 Which of the following statements is true concerning the conditions under which cis-trans isomerism in simple alkenes is possible? Check these statements against the structures in exercise 10.2.

(*a*) Geometric isomerism *can* exist if *either* carbon of the double bond has two identical groups attached to it.

(*b*) Geometric isomerism *cannot* exist if *either* carbon of the double bond has two identical groups attached to it.

In general, physical properties of isomeric alkenes are different but the chemical properties, which depend more on the presence of a double bond than on its precise location, are quite similar. What differences occur, aside from actual products formed, have to do more with the *rates* of reactions than with the *types* of reactions. For this reason, unless attention to cis-trans isomerism is important, alkenes may be represented in simplified, one-line structures. For example, 2-butene may be written $CH_3CH{=}CHCH_3$. (Note that the double bond is never "understood" in condensed structures as single bonds are. It is always written into even the most condensed structures.)

NOMENCLATURE

Common Names. The common names of alkenes have the same ending, "-ylene." An alkene with the carbon skeleton of the three-carbon alkane, propane, is called "propylene." The "prop-" prefix indicates a total of three carbons; the "-ylene" suffix places it in the ethylene or alkene family. The various butylenes (cf. p. 274) are differentiated arbitrarily by the prefixes α, β, and "iso-." In the homologous series the common system loses most of its practicality above the butylenes, and the IUPAC system is used.[1]

IUPAC Names. The IUPAC rules for naming the alkenes are as follows.

1. The characteristic name ending is "-ene."

2. The longest continuous chain *that contains the double bond* is selected. This chain is named by selecting the alkane with the *identical* chain length and changing the suffix in its name from "-ane" to "-ene."

3. The chain is numbered to give the *first* carbon of the double bond the lowest possible number (e.g., $CH_3CH_2CH{=}CH_2$ is 1-butene, not 1,2-butene and not 3-butene).

4. The locations of the groups attached to the main chain are identified by numbers. The word parts are then assembled in a manner analogous to that in alkane nomenclature.

The following examples illustrate correct application of these rules. Study them, noting especially the placing of commas and hyphens. Because rule 3 gives precedence to the double bond, rather than to the side chains, positions bearing side chains will sometimes have larger numbers than they would if an alternate numbering were used. Common names are given in parentheses.

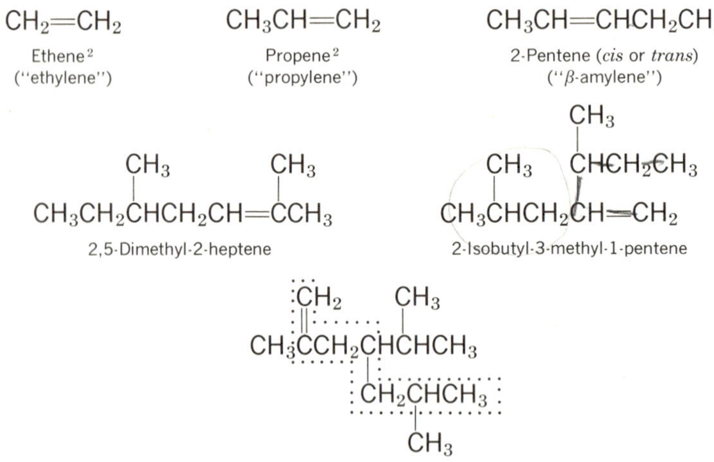
$CH_2{=}CH_2$
Ethene[2]
("ethylene")
$CH_3CH{=}CH_2$
Propene[2]
("propylene")
$CH_3CH{=}CHCH_2CH_3$
2-Pentene (*cis* or *trans*)
("β-amylene")
$CH_3CH_2CH(CH_3)CH_2CH{=}C(CH_3)CH_3$
2,5-Dimethyl-2-heptene
$CH_3CH(CH_3)CH_2C(=CH_2)CH(CH_3)CH_2CH_3$
2-Isobutyl-3-methyl-1-pentene
$CH_3C(=CH_2)CH_2CH(CHCH_3CH_3)CH_2CH(CH_3)CH_3$
4-Isopropyl-2,6-dimethyl-1-heptene
(Dotted lines enclose the longest chain having the double bond.)

[1] A third system of nomenclature is known and used, but we shall not employ it in this text.

[2] Whenever the double bond cannot be located differently to make an isomer, its position need not be designated by a number.

Some important unsaturated groups that occur many times in organic systems have common names, for example,

$$CH_2{=}CH{-} \qquad CH_2{=}CHCH_2{-}$$

Vinyl Allyl

Thus the compound $CH_2{=}CH{-}Cl$ is nearly always called *vinyl chloride,* although its formal name is chloroethene; the compound $CH_2{=}CHCH_2{-}Br$ is almost always called *allyl bromide,* although its formal name is 3-bromo-1-propene.

Exercise 10.4 Write condensed structures for each of the following.

(*a*) 4-methyl-2-pentene (*b*) 3-*n*-propyl-1-heptene
(*c*) 3,3-dimethyl-4-chloro-1-butene (*d*) 2,3-dimethyl-2-butene
(*e*) allyl iodide (*f*) vinyl bromide

Exercise 10.5 Write IUPAC names and the common names if they have already been mentioned for each of the following. Use cis or trans designations if they are applicable.

(*a*) $(CH_3)_2C{=}CH_2$

(*b*) $CH_3{-}CH(CH_3){-}CH_2{-}C(CH_3){-}CH_2{-}CH(CH_3){-}CH_3$, with $C{=}C(CH_3)CH_2{-}CH_3$ (double bond from the central C down to C bearing CH_3 and $CH_2{-}CH_3$)

(*c*) $CH_3{-}CH{=}CH{-}Cl$

(*d*) $Br{-}CH_2{-}CH{=}CH_2$

(*e*) $CH_3{-}CH(CH_2{-}CH_3){-}CH_2{-}CH{=}CH_2$

(*f*) If [cyclohexene ring] is called cyclohexene, what is its *full* structure? What is the structure of cyclopentene?

(*g*) If [ring with CH_3 at position 3; ring positions numbered 1–6, double bond between 1 and 2] is 3-methylcyclohexene,[3] what is [cyclohexene ring with CH_3 on each of the two carbons adjacent to the double bond side] ?

PHYSICAL PROPERTIES OF ALKENES

The alkenes, as a class, closely resemble the alkanes in physical properties; this generalization is the most useful one we can carry forward to our study of other families. Alkene molecules, which in the higher homologs are mostly alkane-like anyway, are relatively nonpolar. Alkenes are insoluble in water, but they dissolve in the typical nonpolar solvents such as benzene, ether, carbon tetrachloride, chloroform, and ligroin (a mixture of liquid alkanes).

[3] The double bond in an acyclic system is *always* regarded as *starting from* position 1. Numbering then proceeds through the double bond around the ring in the direction that will give the lower set of numbers to the substituents.

CHEMICAL PROPERTIES OF ALKENES

There are basically two sites in alkenes that are attacked by chemicals. One is the double bond itself, the other the position on the chain directly attached to the double bond, the *allylic position.* In this chapter we study reactions at the double bond itself.

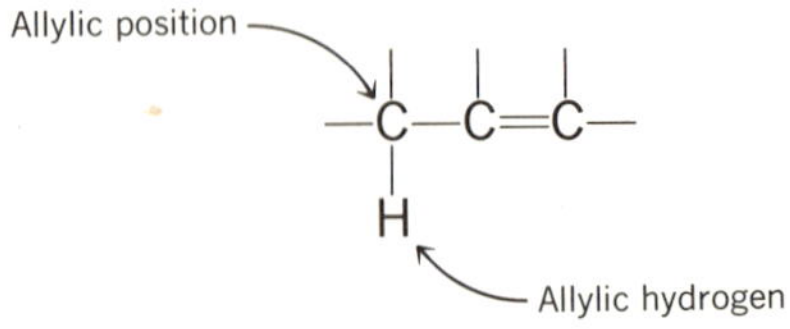

$$\underset{\substack{| \\ H}}{\overset{|}{-C}}-\overset{|}{C}=\overset{|}{C}-$$

Allylic position (arrow to the carbon); Allylic hydrogen (arrow to H)

Addition Reactions. In marked contrast to the alkanes, alkenes undergo many reactions, nearly all involving attack at the double bond. In further contrast to alkanes, these reactions are not substitutions. They are a new type called addition reactions, that is, two molecules come together and react—that is, join—without loss of any fragment from either. (We recall that chlorination of methane, a substitution reaction, produces a fragment, H—Cl, with the hydrogen coming from the alkane.) The following are typical addition reactions of alkenes.

1. **Halogenation,** the addition of chlorine or bromine (only these two halogens will add to a double bond). In general:

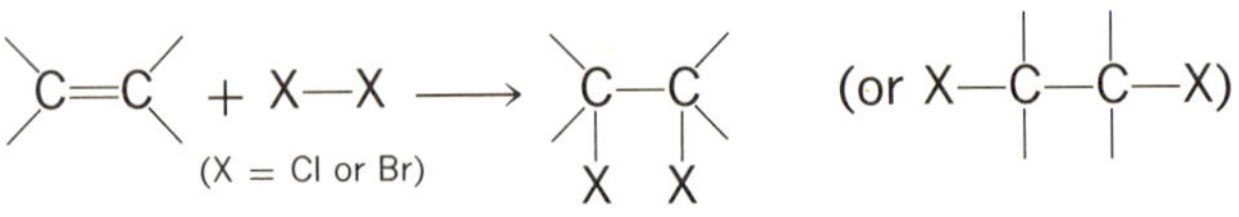

$$>C=C< + X-X \xrightarrow[(X\,=\,Cl\ \text{or}\ Br)]{} -\underset{X}{\overset{|}{C}}-\underset{X}{\overset{|}{C}}- \qquad \left(\text{or } X-\underset{|}{\overset{|}{C}}-\underset{|}{\overset{|}{C}}-X\right)$$

Specific examples:

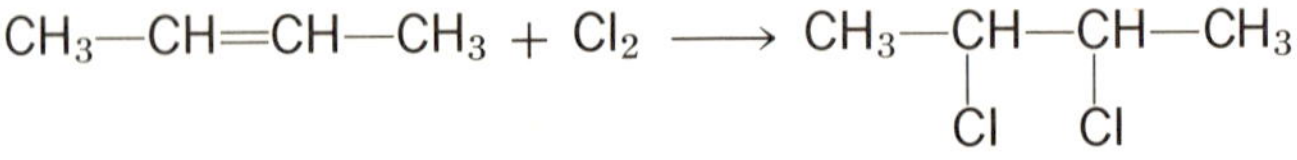

$$\underset{\text{2-Butene}}{CH_3-CH=CH-CH_3} + Cl_2 \longrightarrow \underset{\text{2,3-Dichlorobutane (81\%)}^4}{CH_3-\underset{Cl}{\underset{|}{C}H}-\underset{Cl}{\underset{|}{C}H}-CH_3}$$

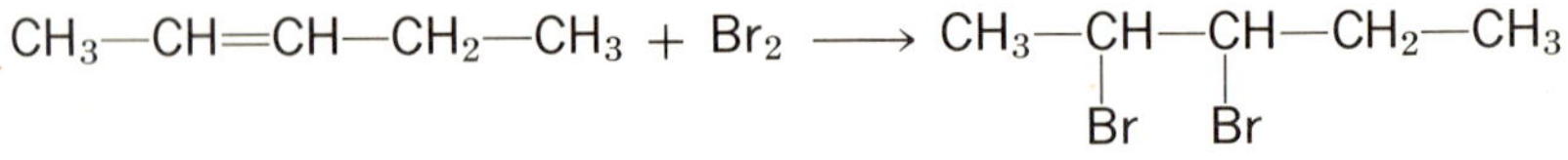

$$\underset{\text{2-Pentene}}{CH_3-CH=CH-CH_2-CH_3} + Br_2 \longrightarrow \underset{\text{2,3-Dibromopentane (94\%)}}{CH_3-\underset{Br}{\underset{|}{C}H}-\underset{Br}{\underset{|}{C}H}-CH_2-CH_3}$$

Usually this reaction is carried out at room temperature, or below, and the reagents are dissolved in some inert solvent such as carbon tetrachloride. If

[4] Throughout the chapters on organic chemistry, most reactions are shown with a figure in parentheses representing the mole percent yield. Side reactions are almost inevitable in organic synthesis. Moreover, in isolating the product, losses occur. Many reactions do not go to completion; unchanged starting materials remain and must be separated from the product. The finally isolated, pure product is nearly always less, on a mole basis, than would be expected if every mole of starting material form the desired product. The percentages cited in nearly all examples in this text are yields actually reported and then tabulated in *Synthetic Organic Chemistry* by R. B. Wagner and H. D. Zook (John Wiley and Sons, New York, 1953).

chlorine is used, the gas is bubbled into the solution. Bromine, a corrosive liquid that must be handled with great care, is usually added as a dilute solution in carbon tetrachloride. This solution of bromine added to an unknown provides a convenient test tube test for the presence of a double bond, for the dibromo compounds that form are colorless or pale yellow, in contrast to the deep-brown and almost opaque bromine solution. If an unknown hydrocarbon believed to be either an alkene or an alkane is tested, it could react immediately at room temperature, with the characteristic bromine color disappearing rapidly. If, simultaneously, fumes of hydrogen bromide gas are *not* evolved, the unknown is probably an alkene, for hydrogen bromide does form if bromine reacts with an alkane by a substitution process.

2. **Hydrogenation,** the addition of the elements of hydrogen, H—H. In general:

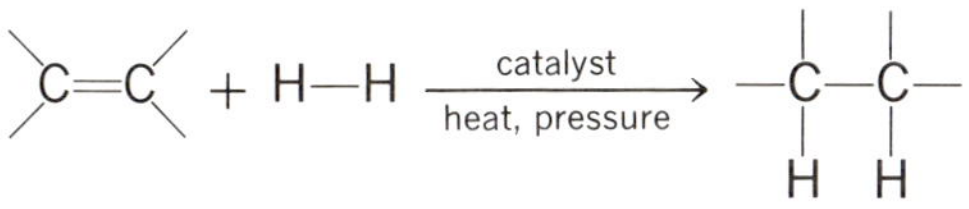

$$\gt C{=}C\lt + H{-}H \xrightarrow[\text{heat, pressure}]{\text{catalyst}} -\overset{|}{\underset{H}{C}}-\overset{|}{\underset{H}{C}}-$$

Specific examples:[5]

$$\underset{\text{Ethylene}}{CH_2{=}CH_2} + H_2 \xrightarrow[\text{heat, pressure}]{\text{Ni}} \underset{\text{Ethane}}{CH_3{-}CH_3}$$

$$\underset{\text{Isobutylene}}{(CH_3)_2C{=}CH_2} + H_2 \xrightarrow[\text{heat, pressure}]{\text{Ni}} \underset{\text{Isobutane}}{(CH_3)_2CH{-}CH_3}$$

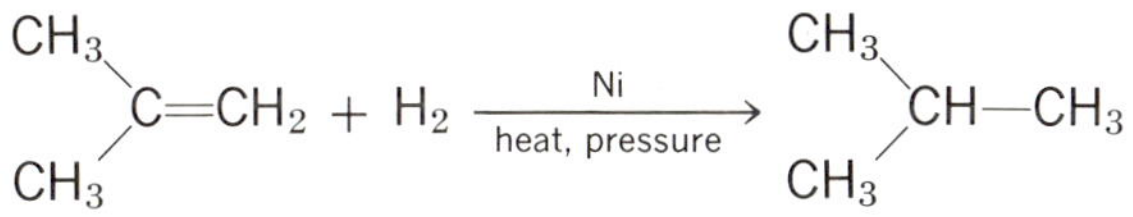

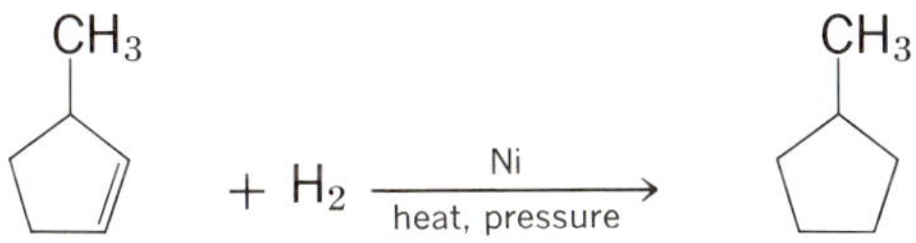

$$\underset{\text{3-Methylcylopentene}}{\text{(cyclopentene ring with } CH_3\text{)}} + H_2 \xrightarrow[\text{heat, pressure}]{\text{Ni}} \underset{\text{Methylcyclopentane}}{\text{(cyclopentane ring with } CH_3\text{)}}$$

The alkene is placed in a heavy-walled bottle or a steel cylinder together with the catalyst, a metal such as nickel, platinum, or palladium in a specially prepared, powdered form. Hydrogen gas is admitted to the vessel under pressure; the bottle is sometimes heated, and it may be shaken to insure intimate contact of the reactants with the metallic catalyst. Figure 10.5 shows a typical apparatus used when the hydrogen pressure is relatively low, less than 100 lb/in.^{2}.

One important commercial application of this reaction is the hydrogenation of vegetable oils. Their molecules contain alkene links which, although the number per molecule varies, qualify as "polyunsaturated." Animal fats, structurally, are almost identical with vegetable oils, but they have fewer double bonds per molecule. By controlled hydrogenation of some of the double bonds in vegetable oils, they be-

[5] The examples have been made up to illustrate the reaction. Reported examples involve more complicated alkenes than those shown here.

3
4
1
2

Figure 10.5 Typical low-pressure hydrogenation apparatus. The unsaturated compound, together with a catalyst and an inert solvent, is placed in a thick-walled glass bottle positioned in a heavy-mesh, protective metal screen and clamped on a rocker, 1. Rocking action, powered by the motor, 2, mixes the contents of the flask while hydrogen gas under pressure is led in by a hose, 3, from a storage tank, 4. (Courtesy of Parr Instrument Company.)

come, like the animal fats, solids at room temperature. This reaction is discussed further in Chapter 17.

3. **Addition of hydrogen chloride or hydrogen bromide.** Molecules of hydrogen or of any of the halogens are symmetrical, but those of hydrogen chloride or hydrogen bromide are not. If a molecule of either acid adds itself to an unsymmetrically substituted double bond, two directions for the addition are possible. Propylene, for example, may react with hydrogen chloride to give either n-propyl chloride or isopropyl chloride or a mixture of both:

$$\underset{\text{Propylene}}{CH_3{-}CH{=}CH_2} + H{-}Cl \longrightarrow \underset{n\text{-Propyl chloride}}{CH_3{-}\underset{\displaystyle H}{\overset{|}{C}H}{-}\underset{\displaystyle Cl}{\overset{|}{C}H_2}} \qquad (\text{or } CH_3CH_2CH_2Cl)$$

$$\text{or} \quad CH_3{-}CH{=}CH_2 + H{-}Cl \longrightarrow \underset{\text{Isopropyl chloride}}{CH_3{-}\underset{\displaystyle Cl}{\overset{|}{C}H}{-}\underset{\displaystyle H}{\overset{|}{C}H_2}} \qquad (\text{or } CH_3\underset{\displaystyle Cl}{\overset{|}{C}H}CH_3)$$

In either case, the double bond becomes a single bond, and the two parts of the hydrogen chloride molecule become attached to the carbons at each end of the former double bond. The product that actually forms is largely isopropyl chloride. Essentially no *n*-propyl chloride is produced.

The several other examples of this reaction to be shown display a consistency first noted by a Russian chemist, Markovnikov.[6] According to Markovnikov's rule, when an unsymmetrical reagent, H—G, adds across an unsymmetrical carbon-carbon double bond of a simple alkene, the hydrogen of H—G attaches to the carbon of the double bond that is already bonded to the greater number of hydrogens. ("Them that has, gets.") For purposes of this rule, "unsymmetrical carbon-carbon double bond" means one in which the carbons bear unequal numbers of hydrogens.

Specific examples:

$$\underset{\text{Ethylene}}{CH_2{=}CH_2} + H{-}Br \longrightarrow \underset{\text{Ethyl bromide}}{CH_3CH_2Br} \qquad \text{(only possible product)}$$

$$\underset{\text{Isobutylene}}{CH_3{-}\underset{\displaystyle CH_3}{\underset{|}{C}}{=}CH_2} + H{-}Cl \longrightarrow \underset{t\text{-Butyl chloride}}{CH_3{-}\overset{\displaystyle Cl}{\overset{|}{\underset{\displaystyle CH_3}{\underset{|}{C}}}}{-}CH_3} \qquad (\text{not } CH_3{-}\underset{\displaystyle CH_3}{\underset{|}{CH}}{-}CH_2{-}Cl)$$

$$\underset{\text{2-Butene}}{CH_3{-}CH{=}CH{-}CH_3} + H{-}Cl \longrightarrow$$

$$\underset{sec\text{-Butyl chloride}}{CH_3{-}CH_2{-}\underset{\displaystyle Cl}{\underset{|}{CH}}{-}CH_3} \qquad \text{(only possible product)}$$

$$\underset{\text{2-Pentene}}{CH_3{-}CH_2{-}CH{=}CH{-}CH_3} + H{-}Br \longrightarrow$$

$$\underset{\text{3-Bromopentane}}{CH_3{-}CH_2{-}\underset{\displaystyle Br}{\underset{|}{CH}}{-}CH_2{-}CH_3} + \underset{\text{2-Bromopentane}}{CH_3{-}CH_2{-}CH_2{-}\underset{\displaystyle Br}{\underset{|}{CH}}{-}CH_3}$$

The reaction is carried out by bubbling the hydrogen halide in gaseous form into the alkene, which may or may not be dissolved in some solvent inert to the reagents (e.g., carbon tetrachloride).[7]

For preparing alkyl chlorides and bromides, this reaction is superior to direct halogenation of an alkane. The latter gives mixtures of mono-, di-, and higher halogenated materials, and the reaction also produces mixtures of isomers.

Alkyl chlorides and bromides (and iodides) are extremely important intermediates in the synthesis of organic compounds. They can also be made from alcohols,

[6] Vladimir Vasil'evich Markovnikov (1838–1904).

[7] When hydrogen bromide is used, it is important that no peroxides, compounds of the type R—O—O—H or R—O—O—R, be present. They catalyze a different mechanism, which we shall not discuss, with the result that in their presence hydrogen bromide (and only this hydrogen halide) adds itself against the Markovnikov rule to unsymmetrical alkenes. Traces of peroxides form when organic chemicals remain exposed to atmospheric oxygen for long periods.

which is often more convenient, but the addition of HX to a double bond illustrates a very important reaction of alkenes.

Exercise 10.6 Write the names and structures for the product(s) that would form under the conditions shown. Assume that peroxides are absent.

(*a*) $CH_2{=}CHCH_2CH_3 + HCl \longrightarrow$

(*b*) $CH_2{=}\underset{\displaystyle CH_3}{\underset{|}{C}}{-}CH_3 + HBr \longrightarrow$

(*c*) $CH_3{-}CH{=}\overset{\displaystyle CH_3}{\overset{|}{C}}{-}C_6H_{11} + HCl \longrightarrow$ (cyclohexyl ring attached)

(*d*) 3-methylcyclohexene (CH_3 on ring) $+ HBr \longrightarrow$

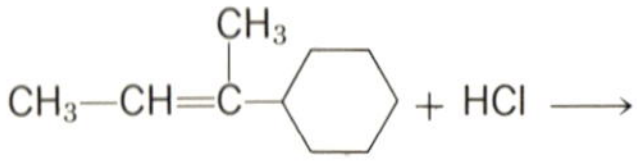

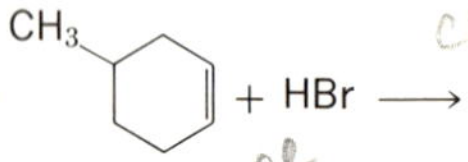

(*e*) $C_6H_{11}{-}CH{=}CH{-}C_6H_{11} + HCl \longrightarrow$ (two cyclohexyl rings)

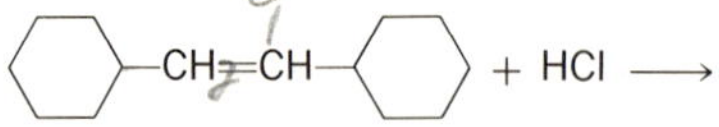

Exercise 10.7 Write the structures and names of alkenes that are best used to prepare each of the following halides. Avoid selecting those yielding isomeric halides that are difficult to separate. If the halide cannot be prepared by the action of HCl or HBr on an alkene (given the limitations of Markovnikov's rule), state this fact.

(*a*) $CH_3CH_2\underset{\displaystyle Cl}{\underset{|}{C}H}CH_3$ (*b*) $CH_3CH_2CH_2Cl$ (*c*) cyclopentyl–Br

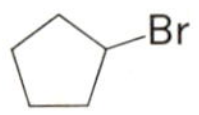

(*d*) $CH_3{-}\underset{\displaystyle Cl}{\underset{|}{\overset{\displaystyle CH_3}{\overset{|}{C}}}}{-}CH_2{-}CH_3$ (*e*) $Cl{-}CH_2{-}\overset{\displaystyle CH_3}{\overset{|}{C}H}{-}CH_2{-}CH_3$

(*f*) $CH_3{-}\underset{\displaystyle Br}{\underset{|}{C}H}{-}CH_2{-}CH_2{-}CH_3$

---

4. **Hydration,** the addition of the elements of water, H—OH. In general:

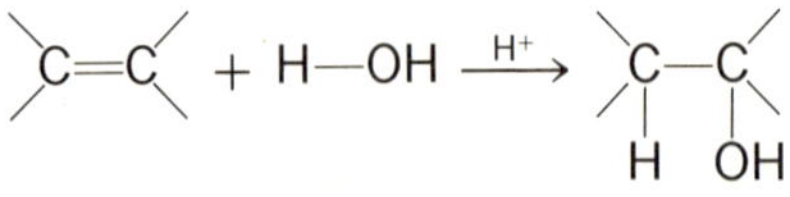

$$\underset{\text{An alkene}}{\gt C{=}C\lt} + H{-}OH \xrightarrow{H^+} \underset{\text{An alcohol}}{\gt\underset{H}{\underset{|}{C}}{-}\underset{OH}{\underset{|}{C}}\lt}$$

Specific examples:

$$\underset{\text{Ethylene}}{CH_2{=}CH_2} + H{-}OH \xrightarrow[240°C]{10\%\ H_2SO_4} \underset{\text{Ethyl alcohol}}{CH_3{-}CH_2{-}OH}$$

$$(CH_3)_2C{=}CH_2 + H{-}OH \xrightarrow[25°C]{10\%\ H_2SO_4} (CH_3)_2C(OH){-}CH_3 \quad (\text{not } (CH_3)_2CH{-}CH_2{-}OH)$$

Isobutylene — *t*-Butyl alcohol

Markovnikov's rule applies to this reaction in which H—G is now H—OH. The H in H—OH goes to the carbon of the double bond that has the greater number of hydrogens; the —OH goes to the other carbon.

Acid catalysis is the key condition. In fact, the reaction can be completed in two discrete steps if we use concentrated sulfuric acid, another unsymmetrical reagent whose molecules add themselves to double bonds in accordance with Markovnikov's rule:

$$CH_3{-}CH{=}CH_2 + H{-}O{-}S(\rightarrow O)_2{-}O{-}H \xrightarrow{0°C} CH_3{-}CH(O{-}S(\rightarrow O)_2{-}O{-}H){-}CH_3$$

Propylene — Sulfuric acid — Isopropyl hydrogen sulfate

If the product, a member of the class of alkyl hydrogen sulfates, is diluted with water, the sulfate group is replaced by —OH and the product is the alcohol:

$$R{-}O{-}S(\rightarrow O)_2{-}O{-}H + H{-}OH \longrightarrow R{-}O{-}H + H{-}O{-}S(\rightarrow O)_2{-}O{-}H$$

An alkyl hydrogen sulfate — An alcohol — Sulfuric acid

The addition of water to double bonds is one of the important ways of making alcohols, R—OH. In living organisms this reaction (catalyzed by enzymes) occurs frequently, as we shall see in our study of the metabolism of lipids and the citric acid cycle. The hydration of alkenes is also important commercially. Ethyl alcohol is made from ethylene which is abundantly available from the cracking of petroleum and has many other uses. By 1970 United States production of ethyl alcohol from ethylene is expected to be over 9 million tons per year.

5. **Polymerization.** In general.

$$n\ \gt C{=}C\lt \xrightarrow{\text{initiator}} \left(\begin{array}{cc} | & | \\ C & C \\ | & | \end{array}\right)_n$$

Alkene — Polyalkene

Specific examples:

$$nCH_2{=}CH_2 \xrightarrow{\text{initiator}} \text{-(}CH_2{-}CH_2\text{)}_n\text{-}$$

Ethylene — Polyethylene (n is several hundred—up to several thousand)

$$nCH(CH_3){=}CH_2 \xrightarrow{\text{initiator}} \left(\begin{array}{c} CH_3 \\ | \\ -CH{-}CH_2- \end{array}\right)_n$$

Propylene — Polypropylene

The starting alkene is called a monomer, the product a polymer (Greek: *polys,* many; *meros,* part). Polymerization usually requires the presence of a chemical called an initiator which performs the services of a catalyst but is consumed by the reaction. Oxygen was the common initiator before substances that permit better control over the properties of the polymer were discovered. With oxygen a high temperature, 100°C, and high pressure, 15,000 lb/in.^{2}, were necessary for the polymerization of ethylene. The product was soft and pliable but unsuitable for spinning into fibers. In 1963 the Nobel prize in chemistry went to Karl Ziegler (Germany) and Giulio Natta (Italy) for their discoveries of initiators that permit polymerizations of alkenes at much lower temperatures and pressures. Ziegler's initiator, a trialkyl derivative of aluminum, R_3Al, plus titanium chloride, produces a polyethylene that does not soften until temperatures well beyond those needed for sterilization of tubes and bottles are reached.

Natta's work produced the symmetrically polymerized polypropylene illustrated in Figure 10.6. The electron clouds of the side chain methyl groups repel each other, causing the chain to coil into a helix. Polymers whose molecules are very long and very symmetrical give promise of being made into serviceable fibers. Natta's symmetrical polypropylene is used more and more to make carpeting for installation around swimming pools, in open entranceways, basements, bathrooms, and even kitchens. One great advantage of both polyethylene and polypropylene is that they are alkanes and have all the chemical unreactivity of this family. They are also thermoplastic polymers; when they are heated, they soften and flow and may be extruded into sheets, molded articles, and tubes. Figures 10.7, 10.8, and 10.9 illustrate some applications of polyethylene and polypropylene. By 1970 the estimated United States production of low- and high-density polyethylene will be 5 billion pounds.

An interesting relative of polyethylene is Teflon, a polymer in which all the hydrogens in polyethylene have been replaced by fluorines. The monomer is tetrafluorethylene, $F_2C{=}CF_2$. Teflon is one of the most chemically inert of all organic

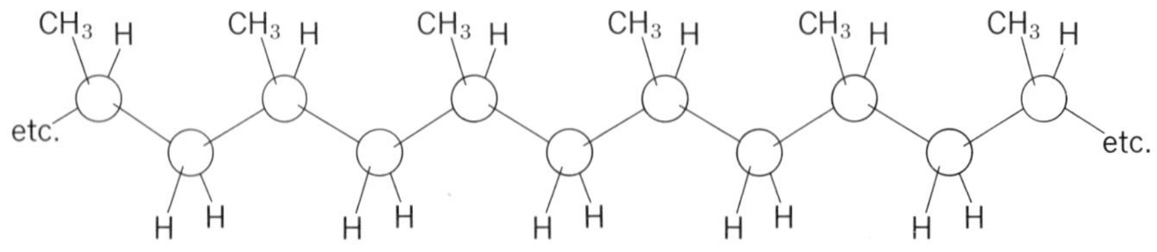

CH3 H CH3 H CH3 H CH3 H CH3 H CH3 H
etc. etc.
H H H H H H H H H H

Figure 10.6 Symmetrical (isotactic) polypropylene.

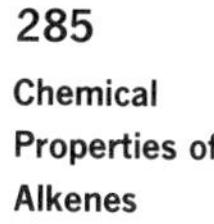

Figure 10.7 Polyethylene has a new use in agriculture, serving as a mulch to protect plants. This photograph was taken at LaCosta, California, in a strawberry field. Because the plants grow through holes in the mulch, weed problems are completely eliminated. The material is being used extensively in Hawaii. (Courtesy of Monsanto Company.)

Figure 10.8 Low-pressure pipe made of high-density polyalkene resins is chosen for an increasing number of applications. It resists rot and corrosion, and it is relatively easy to install. To connect the sections of this 6-inch pipe for natural gas, a small, motor-powered device operated by two men is all that is needed. To install the same pipe in steel would require a much larger crew and much heavier equipment. (Courtesy of Phillips Petroleum Company.)

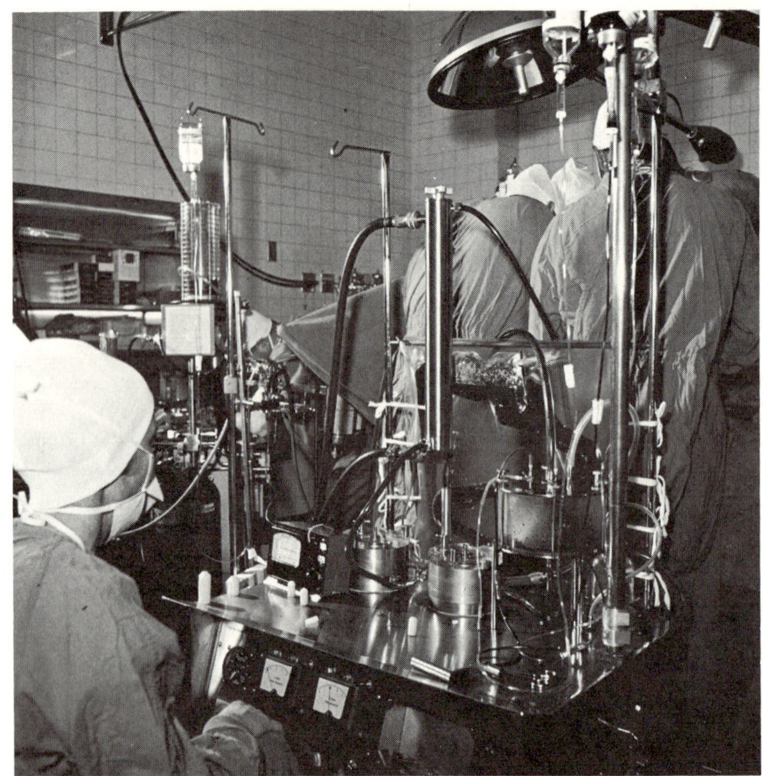

Figure 10.9 The tubing and containers for blood in this Miniprime heart-lung machine are fashioned from polyvinyl chloride, $\text{-(}CH_2CHCl\text{)}_n\text{-}$. The scene is a Los Angeles hospital during open-heart surgery. The machine is manufactured by Travenol Division, Baxter Laboratories. (Courtesy of Monsanto Company.)

substances, for the only chemicals known to attack it are molten sodium and potassium. In addition to the well-known Teflon coating for pots and pans, the polymer is used in electrical insulations and antifriction devices.

Another important monomer related to ethylene is vinyl chloride. In the presence of peroxide catalysts it polymerizes to a brittle resin:

$$n\mathrm{CH_2{=}CH(Cl)} \xrightarrow{\text{peroxide}} \mathrm{-(CH_2{-}CH(Cl))_n-}$$

Vinyl chloride → Polyvinyl chloride (PVC)

Its brittleness can be overcome and a softer product obtained if during the polymerization we add certain compounds known as *plasticizers;* cresyl phosphate and alkyl phthalates are examples. Polyvinyl chloride is in this way made rubberlike or leatherlike so that it can be widely used in floor tiles, raincoats, tubing, electrical insulation, phonograph records, and protective coatings.

The dichloro derivative of ethylene, vinylidene chloride, is the monomer for saran, used for seat covers and packaging film:

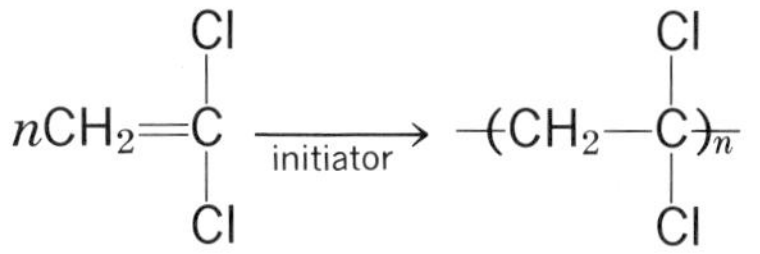

$$n\mathrm{CH_2{=}CCl_2} \xrightarrow{\text{initiator}} \mathrm{-\!(CH_2{-}CCl_2)\!-}_n$$

Vinylidene chloride — Saran

Orlon, a plastic that can be made into fibers for fabrics, is the polymer of a cyano derivative of ethylene called acrylonitrile:

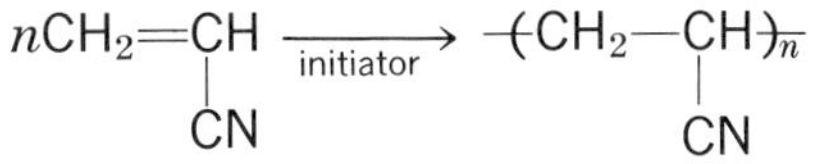

$$n\mathrm{CH_2{=}CH(CN)} \xrightarrow{\text{initiator}} \mathrm{-\!(CH_2{-}CH(CN))\!-}_n$$

Acrylonitrile — Orlon

Natural rubber is a polymer of isoprene, a hydrocarbon with two double bonds:

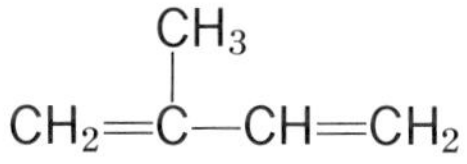

$$\mathrm{CH_2{=}C(CH_3){-}CH{=}CH_2}$$

Isoprene

To understand its relation to the monomer unit, we may imagine that it forms in the following fashion.

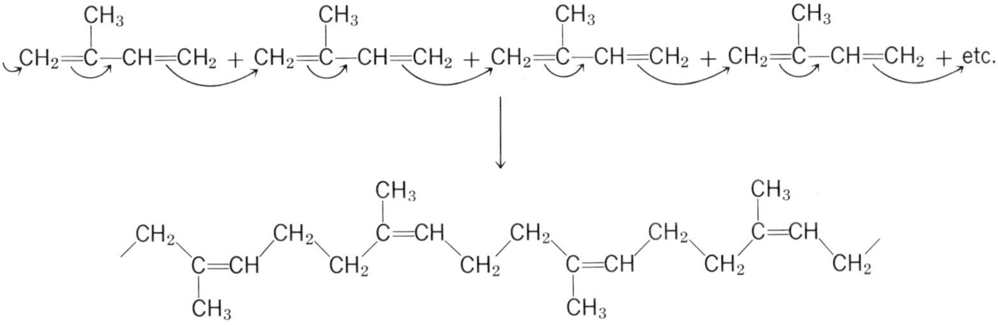

$$\mathrm{CH_2{=}C(CH_3){-}CH{=}CH_2 + CH_2{=}C(CH_3){-}CH{=}CH_2 + CH_2{=}C(CH_3){-}CH{=}CH_2 + CH_2{=}C(CH_3){-}CH{=}CH_2 + etc.}$$

$$\downarrow$$

$$\mathrm{{-}CH_2{-}C(CH_3){=}CH{-}CH_2{-}CH_2{-}C(CH_3){=}CH{-}CH_2{-}CH_2{-}C(CH_3){=}CH{-}CH_2{-}CH_2{-}C(CH_3){=}CH{-}CH_2{-}}$$

Natural rubber

We note that the main chain emerges cis from each double bond, making this cis-1,4-polyisoprene. The polymer in which the main chain emerges trans from each double bond is gutta-percha. Ziegler's catalysts or lithium metal bring about the all-cis polymerization of isoprene, and thus a synthetic source of natural rubber is available. A cheaper source of a rubber substitute is 1,3-butadiene, which can be polymerized to polybutadiene. Industrial chemists have extensively studied the *copolymerization* of dienes with monoenes (alkenes) to find polymers that have fewer double bonds but are still flexible, elastic, and resistant to solvents and abrasion. One of the most important of these polymers is the copolymer of butadiene and styrene, Buna S rubber. In this substance the polymer molecules are not all identical, but the following illustrates some of their features, without regard to the geometry or the exact proportions of monomers:

$$CH_2{=}CHCH{=}CH_2 \qquad C_6H_5CH{=}CH_2$$

1,3-Butadiene — Styrene

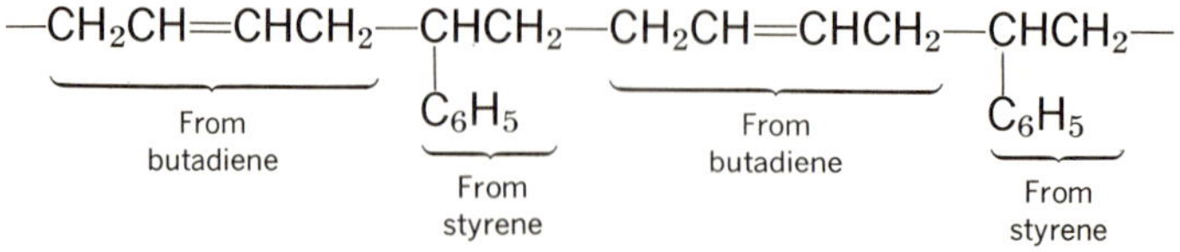

$$\underbrace{-CH_2CH{=}CHCH_2}_{\text{From butadiene}}-\underbrace{\underset{|}{C}HCH_2}_{}-\underbrace{CH_2CH{=}CHCH_2}_{\text{From butadiene}}-CHCH_2-$$

(each CHCH₂ unit bears a C_6H_5 group — from styrene)

Buna S rubber

This product is also called GRS (Government Rubber Styrene)[8] and "cold rubber."

Another synthetic rubber, *butyl rubber,* is made by polymerizing isobutylene in the presence of a small amount of isoprene. Polyisobutylene would resemble an alkane polymer, with a structure similar to that of polypropylene but containing more methyl groups. By mixing in some isoprene, however, the polymer will have a few double bonds. These sites can enter into reactions that produce cross-linking between chains to make, in effect, an interlacing network of chains and a product that wears well and is resistant to chemicals.

MECHANISM OF ADDITION REACTIONS OF ALKENES

We already know that addition reactions happen, that they follow Markovnikov's rule, and that acids or acid catalysts are involved. We must now seek one mechanism to tie these facts together and to explain others still to be introduced.

Let us consider the addition of hydrogen chloride. We already know it as a proton donor, and a proton is an electron-poor reagent. In the terminology of the field, it is said to be an *electrophilic species* (i.e., "electron loving" from *philos,* Greek for loving). The pi-electron cloud at a double bond is "electron-rich"; we should therefore expect some interaction, and we believe it to happen as follows:

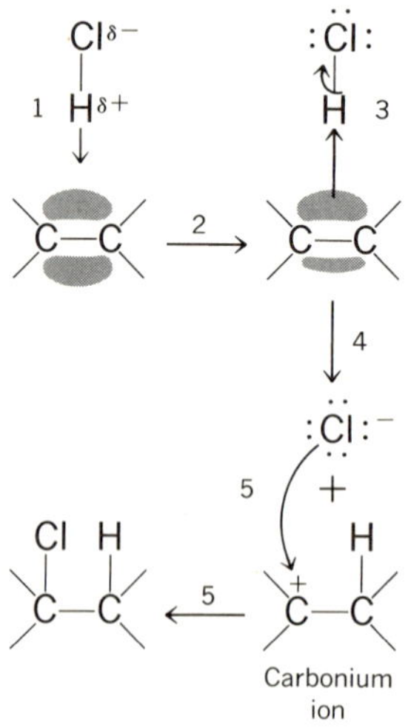

$Cl^{\delta-}$—$H^{\delta+}$ (1) → C=C pi cloud —2→ :Cl:–H (3) bonded to C—C —4→ :Cl:⁻ + H–C—$\overset{+}{C}$ (Carbonium ion) —5→ Cl–C—C–H

[8] After a government project during World War II.

In step 1 we imagine a molecule of an alkene to be on collision course with a molecule of hydrogen chloride. If the collision is violent enough and if it is properly oriented, it will lead (step 2) to a situation in which the pair of pi electrons begins to form a bond to the hydrogen while the chlorine (step 3) begins to disengage from it. The result (step 4) is a pair of ions, one of which has a carbon with a positive charge. Such a species is called a *carbonium ion;* the carbon bearing the positive charge has only six electrons in its valence shell. Carbonium ions have high internal energies and are highly reactive toward electron-rich species. They have only fleeting existences, but they are real and undergo a variety of reactions, *all of them to restore an octet to the carbon.* Hence, in our example, step 5, a chloride ion is very rapidly attached to the site of the positive charge in the carbonium ion to give the final product. To summarize,

(1) H—Cl + >C=C< ⟶ H—C—C$^+$ + Cl$^-$

(A strong σ-bond breaks.) (A weaker π-bond breaks.) (A strong σ-bond, H—C, forms, and a high-energy carbonium ion is also produced.) (A stable chloride ion forms.)

(2) H—C—C$^+$< + Cl$^-$ ⟶ H—C—C—Cl

A carbonium ion (A strong σ-bond, C—Cl, forms.)

Qualitatively, the energy cost to break a strong and a weak bond (step 1) is more than repaid by the energy lowering that accompanies the formation of two strong bonds.

What has thus far been said qualitatively accounts for two facts: first, the reaction does occur, and second, an acidic, proton-donating reagent rather than a basic, electron-rich species is the type that will attack a double bond. What about the direction or the orientation of the addition by Markovnikov's rule?

Consider the addition of hydrogen chloride to propylene to form isopropyl chloride, not *n*-propyl chloride. We have learned that the first general step in the reaction generates a carbonium ion. Two isomeric carbonium ions could form from propylene:

(1) $CH_3CH{=}CH_2 + H{-}Cl \longrightarrow CH_3\overset{+}{C}HCH_3 + Cl^-$

Isopropylcarbonium ion
1

(2) $CH_3CH{=}CH_2 + H{-}Cl \longrightarrow CH_3CH_2CH_2^+ + Cl^-$

n-Propylcarbonium ion
2

In the final step of the overall addition—combination of the carbonium ion with the chloride ion—only the isopropylcarbonium ion, **1,** can lead to isopropyl chloride:

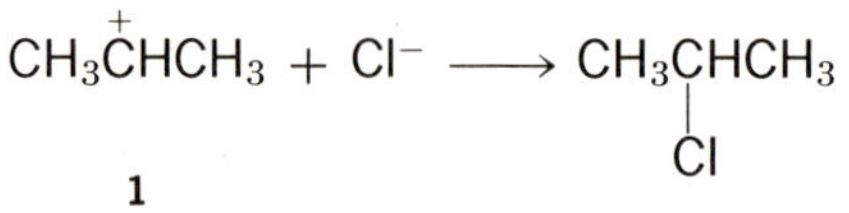

$$\underset{\mathbf{1}}{CH_3\overset{+}{C}HCH_3} + Cl^- \longrightarrow CH_3\underset{\underset{Cl}{|}}{C}HCH_3$$

The result means that **1** rather than **2** is the intermediate. For some reason **1** must form so much more rapidly than **2** that, for all practical purposes, it is the only carbonium ion involved. Perhaps it forms more rapidly because the energy cost of making **1** is far less than that for making **2.** This is, in fact, what a great deal of evidence has led chemists to conclude. In short, a carbonium ion such as **1** is more stable than one such as **2.** We seek now a possible reason for this stability.

Carbonium ions can be classified into three types:

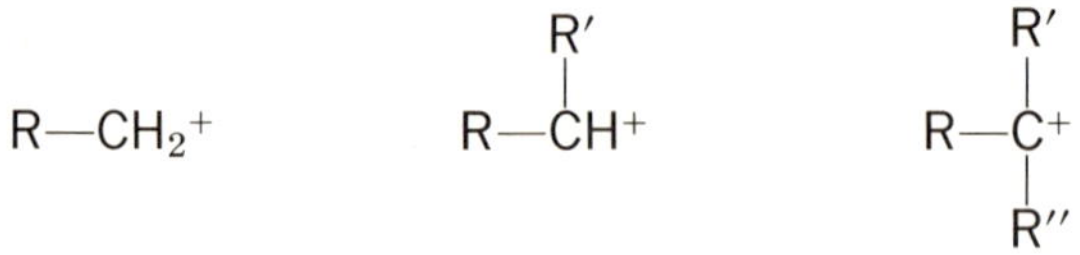

$$R{-}CH_2^+ \qquad R{-}\overset{\overset{R'}{|}}{C}H^+ \qquad R{-}\overset{\overset{R'}{|}}{\underset{\underset{R''}{|}}{C^+}}$$

Primary or 1° Secondary or 2° Tertiary or 3°

The classification relates to the condition of the carbon bearing the positive charge. If only one carbon is attached to it *directly,* it is a primary (1°) carbonium ion.[9] If two carbons are directly attached, it is a secondary (2°) carbonium ion. In a tertiary (3°) carbonium ion three other carbons are directly attached to the carbon bearing the positive charge.

Exercise 10.8 Classify the following carbonium ions as primary, secondary, and tertiary.

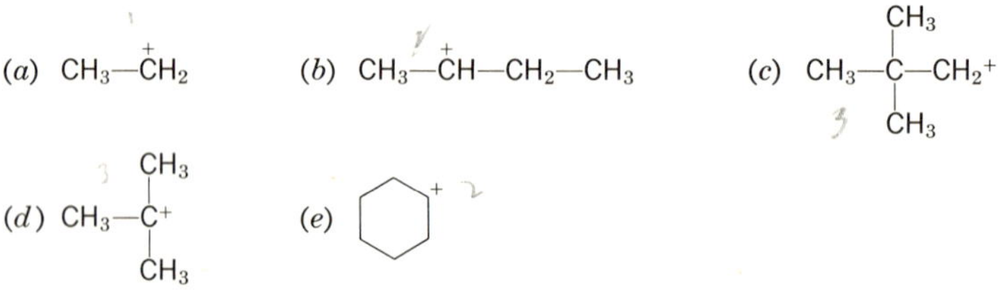

(*a*) $CH_3{-}\overset{+}{C}H_2$ (*b*) $CH_3{-}\overset{+}{C}H{-}CH_2{-}CH_3$ (*c*) $CH_3{-}\overset{\overset{CH_3}{|}}{\underset{\underset{CH_3}{|}}{C}}{-}CH_2^+$

(*d*) $CH_3{-}\overset{\overset{CH_3}{|}}{\underset{\underset{CH_3}{|}}{C^+}}$ (*e*) cyclohexyl cation (ring with +)

In general, *alkyl groups have an electron-releasing inductive effect.* Some evidence for this effect will come later, in Chapter 12, but for our present purposes we make the statement without proof. In a primary carbonium ion only one alkyl group is attached to the positively charged carbon; in the tertiary carbonium ion the number is three. To the extent that alkyl groups induce an electron release toward the positive charge, this charge is partially neutralized. To the extent that it is neutralized, it is stabilized. Three alkyl groups releasing electron density toward a positive site would stabilize it more than two could. And of course two alkyl groups

[9] In a context such as this, 1° is not pronounced "one degree" but rather "primary," 2° is secondary," and 3° is "tertiary."

would be better than one. Thus 3° carbonium ions are more stable than 2°, and these are more stable than 1°:

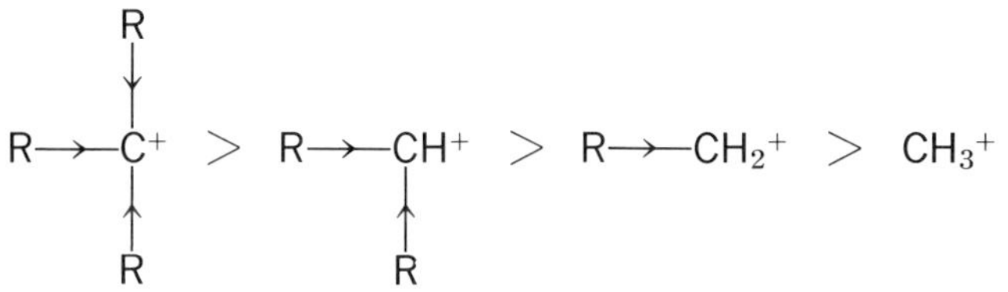

$$R\!\rightarrow\!\overset{\overset{R}{\downarrow}}{\underset{\underset{R}{\uparrow}}{C^+}} > R\!\rightarrow\!\underset{\underset{R}{\uparrow}}{CH^+} > R\!\rightarrow\!CH_2^+ > CH_3^+$$

(order of stability of some types of carbonium ions)

or

Order of stability of carbonium ions: $3° > 2° > 1° > CH_3^+$

Thus propylene reacts with hydrogen chloride to give a secondary carbonium ion (isopropyl) rather than a primary carbonium ion (*n*-propyl) as the intermediate. The more stable ion forms.

Exercise 10.9 Write the condensed structures for the two carbonium ions that can conceivably form if a proton becomes attached to each of the following alkenes. Circle the one that is preferred. Where both are reasonable, state that they are. Write the structures of the alkyl chlorides that will form by the addition of hydrogen chloride to each.

(*a*) $CH_3—CH_2—CH{=}CH_2$ (*b*) $CH_3—\overset{\overset{CH_3}{|}}{C}{=}CH_2$ (*c*) 1-methylcyclohexene (ring structure with CH_3 on the double-bonded carbon)

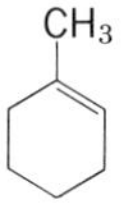

(*d*) $CH_3—CH{=}CH—CH_3$ (*e*) $CH_3—CH{=}CH—CH_2—CH_3$

(*f*) Addition of water to 2-pentene (part *e*) gives a mixture of alcohols. What are they? Why is formation of a mixture to be expected here but not from propylene?

Markovnikov's rule would of course be true however we tried to explain it. It is an example of a purely empirical correlation; that is, it depends only on the *observation* of events, however they may be explained. We have seen, however, that this rule can be understood in terms of general theories of physics and chemistry.

The acid-catalyzed addition of water proceeds in much the same way as the addition of hydrogen chloride:

Step 1. 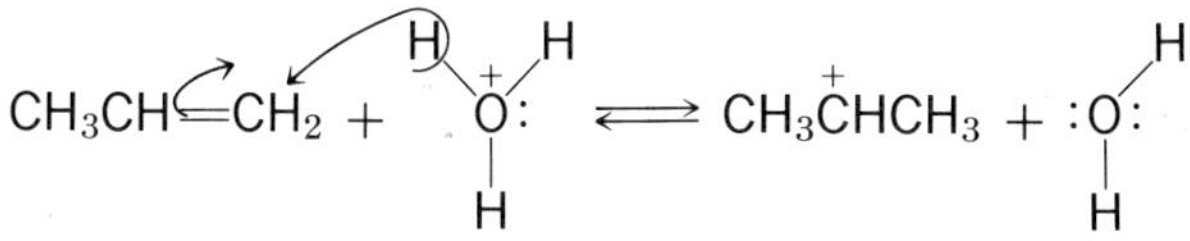 $CH_3CH{=}CH_2 + H—\overset{+}{\underset{\underset{H}{|}}{O}}:—H \rightleftharpoons CH_3\overset{+}{C}HCH_3 + :\underset{\underset{H}{|}}{O}:—H$

Propylene Catalyst Isopropyl-carbonium ion

$$\text{Step 2.}\quad CH_3\overset{+}{C}HCH_3 + :\ddot{O}(H)_2 \rightleftharpoons CH_3CH(\overset{+}{\ddot{O}}H_2)CH_3$$

Isopropyl-carbonium ion | Water | Isopropyl alcohol in its protonated form

$$\text{Step 3.}\quad CH_3CH(\overset{+}{\ddot{O}}H_2)CH_3 + :\ddot{O}(H)_2 \rightleftharpoons CH_3CH(\ddot{O}H)CH_3 + H_3\overset{+}{O}:$$

Isopropyl alcohol | Recovered catalyst

The first step is the generation of a carbonium ion, the more stable of the two possible, which acts like a Lewis acid toward a water molecule (a Lewis base). The alcohol first appears in its protonated form, and then proton transfer gives the final product.

With the carbonium ion we have introduced one of the most important intermediates in organic reactions that occur in solution. One of its many properties is that, being a Lewis acid, it is vulnerable to attack by any electron-rich species that happens to be close by. When water is the most abundant of such neighbors, alcohols will form. Another property of the carbonium ion that must be emphasized is its relative instability. With very few exceptions salts cannot be prepared with an intact carbonium ion. Usually in the circumstances of its formation some electron-rich particle can give it the share of an electron-pair needed to complete its octet.

ALKYNES

The structural feature characteristic of an alkyne—a carbon-carbon triple bond—does occur among natural products, but it is not important to our study of biochemistry in this text. For that reason we note only its structural features and mention that many of its reactions are of the addition type studied for the alkenes.

STRUCTURE

The simplest alkyne is commonly called acetylene:

$$H—C{\equiv}C—H$$

Acetylene
(ethyne)

In it each carbon is bonded to only two other nuclei; in one sense the carbon is divalent. According to one theoretical approach, of the three bonds in a triple bond one is a sigma bond and two are pi bonds (Figure 10.10).

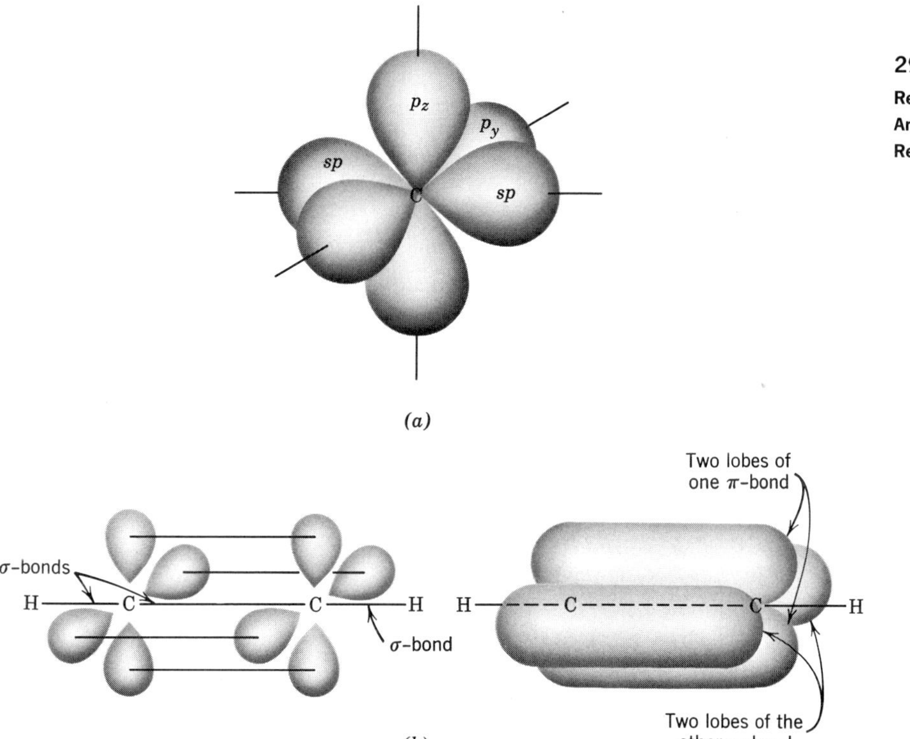
p_z
p_y
sp
C
sp
(a)
Two lobes of one π-bond
σ-bonds
H
C
C
H
σ-bond
H
C
C
H
Two lobes of the other π-bond
(b)

Figure 10.10 The triple bond in acetylene, one theoretical view. (*a*) Orbitals of carbon in its valence state when it is attached to only two other groups, as in an acetylene. (*b*) Two representations of the two π-bonds in acetylene. The carbon-carbon σ-bond results from *sp–sp* overlap. The carbon-hydrogen σ-bonds come from *sp-s* overlap.

REFERENCES AND ANNOTATED READING LIST

BOOKS

J. D. Roberts and M. C. Caserio. *Basic Principles of Organic Chemistry.* W. A. Benjamin, New York, 1964. Chapter 29, "Polymers," is a survey of the major kinds of polymers, with emphasis on how their properties are related to their structures.

C. R. Noller. *Chemistry of Organic Compounds,* third edition. W. B. Saunders Company, Philadelphia, 1965. Chapter 34, "Polyenes, Rubber, and Synthetic Rubbers," provides not only information about the structures and chemistry of the substances named but also interesting historical material.

ARTICLES

H. L. Fisher. "Rubber." *Scientific American,* November 1956, page 75. A brief history of rubber is followed by a general discussion of polymers that serve as synthetic rubber.

G. Oster. "Polyethylene." *Scientific American,* September 1957, page 139. Polyethylene plastic is more than an aggregation of linear molecules. Much branching and cross-linking occur. This article discusses how such structural features arise and how they affect the properties and enhance the usefulness of the plastic. (The entire September 1957 issue of *Scientific American* is devoted to "Giant Molecules.")

PROBLEMS AND EXERCISES

1. Write an illustrated discussion of each of the following terms.
   (*a*) sp^2 hybrid orbital (*b*) pi bond (*c*) cis isomer
   (*d*) trans isomer (*e*) hydrogenation (*f*) unsymmetrical double bond
   (*g*) Markovnikov's rule (*h*) initiator (*i*) polymerization
   (*j*) electrophilic species (*k*) carbonium ion (*l*) inductive effect
2. List the families of organic compounds for which we have learned a method of synthesis in this chapter, and write equations illustrating these reactions.
3. Write IUPAC names for each of the following:

   (*a*) $CH_3(CH_2)_7CH{=}CH_2$

   (*b*) $Cl{-}CH{=}CHCHCH_3$ (with CH_3 on the CH next to CH_3)

   (*c*) $CH_3CH_2CH_2CCH_2CHCHCH_3$ (with CH_2 on the first C of the CCH_2 unit; CH_3 above and CH_3 below the two CH carbons)

   (*d*) $CH_3CCH{=}CH$ (with CH_3 above and CH_3 below the C after the first CH_3; CH_3 on the final CH)

4. Write equations for the reaction of isobutylene with each of the following reagents: (*a*) cold, concentrated H_2SO_4, (*b*) H_2 (Ni, heat, pressure), (*c*) H_2O (H^+ catalysis), (*d*) H—Cl, (*e*) H—Br.
5. Write equations for the reaction of 1-methylcyclopentene with each of the reagents of exercise 4.
6. When 1-butene reacts with hydrogen chloride, the product is 2-chlorobutane. The isomer, 1-chlorobutane, does not form. Explain.
7. Ethane is insoluble in concentrated sulfuric acid. Ethylene dissolves readily. Write an equation to show how ethylene is converted into a substance polar enough to dissolve in the highly polar concentrated sulfuric acid.
8. Write out enough of the structures of each of the following substances to show what their molecules are like. Do not use parentheses to condense these structures.
   (*a*) polyethylene (*b*) polypropylene (*c*) Teflon
   (*d*) PVC (*e*) polybutadiene (*f*) polystyrene
9. Describe a simple chemical test that could be used to distinguish between 2-butene and butane. Describe what you would do and see.

CHAPTER ELEVEN

Aromatic Compounds. Theory of Resonance

AROMATIC AND ALIPHATIC SUBSTANCES

The sixteenth century brought a reorientation of attitudes toward the proper ways for studying and using common substances. The search for the philosopher's stone with which gold might be prepared from baser things was abandoned under the pressure of repeated failures. Almost any ordinary substance was felt to be a proper object of study. The desire to find medicinal agents was certainly one driving motive.

By the seventeenth century a vast amount of information was already available. Furthermore, several techniques—distillation, extraction, pyrolysis, etc.—had been considerably improved. From this period Rudolph Glauber is one figure who stands out. Among his many accomplishments was a study of the action of heat on coal in which he prepared a "pleasant-smelling oil and valuable healing balsams." By treating various plants, Glauber learned the preparation of many aromatic oils. The fragrances of these oils were prized characteristics and made them much sought after.

Johann Rudolph Glauber (1604–1668). Born in Karlsdadt, the son of a barber and orphaned while a boy, Glauber was unusually prolific in technical accomplishments. In the many volumes of works he produced are described ways to make sulfuric acid, nitric acid, hydrochloric acid, many salts, including "Glauber's salt" or sodium sulfate decahydrate, and many organic substances, including aromatic oils and alkaloids. Robert Boyle is usually credited with being the father of modern chemistry. To some historians Glauber might justifiably be called a "German Boyle"; he was clearly a pioneer of organic chemistry.

Michael Faraday (1791–1867). The son of a blacksmith, his formal education consisting of but a few years of elementary school, Faraday ranks as a giant in chemistry and physics. Faraday's laws of electrolysis, the introduction of the field concept in magnetism, the conversion of magnetism into electricity—these accomplishments are the chief basis for his reputation. His discovery of benzene, his investigation of many of its chemical reactions, his inventions for liquefying gaseous chemicals, his introduction of the terms *electrolysis, electrolyte, ion, cation, anion, cathode,* and *anode,* and his contributions to the developing ideas concerning equivalent weights ensure him a high place in the history of chemistry.

We cannot know with certainty what Glauber's pleasant-smelling oil was. It could have been benzene or a mixture of it and related compounds. His healing balsams were probably phenolic compounds. The efforts of scientists during the next three centuries led to the discovery of a feature common to many of the pleasant-smelling and aromatic oils. In 1825 the Portable Gas Company of London turned over to a British scientist, Michael Faraday, an oily liquid that separated out when they compressed illuminating gas for storage and transportation in cylinders. Faraday purified this oil and analyzed it and, because of his care and accuracy, he is credited with the discovery of benzene. The name, however, was given to the oil a few years later by another chemist.

Several of the aromatic, pleasant-smelling oils obtainable from plant sources were much later found to consist of molecules similar in some ways to those of benzene. With time the term *aromatic compound* was applied to any substance whose molecules contained the benzene structural skeleton, even if the substance had an odor not especially pleasing. In fact, a huge number of aliphatic substances (e.g., esters, terpenes, and essential oils) have pleasant fragrances but do not contain the structural skeleton of benzene.

Although the terms *aliphatic* and *aromatic* originated in relation to natural sources of these materials, they are used by modern chemists to divide organic compounds into two great classes on a structural basis. Aromatic compounds are those that have the benzene structural skeleton or typically show the types of reactions peculiar to this system. Aliphatic compounds are the others we have been studying, the alkanes, alkenes, alkynes, and their cyclic relatives, in fact, any open-chain system or cyclic relative of whatever family.

STRUCTURE OF BENZENE

Faraday discovered benzene in 1825 before there were successful theories about the ways atoms are joined together in molecules. The pioneers of structural theory in organic chemistry were Kekulé, Couper, Butlerov, and Crum Brown, and their work was still thirty to thirty-five years in the future. They gave us our present symbols, those we call Lewis structures because Lewis identified the straight lines in structural formulas with shared pairs of electrons. Benzene, however, remained a problem in structural theory until about 1931.

August Kekulé (1829–1896). Born in Darmstadt, the son of a grand-ducal Hessian head councillor in the War Office, Kekulé had the advantage of a gymnasium (roughly, high school) and university education. In 1857 he proposed the tetravalence of carbon, and the following year he explained the great diversity and number of carbon-containing compounds in terms of carbon's ability to form strong bonds to its own kind while also being able to bond to other atoms. A great lecturer as well as a brilliant scientist, he invented ball and stick models to present his theories of valence and structure.

Archibald Scott Couper (1831–1892). At the same time, and independent of Kekulé, Couper proposed essentially the same theory of carbon chain formation. He had his theory ready for presentation to the French Academy of Science early in 1858, but he was not a member and Academy rules permitted only members to present papers. Couper asked a member Charles Wurtz, to present his, but Professor Wurtz procrastinated. In the meantime, Kekulé published his theory. Couper, understandably upset, apparently berated Wurtz who responded by expelling Couper from his laboratory. Couper then persuaded Jean Baptiste Dumas to present his paper to the French Academy in June of 1858, but Kekulé attacked it, insisting on being credited with earlier conception of a more significant theory. Couper suffered a nervous breakdown and for the rest of his life (thirty-four more years) he was incapable of significant intellectual work. Kekulé's claims were accepted and Couper was ignored by the scientific world of his day. Only much later did scientists realize that he deserved far more credit than he received.

Alexander Mikhailovich Butlerov (1828–1886). Born near Kazan, son of a retired army officer, Butlerov's early specialization in butterflies gave way to an interest in chemistry. By 1857 he was a professor of chemistry at the University of Kazan. Foreign travels took him on visits to both Kekulé and Couper shortly before their publications on organic structures. Butlerov further extended these theories, and he was the first chemist to realize that each compound must have its own unique structure. In fact, he invented the term "chemical structure," and his was the first textbook to use structural ideas throughout. He is also remembered for contributions to our understanding of isomerism and tautomerism. Vladimer Markovnikov was his student.

Alexander Crum Brown (1838–1922). Of Scottish parentage, like Couper, Crum Brown was a student of Bunsen and Kolbe. He was apparently the first to designate every valence in an organic structural formula by a separate line, the line G. N. Lewis decades later identified with a shared pair of electrons.

The benzene problem was one of theory making. Structural theory that worked so well for an enormous number of aliphatic compounds simply could not be applied to benzene. Certain known facts about benzene did not correlate well with any simple, Lewis-type structure. Kekulé and Couper worried enough about the benzene problem to devise a theory which, at least for a time, seemed to bring it into line with traditional theory. To understand what they did and why it was not enough, we must survey some of the facts about benzene that show the nature and extent of the benzene problem.

1. The molecular formula of benzene is C_6H_6. With six carbons holding only six hydrogens, the molecule should be highly unsaturated. Addition reactions such as those described for the alkenes should be the rule, but instead they are the rare exception.

2. Benzene undergoes substitution reactions which are catalyzed by acids. When monosubstituted benzenes are made (of the general type $C_6H_5—G$; G = —Br, —Cl, —NO_2, etc.), no isomers form, which means that all six hydrogens in benzene must be equivalent. The principal aromatic substitution reactions are the following.

Nitration:

$$C_6H_6 + HO—NO_2 \xrightarrow[50\text{–}55°C]{H_2SO_4\text{ (concd)}} C_6H_5—NO_2 + H_2O$$

Nitric acid

Nitrobenzene (85%) (bp 211°C)

Under these conditions alkenes undergo extensive oxidation and decomposition.

Sulfonation:

$$C_6H_6 + SO_3 \xrightarrow[\text{room temperature}]{H_2SO_4\text{ (concd)}} C_6H_5—\underset{\underset{O}{\downarrow}}{\overset{\overset{O}{\uparrow}}{S}}—O—H$$

Benzenesulfonic acid (56%) (mp 525°C)

Sulfuric acid *adds* to double bonds in alkenes to form alkyl hydrogen sulfates.

Halogenation:

$$C_6H_6 + Cl_2 \xrightarrow[FeCl_3]{\text{Fe or}} C_6H_5—Cl + H—Cl$$

Chlorobenzene (90%) (bp 132°C)

$$C_6H_6 + Br_2 \xrightarrow{Fe} C_6H_5—Br + H—Br$$

Bromobenzene (59%) (bp 155°C)

$$C_6H_6 + I_2 \xrightarrow{HNO_3} C_6H_5—I + HI \quad \text{(oxidized as it forms by nitric acid)}$$

Iodobenzene (87%) (bp 189°C)

Chlorine and bromine add to an alkene with no catalyst required. Iodine is generally unreactive toward a carbon-carbon double bond. Iodination of benzene occurs only if H—I is removed as soon as it forms. That is the function of the nitric acid.

Benzene is not the only compound reacting in these ways, that is, undergoing substitution reactions for the most part rather than addition reactions.

Reactions such as the three just given are characteristic of aromatic compounds in general. Benzene is unusually stable toward oxidizing agents, which cause extensive changes to alkenes.

3. When a monosubstituted benzene, C_6H_5—G, is made to form a disubstituted product, either $C_6H_4G_2$ or G_6H_4GZ, three isomeric substances can be isolated.

The Kekulé Structure of Benzene. Kekulé proposed structure **1** for benzene as

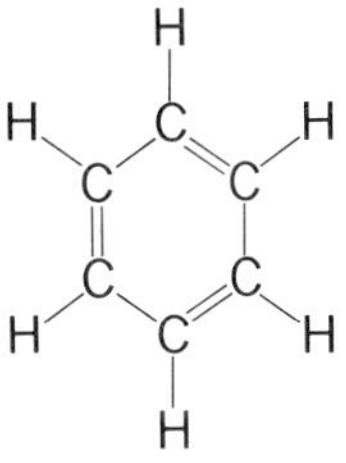

1 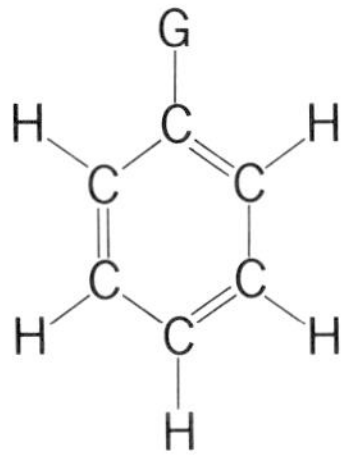 **2**

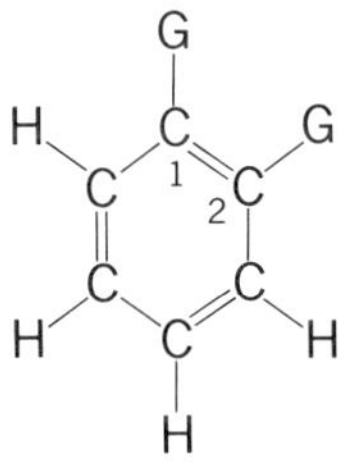

3 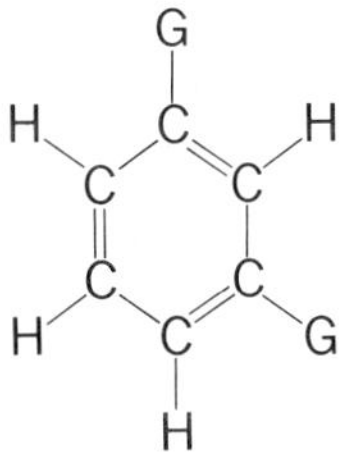 **4** 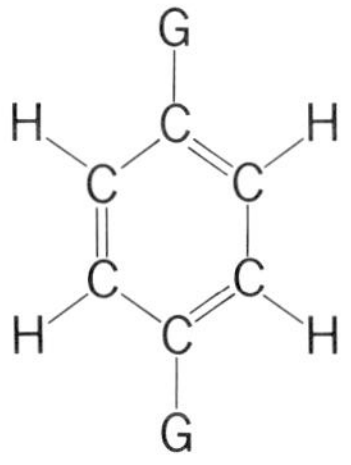 **5** 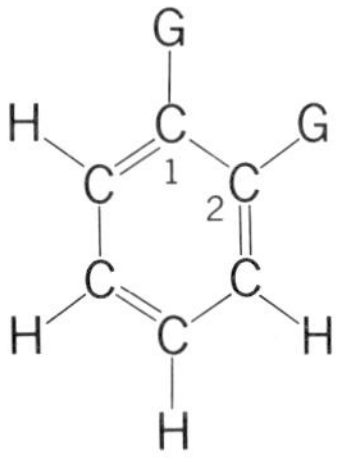 **6**

being the most consistent with the evidence. All six hydrogens are equivalent, and the structure of a monosubstitution product would be of type **2.** Possible disubstitution products would be **3, 4,** and **5.** But it was rightly pointed out by Kekulé's critics that if his structure for benzene were correct, the 1,2-disubstituted product **3** should have an isomer **6** which differed from it only in the location of the double bonds. In **3** the carbons holding the two groups are joined by a double bond; in **6** they are joined by a single bond. Kekulé therefore modified his theory by saying that the double bonds rapidly shift back and forth so that **3** and **6** are isomers in such rapid and mobile equilibrium that they can never be separated. Kekulé's explanation was accepted for decades, but it did not fully solve the problem, for the remarkable nonalkene-like properties of a structure written with three double bonds remained unexplained. One important addition reaction of benzene, hydrogenation to cyclohexane, furnishes data important in understanding benzene.[1]

Hydrogenation of Benzene. Heats of Hydrogenation. Hydrogenation of a double bond is an exothermic process, and it is possible to measure the heat evolved. For carbon-carbon double bonds in most monoalkenes, between 27 and 32 kcal/mole

[1] In the presence of ultraviolet light, benzene will also add chlorine to form "benzene hexachloride" or BHC. This compound is really 1,2,3,4,5,6-hexachlorocyclohexane and consists of several geometric isomers. One of the isomers is a potent insecticide, and for this reason BHC is an important commercial chemical sold under the names gammexane and lindane.

are evolved. Consider, then, the following specific cases. In each the final product is the same, cyclohexane. The heat of hydrogenation of cyclohexene, −28.6 kcal/mole, is taken as the value for 1 mole of double bond in a six-membered ring. 1,3-cyclohexadiene with two double bonds should release 2 × 28.6 = 57.2 kcal/mole. Benzene, if it had three ordinary double bonds, should release 3 × 28.6 = 85.8 kcal/mole.

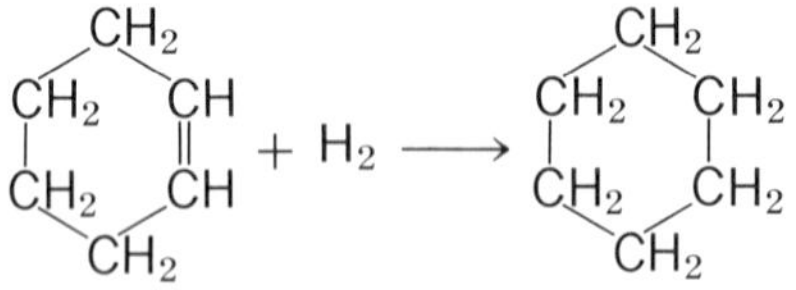

Cyclohexene $+ H_2 \longrightarrow$ Cyclohexane $\quad \Delta H_{obsd} = -28.6$ kcal/mole

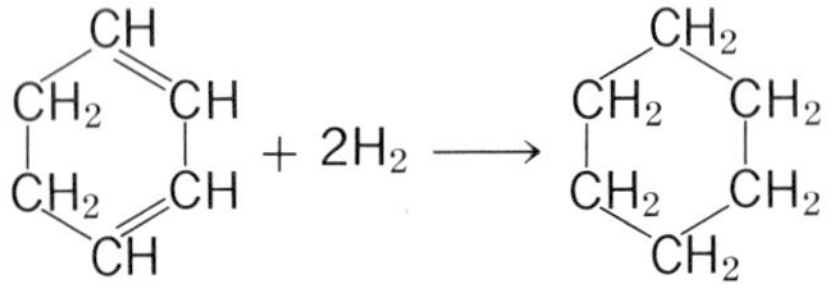

1,3-Cyclohexadiene $+ 2H_2 \longrightarrow$ Cyclohexane $\quad \Delta H_{obsd} = -55.4$ kcal/mole $\quad \Delta H_{calcd} = -57.2$ kcal/mole

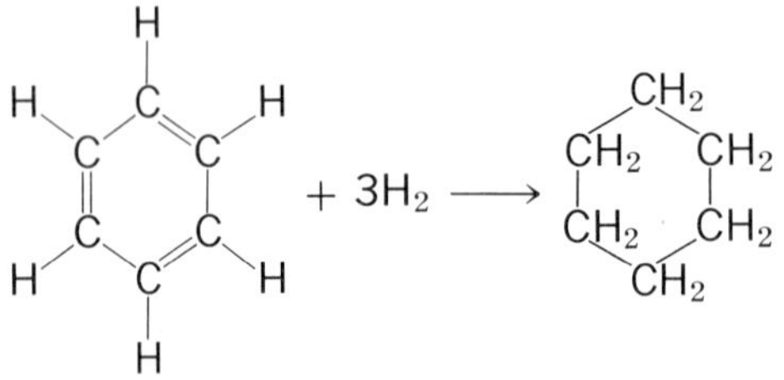

"1,3,5-Cyclohexatriene" (benzene) $+ 3H_2 \longrightarrow$ Cyclohexane $\quad \Delta H_{obsd} = -49.8$ kcal/mole $\quad \Delta H_{calcd} = -85.8$ kcal/mole

Cyclohexene is rapidly hydrogenated at room temperature over nickel at only 20 lb/in.^{2} of pressure. Benzene requires a pressure several hundred times as great and a much higher temperature. Morever, a great deal less energy is liberated than we might expect on the basis of its having three ordinary double bonds. Since the products in the three examples are identical, the evolution of 36 kcal/mole fewer units of energy for benzene must mean that it has this much less internal energy initially—that is, this much less energy than calculated for the Kekulé structure with its three ordinary double bonds. If it has less internal energy than the expected structural system, it must be more stable than that system. Benzene is obviously not well represented by the Kekulé structures. Measured bond distances in benzene further discredit the Kekulé formula.

Bond Lengths in Benzene. If benzene molecules had either Kekulé structure with alternating double and single bonds, X-ray diffraction analysis should reveal two different values for carbon-carbon bond distances, a shorter one for the double bonds and a longer one for the single bonds. Only one is observed, however, 1.39 Å.

This result shows that each carbon-carbon bond is equivalent, and that it is intermediate between a single bond (1.54 Å) and a double bond (1.34 Å). Again, the Kekulé structure is inadequate.

THE THEORY OF RESONANCE

It is clear that a substance such as benzene makes some modification of classical structural theory necessary. Of the approaches available, one is called resonance theory. Organic chemists have found it to be very useful for theorizing about properties of aromatic compounds. According to this theory, whenever two or more structures with the same relative locations of nuclei but different distributions of electrons can be drawn for a molecule, a *condition of resonance* is said to exist.

Two Kekulé structures of benzene, **7** and **8,** differing only in the locations of electrons can be drawn. Two carbons are numbered to show that although **7** and **8** are equivalent, they are not identical. The C_1—C_2 bond in **7** is single, in **8** double.

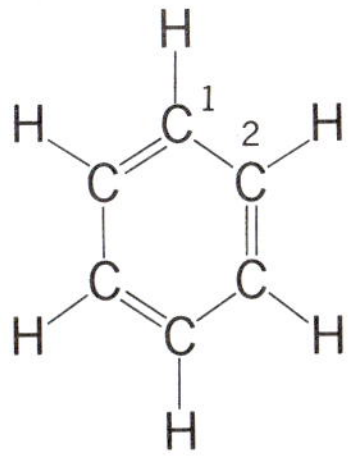

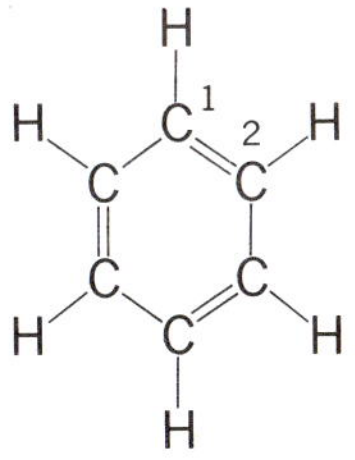

or

or

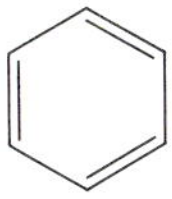

7

8

Chemists have learned through experience that whenever it is possible to draw two or more such structures differing only in the locations of electrons, neither structure adequately represents the molecule. A good approximation of the real molecular structure, however, can be made by thinking of it as a *hybrid,* a resonance hybrid, of the classical Kekulé or Lewis structures. These, instead of being regarded as real, are said to be *contributing structures.* They "contribute" to the hybrid. That is, they contribute to our mental picture of what the real molecule is probably like. As a good rule of thumb, the more nearly equivalent such contributing structures are, the less likely does any one of them represent the true structural state of affairs and the more seriously must we take the hybrid we mentally put together from the contributors. How the hybrid structure for benzene might be put together and represented on paper is described in Figure 11.1. Note especially the new symbol, a double-headed arrow: $\longleftrightarrow$ .We cannot use two arrows pointing in opposite directions because that is our symbol for equilibrium between two or more real substances. The double-headed arrow is our signal that the structures at both ends are contributors to a resonance hybrid. Moreover, we shall usually

Step 1. The contributing structures must be drawn before any hybrid can be fashioned. The trick is to be able to write one Kekulé or Lewis structure and then develop the others from it.

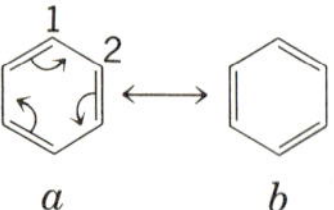
1
2
a
b

The curved arrows in the first indicate how pairs of electrons are relocated to write the second.

Step 2. To start the hybrid, examine the contributors and write down everything that is the same in all.

Step 3. Next note the differences. In structure *a* the C_1—C_2 bond is single, in *b* double. Make it a partial double bond in the hybrid by drawing a dotted line (instead of a solid line) for the second of the two lines.

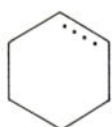

Step 4. Continue this dotted line as in step 3 around each pair of adjacent positions in structures *a* and *b*; the resulting hybrid for benzene is as follows:

Figure 11.1 Developing the hybrid structure of benzene from the contributing structures.

follow the practice of having the structure on the right develop from the one on the left by the relocation of electrons, indicated in the first structure by curved arrows. If several contributors are drawn, the development will be step by step from left to right.

Resonance Energy. When resonance occurs, the substance has less internal energy and is more stable than would be calculated or estimated on the basis of its being any single one of its contributors. The heats of hydrogenation have illustrated this point for benzene, and numerous other examples could be cited. The difference between the predicted and the actual internal energy is called the *resonance energy* of the substance. The substance does *not* have this energy because it does not have the structure used to make the calculation, the best structure that could be written if classical rules of valence were employed. It has instead some nonclassical, hybrid structure.

The more nearly equivalent the contributing structures, the greater this resonance energy of stabilization. The generally accepted value of resonance energy for benzene in which the two contributors are equivalent is 38 kcal/mole—very close to the figure of 36 kcal/mole obtained from heats of hydrogenation. By "equivalent" is meant having equal or nearly equal internal energies—if each contributor had a real existence. We must remember that drawing potential contributing structures is purely a pencil and paper operation. Just because a structure can be drawn does not mean that it is an important contributor. Starting with one Kekulé structure for benzene, we could, on paper, develop the following:

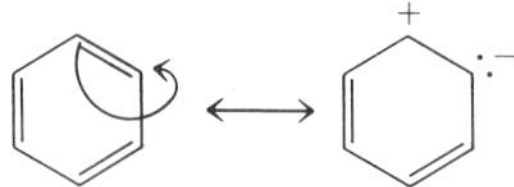

But the second structure is ignored because, if it could exist, it would be much less stable than the first. The separation of a plus and a minus charge is an energy-demanding process. Like charges attract. Thus, although the second structure differs from the first only in the distribution of electrons, it makes no significant contribution to the hybrid, in comparison with the Kekulé structures.

We shall not develop resonance theory any further at this point. The student is not expected to become skilled in writing contributing structures and then selecting those that would contribute most to the hybrid. It is hoped, however, that he will be able to follow arguments based on the use of resonance theory. Before continuing, we shall summarize what we have done and not done in this introduction to the theory of resonance.

The minimum expectation is that the problem making *some* theory necessary be recognized. If a single structure is to mean something, we do not want it to mean everything. If when we draw a carbon-carbon double bond we want it to mean (in part) "addition reactions," we certainly do not want it to mean "no addition reactions" at the same time. But to represent benzene as one or the other of the Kekulé structures confronts us with precisely such a contradiction. Resonance theory is one attempt to solve the contradiction by allowing us to use already familiar symbols, the classical Kekulé or Lewis structures, as a means for making an educated guess at what the actual structure probably is. From the contributors we mentally, or on paper, construct a nonclassical hybrid as a better structure for the substance.

ORBITAL MODEL OF BENZENE

Where resonance occurs, a stabilizing factor is present, and the molecular orbital model for benzene provides insights concerning it. Since the carbons in benzene are individually attached to three other atoms, sp^2 hybridization may be assumed for them. A planar system with bond angles of 120° would then be expected, and this is precisely the internal angle of a regular hexagon. With overlapping of unhybridized p_z orbitals (Figure 11.2), benzene acquires a large circular molecular orbital instead of three isolated, localized double bonds. Instead of the pi electrons being localized as implied by a Kekulé structure, they are considerably delocalized throughout a much larger volume of space. It is believed that this delocalization of electrons into a cyclic molecular orbital largely accounts for the "extra" stability of the benzene molecule. In fact, the term *delocalization energy* is sometimes used synonomously with resonance energy.

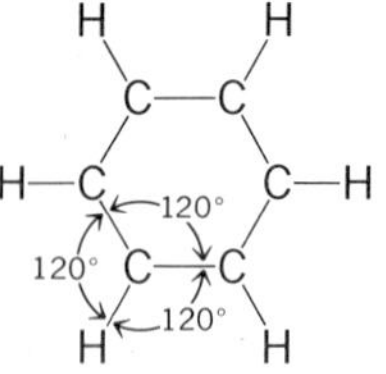
H
H
C—C
H—C
C—H
120°
120°
C—C
120°
H
H

(*a*) σ-Bond network in benzene

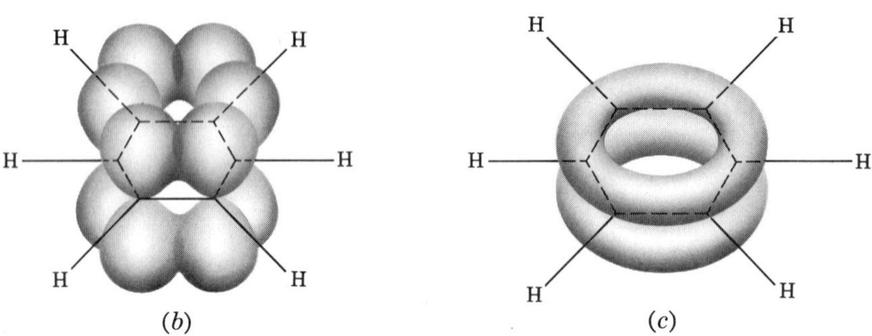
H
H
H
H
H
H
(*b*)
H
H
H
H
H
H
(*c*)

Figure 11.2 Orbital model for benzene. The fundamental framework of the molecule is provided by σ-bonds shown in part *a*. At each carbon is located an unhybridized p_z orbital which can overlap with p_z orbitals on both neighboring carbons as shown in part *b*. The result is the double doughnut-shaped molecular orbital shown in part *c*. This is actually only one of the new molecular orbitals that forms. For our purposes we may treat the shape shown in part *c* as a molecular shell containing three subshells with two electrons in each.

This orbital theory gives us further insight into the structure of benzene, but only with difficulty can we use this theory to understand the reactions of aromatic systems. Resonance theory, on the other hand, permits us to employ conventional structures for deducing the hybrid and in the long run is easier to apply. It is common practice, however, to write the structure of benzene and its derivatives, not with three double bonds showing but rather with a dotted or solid circle in the hexagon as indicated by structures **9** and **10.** We follow this practice except when discussing the mechanism of an aromatic substitution reaction.

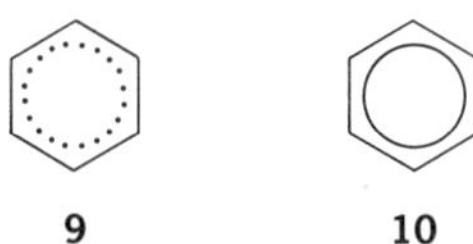

9 **10**

NAMING BENZENE DERIVATIVES

Monosubstituted Compounds. For a few compounds the nomenclature has some system. Prefixes naming the substituent are joined to the word "benzene":

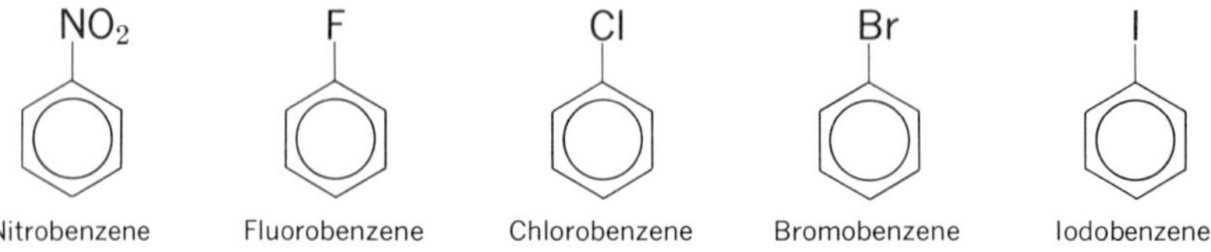
NO2
F
Cl
Br
I
Nitrobenzene
Fluorobenzene
Chlorobenzene
Bromobenzene
Iodobenzene

Other derivatives have common names that are always used, even though systematic names are possible:

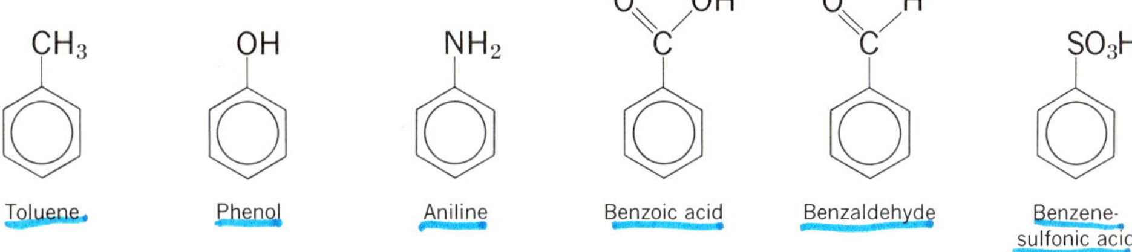
CH3
OH
NH2
O OH
C
O H
C
SO3H

Toluene Phenol Aniline Benzoic acid Benzaldehyde Benzene-sulfonic acid

Disubstituted Compounds. When two or more groups are attached to the same benzene ring, both their nature and their relative locations must be specified in the name. The prefixes "ortho," "meta," and "para" are used to distinguish, respectively, the 1,2-, the 1,3-, and the 1,4- relations. These prefixes are usually abbreviated *o*-, *m*-, and *p*-; in the following examples both substituent groups are the same.

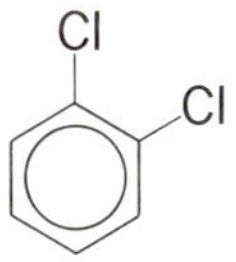

Cl Cl Cl Cl Cl Cl

o-Dichlorobenzene ortho | *m*-Dichlorobenzene meta | *p*-Dichlorobenzene para

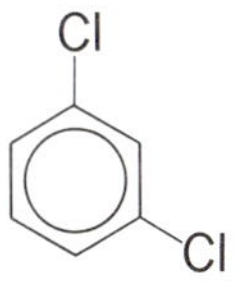

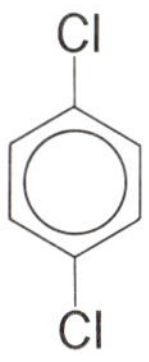

If one of the two groups, alone on the benzene ring, would give it a common name, this name is used and the second group is designated as a substituent. The following examples illustrate this point.

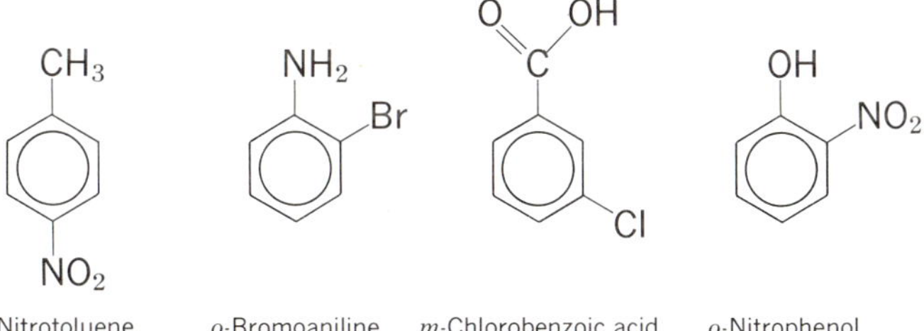
CH3
NO2
NH2
Br
O OH
C
Cl
OH
NO2

p-Nitrotoluene *o*-Bromoaniline *m*-Chlorobenzoic acid *o*-Nitrophenol

If neither group would be associated with a common name were it alone on the benzene ring, both groups are named and located in the name. For example,

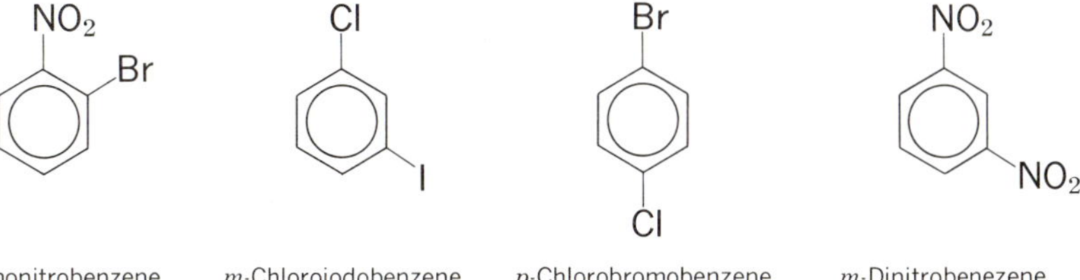
NO2
Br
Cl
I
Br
Cl
NO2
NO2

o-Bromonitrobenzene *m*-Chloroiodobenzene *p*-Chlorobromobenzene *m*-Dinitrobenezene

Polysubstituted Benzenes. If three or more groups are on a benzene ring, the ring positions must be numbered. When one group is associated with a common name, its position number is 1. For example,

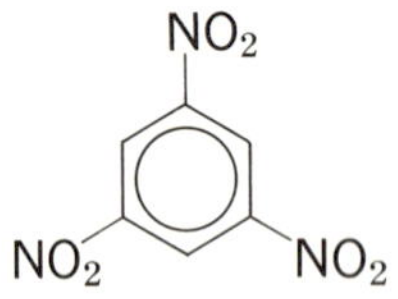

1,3,5-Trinitrobenzene (TNB)

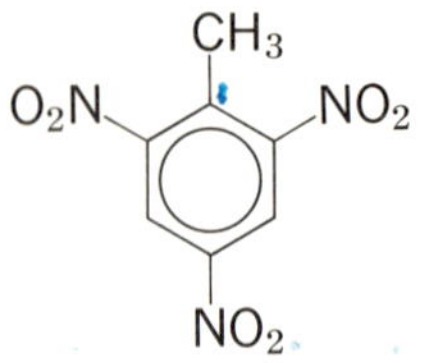

2,4,6-Trinitrotoluene (TNT)

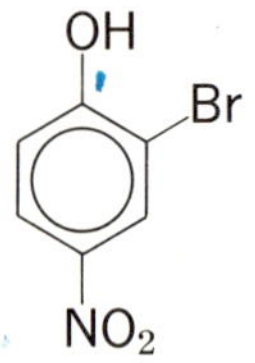

2-Bromo-4-nitrophenol

MECHANISM OF ELECTROPHILIC AROMATIC SUBSTITUTION

The commonly accepted mechanism for nearly all electrophilic aromatic substitutions involves a succession of three steps.

Step 1. Generation of an electrophilic or electron-poor species. The catalyst does its work during this step.

Step 2. Attack by the electrophilic species on the benzene ring. The benzene ring, with its huge pi-electron cloud, is very attractive to the electron-poor electrophilic species. During this step the benzene ring temporarily loses its aromaticity, its circular pi-electron system. A carbonium ion forms which, though resonance-stabilized, is still of relatively high internal energy, like all carbonium ions.

Step 3. The carbonium ion loses a proton from a ring position, and the ring recovers its aromatic character and its pi-electron system.

The Mechanism of the Nitration of Benzene. These steps may be illustrated by the nitration of benzene. (For other mechanisms any textbook of organic chemistry may be consulted.) When concentrated nitric acid and concentrated sulfuric acid are mixed, freezing point depression studies (cf. p. 155) and other analyses show that the following equilibrium is established. This is step 1 in the nitration of benzene. The nitronium ion which is formed is a powerful electrophilic species.

Step 1.

$$HO{-}NO_2 + H_2SO_4 \rightleftharpoons NO_2^+ + H_3O^+ + 2HSO_4^-$$

Nitric acid   Sulfuric acid   Nitronium ion

Step 2. Attack by the electrophilic nitronium ion on a molecule of benzene.

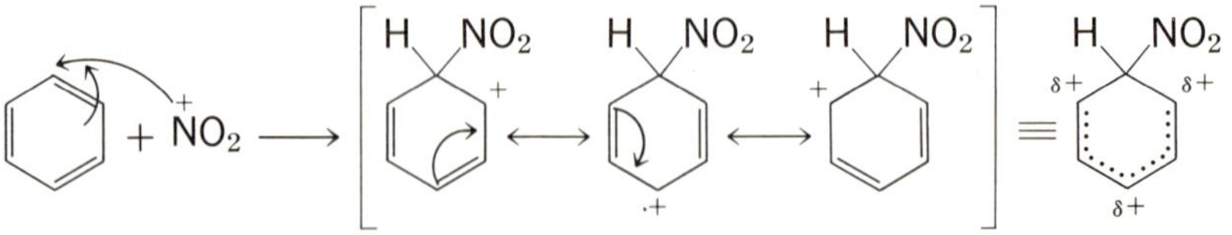

$$+ \overset{+}{N}O_2 \longrightarrow$$

Contributing structures for resonance-stabilized carbonium ion

Hybrid ion (The three $\delta+$ total $1+$.)

Step 3. Loss of a proton, recovery of benzene system, and formation of product.

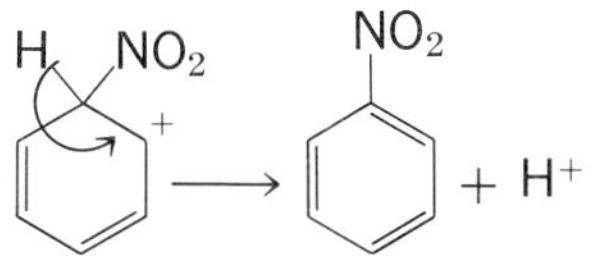

$$\text{(cyclohexadienyl cation with H and } NO_2\text{)} \longrightarrow C_6H_5NO_2 + H^+$$

Nitrobenzene

ARENES. ALKYL BENZENES

The names and structures of several alkyl benzenes or "arenes" are given in Table 11.1. Petroleum and coal tar are important sources. Because they combine an alkane-like group with a benzene system, arenes exhibit properties of both systems.

Chemical Properties

1. **Halogenation.** Under certain conditions (sunlight or high temperature) alkyl benzenes undergo halogenation at the side chain, the alkyl group. If an iron (or iron halide) catalyst is used, substitution into the ring occurs. Nitration and sulfonation also occur at the ring.

2. **Oxidation.** Benzene is exceptionally stable toward strong oxidizing agents and so are alkanes, but strong oxidizing agents will attack an arene at the side chain however long it is. For example,

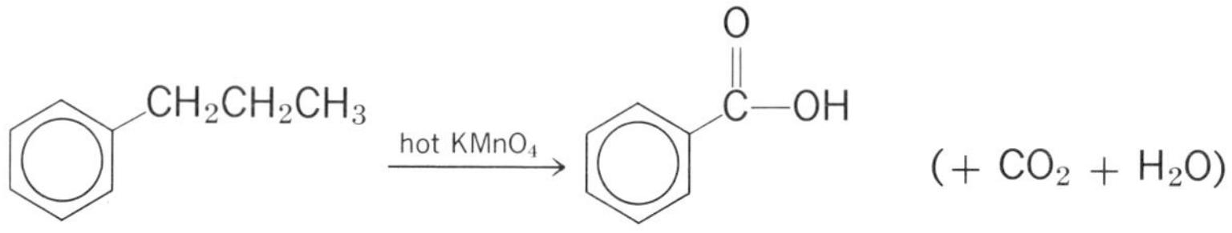

$$C_6H_5CH_2CH_2CH_3 \xrightarrow{\text{hot } KMnO_4} C_6H_5\overset{O}{\overset{\|}{C}}\text{—}OH \quad (+\ CO_2 + H_2O)$$

n-Propylbenzene → Benzoic acid

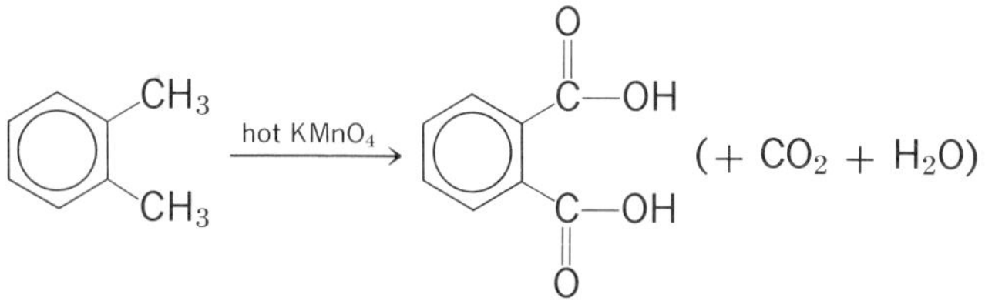

$$C_6H_4(CH_3)_2 \xrightarrow{\text{hot } KMnO_4} C_6H_4\left(\overset{O}{\overset{\|}{C}}\text{—}OH\right)_2 \quad (+\ CO_2 + H_2O)$$

o-Xylene → Phthalic acid

This process is an important source of aromatic carboxylic acids, and reagents such as potassium permanganate or potassium dichromate may be used for the reaction. In this example the properties of one system are modified by the presence of another. An alkane-like group on a benzene ring is much more susceptible to oxidation than an alkane is.

The benzene ring is not this stable in all situations. If an amino (—NH_2) or a hydroxyl (—OH) group is attached to the ring, it is quite susceptible to oxidation.

Table 11.1 **Arenes**

Name	Structure	Mp (°C)	Bp (°C)	Specific Gravity (at 20°C)
Benzene	C_6H_6	5.5	80	0.879
Toluene	$C_6H_5CH_3$	−95	111	0.866
o-Xylene	$C_6H_4(CH_3)_2$	−25	144	0.897
m-Xylene	$C_6H_4(CH_3)_2$	−48	139	0.881
p-Xylene	$C_6H_4(CH_3)_2$	−13	138	0.854
Ethylbenzene	$C_6H_5CH_2CH_3$	−95	136	0.867
Cumene	$C_6H_5CH(CH_3)_2$	−81	152	0.862
p-Cymene	$CH_3C_6H_4CH(CH_3)_2$	−70	177	0.857
Biphenyl	$C_6H_5{-}C_6H_5$	70	255	—

Atmospheric oxygen slowly attacks anilines and phenols, converting them to deeply colored, complex substances. The first synthetic dye was prepared in 1856 by William Henry Perkin who, in an attempt to convert aniline to quinine, oxidized it with potassium dichromate. From the black product he was able to extract a blue dye. In intermediary metabolism in human beings, phenolic substances obtained from proteins are converted into melanins, the dyes responsible for pigmentation of the skin.

POLYNUCLEAR AROMATIC HYDROCARBONS

Compounds exhibit aromatic properties if their molecules consist of flat, cyclic systems with multiple bonds that alternate with single bonds when Kekulé structures are written for them. Several hydrocarbons are in this category, and their structures appear to consist of fused or condensed benzene rings. Examples are given in Figure 11.3. Several of these compounds have color, a rare property among hydrocarbons. We shall encounter other examples illustrating that color is associated with an extensive pi-electron system in which (in Kekulé formulas) double bonds alternate with single bonds.

Naphthalene is a common moth repellant, but its use has declined since the introduction of *p*-dichlorobenzene. Some of these condensed aromatic hydrocarbons occur in the tars of tobacco smoke and in the charred residues on the surfaces of charcoal-broiled steaks. One of them, benzopyrene, is a known carcinogen (cancer inducer).

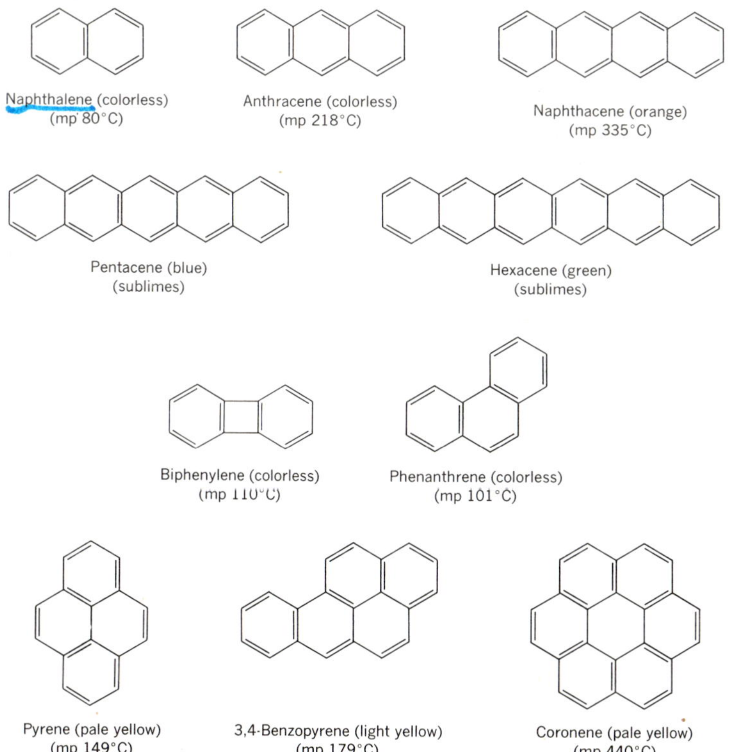
Naphthalene (colorless)
(mp 80°C)
Anthracene (colorless)
(mp 218°C)
Naphthacene (orange)
(mp 335°C)
Pentacene (blue)
(sublimes)
Hexacene (green)
(sublimes)
Biphenylene (colorless)
(mp 110°C)
Phenanthrene (colorless)
(mp 101°C)
Pyrene (pale yellow)
(mp 149°C)
3,4-Benzopyrene (light yellow)
(mp 179°C)
Coronene (pale yellow)
(mp 440°C)

Figure 11.3 Some polynuclear aromatic hydrocarbons.

REFERENCES AND ANNOTATED READING LIST

BOOKS

O. T. Benfey. *From Vital Force to Structural Formulas.* Houghton Mifflin Company, Boston, 1964. Chapters 8 and 9 discuss the theories of Kekulé and Couper with quotations from their writings. (Paperback.)

O. T. Benfey, editor. *Classics in the Theory of Chemical Combination.* Dover Publications, New York, 1963. Kekulé's famous paper and two by Couper are included, Kekulé's in English translation. (Paperback.)

R. T. Morrison and R. N. Boyd. *Organic Chemistry,* second edition. Allyn and Bacon, Boston, 1966. Chapter 10, "Benzene. Resonance. Aromatic Character" is a good reference among currently available textbooks.

ARTICLES

L. P. Lessing. "Coal." *Scientific American,* July 1955, page 193. This article describes what aromatic compounds are obtained from coal, an important source, and how they are obtained.

B. E. Schaar. "Aniline Dyes." *Chemistry,* January 1966, page 12. William Henry Perkin was seventeen years old when he accidentally discovered the first aniline dye. Schaar describes what Perkin did and the impact of his work.

PROBLEMS AND EXERCISES

1. Write structures of each of the following compounds
   (*a*) *m*-nitrobenzoic acid (*b*) 2,4,6-tribromoaniline (*c*) *o*-chlorophenol
   (*d*) *p*-toluenesulfonic acid (*e*) *m*-iodobenzaldehyde

2. Write equations for the steps in the bromination of benzene, catalyzed by $FeBr_3$, if the first step is $Br_2 + FeBr_3 \longrightarrow FeBr_4^- + Br^+$.
3. Naphthalene has three important contributing structures. One of them is the following:

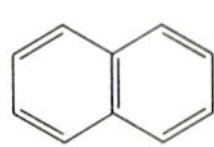

   Write the structures of the other two.

CHAPTER TWELVE

Alcohols

STRUCTURE

Compounds whose molecules contain a hydroxyl group, —OH, attached to a saturated carbon are called *alcohols*. If this carbon, called the carbinol carbon, has a double (or triple) bond going to another group, the substance is classified in

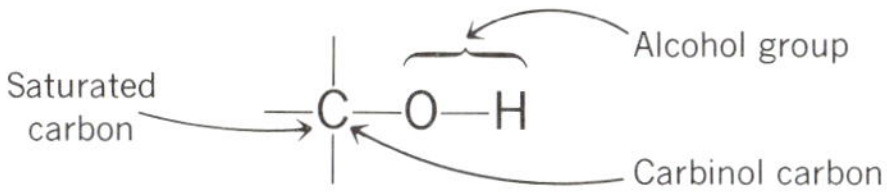

Alcohol group
Saturated carbon
—C—O—H
Carbinol carbon

Essential features of all alcohols

some other way. Several common alcohols are listed in Table 12.1. If the hydroxyl group is attached directly to a benzene ring, the molecule is classified as a *phenol*, which also happens to be the name of a specific compound.

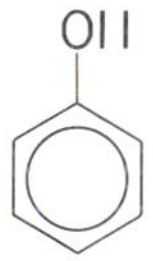

OH

Phenol

If the hydroxyl group is attached to one carbon of a carbon-carbon double bond, the molecule is classified as an *enol*, "ene" to indicate the alk*ene* function, + "-ol" to designate the hydroxyl group. Most enols are unstable. They rearrange, intramolecularly (within the molecule), to give isomeric molecules containing a *carbonyl* group, >C=O, a functional group to be studied in later chapters.

Table 12.1 **Some Monohydric Alcohols**

	Name	Structure	Mp (°C)	Bp (°C)	Density (in g/cc at 20°C)	Solubility (in g/100 g water, 20°C)
C_1	Methyl alcohol	CH_3OH	−98	64.5	0.791	soluble
C_2	Ethyl alcohol	CH_3CH_2OH	−115	78.3	0.789	soluble
C_3	*n*-Propyl alcohol	$CH_3CH_2CH_2OH$	−127	97.2	0.803	soluble
	Isopropyl alcohol	$(CH_3)_2CHOH$	−86	82.4	0.786	soluble
C_4	*n*-Butyl alcohol	$CH_3CH_2CH_2CH_2OH$	−90	118	0.810	8.0
	Isobutyl alcohol	$(CH_3)_2CHCH_2OH$	−108	108	0.802	10.0
	sec-Butyl alcohol	$CH_3CH_2CH(OH)CH_3$	−115	99.5	0.806	12.5
	t-Butyl alcohol	$(CH_3)_3COH$	25.6	82.5	0.786	soluble
C_5 and higher	*n*-Amyl alcohol (1-pentanol)	$CH_3(CH_2)_3CH_2OH$	−79	138	0.814	2.2
	t-Amyl alcohol (2-methyl-2-butanol)	$(CH_3)_2C(OH)CH_2CH_3$	−8.6	102	0.806	12.2
	Isoamyl alcohol (3-methyl-1-butanol)	$(CH_3)_2CHCH_2CH_2OH$		132	0.807	2.9
	Neopentyl alcohol (2,2-dimethyl-2-propanol)	$(CH_3)_3CCH_2OH$	52	113		3.7
	1-Hexanol	$CH_3(CH_2)_4CH_2OH$	−52	157	0.820	0.6
	1-Heptanol	$CH_3(CH_2)_5CH_2OH$	−34	176	0.822	0.1
	1-Octanol	$CH_3(CH_2)_6CH_2OH$	−15	195	0.826	0.04
	1-Decanol	$CH_3(CH_2)_8CH_2OH$	7	229	0.829	0.004
	Cyclopentanol	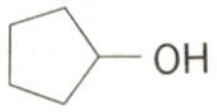 (cyclopentyl)—OH		140	0.949	slightly soluble
	Cyclohexanol	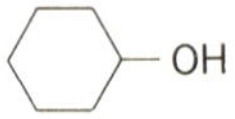 (cyclohexyl)—OH	23	161	0.962	3.6
	Allyl alcohol	$CH_2{=}CHCH_2OH$		97	0.852	soluble
	Benzyl alcohol	$C_6H_5CH_2OH$	−15	205	1.046	4

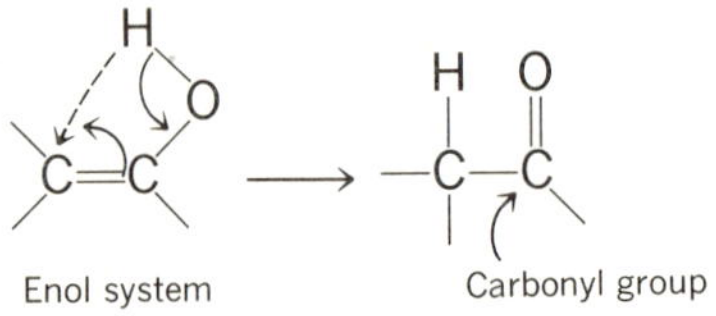

Enol system → Carbonyl group

Subclasses of Alcohols. It is frequently useful to classify an alcohol as primary (1°), secondary (2°), or tertiary (3°) according to the condition of the carbinol carbon (cf. p. 311).

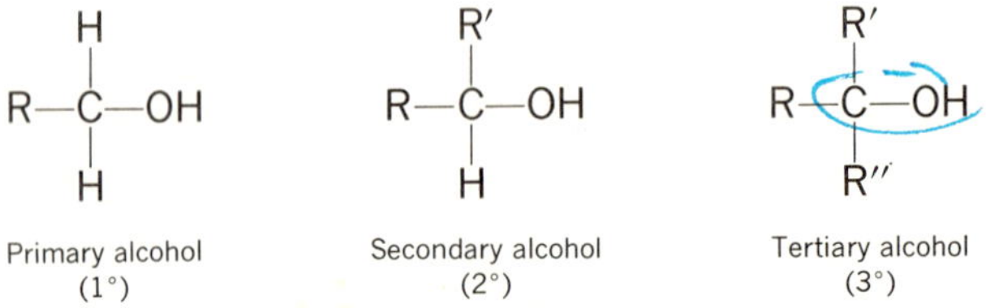

$$R{-}CH_2{-}OH \qquad R{-}CHR'{-}OH \qquad R{-}CR'R''{-}OH$$

Primary alcohol (1°) Secondary alcohol (2°) Tertiary alcohol (3°)

The single and double primes on R (see p. 258) indicate that the alkyl groups may or may not be identical.

The variations of chemical properties from subclass to subclass are not too great, and they often have to do with rates of reactions rather than the kinds of final products that form.

NOMENCLATURE

Of the systems of nomenclature commonly used by chemists for this family, we shall study two, the common names and those based on IUPAC rules.

Common Names. When the alkyl group to which the hydroxyl is attached has a common name, the corresponding alcohol is designated simply by writing the word "alcohol" after this name.

CH_3OH — Methyl alcohol

$(CH_3)_3COH$ — *t*-Butyl alcohol

$C_6H_5CH_2OH$ — Benzyl alcohol

CH_3CH_2OH — Ethyl alcohol

$CH_3CH(CH_3)OH$ — Isopropyl alcohol

$CH_3CH(CH_3)CH_2OH$ — Isobutyl alcohol

$CH_2{=}CH{-}CH_2OH$ — Allyl alcohol

IUPAC Nomenclature Rules for Alcohols

1. For the "parent" structure, select the longest continuous chain of carbons *that includes the hydroxyl group*. Determine the name of the alkane that corresponds to this carbon chain and replace the terminal "-e" by "-ol."

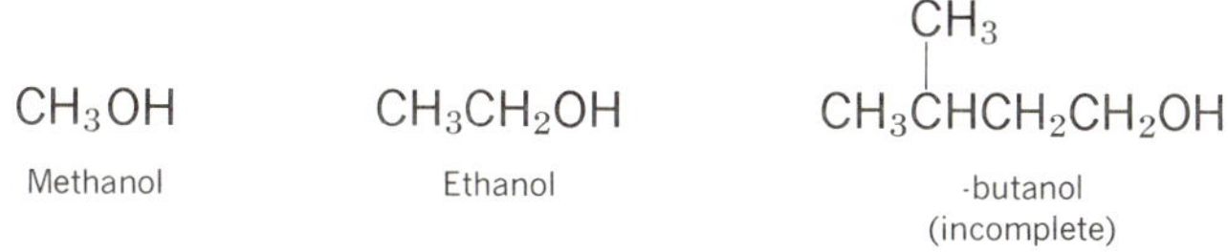

Methanol Ethanol -butanol (incomplete)

2. When isomerism is possible, designate the location of the hydroxyl group on this chain by numbering the carbons from whichever end will give this location the lower number.

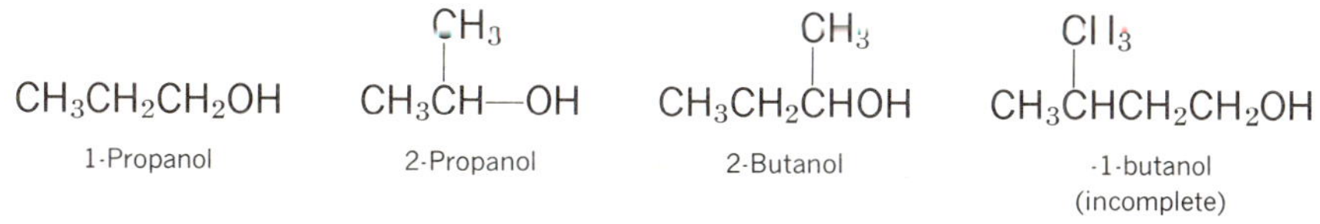

1-Propanol 2-Propanol 2-Butanol -1-butanol (incomplete)

3. Determine the names and the location numbers of any hydrocarbon-like groups (e.g., an alkyl, an aryl) attached to the chain, and assemble these before the parent name developed thus far. Study the examples carefully to determine how commas and hyphens are used to make the final name one word.

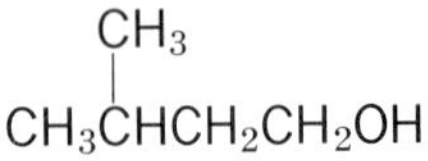

$$\begin{array}{c}CH_3\\ |\\ CH_3CHCH_2CH_2OH\end{array}$$

3-Methyl-1-butanol
(complete)

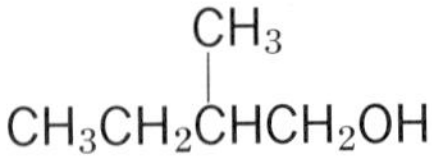

$$\begin{array}{c}CH_3\\ |\\ CH_3CH_2CHCH_2OH\end{array}$$

2-Methyl-1-butanol

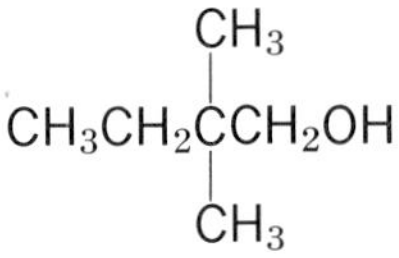

$$\begin{array}{c}CH_3\\ |\\ CH_3CH_2CCH_2OH\\ |\\ CH_3\end{array}$$

2,2-Dimethyl-1-butanol

$$\begin{array}{c}CH_3\\ |\\ CH_3CH \quad OH\\ | \qquad |\\ CH_3CH_2CH_2CH_2C—CCH_3\\ | \qquad |\\ CH_3 \quad CH_3\end{array}$$

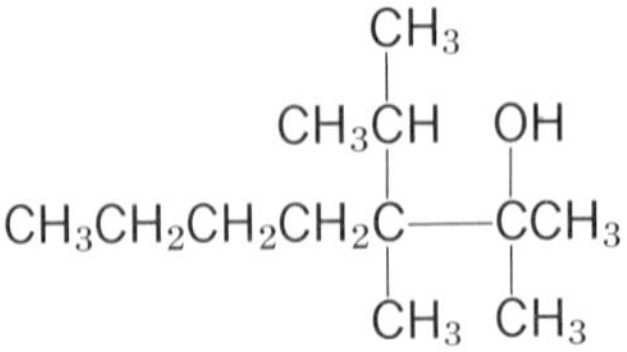

3-Isopropyl-2,3-dimethyl-2-heptanol

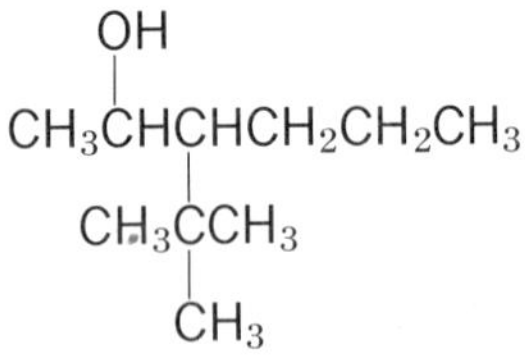

$$\begin{array}{c}OH\\ |\\ CH_3CHCHCH_2CH_2CH_3\\ |\\ CH_3CCH_3\\ |\\ CH_3\end{array}$$

3-*t*-Butyl-2-hexanol

$$\begin{array}{c}CH_2OH\\ |\\ CH_3CH_2CH_2CHCH_2CH_3\end{array}$$

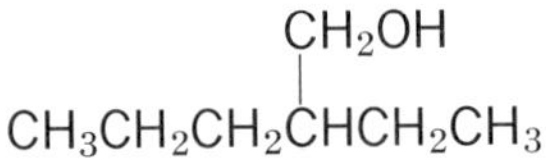

2-Ethyl-1-pentanol
(Note the application of rule 1 here.)

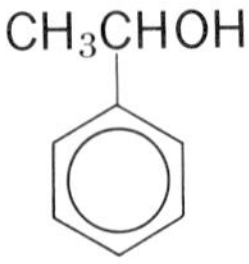

$$CH_3CHOH$$ (attached to a benzene ring)

1-Phenylethanol

4. If other atoms or groups are attached to or incorporated into the main chain of the alcohol molecule, first apply rules 1 through 3, and then work other groups into the name.

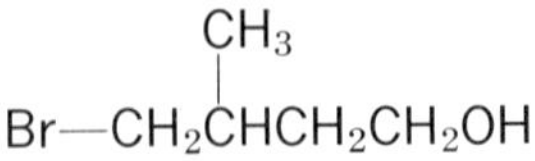

$$\begin{array}{c}CH_3\\ |\\ Br—CH_2CHCH_2CH_2OH\end{array}$$

4-Bromo-3-methyl-1-butanol
(*not* 1-bromo-2-methyl-4-butanol)

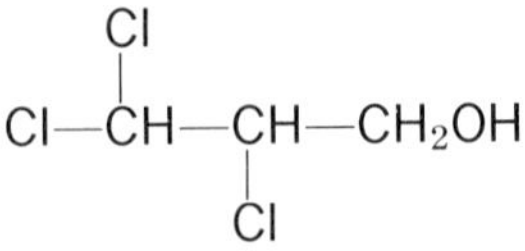

$$\begin{array}{c}Cl\\ |\\ Cl—CH—CH—CH_2OH\\ |\\ Cl\end{array}$$

2,3,3-Trichloro-1-propanol

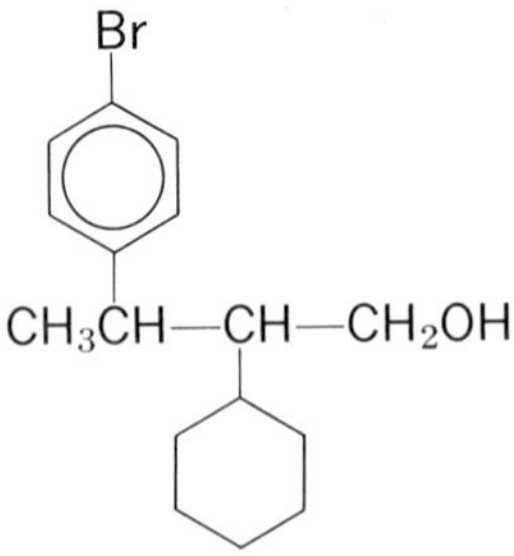

$$CH_3CH—CH—CH_2OH$$ (with 4-bromophenyl on C-3 and cyclohexyl on C-2)

2-Cyclohexyl-3-(4-bromophenyl)-1-butanol
(Note the use of parentheses to avoid confusing the numbering of the parent chain and the numbering of the group.)

Polyhydric Alcohols. Compounds containing two or more hydroxyl groups per molecule are quite common, although the hydroxyl groups must be on separate carbons as stated earlier. *Dihydric alcohols* or *glycols* contain two hydroxyls per molecule. *Trihydric alcohols* contain three hydroxyl groups per molecule. Glycerol

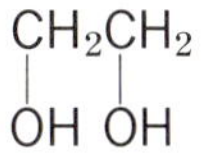

$$\underset{\text{OH}\ \ \text{OH}}{\text{CH}_2\text{CH}_2}$$

Ethylene glycol (1,2-ethanediol, bp 197°C)

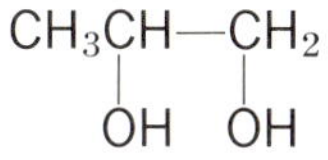

$$\text{CH}_3\underset{|}{\underset{\text{OH}}{\text{CH}}}—\underset{|}{\underset{\text{OH}}{\text{CH}_2}}$$

Propylene glycol (1,2-propanediol, bp 189°C)

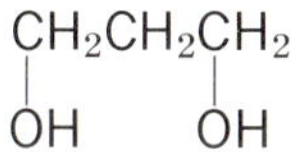

$$\underset{\text{OH}}{\text{CH}_2}\text{CH}_2\underset{\text{OH}}{\text{CH}_2}$$

Trimethylene glycol (1,3-propanediol, bp 214°C)

is the most common example. The common sugars and polysaccharides consist of polyhydric alcohols, with other groups also present (cf. the structure of glucose).

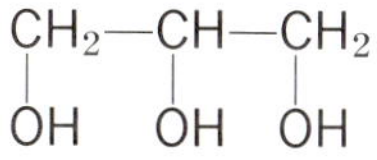

$$\underset{\text{OH}}{\text{CH}_2}—\underset{\text{OH}}{\text{CH}}—\underset{\text{OH}}{\text{CH}_2}$$

Glycerol (1,2,3-propanetriol, bp 290°C)

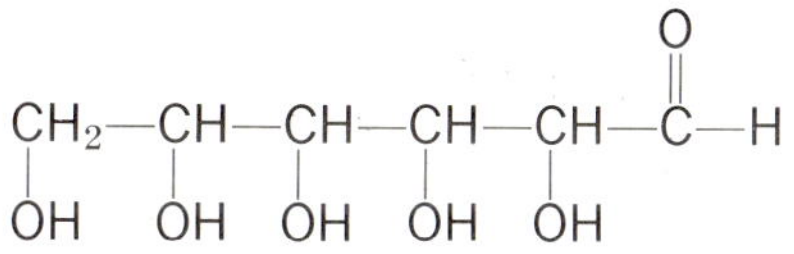

$$\underset{\text{OH}}{\text{CH}_2}—\underset{\text{OH}}{\text{CH}}—\underset{\text{OH}}{\text{CH}}—\underset{\text{OH}}{\text{CH}}—\underset{\text{OH}}{\text{CH}}—\overset{\text{O}}{\overset{\|}{\text{C}}}—\text{H}$$

Glucose (open form, mp 146–150°C)

We know of virtually no compounds that contain both the hydroxyl group and another hydroxyl group, a halo substituent (F, Cl, Br, or I), or an amino group, $—NH_2$, attached to the same carbinol carbon. These systems are apparently too unstable to be isolable, but they are known to exist in solutions under suitable circumstances. Alcohols are therefore largely limited to those in which the carbinol carbon, besides containing the hydroxyl group, holds at the same time only another carbon (or carbons), hydrogens, or both.

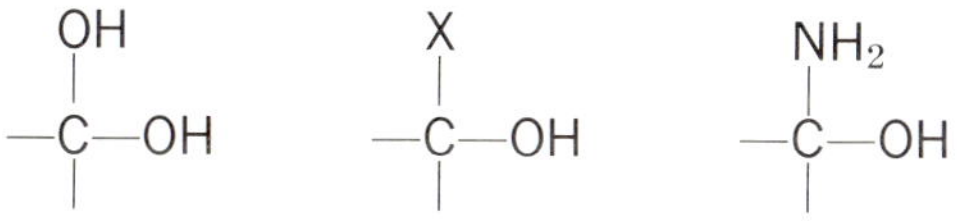

$$—\overset{\text{OH}}{\overset{|}{\underset{|}{\text{C}}}}—\text{OH} \qquad —\overset{\text{X}}{\overset{|}{\underset{|}{\text{C}}}}—\text{OH} \qquad —\overset{\text{NH}_2}{\overset{|}{\underset{|}{\text{C}}}}—\text{OH}$$

Unstable systems

PHYSICAL PROPERTIES. HYDROGEN BONDING. ASSOCIATION

Boiling Points. Figure 12.1 shows that the boiling points of the *normal,* saturated alcohols increase regularly as the formula weights increase. The effect of chain branching is indicated in Figure 12.2. In Figure 12.1 the boiling point curve of the homologous series of alkanes starts out much lower than that of the alcohols, but later the curves of the two families draw closer together. The great difference at the lower end of the two series is striking, however. Ethane and methyl alcohol have comparable formula weights (30 versus 32), yet ethane boils 153° below methyl alcohol (−88.6°C versus 64.5°C). Clearly it takes far more energy to separate molecules of the alcohol than to separate those of the alkane. Even more striking is the fact that water, of even lower formula weight, boils at a higher temperature than both (100°C). This irregularity is also seen in the trends for the boiling points of

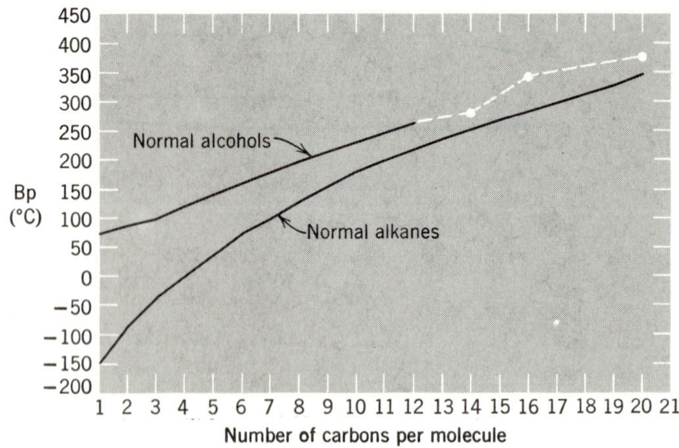
Normal alcohols
Normal alkanes
Bp (°C)
Number of carbons per molecule

Figure 12.1 Variation of boiling point with total carbon content in the homologous series of *normal* alkanes and normal primary alcohols. As the alcohols become increasingly alkane-like in structure, their properties also resemble those of the alkanes, as illustrated here for the boiling points.

simple hydrides (Figure 12.3). Water, hydrogen fluoride, and ammonia are irregular; methane is not.

The data in Figure 12.4 indicate that the high polarity of alcohol molecules as compared to that of alkanes is probably not the whole answer to the abnormally high boiling points of alcohols. Substances of higher polarities that boil at a lower temperature than alcohols of about the same formula weight are known.

Hydrogen Bonds. The irregularities just discussed and illustrated in Figures 12.3 and 12.4 occur in molecules in which a hydrogen atom is bonded to a strongly

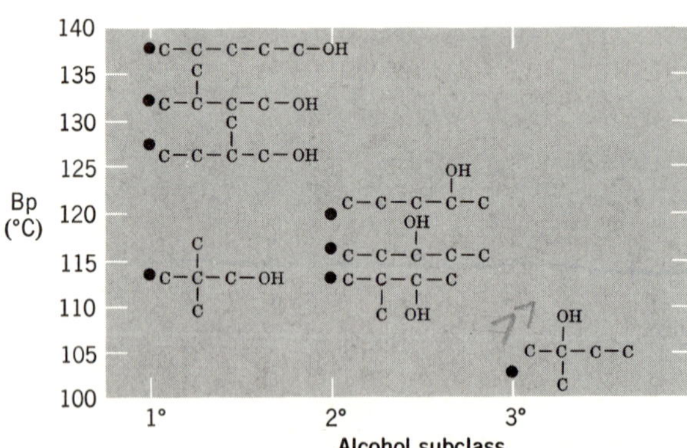
Bp (°C)
Alcohol subclass

Figure 12.2 The more compact a molecule, the lower the boiling point of the substance, all other factors being equal. The boiling points of the isomeric five-carbon alcohols illustrate this rule of thumb. The more compact the molecule, the weaker the forces of attraction between molecules, because compactness and symmetry usually mean lower polarity.

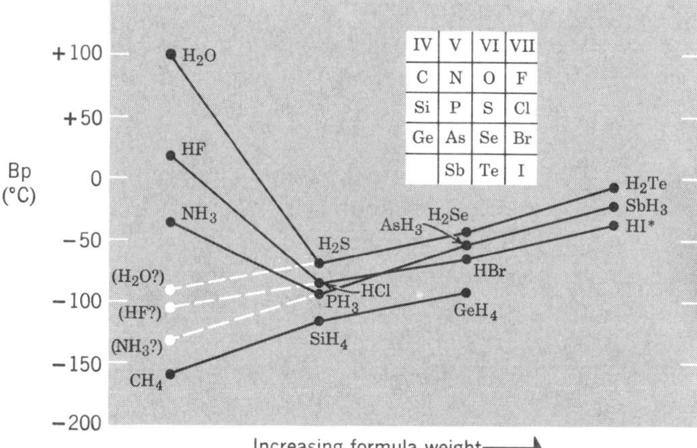
Bp (°C)
+100
+50
0
−50
−100
−150
−200
IV V VI VII
C N O F
Si P S Cl
Ge As Se Br
Sb Te I
H_2O
HF
NH_3
$(H_2O?)$
(HF?)
$(NH_3?)$
CH_4
H_2S
PH_3
HCl
SiH_4
AsH_3
H_2Se
HBr
GeH_4
H_2Te
SbH_3
HI*
Increasing formula weight

Figure 12.3 Hydrides—variation of boiling points with formula weights. The sequence from CH_4 to GeH_4 is the normal trend, that is, boiling points increase with increasing formula weight. The hydrides of group V elements (NH_3, PH_3, AsH_3, and SbH_3), the group VII elements (HF, HCl, HBr, and HI), and the group VI elements (H_2O, H_2S, H_2Se, and H_2Te) all show abrupt breaks in this trend. The boiling points of NH_3, HF, and H_2O are considerably higher than would be predicted on the basis of the rule of thumb that substances consisting of lighter molecules will have lower boiling points. The concept of hydrogen bonding explains these irregularities.

* The boiling point of HI is taken under pressure to prevent decomposition.

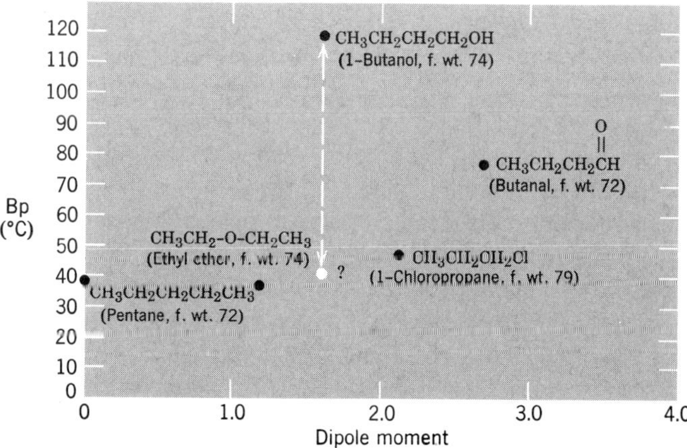
Bp (°C)
120
110
100
90
80
70
60
50
40
30
20
10
0
$CH_3CH_2CH_2CH_2OH$
(1-Butanol, f. wt. 74)
$CH_3CH_2CH_2CH$ (=O)
(Butanal, f. wt. 72)
CH_3CH_2-O-CH_2CH_3
(Ethyl ether, f. wt. 74)
?
$CH_3CH_2CH_2Cl$
(1-Chloropropane, f. wt. 79)
$CH_3CH_2CH_2CH_2CH_3$
(Pentane, f. wt. 72)
0
1.0
2.0
3.0
4.0
Dipole moment

Figure 12.4 Boiling point versus dipole moment, formula weights being approximately equal. Assuming that the dipole moment is a measure of polarity, we might expect the boiling point to increase as the polarity increases. The data plotted here do not support this simple correlation. 1-Butanol has a higher boiling point than would be predicted from polarity alone; only 1-butanol can participate in hydrogen bonding.

electronegative atom. The most electronegative element of all is fluorine; oxygen is next (cf. the alcohols and water); nitrogen and chlorine nearly tie for third. When fluorine, oxygen, and nitrogen are bonded to hydrogen (of much lower electronegativity), these atoms will surely bear rather large partial negative charges, and the hydrogen will have a correspondingly large partial positive charge. Moreover, a hydrogen substituent is very small, and it apparently can with little difficulty locate itself rather close to a partial negative charge on a neighbor molecule. When the distance between partial (and opposite) charges becomes very small, the force of attraction between the two can become sizable. It is not as great as the attractive force between two opposite, *full* charges (ionic bond), nor is it as much as is involved in the sharing of a pair of electrons between two nuclei (covalent bonding). But the force is large enough, nevertheless, for us to say that a bond exists. We encountered it before (p. 139) and called it a *hydrogen bond.* It resembles an ionic bond, for it is a force of attraction between separate, oppositely charged particles. The ionic bond exists between *fully* charged and separate ions; the hydrogen bond occurs between *partially* charged sites, usually on separate molecules. The hydrogen bond (Figure 12.5) can be looked upon as a bridge between two highly electronegative atoms (F, O, or N). It is attached to one by a full covalent bond, and it is attracted to the other by a partly ionic bond, between sites of partial and opposite charge.

Solubility. Substances in which molecules are attracted to each other by means of hydrogen bonds are said to be *associated.* Such substances often tend to be soluble in each other; ammonia in water and water in methyl alcohol are examples. In order for a nonassociated substance like methane to dissolve in, say, water, hydrogen bonds between water molecules would have to be broken and reduced in number. The energy cost of this would not be repaid by the formation of new hydrogen bonds to the foreign material. Methyl alcohol, in contrast to methane, however, can slip into the hydrogen bond network of water (or vice versa) at much lower energy cost (Figure 12.6). For this reason the low-formula-weight alcohols are soluble in water in all proportions (cf. Table 12.1). As the homologous series of alcohols is ascended, however, the larger and larger hydrocarbon "tails" tend more and more to disturb the hydrogen-bonding opportunities between the solvent and the solute. As the solubility data of Table 12.1 reveal, when straight-chain alcohols have five or six carbons or more, they are virtually insoluble in water.

Since all alcohols are also partly alkane-like it is also possible for them to dissolve in the typical so-called "hydrocarbon solvents"—benzene, ether, carbon tetrachloride, chloroform, gasoline, petroleum ether, and methylene chloride. To generalize, water tends to dissolve water-like molecules, nonpolar solvents to dissolve hydrocarbon-like substances. Even more generally speaking, polar solvents tend to dissolve polar solutes, nonpolar solvents to dissolve nonpolar (or moderately polar) solutes. The rule of thumb among chemists is "likes dissolve likes."

Our discussion of solubilities in water has been rather extensive. Since nearly all important reactions in living cells occur between organic molecules in aqueous fluids, we must note very carefully any structural features that help or hinder

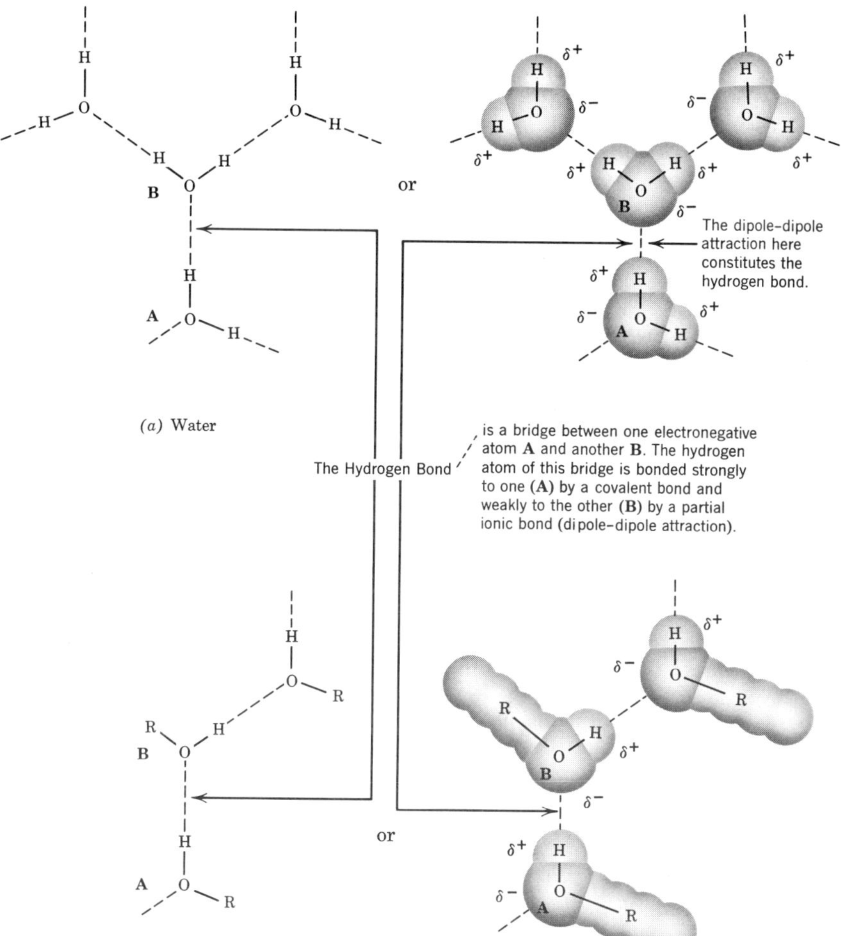
(a) Water
or
The dipole–dipole attraction here constitutes the hydrogen bond.
The Hydrogen Bond
is a bridge between one electronegative atom A and another B. The hydrogen atom of this bridge is bonded strongly to one (A) by a covalent bond and weakly to the other (B) by a partial ionic bond (dipole–dipole attraction).
or

Figure 12.5 Hydrogen bonding in alcohols and water.

the solubility of a compound in an aqueous medium. The —OH group is obviously a very important water-solubilizing group.

Exercise 12.1 Compare the boiling points of *n*-propyl alcohol, propylene glycol, and glycerol (cf. p. 315) with the boiling points of alkanes of approximately the same formula weights (cf. p. 255). Include in your list of data the boiling points of the monohydric alcohols that have formula weights approximately the same as those of propylene glycol and glycerol. Explain the trends in boiling points in going from monohydric to dihydric to trihydric alcohols.

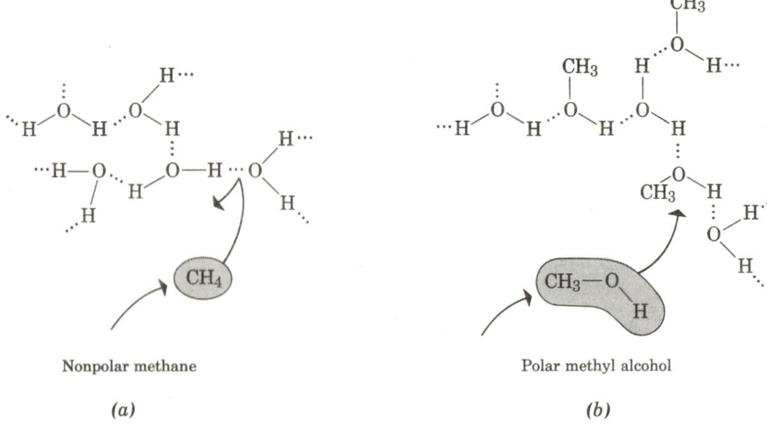
CH₄
Nonpolar methane
(a)
CH₃—O
H
Polar methyl alcohol
(b)

Figure 12.6 Understanding the solubility of short-chain alcohols in water. (*a*) A nonpolar alkane molecule cannot break into the hydrogen-bonded sequence in water. It cannot replace the hydrogen bonds that would have to be broken to let it in. Therefore it is insoluble in water.

(*b*) A short-chain alcohol molecule, capable of hydrogen bonding, can slip into the sequences in water. It can replace at least some of the hydrogen bonds that must be broken to let it in. In any liquid, of course, the molecules shift from place to place. The fixedness of this figure is not meant to indicate any inflexibility. The drawing may be likened to a high-speed flash photograph that has caught the action of the instant.

CHEMICAL PROPERTIES OF ALCOHOLS. GENERAL PRINCIPLES

In our study of the chemistry of alkanes we learned that the carbon-carbon single bond and the carbon-hydrogen single bond in an alkane environment are extremely resistant to chemical attack except by a few substances. Ionic reagents such as concentrated sulfuric acid and sodium hydroxide, active metals such as sodium or potassium, and electrically neutral substances such as water do not attack alkanes—at least not at ordinary temperatures and pressures. One way of understanding this phenomenon is to note that the electronegativities of carbon and hydrogen are quite close together (much closer than those of oxygen and hydrogen or of nitrogen and hydrogen). And, of course, the two atoms at the ends of a carbon-carbon single bond do not differ in relative electronegativities. In short, neither carbon-carbon nor carbon-hydrogen single bonds are very polar. Hence molecules or ions of polar or ionic reagents find little in an alkane molecule to which to be electrically attracted. Random collisions occur between the particles, but little chemical interaction takes place. At very high temperatures the situation may change, for under such circumstances the energies of collisions between potential reactants are much more frequently powerful enough to lead to bond-breaking and bond-making events.

Molecules of alcohols, compared with those of alkanes, have two new bonds, carbon to oxygen and oxygen to hydrogen. Because oxygen has a relatively high electronegativity, both these bonds will be polar. In addition, although alkane molecules have no unshared pairs of electrons anywhere, alcohol molecules do have them—on the oxygen. We should expect, therefore, that polar or ionic reagents will

$$\overset{\delta+}{\text{C}} \longrightarrow \ddot{\underset{\cdot\cdot}{\text{O}}}_{\delta-} \longleftarrow \text{H}^{\delta+}$$

State of polarization of bonds in an alcohol at the functional group

be much more successful in interacting with alcohols than with alkanes. Three types of reactions are observed: (1) reactions involving rupture of the O—H bond, (2) reactions involving the unshared electrons on the oxygen, and (3) reactions involving rupture of the C—O bond.

REACTIONS OF ALCOHOLS INVOLVING THE O—H BOND

Like water, alcohols are weak proton donors. They are weak acids in either the Arrhenius or the Brønsted senses:

$$\underset{\text{Ethanol}}{CH_3CH_2OH} \rightleftarrows \underset{\substack{\text{Ethoxide ion}\\\text{(anion of ethanol)}}}{CH_3CH_2O^-} + H^+ \qquad K_a \approx 10^{-18}$$

The anions of alcohols are known as *alkoxide ions.* They are powerful proton acceptors, as the equilibrium just indicated for ethanol implies.

Exercise 12.2 In the liquid state the order of acidity among the subclasses of the alcohols is

$$\underset{\substack{1°\\\text{stronger}}}{RCH_2OH} > \underset{2°}{R_2CHOH} > \underset{\substack{3°\\\text{weaker}}}{R_3COH}$$

decreasing acidity of alcohols ⟶

What is the *order of basicity* of the corresponding alkoxide ions: RCH_2O^-, R_2CHO^-, and R_3CO^-? *Hint.* What inductive effect of alkyl groups tends to increase the electron density on the oxygen of an alkoxide ion? Or, alternatively, the oxygen of which alkoxide ion, *t*-butoxide, or methoxide is less crowded by alkyl groups and is therefore more open to having its negative charge stabilized by solvation?

Like water, alcohols react with the more active metals such as sodium and potassium. In general:

$$H_2O + Na \longrightarrow HO^-Na^+ + \tfrac{1}{2}H_2 \qquad \text{(violent reaction)}$$

$$ROH + Na \longrightarrow RO^-Na^+ + \tfrac{1}{2}H_2 \qquad \text{(moderate reaction)}$$

Specific examples:

$$\underset{\text{Methanol}}{CH_3OH} + Na \longrightarrow \underset{\text{Sodium methoxide}}{CH_3O^-Na^+} + \tfrac{1}{2}H_2 \qquad \text{(moderate reaction)}$$

$$\underset{t\text{-Butyl alcohol}}{(CH_3)_3COH} + Na \longrightarrow \underset{\text{Sodium }t\text{-butoxide}}{(CH_3)_3CO^-Na^+} + \tfrac{1}{2}H_2 \qquad \text{(extremely slow reaction)}$$

Exercise 12.3 What correlation is there between the vigor of this reaction as described in the specific examples and the relative acidities of the various subclasses of alcohols and water?

When chemists need a powerful base in a nonaqueous system, they frequently choose the sodium or potassium alkoxides dissolved in the alcohol used to make them.

REACTIONS OF ALCOHOLS CONFINED TO THE UNSHARED PAIRS OF ELECTRONS ON OXYGEN

The reaction of hydrogen bromide or other strong acids with water has its counterpart in the alcohol family. In general, with water:

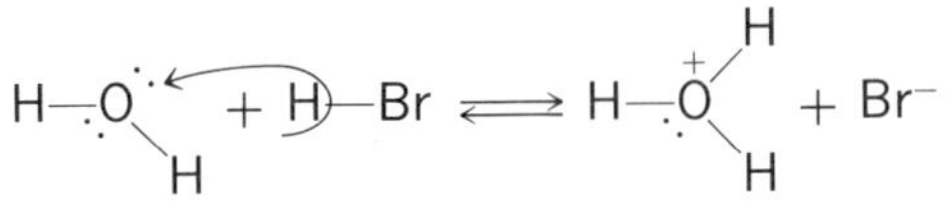

$$H-\ddot{O}-H + H-Br \rightleftharpoons H-\overset{+}{\ddot{O}}(H)-H + Br^-$$

Hydronium ion

with alcohols:

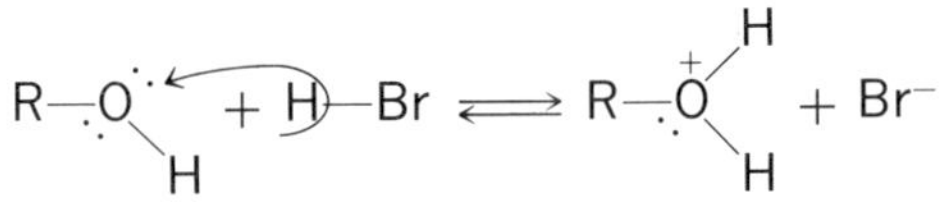

$$R-\ddot{O}-H + H-Br \rightleftharpoons R-\overset{+}{\ddot{O}}(H)-H + Br^-$$

Alkyloxonium ion

Specific examples:

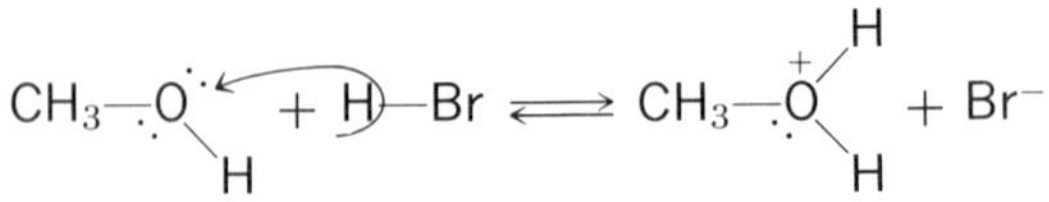

$$CH_3-\ddot{O}-H + H-Br \rightleftharpoons CH_3-\overset{+}{\ddot{O}}(H)-H + Br^-$$

Methyloxonium bromide
(in solution only)

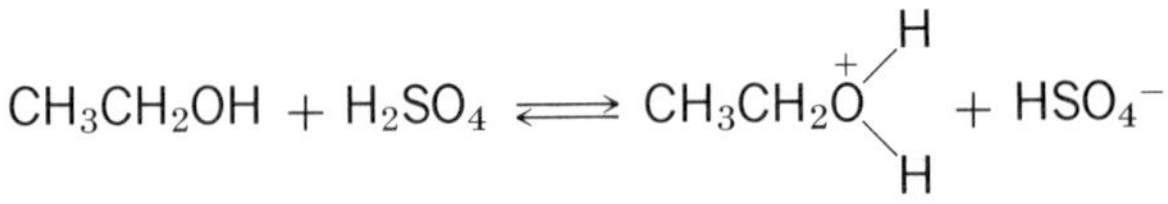

$$CH_3CH_2OH + H_2SO_4 \rightleftharpoons CH_3CH_2\overset{+}{O}(H)-H + HSO_4^-$$

Ethyloxonium hydrogen sulfate
(in solution only)

The alkyloxonium salts that form exist only in solution. They are important because the oxygen of the alcohol acquires a positive charge when it picks up a proton. As a result, the pair of electrons this oxygen shares with the carbinol carbon is much more strongly attracted toward the oxygen than before. In fact, the alcohol is on the verge of ionizing as follows:

$$R{:}\overset{+}{O}(H)_2 \rightleftharpoons R^+ + {:}\ddot{O}(H)_2$$

Alkyloxonium ion (a protonated alcohol) — Alkyl-carbonium ion — Water

Whether it will ionize depends on several factors to be discussed, but the possibility leads us to a study of the third type of reactions undergone by alcohols.

REACTIONS OF ALCOHOLS INVOLVING RUPTURE OF THE C—O BOND

Nucleophilic Substitution Reactions. General Principles. As mentioned earlier, the C—O bond is strong enough to resist ionization as follows:

$$R\text{—}OH \not\rightarrow R^+ + OH^-$$

(does not happen)

However, if some acid interacts with the oxygen to put a positive charge on it, $R\text{—}\overset{+}{O}H_2$, the C—O bond is considerably weakened. Whether or not it breaks to form a carbonium ion and water ($R\text{—}\overset{+}{O}H_2 \longrightarrow R^+ + OH_2$) depends on two factors. (1) Can the solvent stabilize the carbonium ion somewhat by solvating it? (2) Are there structural features in the developing carbonium ion that tend to stabilize it?

With respect to the first question, the presence of water or water-like molecules (e.g., low-formula-weight alcohols) will help the formation of R^+ because they can solvate this ion. As for the second question, we have already learned that 3° carbonium ions are more stable than 2° which are more stable than 1° carbonium ions.

We might therefore expect that C—O bonds in alcohols will most readily break (1) in the presence of strong acids, (2) in a solvent of good ionizing power, and (3) among those alcohols (3°) capable of yielding the most stable carbonium ions. In a solvent of poor ionizing power and among those alcohols (1°) incapable of forming the more stable carbonium ions, ionization will not occur. The *leaving group,* however, which is water for ROH_2^+, can still be "kicked out" by some species

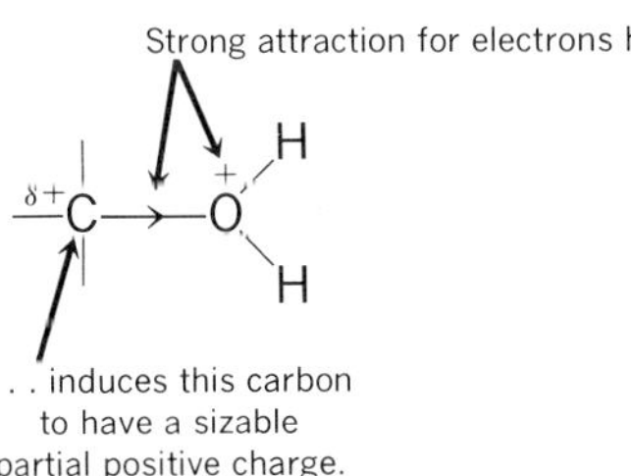
Strong attraction for electrons here . . .
δ+
C
O
H
H
+
. . . induces this carbon to have a sizable partial positive charge.

attracted to the developing partial positive charge on the carbinol carbon. Such a species, to act this way, must be electron-rich. It could be electrically neutral provided it had unshared pairs of electrons (e.g., $:NH_3$, or $R\text{—}\ddot{O}\text{—}H$), or it could be negatively charged (e.g., halide ion). Such electron-rich particles capable of bonding to carbon are said to be *nucleophilic reagents* or, simply, *nucleophiles* ("-phile" from *philos,* Greek for loving; "nucleo-," meaning nucleus—in this instance an ambiguous way of talking about a site with only a *partial* positive charge, such as a carbon at the end of a bond polarized so as to place a partial positive charge on it).

Strong nucleophiles are often, but not always, strong Brønsted bases or proton acceptors.[1] This property makes possible another important reaction of alcohols, elimination of the elements of water to form alkenes. After we have discussed the substitution reactions of alcohols, this elimination reaction will be integrated with general principles covering both types.

What we are about to do, then, is to study in some depth two of the most important types of reactions that organic compounds can undergo, nucleophilic substitution reactions and elimination reactions. Descriptions of these events answer the question "What happens?" The theoretical parts answer the question "How does it happen?" The principles we shall study have wide applicability. Careful and patient attention to them now will be very helpful to our future study. In most courses in organic chemistry these principles are examined during a study of alkyl halides, but this class of organic compounds is of minor significance in biochemistry. We shall study the principles in connection with the alcohols to which they also apply and which are very important to our future work. (Both for convenience and completeness, however, some of the chemistry of alkyl halides is included.)

Formation of Alkyl Halides by Reaction of Alcohols with Hydrogen Halides. In general:

$$\underset{\substack{\text{Alcohol}\\(\text{R = alkyl, }\textit{not}\\\text{aryl)}}}{\text{R—OH}} + \underset{\substack{\text{Hydrogen}\\\text{halide}\\(\text{X = Cl, Br, I})}}{\text{H—X}} \longrightarrow \underset{\substack{\text{Alkyl}\\\text{halide}}}{\text{R—X}} + H_2O$$

Specific examples:

$$\underset{n\text{-Butyl alcohol}}{CH_3CH_2CH_2CH_2OH} + \text{concd HBr} \xrightarrow[\text{heat}]{} \underset{n\text{-Butyl bromide (95\%)}}{CH_3CH_2CH_2CH_2Br} + H_2O$$

$$\underset{\text{1-Pentanol}}{CH_3CH_2CH_2CH_2CH_2OH} + \text{concd HBr} \xrightarrow[\text{heat}]{} \underset{\text{1-Bromopentane (78\%)}}{CH_3CH_2CH_2CH_2CH_2Br} + H_2O$$

$$\underset{t\text{-Butyl alcohol}}{CH_3-\overset{\overset{CH_3}{|}}{\underset{\underset{CH_3}{|}}{C}}-OH} + \text{concd HCl} \xrightarrow[\text{room temperature}]{} \underset{t\text{-Butyl chloride (88\%)}}{CH_3-\overset{\overset{CH_3}{|}}{\underset{\underset{CH_3}{|}}{C}}-Cl} + H_2O$$

[1]Basicity in this sense refers to the position of *equilibrium* in an acid-base interaction involving proton donors and acceptors. Strong bases are those capable of coordinating strongly with *protons.* Nucleophilicity refers to the relative *rate* at which the nucleophile can coordinate or bond to *carbon* while (or after) some leaving group departs. Bromide ion (or, better, iodide ion) is a strong nucleophile but a weak base. A more detailed discussion of these factors is not intended, but the instructor may wish to use these statements as a starting point.

The presence of a strong acid is essential for these reactions. Salts such as sodium bromide or potassium chloride, also suppliers of halide ions, do not react with alcohols.

The order of reactivity of the subclasses of alcohols is

$$3° > 2° > 1° > CH_3OH$$

which is the order of stability of the corresponding carbonium ions. These two facts, the need for acid and the order of reactivity, are understood in terms of the most probable mechanism for the reaction. Tertiary alcohols normally react by one mechanism, primary alcohols by another. For tertiary alcohols the following steps are believed to occur.

Step 1. The alcohol is protonated by the acid.

$$R-\ddot{\underset{\cdot\cdot}{O}}-H + H-X \rightleftharpoons R-\overset{+}{O}{:}H_2 + X^- \quad \text{(rapid formation of equilibrium)}$$

Step 2. The C—O bond is now much weaker and breaks, forming a carbonium ion.

$$R-\overset{+}{O}{:}H_2 \rightleftharpoons \underset{\text{Carbonium ion}}{R^+} + H_2O \quad \text{(slow rate of ionization)}$$

Step 3. The carbonium ion combines with a halide ion.

$$R^+ + X^- \rightleftharpoons R-X \quad \text{(rapid)}$$

The slow step, step 2, is the bottleneck in the reaction. Chemists call it the *rate-determinining step* because the final product cannot form any faster than carbonium ions can be generated. It does not matter how concentrated the solution is in halide ions; they do not enter the reaction until step 2 provides carbonium ions. Since this slow step involves something happening to only one molecule, the entire reaction is called a *unimolecular* process. Just how the C—O bond breaks may be visualized as follows. This bond, as any bond, can be likened to a tiny spring.[2] The groups joined by the bond vibrate—stretch in and out and back and forth—at the ends of the spring. But any spring has its limit. If sufficient energy is provided, it may become stretched beyond this limit and break. In any large collection of alkyloxonium ions, $R-OH_2^+$, a certain fraction of them will at any moment receive enough energy of activation from collisions and absorption to break the C—O spring, separating into the fragments R^+ and H_2O. Since the organic ion, R^+, is not preformed in the alcohol, R—OH, and since no carbonion ion, not even a 3° one, is very stable, it takes energy and therefore time for its formation. A progress-of-reaction diagram, Figure 12.7, explains this further.

[2]This is a model for a bond that is useful in other situations too, for example, in understanding certain aspects of infrared spectroscopy.

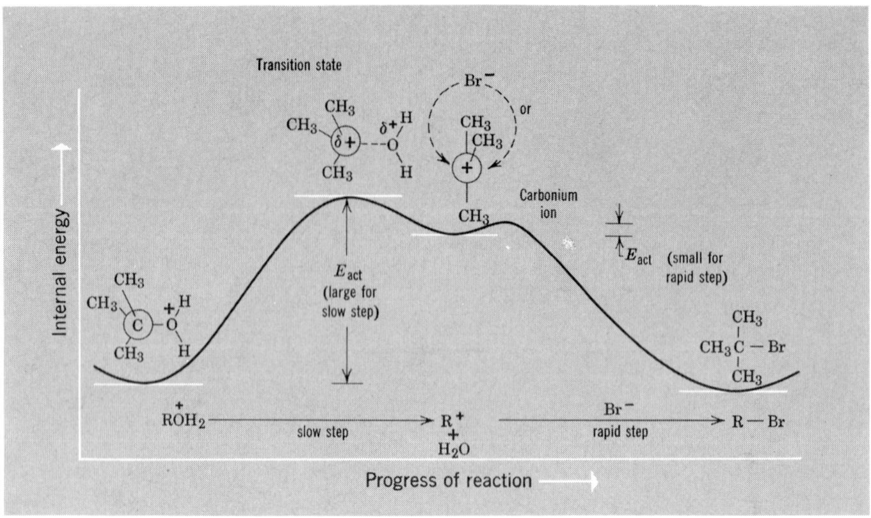
Transition state
Internal energy
Carbonium ion
E_{act} (large for slow step)
E_{act} (small for rapid step)
$\overset{+}{ROH_2}$
slow step
R^+ + H_2O
Br^-
rapid step
R—Br
Progress of reaction

Figure 12.7 The S_N1 mechanism. The rate-determining step (slow step) is the ionization of the bond between carbon and the group leaving, which is here a water molecule. The E_{act} for this is relatively large, which means that the rate is slow in relation to other steps. Once the carbonium ion forms, it rapidly combines with any nucleophile nearby—bromide ion in the example.

The order of reactivity of alcohols is understandable in terms of this mechanism. The most reactive alcohols are those that can form the most stable carbonium ions. The mechanism has a useful shorthand symbol, S_N1, with S standing for substitution, N for nucleophilic, and 1 for unimolecular. This symbol could also be read "carbonium ion mechanism" or "order of reactivity will be the same as order of stability of carbonium ions" or "this reaction pathway will be favored when the reaction occurs in solvents of good ionizing power." The one symbol S_N1 carries all this meaning.

Primary alcohols would yield primary carbonium ions if they reacted by an S_N1 mechanism. Probably because these are the least stable of carbonium ions, primary alcohols normally react by another mechanism, one not involving a carbonium ion. Step 1 in this new mechanism is the same as before.

Step 1.

$$R{-}CH_2{-}OH + H{-}X \rightleftharpoons R{-}CH_2{-}\overset{+}{O}H_2 + X^- \qquad \text{(rapid)}$$

Step 2. The nucleophile, X^-, attacks the carbon holding the oxygen.

$$X^- + RCH_2{-}\overset{+}{O}H_2 \longrightarrow X{-}CH_2R + H_2O \qquad \text{(slow)}$$

Alkyl halide

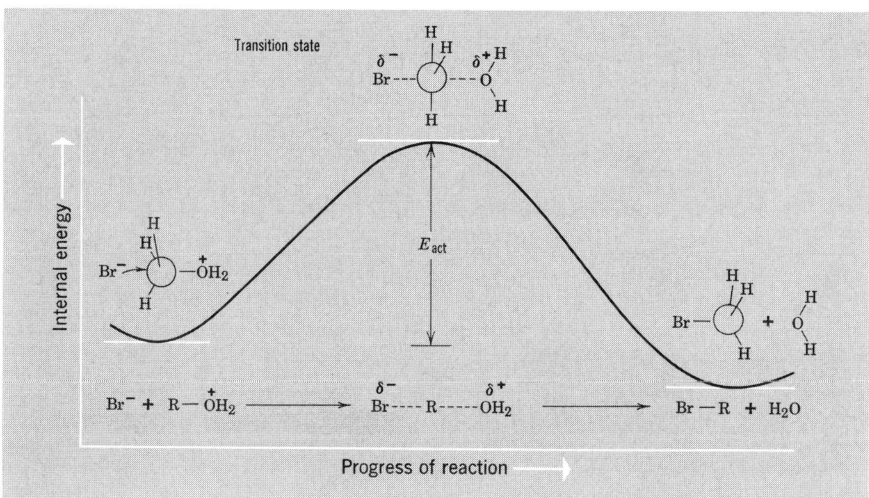
Transition state
Internal energy
E_{act}
$Br^- + R-\overset{+}{O}H_2 \longrightarrow \overset{\delta-}{Br}\text{----}R\text{-----}\overset{\delta+}{O}H_2 \longrightarrow Br-R + H_2O$
Progress of reaction

Figure 12.8 The S_N2 mechanism. The rate-determining step (slow step) produces a transition state in which carbon is temporarily holding five groups. The bulkier these are, the higher E_{act} is and the slower the rate will be. Note that as the reaction proceeds the bonds from the central carbon are turned inside out like an umbrella in a gale (inversion of configuration).

This is the rate-determining step, and because two particles participate directly in it, X^- and $RCH_2\overset{+}{O}H_2$, the overall process is called *bimolecular*. Experimentally, the rate at which alkyl halide forms will be doubled by doubling the concentration of either X^- or $RCH_2\overset{+}{O}H_2$. Or the rate can be cut in half by cutting either concentration in half. The shorthand symbol for this mechanism is S_N2, with S for substitution, N for nucleophilic, and 2 for bimolecular. Also implied by this symbol is "no carbonium ion," as well as other aspects to be discussed.

The S_N2 mechanism is discussed further in Figure 12.8. The transition state for the reaction is an unstable species in which five groups are for a very brief time partially or fully attached to one carbon. There is considerable crowding in this

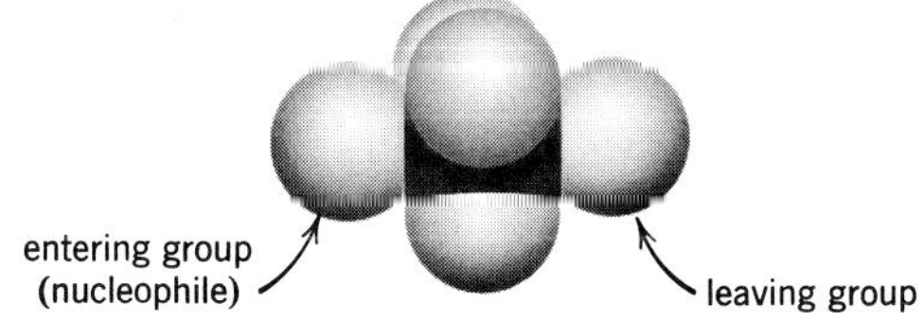
entering group
(nucleophile)
leaving group

The larger these three groups around the middle, the more energetic a collision between the nucleophile and the carbon must be to be successful.

particle, which helps us understand how it is that 1° alcohols have a much greater tendency than 3° alcohols to react by an S_N2 process. As Figure 12.9 shows, the carbon that must be attacked by the nucleophile is much more open to this attack in a 1° alcohol than in a 3° alcohol. Furthermore, one factor favoring attack at this carbon is the partial positive charge there. In a 3° alcohol this partial charge is reduced by the inductive effect of the alkyl groups attached to the 3° carbon, but in the 1° alcohol no such reduction occurs. Therefore the 1° alcohols have a greater

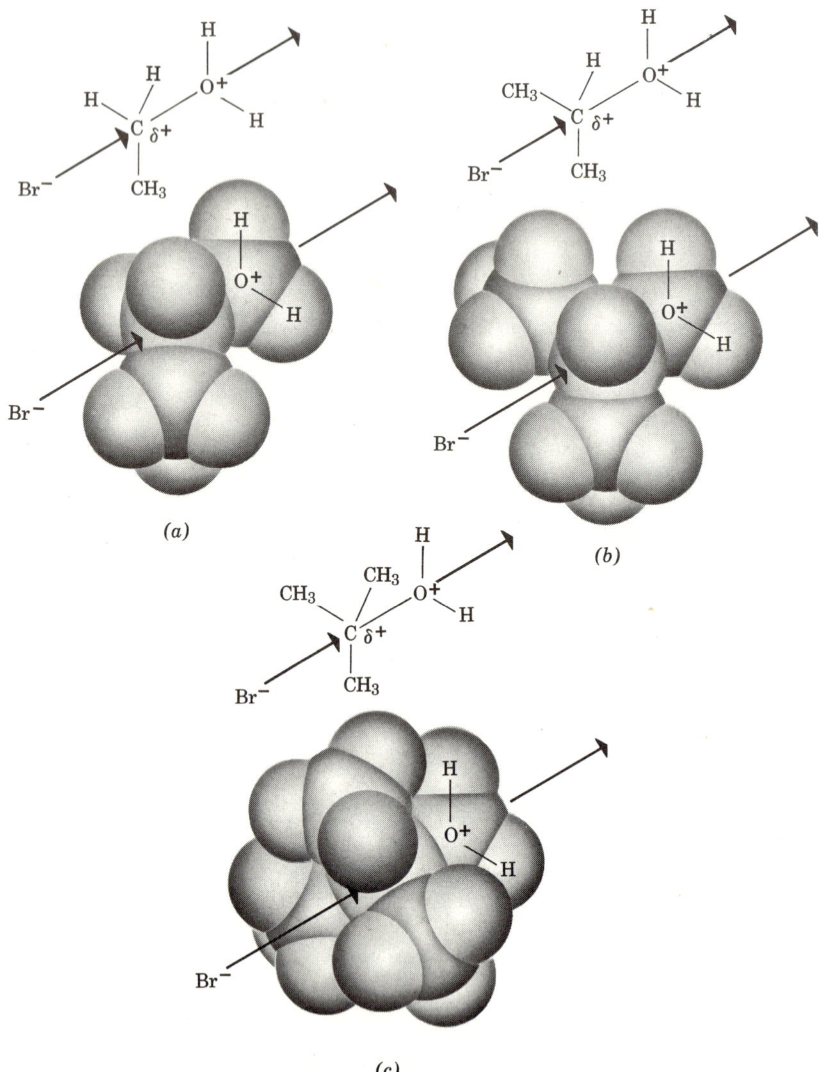
H
H
H
O^+
C $\delta+$
H
Br^-
CH_3
H
O^+
H
Br^-
(a)
H
H
CH_3
O^+
H
C $\delta+$
Br^-
CH_3
H
O^+
H
Br^-
(b)
H
CH_3
CH_3
O^+
H
C $\delta+$
Br^-
CH_3
H
O^+
H
Br^-
(c)

Figure 12.9 The S_N2 mechanism, steric factors. Attack by a bromide ion on the protonated form of a 1° alcohol, part *a*, encounters relatively little hindrance from groups already attached to the carbinol carbon. For a 2° alcohol, part *b*, there are two alkyl groups on the carbinol carbon, and they cause some hindrance. With the 3° alcohol, part *c*, this steric hindrance is at its maximum; such alcohols usually react by an S_N1 process.

tendency than the 3° alcohols to react by an S_N2 mechanism for three reasons: (1) the 1° carbonium ion is less stable than the 3°; (2) the partial positive charge on the carbon of the C—O system is greater in the 1° alcohol than in the 3°; (3) this carbon in the 1° alcohol is more open to attack than that in the 3°. The last factor is called a *steric factor*—it has to do with spatial requirements for certain reaction pathways.

Since ionization is not a step in the S_N2 mechanism, the solvent need not have good solvating power.

The symbol S_N2 therefore carries all the following meanings:

—Substitution, nucleophilic, bimolecular.
—Order of reactivity: $CH_3OH > 1° > 2° > 3°$ (just the reverse of S_N1).
—Carbonium ion not involved, but rather a transition state with the attacked carbon temporarily pentavalent.
—The configuration of the groups at the attacked carbon being turned inside out or inverted (cf. Figure 12.8).

Since 2° alcohols are capable of forming carbonium ions of intermediate stability, and since they are not as "crowded" at the carbinol carbon as in a 3° alcohol, some molecules may react by an S_N1 process and others in the same mixture by an S_N2 process.

Conversion of Alcohols into Alkenes. Elimination Reactions. In general:

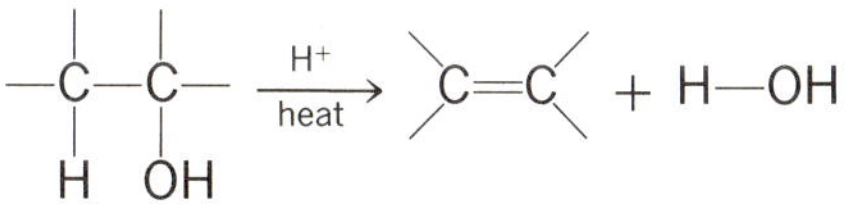

$$-\underset{\substack{|\\ \text{H}}}{\overset{|}{\text{C}}}-\underset{\substack{|\\ \text{OH}}}{\overset{|}{\text{C}}}- \xrightarrow[\text{heat}]{\text{H}^+} \gt\text{C}=\text{C}\lt + \text{H—OH}$$

Alcohol　　　　Alkene

Specific examples:

$$CH_3CH_2—OH \xrightarrow[170\text{–}180°C]{\text{concd } H_2SO_4} CH_2=CH_2 + H_2O$$

Ethyl alcohol　　　　Ethylene

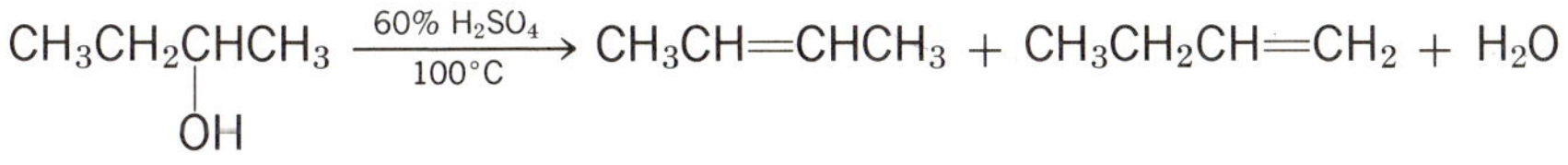

$$CH_3CH_2\underset{\substack{|\\ OH}}{C}HCH_3 \xrightarrow[100°C]{60\%\ H_2SO_4} CH_3CH=CHCH_3 + CH_3CH_2CH=CH_2 + H_2O$$

sec-Butyl alcohol　　　　2-Butene (principal product)　　　　1-Butene

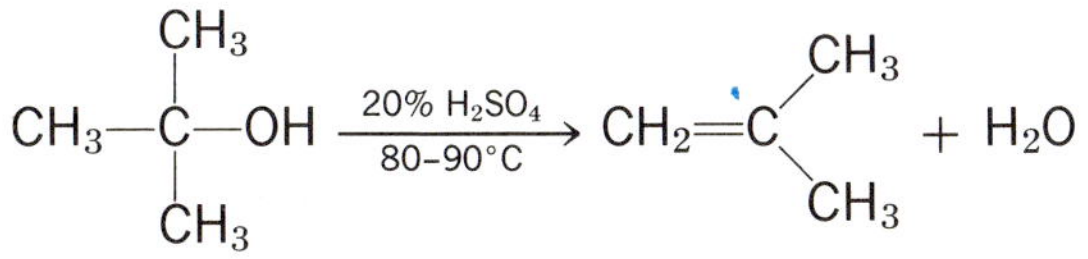

$$CH_3-\underset{\substack{|\\ CH_3}}{\overset{\substack{CH_3\\ |}}{C}}-OH \xrightarrow[80\text{–}90°C]{20\%\ H_2SO_4} CH_2=C\begin{matrix}\diagup CH_3\\ \diagdown CH_3\end{matrix} + H_2O$$

t-Butyl alcohol　　　　Isobutylene

In the general example we see that the elements of water, H and OH, are taken from *adjacent* carbons. In the specific examples it is much easier to form the alkene from *t*-butyl alcohol than from ethyl alcohol. The examples illustrate something that is generally true about this reaction: the order of ease of dehydration of alcohols to form alkenes is

$$3° > 2° > 1°$$

Exercise 12.4 Methyl alcohol is not included in this order. Why?

Two more facts are illustrated by the examples. An acid catalyst is needed. Furthermore, whenever it is possible for more than one alkene to form, the one more highly branched will be the major product (cf. dehydration of *sec*-butyl alcohol). By "more highly branched" is meant one in which the greater number of bonds to other carbons go out from the C=C system. 2-Butene has two branches by this definition, 1-butene only one.

(We shall mention one additional fact. During the dehydration of some alcohols, the intermediate carbonium ions rearrange, forming alkenes with structures that could not have been predicted from the general example given at the beginning of this section. We shall not encounter carbonium-ion rearrangements in our study of biochemistry, and nothing further will be said about them.)

The facts of the dehydration reaction are correlated by the mechanisms, two of which are recognized as important. One is very similar to the S_N1 mechanism and is designated E_1, the E for elimination, the 1 for unimolecular. The other resembles the S_N2 and is called E_2 or elimination, bimolecular.

The E_1 Mechanism. The first two steps are identical with the first two steps in an S_N1 process, and they may be illustrated by the dehydration of *t*-butyl alcohol.

Step 1. Protonation of the alcohol, a rapid acid-base equilibrium. Protona-

$$\underset{t\text{-Butyl alcohol}}{(CH_3)_3COH} + H_2SO_4 \rightleftharpoons (CH_3)_3C-\overset{+}{O}H_2 + HSO_4^-$$

tion of the oxygen places on it a positive charge, which attracts electrons to it from the C—O bond. A molecule of water is ready to leave.

Step 2. Ionization at the C—O bond, carbonium-ion formation, the rate-determining step. Since primary carbonium ions are the least stable types, it is under-

$$(CH_3)_3C-\overset{+}{O}H_2 \longrightarrow \underset{\text{Transition state}}{\left[(CH_3)_3\overset{\delta+}{C}\cdots\overset{\delta+}{O}H_2\right]} \longrightarrow \underset{t\text{-Butylcarbonium ion}}{(CH_3)_3C^+} + H_2O$$

standable why 1° alcohols dehydrate much more slowly than 3° alcohols. Having formed, the carbonium ion can now be attacked by some nucleophile, but this would lead to a by-product, not to an alkene. It is alkene formation that we seek to explain here.

Step 3. Removal of a proton from the carbonium ion. Monohydrogen sulfate ion (HSO_4^-) is shown acting as the proton acceptor, although a water molecule or another alcohol molecule could serve the same function. As the proton is pulled off, the pair of electrons holding it to the carbon pivots toward the positively charged carbon and the pi bond forms.

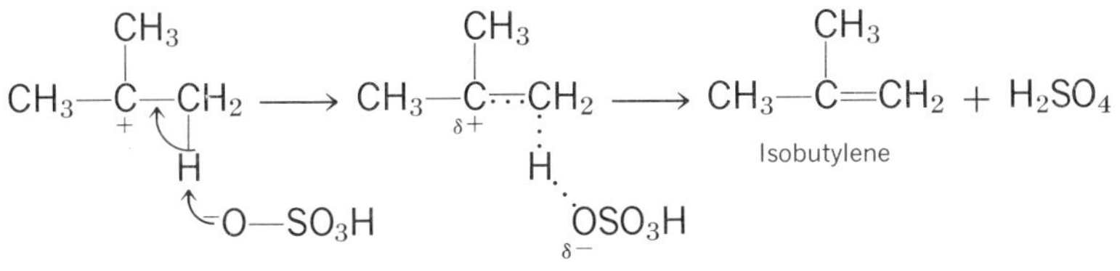

$$CH_3-\overset{CH_3}{\underset{+}{C}}-CH_2(H \leftarrow {}^-O-SO_3H) \longrightarrow CH_3-\overset{CH_3}{\underset{\delta+}{C}}\cdots CH_2(H\cdots \overset{\delta-}{O}SO_3H) \longrightarrow CH_3-\overset{CH_3}{C}=CH_2 + H_2SO_4$$

Second transition state of the E_1 mechanism

Isobutylene

The E_2 Mechanism. Illustrated with Ethyl Alcohol

Step 1. Rapidly established acid-base equilibrium, protonation of the alcohol. This step is identical with the first step in either the E_1 or the S_N1 reactions. If the temperature of the medium is unusually high, enough energy may be available for the reaction of a 1° alcohol to continue by means of a carbonium-ion mechanism, but at lower temperatures (below the boiling point of the mixture) the next step is bimolecular.

$$CH_3CH_2OH + H_2SO_4 \rightleftharpoons CH_3CH_2-\overset{+}{O}H_2 + HSO_4^-$$

Step 2. A proton acceptor strikes the hydrogen that is to be pulled off *while* the molecule of water leaves and the pi bond forms. It is said to be a *concerted* process.

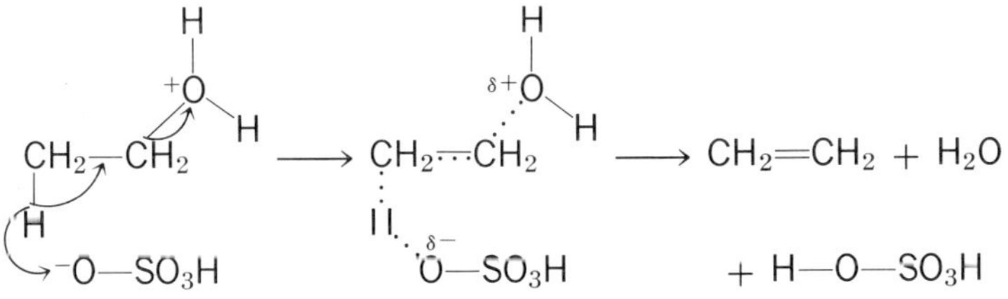

$$\underset{\text{Bisulfate ion}}{H-CH_2-CH_2-\overset{+}{O}H_2 + {}^-O-SO_3H} \longrightarrow \underset{\text{Transition state of the } E_2 \text{ mechanism}}{(\overset{\delta-}{O}SO_3H\cdots H\cdots)CH_2\cdots CH_2\cdots\overset{\delta+}{O}H_2} \longrightarrow CH_2=CH_2 + H_2O + H-O-SO_3H$$

Bisulfate ion

Transition state of the E_2 mechanism

Sulfuric acid catalyst is recovered

These mechanisms explain the necessity for an acid catalyst to protonate the alcohol, weaken the C—O bond, and create a stable leaving group, water. They also explain the order of ease of dehydration of alcohols. We have yet to explain the formation of the most highly branched alkene.

Relative Stabilities of Alkenes. The following reactions, including the heats of hydrogenation, illustrate a general trend. The more highly branched the alkene, the more stable it is. Since the same product forms in each case, if the 2-butene isomers give off less energy than 1-butene, they must initially have had less energy. Hence

$$\underset{\text{1-Butene}}{CH_3CH_2CH{=}CH_2} + H_2 \xrightarrow{\text{catalyst, etc.}} \underset{\text{Butane}}{CH_3CH_2CH_2CH_3} \quad \Delta H = -30.3 \text{ kcal/mole}$$

$$\underset{\substack{\text{2-Butene} \\ \text{(cis or trans)}}}{CH_3CH{=}CHCH_3} + H_2 \longrightarrow \underset{\text{Butane}}{CH_3CH_2CH_2CH_3} \quad \begin{array}{l} \Delta H = -27.6 \text{ kcal/mole} \\ \text{(from the trans alkene)} \\ \Delta H = -28.6 \text{ kcal/mole} \\ \text{(from the cis alkene)} \end{array}$$

they must be more stable. They are also the more highly branched alkenes, which does not explain *why* they are more stable but simply provides some evidence that they are.

In both the E_1 and the E_2 mechanisms a transition state with a partial double bond emerges. We may assume that whatever stabilizes a full double bond will also stabilize a partial one, which means that the transition state leading to the more highly branched alkene is of lower internal energy and more stable than the one leading to the less highly branched isomer. Consequently, the energy of activation needed for the more stable alkene is less and the rate of its formation is faster than that for the less stable alkene. In the dehydration of *sec*-butyl alcohol, shown in the specific examples (p. 329), 2-butene is the major product because its rate of formation is faster than that for 1-butene, And this rate is faster because the energy of activation leading to it is less. We have just explained why this energy of activation is lower. Whenever there are competing reactions—and they are the rule rather than the exception in organic chemistry—the product forming at the faster rate will obviously be the major product. Figure 12.10 illustrates the dehydration of *sec*-butyl alcohol by an E_2 mechanism.

Competing reactions are usually the rule in the chemical systems of living cells. Which of several possible reactions takes place is determined not only by the availability of the chemicals and by how relatively favorable the Gibbs free energy changes may be, but also by the presence of a catalyst (enzyme) that somehow lowers the energy of activation for one of the possible reactions and not for any of the others.

OXIDATION OF ALCOHOLS

Primary and secondary alcohols may be oxidized by a variety of reagents to compounds whose molecules contain a carbonyl group, one of the most widely

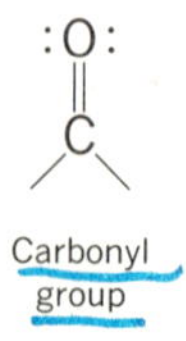
:O:
C

Carbonyl group

occurring functional groups in organic chemistry. A variety of families whose names and properties are studied in later chapters are characterized by it. We shall

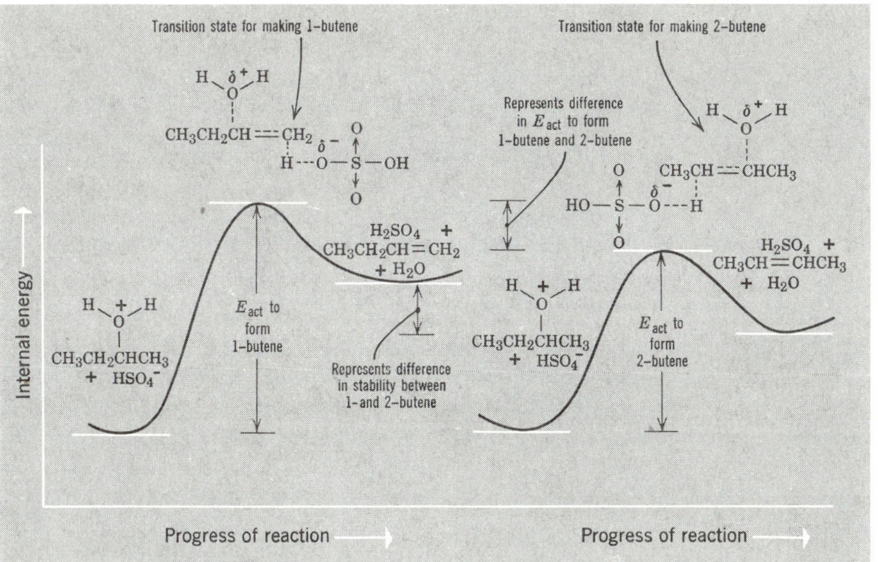
Transition state for making 1-butene
Transition state for making 2-butene
Represents difference in E_{act} to form 1-butene and 2-butene
H_2SO_4 + $CH_3CH_2CH{=}CH_2$ + H_2O
H_2SO_4 + $CH_3CH{=}CHCH_3$ + H_2O
$CH_3CH_2CHCH_3$ + HSO_4^-
E_{act} to form 1-butene
E_{act} to form 2-butene
Represents difference in stability between 1-and 2-butene
Internal energy
Progress of reaction
Progress of reaction

Figure 12.10 The E_2 mechanism, competing reactions. In the acid-catalyzed dehydration of 2-butanol, 2-butene rather than 1-butene is the major product. 2-Butene is known to be slightly more stable than 1-butene, and whatever provides this extra stability probably stabilizes the transition state leading to it. As shown here, the E_{act} for forming 2-butene is therefore probably less than that for forming 1-butene; and this automatically means that 2-butene forms at a faster rate than its isomer.

concentrate here only on the fact that alcohols are a potential source of the following types:

$$R{-}\overset{\overset{\displaystyle O}{\|}}{C}{-}H \qquad R{-}\overset{\overset{\displaystyle O}{\|}}{C}{-}O{-}H \qquad R{-}\overset{\overset{\displaystyle O}{\|}}{C}{-}R'$$

Aldehydes — Carboxylic acids — Ketones

When an alcohol is oxidized, the "elements" of hydrogen, ($H\!:^-$ + H^+),[3] are lost:

[3] In Chapter 4 we learned that oxidation is defined as loss of electrons and that reduction is a gain of electrons. Complete loss and complete gain (as in an electron transfer) represent two extremes. In organic chemistry it is not always obvious what specific molecule or part of it has lost or gained electrons. Between these extremes we could speak of losing or gaining *control* of some small or large amount of electron density, but this approach is strewn with stumbling blocks. As working definitions we use the following. When an organic molecule has an increase in its oxygen content or a decrease in its hydrogen content, we say that it has been oxidized. Conversely, if it has lost oxygen or gained hydrogen, it has been reduced. Thus removing the elements of hydrogen from an alcohol is an oxidation and is also rather clearly a loss of electrons, for the expression "elements of hydrogen," a contraction of "elements of the element hydrogen, H_2," refers to the building blocks of a hydrogen molecule, namely two protons and two electrons. These units are pulled from the alcohol molecule. The reagent doing this, the oxidizing agent, accepts at least the electrons and is thereby reduced.

$$\underset{\text{1° or 2° alcohol}}{\overset{|}{\underset{H}{-C}}-O-H} \longrightarrow \underset{\text{Aldehyde or ketone}}{>C=O} + \underset{\text{"Elements" of hydrogen}}{(H:^- + H^+)}$$

It is only a matter of convenience to write the elements of hydrogen as shown. In most oxidations the oxidizing agent converts them to water, but in one procedure hydrogen gas, H_2, is produced. In higher animals the elements of hydrogen are passed along a series of enzymes until they eventually combine with oxygen taken in at the lungs and form water. It is interesting that the word "aldehyde" is a contraction of "*al*cohol-*dehyd*rogenation product."

Oxidation of 1° Alcohols. Synthesis of Aldehydes and Carboxylic Acids. Primary alcohols are oxidized first to aldehydes, but aldehydes are even more readily oxidized than alcohols, with the product of this second oxidation a carboxylic acid. As soon as some aldehyde forms, it competes for unchanged oxidizing agent with unchanged 1° alcohol, and it wins, provided it is not removed from the solution. Primary alcohols are therefore an excellent source of carboxylic acids. The oxidation can be stopped at the aldehyde stage by carefully choosing the oxidizing agent or by removing the aldehyde as soon as it forms. Since the aldehydes of lower formula weight usually have much lower boiling points than the parent alcohols, they can be distilled continuously from the mixture. In general (the equations are not balanced):

$$\underset{\text{1° alcohol}}{RCH_2OH} \xrightarrow[200\text{–}300°C]{Cu} \underset{\text{Aldehyde}}{R-\overset{\overset{O}{\|}}{C}-H} + H_2$$

$$RCH_2OH + \underset{\text{Potassium permanganate}}{KMnO_4} \longrightarrow R-\overset{\overset{O}{\|}}{C}-H \xrightarrow{KMnO_4} \underset{\text{Salt of carboxylic acid}}{R-\overset{\overset{O}{\|}}{C}-O^-K^+} + \underset{\text{Manganese dioxide (brown sludge)}}{MnO_2} + KOH$$

$$\downarrow H_3O^+$$

$$\underset{\text{Carboxylic acid}}{R-\overset{\overset{O}{\|}}{C}-O-H} + H_2O + K^+$$

$$RCH_2OH + \underset{\text{Sodium dichromate (bright orange)}}{Na_2Cr_2O_7} \xrightarrow{H^+} R-\overset{\overset{O}{\|}}{C}-H \xrightarrow{Cr_2O_7^{2-}} R-\overset{\overset{O}{\|}}{C}-OH + \underset{\text{Chromium ion (bright green)}}{Cr^{3+}}$$

Specific examples (equations are not balanced, sufficient oxidizing agent is assumed available):

Aldehydes from 1° alcohols:

$$\underset{\substack{\text{Ethyl alcohol}\\ \text{(bp 78°C)}}}{CH_3CH_2OH} \xrightarrow{(O)} \underset{\substack{\text{Acetaldehyde}\\ \text{(bp 20°C)}}}{CH_3\overset{\overset{\displaystyle O}{\|}}{C}H} \qquad (O) = Cr_2O_7^{2-},\ H^+;\ 74\%\ \text{yield}$$

$$\underset{\substack{n\text{-Propyl alcohol}\\ \text{(bp 97°C)}}}{CH_3CH_2CH_2OH} \xrightarrow{(O)} \underset{\substack{\text{Propionaldehyde}\\ \text{(bp 55°C)}}}{CH_3CH_2\overset{\overset{\displaystyle O}{\|}}{C}H} \qquad \begin{aligned}(O) &= \text{Cu, high temperature; 67\% yield}\\ &= Cr_2O_7^{2-},\ H^+;\ 49\%\ \text{yield}\end{aligned}$$

$$\underset{\substack{n\text{-Butyl alcohol}\\ \text{(bp 118°C)}}}{CH_3CH_2CH_2CH_2OH} \xrightarrow{(O)} \underset{\substack{n\text{-Butyraldehyde}\\ \text{(bp 82°C)}}}{CH_3CH_2CH_2\overset{\overset{\displaystyle O}{\|}}{C}H} \qquad \begin{aligned}(O) &= \text{Cu, heat; 62\% yield}\\ &= Cr_2O_7^{2-},\ H^+;\ 72\%\ \text{yield}\end{aligned}$$

$$\underset{\substack{\text{1-Nonanol}\\ \text{(bp 212°C)}}}{CH_3(CH_2)_7CH_2OH} \xrightarrow{(O)} \underset{\substack{\text{Nonanal}\\ \text{(bp 185°C)}}}{CH_3(CH_2)_7\overset{\overset{\displaystyle O}{\|}}{C}H} \qquad (0) = \text{Cu, heat; 90\% yield}$$

Carboxylic acids from 1° alcohols:

$$\underset{n\text{-Propyl alcohol}}{CH_3CH_2CH_2OH} \xrightarrow{Cr_2O_7^{2-},\ H^+} \underset{\text{Propionic acid}}{CH_3CH_2\overset{\overset{\displaystyle O}{\|}}{C}OH} \qquad (65\%\ \text{yield})$$

$$\underset{\text{2-Ethyl-1-hexanol}}{CH_3CH_2CH_2CH_2\overset{\overset{\displaystyle CH_3CH_2}{|}}{C}HCH_2OH} \xrightarrow[\text{followed by } H^+]{MnO_4^-,\ OH^-} \underset{\text{2-Ethylhexanoic acid}}{CH_3CH_2CH_2CH_2\overset{\overset{\displaystyle CH_3CH_2}{|}}{C}H-\overset{\overset{\displaystyle O}{\|}}{C}OH} \qquad (74\%\ \text{yield})$$

$$\underset{\text{6-Methyl-1-octanol}}{CH_3CH_2\underset{\underset{\displaystyle CH_3}{|}}{C}H(CH_2)_4CH_2OH} \xrightarrow[\text{followed by } H^+]{MnO_4^-,\ OH^-} \underset{\text{6-Methyloctanoic acid}}{CH_3CH_2\underset{\underset{\displaystyle CH_3}{|}}{C}H(CH_2)_4\overset{\overset{\displaystyle O}{\|}}{C}OH} \qquad (66\%\ \text{yield})$$

Ketones from 2° Alcohols. The ketone system is characterized by a carbonyl

$$\underset{\text{Ketone system}}{C-\overset{\overset{\displaystyle O}{\|}}{C}-C}$$

group flanked on both sides by carbons, and it is resistant to further oxidation except under vigorous conditions.

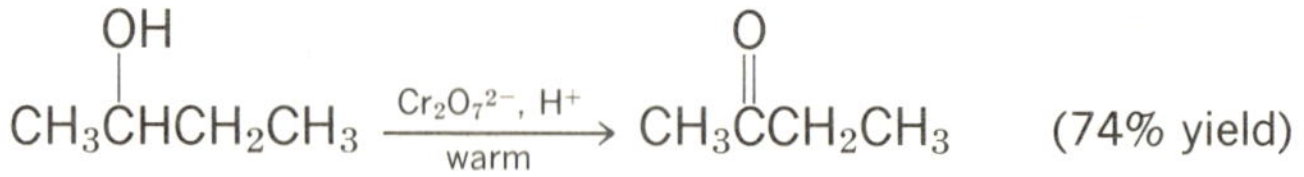

$$\underset{\text{2-Butanol}}{CH_3\overset{OH}{\overset{|}{C}}HCH_2CH_3} \xrightarrow[\text{warm}]{Cr_2O_7^{2-},\ H^+} \underset{\text{2-Butanone}}{CH_3\overset{O}{\overset{\|}{C}}CH_2CH_3} \qquad \text{(74\% yield)}$$

$$\text{Cyclohexanol} \xrightarrow[\text{warm}]{Cr_2O_7^{2-},\ H^+} \text{Cyclohexanone} \qquad \text{(85\% yield)}$$

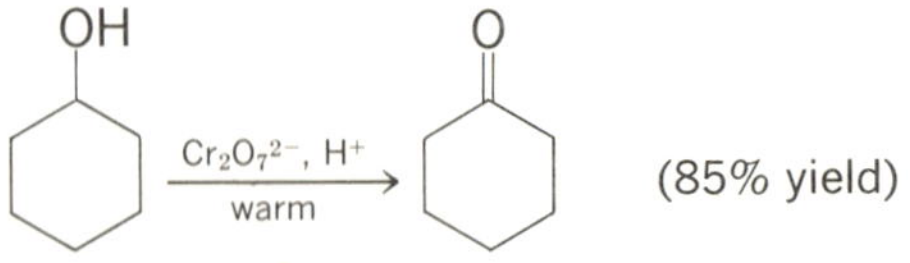

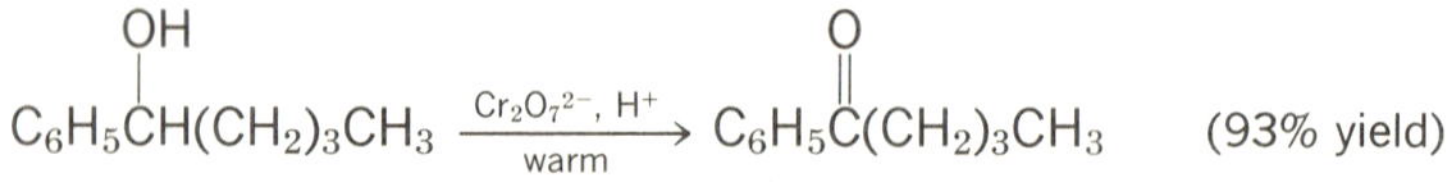

$$\underset{\text{1-Phenyl-1-pentanol}}{C_6H_5\overset{OH}{\overset{|}{C}}H(CH_2)_3CH_3} \xrightarrow[\text{warm}]{Cr_2O_7^{2-},\ H^+} \underset{\text{Phenyl } n\text{-butyl ketone}}{C_6H_5\overset{O}{\overset{\|}{C}}(CH_2)_3CH_3} \qquad \text{(93\% yield)}$$

Examples from molecular biology. During the degradation of a long-chain carboxylic acid (delivered to the system by fats and oils), the following reaction takes place.

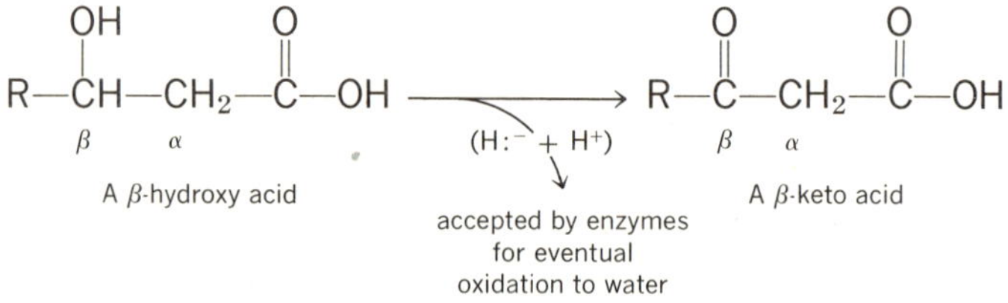

$$\underset{\text{A }\beta\text{-hydroxy acid}}{R-\underset{\beta}{\overset{OH}{\overset{|}{C}H}}-\underset{\alpha}{CH_2}-\overset{O}{\overset{\|}{C}}-OH} \xrightarrow[(H:^- + H^+)]{} \underset{\text{A }\beta\text{-keto acid}}{R-\underset{\beta}{\overset{O}{\overset{\|}{C}}}-\underset{\alpha}{CH_2}-\overset{O}{\overset{\|}{C}}-OH}$$

$(H:^- + H^+)$ accepted by enzymes for eventual oxidation to water

Two other instances of the oxidation of a 2° alcohol occur during the citric acid cycle, a series of reactions by which acetic acid units ($CH_3\overset{O}{\overset{\|}{C}}-OH$) are degraded to carbon dioxide and water. The details are discussed more fully in Chapter 21, but two of its steps (shown here for purposes of illustration—they need not be memorized) are as follows.

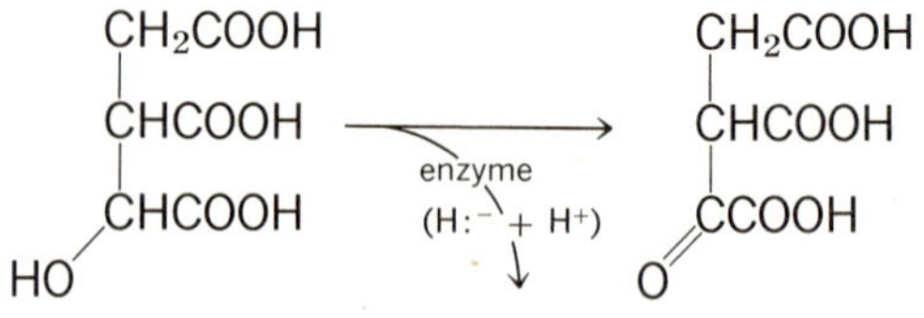

Isocitric acid ($HOOCCH_2-CH(COOH)-CH(OH)COOH$) $\xrightarrow[\text{enzyme}]{}$ Oxalosuccinic acid ($HOOCCH_2-CH(COOH)-C(=O)COOH$) + $(H:^- + H^+)$

Malic acid ($HO-CHCOOH-CH_2COOH$) $\xrightarrow[\text{enzyme}]{}$ Oxaloacetic acid ($O{=}CCOOH-CH_2COOH$) + $(H:^- + H^+)$

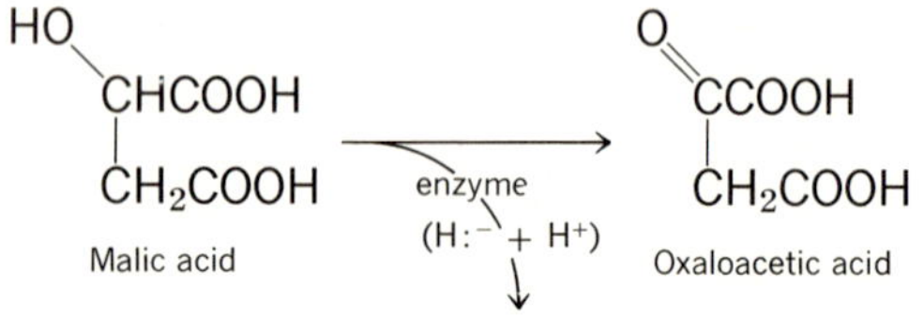

Tertiary alcohols are not oxidized if the medium is alkaline. Under acidic conditions tertiary alcohols easily dehydrate to alkenes, and these are readily attacked by oxidizing agents.

SYNTHESES OF ALCOHOLS

The alcohol function may be introduced into a molecule in the following ways. This list of methods is by no means complete, for only those relevant to our study of the molecular basis of life are mentioned.

1. **Reduction of a carbonyl group.** Reduction of an aldehyde produces a 1° alcohol.[4]

$$R-\overset{\overset{\displaystyle O}{\|}}{C}-H \xrightarrow{(2H)} \underset{1^\circ\text{ alcohol}}{RCH_2OH}$$

Reduction of a ketone produces a 2° alcohol.

$$R-\overset{\overset{\displaystyle O}{\|}}{C}-R' \xrightarrow{(2H)} \underset{2^\circ\text{ alcohol}}{R\overset{\overset{\displaystyle OH}{|}}{C}HR'}$$

Hydrogen over a metal catalyst such as powdered nickel or platinum (catalytic reduction), lithium aluminum hydride ($LiAlH_4$), and sodium borohydride ($NaBH_4$) are a few of the reducing agents available. The reduction of a keto group to a 2° alcohol has its counterpart in molecular biology:

$$\underset{\text{Pyruvic acid}}{CH_3-\overset{\overset{\displaystyle O}{\|}}{C}-\overset{\overset{\displaystyle O}{\|}}{C}-OH} \xrightarrow[\text{enzyme}]{\substack{\text{enzyme donor of hydrogen}\\(H:^- + H^+)}} \underset{\text{Lactic acid}}{CH_3-\overset{\overset{\displaystyle OH}{|}}{C}H-\overset{\overset{\displaystyle O}{\|}}{C}-OH}$$

Pyruvic acid is an intermediate in the metabolism of glucose. The reaction is reversible. The source of hydrogen is another organic compound, an enzyme, which is itself dehydrogenated in the reaction.

2. **The addition of water to a carbon-carbon double bond.** This method was discussed in Chapter 10. The reaction occurs in molecular biology, for example, during the metabolism of a carboxylic acid (fatty acid) when the elements of water add to the double bond of an unsaturated acid:

$$\underset{\text{Unsaturated fatty acid}}{R-CH{=}CH-\overset{\overset{\displaystyle O}{\|}}{C}-OH} + H_2O \xrightarrow[\text{enzyme}]{} \underset{\text{A }\beta\text{-hydroxy acid}}{R-\underset{\beta}{\overset{\overset{\displaystyle OH}{|}}{C}H}-\underset{\alpha}{\overset{\overset{\displaystyle H}{|}}{C}H}-\overset{\overset{\displaystyle O}{\|}}{C}-OH}$$

[4] Just as (O) was a symbol for an oxidizing agent, (2H) stands for a reducing agent that will accomplish the reaction.

It also occurs at two places in the citric acid cycle, in the conversions of aconitic acid to isocitric acid and fumaric acid to malic acid; they are shown here only for purposes of illustrating how *in vitro* reactions have counterparts *in vivo*.[5]

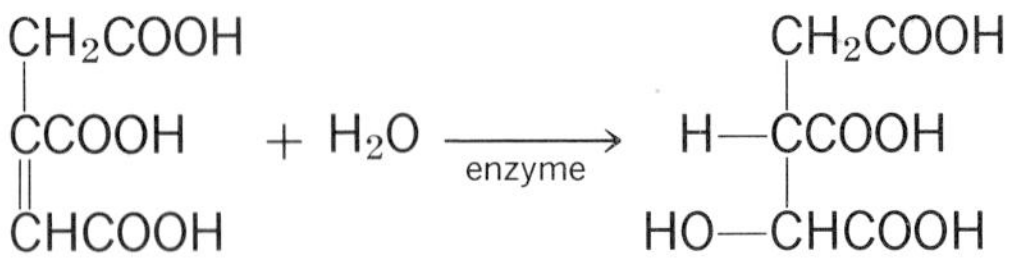

$$\begin{array}{l} CH_2COOH \\ | \\ CCOOH \\ \| \\ CHCOOH \end{array} + H_2O \xrightarrow[\text{enzyme}]{} \begin{array}{r} CH_2COOH \\ | \\ H—CCOOH \\ | \\ HO—CHCOOH \end{array}$$

Aconitic acid — Isocitric acid

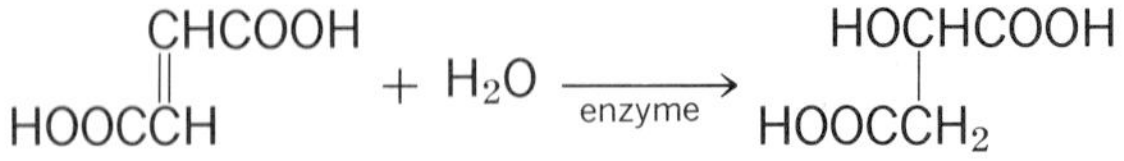

$$\begin{array}{r} CHCOOH \\ \| \\ HOOCCH \end{array} + H_2O \xrightarrow[\text{enzyme}]{} \begin{array}{r} HOCHCOOH \\ | \\ HOOCCH_2 \end{array}$$

Fumaric acid — Malic acid

IMPORTANT INDIVIDUAL ALCOHOLS

Methyl alcohol, $CH_3—OH$, wood alcohol, methanol. Carbon monoxide and hydrogen, under a pressure of 200 to 300 atmospheres and a temperature of 300 to 400°C and in the presence of a mixed metal oxide catalyst, combine directly to form methyl alcohol:

$$2H_2 + CO \xrightarrow[\substack{300\text{–}400°C \\ ZnO—Cr_2O_3}]{200\text{–}300 \text{ atm}} CH_3OH$$

Methyl alcohol made this way is much cheaper than that available from the heating of wood, but archaic laws keep the latter method alive; federally approved formulas for denatured alcohol specify methyl alcohol prepared from wood. The annual United States production of synthetic methyl alcohol is about 3.3 billion pounds, or about 500 million gallons. Both the carbon monoxide and hydrogen needed for its synthesis can be produced from the partial oxidation of methane, obtained from petroleum reserves:

$$2CH_4 + O_2 \longrightarrow 2CO + 4H_2$$

Methyl alcohol is a dangerous poison, causing blindness or death. About half the annual production goes for making formaldehyde, an important raw material for plastics. Methyl alcohol is also used as a temporary antifreeze, as a component of jet fuel, and as a laboratory and commercial solvent.

Ethyl alcohol, $CH_3CH_2—OH$, grain alcohol; ethanol. Ethyl alcohol is made by the fermentation of sugars (discussed in Chapters 21 and 22) or by the hydration of ethylene. The source of ethylene for the second process is the cracking of ethane and other components of petroleum. Ethylene is absorbed in concentrated sulfuric acid at about 100°C, and a mixture of ethyl hydrogen sulfate and diethyl sulfate forms. When this mixture is diluted with water, the sulfates are hydrolyzed and ethyl alcohol is formed. In the year 1967, 700 million gallons of ethyl alcohol were

[5] *In vitro* means "in a glass vessel." The expression applies to reactions and processes performed with laboratory glassware. *In vivo* refers to processes in living systems.

produced in the United States, and of this 58% was from ethylene and only 15% from sources of sugar (grains, fruits, molasses). Another method, about 18% of the production, was the direct, vapor phase hydration of ethylene.

Industrially, ethyl alcohol is used as a solvent and in the compounding of pharmaceuticals, perfumes, lotions, tonics, and rubbing compounds. For these purposes it is adulterated (*denatured*) by poisons that are difficult to remove and that make it thoroughly undesirable for drinking purposes. Nearly all governments derive considerable revenue by taxing potable alcohol. Tax-free alcohol in the United States costs about 70 cents a gallon; the tax on a gallon of potable alcohol is slightly over 20 dollars. In days of old whiskey was tested by pouring it on gunpowder and seeing whether the gunpowder would ignite after the alcohol burned away. If it did, the tester had *proof* that the whiskey did not contain too much water. This was the origin of the term used in connection with alcohol-water solutions. Pure alcohol is 200 proof. Alcohol gives the illusion that it is a stimulant when taken internally through its first effect, which is to depress activity in the uppermost level of the brain, the center of judgment, inhibition, and restraint.

Isopropyl alcohol, $CH_3\underset{\displaystyle OH}{\underset{|}{C}}HCH_3$, 2-propanol. This alcohol, twice as toxic as ethyl alcohol, is a common substitute for it as a rubbing compound.

Ethylene glycol, $HOCH_2CH_2OH$; **propylene glycol,** $CH_3\underset{\displaystyle OH}{\underset{|}{C}}HCH_2OH$. Both liquids have high boiling points, are soluble in water in all proportions, and are used chiefly as permanent antifreezes and coolants in refrigerator systems. The 1966 production of ethylene glycol in the United States was 2.1 billion pounds, most of it made by the hydrolysis of ethylene oxide:

$$\underset{\text{Ethylene oxide (made from ethylene)}}{CH_2\!-\!CH_2 \text{ (bridged by O)}} + H_2O \xrightarrow[60^\circ C]{H^+} \underset{\text{Ethylene glycol}}{HOCH_2CH_2OH}$$

Production of propylene glycol, largely from the hydrolysis of propylene oxide, was about 260 million pounds in the United States in 1966.

Glycerol, $\underset{\displaystyle OH}{\underset{|}{CH_2}}\!-\!\underset{\displaystyle OH}{\underset{|}{CH}}\!-\!\underset{\displaystyle OH}{\underset{|}{CH_2}}$, glycerin. This colorless, syrupy liquid with a sweet taste is freely soluble in water and insoluble in nonpolar solvents. A by-product in the manufacture of soap, over half the annual United States production of about 300 million pounds is made from one of two synthetic processes which start from propylene:

$$\underset{\text{Propylene}}{CH_2\!=\!CH\!-\!CH_3} + Cl_2 \xrightarrow[500\text{–}600^\circ C]{}$$

$$\underset{\text{Allyl chloride}}{CH_2\!=\!CH\!-\!CH_2\!-\!Cl} \xrightarrow[NaOH]{Na_2CO_3} \underset{\text{Allyl alcohol}}{CH_2\!=\!CH\!-\!CH_2\!-\!OH}$$

Then

$$CH_2{=}CH{-}CH_2{-}OH + Cl_2 + H_2O \longrightarrow \underset{Cl}{\underset{|}{CH_2}}{-}\underset{OH}{\underset{|}{CH}}{-}\underset{OH}{\underset{|}{CH_2}} \xrightarrow[NaOH]{Na_2CO_3} \underset{OH}{\underset{|}{CH_2}}{-}\underset{OH}{\underset{|}{CH}}{-}\underset{OH}{\underset{|}{CH_2}}$$

Glycerol

or

$$CH_2{=}CH{-}CH_2{-}OH + H_2O_2 \xrightarrow[65°C]{WO_3} \underset{OH}{\underset{|}{CH_2}}{-}\underset{OH}{\underset{|}{CH}}{-}\underset{OH}{\underset{|}{CH_2}}$$

Glycerol

Glycerol is used in the synthesis of glyptal resins (p. 430), as a humectant for tobacco, as a softening agent for cellophane, and in the compounding of cosmetics and drugs. It is also a raw material for the manufacture of nitroglycerin, a powerful explosive, and it is an important intermediate in the metabolism of lipids.

REFERENCES AND ANNOTATED READING LIST

C. R. Noller. *Chemistry of Organic Compounds,* third edition. W. B. Saunders Company, Philadelphia, 1965. This textbook for a first-year course in organic chemistry is a rich source of information, not only about the subject in general but also about the history, manufacturing processes, and uses of individual compounds.

R. T. Morrison and R. N. Boyd, *Organic Chemistry,* second edition. Allyn and Bacon, Boston, 1966. This textbook contains two chapters on alcohols.

PROBLEMS AND EXERCISES

1. Write condensed structural formulas for each of the following compounds.

   (*a*) allyl alcohol (*b*) benzyl alcohol (*c*) isopropyl alcohol
   (*d*) *sec*-butyl alcohol (*e*) *n*-octyl alcohol (*f*) *n*-propyl alcohol
   (*g*) glycerol (*h*) propylene glycol (*i*) *t*-butyl alcohol
   (*j*) *p*-nitrophenol (*k*) ethylene glycol (*l*) formaldehyde

2. Write names according to the IUPAC rules for the following compounds.

   (*a*) $CH_3CH_2CH_2CH_2CH_2CH_2OH$

   (*b*) $CH_3\overset{CH_3}{\overset{|}{C}}HCH_2OH$

   (*c*) $CH_3{-}\underset{CH_3}{\underset{|}{\overset{CH_3}{\overset{|}{C}}}}{-}OH$

   (*d*) $CH_3CH_2\underset{CH(CH_3)_2}{\underset{|}{\overset{CH_3}{\overset{|}{C}}}}CH_2\overset{OH}{\overset{|}{C}}HCH_2\overset{CH_2CH_2CH_3}{\overset{|}{C}}H\underset{CH_3}{\underset{|}{C}}HCH_3$

   (*e*) $CH_3\underset{CH_3}{\underset{|}{C}}H{-}OH$

3. Arrange the following in order of their increasing boiling points and explain the reasoning behind your choices.

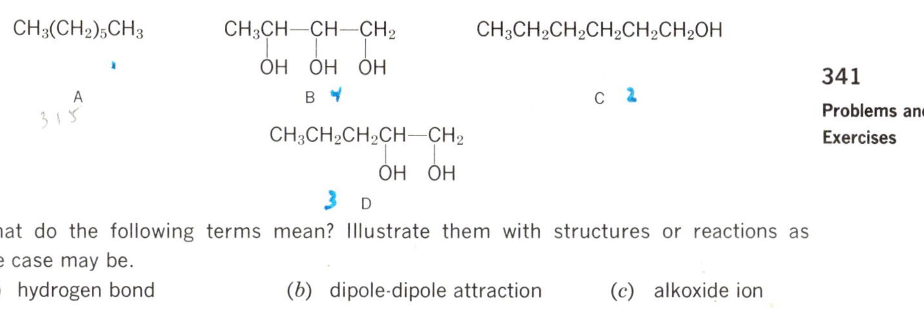
CH3(CH2)5CH3
A
CH3CH—CH—CH2
OH OH OH
B
CH3CH2CH2CH2CH2CH2OH
C
CH3CH2CH2CH—CH2
OH OH
D

4. What do the following terms mean? Illustrate them with structures or reactions as the case may be.
   (*a*) hydrogen bond (*b*) dipole-dipole attraction (*c*) alkoxide ion
   (*d*) carbinol carbon (*e*) S_N1 (*f*) alkyloxonium ion
   (*g*) methylcarbonium ion (*h*) elements of hydrogen (*i*) S_N2
   (*j*) nucleophile (*k*) rate-determining step (*l*) E_1
   (*m*) E_2 (*n*) steric factor
5. Write an equation (not necessarily balanced) for the reaction of *n*-propyl alcohol with each of the following reagents.
   (*a*) potassium (*b*) sodium dichromate, H^+ (*c*) copper (high temperature)
   (*d*) concentrated HBr (*e*) concentrated H_2SO_4, heat
6. Outline steps for the conversions indicated. More than one step will be needed for each.
   (*a*) synthesis of propane from isopropyl alcohol
   (*b*) synthesis of $CH_3\overset{O}{\overset{\|}{C}}CH_2CH_3$ from 2-butene
7. Write the structure of an alcohol from which each of the following compounds could be made; specify a reagent for carrying out the synthesis.
   (*a*) 3-pentanone, $CH_3CH_2\overset{O}{\overset{\|}{C}}CH_2CH_3$ (*b*) $C_6H_5—\overset{O}{\overset{\|}{C}}—H$ (*c*) cyclohexene
   (*d*) $(CH_3)_3C—Cl$ (*e*) isobutylene
8. Write all the steps in the mechanism of the dehydration of 1-propanol assuming that the acid-catalyzed process is E_2. Be sure to write representations of the transition state(s) and then construct a progress of reaction diagram.
9. Write all the steps in the mechanism of the conversion of benzyl alcohol to benzyl bromide by the action of concentrated hydrobromic acid. Assume that the rate of this reaction is not affected by the concentration of bromide ion.
10. Write all the steps in the mechanism of the dehydration of 2-methyl-2-butanol to 2-methyl-2-butene, assuming that the process is E_1. Draw a progress of reaction diagram.
11. Follow the instructions for problem 10, except take the reaction to the formation of the less stable alkene, 2-methyl-1-butene. Draw a progress of reaction diagram *on the same scale* used for problem 10.

CHAPTER THIRTEEN

Optical Isomerism. Organohalogen Compounds. Ethers. Amines. Mercaptans

Even though two (or more) compounds have the same general structure, they may be different in a very subtle way, a way even more subtle than cis-trans isomerism, which is merely a geometric difference. This new isomerism, optical isomerism, occurs widely in nature, and to interpret some of nature's workings we must understand it. We shall find examples of optical isomerism and its implications among the organic compounds and reactions yet to be studied.

Several families of organic compounds have reactions that are of particular importance in biochemistry but need not be studied in great depth. The alkyl halides and other organohalogen compounds, although seldom encountered in the body, nevertheless provide a basis for illustrating some important principles. Another family, the ethers, is also rarely represented in the molecules of living things, but when it occurs we need to know what to expect of its functional group. Mercaptans are sulfur analogs of alcohols, and the —SH group is important in protein chemistry. Finally, the amines, the organic derivatives of ammonia, are essential to a study of protein chemistry. Because only a few of their reactions have a bearing on the molecular basis of life, our study of them will be brief. Rather than creating separate, very short chapters for each of these topics, we have gathered them together in this one.

OPTICAL ISOMERISM

TYPES OF ISOMERISM

We have already encountered several ways in which substances can have the same molecular formula and still be different. The more obvious examples were *chain isomers, position isomers,* and *functional isomers.*

Chain isomers: —C—C—C—C— (Butane) and isobutane, in which the middle carbon of —C—C—C— carries a fourth carbon (Isobutane)

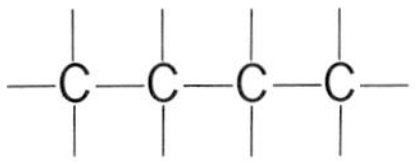

Butane 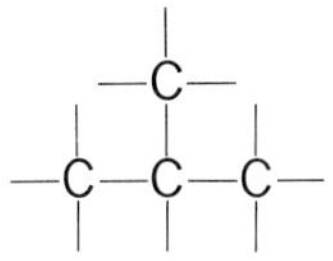 Isobutane

Position isomers: $CH_3CH_2CH_2Cl$ and CH_3CHCH_3 with Cl on the middle carbon

1-Chloropropane 2-Chloropropane

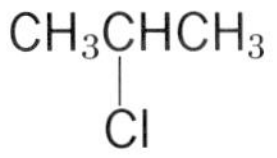

Functional group isomers: CH_3CH_2OH and $CH_3—O—CH_3$

Ethyl alcohol Methyl ether

Less obvious was stereoisomerism, and we have studied only one of the two types, cis-trans or geometrical. For two compounds to be related as stereoisomers, they must have not only the same molecular formula but also the same nucleus-to-nucleus sequence. If they are then different because one molecular part does not rotate freely about a bond, this particular stereoisomerism is cis-trans. It commonly occurs among the alkenes and with cyclic compounds.

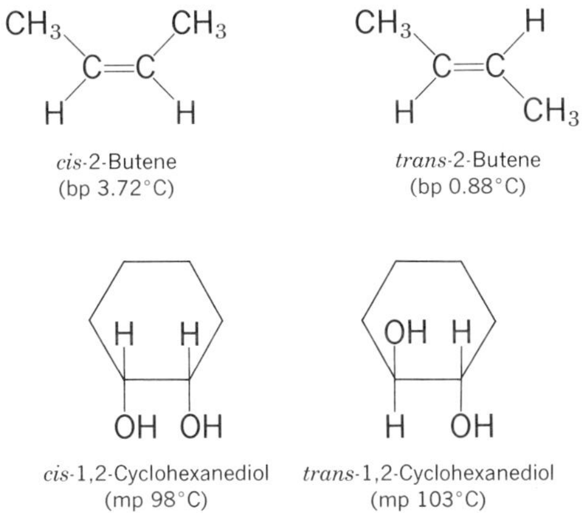
CH3 CH3
C=C
H H
CH3 H
C=C
H CH3
H H
OH OH
OH H
H OH

cis-2-Butene (bp 3.72°C) *trans*-2-Butene (bp 0.88°C)

cis-1,2-Cyclohexanediol (mp 98°C) *trans*-1,2-Cyclohexanediol (mp 103°C)

The second kind of stereoisomerism has to do with the lack of symmetry in the molecule.

"ONE SUBSTANCE, ONE STRUCTURE"

Chemists have long worked with the principle "one substance, one structure." If two substances are identical in their physical and chemical properties, they must be identical at the level of their individual molecules. If two substances differ in even one way, their individual molecules must also differ in some way.

Consider now a substance known as asparagine (ăs-păr′-à-jĭn), a white solid with a bitter taste, first isolated in 1806 from the juice of asparagus. Its molecular formula is $C_4H_8N_2O_3$, and its structure is now known to be **1.** Eighty years later, in

1886, a chemist isolated from sprouting vetch[1] a substance of the same molecular

$$NH_2{-}\overset{\overset{\displaystyle O}{\|}}{C}{-}CH_2{-}\underset{\underset{\displaystyle NH_2}{|}}{CH}{-}\overset{\overset{\displaystyle O}{\|}}{C}{-}OH$$

1

Asparagine

formula and structure, but it had a sweet taste. To distinguish these two substances from each other, we call the one isolated from asparagus L-asparagine and the one from vetch sprouts D-asparagine. The small capitals D and L, although arbitrary here, acquire significance later.

These two samples of asparagine have the same solubilities in solvents. Toward ordinary chemicals they have identical reactions with identical rates. Their solutions have identical spectra (which are comparable to identical fingerprints). From all the data the same structure has to be assigned the molecules in both samples, namely structure **1.** But they have this rather dramatic difference in taste. (Ordinarily chemists do not taste chemicals, but the fact remains that this difference in taste was discovered decades ago.)

Taste is a chemical sense. Chemical reactions between the substance being tasted and chemicals in the taste buds result in patterns of signals sent to the brain. We label these signals by words such as sweet, sour, bitter, salty, etc. The two samples of asparagine therefore show one difference in chemical property. (One physical property, to be described later, is also different.) Under the one-substance, one-structure doctrine, the molecules of D- and L-asparagine must be structurally different in some way, even though they both "answer" to structure **1.** Although all the forms of isomerism thus far discussed are ruled out by the physical data alone, another type must be possible.

Nature is rich in substances such as asparagine. Monosodium glutamate is a common, widely used flavor-enhancing agent. Yet two forms, identical in the same

$$HO{-}\overset{\overset{\displaystyle O}{\|}}{C}{-}CH_2CH_2\underset{\underset{\displaystyle NH_2}{|}}{CH}\overset{\overset{\displaystyle O}{\|}}{C}{-}O^-Na^+$$

Monosodium glutamate

$$HO{-}\overset{\overset{\displaystyle O}{\|}}{C}{-}\underset{\underset{\displaystyle OH}{|}}{CH}{-}\underset{\underset{\displaystyle OH}{|}}{CH}{-}\overset{\overset{\displaystyle O}{\|}}{C}{-}OH$$

Tartaric acid

ways that the two forms of asparagine are identical, are known. One is a flavor enhancer, the other not. Tartaric acid is another example. One form is fermented through the help of the enzymes in the organism *Penicillium glaucum,* another not. Chloromycetin, an important antibiotic of very complicated structure, also exists in two forms. One of the forms is effective as an antibiotic, the other not. These examples are only a few of the great many in molecular biology. Organic chemistry has still more. To understand how it might be possible for these differences to exist,

[1] Vetch is a member of a genus of herbs, some of which are useful as fodder.

we must understand the most fundamental way of deciding whether two molecular structures (in spatial models) represent identical compounds.

MOLECULAR SYMMETRY AND ASYMMETRY

Two partial ball and stick models of a molecule of asparagine are shown in Figure 13.1*a*. Examine each carefully to assure yourself that each is a representation of structure **1.** For convenience, each model in part *a* of the figure has been simplified as shown in part *b*. In each the same four groups are attached to a carbon, yet the two structures are not identical. For two structures to be identical it must be possible, mentally, to complete an operation known as superimposition, illustrated in Figure 13.1*c*. This is the fundamental criterion of identity for two molecular structures. Each must have a conformation by which the two become superimposable.

In asparagine the two nonidentical structures that cannot be superimposed are related in a significant way—as an object to its mirror image. Consider now what is characteristic of an object whose mirror image is superimposable. Imagine a perfect cube before a mirror, and imagine that the image in the mirror can be taken out and a model made of it. The two cubes would be superimposable and therefore essentially identical. The same "thought experiment" could be performed with a sphere, a pyramid, or any regular solid. By "regular" of course is meant symmetrical. Although we may not think about it this way, when we use the word symmetrical to designate an object, we are saying in effect that the object and its mirror image are superimposable. The two forms of asparagine, have the same molecular formula, the same skeletons, the same functional groups located at the same places on the chain. Yet they are not in a cis-trans relation; simple, free rotation cannot convert one into a form superimposable with the other. The two asparagines differ structurally only in the *relative* directions by which the four groups attached to the carbon shown in Figure 13.1 project into space. Chemists use the word *configuration* to denote a relative arrangement of nuclei in three-dimensional space. The two forms of asparagine could therefore be called configurational isomers, but instead, for historical reasons which will soon be apparent, the term *optical isomers* is used almost exclusively, There are several types of optical isomers, and the two asparagines illustrate one.

Enantiomers. Isomers which, like the two asparagines, are related as object and mirror image that cannot be superimposed are called *enantiomers* (Greek: *enantios*, opposite; *meros*, part). To deal as swiftly as possible with one source of misunderstanding, *any* object will have a mirror image. Only when the object and mirror image cannot be superimposed are they said to be enantiomers. With this definition of enantiomers, optical isomers in general can be better defined. Optical isomers are members of a set of stereoisomers that includes at least two related as enantiomers.

Physical properties are identical for two enantiomers. This *must* be so. Within two enantiomers intranuclear distances and bond angles are identical, and polarities must therefore be identical. Melting points, boiling points, densities, solubilities, spectra, and of course molecular weights have to be the same for two enantiomers.

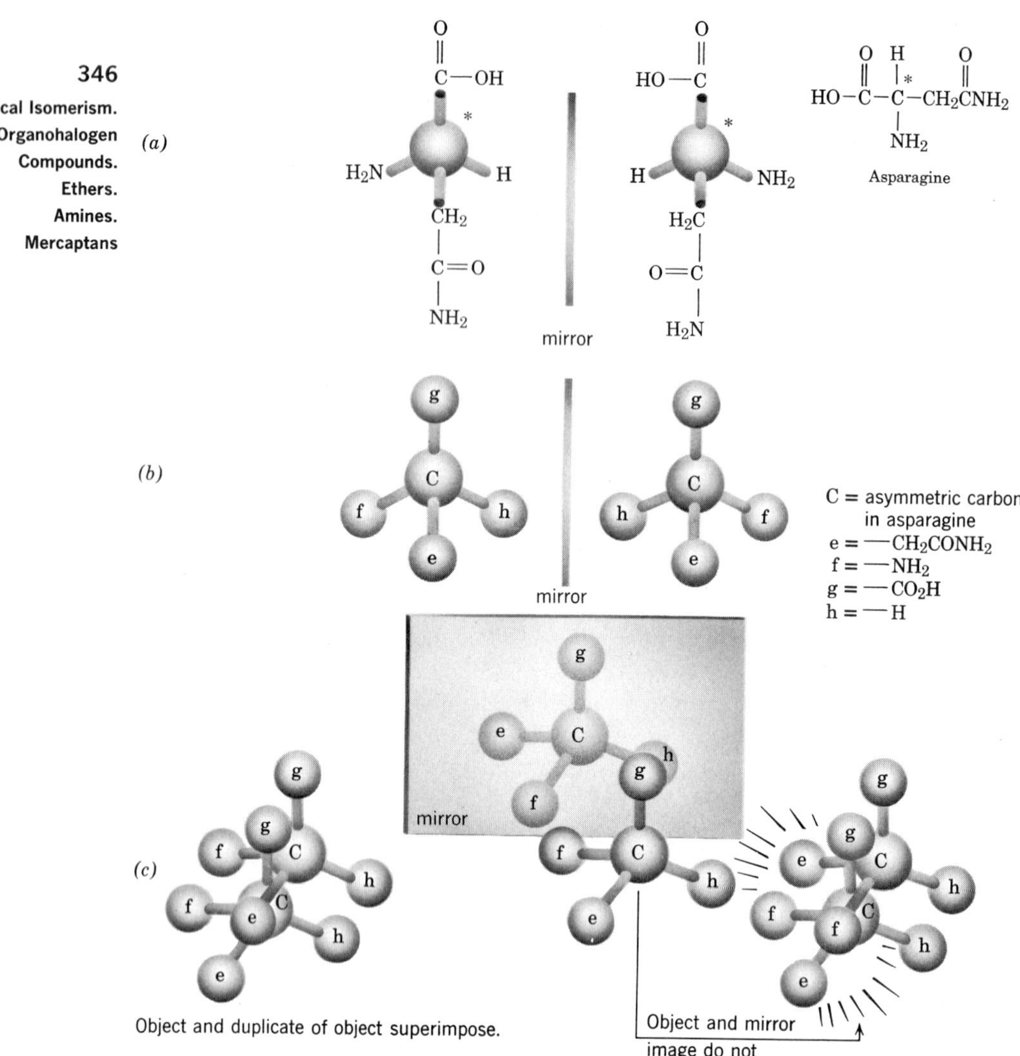
(a)
O
C—OH
H2N
H
CH2
C=O
NH2
mirror
O
HO—C
H
NH2
H2C
O=C
H2N
O H O
HO–C–C–CH2CNH2
NH2
Asparagine
(b)
g
C
f
h
e
mirror
C = asymmetric carbon in asparagine
e = —CH2CONH2
f = —NH2
g = —CO2H
h = —H
(c)
mirror
Object and duplicate of object superimpose.
Object and mirror image do not superimpose.

Figure 13.1 The enantiomers of asparagine are shown in part *a*. When the groups attached to an asymmetric carbon do not themselves contribute further to asymmetry in the molecule, they may be more simply represented as in part *b*. Checking superimposability requires the imaginary operation begun in part *c* but prevented by the impenetrability of wood.

In thinking about intranuclear distances and angles in enantiomers, let us examine our two hands, disregarding differences in fingerprints and wrinkles. The left hand and the right hand can serve as enantiomers because they are related as object and mirror image that cannot be superimposed. If we try to superimpose them with both palms in the same position, facing toward us, the fingers and thumb in one hand do not line up with the corresponding fingers and thumb in the other hand. If we try to superimpose them with the fingers and thumbs lined up, the palm of the first hand, when we imagine it going "through" the other, comes out on the back side of the second. Yet in each hand each finger has the same position in relation to the others. The only difference is that with the two hands held palms facing us, we "read" clockwise from thumb to little finger in the left hand and counterclockwise in the right hand.

Chemical properties of enantiomers are also identical, provided other reagents are symmetrical in the sense used here. If the reaction is with an unsymmetrical reagent, we observe striking differences in properties. A widely used illustration involves a pair of gloves. Because two hands are related as enantiomers, so too are a pair of ordinary gloves:

$$\text{Right hand} + \text{right glove} \longrightarrow \text{gloved right hand}$$
$$\text{Right hand} + \text{left glove} \longrightarrow \text{no fit}$$

Taking the gloves as the unsymmetrical "reagents," we see that the ease and comfort with which each glove can "react" with each hand is certainly different. By contrast, if we use a pair of, say, ski poles which are superimposable and symmetrical as the reagent system, no differences can be noted in the way these "react" with either hand.

Deciding Whether a Molecule Is Symmetrical. Any structure that is identical with its mirror image is symmetrical in some way. Three ways are recognized, but for our purposes we need only one, the existence of a *plane of symmetry*.[2] When optical isomerism is a possibility, we search the structures in question for a plane of symmetry. If we can discover that it is either present or absent, our work is greatly simplified, for we can dispense with a tedious task, making or drawing a three-dimensional model, reflecting it in an imaginary mirror, constructing a model of the mirror image, and finally checking the superimposability of the object with the mirror image. If we find a plane of symmetry, we are guaranteed that the object and mirror images are identical (superimposable). On the other hand, if we cannot discover this (or any) symmetry in a structure, in at least one conformation of it, we are guaranteed that the object and mirror images are not superimposable. It is apparent that ways to discover the symmetry element will save considerable time in trying to decide whether the system supports optical isomerism.

A plane of symmetry is an imaginary plane that divides an object into two halves that are mirror images of each other. In Figure 13.2 some simple molecules that have planes of symmetry are shown. Some of them have several planes, but we require only one to guarantee that object and mirror image are superimposable and therefore identical. (Figure 13.2*c* has only one.) Only when there is no element of symmetry do the interesting possibilities seen in asparagine, glutamic acid, tartartic acid, and chloromycetin materialize.

The Asymmetric Carbon—A Shortcut. Chemists have noted that in nearly all examples of optical isomerism the molecules possess at least one carbon to which are attached four *different* atoms or groups. In asparagine (Figure 13.1) one and

[2] For the record, the other two elements of symmetry are a center of symmetry and a fourfold alternating (or mirror) axis of symmetry.

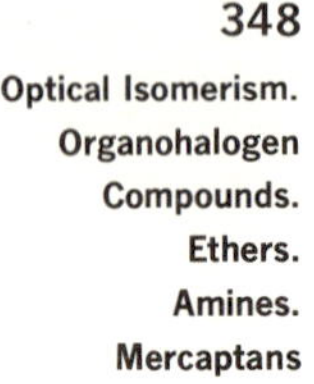

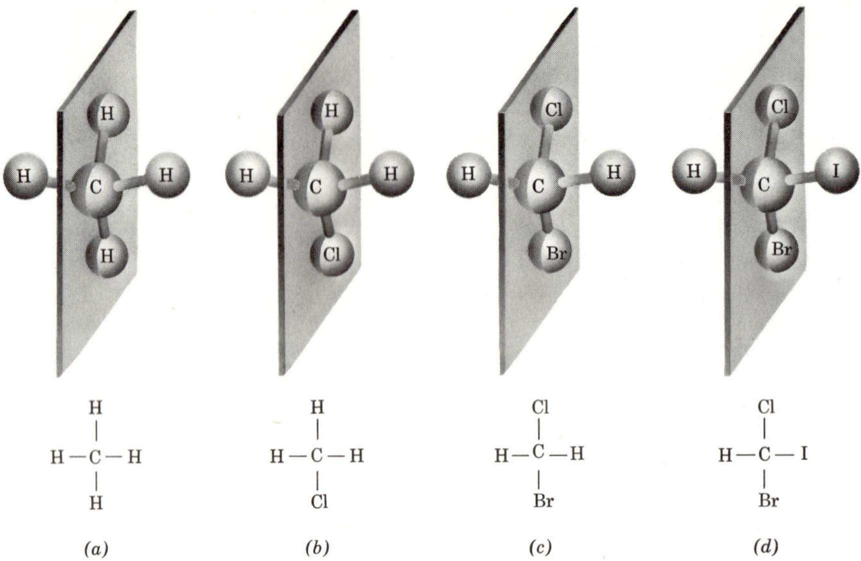
H
H C H
H
H C H
Cl
Cl
H C H
Br
Cl
H C I
Br
(a)
(b)
(c)
(d)

Figure 13.2 The plane of symmetry. (*a*) Methane, (*b*) methyl chloride, and (*c*) chlorobromomethane each possess at least one plane of symmetry. This plane, serving as an imaginary mirror, splits each molecule into two halves, one being the mirror image of the other. No plane of symmetry exists for molecule *d*, which is asymmetric. If a mirror image were made of it, the object and it would not be superimposable.

only one carbon has attached to it four *different* substituents: —H, $—NH_2$, —COOH, and $NH_2COCH_2—$. Such a carbon is defined as an *asymmetric* carbon. Its presence in a structure nearly always guarantees that there will be no element of symmetry in any conformation of the molecule. Therefore molecules with an asymmetric carbon will nearly always be asymmetric themselves and will belong to a set of optical isomers.

The qualification "nearly always" used repeatedly in these statements is an important one. Many substances with no asymmetric carbons have asymmetric molecules (Figure 13.3), and a few substances with molecules containing asymmetric carbons are yet symmetrical in some way (Figure 13.4). Some of these substances will be studied in more detail later. In general, if a molecule has *only one* asymmetric carbon, it will be asymmetric as a whole and it will belong to a set of optical isomers. We need not qualify this statement with "nearly always."

When it works, the concept of an asymmetric carbon is extremely useful, for we can usually consult an ordinary structural formula in trying to discover whether optical isomerism is a possibility. Thus we have a shortcut to replace the more difficult effort of seeking an element of symmetry, a method which in turn is a great improvement over making or drawing two or more structures. It must be remembered that the shortcut is fallible (cf. Figures 13.3 and 13.4), but with a little experience the exceptions are easily recognized. Moreover, they do not occur frequently.

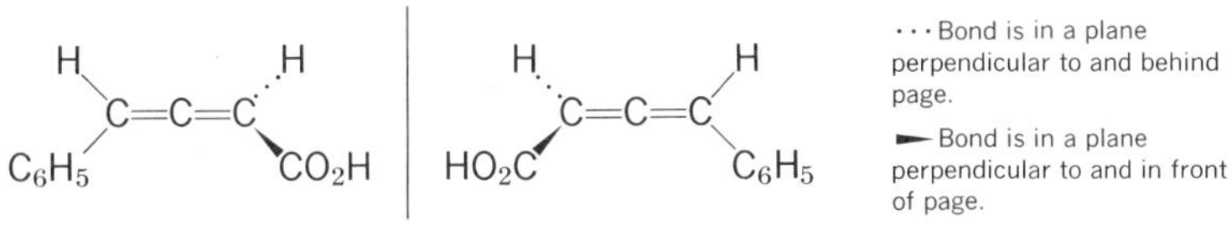

H C=C=C H C_6H_5 CO_2H | H C=C=C H HO_2C C_6H_5

· · · Bond is in a plane perpendicular to and behind page.

▬ Bond is in a plane perpendicular to and in front of page.

Enantiomeric allenes

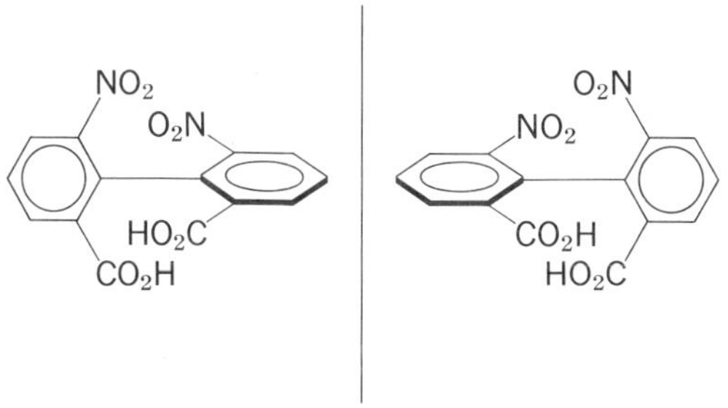

NO_2 O_2N HO_2C CO_2H | O_2N NO_2 CO_2H HO_2C

Enantiomeric biphenyls

Figure 13.3 Asymmetric molecules having no asymmetric carbon atoms. In the allenes the rigid geometry about the ends of the double bonds makes asymmetry possible with the appropriate end groups. In the biphenyls the bulkiness of the groups located at the ortho positions prevents free rotation about the single bond joining the two benzene rings. With the appropriate choice and distribution of these ortho substituents, the molecules can be asymmetric.

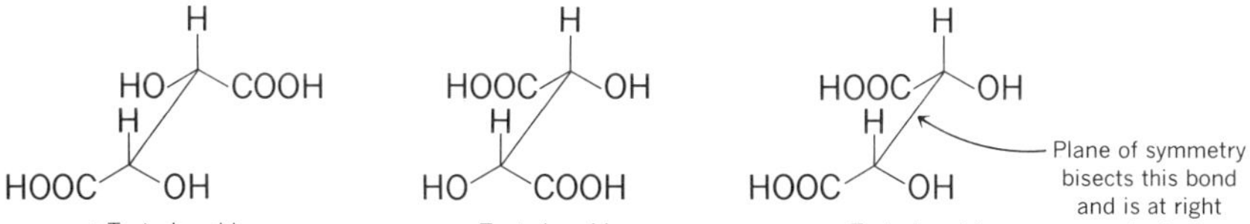

H HO COOH H HOOC OH
D-Tartaric acid (levorotatory, rare)

H HOOC OH H HO COOH
L-Tartaric acid (dextrorotatory, common)

H HOOC OH H HOOC OH
meso-Tartaric acid

Plane of symmetry bisects this bond and is at right angles to it.

$$\mathrm{HO{-}\overset{\overset{\displaystyle O}{\|}}{C}{-}\underset{\underset{\displaystyle OH}{|}}{\overset{*}{C}H}{-}\underset{\underset{\displaystyle OH}{|}}{\overset{*}{C}H}{-}\overset{\overset{\displaystyle O}{\|}}{C}{-}OH}$$

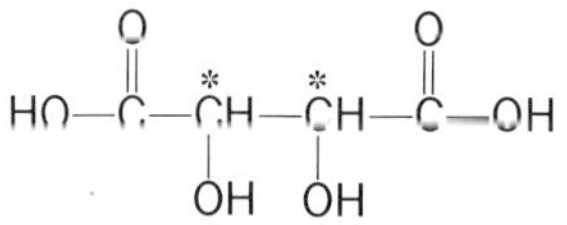

Tartaric acid

Figure 13.4 The presence of an asymmetric carbon does not always guarantee that the molecule will be asymmetric. For tartaric acid there are two enantiomeric forms, D- and L-tartaric acid, related as object and mirror image that cannot be superimposed. Another form of tartaric acid, the meso form, has a plane of symmetry (Greek *meso,* in the middle). Meso forms have molecules possessing two (or more) asymmetric carbons, to each of which are attached the same set of four different groups.

Exercise 13.1 Examine the given structures. Place an asterisk above each carbon to which four different groups are attached. If ball and stick sets or their equivalent are available to you, make molecular models of each structure and try to discover planes of symmetry. Make molecular models of the mirror images and check to see whether the objects and mirror images are superimposable.

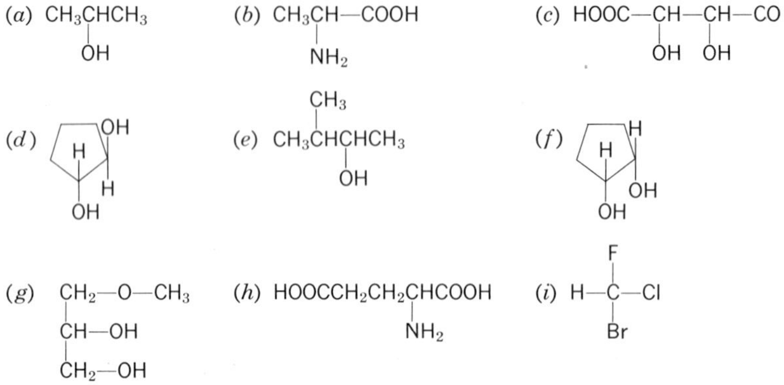

(*a*) $CH_3CH(OH)CH_3$

(*b*) $CH_3CH(NH_2)—COOH$

(*c*) $HOOC—CH(OH)—CH(OH)—CO$

(*d*) cyclopentane ring bearing H and OH on one carbon, and H and OH on the adjacent carbon

(*e*) $CH_3CH(CH_3)CH(OH)CH_3$

(*f*) cyclopentane ring bearing H and OH on one carbon, and H and OH on the adjacent carbon

(*g*) $CH_2—O—CH_3$ / $CH—OH$ / $CH_2—OH$

(*h*) $HOOCCH_2CH_2CH(NH_2)COOH$

(*i*) $H—C(F)(Br)—Cl$

OPTICAL ACTIVITY

Taste happens to be one way in which the asparagine enantiomers differ, but for general use the taste test is unreliable and physically dangerous. We must be able to detect enantiomers in some other way. Fortunately, one physical measurement, the effect of an enantiomer on plane-polarized light, is available.

In Chapter 1 light was described as electromagnetic radiation involving oscillations in the strengths of electric and magnetic fields set up by the light source. In ordinary light these oscillations or vibrations occur equally in all directions about the line that may be used to define the path of the light ray.

Certain materials, for example, the Polaroid film used in some sunglasses, affect ordinary light in a special manner. They interact with the oscillating electrical field in such a way that the emerging light vibrates in one plane. Such light is called *plane-polarized light* (Figure 13.5). If we look at some lighted object through a piece of Polaroid film and then place in front of the first film (or behind it) a second film, we can by rotating the two films in relation to each other find one orientation in which the lighted object has maximum brightness and one in which it can no longer be seen. To go from the first orientation to the second, one film is simply rotated 90° with respect to the other. It is as though, to use a rather unsatisfactory analogy, the film acts as a lattice fence forcing any emerging light to vibrate in the plane of the long axis of the slats. If, when this light reaches the second film or lattice fence, the second slats are related perpendicular to those of the first fence, the light cannot go through. At intermediate angles fractional amounts of light can go through the second lens or fence. Even though this analogy is not completely valid, we can take solace in the fact that a complete understand-

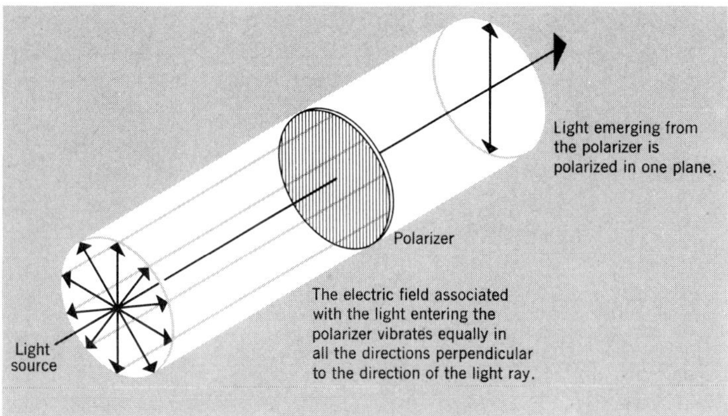
Light emerging from the polarizer is polarized in one plane.
Polarizer
The electric field associated with the light entering the polarizer vibrates equally in all the directions perpendicular to the direction of the light ray.
Light source

Figure 13.5 Light becomes polarized into vibrations in one plane when it passes through certain materials generally called polarizers. Polaroid film is an example.

ing of polarized light is unnecessary to our comprehension of the structural and chemical aspects of optical isomerism. Very few chemists are deeply versed in the mathematicophysical theories of plane-polarized light.

Optical Rotation. If a sample of an organic substance in which the molecules of one enantiomer predominate, like D-asparagine, is dissolved in an ordinary solvent, like water, and if this solution is placed in the path of plane-polarized light, the plane of polarization will be rotated. Any substance that will do this to plane-polarized light—rotate the plane—is said to be *optically active*.

The Polarimeter. The instrument used to detect optical activity and to measure the number of degrees of rotation of plane-polarized light is called a polarimeter. Figure 13.6 illustrates its principal working parts, which consist of a polarizer, a tube for holding solutions in the light path, an analyzer (actually, just another polarizing device), and a scale for measuring degrees of rotation. As shown in Figure 13.6*a*, when polarizer and analyzer are "parallel" and the tube contains no optically active material, the polarized light goes through to the observer. If, now, a solution containing an optically active material is placed in the light path, the plane-polarized light encounters asymmetric molecules, and its plane of oscillation is rotated. The polarized light that leaves the solution is no longer "parallel" with the analyzer (cf. Figures 13.6*a* and *c*), and the intensity of the observed light is reduced. To restore the original intensity, the analyzer is rotated to the right or to the left until it is again parallel with the light emerging from the tube. As the operator looks toward the light source, he will, if he rotates the analyzer to the right, record the degrees as positive, and the optically active substance will be called *dextrorotatory*. If the rotation is counterclockwise, the degrees will be recorded as negative, and the substance will be called *levorotatory*. (The direction requiring the fewer degrees to restore the original brightness is taken.) In the example of Figure 13.6*c*, the record will read $\alpha = -40°$, where α stands for the *optical rotation*, the observed number of degrees of rotation.

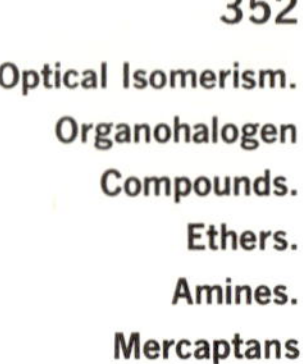
Analyzer
(can be rotated)
+
0°
−
Observer
+90°
−90°
180°
Degree scale
(fixed)
As the arrows indicate, the optically active substance in solution in the tube is causing the plane of the polarized light to rotate.
The plane of polarization of the emerging light is not the same as that of the entering polarized light.
Polarimeter tube
Polarizer (fixed)
Light source
(a)
–Polarizer and analyzer are parallel.
–No optically active substance is present.
–Polarized light can get through analyzer.
(b)
–Polarizer and analyzer are crossed.
–No optically active substance is present.
–No polarized light can emerge from analyzer.
−40°
(c)
–Substance between polarizer and analyzer is optically active.
–Analyzer has been rotated to the left (from observer's point of view) to permit rotated polarized light through (substance is levorotatory).
Polarizers
Analyzers
Observer

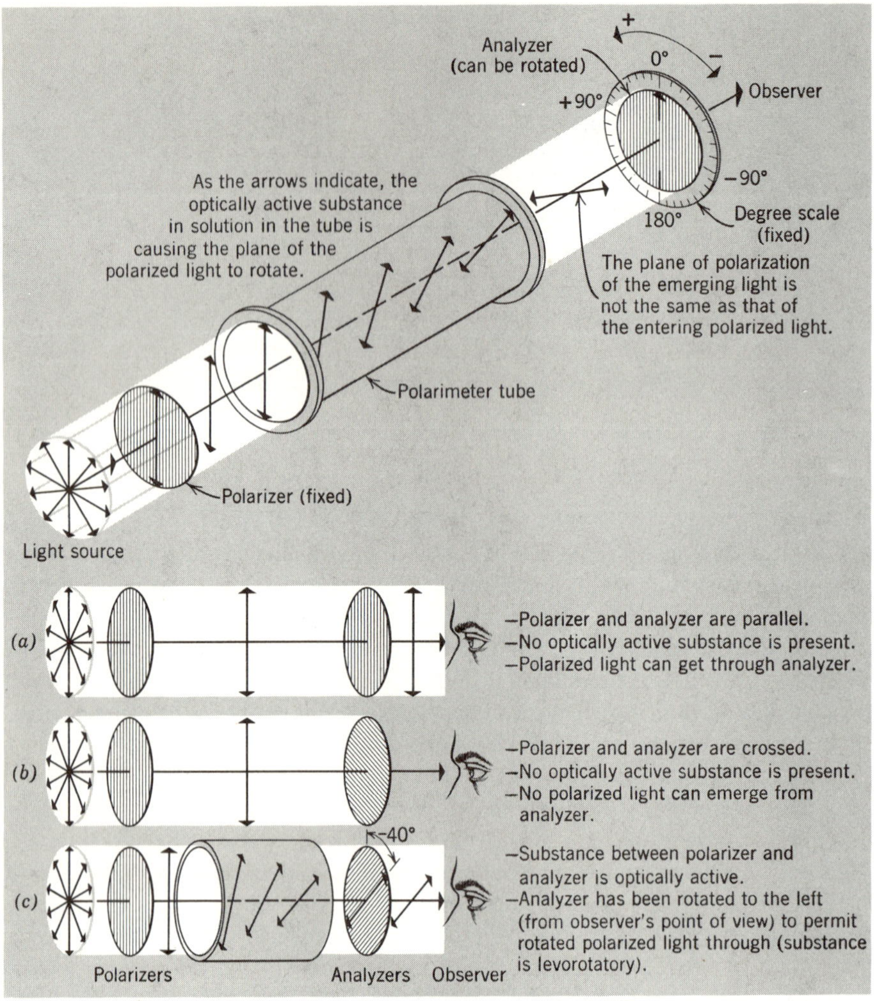

Figure 13.6 Principal working parts of a polarimeter and the measurement of optical rotation.

Since the optical rotation varies with both the temperature of the solution and the frequency of the light, the record actually notes both data, for example, $\alpha_D^{20} = -40°$. The subscript D stands for the D-line of the spectrum of sodium, which is one particular frequency that electromagnetic radiation can have, and 20 means that the solution has a temperature of 20°C.

Specific Rotation. The extent of the optical rotation also depends on the population of asymmetric molecules that the light beam encounters. If it travels through a longer tube, the plane of oscillation of the light will be rotated more. Or if the tube length is held constant but the concentration of the solution is increased, the degrees of rotation also increase. To incorporate such effects into the record, the observer converts the observed value of α into a ratio expressing the number of degrees per unit concentration per unit of path length. The *specific rotation* is one way of expressing this ratio. Its symbol is $[\alpha]_{\lambda}^{t°}$, or with temperature and frequency specified $[\alpha]_{\mathrm{D}}^{20}$. By definition,

$$[\alpha]_{\lambda}^{t°} = \frac{100\alpha}{cl}$$

where c = concentration in grams per 100 cc of solution
l = length of the light path in the solution, measured in decimeters (1 decimeter = 10 cm)
α = observed rotation in degrees (plus or minus)
λ = wavelength of light
$t°$ = temperature of the solution in degrees centigrade

The specific rotation of a compound is an important physical constant, comparable to its melting point, boiling point, etc. It is one more physical characteristic that a chemist can use to identify a substance. Moreover, by measuring the actual rotation α of a solution of a substance of known specific rotation in a tube of fixed length, we can calculate the concentration of the solution. That is, with $[\alpha]$, α, and l known, the equation readily permits calculation for c.

Table 13.1 lists specific rotations for several materials, Table 13.2 physical constants for some sets of optical isomers. *Pairs of enantiomers differ physically only in the sign or direction in which each rotates the plane of plane-polarized light.* All other physical constants are the same, including even the number of degrees of rotation.

Table 13.1. Specific Rotations of Various Substances

Physical State	Substance	Specific Rotation (sodium D light at 20°C in solvent specified)
Solutions*	asparagine	+5.41 (water)
	albumin (a protein)	−25 to −38 (water)
	cholesterol	−31.61 (chloroform)
	glucose	+52.5 (aged solution in water)
	sucrose (table sugar)	+66.4 (water)
	3-methyl-2-butanol ("active amyl alcohol")	+5.34 (ethyl alcohol)
	quinine sulfate	−214 (water, at 17°C)
Pure liquids*	turpentine	−37
	cedar oil	−30 to −40
	citron oil	+62 (at 15°C)
	nicotine	−162
Pure solids†	quartz	+21.7
	cinnabar (HgS)	+32.5
	sodium chlorate	+3.13

* Specific rotation in degrees per decimeter.
† Specific rotation in degrees per millimeter of path length in the crystal.

Table 13.2 **Physical Constants of Optical Isomers**

Set	Members of Set	Mp (°C)	$[\alpha]_D^{20}$ (degrees)	Miscellaneous
Mandelic acids $C_6H_5CHCO_2H$ (with OH on the CH)	(+) mandelic acid	132.8	+155.5	Solubility: 8.54 g/100 cc water (at 20°C)
	(−) mandelic acid	132.8	−155.4	Solubility: 8.64 g/100 cc water (at 20°C)
Asparagine	(+) asparagine	234.5 (decomposes)	+5.41	$d_4^{15} = 1.534$ g/cc
	(−) asparagine	235 (decomposes)	−5.41	$d_4^{15} = 1.54$ g/cc
Tartaric acids	(+) tartaric acid	170	+11.98	$d_4^{15} = 1.760$ g/cc
	(−) tartaric acid	170	−11.98	$d_4^{15} = 1.760$ g/cc
	meso-tartaric acid	140	0	$d_4^{15} = 1.666$ g/cc
	racemic tartaric acid*	205	0	$d_4^{15} = 1.687$ g/cc

* This is a 50:50 "mixture" of the plus and minus forms of tartaric acid, which in this case happens to form what is known as a racemic compound.

OTHER KINDS OF OPTICAL ISOMERS

Meso Compounds. In some sets of optical isomers, for example, the tartaric acids of Figure 13.4, one member is found to be optically inactive, even though it possesses asymmetric carbons. We are reminded that the fundamental criterion for asymmetry in a molecule is not the asymmetric carbon but rather the impossibility of superimposing object and mirror images. The isomer that belongs to a set of optically active substances but is optically inactive itself is called the meso isomer. The mirror image of *meso*-tartaric acid would be superimposable on the object. The acid also possesses within the molecule a plane of symmetry. Behind this plane is a reflection of what is in front. Hence, as one-half of the molecule tends to rotate polarized light, say, to the left, the other half cancels this with an equal rightward rotation. The net effect is optical inactivity.

Diastereomers. Optical isomers that are not related as object to mirror image are called diastereomers. Figure 13.7 illustrates examples, and we shall encounter them again in carbohydrate chemistry. For two molecules to qualify as diastereomers, they must have identical molecular formulas, they must have identical nucleus-to-nucleus sequences (without regard to geometry), they must be members of a set of optical isomers, but they may not be related as object to mirror image. Generally, they will have different physical properties and sometimes different chemical properties.

Racemic Mixtures. A substance that is composed of a 50:50 mixture of enantiomers is called a racemic mixture. One enantiomer cancels the effects of the other on polarized light, and a racemic mixture is optically inactive, emphazing the point that the phenomenon of optical activity relates to a measurement we can make on a substance. The explanation of the phenomenon, in terms of asymmetric molecules, relates to our theory about it.

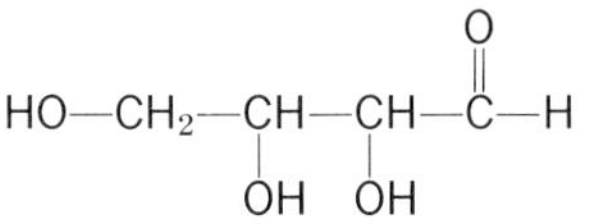

$$HO-CH_2-\underset{OH}{CH}-\underset{OH}{CH}-\overset{O}{\overset{\|}{C}}-H$$

I

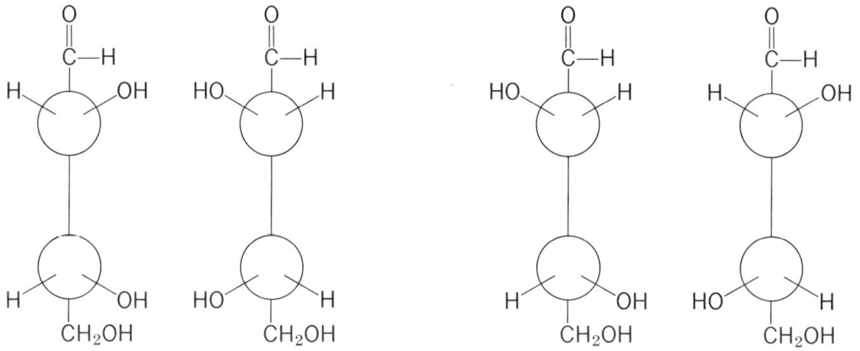

D-Erythrose $[\alpha]_D^{20}$ −14.8°

L-Erythrose +14.8°

D-Threose $[\alpha]_D^{20}$ +19.6°

L-Threose −19.6°

Figure 13.7 Optical isomers that illustrate pairs of enantiomers and diastereomers. The pairs of enantiomers of I are D- and L-erythrose and D- and L-threose. Any member of the first pair is a diastereomer of any member of the second. Compounds related as diastereomers are optical isomers that are not related as object and mirror image. As the values of specific rotation indicate, diastereomers may have different physical properties. Both erythrose and threose are carbohydrates.

A racemic mixture must not be confused with a meso compound. In a meso compound optical inactivity is the result of an internal plane of symmetry. In a racemic mixture optical inactivity is the result of both enantiomers' being present in a 1:1 ratio. Racemic comes from racemic acid, the name once given to a form of tartaric acid that Pasteur (p. 574) demonstrated consisted of a 50:50 mixture of enantiomers.

CONFIGURATIONAL CHANGES IN CHEMICAL REACTIONS

n-Butane has no asymmetric carbon, and it has a plane of symmetry. A molecule of bromine, Br_2, is also symmetrical. Yet if bromine can be made to react with *n*-butane at the second carbon, this carbon becomes asymmetric in the product.

$$\underset{n\text{-Butane}}{CH_3CH_2CH_2CH_3} + Br_2 \longrightarrow \underset{sec\text{-Butyl bromide}}{CH_3CH_2\underset{Br}{CH}CH_3} + HBr$$

The question is whether the substance actually isolated as a product of this reaction and identified as *sec*-butyl bromide is optically active. The answer is *no*. We must distinguish between a *substance* that is optically active and one that has asymmetric molecules. Optical activity is something we can measure with a polarimeter, completely without regard for an interpretation of the observation. The reaction produces an asymmetric carbon but in such a way that the product is a 50:50 mixture of enantiomers, a racemic or optically inactive mixture. We can understand the reason for this outcome by examining Figure 13.8. In *n*-butane, at the site where bromine will eventually become affixed, are two equivalent hydrogens. In a substitution occurring to any one molecule there is an equal chance that each hydrogen will be the one replaced. In a collection approaching Avogadro's number in size, the statistics are such that production of a racemic mixture is a certainty.

It is, in general, true that optically active *substances* cannot be synthesized from optically inactive reagents. Asymmetric centers can easily be made, but the final substance will be either a racemic mixture or a meso compound.

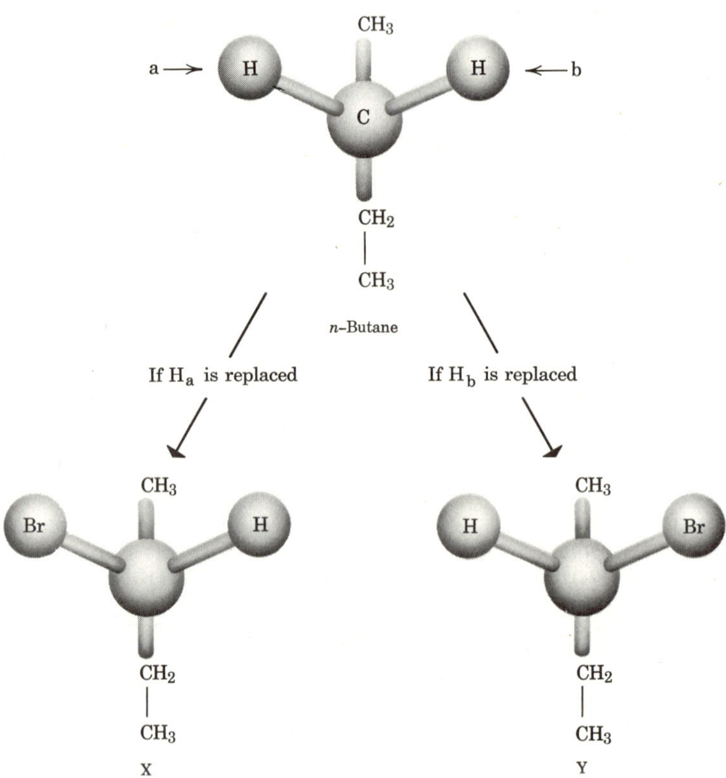
CH_3
a
H
H
b
C
CH_2
CH_3
n–Butane
If H_a is replaced
If H_b is replaced
CH_3
Br
H
CH_2
CH_3
X
CH_3
H
Br
CH_2
CH_3
Y

Figure 13.8 Illustrating a general fact: optically active substances cannot be prepared by the action of an optically inactive reagent on an optically inactive compound. In *n*-butane the two hydrogens that are candidates for replacement, if *sec*-butyl bromide is to form, have the same chance to be replaced. Therefore a 50:50 mixture of the two enantiomers of *sec*-butyl bromide, X and Y, is produced. Although the enantiomers would individually be optically active, the racemic mixture of them that actually forms is not.

ORGANOHALOGEN COMPOUNDS

TYPES OF ORGANOHALOGEN COMPOUNDS. STRUCTURE. NOMENCLATURE

The monohalogen derivatives of hydrocarbons of widest use in organic chemistry are the monochloro, monobromo, and monoiodo derivatives of alkanes and aromatic hydrocarbons. Several examples are given in Table 13.3. The fluoro derivatives are virtually in a class by themselves, partly because they are more difficult to make. They are, however, of great industrial and commercial importance, and some of the more widely used examples are listed in Table 13.4. Alkyl halides are not found in the body, but they illustrate principles important in living systems.

Table 13.3 Some Classes of Organohalogen Compounds

Halides	Example	Structure	Bp (°C)
Alkyl halides	methyl bromide	$CH_3—Br$	4
	isopropyl chloride	$(CH_3)_2CH—Cl$	35
	t-butyl chloride	$(CH_3)_3C—Cl$	51
	cyclohexyl chloride	$C_6H_{11}—Cl$ (cyclohexane ring—Cl)	164
	allyl chloride	$CH_2{=}CHCH_2—Cl$	45
	β-phenylethyl bromide	$C_6H_5CH_2CH_2—Br$	217
	benzyl chloride	$C_6H_5CH_2—Cl$	179
Vinyl halides			
	vinyl bromide	$CH_2{=}CH—Br$	−14
	2-chloropropene	$CH_2{=}C(Cl)—CH_3$	23
Aryl halides			
	bromobenzene	$C_6H_5—Br$ (benzene ring—Br)	155
	o-chlorotoluene	benzene ring with CH_3 and ortho Cl	159
Polyhalogen compounds			
	methylene chloride	CH_2Cl_2	40
	chloroform	$CHCl_3$	61
	carbon tetrachloride	CCl_4	77
	iodoform	CHI_3	119 (mp)
	trichloroethylene (Triclene)	$Cl—CH{=}CCl_2$	87

Table 13.4 Important Organofluorine Compounds

Name	Structure	Uses
Freon 11	CCl_3F (bp 24°C)	propellant; refrigerant
Freon 12	CCl_2F_2 (bp −30°C)	first fluorocarbon aerosol propellant; refrigerant
Freon 114	$CClF_2CClF_2$ (bp 4°C)	refrigerant (very stable, odorless); propellant for shaving creams, colognes
Bromotrifluoromethane	$CBrF_3$ (bp −58°C)	fire extinguishers (low toxicity)
Perfluorocyclobutane	$F_2C—CF_2$ / $F_2C—CF_2$ (four-membered ring) (bp −4°C)	propellant for aerosol cans of food products (e.g., whipped cream)
Teflon	$-(CF_2-CF_2)_n-$	cookware; bearings
Tedlar (polyvinyl fluoride)	$-(CH_2-CHF)_n-$	weatherproofing material
Scotchgard	Organofluorine substance (structure not publicly known)	grease and oil repellent for fabrics and paper (see photograph)

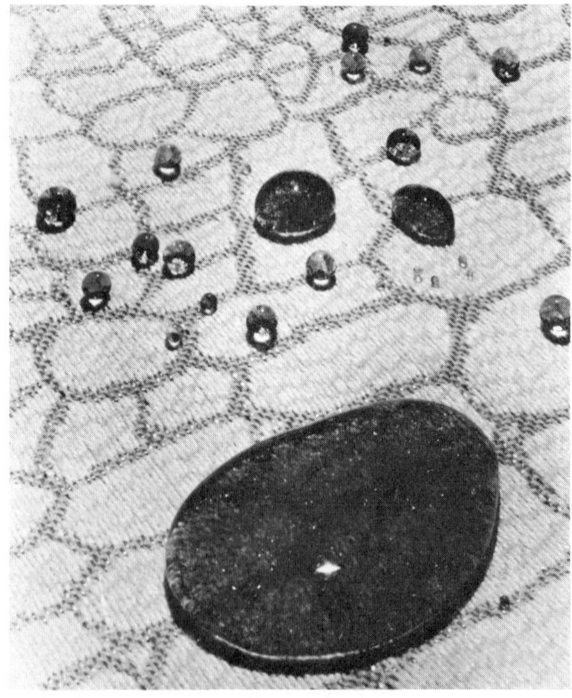

This closeup shows how oil and water bead on a fabric treated with Scotchgard fluorocarbon repeller, which does not close the pores of the fabric. (Courtesy of the 3M Company.)

SYNTHESIS

In Chapter 12 we learned that alkyl halides may be made by the action of concentrated H—X on an alcohol. Earlier we learned that H—X can also be made to add to an alkene linkage. Other reagents that will convert alcohols to halides include thionyl chloride ($SOCl_2$), phosphorus tribromide (PBr_3), and a mixture of potassium iodide and phosphoric acid:

$$ROH + SOCl_2 \longrightarrow R{-}Cl + SO_2 + HCl$$
$$3ROH + PBr_3 \longrightarrow 3R{-}Br + H_3PO_3$$
$$ROH + KI + H_3PO_4 \longrightarrow R{-}I + KH_2PO_4 + H_2O$$

NUCLEOPHILIC SUBSTITUTION REACTIONS OF ALKYL HALIDES

Several types of nucleophilic substitution reactions are tabulated in Table 13.5. For the most part the nucleophiles are also bases. Therefore an important side reaction is the elimination of the elements of H—X and the formation of an alkene (see the next section). Tertiary halides are particularly susceptible to this reaction, making it very difficult to use them successfully in a nucleophilic substitution reaction.

Stereochemistry of Nucleophilic Substitutions. If the carbon holding the halogen in an alkyl halide is asymmetric, what will be the fate of this center of asymmetry when it experiences nucleophilic substitution? If the mechanism is S_N1, the product will be a racemic mixture, or very nearly one, because the formation of a carbonium ion at a carbon that is initially asymmetric destroys that asymmetry. The carbon in the ion has only three groups attached to it. The nucleophile can approach the positive charge from either side of the ion, and the product will be a racemic mixture (Figure 13.9).

Table 13.5 Some Nucleophilic Substitution Reactions of Alkyl Halides*

Nucleophilic Agent	$R{-}X + Y{:} \longrightarrow R{-}Y + X{:}$
Negatively charged nucleophiles	
Hydroxide ion, OH^-	$R{-}X + HO^- \longrightarrow R{-}OH + X^-$ Alcohols
Alkoxide ion, RO^-	$+ RO^- \longrightarrow R{-}O{-}R + X^-$ Ethers
Cyanide ion, ^-CN	$+ {}^-CN \longrightarrow R{-}CN + X^-$ Nitriles
Amide ion, $^-NH_2$	$+ {}^-NH_2 \longrightarrow R{-}NH_2 + X^-$ Amines
Neutral nucleophiles	
Ammonia, $:NH_3$	$+ :NH_3 \longrightarrow R{-}\ddot{N}H_2 + (HX)$ Amines
Water, $:\ddot{O}H_2$	$+ H_2\ddot{O}: \longrightarrow R{-}\ddot{O}H + (HX)$ Alcohols

* The alkyl halide may be 1° or 2°, alkyl or cycloalkyl, allylic or benzylic. It may not be vinyl or aryl, which are usually inert toward nucleophilic substitution.

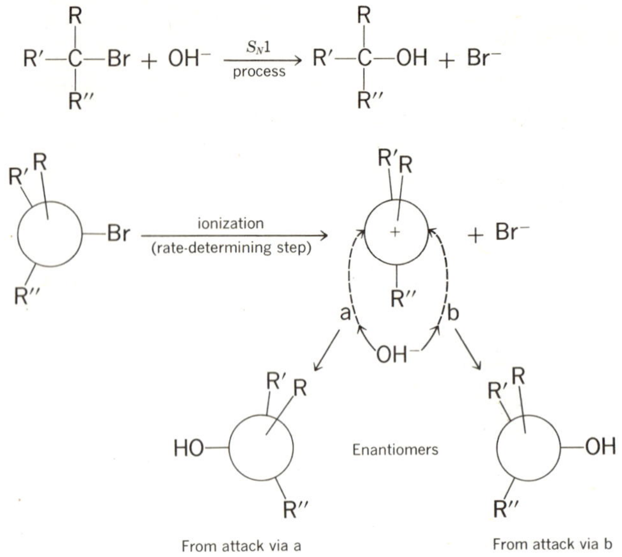

$$R'-\overset{\displaystyle R}{\underset{\displaystyle R''}{C}}-Br + OH^- \xrightarrow[\text{process}]{S_N1} R'-\overset{\displaystyle R}{\underset{\displaystyle R''}{C}}-OH + Br^-$$

R', R, R''
Br
ionization
(rate-determining step)
+ Br⁻
a
b
OH⁻
HO
OH
Enantiomers
From attack via a
From attack via b

Figure 13.9 Stereochemistry in an S_N1 reaction. If an optically active bromide is converted to the corresponding alcohol under conditions that assure an S_N1 mechanism, asymmetry will be lost at the carbonium-ion stage. Approach to the positively charged carbon by the nucleophile (OH^- in our example) is equally likely from either side. The enantiomers of the product will form in a ratio of 1:1, and the product will be optically inactive. If the departing bromide ion stays for a while close by one side of the positively charged carbon, the nucleophile may more frequently attack the other side, and the product will have some optical activity. Only when the opportunities for approach to either side are the same will a fully racemic mixture form.

If the mechanism is S_N2, the configuration of the asymmetric carbon will be inverted. The product will be optically active, but its configuration will not be the same as in the starting material (Figure 13.10).

In the last chapter we learned relative orders of reactivity of alcohols in S_N1 and S_N2 mechanisms. These same orders apply to the nucleophilic substitution reactions of alkyl halides and for the same reasons. Aryl and vinyl halides are exceptionally inert toward nucleophilic substitution.

HO⁻
R, H, R'
Br
δ− HO---
---Br δ−
HO
+ Br⁻
Transition state

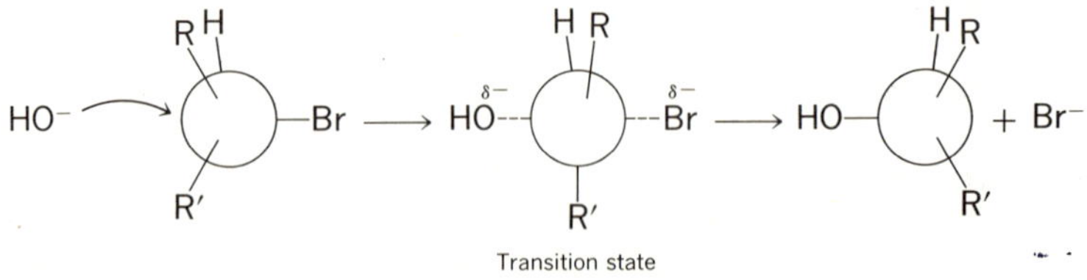

Figure 13.10 Stereochemistry in an S_N2 reaction. If an optically active bromide is converted to the corresponding alcohol under conditions that assure an S_N2 mechanism, asymmetry will be preserved, but the configuration of the asymmetric carbon will be inverted. The product will be optically active.

ELIMINATION REACTIONS OF ALKYL HALIDES

A carbon-carbon double bond can be introduced into a molecule by splitting the elements of H—X out of an alkyl halide:

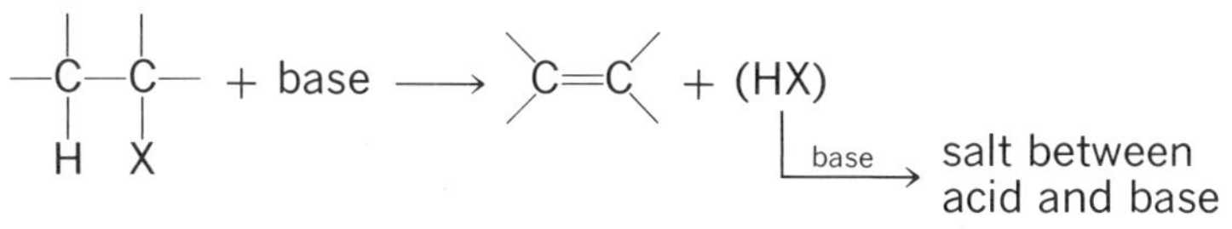

$$\underset{\text{H}\quad\text{X}}{-\overset{|}{\underset{|}{C}}-\overset{|}{\underset{|}{C}}-} + \text{base} \longrightarrow \;\rangle C{=}C\langle\; + (HX) \xrightarrow{\text{base}} \text{salt between acid and base}$$

The elements lost, H—X, must be on adjacent carbons, as shown. A common reagent for accomplishing this splitting is a solution of potassium hydroxide in alcohol. Since hydroxide ion is also a nucleophile, some molecules of halide can lose H—X to form an alkene, while others, undergoing different kinds of collisions, can take part in a substitution reaction to form R—OH. By suitable selection of conditions, however, chemists have been able to make one or the other of these two, elimination or substitution, dominate the course of the reaction. In general:

—Elimination is favored by high concentration of hydroxide ion in *alcohol* solvent at relatively high temperatures.
—Substitution is favored by low concentration of hydroxide ion in *water* at lower temperatures.

In more advanced courses these observations are explained in terms of general principles. The student is not expected to memorize these facts. They are mentioned simply to show that variations in experimental conditions (concentrations, solvents, temperatures for example) can greatly affect the course of a reaction.

The experimental work of many chemists is devoted to studies of these factors, not only what works (practical) but how it works (theoretical). If a particular synthesis is to be put on stream in an around-the-clock industrial operation employing many people and producing organic chemicals in tank car quantities for sale in a competitive business, an experimental factor that increases the yield by even a few percent can make a great difference in the profit-loss columns.

ETHERS

STRUCTURE AND NOMENCLATURE

The ethers are substances whose molecules are of the general formulas R—O—R, R—O—Ar, and Ar—O—Ar, with R any alkyl group and Ar any aromatic group. They are generally designated by first naming the two groups attached to oxygen and then adding the separate word *ether,* as illustrated by the following examples. When the two groups attached to oxygen are identical, the ether is said to be *symmetrical;* otherwise it is said to be *unsymmetrical.* The IUPAC has rules for naming more complicated systems, but we shall not study them.

$CH_3—O—CH_3$
Dimethyl ether[3]

$CH_3—O—CH_2CH_3$
Ethyl methyl ether

$CH_3CH_2—O—CH_2CH_3$
Diethyl ether (anesthetic ether)

$CH_2{=}CH—O—CH{=}CH_2$
Divinyl ether (vinethene, another anesthetic)

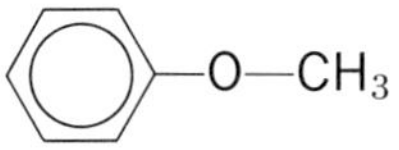

$C_6H_5—O—CH_3$
Methyl phenyl ether (anisole)

SYNTHESIS OF ETHER

Ethers can be made in two important ways—through the action of the salt of an alcohol on an alkyl halide, or through the action of heat and an acid catalyst on an alcohol. The first is important in organic synthesis in research; the second is important commercially for making certain ethers.

The Williamson Ether Synthesis. In general:

First: $$R—O—H + Na \longrightarrow \underset{\text{A sodium alkoxide}}{R—O^-Na^+} + \tfrac{1}{2}H_2$$

Then: $$R—O^- + R'—X \longrightarrow R—O—R' + X^-$$

Specific examples:

$$\underset{\text{Methyl iodide}}{CH_3—I} + \underset{n\text{-Butoxide ion}}{^-O—CH_2CH_2CH_2CH_3} \longrightarrow \underset{\text{Methyl } n\text{-butyl ether (71\%)}}{CH_3—O—CH_2CH_2CH_2CH_3} + I^-$$

$$\underset{\text{Ethyl iodide}}{CH_3CH—I} + {^-O—CH_2CH_2CH_2CH_3} \longrightarrow \underset{\text{Ethyl } n\text{-butyl ether (71\%)}}{CH_3CH_2—O—CH_2CH_2CH_2CH_3} + I^-$$

$$\underset{\text{Ethyl iodide}}{CH_3CH_2—I} + \underset{n\text{-Pentoxide ion}}{^-O—CH_2CH_2CH_2CH_2CH_3} \longrightarrow \underset{\text{Ethyl } n\text{-pentyl ether (47\%)}}{CH_3CH_2—O—CH_2CH_2CH_2CH_2CH_3} + I^-$$

Because alkoxide ions are powerful bases as well as nucleophiles, 3° halides will undergo dehydrohalogenation, not ether formation. The Williamson synthesis is very successful when the halide is primary and then the mechanism is usually S_N2.

Ethers from Dehydration of Alcohols. In general:

[3] The names of symmetrical ethers frequently do not contain the prefix "di-"; thus dimethyl ether can be named methyl ether. Diethyl ether is often called simply ethyl ether and sometimes, as here, it can be called just ether.

$$2ROH \xrightarrow{H^+} R—O—R + H_2O$$

or

$$R—O—H + H—O—R \xrightarrow{H^+} R—O—R + H_2O$$

Specific example:

$$\underset{\text{Ethanol}}{2CH_3CH_2OH} \xrightarrow[H^+]{140°C} \underset{\text{Diethyl ether}}{CH_3CH_2—O—CH_2CH_3} + H_2O$$

Mechanism. Type S_N2.

Step 1. $\underset{\text{Ethyl alcohol}}{CH_3CH_2—OH} + H_2SO_4 \rightleftharpoons \underset{\text{Ethyloxonium ion}}{CH_3CH_2—\overset{+}{O}H_2} + HSO_4^-$ (rapid)

Step 2.

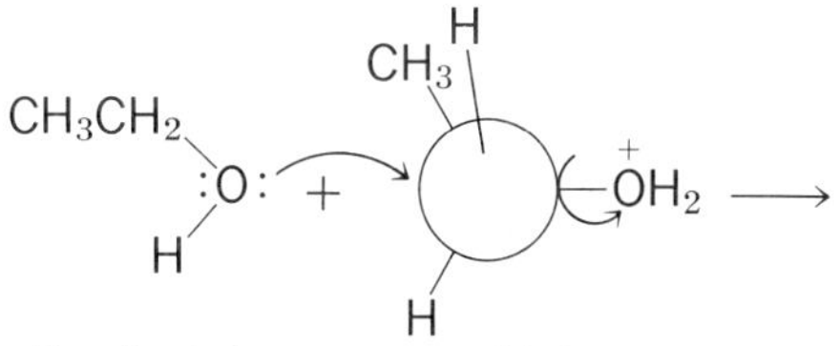

$CH_3CH_2(H)\ddot{O}: + CH_3CH_2—\overset{+}{O}H_2 \longrightarrow$

Unprotonated ethyl alcohol, the nucleophile — Ethyloxonium ion

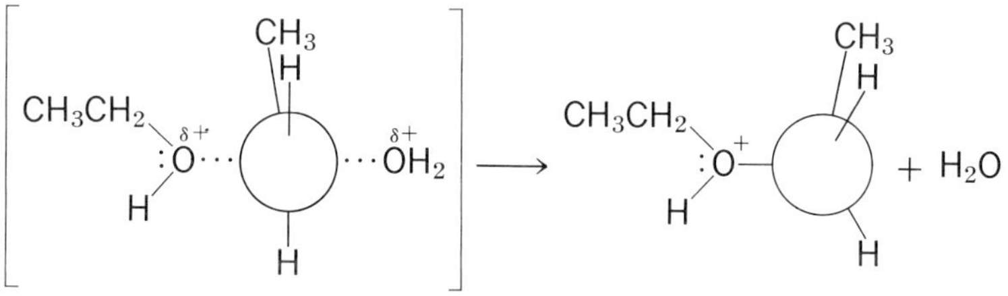

$\left[CH_3CH_2(H)\overset{\delta+}{O}\cdots C \cdots \overset{\delta+}{O}H_2 \right] \longrightarrow CH_3CH_2(H)\overset{+}{O}—CH_2CH_3 + H_2O$

Transition state — Protonated form of diethyl ether, diethyloxonium ion

Step 3.

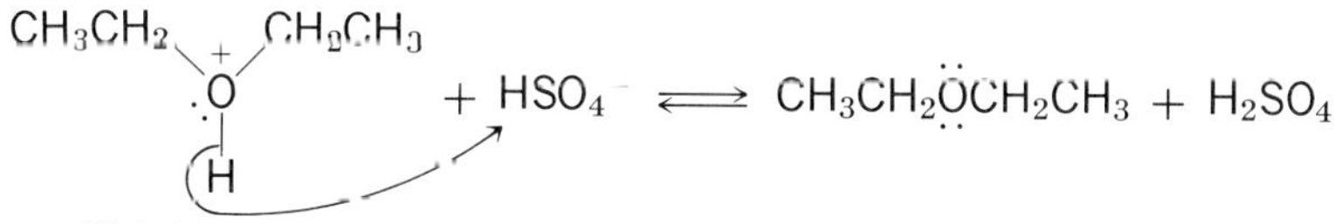

$$\underset{\text{Diethyloxonium ion}}{CH_3CH_2\overset{+}{O}(H)CH_2CH_3} + HSO_4^- \rightleftharpoons \underset{\text{Diethyl ether}}{CH_3CH_2\ddot{O}CH_2CH_3} + H_2SO_4$$

Step 1 is simply an acid-base equilibrium involving ethanol as the stronger proton acceptor and sulfuric acid as the stronger proton donor. Acid-base equilibria of this type are rapidly established. Step 2 is the rate-determining or slow step—the attack by the nucleophile ethanol on the carbinol carbon of the ethyloxonium ion. Because its oxygen is protonated and positively charged, this carbinol carbon bears a significant partial positive charge. The product of this step is the diethyloxonium ion. Step 3 is, again, a rapidly established acid-base equilibrium, wherein bisulfate ion is shown as the proton acceptor. Actually, this is only one of the proton acceptors present. Molecules of water or ethanol could serve here too.

The formation of ethers by acid-catalyzed dehydration of alcohols depends on careful control of conditions. If the temperature is allowed to climb too high, an alternative reaction, the *internal* dehydration of an alcohol to produce an alkene, becomes significant. In fact, the internal dehydration of alcohols is one of the important ways of introducing a carbon-carbon double bond.

We shall later encounter ether formation in a special reaction of an aldehyde with an alcohol; our study of the mechanism whereby ethers may form from alcohols will be important at that time.

PHYSICAL PROPERTIES OF ETHERS

Several representative ethers are listed in Table 13.6 together with certain of their physical constants. Ether molecules are slightly polar. Since they possess no O—H groups, they are incapable of the hydrogen bonding between two molecules that alcohols engage in. Hence the boiling points of ethers and alkanes of comparable formula weights are much the same. Diethyl ether (formula weight 74) boils at 34.6°C; *n*-pentane (formula weight 72) boils at 36°C; *n*-butyl alcohol (formula weight 74) boils much higher, at 117°C.

Since ether molecules do contain an oxygen, they can be involved in hydrogen bonding with water molecules. Even though an ether molecule cannot "donate" a hydrogen bridge, it can, of course, "accept" one (cf. Figure 13.11). The solubilities in water of ethers and alcohols of comparable formula weight are therefore found to be similar. For example, both *n*-butyl alcohol and diethyl ether dissolve to the extent of about 8 g in 100 cc of water.

Neither boiling points nor the solubilities are to be memorized, for they are cited only to help indicate how structural factors are related to these physical properties.

CHEMICAL PROPERTIES OF ETHERS

Although ethers undergo a few reaotions, none is very important to our later study of biochemistry. We simply note the fact that the ether linkage is chemically

Table 13.6 Some Ethers

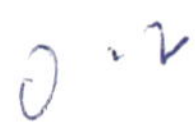

Name	Structure	Mp (°C)	Bp (°C)	Solubility in Water
Dimethyl ether	$CH_3—O—CH_3$	−138	−23	37 volumes gas dissolve in 1 volume H_2O at 18°C
Methyl ethyl ether	$CH_3—O—CH_2CH_3$	−116	11	—
Diethyl ether	$CH_3CH_2—O—CH_2CH_3$	−116	34.5	8 g/100 cc (16°C)
Di-*n*-propyl ether	$CH_3CH_2CH_2—O—CH_2CH_2CH_3$	−122	91	slightly soluble
Methyl phenyl ether	$CH_3—O—C_6H_5$	−38	155	insoluble
Diphenyl ether	$C_6H_5—O—C_6H_5$	28	259	insoluble

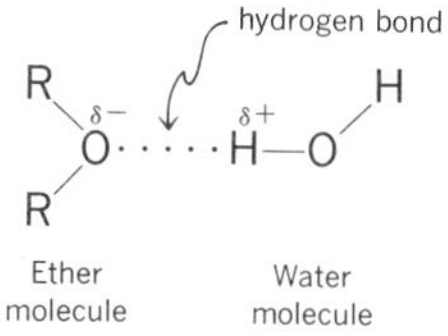

hydrogen bond

R—$\overset{\delta-}{O}$(—R) · · · · · $\overset{\delta+}{H}$—O—H

Ether molecule   Water molecule

Figure 13.11 The oxygen of an ether molecule can serve as a hydrogen bond acceptor, which makes ethers somewhat more soluble in water than alkanes.

quite stable. At room temperature strong aqueous bases do not react with ethers. Concentrated, strong acids, for example, hydrobromic and hydriodic acid, will cleave ethers:

$$R{-}O{-}R + 2H{-}Br \xrightarrow[\text{heat}]{} 2R{-}Br + H_2O$$

Such conditions are obviously not found in living organisms. Only a special kind of ether (an acetal) is easily cleaved, a fact essential to understanding the digestion of sugars (cf. p. 451).

HAZARDS

Most liquid, aliphatic ethers (R—O—R), when exposed to air, react slowly to form unstable peroxides which even in small concentrations can be quite dangerous. They can explode violently when exposed to air, setting fire to the ether. The presence of peroxides can be checked easily by shaking the ether with an aqueous solution of iron(II) ammonium sulfate plus potassium thiocyanate. If peroxides are present, a red color is produced.

Anesthetic ether, since it has such a low boiling point, readily volatilizes in an open container. This mixture of air and ether vapor is highly explosive. When diethyl ether is used as an anesthetic, precautions must be taken to ensure that no stray spark will set off the reaction.

IMPORTANT INDIVIDUAL ETHERS

Diethyl Ether ("Ether"). Diethyl ether, a colorless, volatile liquid with a pungent, somewhat irritating odor, is a depressant for the central nervous system and is at the same time somewhat of a stimulant for the sympathetic system. It exerts an effect on nearly all tissues of the body, but it is still one of the safest anesthetics for human use.

Divinyl Ether ("Vinethene"). This ether, another anesthetic, is more rapid in its action than diethyl ether. It too forms an explosive mixture with air.

MERCAPTANS AND RELATED COMPOUNDS

STRUCTURE AND NOMENCLATURE. OCCURRENCE

Mercaptans, sulfur analogs of alcohols, are sometimes called thioalcohols.[4] Table 13.7 lists a few representative examples together with names and certain

[4] Mercaptan is derived from the Latin term *mercurium captans,* meaning "seizing mercury." Mercuric ions react and form a precipitate (a mercuric salt) with mercaptans.

Table 13.7 Some Mercaptans

Name	Structure	Mp (°C)	Bp (°C)	Solubility in Water
Methyl mercaptan	CH_3SH	−123	6	soluble
Ethyl mercaptan	CH_3CH_2SH	−144	37	slightly soluble
n-Propyl mercaptan	$CH_3CH_2CH_2SH$	−113	67	slightly soluble
Isopropyl mercaptan	$(CH_3)_2CHSH$	−131	58	slightly soluble
n-Butyl mercaptan	$CH_3CH_2CH_2CH_2SH$	−116	98	slightly soluble
Cysteine (an amino acid)	$HSCH_2CH(NH_2)CO_2H$	258 (decomposes)	—	moderately soluble

physical properties. The —SH group that characterizes mercaptans is found among many of the protein molecules in the chemical inventories of living cells. The low-formula-weight mercaptans, which are therefore the more volatile, possess some of the most disagreeable odors of all organic materials. *n*-Butyl mercaptan is the principle constituent of the well-known fluid that is the defensive weapon of the skunk.

OXIDATION OF MERCAPTANS TO DISULFIDES

Mercaptans are reducing agents easily oxidized by hydrogen peroxide or sodium hypochlorite to disulfides:

$$2R\text{—}S\text{—}H + H_2O_2 \longrightarrow R\text{—}S\text{—}S\text{—}R + H_2O$$

Disulfides are easily reduced to mercaptans:

$$R\text{—}S\text{—}S\text{—}R + (2H) \longrightarrow 2R\text{—}S\text{—}H$$

These two reactions, interconversions of mercaptan and disulfide groups, are quite important in protein chemistry.

AMINES

STRUCTURE AND NOMENCLATURE

Amines are alkyl or aryl derivatives of ammonia. One, two, or all three of the hydrogens on a molecule of ammonia may be replaced by such groups. For example,

CH_3NH_2	$(CH_3)_2NH$	$(CH_3)_3N$	$CH_3NHCH_2CH_3$
Methylamine	Dimethylamine	Trimethylamine	Methylethylamine

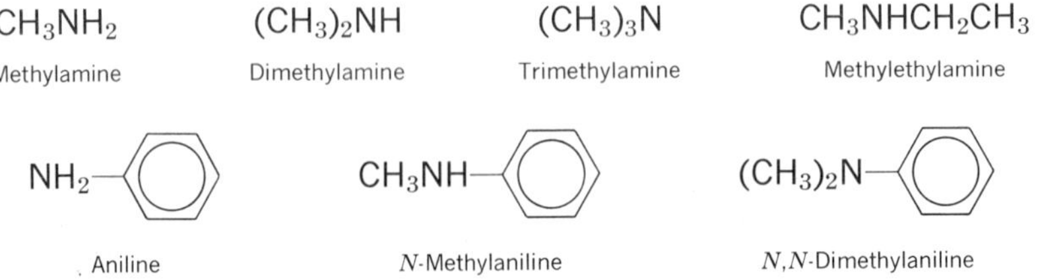
NH_2
CH_3NH
$(CH_3)_2N$

Aniline *N*-Methylaniline *N*,*N*-Dimethylaniline

Aliphatic amines are those in which the carbon(s) attached directly to the nitrogen has (or have) only single bonds to other groups. Thus the compound $\mathrm{R{-}\overset{O}{\overset{\|}{C}}{-}NH_2}$ is not an amine, but the compound $\mathrm{R{-}\overset{O}{\overset{\|}{C}}{-}CH_2{-}NH_2}$ has an amine function. If even one of the groups attached directly to the nitrogen is aromatic, the amine is classified as an *aromatic amine*. This subclassification is useful because aromatic amines have benzene rings highly reactive toward aromatic substitution reactions and because they are much weaker bases than aliphatic amines. In other words, the differences between the two are enough to make two subclasses convenient. They also share several types of reactions.

Another way of subclassifying amines is by the designations 1° (primary), 2° (secondary), and 3° (tertiary) amines. The numerals denote the number of groups attached to the *nitrogen:*

$\mathrm{R{-}\overset{H}{\overset{\vert}{N}}{-}H}$	$\mathrm{R{-}\overset{H}{\overset{\vert}{N}}{-}R'}$	$\mathrm{R{-}\overset{R''}{\overset{\vert}{N}}{-}R'}$	$\mathrm{R{-}\underset{R'''}{\underset{\vert}{\overset{R''}{\overset{\vert}{N^+}}}}{-}R}$
Primary amines 1°	Secondary amines 2°	Tertiary amines 3°	Quaternary ammonium ions 4°

Because of structural similarity to amines, tetraalkyl derivatives of the ammonium ion, NH_4^+, called *quaternary ammonium ions,* R_4N^+, are also included here.

Heterocyclic Amines. Many compounds of biochemical importance consist, at least in part, of ring systems in which one of the ring atoms is a nitrogen. Some of the important heterocyclic amines are

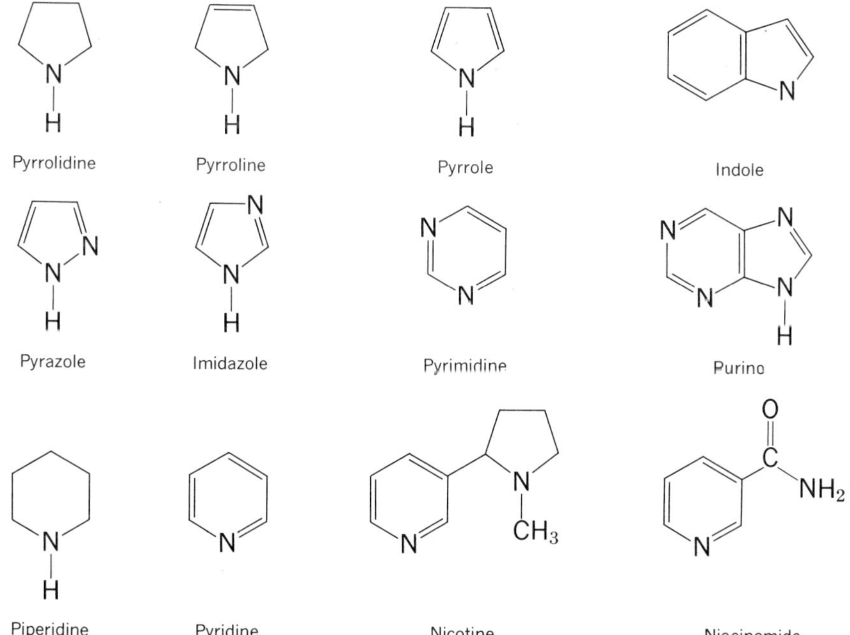
N
H
Pyrrolidine
Pyrroline
Pyrrole
Indole
Pyrazole
Imidazole
Pyrimidine
Purine
Piperidine
Pyridine
Nicotine
CH3
Niacinamide
O
C
NH2

The pyrimidine and purine systems are particularly important in biochemical genetics, the chemistry of heredity, but little more need be said about these heterocyclic amines at this point. It can be noted, however, that the saturated systems, pyrrolidine and piperidine, behave like ordinary 2° amines. Pyridine is a weakly basic 3° amine with a benzene-like ring. In fact, it has many aromatic properties.

Nomenclature. In the *common system* of nomenclature, aliphatic amines are named by designating the alkyl groups attached to the nitrogen and following this series by the word part "-amine."

$$(CH_3)_2CHCH_2{-}NH_2$$

Isobutylamine

$$(CH_3)_2CH{-}NH{-}CH_2CH_3$$

Ethylisopropylamine

$$CH_3CH_2CH_2{-}\underset{}{\overset{\displaystyle CH_3\\ |}{N}}{-}CH_2CH_2CH_2CH_3$$

Methyl-*n*-propyl-*n*-butylamine

Where this system breaks down, the IUPAC system can be employed. In this system the $-NH_2$ group is named the *amino* group, or *N*-alkyl or *N*,*N*-dialkylamino, and the specific groups are of course named.

$$CH_3{-}\underset{\displaystyle |\\ NH_2}{CH}{-}CH_2{-}OH$$

2-Amino-1-propanol

$$CH_3CH_2NH{-}CH_2CH_2CH_2CH_2CH_2CH_2CH_2CH_3$$

1-(*N*-Ethylamino)octane (That is, at the 1-position in octane an *N*-ethylamino group is attached.)

The capital *N* designation means that the group immediately following it is attached to nitrogen. (See also the earlier examples of *N*-methylaniline and *N*,*N*-dimethylaniline.)

The simplest aromatic amine is always called *aniline*.

Exercise 13.2 Write structures for the following compounds.

(*a*) triethylamine
(*b*) dimethylethylamine
(*c*) *t*-butyl-*sec*-butyl-*n*-butylamine
(*d*) *p*-nitroaniline
(*e*) *p*-aminophenol
(*f*) 1,3-diamino-2-pentanol
(*g*) cyclohexylamine
(*h*) diphenylamine
(*i*) 2-(*N*-methyl-*N*-ethylamino)butane
(*j*) benzylmethylamine

Exercise 13.3 Classify the amines of the preceding exercise according to (*a*) aromatic versus aliphatic subclasses, (*b*) 1°, 2°, or 3° subclasses.

Exercise 13.4 Write names for each of the following according to the common system whenever possible, otherwise according to the IUPAC system.

(*a*) $CH_3CH_2N(CH_2CH_2CH_2CH_3)_2$
(*b*) $NH_2CH_2CH_2NH_2$
(*c*) $C_6H_5{-}N(CH_2CH_3)_2$
(*d*) $C_5H_9{-}NH_2$ (cyclopentane ring bearing NH_2)
(*e*) $(CH_3)_4N^+Br^-$

PHYSICAL PROPERTIES

In Table 13.8 several representative amines are listed together with important physical properties. Molecules of amines are moderately polar. Hydrogens on nitrogen in amines can form hydrogen bonds to nitrogens on neighboring amines or to oxygens on neighboring molecules of water if that is present. As a result, when compounds of similar formula weights are compared, amines have higher boiling points than alkanes but lower boiling points than alcohols.

	F.Wt.	Bp (°C)
CH_3CH_3	30	−89
CH_3NH_2	31	− 6
CH_3OH	32	65

Like the —OH group in alcohols, the $—NH_2$ group in amines helps make them soluble in water (Figure 13.12). Amines are also soluble in less polar solvents.

The lower-formula-weight amines (e.g., methylamine, ethylamine) have odors very much like that of ammonia. At higher formula weights the odors become "fishy." Aromatic amines have moderately pleasant, pungent odors, but they are at the same time very toxic. They are absorbed directly through the skin with sometimes fatal results. Aromatic amines are also easily oxidized by atmospheric oxygen. Dark-colored substances form during this process, and although highly purified aromatic amines are usually colorless, after standing for a time on the shelf they are discolored or sometimes black.

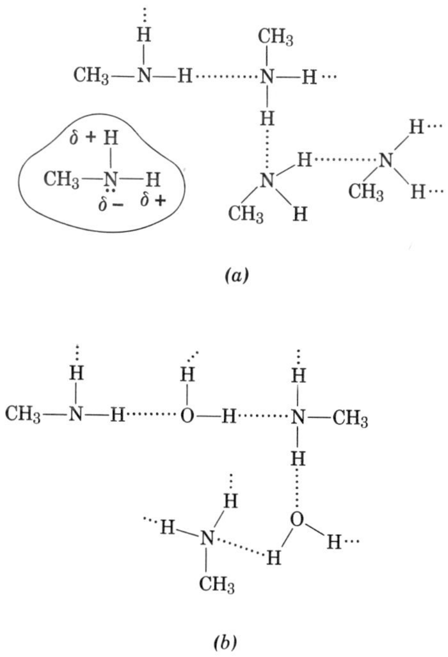

(a)

(b)

Figure 13.12 Amines as hydrogen bond donors and acceptors. (*a*) Hydrogen bonds (dotted lines) can exist between molecules of an amine. (*b*) Low-formula-weight amines are soluble in water because they can slip into the hydrogen-bonding network of water.

Table 13.8 Amines

Name	Structure	Mp (°C)	Bp (°C)	Solubility in H_2O	K_b (at 25°C)
Ammonia	NH_3	−78	−33	very soluble	1.8×10^{-5}
Methylamine	CH_3NH_2	−94	−6	very soluble	4.4×10^{-4}
Dimethylamine	$(CH_3)_2NH$	−96	7	very soluble	5.2×10^{-4}
Trimethylamine	$(CH_3)_3N$	−117	3.5	very soluble	0.5×10^{-4}
Ethylamine	$CH_3CH_2NH_2$	−84	17	very soluble	5.6×10^{-4}
Diethylamine	$(CH_3CH_2)_2NH$	−48	56	very soluble	9.6×10^{-4}
Triethylamine	$(CH_3CH_2)_3N$	−115	90	14 g/100 cc	5.7×10^{-4}
n-Propylamine	$CH_3CH_2CH_2NH_2$	−83	49	very soluble	4.7×10^{-4}
Di-*n*-propylamine	$(C_3H_7)_2NH$	−40	110	very soluble	9.5×10^{-4}
Tri-*n*-propylamine	$(C_3H_7)_3N$	−90	156	slightly soluble	4.4×10^{-4}
Isopropylamine	$(CH_3)_2CHNH_2$	−101	33	very soluble	5.3×10^{-4}
n-Butylamine	$CH_3CH_2CH_2CH_2NH_2$	−51	78	very soluble	4.1×10^{-4}
Isobutylamine	$(CH_3)_2CHCH_2NH_2$	−86	68	very soluble	3.1×10^{-4}
sec-Butylamine	$CH_3CH_2CH(CH_3)NH_2$	<-72	63	very soluble	3.6×10^{-4}
t-Butylamine	$(CH_3)_3CNH_2$	−68	45	very soluble	2.8×10^{-4}
Ethylenediamine	$NH_2CH_2CH_2NH_2$	9	117	very soluble	8.5×10^{-4}
Tetramethylammonium hydroxide	$(CH_3)_4N^+OH^-$	130–135 (decomposes)	—	very soluble	very strong base
Aromatic Amines					
Aniline	C_6H_5–NH_2	−6	184	3.7 g/100 cc	3.8×10^{-10}
N-Methylaniline	C_6H_5–$NHCH_3$	−57	196	slightly soluble	5×10^{-10}
N,*N*-Dimethylaniline	C_6H_5–$N(CH_3)_2$	3	194	slightly soluble	11.5×10^{-10}
Sulfanilamide	NH_2–C_6H_4–SO_2NH_2	163	—	0.4 g/100 cc	

BASICITY OF AMINES

Ammonia is a stronger base than water, for it holds a proton better than a water molecule can.

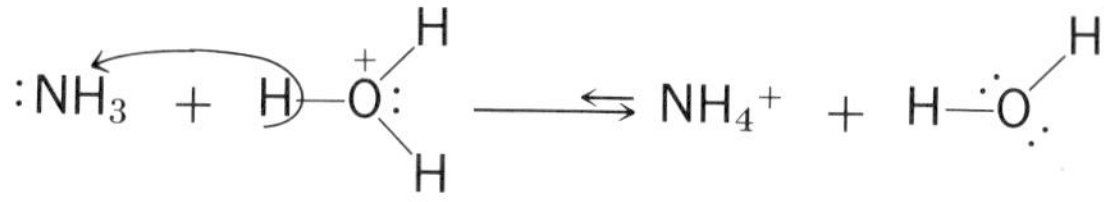

$$:NH_3 + H\!-\!\overset{+}{O}H_2 \rightleftharpoons NH_4^+ + H\!-\!\ddot{O}\!-\!H$$

Ammonia (stronger base) | Hydronium ion (stronger acid) | Ammonium ion (weaker acid) | Water (weaker base)

The ammonium ion, as it occurs in ammonium salts, will give up its extra proton to hydroxide ion. Hence ammonia is not as strong a base as the hydroxide ion.

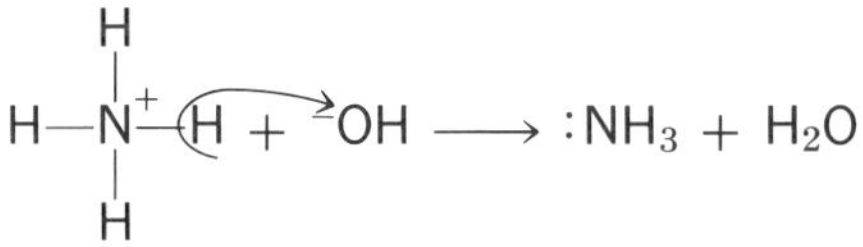

$$H_3\overset{+}{N}\!-\!H + {}^{-}OH \longrightarrow :NH_3 + H_2O$$

Stronger acid | Stronger base | Weaker base | Weaker acid

Much of the chemistry of amines can be understood by comparing them to ammonia and to ammonium ions. Amines are simply close relatives of ammonia.

$$R\!-\!\ddot{N}H_2 + H_3O^+ \rightleftharpoons R\!-\!NH_3^+ + H_2O$$

1° amine Stronger base | Stronger acid | Weaker acid | Weaker base

$$R_2\ddot{N}H + H_3O^+ \rightleftharpoons R_2NH_2^+ + H_2O$$

2° amine

$$R_3N: + H_3O^+ \rightleftharpoons R_3NH^+ + H_2O$$

3° amine

$$R_4N^+ + H_3O^+ \not\longrightarrow \text{no reaction}$$

4° ammonium ion

Since the 4° ammonium ion has no unshared pair of electrons, and since like the hydronium ion it is positively charged, no reaction can occur with acid. *The basicity of amines and of ammonia is correlated with the unshared pair of electrons on nitrogen.* Any structural factor in an amine that makes this pair more "available" to a proton than it is in ammonia will make the amine a better proton acceptor, that is, a better or stronger base than ammonia. Any structural factor in an amine that makes this unshared pair less available than it is in ammonia will make the amine a weaker proton acceptor or a weaker base.

$$G \rightarrow \ddot{N}H_2$$

If G is electron-donating, the amine is a stronger base than ammonia.

$$G \leftarrow \ddot{N}H_2$$

If G is electron-withdrawing, the amine is a weaker base than ammonia.

Relative basicities of amines can be seen by comparing the basicity constants for their reaction with water.

$$R\ddot{N}H_2 + H_2O \rightleftharpoons RNH_3^+ + OH^-$$

$$K_b = \frac{[RNH_3^+][OH^-]}{[RNH_2]}$$

The higher the K_b, the stronger the basicity of the amine. The K_b values for several amines are listed in Table 13.8, and it is apparent that the aliphatic amines for the most part are *stronger* bases than ammonia.

Basicity and Structure. The difference, structurally, between ammonia and an aliphatic amine such as methylamine is, of course, the presence of the methyl group. We have learned that alkyl groups have an electron-releasing effect. Being directed toward nitrogen in methylamine, the unshared pair of electrons on nitrogen is made more available for sharing with a proton and accepting it. Once the proton is accepted, the same electron-releasing effect of the methyl group pushes electrons toward a positively charged site, stabilizing it.

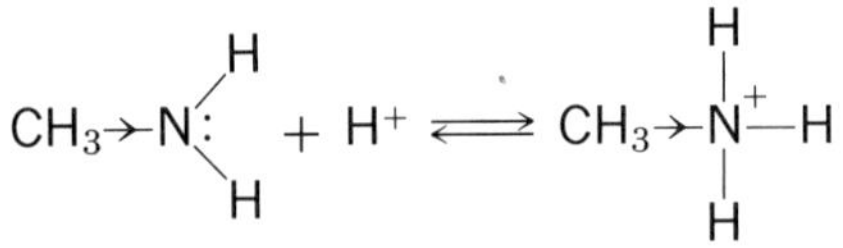

$$CH_3 \rightarrow \ddot{N}H_2 + H^+ \rightleftharpoons CH_3 \rightarrow \overset{+}{N}H_3$$

Methyl releases electron density, making the unshared pair of electrons more available. Hence methylamine is a stronger base than ammonia.

Methyl releases electron density, tending to stabilize the positive charge. Hence methylamine is a stronger base than ammonia.

According to the K_b values in Table 13.8, aromatic amines are weaker bases than ammonia or aliphatic amines by factors of about one million (10^{-4} versus 10^{-10}). Apparently the unshared pair on nitrogen in an aniline is much less available for accepting a proton than that in an aliphatic amine. Resonance theory helps to make this decreased availability understandable. The principal contributing resonance forms for aniline are structures **A** through **E.** Of these, **B, C,** and **D**

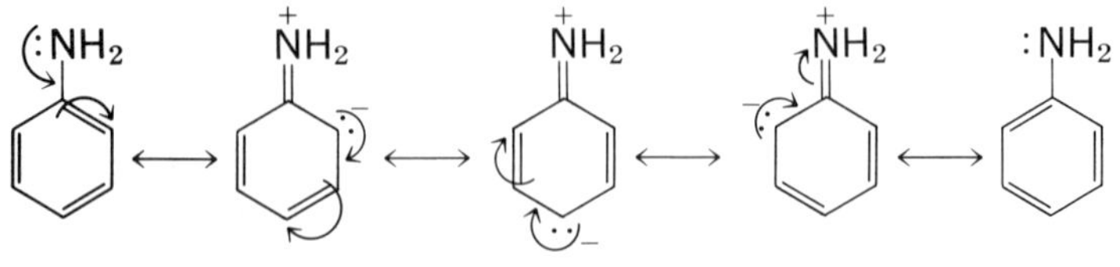

:NH₂ ⟷ $\overset{+}{N}H_2$ ⟷ $\overset{+}{N}H_2$ ⟷ $\overset{+}{N}H_2$ ⟷ :NH₂

A **B** **C** **D** **E**

involve both a separation of unlike charges (energy-demanding) and a disruption of the alternating double-bond–single-bond network of the benzene ring (also energy-demanding). We therefore conclude that these three cannot contribute significantly to the true state of the aniline molecule. Forms **A** and **E** must make the principal contributions. However, *to the extent that* **B, C,** *and* **D** *make any contribution,* the unshared pair of electrons on the nitrogen is *at least somewhat delocalized into the ring* The pair is to that extent less available for accepting a proton, and aniline is to that extent a weaker base than ammonia or aliphatic amines.

AMINE SALTS

Table 13.9 contains a list of a few representative salts of amines. These are salts in the true sense, that is, they are typical ionic compounds. All are solids, all have relatively high melting points, especially compared with the melting points of their corresponding amines, and all are like ammonium salts generally soluble in water and insoluble in nonpolar compounds such as carbon tetrachloride.

In amine salts the positive ion can be called a *protonated amine;* such positively charged systems are of considerable importance in protein chemistry as proton donors. Hence they can act in cells to neutralize bases and help control the pH of body fluids. They also help make the substance soluble in water.

CHEMICAL PROPERTIES OF AMINES

Several of the reactions of amines with organic compounds will be studied in later sections on these materials. Anticipating these future discussions, we call attention to the most important way of regarding an amine molecule. The unshared pair of electrons makes it not only a basic substance (that is, a proton-seeking molecule) but also a nucleophilic substance. Whenever an organic molecule contains a partial positive charge on carbon, it must be regarded as a likely site for attack by an amine.

Reaction of Amines with Alkyl Halides. This reaction is an example of an amine attacking an organic molecule with a partial positive charge on carbon. In general:

$$\underset{1^\circ \text{ amine}}{R{-}\ddot{N}H_2} + R'{-}X \longrightarrow R{-}\overset{+}{N}H_2{-}R' + X^-$$

$$R{-}\overset{+}{N}H_2{-}R' \xrightarrow{OH^-} \underset{2^\circ \text{ amine}}{R{-}\ddot{N}H{-}R'} + H_2O$$

Table 13.9 Amine Salts

Name	Structure	Mp (°C)
Methylammonium chloride	$CH_3NH_3^+Cl^-$	232
Dimethylammonium chloride	$(CH_3)_2NH_2^+Cl^-$	171
Dimethylammonium bromide	$(CH_3)_2NH_2^+Br^-$	134
Dimethylammonium iodide	$(CH_3)_2NH_2^+I^-$	155
Tetramethylammonium hydroxide (base as strong as KOH)	$(CH_3)_4N^+OH^-$	130–135 (decomposes)

Ordinarily, sodium hydroxide is added after the initial reaction is over to liberate the amine.

$$\underset{\text{2° amine}}{R-\ddot{N}(R')-H} + R''-X \xrightarrow[\text{hydroxide ion}]{\text{followed by}} \underset{\text{3° amine}}{R-\ddot{N}(R')-R''}$$

$$\underset{\text{3° amine}}{R-\ddot{N}(R')-R''} + R^*-X \longrightarrow \underset{\text{4° ammonium salt}}{R-\overset{+}{N}(R^*)(R')-R''\ \ X^-}$$

When a 1° amine is allowed to react with an alkyl halide, a mixture consisting principally of 2° and 3° amines plus smaller amounts of 4° ammonium salt and unchanged 1° amine results. The relative amounts of these products can be controlled by adjusting the proportions of reactants. A large excess of 1° amine over alkyl halide, for example, will favor the 2° amine with only traces of higher alkylated amines. (Why?)

Specific examples:

$$C_6H_5-NH_2 + Cl-CH_2-C_6H_5 \longrightarrow C_6H_5-NH-CH_2-C_6H_5$$

Aniline 4 parts : Benzyl chloride 1 part → Phenylbenzylamine (96% yield)

$$(CH_3CH_2)_2NH + Br-CH(CH_3)_2 \longrightarrow (CH_3CH_2)_2NCH(CH_3)_2$$

Diethylamine 1.3 parts : Isopropyl bromide 1 part → Diethylisopropylamine (60%)

Mechanism: In the alkyl halide the bond is polarized as shown. The carbon holding the halogen has a partial positive charge, a likely site for an electron-rich reagent (the amine) to attack. The attack is therefore usually of the S_N2 type:

$$R-\ddot{N}H_2 + \overset{\delta+}{C}H_2(R')-\overset{\delta-}{X} \longrightarrow R-\overset{+}{N}H_2-CH_2-R' + X^- \xrightarrow[\text{(added later)}]{^-OH} R-\ddot{N}H-CH_2-R' + H-OH$$

IMPORTANT INDIVIDUAL AMINES

Invert Soaps. $CH_3(CH_2)_n\overset{+}{N}(CH_3)_3Cl^-$. Quaternary ammonium salts in which one of the alkyl groups is long (e.g., $n = 15$) have detergent and germicidal properties. In ordinary soaps the detergent action is associated with a negatively charged organic ion. In an invert soap a positive ion has the detergent property. (Detergent action is discussed on page 464.)

Aniline. $C_6H_5NH_2$. The annual production of aniline in the United States is well over 100,000 tons. Virtually all of it is used in the manufacture of aniline dyes, pharmaceuticals, and chemicals for the plastics industry. In one industrial synthesis benzene is first converted to nitrobenzene, which is in turn reduced by the action of iron and hydrochloric acid.

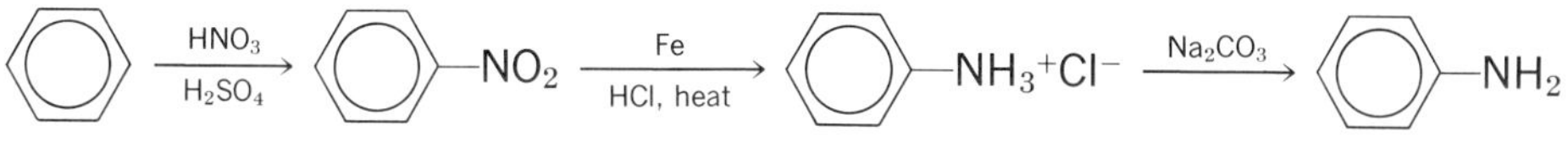

$$C_6H_6 \xrightarrow[H_2SO_4]{HNO_3} C_6H_5\text{—}NO_2 \xrightarrow[\text{HCl, heat}]{Fe} C_6H_5\text{—}NH_3^+Cl^- \xrightarrow{Na_2CO_3} C_6H_5\text{—}NH_2$$

Benzene — Nitrobenzene — Anilinium chloride — Aniline

REFERENCES AND ANNOTATED READING LIST

R. T. Morrison and R. N. Boyd. *Organic Chemistry*, second edition. Allyn and Bacon, Boston, 1966. The following chapters in this standard first-year text for organic chemistry make a helpful supplement going beyond the material we have covered but not in a way that is difficult to handle: Chapter 3, "Stereochemistry I"; Chapter 7, "Stereochemistry II"; Chapter 14, "Alkyl Halides. Nucleophilic Aliphatic Substitution."

F. G. Bordwell. *Organic Chemistry*. The Macmillan Company, New York, 1963. Chapter 17, "Stereoisomerism," includes some interesting information about the discovery of optical activity.

In addition to these sources, any standard textbook on organic chemistry will have one or more chapters devoted to optical isomerism.

PROBLEMS AND EXERCISES

1. Write definitions for each of the following terms and when appropriate provide suitable illustrations.

   (*a*) plane of symmetry
   (*b*) optical isomers
   (*c*) enantiomers
   (*d*) superimposability
   (*e*) asymmetric carbon
   (*f*) optical activity
   (*g*) polarimeter
   (*h*) dextrorotatory
   (*i*) levorotatory
   (*j*) optical rotation (and symbol)
   (*k*) specific rotation
   (*l*) meso compound
   (*m*) diastereomers
   (*n*) racemic mixture
   (*o*) inversion of configuration
   (*p*) Williamson reaction
   (*q*) 2° amine
   (*r*) heterocyclic amine
   (*s*) basicity constant

2. Examine the following structures and predict whether optical isomerism is a possibility.

   (*a*) $CH_3C(CH_3)(OH)CH_2CH_3$ (i.e., $CH_3\overset{CH_3}{\underset{OH}{C}}CH_2CH_3$)

   (*b*) $CH_3CH(CH_3)CH(OH)CH_3$ (i.e., $CH_3\overset{CH_3}{CH}\underset{OH}{CH}CH_3$)

   (*c*) $HOC(=O)CH(Br)CH_3$ (i.e., $HO\overset{O}{\overset{\|}{C}}\underset{Br}{CH}CH_3$)

(*d*) $HO\overset{O}{\overset{\|}{C}}CH_2\underset{Br}{\underset{|}{C}}H_2$ (*e*) *cis*-1,2-cyclopentanediol (*f*) *trans*-1,2-cyclopentanediol

3. Explain why enantiomers *should* have identical physical properties (except for the sign of rotation of plane-polarized light).
4. Explain why diastereomers *should not* be expected to have identical physical properties (except coincidentally).
5. A solution of sucrose in water at 25°C in a tube 10 cm long gives an observed rotation of +2.0°. The specific rotation of sucrose in this solvent at this temperature is +66.4°. What is the concentration of the sucrose solution in grams per 100 cc?
6. Complete the following equations by supplying the structures of the principal organic products. Do not balance the equations. If no reaction is to be expected, write "No reaction."

   (*a*) $CH_3CH_2OH + SOCl_2 \longrightarrow$

   (*b*) $CH_3Br + NaCN \longrightarrow$

   (*c*) $2CH_3CH_2SH + H_2O_2 \longrightarrow$

   (*d*) $CH_3CH_2OCH_2CH_3 + Na \longrightarrow$

   (*e*) $CH_3CH_2CH_2OH + PBr_3 \longrightarrow$

   (*f*) $(CH_3)_2CHBr + NH_3 \text{ (excess)} \xrightarrow[\text{[followed by NaOH}(aq)]]{}$

   (*g*) $(CH_3)_3CCl + Na^{+-}OCH_3 \xrightarrow[HOCH_3]{}$

   (*h*) $CH_3CH_2CH_2CH_2Cl + Na^{+-}OCH_3 \xrightarrow[HOCH_3]{}$

   (*i*) $(CH_3)_2CHCH_2—S—S—CH_2CH(CH_3)_2 + (2H) \longrightarrow$

   (*j*) $CH_3NH_2 + HCl(aq) \xrightarrow[\text{room temperature}]{}$

   (*k*) $CH_3CH_2NH_3^+Cl^- + NaOH(aq) \longrightarrow$

   (*l*) $CH_3NH_2 + C_6H_5CH_2Cl \xrightarrow[\text{[followed by NaOH}(aq)]]{}$

7. The number of one-step organic reactions is growing rather rapidly during the course of our study. If they are to be of use to us in the biochemistry section, some way of organizing them is needed. Many students have found it helpful to make cards. For example, on one side of a 5 × 8-in. card the following question might appear: "What are all the ways we have learned for introducing the hydroxyl group into an organic molecule in one step from anything?" The reverse side of the card might look like this:

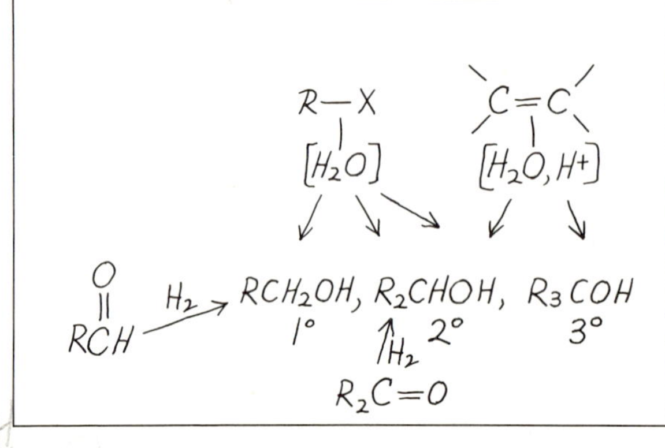
R—X
[H₂O]
C=C
[H₂O, H+]
O
RCH
H₂
RCH₂OH, R₂CHOH, R₃COH
1°
2°
3°
H₂
R₂C=O

Prepare similar cards for the following functional groups: carbon-carbon double bond, aldehyde, ketone, carboxylic acid, aliphatic amine, ether, and alkyl halide. Keep these cards current as additional methods are discussed in later chapters.

8. Students have found collecting the one-step reactions of each functional group useful. Thus on one side of a 5 × 8-in. card might appear the question "What reactions that alcohols undergo have we studied?" On the reverse side might appear something like this:

R—X
concd HX, $SOCl_2$
PBr_3, KI, H_3PO_4
RO^-Na^+
Na
H_2SO_4 Δ
C=C
ROH
H_2SO_4 Δ
R—O—R
(O)
aldehyde, or acid, or ketone

Prepare similar cards for the following functional groups: carbon-carbon double bond, aldehyde, ketone, carboxylic acid, aliphatic amines (all subclasses), aromatic amines, alcohols (all subclasses), alkyl halides, alkanes, and benzene. Keep these cards current as new reactions are described in later chapters.

9. Still another useful way of studying organic reactions is to look at them by reagent. Thus on one side of a 5 × 8-in. card might appear the question "What are the functional groups that react with water with (*a*) no added catalyst, (*b*) acid catalyst, (*c*) base catalyst?" Go back through all the functional groups we have studied and name those that answer the question. Write a typical reaction to illustrate each.

Prepare similar cards for the following reagents: water; water plus acid catalyst; water plus base catalyst; aqueous acid [e.g., HCl(*aq*) as in the reaction with an amine]; aqueous base (e.g., NaOH, room temperature); oxidizing agents; reducing agents; sodium; ammonia; concentrated sulfuric acid (and heat); concentrated HX (and heat); hydrogen; chlorine; and bromine. Keep these cards current, making new ones as new reagents are described, throughout the rest of our study of organic chemistry.

10. Write structures for each of the following compounds.

(*a*) diisopropyl ether	(*b*) *n*-butyl mercaptan
(*c*) *t*-butyl bromide	(*d*) *p*-nitroaniline
(*e*) methylethyl-*n*-propylamine	(*f*) ethyl phenyl ether
(*g*) tetramethylammonium iodide	(*h*) methyl isobutyl ether
(*i*) an optically active ether of formula $C_5H_{12}O$	(*j*) ethylammonium chloride

CHAPTER FOURTEEN

Aldehydes and Ketones

THE CARBONYL GROUP

The carbonyl group consists of a carbon-to-oxygen double bond. The oxygen atom of this group, being more electronegative than carbon, carries a partial negative charge, which leaves the carbonyl carbon with a partial positive charge. These two simple considerations will be of great help to us in understanding several of the reactions of this group.

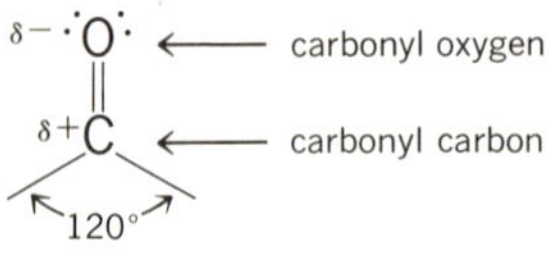

$\delta-$ O ← carbonyl oxygen

$\delta+$ C ← carbonyl carbon

120°

Carbonyl group

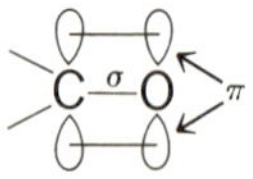

C —σ— O π

The two carbon-oxygen bonds in the carbonyl group consist of one σ-bond and one π-bond.

FAMILIES OF ORGANIC COMPOUNDS WITH CARBONYL GROUPS

Several important families of organic compounds consist of molecules containing the carbonyl group, each family being characterized by the nature of the two groups attached to the carbonyl carbon. The essential characteristics of these families, outlined in Table 14.1 and discussed in the following paragraphs,[1] should be memorized.

Aldehydes. (H)R—C(=O)—H. To be classified as an aldehyde, the molecules of the substance must have a hydrogen attached to the carbonyl carbon. The other group

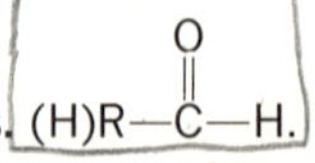

[1] In the examples given, R— is not limited just to alkyl groups. It may also be aryl, or any other group in which the bond to the carbonyl group is from a *carbon*, saturated or otherwise.

must involve a bond from the carbonyl carbon to another *carbon* (or a hydrogen in formaldehyde, $\mathrm{H{-}\overset{\overset{O}{\|}}{C}{-}H}$). The *aldehyde group,* then, is $\mathrm{{-}\overset{\overset{O}{\|}}{C}{-}H}$. (Specific examples are listed in Table 14.3, page 390.)

Ketones. $\mathrm{R{-}\overset{\overset{O}{\|}}{C}{-}R'}$. The molecules in ketones must have a carbonyl group flanked on both sides of the carbonyl carbon by bonds to other *carbons.* (Specific examples are listed in Table 14.4, page 391.) When it occurs in ketones, the carbonyl group is frequently called the *keto group.*

Carboxylic Acids. $\mathrm{(H)R{-}\overset{\overset{O}{\|}}{C}{-}OH}$. The molecules in substances classified as carboxylic acids must have a hydroxyl group attached to the carbonyl carbon. The

Table 14.1 The Carbonyl Group in Families of Organic Compounds

Family Name	Generic Family Structure*
Aldehydes	$\mathrm{(H)R{-}\overset{\overset{O}{\|}}{C}{-}H}$
Ketones	$\mathrm{R{-}\overset{\overset{O}{\|}}{C}{-}R'}$
Carboxylic acids	$\mathrm{(H)R{-}\overset{\overset{O}{\|}}{C}{-}OH}$
Derivatives of carboxylic acids	
Acid chlorides	$\mathrm{R{-}\overset{\overset{O}{\|}}{C}{-}Cl}$
Anhydrides	$\mathrm{R{-}\overset{\overset{O}{\|}}{C}{-}O{-}\overset{\overset{O}{\|}}{C}{-}R'}$
Esters	$\mathrm{(H)R{-}\overset{\overset{O}{\|}}{C}{-}O{-}R'}$
Amides	$\mathrm{(H)R{-}\overset{\overset{O}{\|}}{C}{-}NH_2}$ (simple amides)
	$\mathrm{(H)R{-}\overset{\overset{O}{\|}}{C}{-}NH{-}R'}$ (*N*-alkyl amides)
	$\mathrm{(H)R{-}\overset{\overset{O}{\|}}{C}{-}\underset{\underset{R''}{\|}}{N}{-}R'}$ (*N,N*-dialkyl amides)

* When the symbol for hydrogen is placed in parentheses before (or after) R [e.g., (H)R— or —R(H)], the specific R— group may be either an alkyl group or hydrogen. In these compounds, R— may also be aryl. When R— is primed (e.g., R′— or R″—), the two or more R— groups in a given symbol may be different.

other bond from the carbonyl carbon must be to another *carbon* (or to hydrogen in formic acid, $H{-}\overset{\overset{\large O}{\|}}{C}{-}OH$). The group, $-\overset{\overset{\large O}{\|}}{C}{-}OH$, is called the *carboxylic acid group* or *carboxyl group.* It is frequently condensed to —COOH or to $-CO_2H$, but when this is done the reader must remember the presence of the carbon-oxygen double bond and the hydroxyl group. Although they possess hydroxyl groups, carboxylic acids are definitely not alcohols. The carbonyl group so changes the properties of the —OH attached to it (and vice versa) that it is more convenient to designate a new class. (Specific examples of carboxylic acids are listed in Table 15.1, page 406.)

Carboxylic Acid Chlorides. Acid Chlorides. $R{-}\overset{\overset{\large O}{\|}}{C}{-}Cl$. These molecules have carbonyl-to-chlorine bonds, and the group, $-\overset{\overset{\large O}{\|}}{C}{-}Cl$, is called the *acid chloride function.* (Specific examples are given in Table 15.5, page 420.)

Carboxylic Acid Anhydrides. Anhydrides. $R{-}\overset{\overset{\large O}{\|}}{C}{-}O{-}\overset{\overset{\large O}{\|}}{C}{-}R$. In these molecules an oxygen atom is flanked on both sides by carbonyl groups. They are called anhydrides ("not hydrated") because they can be considered to result from the loss of water between two molecules of a carboxylic acid:

$$R{-}\overset{\overset{\large O}{\|}}{C}{-}O{-}H + H{-}O{-}\overset{\overset{\large O}{\|}}{C}{-}R \longrightarrow R{-}\overset{\overset{\large O}{\|}}{C}{-}O{-}\overset{\overset{\large O}{\|}}{C}{-}R + H{-}OH$$

In both acid chlorides and anhydrides the R— groups may not be replaced by hydrogens. Such substances, for example, $H{-}\overset{\overset{\large O}{\|}}{C}{-}Cl$, are not known, for they are too unstable to exist. (A few examples of anhydrides are given in Table 15.5, page 420.)

Carboxylic Acid Esters. Esters. $(H)R{-}\overset{\overset{\large O}{\|}}{C}{-}O{-}R'$ or $(H)RCO_2R'$. The *ester group* has the features shown, and the bond drawn with a heavier line is called the *ester linkage.* To be an ester the molecule must have a carbonyl-oxygen-carbon network. (Several are listed in Tables 15.6 and 15.7, pages 424 and 425.)

$$-\overset{\overset{\large O}{\|}}{C}\boldsymbol{-}O{-}\overset{|}{\underset{|}{C}}{-}$$

ester linkage

Ester group

Carboxylic Acid Amides. Amides. $(H)R{-}\overset{\overset{\large O}{\|}}{C}{-}\overset{\overset{\large R''(H)}{|}}{N}{-}R'(H)$. So-called "simple amides" are ammonia derivatives of carboxylic acids, for they may be regarded as having formed by the splitting out of water between an acid and ammonia:

$$R{-}\overset{O}{\overset{\|}{C}}{-}OH + H{-}\overset{H}{\overset{|}{N}}{-}H \xrightarrow[-H_2O]{} R{-}\overset{O}{\overset{\|}{C}}{-}NH_2 \quad (\text{or } RCONH_2)$$

Simple amides

Other amides (the so-called N-alkyl or N,N-dialkyl amides, where N denotes attachment to the nitrogen of a simple amide) are also well-known types of compounds:

$$R{-}\overset{O}{\overset{\|}{C}}{-}\overset{H}{\overset{|}{N}}{-}R' \quad \text{or} \quad R{-}\overset{O}{\overset{\|}{C}}{-}NHR' \quad (\text{or } RCONHR')$$

$$R{-}\overset{O}{\overset{\|}{C}}{-}\overset{R'}{\overset{|}{N}}{-}R'' \quad \text{or} \quad R{-}\overset{O}{\overset{\|}{C}}{-}NR'R'' \quad (\text{or } RCONR'R'')$$

To be classified as an amide, the molecule must have a carbonyl-to-nitrogen bond, called the *amide linkage.* In proteins the name peptide bond is synonymous with amide linkage. (Several examples of amides are given in Table 15.8, page 432.)

$$-\overset{O}{\overset{\|}{C}}{-}N\langle$$

amide linkage

The amide group

Exercise 14.1 The ability to recognize quickly the family to which a structure belongs is important if we are to have any degree of success in correlating structures and properties. Examine each of the following structural formulas and place each in the correct family. In some more than one functional group is present. List such examples in all classes to which they belong.

Examples.

$CH_3\overset{O}{\overset{\|}{C}}OH$	Carboxylic acid	$CH_3\overset{O}{\overset{\|}{C}}CH_2\overset{O}{\overset{\|}{C}}CH_3$	Two keto groups
$C_6H_5\overset{O}{\overset{\|}{C}}H$	Aldehyde	$CH_3\overset{O}{\overset{\|}{C}}O\overset{O}{\overset{\|}{C}}CH_3$	Anhydride
$CH_3\underset{OH}{\underset{\|}{C}H}\overset{O}{\overset{\|}{C}}OH$	Carboxylic acid and alcohol	$H\overset{O}{\overset{\|}{C}}OH$	Carboxylic acid (This acid is not in the aldehyde class, but it does have some aldehyde properties.)

(*a*) $HO\overset{O}{\overset{\|}{C}}CH_3$ (*b*) $H\overset{O}{\overset{\|}{C}}C_6H_5$ A (*c*) $CH_3\overset{O}{\overset{\|}{C}}Cl$

(*d*) $H\overset{O}{\overset{\|}{C}}NH_2$ (*e*) $CH_3O\overset{O}{\overset{\|}{C}}CH_3$ (*f*) $CH_3\overset{O}{\overset{\|}{C}}CH_3$ K

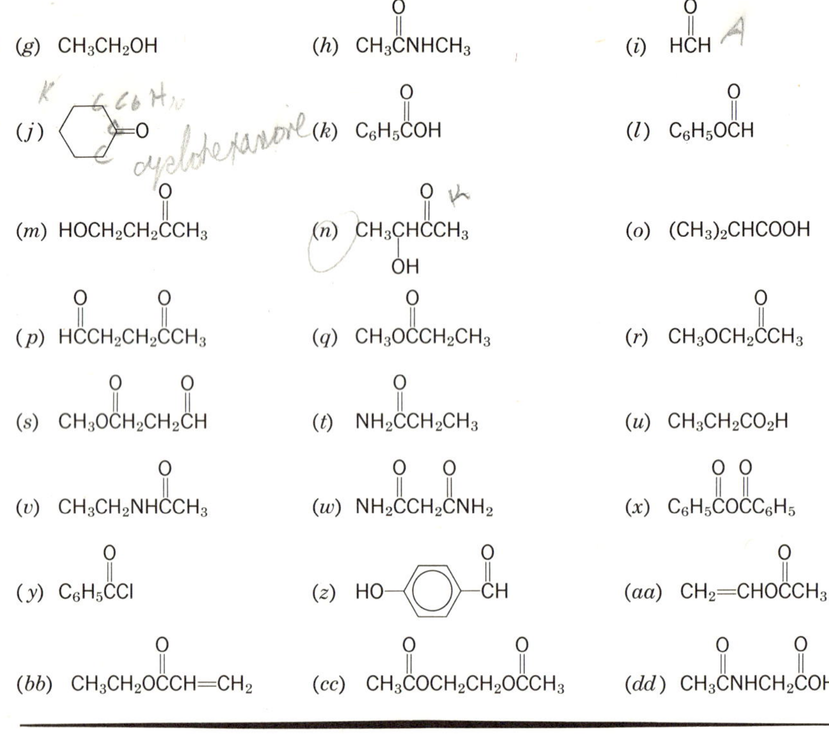

NOMENCLATURE OF ORGANIC COMPOUNDS WITH CARBONYL GROUPS

Although discussions of families with carbonyl groups will be spread out over several chapters, it will be convenient for our work to assemble in one place the rules for naming them. Our major concern is knowing how to write a structure from a name.

All families have both a common and an IUPAC system of nomenclature. We shall concentrate mainly on the common system, for the IUPAC names are very seldom applied to the simpler members of the families.

Common names for aldehydes and for the carboxylic acids and their derivatives share common prefix portions. These are summarized in Table 14.2.

Common Names. Considering first the carboxylic acids and their derivatives, we define the carbonyl portion or acid portion of the compound as that part in which

$$\underbrace{\text{Cl—}\overset{\overset{\text{O}}{\|}}{\text{C}}\text{—CH}_2\text{CH}_3}_{\text{carbonyl portion}} \qquad \underbrace{\text{CH}_3\overset{\overset{\text{O}}{\|}}{\text{C}}}_{\text{carbonyl portion}}\text{—O—CH}_2\text{CH}_3 \qquad \text{CH}_3\text{CH}_2\text{O}\underbrace{\overset{\overset{\text{O}}{\|}}{\text{C}}\text{CH}_2\text{CH}_3}_{\text{carbonyl portion}}$$

the carbon skeleton has the carbonyl group. It is for this part of the molecule that the prefix portions given in Table 14.2 are characteristic. In learning these word

Table 14.2 Common Names for Aldehydes, Acids, and Acid Derivatives

Class	Characteristic Suffix for the Name	Characteristic Prefix in the Carbonyl Portion of the Name				
		C_1 form-	C_2 acet-	C_3 propion-	C_4 *n*-butyr-*	C_4 isobutyr-
Aldehydes	-aldehyde	$H\overset{\overset{O}{\|}}{C}H$ formaldehyde	$CH_3\overset{\overset{O}{\|}}{C}H$ acetaldehyde	$CH_3CH_2\overset{\overset{O}{\|}}{C}H$ propionaldehyde	$CH_3CH_2CH_2\overset{\overset{O}{\|}}{C}H$ *n*-butyraldehyde	$(CH_3)_2CH\overset{\overset{O}{\|}}{C}H$ isobutyraldehyde
Carboxylic acids	-ic acid	HCO_2H formic acid	CH_3CO_2H acetic acid	$CH_3CH_2CO_2H$ propionic acid	$CH_3CH_2CH_2CO_2H$ *n*-butyric acid	$(CH_3)_2CHCO_2H$ isobutyric acid
Carboxylic acid salts	-ate (with the name of the positive ion written first as a separate name)	$HCO_2{}^-Na^+$ sodium formate	$CH_3CO_2{}^-Na^+$ sodium acetate	$CH_3CH_2CO_2{}^-K^+$ potassium propionate	$CH_3CH_2CH_2CO^-NH_4{}^+$ ammonium *n*-butyrate	$(CH_3)_2CHCO_2{}^-Na^+$ sodium isobutyrate
Acid chlorides	-yl chloride	(unstable)	$CH_3\overset{\overset{O}{\|}}{C}Cl$ acetyl chloride	$CH_3CH_2\overset{\overset{O}{\|}}{C}Cl$ propionyl chloride	$CH_3CH_2CH_2\overset{\overset{O}{\|}}{C}Cl$ *n*-butyryl chloride	$(CH_3)_2CH\overset{\overset{O}{\|}}{C}Cl$ isobutyryl chloride
Anhydrides	-ic anhydride	(unstable)	$CH_3\overset{\overset{O}{\|}}{C}O\overset{\overset{O}{\|}}{C}CH_3$ acetic anhydride	$CH_3CH_2\overset{\overset{O}{\|}}{C}O\overset{\overset{O}{\|}}{C}CH_2CH_3$ propionic anhydride	$(CH_3CH_2CH_2\overset{\overset{O}{\|}}{C})_2O$ *n*-butyric anhydride	$((CH_3)_2CH\overset{\overset{O}{\|}}{C})_2O$ isobutyric anhydride
Esters	-ate (with the name of the alkyl group on oxygen written first as a separate word)	$H\overset{\overset{O}{\|}}{C}OCH_2CH_3$ ethyl formate	$CH_3\overset{\overset{O}{\|}}{C}OCH(CH_3)_2$ isopropyl acetate	$CH_3CH_2\overset{\overset{O}{\|}}{C}OCH_2CH_3$ ethyl propionate	$CH_3CH_2CH_2\overset{\overset{O}{\|}}{C}OCH_3$ methyl *n*-butyrate	$(CH_3)_2CH\overset{\overset{O}{\|}}{C}OCH(CH_3)_2$ isopropyl isobutyrate
Simple amides	-amide	$H\overset{\overset{O}{\|}}{C}NH_2$ formamide	$CH_3\overset{\overset{O}{\|}}{C}NH_2$ acetamide	$CH_3CH_2\overset{\overset{O}{\|}}{C}NH_2$ propionamide	$CH_3CH_2CH_2\overset{\overset{O}{\|}}{C}NH_2$ *n*-butyramide	$(CH_3)_2CH\overset{\overset{O}{\|}}{C}NH_2$ isobutyramide

* It is also very common to omit *n*- from *n*-butyr- in all these names. Thus "butyr-" without the "*n*-" means the normal isomer.

parts, we must be certain to associate the *total* carbon content *of the acid portion* with the prefixes. In counting carbon content, the carbonyl carbon is included.

Exercise 14.2 In the structures given, circle the carbonyl portions. Write the prefix associated with it.

Examples.

$NH_2\overset{O}{\overset{\|}{C}}CH_3$ acet-

$CH_3CH_2CH_2\overset{O}{\overset{\|}{C}}OCH_2CH_2CH_2CH_3$ *n*-butyr-

Note that the oxygen that has only single bonds and the R— group attached to it are not included.

(*a*) $H\overset{O}{\overset{\|}{C}}OCH_3$ (*b*) $CH_3CH_2\overset{O}{\overset{\|}{C}}O^-Na^+$ (*c*) $CH_3CH_2O\overset{O}{\overset{\|}{C}}H$

(*d*) $CH_3CH_2O\overset{O}{\overset{\|}{C}}CH_3$ (*e*) $CH_3CH_2\overset{O}{\overset{\|}{C}}OCH_3$ (*f*) $(CH_3)_2CH\overset{O}{\overset{\|}{C}}OCH(CH_3)_2$

In esters the name of the group attached by a *single* bond to oxygen is determined first and designated. Then the carbonyl portion must be recognized and named by attaching the suffix "-ate" to the appropriate prefix.

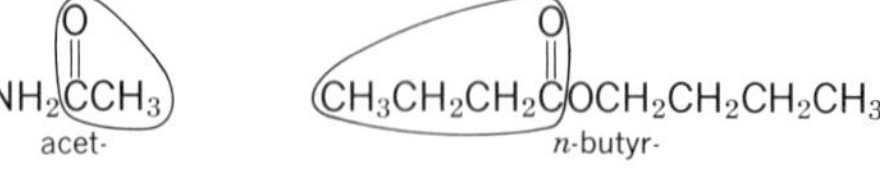
-ate (for ester family)
CH3—C—O—CH3
acet-
methyl
Methyl acetate

Common names for aldehydes are based on common names for their corresponding carboxylic acids. To the prefix for this acid we therefore attach the suffix "-aldehyde."

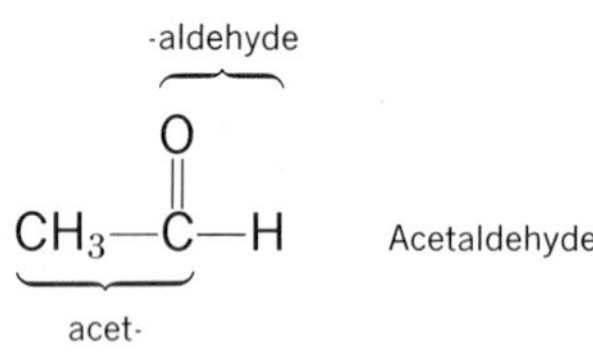
-aldehyde
CH3—C—H
acet-
Acetaldehyde

Exercise 14.3 Complete the following table according to the example given.

Structure	Family	Common Name
$CH_3\overset{O}{\overset{\|}{C}}OCH_3$	ester	methyl acetate

$(CH_3)_2CHOC(=O)CH_3$	______	______
$CH_3CH_2C(=O)NH_2$	______	______
$CH_3OC(=O)CH_2CH_2CH_3$	______	______
$(CH_3)_2CHCH_2OC(=O)CH_2CH_2CH_3$	______	______
$H{-}C(=O)CH_2CH_2CH_3$	Aldehyde	n-butyraldehyde

Common names for slightly more complicated carbonyl compounds are derived by designating with Greek letters the positions in the carbon skeleton as it extends away from the carbonyl group:

$$\underset{\delta}{C}{-}\underset{\gamma}{C}{-}\underset{\beta}{C}{-}\underset{\alpha}{C}{-}\overset{O}{\overset{\|}{C}}{-}$$

The α-position is always the carbon attached directly to the carbonyl. The carbonyl carbon is *not* lettered. (The IUPAC system, discussed later, will be different at this point.) The following examples illustrate how the common system can be extended by using these designations.

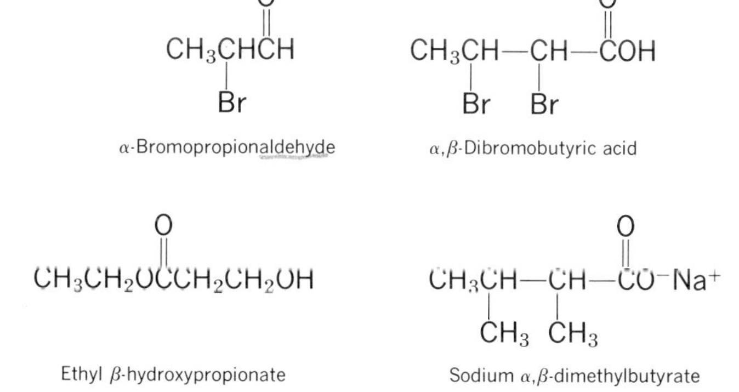

α-Bromopropionaldehyde

α,β-Dibromobutyric acid

Ethyl β-hydroxypropionate

Sodium α,β-dimethylbutyrate

Common names for ketones are devised by giving the name ketone to the carbonyl group and then by designating the groups attached to it. The following examples illustrate the method. The simplest ketone possible, dimethyl ketone, is, however, always called by its trivial name, acetone.

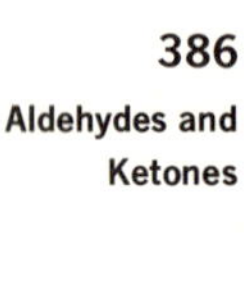

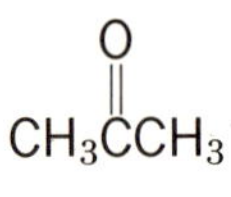

Dimethyl ketone (acetone)

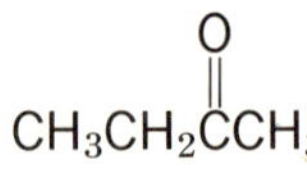

Ethyl methyl ketone

$CH_3CH_2\overset{O}{\overset{\|}{C}}CH_2CH_3$

Diethyl ketone

$(CH_3)_2CH\overset{O}{\overset{\|}{C}}CH_2CH_2CH_2CH_3$

Isopropyl *n*-butyl ketone

$(CH_3)_3C\overset{O}{\overset{\|}{C}}C(CH_3)_3$

Di-*t*-butyl ketone

$C_6H_5\overset{O}{\overset{\|}{C}}C_6H_5$

Diphenyl ketone

Exercise 14.4 Write condensed structural formulas for each of the following.

(*a*) methyl acetate
(*b*) α-bromobutyramide
(*c*) α,β-dimethylbutyraldehyde
(*d*) sodium β-chloropropionate
(*e*) propionic anhydride
(*f*) *t*-butyl isobutyrate
(*g*) isopropyl propionate
(*h*) ammonium α-hydroxypropionate
(*i*) *sec*-butyl acetate
(*j*) α-chloropropionyl chloride
(*k*) α,β-dichloroisobutyraldehyde
(*l*) *n*-butyl methyl ketone
(*m*) potassium formate
(*n*) dibenzyl ketone
(*o*) phenyl *p*-bromophenyl ketone
(*p*) ethyl β-hydroxybutyrate
(*q*) *n*-hexyl formate
(*r*) α,β-dihydroxybutyraldehyde
(*s*) α-methylbutyryl chloride
(*t*) *n*-butyl butyrate

Exercise 14.5 If the structure of benzoic acid is $C_6H_5\overset{O}{\overset{\|}{C}}\text{—}OH$, write the structures of

(*a*) benzoyl chloride
(*b*) benzaldehyde
(*c*) phenyl benzoate
(*d*) ethyl *p*-nitrobenzoate
(*e*) benzamide
(*f*) benzoic anhydride

Exercise 14.6 Write suitable common names for each of the following.

(*a*) $(CH_3)_2CH\overset{O}{\overset{\|}{C}}CH_2CH_3$

(*b*) $CH_3\underset{OH}{\underset{|}{C}H}\overset{O}{\overset{\|}{C}}NH_2$

(*c*) $H\overset{O}{\overset{\|}{C}}OCH_2CH_2CH_2CH_3$

(*d*) $CH_3CH_2CH_2\overset{O}{\overset{\|}{C}}CH_2CH_2CH_3$

(*e*) $NH_2\overset{O}{\overset{\|}{C}}CH_2\underset{Br}{\underset{|}{C}H}CH_3$

(*f*) $CH_3\underset{Cl}{\underset{|}{C}H}\text{—}\underset{Cl}{\underset{|}{C}H}\overset{O}{\overset{\|}{C}}H$

(*g*) $H\text{—}\overset{O}{\overset{\|}{C}}\text{—}H$

(*h*) $CH_3O\overset{O}{\overset{\|}{C}}CH_2CH_3$

(*i*) $CH_3\overset{O}{\overset{\|}{C}}OCH_2CH_3$

(*j*) $Br\text{—}C_6H_4\text{—}\overset{O}{\overset{\|}{C}}\text{—}NH_2$

IUPAC Nomenclature. For aldehydes, ketones, and carboxylic acids, the longest continuous sequence of carbons, including the carbonyl carbon, is taken as the base. This chain is numbered from the end that will give the carbonyl carbon the lower number, regardless of the other groups that may be present.

For aliphatic or cycloaliphatic systems, the name of the saturated hydrocarbon corresponding to the selected longest chain or largest ring is determined. This name is then altered in a way characteristic for the kind of carbonyl compound involved.

For aldehydes, the "-e" at the end of the alkane name is changed to "-al." The "-al" function requires no number to designate its location because it can occur only at the end of a chain and the 1-position for the aldehyde carbon is understood.

$$\mathrm{H{-}\overset{O}{\overset{\|}{C}}{-}H} \qquad \mathrm{CH_3\overset{O}{\overset{\|}{C}}{-}H} \qquad \mathrm{CH_3CH_2\overset{O}{\overset{\|}{C}}{-}H}$$

Methanal (formaldehyde) | Ethanal (acetaldehyde) | Propanal (propionaldehyde)

For ketones, the "-e" ending of the alkane corresponding to the longest chain in the ketone is changed to "-one." Whenever it is possible for the keto group to be located in different positions, the number of its carbon must be designated.

$$\mathrm{CH_3{-}\overset{O}{\overset{\|}{C}}{-}CH_3} \qquad \mathrm{CH_3{-}\overset{O}{\overset{\|}{C}}{-}CH_2CH_3} \qquad \mathrm{CH_3{-}\overset{O}{\overset{\|}{C}}{-}CH_2CH_2CH_3}$$

Propanone (acetone) | Butanone (ethyl methyl ketone) | 2-Pentanone (methyl *n*-propyl ketone)

$$\mathrm{CH_3CH_2{-}\overset{O}{\overset{\|}{C}}{-}CH_2CH_3}$$

3-Pentanone (diethyl ketone)

For carboxylic acids, the "-e" ending of the alkane corresponding to the longest chain in the acid is changed to "-oic acid". As with the aldehydes, the location of the carbonyl carbon is not specified because it can occur only at the end of

$$\mathrm{H{-}\overset{O}{\overset{\|}{C}}{-}OH} \qquad \mathrm{CH_3{-}\overset{O}{\overset{\|}{C}}{-}OH} \qquad \mathrm{CH_3CH_2CH_2{-}\overset{O}{\overset{\|}{C}}{-}OH}$$

Methanoic acid (formic acid) | Ethanoic acid (acetic acid) | Butanoic acid (butyric acid)

$$\mathrm{CH_3{-}\overset{CH_3}{\overset{|}{C}}H{-}\overset{O}{\overset{\|}{C}}{-}OH}$$

2-Methylpropanoic acid (isobutyric acid)

the chain and its 1-position is understood. Substituents on the basic chain are named and located in the usual way, as illustrated by 2-methylpropanoic acid.

The IUPAC names for acid derivatives (salts, acid chlorides, anhydrides, esters, and amides) are based on the name of the acid itself. To name any of these derivatives, first write the name of the parent acid. Then drop the "-ic acid" suffix portion and replace it by the suffix specified for the derivative in the second column of Table 14.2. These word parts are the same in both the common and the IUPAC system. For example,

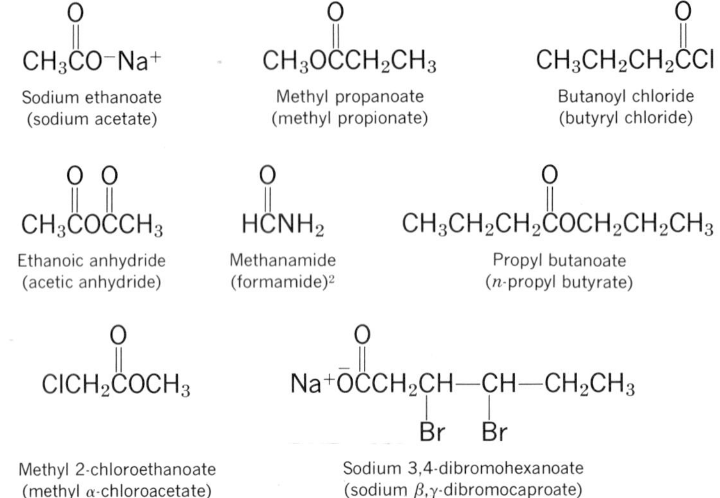

Exercise 14.7 Write condensed structural formulas for each of the following names.

(*a*) ammonium pentanoate
(*b*) isopropyl propanoate
(*c*) butanoic anhydride
(*d*) 3,4-dimethylhexanal
(*e*) 2,4-dibromo-3-pentanone
(*f*) decanamide
(*g*) methyl 2,3-dimethylhexanoate
(*h*) butyl methanoate
(*i*) 2-hydroxypropanal
(*j*) 1-phenyl-1-propanone
(*k*) butanoyl chloride
(*l*) sodium 2-methylpropanoate
(*m*) 2,3-dibromopentanal
(*n*) *t*-butyl pentanoate
(*o*) cyclohexanone
(*p*) 2,5-dichlorocyclopentanone
(*q*) *N*,2-dimethylpropanamide

Exercise 14.8 Write IUPAC and common names for each of the following.

(*a*) $CH_3CH_2\overset{\overset{\Large O}{\|}}{C}H$

(*b*) $Na^+\overset{-}{O}\overset{\overset{\Large O}{\|}}{C}CH_2CH_3$

(*c*) $CH_3CH_2O\overset{\overset{\Large O}{\|}}{C}CH_3$

(*d*) $CH_3CH_2\overset{\overset{\Large O}{\|}}{C}CH_2CH_2CH_3$

(*e*) $CH_3\overset{\overset{\Large O}{\|}}{C}O\overset{\overset{\Large O}{\|}}{C}CH_3$

(*f*) $H\overset{\overset{\Large O}{\|}}{C}OH$

(*g*) $NH_2\overset{\overset{\Large O}{\|}}{C}\underset{\underset{\Large OH}{|}}{C}HCH_2CH_3$

(*h*) $(CH_3)_2CH\overset{\overset{\Large O}{\|}}{C}OCH(CH_3)_2$

(*i*) $CH_3CH_2CH_2O\overset{\overset{\Large O}{\|}}{C}CH_3$

(*j*) $(CH_3)_2CH\overset{\overset{\Large O}{\|}}{C}H$

[2] Note that the "o" is also dropped. The name is not methanoamide.

Exercise 14.9 If the common name for $CH_3CH_2CH_2CH_2\overset{O}{\overset{\|}{C}}OH$ is valeric acid, write common and IUPAC names for the following.

(*a*) $CH_3CH_2CH_2CH_2\overset{O}{\overset{\|}{C}}H$ (*b*) $CH_3O\overset{O}{\overset{\|}{C}}CH_2CH_2CH_2CH_3$

(*c*) $Cl\overset{O}{\overset{\|}{C}}CH_2CH_2CH_2CH_3$ (*d*) $CH_3CH_2CH_2CH_2\overset{O}{\overset{\|}{C}}NH_2$

(*e*) $CH_3CH_2CH_2CH_2\overset{O}{\overset{\|}{C}}O^-Na^+$

PROPERTIES OF ALDEHYDES AND KETONES

PHYSICAL PROPERTIES

Aldehydes and ketones are moderately polar compounds. An approximate idea of how their polarity relates to that of other families may be obtained from the following boiling point data for compounds of closely similar formula weights, given in parentheses.

Propane	$CH_3CH_2CH_3$	(44)	bp −45°C
Dimethyl ether	CH_3OCH_3	(46)	bp −25
Methyl chloride	CH_3Cl	(50)	bp −24
Ethylamine	$CH_3CH_2NH_2$	(45)	bp 17
Acetaldehyde	$CH_3CH{=}O$	(44)	bp 21
Ethyl alcohol	CH_3CH_2OH	(46)	bp 78.5

Low-formula-weight aldehydes and ketones are soluble in water, but by the time the molecule has five carbons, this solubility has become quite low. The carbonyl group cannot act as a hydrogen bond donor, but the carbonyl oxygen, of course, is a hydrogen bond acceptor. Trends in typical properties are indicated by the data in Table 14.3 and 14.4.

OXIDATION

At room temperature molecular oxygen (from air) is not considered a vigorous oxidizing agent. Yet the aldehyde group is sensitive enough to oxidation that aldehydes cannot be stored exposed to air or they will slowly convert to their corresponding carboxylic acids. In sharp contrast, ketones are ordinarily very stable to further oxidation. In fact, this is the principal difference between aldehydes and ketones. Otherwise, the two families exhibit very similar chemical properties.

The ease with which the aldehyde group is oxidized makes it possible to detect its presence with reagents that leave other groups unaffected.

Tollens' Test. Silvering Mirrors. Tollens' reagent is a solution of the diammonia complex of the silver ion in dilute ammonium hydroxide. (The silver ion is precipitated as silver oxide in the presence of hydroxide ion unless it is complexed with ammonia

Table 14.3 Aldehydes

Common Name	Structure	Mp (°C)	Bp (°C)	Solubility in Water
Formaldehyde	$CH_2{=}O$	−92	−21	very soluble
Acetaldehyde	$CH_3CH{=}O$	−125	21	very soluble
Propionaldehyde	$CH_3CH_2CH{=}O$	−81	49	16 g/100 cc (25°)
n-Butyraldehyde	$CH_3CH_2CH_2CH{=}O$	−99	76	4 g/100 cc
Isobutyraldehyde	$(CH_3)_2CHCH{=}O$	−66	65	9 g/100 cc
Valeraldehyde	$CH_3CH_2CH_2CH_2CH{=}O$	−92	102	slightly soluble
Caproaldehyde	$CH_3CH_2CH_2CH_2CH_2CH{=}O$	−56	128	very slightly soluble
Acrolein	$CH_2{=}CHCH{=}O$	−87	53	40 g/100 cc
Crotonaldehyde	$CH_3CH{=}CHCH{=}O$	−77	104	moderately soluble
Benzaldehyde	C_6H_5—CH=O	−56	178	0.3 g/100 cc
Salicylaldehyde	C_6H_4(OH)—CH=O (OH ortho)	−10	197	slightly soluble
Vanillin	HO—C_6H_3(OCH_3)—CH=O	81	285	1 g/100 cc
Cinnamaldehyde (*trans*)	C_6H_5—CH=CHCH=O	−8	253	insoluble
Furfural	furan ring (O)—CH=O	−31	162	9 g/100 cc
Glucose* (dextrose)	CH_2(OH)—CH(OH)—CH(OH)—CH(OH)—CH(OH)—CH=O	146 (decomposes)	—	very soluble

* Only the general structural features of the open-chain form of glucose are shown. Additional details are discussed in Chapter 16.

molecules.) In general:

$$\underset{\text{Aldehyde}}{RCH{=}O} + 2Ag(NH_3)_2^+ + 3OH^- \longrightarrow \underset{\text{Carboxylate ion}}{RCO_2^-} + 2Ag\downarrow + 2H_2O + 4NH_3$$

Specific example:

$$\underset{\text{Acetaldehyde}}{CH_3CH{=}O} + 2Ag(NH_3)_2^+ + 3OH^- \longrightarrow \underset{\text{Acetate ion}}{CH_3CO_2^-} + \underset{\text{Silver}}{2Ag\downarrow} + 2H_2O + 4NH_3$$

Any good test for a functional group must provide sound, positive evidence that reaction occurs with that functional group. The positive evidence in the Tollens

Table 14.4 Ketones

Name	Structure	Mp (°C)	Bp (°C)	Solubility in Water
Acetone	CH_3COCH_3	−95	56	very soluble
Methyl ethyl ketone	$CH_3COCH_2CH_3$	−87	80	33 g/100 cc (25°)
2-Pentanone	$CH_3COCH_2CH_2CH_3$	−84	102	6 g/100 cc
3-Pentanone	$CH_3CH_2COCH_2CH_3$	−40	102	5 g/100 cc
2-Hexanone	$CH_3COCH_2CH_2CH_2CH_3$	−57	128	1.6 g/100 cc
3-Hexanone	$CH_3CH_2COCH_2CH_2CH_3$	—	124	1.5 g/100 cc
Cyclopentanone	cyclopentane ring =O	−53	129	slightly soluble
Cyclohexanone	cyclohexane ring =O		156	slightly soluble
Muscone*	CH_3–CH–CH_2–C=O, ring closed by $(CH_2)_{12}$		328	insoluble
Civetone†	CH–$(CH_2)_7$–C=O, CH=CH, ring closed by CH–$(CH_2)_7$	33	342	insoluble
Camphor	bicyclic ring with $C(CH_3)_2$ bridge, =O	176	209	insoluble
Acetophenone	C_6H_5–$COCH_3$	−23	137	insoluble
Fructose (levulose)	$CH_2(OH)$–CH(OH)–CH(OH)–CH(OH)–$C(=O)CH_2OH$	48		soluble

* Active principle of musk, a substance of powerful odor used in trace amounts as a perfume base and obtained from an abdominal gland of the Himalayan male musk deer. The male uses musk to attract the female deer.

† Also a rare perfume base, civetone is the active fragrant principle of civet, a substance found in the African civet cat.

test is the appearance of metallic silver in a previously clear, colorless solution. If the inner wall of the test vessel is clean and grease-free, silver deposits to form a beautiful mirror, a reaction that serves as the basis for silvering mirrors. If the glass surface is not clean, the silver separates as a gray, finely divided, powdery precipitate. Tollens' reagent does not keep well and is prepared just before use.

Benedict's Test and Fehling's Test. Benedict's solution and Fehling's solution are both alkaline and contain the copper(II) ion, Cu^{2+}, stabilized by a complexing agent. This agent is the citrate ion in Benedict's solution and the tartrate ion in Fehling's solution. Without these complexing agents, copper(II) ions would be precipitated as copper(II) hydroxide by the alkali present. Benedict's solution is more frequently used because it stores well. Fehling's solution must be prepared just before use.

With either solution the Cu^{2+} ion is the oxidizing agent. In the presence of certain easily oxidized groups it is reduced to the copper(I) state, but copper(I) ions cannot be solubilized by either citrate or tartrate ions and a precipitate of copper(I) oxide, Cu_2O, forms. What we see in a positive test, then, is a change from the brilliant blue color of the test solution to the bright orange-red color of precipitated copper(I) oxide. The principal systems that give this result are α-hydroxy aldehydes, α-keto aldehydes, and α-hydroxy ketones. α-Hydroxy aldehydes and α-hydroxy ketones

$$\underset{\text{α-Hydroxy aldehyde}}{\mathrm{R\underset{\underset{OH}{|}}{C}H\overset{\overset{O}{\|}}{C}H}} \qquad \underset{\text{α-Keto aldehyde}}{\mathrm{R{-}\overset{\overset{O}{\|}}{C}{-}\overset{\overset{O}{\|}}{C}{-}H}} \qquad \underset{\text{α-Hydroxy ketone}}{\mathrm{R\underset{\underset{OH}{|}}{C}H\overset{\overset{O}{\|}}{C}R}}$$

are common to the sugar families (cf. the last compounds listed in Tables 14.3 and 14.4, glucose and fructose). Benedict's test, or Fehling's test, is a common method for detecting the presence of glucose in urine. Normally, urine does not contain glucose, but in certain conditions, for example, diabetes, it does, A positive test for glucose varies from a bright green color (0.25% glucose), to yellow-orange (1% glucose), to brick red (over 2% glucose).

Some ordinary aliphatic aldehydes react with these reagents, but the changes are complex. A dark, gummy precipitate may form, which apparently is not copper(I) oxide, while the color of the test solution remains blue or becomes any hue from yellow to green to brown. Aromatic aldehydes, in general, do not react.[3]

Strong Oxidizing Agents. Aldehydes are of course readily oxidized to carboxylic acids by permanganate ion, dichromate ion, and other oxidizing agents. Ketones also eventually yield to such reagents, but except in isolated instances the reaction is not particularly useful. Ketones tend to fragment somewhat randomly on both sides of the carbonyl carbon:

$$\mathrm{R{-}\overset{\overset{O}{\|}}{C}{-}R'} \xrightarrow{\text{strong oxidizing agent, heat}} \mathrm{R{-}CO_2H + R'CO_2H} \quad \text{(as well as acids of lower formula weights)}$$

[3] R. Daniels, C. C. Rush, and L. Bauer, *Journal of Chemical Education,* Vol. 37, 1960, page 205.

REDUCTION

Several methods are available for reducing aldehydes and ketones to their corresponding primary and secondary alcohols.

Catalytic Hydrogenation. In general:

$$\underset{\text{Aldehyde}}{R{-}\overset{\overset{\displaystyle O}{\|}}{C}{-}H} + H_2 \xrightarrow[\text{pressure, heat}]{Ni} \underset{1^\circ\text{ alcohol}}{R{-}CH_2{-}OH}$$

$$\underset{\text{Ketone}}{R{-}\overset{\overset{\displaystyle O}{\|}}{C}{-}R'} + H_2 \xrightarrow[\text{pressure, heat}]{Ni} \underset{2^\circ\text{ alcohol}}{R{-}\overset{\overset{\displaystyle OH}{|}}{CH}{-}R'}$$

Specific examples:

$$\underset{n\text{-Butyraldehyde}}{CH_3CH_2CH_2CH{=}O} + H_2 \longrightarrow \underset{n\text{-Butyl alcohol (85\%)}}{CH_3CH_2CH_2CH_2OH}$$

$$\underset{\text{Acetone}}{CH_3\overset{\overset{\displaystyle O}{\|}}{C}CH_3} + H_2 \longrightarrow \underset{\text{Isopropyl alcohol (100\%)}}{CH_3\overset{\overset{\displaystyle OH}{|}}{C}HCH_3}$$

Other Methods. Metallic hydrides such as lithium aluminum hydride ($LiAlH_4$) and sodium borohydride ($NaBH_4$) are excellent but expensive reducing agents for aldehydes and ketones.

ADDITIONS OF HYDROXYLIC COMPOUNDS TO ALDEHYDES AND KETONES

The analogy between the alkene double bond (C=C) and the carbonyl double bond (C=O) does not extend too far. Although both will add hydrogen in the presence of a suitable catalyst, the carbonyl group does not add halogens (Cl_2, etc) or water to form stable addition compounds. Yet the interaction of the carbonyl group with water, and especially with alcohols, is of primary importance to our future study of carbohydrates.

Hydrates. Most aldehydes react in aqueous solutions with water to establish an equilibrium mixture with the hydrated form:

$$R{-}\overset{\overset{\displaystyle O}{\|}}{C}{-}H + H{-}OH \overset{H^+}{\rightleftharpoons} R{-}CH\begin{matrix}\diagup OH\\ \diagdown OH\end{matrix}$$

Formaldehyde is especially prone to form its hydrate. Of course, of all simple aldehydes and ketones, formaldehyde has a carbonyl carbon that is least hindered or crowded by neighboring groups. Because the carbonyl carbon has no alkyl groups with their electron-releasing effects, it bears in formaldehyde the largest partial positive charge of all simple aldehydes and ketones. Water is probably added as follows:

$$H-\overset{\overset{\displaystyle :\ddot{O}:}{\|}}{C}-H + H^+ \underset{\text{Catalyst}}{\rightleftharpoons} H-\overset{\overset{\displaystyle H-\ddot{O}:}{|}}{\underset{+}{C}}-H$$

$$H-\overset{\overset{\displaystyle H-\ddot{O}:}{|}}{\underset{+}{C}H} + H-\ddot{\underset{\cdot\cdot}{O}}-H \rightleftharpoons H-CH(O-H)(\overset{+}{O}(H)-H) \underset{\text{Proton loss}}{\rightleftharpoons} H-CH(OH)_2 + \underset{\text{Recovery of catalyst}}{H^+}$$

Hemiacetals and Acetals. In a similar manner, solutions of most aldehydes in alcohols consist of equilibrium mixtures:

$$\underset{\text{Aldehyde}}{R-\overset{\overset{\displaystyle O}{\|}}{C}-H} + \underset{\text{Alcohol}}{H-O-R'} \overset{H^+}{\rightleftharpoons} \underset{\text{A hemiacetal}}{R-CH(OH)(OR')}$$

(arrow to the CH carbon of the hemiacetal: original carbonyl carbon)

From the alcohol the hydrogen goes to the carbonyl oxygen; the alcohol oxygen goes to the carbonyl carbon. Most hemiacetals cannot be isolated, for efforts to do so cause them to convert back to the original aldehyde and alcohol. In fact, the hemiacetal group is often called a *potential aldehyde group*. The hemiacetals that can be isolated are, however, very important, for they occur among carbohydrate molecules.

The hemiacetal system is defined by

$$>\mathbf{C}<\begin{matrix} O-H \\ O-\underset{|}{\overset{|}{C}}- \end{matrix}$$

One carbon, in boldface, carries both an OH and an OR group. This carbon was the original carbonyl carbon. The hemiacetal is both an alcohol and an ether. Since these two groups occur so close together, however, each modifies the properties of the other. For example, ether linkages are difficult to break, yet in hemiacetals they are seldom strong enough to hold the system together. These compounds usually exist only as part of an equilibrium in solution. The properties of the —OH group are similarly changed by the neighboring ether linkage. Ordinary alcohols interact and form ordinary ethers only under rather extreme conditions (cf. p. 362). As we shall see later, however, the —OH group in a hemiacetal is easily converted into a second ether linkage.

The key to this behavior, at least under certain conditions, is the relative stability of the carbonium ion formed as follows:

$$\text{R—CH(OH)(OR')} \xrightarrow[\text{(H}_2\text{O leaves)}]{\text{H}^+} \left[\text{R—}\overset{+}{\text{C}}\text{H—}\ddot{\text{O}}\text{—R'} \longleftrightarrow \text{R—CH=}\overset{+}{\ddot{\text{O}}}\text{—R}\right] + H_2O$$

The carbonium ion
is a resonance-stabilized
hybrid of these two contributors.

This carbonium ion, in which the positive charge is delocalized between carbon and oxygen, is much more stable than ordinary carbonium ions for which delocalization by resonance cannot occur. Alteratively,

$$\text{R—CH(OH)(OR')} \xrightarrow[\text{(R'OH leaves)}]{\text{H}^+} \left[\text{R—}\overset{+}{\text{C}}\text{H—}\ddot{\text{O}}\text{H} \longleftrightarrow \text{R—CH=}\overset{+}{\ddot{\text{O}}}\text{—H}\right] + \text{H—O—R'}$$

In this example the ether linkage is broken; the driving force is again the relative stability of the carbonium ion.

Hemiacetals can be converted into acetals by the action of additional alcohol in the presence of an acid catalyst such as dry hydrogen chloride:

$$\underset{\text{Hemiacetal}}{\text{R—CH(O—H)(O—R')}} + \underset{\text{Alcohol}}{\text{H—O—R'}} \xrightarrow{\text{H}^+} \underset{\text{Acetal}}{\text{R—CH(O—R')(O—R')}} + H_2O$$

The acetal system is defined as

$$\text{>}\mathbf{C}\text{(O—}\overset{|}{\underset{|}{\text{C}}}\text{—)(O—}\overset{|}{\underset{|}{\text{C}}}\text{—)}$$

The carbon shown in boldface was the original carbonyl carbon. Wherever two ether linkages come to the *same* carbon, the acetal system exists. It occurs widely in carbohydrates and is the key to understanding such substances as sucrose, lactose, maltose, cellulose, starch, and glycogen.

Acetals are stable in neutral and alkaline media. They can be isolated and stored, in great contrast to the hemiacetals. The one reaction of importance to our future studies is their ready *acid-catalyzed* hydrolysis—first to the hemiacetal stage, which of course is immediately in equilibrium with the original alcohol and carbonyl compound. In general:

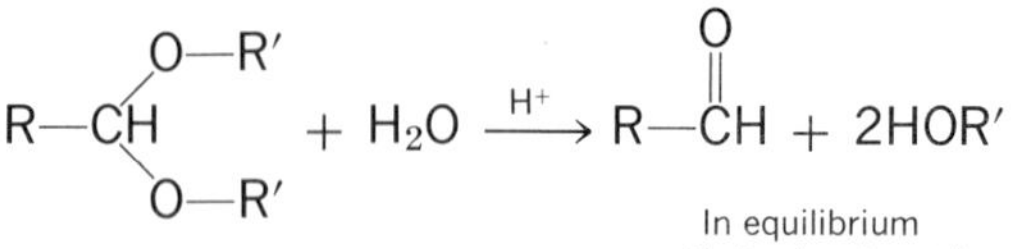

$$R{-}CH(O{-}R')_2 + H_2O \xrightarrow{H^+} R{-}\overset{O}{\overset{\|}{C}}H + 2HOR'$$

In equilibrium with the hemiacetal

Specific example:

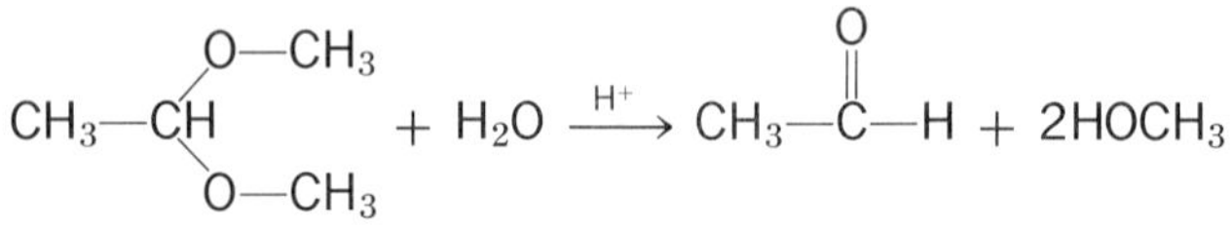

$$CH_3{-}CH(O{-}CH_3)_2 + H_2O \xrightarrow{H^+} CH_3{-}\overset{O}{\overset{\|}{C}}{-}H + 2HOCH_3$$

Acetaldehyde dimethylacetal — Acetaldehyde — Methyl alcohol

They hydrolyze in acid media. These two facts about acetals in general help determine the specific properties of carbohydrates.

Exercise 14.10 Write the structure of the hemiacetal that will exist in equilibrium with each of the following pairs of compounds.

(*a*) $CH_3\overset{O}{\overset{\|}{C}}H + HOCH_2CH_3$ (*b*) $CH_3CH_2\overset{O}{\overset{\|}{C}}H + HOCH_3$

(*c*) $CH_3OH + (CH_3)_2CH\overset{O}{\overset{\|}{C}}H$ (*d*) $CH_3CH{=}CH\overset{O}{\overset{\|}{C}}H + HOCH_2CH{=}CH_2$

(*e*) $C_6H_{11}{-}\overset{O}{\overset{\|}{C}}H + CH_3OH$ (cyclohexyl)

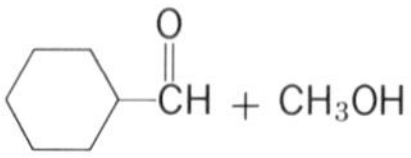

Exercise 14.11 Write the structure of the acetal that will form if the aldehyde in each of the parts of exercise 14.10 combines with two molecules of the alcohol shown with it.

Exercise 14.12 Write the structures of the original aldehyde and the original alcohol that will form if the following acetals are hydrolyzed. One example is not an acetal. Which one is it?

(*a*) $CH_2(O{-}CH_3)_2$ (*b*) $CH_3CH(O{-}CH_2CH_3)_2$

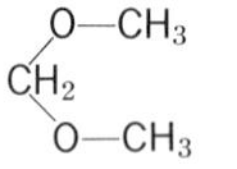

(*c*) $CH_3{-}O{-}CH(O{-}CH_3){-}CH_3$ (*d*) $CH_3{-}O{-}CH_2{-}CH_2{-}O{-}CH_3$

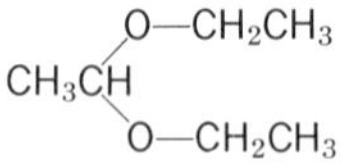

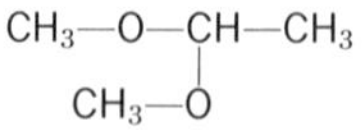

(*e*) $(CH_3O)_2CH{-}C_6H_5$ (phenyl)

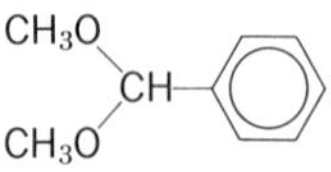

Exercise 14.13 The following compound is one of the forms in which glucose can exist. The carbons are numbered. Which is the hemiacetal carbon (the one that was "originally" a carbonyl carbon)?

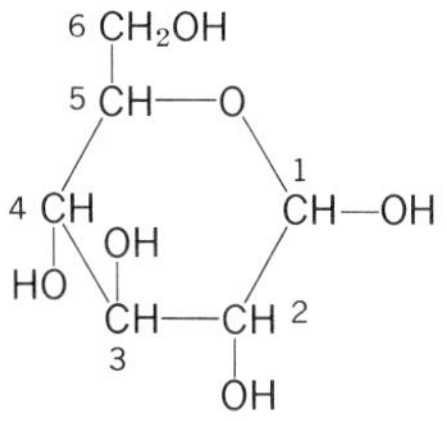

6 CH_2OH (on C5); ring: 5 CH—O—1 CH—OH; 4 CH(HO)—3 CH(OH)—2 CH(OH)

If this hemiacetal were to undergo a regeneration of the original carbonyl group and the original —OH group, what would the structure of the resulting substance be? In this regeneration the new alcohol portion is elsewhere on the same new aldehyde molecule.

Exercise 14.14 Suggest a mechanism for the acid-catalyzed conversion of a hemiacetal to an acetal.

REACTIONS AT THE α-POSITION. THE ALDOL CONDENSATION AND RELATED REACTIONS

When heated in the presence of 10% sodium hydroxide, acetaldehyde reacts to form β-hydroxybutyraldehyde, a reaction called the *aldol condensation.*

$$2CH_3CH{=}O \xrightarrow[\text{heat}]{\text{NaOH}(aq)} CH_3\overset{\overset{\displaystyle OH}{|}}{C}HCH_2CH{=}O$$

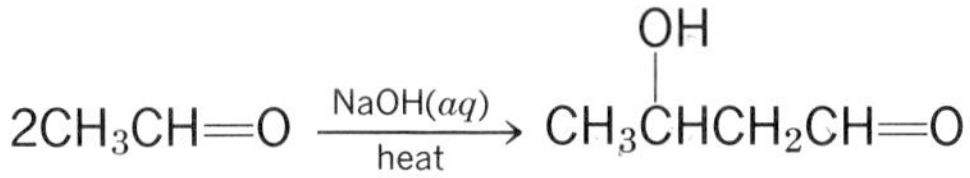

Acetaldehyde → β-Hydroxybutyraldehyde (aldol)

In general:

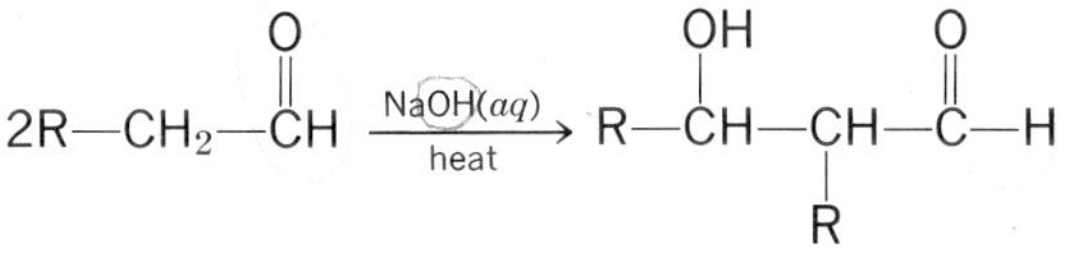

$$2R{-}CH_2{-}\overset{\overset{\displaystyle O}{\|}}{C}H \xrightarrow[\text{heat}]{\text{NaOH}(aq)} R{-}\overset{\overset{\displaystyle OH}{|}}{C}H{-}\underset{\underset{\displaystyle R}{|}}{C}H{-}\overset{\overset{\displaystyle O}{\|}}{C}{-}H$$

This reaction is illustrative of a large number that have the following feature in common:

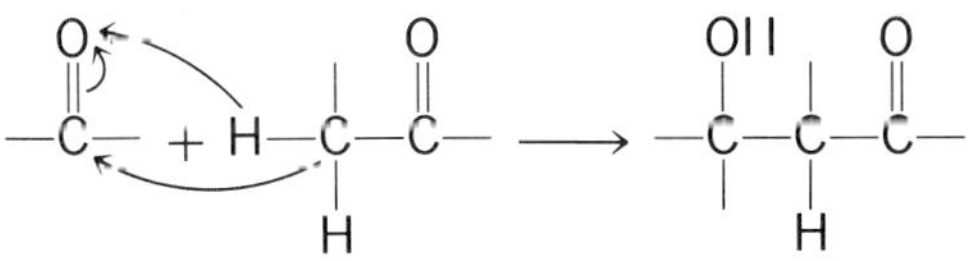

$$-\overset{\overset{\displaystyle O}{\|}}{C}- + H{-}\underset{\underset{\displaystyle H}{|}}{\overset{|}{C}}{-}\overset{\overset{\displaystyle O}{\|}}{C}- \longrightarrow -\overset{\overset{\displaystyle OH}{|}}{\underset{|}{C}}{-}\underset{\underset{\displaystyle H}{|}}{\overset{|}{C}}{-}\overset{\overset{\displaystyle O}{\|}}{C}-$$

Carbonyl compound + Carbonyl compound with two α-hydrogens → A β-hydroxy carbonyl compound

In words, one aldehyde adds across the carbonyl group of another. Most other examples of this type of reaction continue beyond the β-hydroxy carbonyl stage. Dehydration occurs and an unsaturated carbonyl compound forms:

$$\begin{array}{c}\text{OH}\quad\;\text{O}\\ |\quad\;|\quad\;\|\\ -\text{C}-\text{C}-\text{C}-\\ |\quad\;|\quad\;\\ \quad\;\text{H}\end{array} \longrightarrow \begin{array}{c}\quad\quad\;\;\text{O}\\ \quad\quad|\;\;\;\|\\ {>}\text{C}=\text{C}-\text{C}-\\ \beta\quad\;\alpha\end{array} + H_2O$$

A β-hydroxy carbonyl compound — An α,β-unsaturated carbonyl compound

Before examining specific examples, including some from metabolic reactions in living cells, the mechanism for this general reaction will be described.

General Mechanism for Aldol-Type Condensations

Step 1. Acid-base equilibration between the base and a molecule of carbonyl compound. The base takes a proton from the α-position and only from the α-position.

In general:

$$-\underset{\text{H}}{\overset{|}{\text{C}}}-\overset{\text{O}}{\overset{\|}{\text{C}}}- + {}^{-}OH \rightleftharpoons -\overset{|}{\underset{-\,\cdot\cdot}{\text{C}}}-\overset{\text{O}}{\overset{\|}{\text{C}}}- + H-OH$$

Anion

Example:

$$CH_3-\overset{\text{O}}{\overset{\|}{\text{C}}}-H + {}^{-}OH \rightleftharpoons {}^{-}{:}CH_2-\overset{\text{O}}{\overset{\|}{\text{C}}}-H$$

Acetaldehyde — An anion

Step 2. Attack by the anion[4] formed in step 1 at the carbonyl carbon of another carbonyl-containing molecule.

In general:

$$-\overset{\text{O}}{\overset{\|}{\text{C}}}- + {}^{-}{:}\overset{|}{\underset{|}{\text{C}}}-\overset{\text{O}}{\overset{\|}{\text{C}}}- \rightleftharpoons -\overset{\text{O}^-}{\overset{|}{\underset{|}{\text{C}}}}-\overset{|}{\underset{|}{\text{C}}}-\overset{\text{O}}{\overset{\|}{\text{C}}}-$$

A new, more stable anion

Example:

$$CH_3-\overset{\text{O}}{\overset{\|}{\text{C}}}-H + {}^{-}{:}CH_2-\overset{\text{O}}{\overset{\|}{\text{C}}}-H \rightleftharpoons CH_3\overset{\text{O}^-}{\overset{|}{\text{C}}}H-CH_2-\overset{\text{O}}{\overset{\|}{\text{C}}}-H$$

An alkoxide ion

Step 3. Acid-base equilibration between the alkoxide ion formed in step 2 and a proton donor (such as water).

In general:

$$-\overset{\text{O}^-}{\overset{|}{\underset{|}{\text{C}}}}-\overset{|}{\underset{|}{\text{C}}}-\overset{\text{O}}{\overset{\|}{\text{C}}}- + H-OH \rightleftharpoons -\overset{\text{OH}}{\overset{|}{\underset{|}{\text{C}}}}-\overset{|}{\underset{|}{\text{C}}}-\overset{\text{O}}{\overset{\|}{\text{C}}}- + {}^{-}OH$$

[4] The term *anion* denotes any negatively charged ion.

Example:

$$CH_3-\underset{|}{\overset{O^-}{C}}H-CH_2-\overset{O}{\overset{\|}{C}}-H + H-OH \rightleftarrows CH_3-\underset{|}{\overset{OH}{C}}H-CH_2-\overset{O}{\overset{\|}{C}}-H + {}^-OH$$

β-Hydroxybutyraldehyde (aldol)

In step 1 a negatively charged ion forms. Since the charge is associated with a carbon, the particle is called a carbanion. Carbanions derived from alkanes are among the strongest bases known, which is an inverted way of saying that alkanes are among the weakest acids or proton donors that can be formed. A hydrogen attached to a carbon in an alkane does not ionize, and a hydroxide ion cannot pull it off. But when the carbon holding a hydrogen also holds a carbonyl group, the hydrogen can be taken because the resulting carbanion can be stabilized by resonance. The negative charge can be delocalized.

$$HO^- + -\overset{:O:}{\overset{\|}{C}}-\overset{H}{\underset{|}{C}}- \rightleftarrows H_2O + \left[-\overset{:O:}{\overset{\|}{C}}-\underset{|}{\ddot{C}}^- \longleftrightarrow -\overset{:\ddot{O}:^-}{\overset{|}{C}}=\underset{|}{C}- \right] \equiv -\overset{O^{\delta-}}{\overset{\vdots}{C}}\cdots C^{\delta-}\langle$$

System with a hydrogen atom in the α-position to a carbonyl.

Contributors to the structure of the anion; they differ only in positions of electrons, and resonance exists.

Hybrid structure of anion deduced from resonance contributors; the negative charge is delocalized and spread out, thereby stabilizing the system.

$$HO^- + -CH_2-\overset{H}{\underset{|}{C}}- \not\rightleftarrows H_2O + -CH_2-\underset{|}{\ddot{C}}^-$$

System in an alkane

Carbanion (There is no way to relocate electrons and develop contributing structures of an estimated comparable internal energy. This carbanion is not stabilized by resonance, the charge cannot be delocalized.)

In an aldol condensation the carbonyl compound must have an α-hydrogen. The base cannot take a proton from a β-position or one farther down the chain, because anions resulting from such changes cannot be stabilized by resonance with the oxygen.

The anion formed in step 1 of the aldol condensation is a species actively attracted to positively charged sites. The carbonyl carbon of another molecule is such a site. An alkoxide ion (with a carbonyl group elsewhere in the same molecule) is the product. This, of course, is a powerful proton acceptor; in step 3 it takes a proton and the final product, the aldol, forms. Aldol is a specific name for β-hydroxybutyraldehyde and a general name for β-hydroxy aldehydes.

Step 4. The aldol suffers loss of the elements of water.

In general:

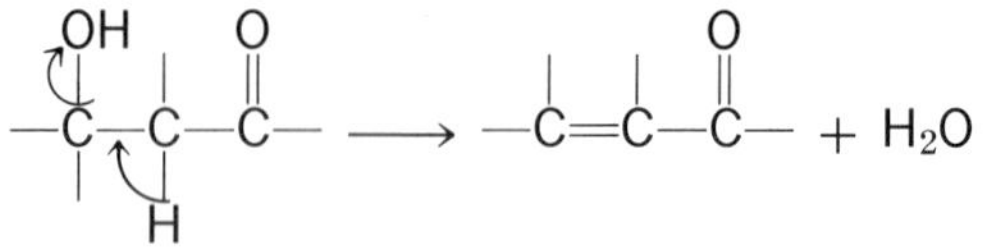

$$-\overset{\overset{\displaystyle OH}{|}}{\underset{|}{C}}-\overset{|}{\underset{\underset{\displaystyle H}{|}}{C}}-\overset{\overset{\displaystyle O}{\|}}{C}- \longrightarrow -\overset{|}{C}=\overset{|}{C}-\overset{\overset{\displaystyle O}{\|}}{C}- + H_2O$$

Example:

$$CH_3-\overset{\overset{\displaystyle OH}{|}}{C}H-CH_2-\overset{\overset{\displaystyle O}{\|}}{C}-H \longrightarrow \underset{\text{Crotonaldehyde}}{CH_3-CH=CH-\overset{\overset{\displaystyle O}{\|}}{C}-H} + H_2O$$

This last step is essentially irreversible, but dehydration is not spontaneous with the aldols formed by condensation between aldehyde molecules. For these an acid catalyst is usually necessary. In many variations of the aldol condensation dehydration is spontaneous, however, and the final isolable products are unsaturated carbonyl compounds.

Exercise 14.15 Write the structure of the product of the aldol condensation involving (*a*) propionaldehyde, (*b*) butyraldehyde, (*c*) phenylacetaldehyde ($C_6H_5CH_2CH{=}O$).

Thus far our discussion of the aldol condensation has concerned *in vitro* reactions. Reactions in cells are under the control of highly specialized catalysts, enzymes, which can accomplish changes that are not possible in ordinary catalysis. One variation of the aldol condensation that takes place in plant and animal cells occurs as a step in the synthesis of glucose units, the building blocks for starches, glycogen, and cellulose.

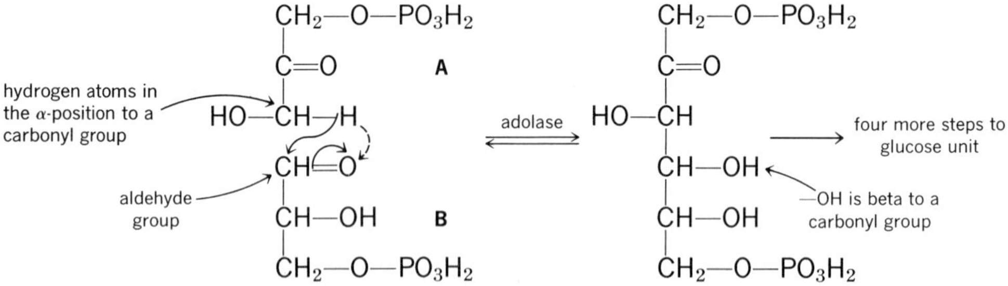

$$\begin{array}{l} CH_2-O-PO_3H_2 \\ | \\ C=O \qquad\qquad \textbf{A} \\ | \\ HO-CH-H \\ \\ CH=O \\ | \\ CH-OH \qquad \textbf{B} \\ | \\ CH_2-O-PO_3H_2 \end{array} \quad \underset{}{\overset{\text{adolase}}{\rightleftharpoons}} \quad \begin{array}{l} CH_2-O-PO_3H_2 \\ | \\ C=O \\ | \\ HO-CH \\ | \\ CH-OH \\ | \\ CH-OH \\ | \\ CH_2-O-PO_3H_2 \end{array} \longrightarrow \text{four more steps to glucose unit}$$

hydrogen atoms in the α-position to a carbonyl group

aldehyde group

—OH is beta to a carbonyl group

A is dihydroxyacetone phosphate
B is glyceraldehyde 3-phosphate

Fructose 1,6-diphosphate

The enzyme in this reaction is called aldolase to denote that it catalyzes an aldol-type condensation. *The aldol condensation is reversible* (step 3 backward to the start of step 1, in our mechanism), and the reverse aldol condensation is a key step in the degradation of a glucose unit which is eventually converted into carbon dioxide, water, and chemical energy plus some heat.

SYNTHESIS OF ALDEHYDES AND KETONES

Aldehydes and ketones can be synthesized in dozens of ways, but only one, the oxidation of alcohols, is important to our study of molecular biology. This method was discussed in Chapter 12.

AN IMPORTANT ALDEHYDE

Formaldehyde. $H-\overset{\overset{\displaystyle O}{\|}}{C}-H$. At room temperature formaldehyde is a gas with a very irritating and distinctive odor. In one industrial method it is prepared by passing a mixture of methyl alcohol and air over a silver catalyst at a temperature of over 600°C. The newly formed formaldehyde together with unchanged methanol is absorbed in water until the solution has a concentration of about 37% formaldehyde. In this form it is marketed as Formalin and is used as a disinfectant and for preserving biological samples and specimens. The largest amounts of Formalin, however, are consumed in the manufacture of various resins such as Bakelite and melamine (plastic dinnerware).

REFERENCES

All the standard textbooks in organic chemistry have one or more chapters on the chemistry of aldehydes and ketones.

PROBLEMS AND EXERCISES

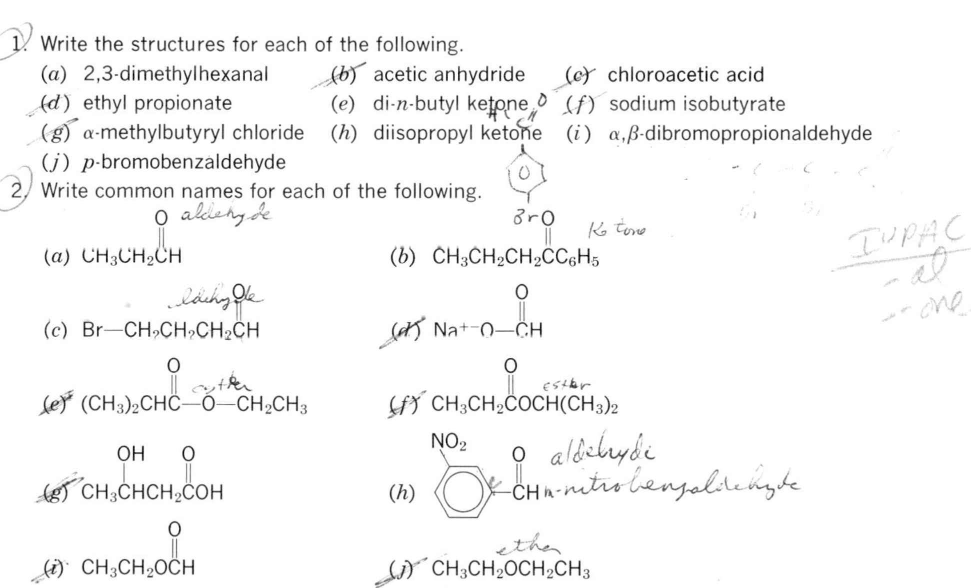

1. Write the structures for each of the following.
   (*a*) 2,3-dimethylhexanal (*b*) acetic anhydride (*c*) chloroacetic acid
   (*d*) ethyl propionate (*e*) di-*n*-butyl ketone (*f*) sodium isobutyrate
   (*g*) α-methylbutyryl chloride (*h*) diisopropyl ketone (*i*) α,β-dibromopropionaldehyde
   (*j*) *p*-bromobenzaldehyde
2. Write common names for each of the following.
   (*a*) $CH_3CH_2\overset{\overset{\displaystyle O}{\|}}{C}H$ (*b*) $CH_3CH_2CH_2\overset{\overset{\displaystyle O}{\|}}{C}C_6H_5$
   (*c*) $Br-CH_2CH_2CH_2\overset{\overset{\displaystyle O}{\|}}{C}H$ (*d*) $Na^{+}\,{}^{-}O-\overset{\overset{\displaystyle O}{\|}}{C}H$
   (*e*) $(CH_3)_2CH\overset{\overset{\displaystyle O}{\|}}{C}-O-CH_2CH_3$ (*f*) $CH_3CH_2\overset{\overset{\displaystyle O}{\|}}{C}OCH(CH_3)_2$
   (*g*) $CH_3\underset{}{\overset{\overset{\displaystyle OH}{|}}{C}}HCH_2\overset{\overset{\displaystyle O}{\|}}{C}OH$ (*h*) NO_2-substituted benzene ring $-\overset{\overset{\displaystyle O}{\|}}{C}H$
   (*i*) $CH_3CH_2O\overset{\overset{\displaystyle O}{\|}}{C}H$ (*j*) $CH_3CH_2OCH_2CH_3$

3. Write the structure(s) of the principal organic product(s) that might reasonably be expected to form if *acetaldehyde* were subjected to each of the following reagents and conditions.

(*a*) H_2, Ni, pressure, heat (*b*) excess CH_3OH, dry HCl
(*c*) CH_3OH (as a solvent) (*d*) 10% NaOH, heat
(*e*) Tollens' reagent (*f*) CrO_3, H^+

4. Repeat question 3 using *acetone* instead of acetaldehyde.

5. Which of the following would be expected to give a positive Benedict or Fehling test?

(*a*) $CH_3\underset{OH}{\underset{|}{C}}H\overset{O}{\overset{\|}{C}}H$ (*b*) $CH_3\overset{O}{\overset{\|}{C}}CH_2\overset{O}{\overset{\|}{C}}CH_3$ (*c*) $CH_3\overset{O}{\overset{\|}{C}}\overset{O}{\overset{\|}{C}}H$ (*d*) $CH_3\underset{OH}{\underset{|}{C}}HCH_2CH_2\overset{O}{\overset{\|}{C}}H$

(*e*) $CH_3\underset{OH}{\underset{|}{C}}HCH_2CH_2\overset{O}{\overset{\|}{C}}CH_3$ (*f*) $CH_3CH_2\underset{OH}{\underset{|}{C}}H\overset{O}{\overset{\|}{C}}CH_3$

6. Shown below is the structure of a hypothetical molecule.

$$\underset{①}{CH_2{=}CH}{-}CH_2{-}\underset{②}{O}{-}CH_2{-}\overset{OH}{\overset{|③}{\underset{\underset{O \,④\, H}{\underset{|}{C}}}{\underset{|}{C}}}}{-}CH_2{-}CH_2{-}\underset{⑤}{\overset{O}{\overset{\|}{C}}}{-}CH_3$$

Predict the chemical reactions it would probably undergo if it were subjected to the reagents and conditions listed. To simplify the writing of the equations illustrating these reactions, isolate the portion of the molecule that would be involved, replace the other non-involved portions by symbols such as G, G′, G″, etc., and then write the equation (see example). You are expected to make reasonable predictions based only on the reactions we have studied thus far. If no reaction is to be predicted, write "None."

Example. Reagent: excess H_2, Ni, heat, pressure. We approach this problem in the following steps.

(*a*) First identify *by name* the functional groups in the molecule. (By doing this you take advantage of the way you mentally store information about the reactions of functional groups. Thus the *name* for —CH═O is aldehyde or aldehyde group. In your mental "file" on this group there should be such statements as "Aldehydes can be reduced to 1° alcohols" or "Aldehydes can be easily oxidized to acids," etc.

In the structure given you should recognize ① alkene, ② ether, ③ 3° alcohol, ④ aldehyde, ③ and ④ α-hydroxy aldehyde, ⑤ ketone.

(*b*) To continue the example, ask yourself how each group would respond to the reagent given.

① Alkenes will add hydrogen catalytically under heat and pressure.
② Ethers do not react with hydrogen.
③ There is no reaction of 3° alcohols with hydrogen.
④ Aldehydes can be reduced to 1° alcohols.
⑤ Ketones can be reduced to 2° alcohols.

(*c*) Third, write the equations. For example,

$$CH_2{=}CH{-}G + H_2 \xrightarrow[\text{heat, pressure}]{\text{Ni}} CH_3{-}CH_2{-}G$$

where G in this reaction is $\quad -CH_2OCH_2\overset{CH=O}{\overset{|}{\underset{OH}{\underset{|}{C}}}}CH_2CH_2\overset{O}{\overset{\|}{C}}CH_3.$

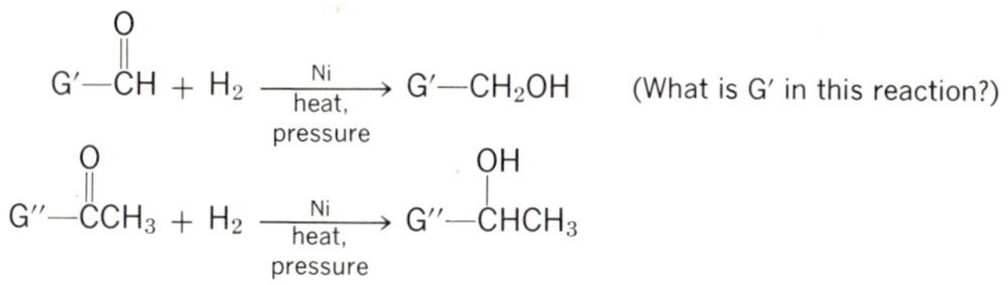

$$G'\!-\!\overset{\overset{\displaystyle O}{\|}}{C}H + H_2 \xrightarrow[\text{heat, pressure}]{\text{Ni}} G'\!-\!CH_2OH$$ (What is G′ in this reaction?)

$$G''\!-\!\overset{\overset{\displaystyle O}{\|}}{C}CH_3 + H_2 \xrightarrow[\text{heat, pressure}]{\text{Ni}} G''\!-\!\overset{\overset{\displaystyle OH}{|}}{C}HCH_3$$

The reagents and conditions for this exercise are

(1) Br_2 in CCl_4 solution, room temperature.

(2) Excess CH_3OH in the presence of dry HCl.

(3) Tollens' reagent.

(4) Benedict's reagent (do not attempt to write an equation).

(5) Dilute sodium hydroxide at room temperature.

7. Repeat the directions given in exercise 6 for the following hypothetical structure and the reagents given.

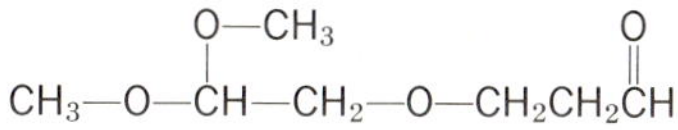

$$CH_3\!-\!O\!-\!\overset{\overset{\displaystyle O\!-\!CH_3}{|}}{C}H\!-\!CH_2\!-\!O\!-\!CH_2CH_2\overset{\overset{\displaystyle O}{\|}}{C}H$$

(*a*) Tollens' reagent (which is a basic solution).

(*b*) Excess water, H^+ catalyst.

(*c*) MnO_4^-, OH^-.

8. One of our goals is to be able to understand chemical events occurring among molecules of biochemical interest. Our chief purpose in studying organic chemistry has been to learn several of these reactions as they occur in simple systems. If we are to apply what we have learned, we must be able to recognize the simple organic reactions even when they take place with complicated structures.

The following reactions have been established for one or more biological systems. (Enzymes and other reagents have been omitted. Only the overall results are shown.) Classify each according to the following types.

A. Oxidation of a 2° alcohol to a ketone.

B. Oxidation of a mercaptan to a disulfide.

C. Oxidation of an aldehyde to a carboxylic acid.

D. Oxidation of a 1° alcohol to an aldehyde.

E. Dehydration of an alcohol.

F. Reduction of a ketone to a 2° alcohol.

G. Addition of water to an alkene linkage.

H. Reduction of an alkene linkage.

I. Reverse aldol condensation.

J. Hemiacetal formation.

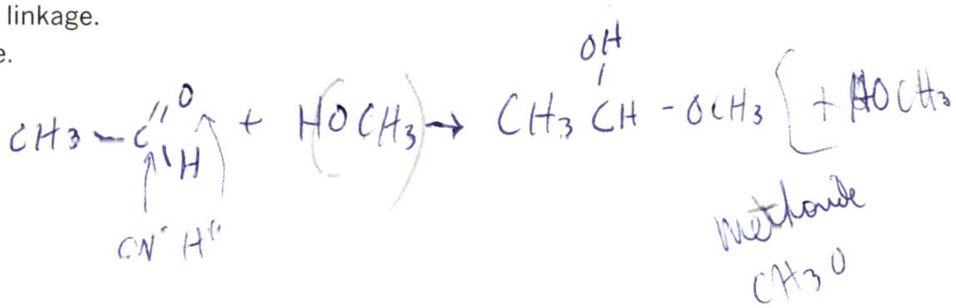

(*a*) $CH_2OH\!-\!C(=O)\!-\!OH \longrightarrow CH(=O)\!-\!C(=O)\!-\!OH$

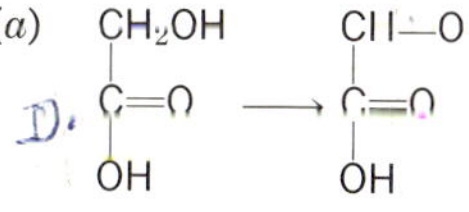

Glycolic acid → Glyoxylic acid

(*b*) $CO_2H\!-\!CH\!=\!CH\!-\!CO_2H \longrightarrow CO_2H\!-\!CH_2\!-\!CH_2\!-\!CO_2H$

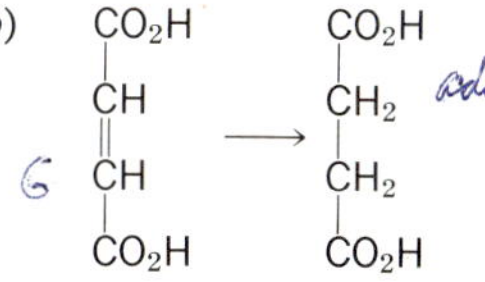

Fumaric acid → Succinic acid

(*c*)

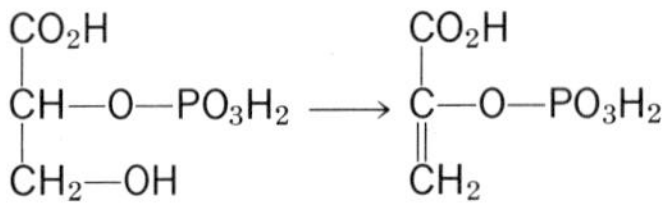

$$\begin{array}{l} CO_2H \\ | \\ CH{-}O{-}PO_3H_2 \\ | \\ CH_2{-}OH \end{array} \longrightarrow \begin{array}{l} CO_2H \\ | \\ C{-}O{-}PO_3H_2 \\ \| \\ CH_2 \end{array}$$

Phosphoglyceric acid → Phosphoenolpyruvic acid

(*d*)

$$\begin{array}{l} SH \qquad SH \\ | \qquad\quad | \\ CH_2CH_2CH(CH_2)_4CO_2H \end{array} \longrightarrow \begin{array}{l} S{-\!\!-\!\!-}S \\ | \qquad\quad | \\ CH_2 \quad CH(CH_2)_4CO_2H \\ \quad \diagdown CH_2 \diagup \end{array}$$

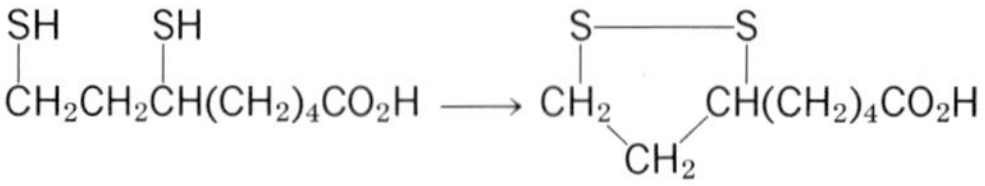

Dihydrolipoic acid → Lipoic acid

(*e*)

$$\begin{array}{l} CHCO_2H \\ \| \\ C{-}CO_2H \\ | \\ CH_2CO_2H \end{array} \longrightarrow \begin{array}{l} HO{-}CHCO_2H \\ \qquad | \\ \qquad CH{-}CO_2H \\ \qquad | \\ \qquad CH_2CO_2H \end{array}$$

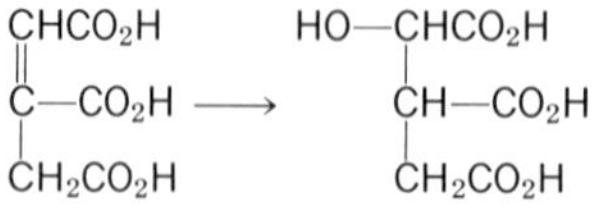

Aconitic acid → Isocitric acid

(*f*)

$$\begin{array}{l} CH_3 \\ | \\ CH{-}OH \\ | \\ CO_2H \end{array} \longrightarrow \begin{array}{l} CH_3 \\ | \\ C{=}O \\ | \\ CO_2H \end{array}$$

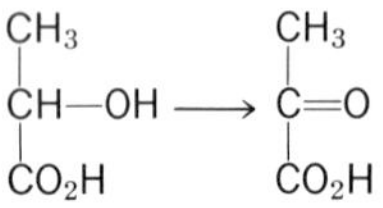

Lactic acid → Pyruvic acid

(*g*)

$$\begin{array}{l} CH{=}O \\ | \\ (CHOH)_4 \\ | \\ CH_2OPO_3H_2 \end{array} \longrightarrow \begin{array}{l} CO_2H \\ | \\ (CHOH)_4 \\ | \\ CH_2OPO_3H_2 \end{array}$$

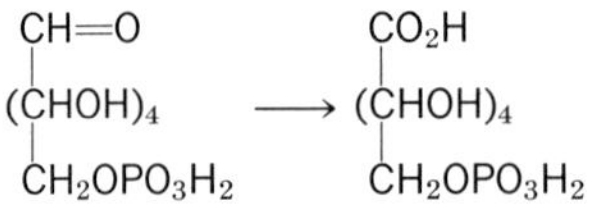

Glucose 6-phosphate → 6-Phosphogluconic acid

(*h*)

$$\begin{array}{l} CH_2OH \\ | \\ C{=}O \\ | \\ CH_2OPO_3H_2 \end{array} \longrightarrow \begin{array}{l} CH_2OH \\ | \\ CHOH \\ | \\ CH_2OPO_3H_2 \end{array}$$

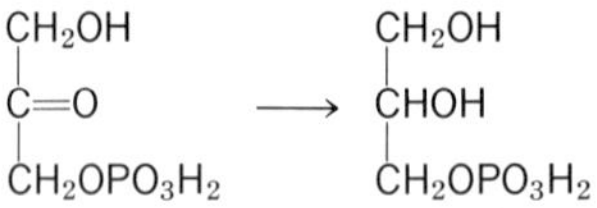

Dihydroxyacetone phosphate → α-Glycerophosphate

(*i*)

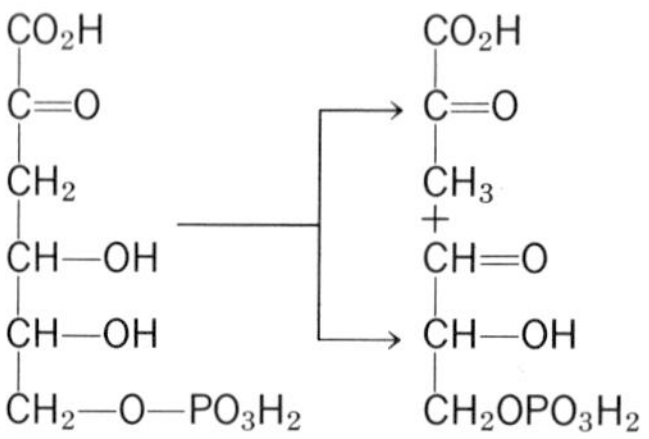

$$\begin{array}{l} CO_2H \\ | \\ C{=}O \\ | \\ CH_2 \\ | \\ CH{-}OH \\ | \\ CH{-}OH \\ | \\ CH_2{-}O{-}PO_3H_2 \end{array} \longrightarrow \begin{array}{l} CO_2H \\ | \\ C{=}O \\ | \\ CH_3 \\ + \\ CH{=}O \\ | \\ CH{-}OH \\ | \\ CH_2OPO_3H_2 \end{array} \begin{array}{l} \\ \\ \text{Pyruvic acid} \\ \\ \\ \\ \\ \\ \text{Glyceraldehyde 3-phosphate} \\ \\ \end{array}$$

2-Keto-3-deoxy-6-phosphogluconic acid

(*j*)

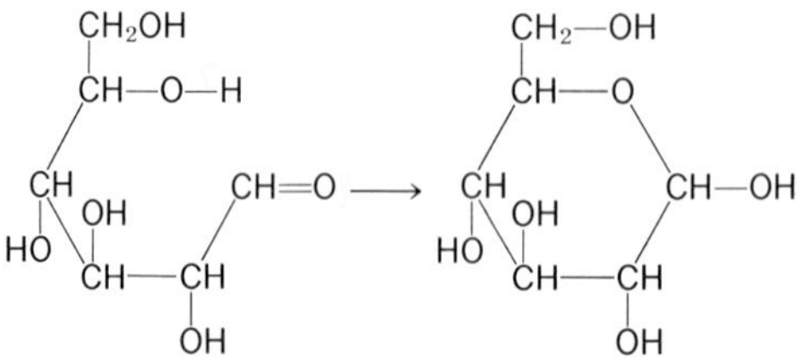

Glucose (open form): $CH_2OH{-}CH(O{-}H){-}CH(OH){-}CH(OH){-}CH(OH){-}CH{=}O$ ⟶ Glucose (closed form): ring $CH(CH_2{-}OH){-}O{-}CH(OH){-}CH(OH){-}CH(OH){-}CH(OH)$ (closing the ring via $CH{-}O$)

Glucose (open form) → Glucose (closed form)

CHAPTER FIFTEEN

Carboxylic Acids and Related Compounds

The carboxylic acid group occurs widely among substances of biochemical and industrial importance. Three closely related types of substances, salts of carboxylic

$$\begin{array}{c} O \\ \| \\ -C-OH \end{array}$$

Carboxyl group
(*carb*onyl + hydr*oxyl*)

acids, esters, and amides are also important. The chemistry of esters is basic to an understanding of simple lipids, and the chemistry of amides provides many of our insights into protein chemistry.

The nomenclature of these substances was described in Chapter 14. The several tables in this chapter provide additional illustrations.

CARBOXYLIC ACIDS

Several representative carboxylic acids are listed in Table 15.1. They are often called fatty acids because many are obtainable from fats and oils. Formic acid is a sharp-smelling, irritating liquid responsible for the sting of certain ants and the nettle. Acetic acid is the constituent of vinegar. The odor of rancid butter is produced by the presence of butyric acid. What is "strong" about valeric acid, from the Latin *valerum,* meaning "to be strong," is its odor. The same is true of caproic, caprylic, and capric acids, from the Latin *caper,* meaning "goat." Smelly goats and locker rooms telegraph their condition by molecules of these oily acids and their corresponding aldehydes. The vapor pressures of the acids with twelve and more carbons are low enough at room temperature that odors are slight.

Table 15.1 Carboxylic Acids

n	Structure	Name	Origin of Name	Mp (°C)	Bp (°C)	Solubility (in g/100 g water at 20°C)	K_a (at 25°C)
Straight-chain saturated acids, $C_nH_{2n}O_2$							
1	HCO_2H	Formic acid (methanoic)	L.*formica,* ant	8	101	∞	1.77×10^{-4} (20°C)
2	CH_3CO_2H	Acetic acid (ethanoic)	L.*acetum,* vinegar	17	118	∞	1.76×10^{-5}
3	$CH_3CH_2CO_2H$	Propionic acid (propanoic)	Gr.*proto,* first *pion,* fat	−21	141	∞	1.34×10^{-5}
4	$CH_3CH_2CH_2CO_2H$	Butyric acid (butanoic)	L.*butyrum,* butter	−6	164	∞	1.54×10^{-5}
5	$CH_3(CH_2)_3CO_2H$	Valeric acid (pentanoic)	L.*valere,* to be strong (valerian root)	−35	186	4.97	1.52×10^{-5}
6	$CH_3(CH_2)_4CO_2H$	Caproic acid (hexanoic)	L.*caper,* goat	−3	205	1.08	1.31×10^{-5}
7	$CH_3(CH_2)_5CO_2H$	Enanthic acid (heptanoic)	Gr.*oenanthe,* vine blossom	−9	223	0.26	1.28×10^{-5}
8	$CH_3(CH_2)_6CO_2H$	Caprylic acid (octanoic)	L.*caper,* goat	16	238	0.07	1.28×10^{-5}
9	$CH_3(CH_2)_7CO_2H$	Pelargonic acid (nonanoic)	pelargonium (geranium family)	15	254	0.03	1.09×10^{-5}
10	$CH_3(CH_2)_8CO_2H$	Capric acid (decanoic)	L.*caper,* goat	32	270	0.015	1.43×10^{-5}
12	$CH_3(CH_2)_{10}CO_2H$	Lauric acid (dodecanoic)	laurel	44	—	0.006	—
14	$CH_3(CH_2)_{12}CO_2H$	Myristic acid (tetradecanoic)	myristica (nutmeg)	54	—	0.002	—
16	$CH_3(CH_2)_{14}CO_2H$	Palmitic acid (hexadecanoic)	palm oil	63	—	0.0007	—
18	$CH_3(CH_2)_{16}CO_2H$	Stearic acid (octadecanoic)	Gr.*stear,* tallow	70	—	0.0003	—
Miscellaneous carboxylic acids							
	$C_6H_5CO_2H$	Benzoic acid	gum benzoin	122	249	0.34 (25°C)	6.46×10^{-5}
	$C_6H_5CH{=}CHCO_2H$	Cinnamic acid (*trans*)	cinnamon	42	—	0.04	3.65×10^{-5}
	$CH_2{=}CHCO_2H$	Acrylic acid	L.*acer,* sharp	13	141	soluble	5.6×10^{-5}
	CO_2H OH	Salicylic acid	L.*salix,* willow	159	211	0.22 (25°C)	1.1×10^{-3} (19°C)

PHYSICAL PROPERTIES

The carboxyl group confers considerable polarity to molecules possessing it. A carboxylic acid has a higher boiling point than an alcohol of the same formula weight,

CH_3CH_2OH	(f.wt. 46), bp 78°C	$CH_3CH_2CH_2OH$	(f.wt. 60), bp 97°C
$H—CO_2H$	(f.wt. 46), bp 101°C	CH_3CO_2H	(f.wt. 60), bp 118°C

partly because pairs of carboxylic acid molecules can be held together by two hydrogen bonds:

$$R—C\begin{matrix} \overset{\delta-}{O}\cdots\overset{\delta+}{H}—O \\ O—\underset{\delta+}{H}\cdots\underset{\delta-}{O} \end{matrix}C—R$$

Because a carboxylic acid has these dimeric forms (Greek: *di-*, two; *meros*, part), the effective formula weight is higher than the structural formula implies, and the boiling point is thus higher than otherwise expected. Other hydrogen-bonded polymers are also present.

The first four members of the homologous series of carboxylic acids are soluble in water, for the carboxyl group can both donate and accept hydrogen bonds to and from water molecules. Even so, as the hydrocarbon chain in the acid lengthens, solubility of the acid in water falls off sharply. Since the long-chain fatty acids (those with twelve carbons and more) are normal products of the digestion of fats and oils, the bloodstream, largely aqueous in nature, contains substances that help to dissolve the relatively insoluble fatty acids and transport them.

ACIDITY OF CARBOXYLIC ACIDS

In an aqueous medium molecules of carboxylic acids interact with water molecules. Collisions that are strong enough occur with a relatively low frequency, and only a small concentration of hydronium ions is present. In general:

$$R—\overset{\overset{O}{\|}}{C}—O—H + H_2O \rightleftharpoons R—\overset{\overset{O}{\|}}{C}—O:^- + H_3O^+$$

Weaker acid — Weaker base — Stronger base — Stronger acid

Specific example:

$$CH_3\overset{\overset{O}{\|}}{C}—O—H + H_2O \rightleftharpoons CH_3\overset{\overset{O}{\|}}{C}—O:^- + H_3O^+$$

Acetic acid — Water — Acetate ion — Hydronium ion

A 1 molar solution of acetic acid is ionized only to about 0.5% at room temperature. Evidently water molecules are not very strong proton acceptors in relation to carboxylic acids, which are relatively weak proton donors. Compared with alcohols, however, which also have the hydroxyl group,[1] carboxylic acids are stronger acids by several orders of magnitude.

Acidity Constants. As we learned in Chapter 8, the higher the value of K_a, the stronger the acid. Hence measured K_a values may be used to compare the strengths of different acids. In Tables 15.1 and 15.2 several such values of K_a, most on the order of 10^{-5} (0.00001), are recorded. In marked contrast, 1° alcohols have K_a values on the order of 10^{-16}. To understand this great difference in acidity for two classes possessing hydroxyl groups, we must look to structural differences.

Structure and Acidity. The ionization of an alcohol molecule produces an alkoxide ion. A carboxylate ion is produced from an acid.

$$\text{R—}\ddot{\underset{..}{\text{O}}}\text{—H} \rightleftharpoons \text{R—}\ddot{\underset{..}{\text{O}}}\text{:}^- + \text{H}^+$$

Alkoxide ion

$$\text{R—}\overset{\overset{\displaystyle \ddot{\text{O}}}{\|}}{\text{C}}\text{—}\ddot{\underset{..}{\text{O}}}\text{—H} \rightleftharpoons \text{R—}\overset{\overset{\displaystyle \ddot{\text{O}}}{\|}}{\text{C}}\text{—}\ddot{\underset{..}{\text{O}}}\text{:}^- + \text{H}^+$$

Carboxylate ion

Table 15.2 Structure and Acidity of Substituted Carboxylic Acids

Structure	K_a (25°C)
α-Haloacetic acids	
F—CH_2CO_2H	219×10^{-5}
Cl—CH_2CO_2H	155×10^{-5}
Br—CH_2CO_2H	138×10^{-5}
I—CH_2CO_2H	75×10^{-5}
H—CH_2CO_2H	1.8×10^{-5}
Chloroacetic acids	
H—CH_2CO_2H	1.8×10^{-5}
Cl—CH_2CO_2H	155×10^{-5}
Cl_2CHCO_2H	$5{,}100 \times 10^{-5}$
Cl_3CCO_2H	$90{,}000 \times 10^{-5}$
Monochlorobutyric acids	
α-chloro $CH_3CH_2\underset{\displaystyle Cl}{\underset{\vert}{C}}HCO_2H$	139×10^{-5}
β-chloro $CH_3\underset{\displaystyle Cl}{\underset{\vert}{C}}HCH_2CO_2H$	8.9×10^{-5}
γ-chloro $\underset{\displaystyle Cl}{\underset{\vert}{C}}H_2CH_2CH_2CO_2H$	3.8×10^{-5}
no chloro $CH_3CH_2CH_2CO_2H$	1.54×10^{-5}

[1] Here, as always, hydroxyl *group* is not to be confused with hydroxide *ion*.

The classical structure for an alkoxide ion adequately represents it. Reasonable resonance structures, for example, cannot be drawn for it. For the carboxylate ion two reasonable structures—differing only in locations of electrons—can be drawn, **1** and **2.**

(hybrid)

1 **2** **3**

According to resonance theory, whenever such a pencil and paper operation is possible, neither structure can be satisfactory for the species. Both **1** and **2** must be used to represent the *one* species present. Blending **1** and **2** into a hybrid produces structure **3.** Since **1** and **2,** although not identical, are still equivalent as far as predicted internal energies are concerned, each must make the same contribution to the hybrid. If they do so, the two carbon-oxygen bonds in **3** must be alike; both carbon-oxygen bond distances must be the same. Spectroscopic examinations do, in fact, confirm this prediction.

The net effect is that, compared with the molecule from which it forms, the carboxylate ion is more stable than an alkoxide ion. The free energy change for ionization of an acid is more favorable than that for ionization of an alcohol, and the carboxylic acid is a stronger acid than the alcohol.

The Inductive Effect and Acidity. The K_a values for α-haloacetic acids given in Table 15.2 reveal that as the halogen substituent changes from —I to —Br to —Cl to —F, the acidity *increases*. This order is the same as the order of the relative electronegativities of the halogens. We have learned that one way to stabilize a

If X is a group that withdraws electrons, the anion is stabilized further; hence the acid is a stronger acid.

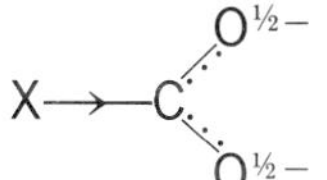

If X tends to release electrons, the anion is destabilized; negative charge is being forced into a region already negatively charged, making the acid weaker.

charge is to disperse it. Electron-withdrawing substituents such as halogens have this kind of inductive effect, and their presence on the α-carbon of acetic acid strengthens the acid. By the same token, an alkyl group, being electron-releasing, weakens the acid. Thus the drop in K_a value between formic acid and acetic acid is reasonable because a methyl group has an electron-releasing effect.

As more and more electron-withdrawing substituents are placed on the α-carbon of acetic acid, its acidity rises sharply. The K_a values for mono-, di-, and trichloroacetic acids (Table 15.2) reveal this. Trichloroacetic acid is almost as strong as a mineral acid and is sometimes used in its place as an acid catalyst.

As the substituent exerting an inductive effect becomes farther and farther removed from the negatively charged end of the carboxylate ion, its inductive influence on that group drops off markedly. The K_a values for the three monochlorobutyric acids in Table 15.2 illustrate this.

SALTS OF CARBOXYLIC ACIDS

The K_a value of an acid is a measure of the ability of the carboxylic acid group to donate a proton to a water molecule. In this role the neutral water molecule acts as a base, a proton acceptor, and a very weak one at that. In contrast, the hydroxide ion is a far stronger base. Aqueous sodium hydroxide, for example, can quantitatively convert carboxylic acids into their salts. In general:

$$R{-}\overset{\overset{\displaystyle O}{\|}}{C}{-}O{-}H + Na^{+}\,OH^{-} \rightleftharpoons R{-}\overset{\overset{\displaystyle O}{\|}}{C}{-}O^{-}Na^{+} + H{-}OH$$

Stronger acid | Stronger base | Weaker base | Weaker acid

Specific examples:

$$CH_3{-}\overset{\overset{\displaystyle O}{\|}}{C}{-}O{-}H + Na^{+}OH^{-} \rightleftharpoons CH_3{-}\overset{\overset{\displaystyle O}{\|}}{C}{-}O^{-}Na^{+} + H{-}OH$$

Acetic acid | Sodium hydroxide | Sodium acetate | Water

$$CH_3(CH_2)_{16}\overset{\overset{\displaystyle O}{\|}}{C}{-}O{-}H + Na^{+}OH^{-} \rightleftharpoons CH_3(CH_2)_{16}\overset{\overset{\displaystyle O}{\|}}{C}{-}O^{-}Na^{+} + H{-}OH$$

Stearic acid (insoluble in water) | Sodium stearate (soluble in water, a soap)

$$C_6H_5\overset{\overset{\displaystyle O}{\|}}{C}{-}O{-}H + Na^{+}OH^{-} \rightleftharpoons C_6H_5\overset{\overset{\displaystyle O}{\|}}{C}{-}O^{-}Na^{+} + H{-}OH$$

Benzoic acid (insoluble in water) | Sodium benzoate (soluble in water)

The salts that form in these reactions are obtained by evaporating the water. All are crystalline solids with relatively high melting points; many decompose before they melt. They are fully ionic substances with the properties to be expected of such materials, solubility in water, insolubility in nonpolar solvents, and relatively high melting points. Table 15.3 lists a few representative examples.

Since they form from substances that are not good proton donors, carboxylate ions (RCO_2^-) must be good proton acceptors. In other words, they are quite good bases, especially toward a good proton donor such as hydronium ion:

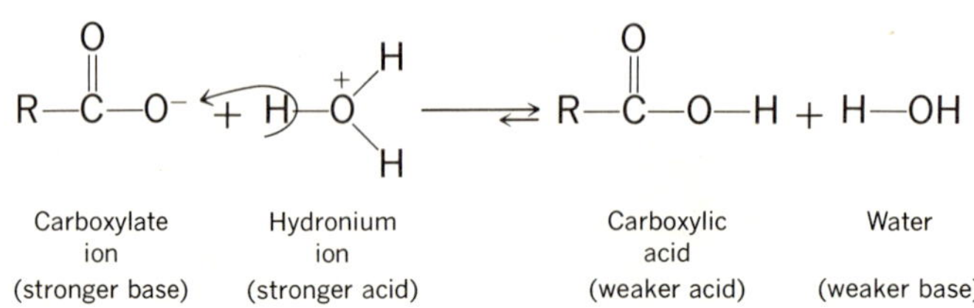
R—C—O⁻ + H—O⁺(H)H ⇌ R—C—O—H + H—OH
Carboxylate ion (stronger base)
Hydronium ion (stronger acid)
Carboxylic acid (weaker acid)
Water (weaker base)

For example,

$$\underset{\substack{\text{Sodium benzoate}\\\text{(soluble in water)}}}{C_6H_5CO_2^-Na^+} + \underset{\substack{\text{Hydrochloric}\\\text{acid}}}{HCl} \xrightarrow[\text{water}]{} \underset{\substack{\text{Benzoic acid}\\\text{(insoluble in water)}}}{C_6H_5CO_2H} + \underset{\text{Sodium chloride}}{Na^+Cl^-}$$

The ability of a negatively charged carboxylate group to accept a proton from a hydronium ion is important in maintaining control of pH in body fluids. Proteins in the bloodstream, for example, possess such groups in their molecules, and they act to buffer the pH of the blood against any downward change. These same groups in molecules of proteins also help to bring them into solution in water. For helping to make an organic compound soluble in water, the $—CO_2^-$ group is one of the best we shall encounter. Quaternary ammonium ions are also very soluble in water, and such groups are also found in proteins.

Hydrolysis of Salts of Carboxylic Acids. Aqueous solutions of salts of carboxylic acids will usually test basic to litmus because of hydrolysis, as discussed on page 231.

REACTIONS OF CARBOXYLIC ACIDS

The changes of the carboxylic acid group of principal concern in this study are of the type

$$R—\overset{\overset{\displaystyle O}{\|}}{C}—OH \xrightarrow[\text{reagent}]{} R—\overset{\overset{\displaystyle O}{\|}}{C}—G$$

where G is —Cl, —OR′, $—NH_2$, —NHR′, or $—NHR'_2$. By action of some reagent the —OH group may be replaced by one of these groups. We shall study these transformations next.

Synthesis of Acid Chlorides. $R—\overset{\overset{\displaystyle O}{\|}}{C}—Cl$. Action of any one of three reagents, thionyl chloride ($SOCl_2$), phosphorus trichloride (PCl_3), or phosphorus pentachloride (PCl_5), on a carboxylic acid converts it to its acid chloride. In general:

Table 15.3 Salts of Carboxylic Acids

Name	Structure	Mp (°C)	Solubility: Water	Solubility: Ether
Sodium formate	$HCO_2^-Na^+$	253	soluble	insoluble
Sodium acetate	$CH_3CO_2^-Na^+$	323	soluble	insoluble
Sodium propionate	$CH_3CH_2CO_2^-Na^+$		soluble	insoluble
Sodium benzoate	$C_6H_5CO_2^-Na^+$		66 g/100 cc	insoluble
Sodium salicylate	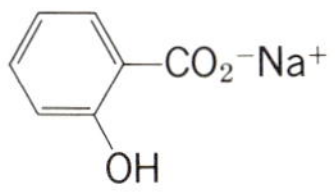 $C_6H_4(OH)CO_2^-Na^+$ (benzene ring with $—CO_2^-Na^+$ and ortho OH)		111 g/100 cc	insoluble

$$R{-}\overset{\overset{\displaystyle O}{\|}}{C}{-}OH \xrightarrow[PCl_5]{SOCl_2 \text{ or } PCl_3 \text{ or}} R{-}\overset{\overset{\displaystyle O}{\|}}{C}{-}Cl$$

Specific examples:

$$\underset{\text{Acetic acid}}{CH_3\overset{\overset{\displaystyle O}{\|}}{C}{-}OH} \xrightarrow{PCl_3} \underset{\text{Acetyl chloride (67\%)}}{CH_3\overset{\overset{\displaystyle O}{\|}}{C}{-}Cl} \quad (+H_3PO_3)$$

$$\underset{\text{Benzoic acid}}{C_6H_5\overset{\overset{\displaystyle O}{\|}}{C}{-}OH} \xrightarrow{SOCl_2} \underset{\text{Benzoyl chloride (91\%)}}{C_6H_5\overset{\overset{\displaystyle O}{\|}}{C}{-}Cl} \quad (+SO_2 + HCl)$$

$$\underset{\text{Caprylic acid}}{CH_3(CH_2)_6\overset{\overset{\displaystyle O}{\|}}{C}{-}OH} \xrightarrow{PCl_5} \underset{\text{Caprylyl chloride (82\%)}}{CH_3(CH_2)_6\overset{\overset{\displaystyle O}{\|}}{C}{-}Cl} \quad (+POCl_3 + HCl)$$

Acid chlorides are the most reactive of the acid derivatives. Although they do not occur in the chemical inventory of living organisms, they are important to our study in other ways. *In vitro* synthesis of esters and amides are frequently accomplished by first converting the less reactive acid to its acid chloride.

Synthesis of Acid Anhydrides. $R{-}\overset{\overset{\displaystyle O}{\|}}{C}{-}O{-}\overset{\overset{\displaystyle O}{\|}}{C}{-}R$. With reactivity somewhat comparable to that of acid chlorides are types of anhydrides which do occur in living cells. The most important aliphatic anhydride is acetic anhydride, and one of the ways it may be prepared is by the reaction of acetic acid with acetyl chloride:

$$CH_3{-}\overset{\overset{\displaystyle O}{\|}}{C}{-}OH + Cl{-}\overset{\overset{\displaystyle O}{\|}}{C}{-}CH_3 \xrightarrow{\text{pyridine}}$$

$$CH_3{-}\overset{\overset{\displaystyle O}{\|}}{C}{-}O{-}\overset{\overset{\displaystyle O}{\|}}{C}{-}CH_3 + HCl \quad \text{(as its pyridine salt)}$$

This method is quite general.

Synthesis of Esters. $R{-}\overset{\overset{\displaystyle O}{\|}}{C}{-}O{-}R'$. Esters may be made in a variety of ways. Direct esterification is presented here. In general:

$$R{-}\overset{\overset{\displaystyle O}{\|}}{C}{-}OH + H{-}O{-}R' \overset{H^+}{\rightleftharpoons} R{-}\overset{\overset{\displaystyle O}{\|}}{C}{-}O{-}R' + H{-}OH$$

Specific examples:

$$\underset{\text{Acetic acid}}{CH_3\overset{\overset{\displaystyle O}{\|}}{C}OH} + \underset{\text{Methyl alcohol}}{HOCH_3} \underset{\text{heat}}{\overset{H^+}{\rightleftharpoons}} \underset{\text{Methyl acetate}}{CH_3\overset{\overset{\displaystyle O}{\|}}{C}OCH_3} + H_2O$$

$$CH_3CH_2\overset{O}{\overset{\|}{C}}OH + HOCH_2CH_3 \underset{heat}{\overset{H^+}{\rightleftharpoons}} CH_3CH_2\overset{O}{\overset{\|}{C}}OCH_2CH_3 + H_2O$$

Propionic acid, Ethyl alcohol, Ethyl propionate

$$C_6H_5\overset{O}{\overset{\|}{C}}OH + CH_3OH \underset{heat}{\overset{H^+}{\rightleftharpoons}} C_6H_5\overset{O}{\overset{\|}{C}}OCH_3 + H_2O$$

Benzoic acid, Methyl alcohol, Methyl benzoate

When a straight-chain acid and a 1° alcohol are mixed in a 1:1 mole ratio and heated, about one-third mole percent of acid and alcohol is still present at the eventual equilibrium; about two-thirds mole percent changes to ester and water. The equilibrium may be shifted in favor of the ester by using a large excess of either alcohol (usually the cheaper starting material) or acid or by removing the water (or ester) as it forms.

Mechanism of Direct Esterification. If methanol enriched in ^{18}O is allowed to react with benzoic acid, the ^{18}O appears in the ester, methyl benzoate, rather than in the water:

$$C_6H_5C(=O)\text{—}O\text{—}H + H\text{—}^{18}O\text{—}CH_3 \xrightarrow{H^+} C_6H_5C(=O)\text{—}^{18}O\text{—}CH_3 + H\text{—}OH$$

This result discloses that direct esterification involves rupture of the C—OH bond in the acid and the H—O bond in the alcohol.

A mineral acid is normally used to catalyze the reaction. Its function is to *protonate* the carboxyl group to increase the size of the positive charge on the carbonyl carbon. In this condition it will more readily undergo nucleophilic attack by an alcohol molecule.

$$R\text{—}\overset{\delta+}{C}(=\overset{\delta-}{O})\text{—}OH + H^+ \underset{}{\overset{\text{protonation}}{\rightleftharpoons}} \left[R\text{—}\overset{+}{C}(\text{—}O\text{—}H)\text{—}O\text{—}H\right]$$

Catalyst

partial plus, mildly reactive toward a nucleophile (i.e., an electron-rich reagent)

full plus, much more reactive toward a nucleophile

After this protonation of the acid, a succession of reactions takes place, all involving equilibria.

$$R\text{—}C(=O)\text{—}OH + H^+ \underset{\text{protonation}}{\rightleftharpoons} \left[R\text{—}\overset{+}{C}(\text{—}OH)\text{—}OH\right] + H\text{—}O\text{—}R' \underset{\text{attack by nucleophile, } R'\ddot{O}H}{\rightleftharpoons}$$

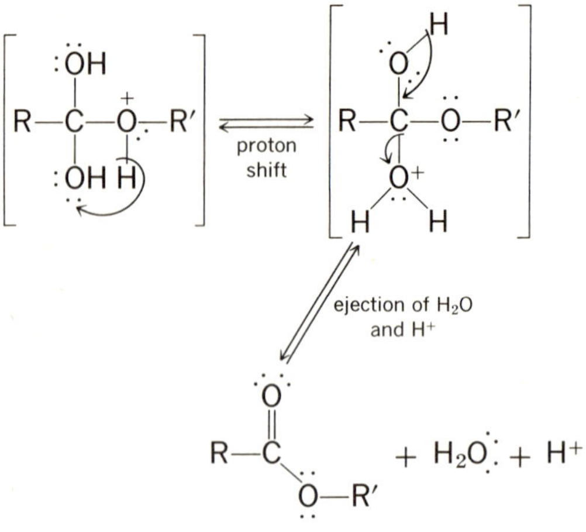

$$\left[\begin{array}{c} :\ddot{O}H \\ | \\ R-C-\overset{+}{O}-R' \\ | \quad | \\ :\ddot{O}H \; H \end{array} \right] \underset{\text{proton shift}}{\rightleftharpoons} \left[\begin{array}{c} \ddot{O}-H \\ | \\ R-C-\ddot{O}-R' \\ | \\ \overset{+}{O}(H)H \end{array} \right]$$

ejection of H_2O and H^+

$$R-\overset{\overset{\displaystyle \ddot{O}}{\|}}{C}-\ddot{O}-R' + H_2\ddot{O}: + H^+$$

The equilibria are driven from left to right, that is, toward ester formation, by an excess of alcohol or by removing one of the products as it forms. Conversely, an ester can be converted to its original acid and alcohol by the action of excess water in the presence of a strong acid catalyst. The mechanism for the hydrolysis of an ester is the exact reverse of esterification.

From the standpoint of predicting correctly what forms in a specific problem of esterification, the following step-by-step procedure is suggested until experience has been gained. A sample question might ask what ester forms between propionic acid and methyl alcohol.

1. Write the correct structures of the acid and the alcohol,

$$\underset{\text{Propionic acid}}{CH_3CH_2\overset{\overset{O}{\|}}{C}-OH} + \underset{\text{Methyl alcohol}}{H-O-CH_3}$$

2. Blacken out or erase the hydrogen on the OH of the alcohol and the hydroxyl group of the acid.

$$CH_3CH_2\overset{\overset{O}{\|}}{C}-\blacksquare OH\blacksquare + \blacksquare H\blacksquare-O-CH_3 \longrightarrow$$

3. Link the remaining fragments; the oxygen of the alcohol (and everything still attached to it) goes to the carbonyl carbon of the acid:

$$CH_3CH_2\overset{\overset{O}{\|}}{C}\leftarrow\!\!-\!\!-O-CH_3 \longrightarrow CH_3CH_2\overset{\overset{O}{\|}}{C}-O-CH_3$$

4. Having determined the correct structure of the ester, write the equation for its formation in a neat and orderly manner. The first three steps are done on scratch paper and later, with practice, mentally.

$$\underset{\text{Propionic acid}}{CH_3CH_2CO_2H} + \underset{\text{Methyl alcohol}}{CH_3OH} \xrightarrow{H^+} \underset{\text{Methyl propionate}}{CH_3CH_2CO_2CH_3} + H_2O$$

A mastery of esterification will be of great value in our future study of the chemistry of lipids, for the simple lipids are merely esters of glycerol, a trihydric alcohol.

Exercise 15.1 Write equations for direct esterification between the compounds in the following pairs.

(*a*) Formic acid and ethyl alcohol.
(*b*) Acetic acid and methyl alcohol.
(*c*) Propionic acid and isobutyl alcohol.
(*d*) Benzoic acid and methyl alcohol.
(*e*) Butyric acid and *n*-butyl alcohol.
(*f*) Acetic acid (three moles) and glycerol (one mole).
(*g*) Salicylic acid and methyl alcohol.
(*h*) Acetic acid and *n*-pentyl alcohol.
(*i*) Valeric acid and isopropyl alcohol.
(*j*) Terephthalic acid and ethyl alcohol (two moles).

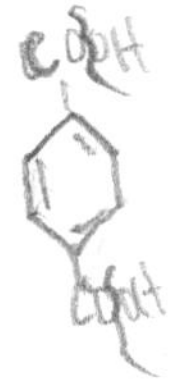

Synthesis of Amides. $R{-}\overset{O}{\overset{\|}{C}}{-}NH_2$, $R{-}\overset{O}{\overset{\|}{C}}{-}NHR'$, $R{-}\overset{O}{\overset{\|}{C}}{-}NR'_2$. Under appropriate conditions ammonia, 1° amines, or 2° amines, reacting with a carboxylic acid, can be made to split out water:

$$\underset{\text{Carboxylic acid}}{R{-}\overset{O}{\overset{\|}{C}}{-}OH} + \underset{NH_3,\ NH_2R',\ \text{or } NHR'_2}{H{-}N\langle} \rightleftharpoons \underset{\text{Amide}}{R{-}\overset{O}{\overset{\|}{C}}{-}N\langle} + H{-}OH$$

In vivo, the reaction is smoothly controlled by enzyme catalysis, as in protein synthesis (cf. p. 472). *In vitro,* to drive the reaction from left to right and obtain reasonable yields of amides, temperatures must be high, there must be an excess of one of the reactants, and/or water must be removed as it forms.

$$\underset{\text{Acetic acid}}{CH_3CO_2H} + \underset{\text{Ammonia}}{NH_3} \rightleftharpoons \underset{\text{Ammonium acetate}}{CH_3CO_2^-\ NH_4^+} \xrightarrow[\text{acetic acid, } 110^\circ C]{\text{excess}} \underset{\text{Acetamide (90\%)}}{CH_3\overset{O}{\overset{\|}{C}}NH_2} + H_2O$$

This method of preparing amides is not often used in the laboratory. Usually the acid is first converted to its acid chloride, which is then allowed to react with ammonia (or 1° or 2° amine). The enhanced reactivity of the acid chloride makes up for the inconvenience of using two steps instead of one (cf. p. 419).

Summary. The principal general reactions of carboxylic acids are the following:

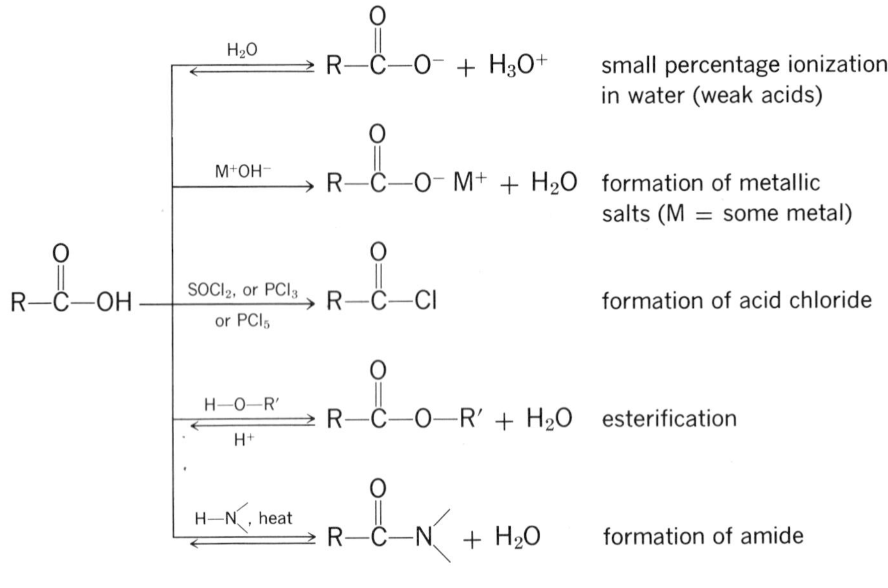
H_2O
R—C(=O)—O⁻ + H_3O^+
small percentage ionization in water (weak acids)
M^+OH^-
R—C(=O)—O⁻ M⁺ + H_2O
formation of metallic salts (M = some metal)
R—C(=O)—OH
$SOCl_2$, or PCl_3 or PCl_5
R—C(=O)—Cl
formation of acid chloride
H—O—R′
H⁺
R—C(=O)—O—R′ + H_2O
esterification
H—N<, heat
R—C(=O)—N< + H_2O
formation of amide

Exercise 15.2 Write the structure of the amide that would be expected to form from the following pairs of compounds. Until you have had a little practice, approach each problem in much the same way suggested for similar problems involving esters. (1) Set down the correct structures of the acid and the amine (or ammonia). (2) Erase or blacken out the —OH on the acid and one —H on the amine. (If the amine is a 3° amine, it has no —H, and we write "No amide forms.") (3) Join the remaining fragments—attach the nitrogen to the carbonyl carbon, copying over all other groups attached to each. (4) Write the equation in the conventional way.

(*a*) Acetic acid and methylamine.
(*b*) Butyric acid and diethylamine.
(*c*) Benzoic acid and ammonia.
(*d*) Propionic acid and trimethylamine.
(*e*) Valeric acid and aniline.

SYNTHESIS OF CARBOXYLIC ACIDS

The carboxylic acid group can be put into an organic molecule in several ways, but for our study of molecular biology only two are important, the oxidation of 1° alcohols and the oxidation of aldehydes. Both have already been studied.

$$\underset{1^\circ\ \text{alcohol}}{R{-}CH_2{-}OH} \xrightarrow{(O)} R{-}\overset{O}{\overset{\|}{C}}{-}OH \quad \text{(page 334)}$$

$$\underset{\text{Aldehyde}}{R{-}\overset{O}{\overset{\|}{C}}{-}H} \xrightarrow{(O)} R{-}\overset{O}{\overset{\|}{C}}{-}OH \quad \text{(page 389)}$$

DICARBOXYLIC ACIDS

Many important, naturally occurring carboxylic acids have two carboxyl groups per molecule; citric acid, a key intermediate in glucose catabolism, has three. Several are listed in Table 15.4. Just like the monocarboxylic acids, these substances form salts and esters. Complications sometimes arise in trying to prepare other derivatives, but we shall be able to ignore them.

IMPORTANT INDIVIDUAL ACID

Acetic Acid. CH_3CO_2H. Acetic acid, the principal constituent (4 to 5%) in vinegar, is made industrially in a variety of ways, notably from acetylene and by the controlled, catalytic oxidation of butane. Among its many uses may be mentioned the manufacture of cellulose acetate and white lead. The acetate unit is one of the most important intermediates in metabolism.

Table 15.4 Dicarboxylic Acids

n	Structure	Name	Origin of Name	Mp (°C)	K_a (25°C) First proton	K_a (25°C) Second proton
Straight-chain, saturated dicarboxylic acids, $C_nH_{2n-2}O_4$						
2	HO_2CCO_2H	oxalic acid (ethanedioic)	L. *oxalis*, sorrel	190 (decomposes)	$5{,}900 \times 10^{-5}$	6.4×10^{-5}
3	$HO_2CCH_2CO_2H$	malonic acid (propanedioic)	L. *malum*, apple	136	149×10^{-5}	0.20×10^{-5}
4	$HO_2C(CH_2)_2CO_2H$	succinic acid (butanedioic)	L. *succinum*, amber	182	6.9×10^{-5}	0.25×10^{-5}
5	$HO_2C(CH_2)_3CO_2H$	glutaric acid (pentanedioic)	L. *gluten*, glue	98	4.6×10^{-5}	0.39×10^{-5}
6	$HO_2C(CH_2)_4CO_2H$	adipic acid (hexanedioic)	L. *adipis*, fat	153	3.7×10^{-5}	0.24×10^{-5}
Miscellaneous dicarboxylic acids						
	$HCCO_2H \Vert HCCO_2H$	maleic acid	L. *malum*, apple	131°	$1{,}420 \times 10^{-5}$	0.08×10^{-5}
	$HO_2CH \Vert HCCO_2H$	fumaric acid	fumitory (a garden annual)	sublimes above 200	93×10^{-5}	3.6×10^{-5}
	$C_6H_4(CO_2H)_2$ (1,2)	phthalic acid	na*phthalene*	231	130×10^{-5}	0.39×10^{-5}
	$C_6H_4(CO_2H)_2$ (1,3)	isophthalic acid		345	29×10^{-5}	0.25×10^{-5}
	$C_6H_4(CO_2H)_2$ (1,4)	terephthalic acid		sublimes	31×10^{-5}	1.5×10^{-5} (16°C)

DERIVATIVES OF ORGANIC ACIDS

When the term derivative is used in the context "derivative of an organic acid," we mean that the compound can be prepared from the acid directly or indirectly and can be converted to the acid again, usually by the action of water. In fact, much of the chemistry of acid derivatives consists of converting one into another or into the parent acid. Even though we must deal with several types of compounds, their reactions are so similar that it would be best to note these general patterns first.

For the remainder of this chapter the most noteworthy aspect of acids and their derivatives will be the polarity of the carbonyl group. The carbonyl carbon bears a partial positive charge and is therefore vulnerable to attack by a reagent that is electron-rich. The common theme running throughout the chemistry of these substances is a substitution illustrated by the following:

$$R{-}C(=O){-}G + :Z \longrightarrow R{-}C(=O){-}Z + :G$$

where G = —OH (as in carboxylic acids)
—Cl (as in acid chlorides)
$-O-\overset{\overset{O}{\|}}{C}-R$ (as in an acid anhydride)
—O—R′ (as in an ester)
—NH_2 (as in amides)
—NHR (as in amides)
—NR_2 (as in amides)
:Z = some nucleophile which may be a neutral species or negatively charged (e.g., OH^- or H_2O, ^-OR or HOR, NH_3, etc.)

These substitution reactions at the unsaturated carbon of the carbonyl group usually proceed much more readily than analogous substitutions at a saturated carbon (as in an alkyl halide or an alcohol or an ether or an amine).

One reason for the enhanced reactivity is the fact that the carbonyl carbon goes from an essentially trivalent state to a tetravalent state as the intermediate forms:

$$R{-}\overset{\delta+}{C}(=\overset{\delta-}{O}){-}G + :Z \longrightarrow R{-}C(-O^-)(G){-}Z \longrightarrow RC(=O){-}Z + :G$$

Carbonyl carbon is trivalent in the sense that it is holding only three groups.

The former carbonyl carbon is tetravalent.

In contrast, substitution at a saturated carbon requires, in the concerted mechanism, a shift from a tetravalent carbon to a carbon that is temporarily pentavalent (attempting to hold five groups):

$$G{-}R + {:}Z \longrightarrow [G\cdots R\cdots Z] \longrightarrow G{:} + R{-}Z$$

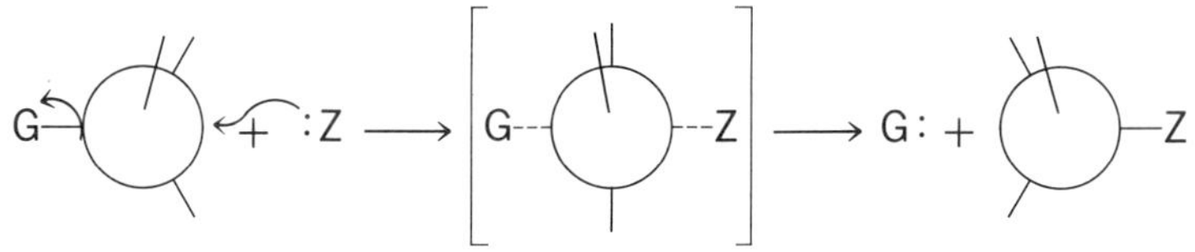

$$G{-}\bigcirc + {:}Z \longrightarrow [G{-}{-}\bigcirc{-}{-}Z] \longrightarrow G{:} + \bigcirc{-}Z$$

The carbon holding —G is tetravalent.

In the transition state the carbon is holding five groups.

The pentasubstituted carbon, because it produces severe crowding of groups, causes this transition state to be of higher internal energy than a transition state from the carbonyl compound involving a temporarily tetrasubstituted carbon. Substitution at the unsaturated carbon of the carbonyl group therefore usually proceeds with a lower energy of activation, resulting of course in a higher frequency of successful collisions and a more rapid rate.

Another factor making acid derivatives $R{-}\overset{\overset{\displaystyle O}{\|}}{C}{-}G$ more reactive than their counterparts in saturated systems, R—G, is the fact that in acid derivatives, two groups are acting to exert electron withdrawal from the carbonyl carbon. In R—G, only the group, G, is acting in this way. Hence the relative size of the partial positive charge on a carbonyl carbon may be expected to be greater than that on the carbon in R—G which is directly holding —G. In a sense, then, the carbon to be attacked in acid derivatives is more attractive to the nucleophile than the carbon in saturated systems, R—G.

We can make the following general comparisons between the reactivities of acid derivatives toward nucleophilic substitution and those of their analogs without carbonyl groups.

$$R{-}\overset{\overset{\displaystyle O}{\|}}{C}{-}Cl > R{-}Cl$$

—Acid chlorides are more reactive than alkyl chlorides.

$$R{-}\overset{\overset{\displaystyle O}{\|}}{C}{-}O{-}R' > R{-}O{-}R'$$

—Esters are more reactive than ethers.

$$R{-}\overset{\overset{\displaystyle O}{\|}}{C}{-}NH_2 > R{-}NH_2$$

—Amides are more reactive than amines.

Within the series of acid derivatives, relative reactivities toward a common nucleophile (e.g., water) are generally in the following order:

$$\text{Acid chlorides} \gtrsim \text{acid anhydrides} > \text{esters} > \text{amides}$$

Most reactive ... Least reactive

——decreasing reactivity——→

Some of the reactions of acid derivatives are catalyzed by mineral acids which render the carbonyl carbon even more positively charged:

$$R-\overset{\delta+}{C}(=\ddot{O}^{\delta-})-G \underset{\text{protonation}}{\overset{H^+}{\rightleftharpoons}} R-\overset{+}{C}(-\ddot{O}-H)-G \xrightarrow[\text{attack by nucleophile}]{:Z} R-C(-\ddot{O}-H)(-G)-Z \longrightarrow R-C(=\ddot{O}:)-Z + :G + H^+$$

ACID CHLORIDES AND ANHYDRIDES

Some examples of these substances are given in Table 15.5. Since the reactions of these two types of derivatives are so similar, they will be considered together. The principal reactions are outlined in the following equations.

Hydrolysis. Conversion to Original Acid

$$\underset{\text{Acid chloride}}{R-\overset{O}{\overset{\|}{C}}-Cl} + H_2O \longrightarrow R-\overset{O}{\overset{\|}{C}}-OH + H-Cl$$

Table 15.5 Acid Chlorides and Anhydrides

Name	Structure	Bp (°C)
(Formyl chloride)	(H—CO—Cl) unknown substance	
Acetyl chloride	CH_3COCl	51
Propionyl chloride	CH_3CH_2COCl	80
Benzoyl chloride	C_6H_5COCl	197
Acetic anhydride	$CH_3\overset{O}{\overset{\|}{C}}-O-\overset{O}{\overset{\|}{C}}CH_3$	136
Benzoic anhydride	$C_6H_5\overset{O}{\overset{\|}{C}}-O-\overset{O}{\overset{\|}{C}}C_6H_5$	42 (mp)
Phthalic anhydride	$C_6H_4(CO)_2O$ (cyclic)	131 (mp)

$$R-\overset{O}{\overset{\|}{C}}-O-\overset{O}{\overset{\|}{C}}-R + H_2O \longrightarrow R-\overset{O}{\overset{\|}{C}}-OH + H-O-\overset{O}{\overset{\|}{C}}-R$$

Anhydride

Alcoholysis. Ester Formation

$$R-\overset{O}{\overset{\|}{C}}-Cl + H-O-R' \longrightarrow R-\overset{O}{\overset{\|}{C}}-O-R' + H-Cl$$

$$R-\overset{O}{\overset{\|}{C}}-O-\overset{O}{\overset{\|}{C}}-R + H-O-R' \longrightarrow R-\overset{O}{\overset{\|}{C}}-O-R' + H-O-\overset{O}{\overset{\|}{C}}-R$$

Ammonolysis. Amide Formation

$$R-\overset{O}{\overset{\|}{C}}-Cl + H-N\langle \text{ (excess)} \longrightarrow R-\overset{O}{\overset{\|}{C}}-N\langle + HCl$$

(as its salt with the excess ammonia or amine)

$H-NH_2$, $H-NHR'$, or $H-NR'_2$

$R-\overset{O}{\overset{\|}{C}}-NH_2$, $R-\overset{O}{\overset{\|}{C}}-NHR'$, or $R-\overset{O}{\overset{\|}{C}}-NR'_2$

$$R-\overset{O}{\overset{\|}{C}}-O-\overset{O}{\overset{\|}{C}}-R + H-N\langle \text{ (excess)} \longrightarrow$$

$$R-\overset{O}{\overset{\|}{C}}-N\langle + HO\overset{O}{\overset{\|}{C}}-R$$

(as a salt; see previous reaction)

Exercise 15.3 Complete the following equations by writing the structures of the products that would be expected to form in each. If no reaction is to be expected, write "No reaction."

(*a*) $CH_3-\overset{O}{\overset{\|}{C}}-Cl + H_2O \longrightarrow$

(*b*) $C_6H_5\overset{O}{\overset{\|}{C}}-Cl + CH_3OH \longrightarrow$

(*c*) $CH_3-\overset{O}{\overset{\|}{C}}-O-\overset{O}{\overset{\|}{C}}-CH_3 + CH_3CH_2OH \longrightarrow$

(*d*) $CH_3CH_2\overset{O}{\overset{\|}{C}}Cl + NH_3 \text{ (excess)} \longrightarrow$

(*e*) $C_6H_5CH_2-\overset{O}{\overset{\|}{C}}-Cl + (CH_3)_2NH \text{ (excess)} \longrightarrow$

(*f*) $CH_3\overset{O}{\overset{\|}{C}}-Cl + (CH_3)_3N \longrightarrow$

Exercise 15.4 Write a mechanism for the reaction of water with an acid chloride, for the reaction of an alcohol with an anhydride, and for the reaction of ammonia with an acid chloride.

The reactions of acid chlorides and anhydrides are quite exothermic, and those shown occur readily. The counterpart of anhydride chemistry of particular importance in biochemistry involves derivatives of phosphoric acid.

Phosphoric acid, H_3PO_4, is commonly represented by the following structure:

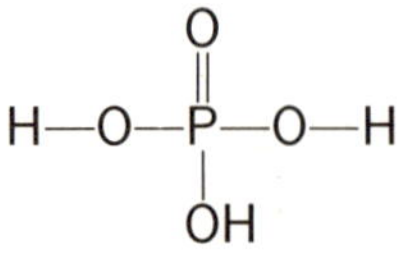

$$H-O-\overset{O}{\overset{\|}{\underset{OH}{\underset{|}{P}}}}-O-H$$

If a molecule of water is split out between two such molecules, a substance known as pyrophosphoric acid, $H_4P_2O_7$, is the result.

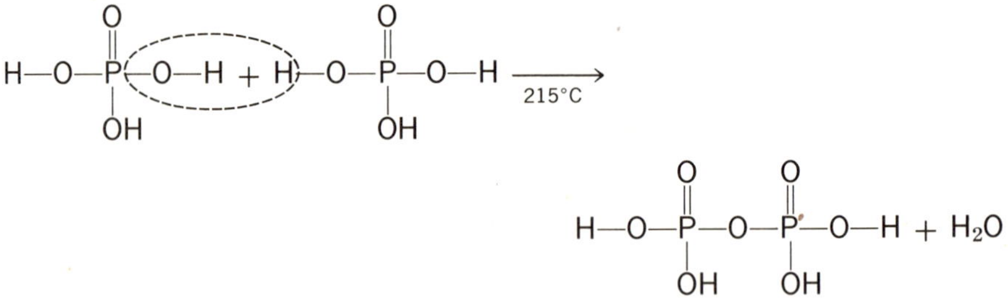

$$H-O-\overset{O}{\overset{\|}{\underset{OH}{\underset{|}{P}}}}-O-H + H-O-\overset{O}{\overset{\|}{\underset{OH}{\underset{|}{P}}}}-O-H \xrightarrow[215°C]{}$$

$$H-O-\overset{O}{\overset{\|}{\underset{OH}{\underset{|}{P}}}}-O-\overset{O}{\overset{\|}{\underset{OH}{\underset{|}{P}}}}-O-H + H_2O$$

Pyrophosphoric acid
(mp 61°C)

This is an acid in which all four hydrogens are replaceable and the successive ionization constants for ionization of the first, second, third, and fourth hydrogens are $K_1 = 1.4 \times 10^{-1}$, $K_2 = 1.1 \times 10^{-2}$, $K_3 = 2.9 \times 10^{-7}$, and $K_4 = 3.6 \times 10^{-9}$. In a sense pyrophosphoric acid is also very much like an anhydride.

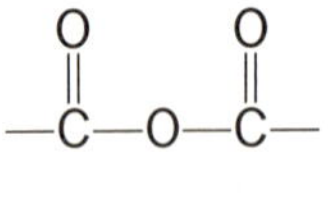

$$-\overset{O}{\overset{\|}{C}}-O-\overset{O}{\overset{\|}{C}}- \qquad -\overset{O}{\overset{\|}{\underset{|}{P}}}-O-\overset{O}{\overset{\|}{\underset{|}{P}}}-$$

Skeleton of a carboxylic acid anhydride

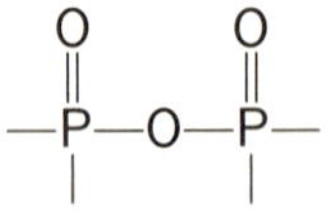

Skeleton of a phosphoric acid anhydride

The system occurs widely in living cells in the form of compounds of the following general type. They are shown in the acid forms, but usually they are present as singly or doubly ionized particles.

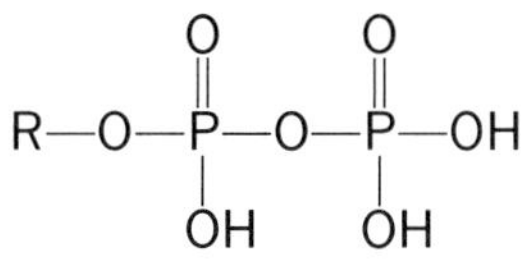

$$\mathrm{R{-}O{-}\overset{\overset{\displaystyle O}{\|}}{\underset{\underset{\displaystyle OH}{|}}{P}}{-}O{-}\overset{\overset{\displaystyle O}{\|}}{\underset{\underset{\displaystyle OH}{|}}{P}}{-}OH}$$

Monoester of pyrophosphoric acid (a diphosphate)[2]

for example:

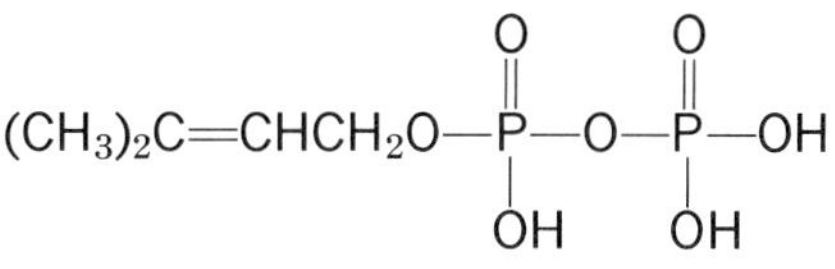

$$\mathrm{(CH_3)_2C{=}CHCH_2O{-}\overset{\overset{\displaystyle O}{\|}}{\underset{\underset{\displaystyle OH}{|}}{P}}{-}O{-}\overset{\overset{\displaystyle O}{\|}}{\underset{\underset{\displaystyle OH}{|}}{P}}{-}OH}$$

Isopentenyl pyrophosphate

The R— group is not necessarily a purely hydrocarbon system; in fact, it usually is not. Pyrophosphate esters are simultaneously esters, acids, and anhydrides. As anhydrides they participate in typical reactions of these compounds, for example, hydrolysis:

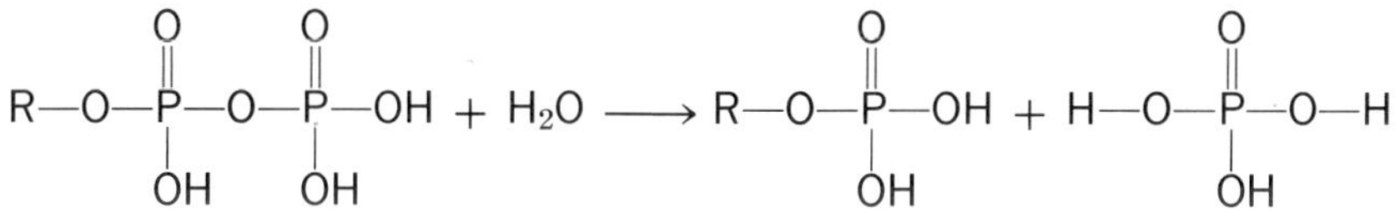

$$\underset{\text{Diphosphate}}{\mathrm{R{-}O{-}\overset{\overset{\displaystyle O}{\|}}{\underset{\underset{\displaystyle OH}{|}}{P}}{-}O{-}\overset{\overset{\displaystyle O}{\|}}{\underset{\underset{\displaystyle OH}{|}}{P}}{-}OH}} + \mathrm{H_2O} \longrightarrow \underset{\text{Monophosphate}}{\mathrm{R{-}O{-}\overset{\overset{\displaystyle O}{\|}}{\underset{\underset{\displaystyle OH}{|}}{P}}{-}OH}} + \underset{\text{Phosphoric acid}}{\mathrm{H{-}O{-}\overset{\overset{\displaystyle O}{\|}}{\underset{\underset{\displaystyle OH}{|}}{P}}{-}O{-}H}}$$

Triphosphate derivatives are also known and are essentially double anhydrides, plus being esters as well as acids. Much more will be studied about these substances, particularly adenosine triphosphate or ATP, in later chapters.

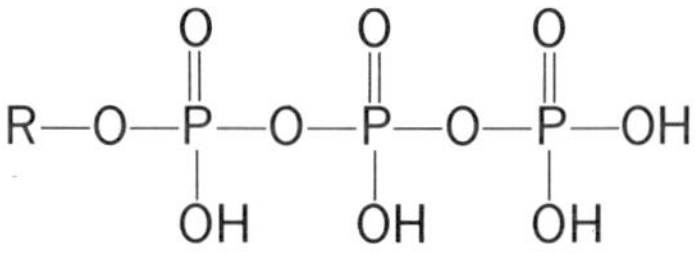

$$\mathrm{R{-}O{-}\overset{\overset{\displaystyle O}{\|}}{\underset{\underset{\displaystyle OH}{|}}{P}}{-}O{-}\overset{\overset{\displaystyle O}{\|}}{\underset{\underset{\displaystyle OH}{|}}{P}}{-}O{-}\overset{\overset{\displaystyle O}{\|}}{\underset{\underset{\displaystyle OH}{|}}{P}}{-}OH}$$

A triphosphate

ESTERS

Physical Properties. Since molecules of an ester cannot be hydrogen bond donors, they behave as less polar compounds than carboxylic acids with respect to solubilities and boiling points. The ethyl esters of the fatty acids shown in Table 15.6 all boil at lower temperatures and have lower solubilities in water than the parent acids. The lower-formula-weight esters possess some very fragrant odors (Table 15.7), in sharp contrast with those of the parent acids.

Chemical Properties. Esters undergo reactions of hydrolysis, alcoholysis, and ammonolysis, but only the reactions involving water will be studied here. In general:

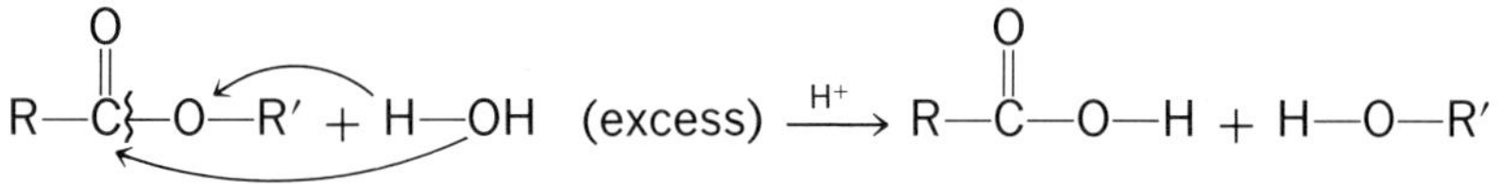

$$\mathrm{R{-}\overset{\overset{\displaystyle O}{\|}}{C}{-}O{-}R' + H{-}OH}\ \text{(excess)} \xrightarrow{\mathrm{H^+}} \mathrm{R{-}\overset{\overset{\displaystyle O}{\|}}{C}{-}O{-}H + H{-}O{-}R'}$$

[2] Organic chemists and biochemists frequently refer to these compounds as diphosphates. Note carefully that in this and similar contexts the term does *not* mean two separate PO_4^{3-} groups.

Table 15.6 Esters of Carboxylic Acids

Name	Structure	Mp (°C)	Bp (°C)	Solubility (in g/100 g water at 20°C)
Ethyl esters of straight-chain carboxylic acids, $RCO_2C_2H_5$				
Ethyl formate	$HCO_2C_2H_5$	−79	54	miscible in all proportions
Ethyl acetate	$CH_3CO_2C_2H_5$	−82	77	7.39 (25°C)
Ethyl propionate	$CH_3CH_2CO_2C_2H_5$	−73	99	1.75
Ethyl butyrate	$CH_3(CH_2)_2CO_2C_2H_5$	−93	120	0.51
Ethyl valerate	$CH_3(CH_2)_3CO_2C_2H_5$	−91	145	0.22
Ethyl caproate	$CH_3(CH_2)_4CO_2C_2H_5$	−68	168	0.063
Ethyl enanthate	$CH_3(CH_2)_5CO_2C_2H_5$	−66	189	0.030
Ethyl caprylate	$CH_3(CH_2)_6CO_2C_2H_5$	−43	208	0.007
Ethyl pelargonate	$CH_3(CH_2)_7CO_2C_2H_5$	−45	222	0.003
Ethyl caproate	$CH_3(CH_2)_8CO_2C_2H_5$	−20	245	0.0015
Esters of acetic acid, CH_3CO_2R				
Methyl acetate	$CH_3CO_2CH_3$	−99	57	24.4
Ethyl acetate	$CH_3CO_2CH_2CH_3$	−82	77	7.39 (25°C)
n-Propyl acetate	$CH_3CO_2CH_2CH_2CH_3$	−93	102	1.89
n-Butyl acetate	$CH_3CO_2CH_2CH_2CH_2CH_3$	−78	125	1.0 (22°C)
Miscellaneous esters				
Methyl acrylate	$CH_2{=}CHCO_2CH_3$		80	5.2
Methyl benzoate	$C_6H_5CO_2CH_3$	−12	199	insoluble
Methyl salicylate	benzene ring with CO_2CH_3 and ortho OH	−9	223	insoluble
Acetylsalicylic acid	benzene ring with CO_2H and ortho $OCCH_3$ (C=O)	135		
"Waxes"	$CH_3(CH_2)_nCO_2(CH_2)_nCH_3$	n = 23–33: carnauba wax = 25–27: beeswax = 14–15: spermaceti		

Specific examples:

$$CH_3\overset{O}{\overset{\|}{C}}{-}OCH_2CH_3 + H_2O \longrightarrow CH_3\overset{O}{\overset{\|}{C}}OH + CH_3CH_2OH$$

Ethyl acetate — Acetic acid — Ethyl alcohol

$$C_6H_5\overset{O}{\overset{\|}{C}}{-}OCH_3 + H_2O \longrightarrow C_6H_5\overset{O}{\overset{\|}{C}}OH + CH_3OH$$

Methyl benzoate — Benzoic acid — Methyl alcohol

Table 15.7 Fragrances or Flavors Associated with Some Esters

Name	Structure	Source or Flavor
Ethyl formate	$HCO_2CH_2CH_5$	rum
Isobutyl formate	$HCO_2CH_2CH(CH_3)_2$	raspberries
n-Pentyl acetate (*n*-amyl acetate)	$CH_3CO_2CH_2CH_2CH_2CH_2CH_3$	bananas
Isopentyl acetate (isoamyl acetate)	$CH_3CO_2CH_2CH_2CH(CH_3)_2$	pears
n-Octyl acetate	$CH_3CO_2(CH_2)_7CH_3$	oranges
Ethyl butyrate	$CH_3CH_2CH_2CO_2CH_2CH_5$	pineapples
n-Pentyl butyrate	$CH_3CH_2CH_2CO_2(CH_2)_4CH_3$	apricots
Methyl salicylate	benzene ring bearing $-CO_2CH_3$ with OH on the adjacent (ortho) carbon	oil of wintergreen

Saponification. This important reaction of esters is a slight variation of ester hydrolysis. Ester hydrolysis occurs in the presence of an acid catalyst (or an enzyme); saponification in the presence of aqueous sodium or potassium hydroxide. Essentially the same products form, except that saponification produces not the free acid but its salt. In both hydrolysis and saponification the alcohol portion is liberated as the free alcohol. In general:

$$R-\overset{\overset{\displaystyle O}{\|}}{C}-O-R' + NaOH \xrightarrow[\text{heat}]{H_2O} R-\overset{\overset{\displaystyle O}{\|}}{C}-O^-Na^+ + HOR'$$

Specific examples (compare these with the examples of ester hydrolysis):

$$\underset{\text{Ethyl acetate}}{CH_3\overset{\overset{\displaystyle O}{\|}}{C}OCH_2CH_3} + NaOH \xrightarrow[\text{heat}]{H_2O} \underset{\text{Sodium acetate}}{CH_3\overset{\overset{\displaystyle O}{\|}}{C}O^-Na^+} + \underset{\text{Ethyl alcohol}}{CH_3CH_2OH}$$

$$\underset{\text{Methyl benzoate}}{C_6H_5\overset{\overset{\displaystyle O}{\|}}{C}-O-CH_3} + NaOH \xrightarrow[\text{heat}]{H_2O} \underset{\text{Sodium benzoate}}{C_6H_5\overset{\overset{\displaystyle O}{\|}}{C}-O^-Na^+} + \underset{\text{Methyl alcohol}}{CH_3OH}$$

The term *saponification* comes from the Latin *sapo* and *onis*, "soap" and "-fy"—that is, to make soap. Ordinary soap is a mixture of sodium salts of long-chain carboxylic acids.

Claisen Ester Condensation. Esters can also be made to react with themselves, undergoing a reaction very similar to the aldol condensation. In esters as in aldehydes and ketones, the carbonyl carbon is not the only reactive site, for the α-position is also reactive. If a hydrogen is attached there, a strong proton acceptor may take it; for example,

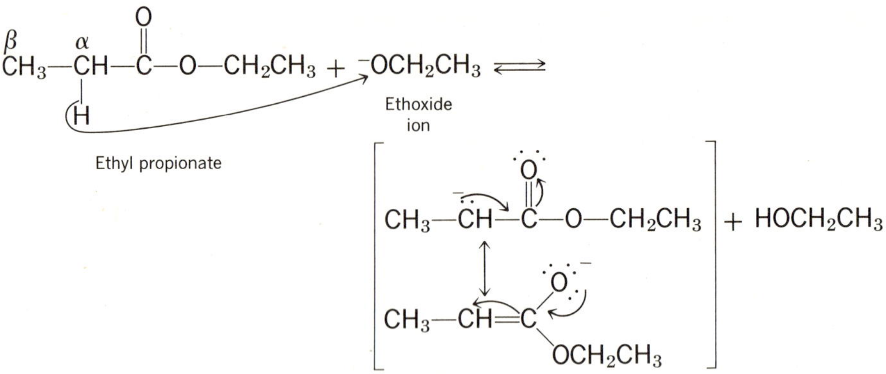

$$\overset{\beta}{CH_3}-\underset{\underset{H}{|}}{\overset{\alpha}{CH}}-\overset{\overset{O}{\|}}{C}-O-CH_2CH_3 + {}^{-}OCH_2CH_3 \rightleftharpoons$$

Ethyl propionate

Ethoxide ion

$$\left[CH_3-\overset{-}{\ddot{C}H}-\overset{\overset{\ddot{O}}{\|}}{C}-O-CH_2CH_3 \longleftrightarrow CH_3-CH{=}C\begin{smallmatrix}\ddot{O}:^{-} \\ OCH_2CH_3\end{smallmatrix} \right] + HOCH_2CH_3$$

Anion of ethyl propionate

Resonance stabilization of the anion helps to explain its formation.[3] Even so, it is highly reactive. It can stabilize itself by abstracting a proton from a molecule of ethyl alcohol, regenerating the starting materials. (Hence the equilibrium arrows.) However, the anion of the ester is a new electron-rich species, a nucleophile. What other event might we *expect* to occur? If this anion encounters a molecule of unchanged ester, it could easily be attracted to the relatively electron-poor carbonyl carbon:

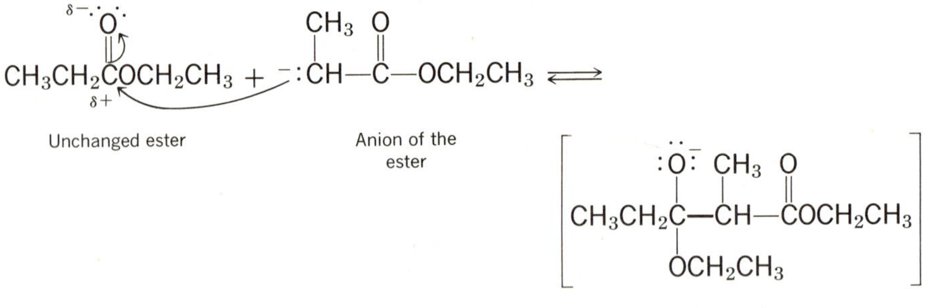

$$CH_3CH_2\overset{\overset{\delta-\ddot{O}}{\|}}{\underset{\delta+}{C}}OCH_2CH_3 + {}^{-}:\overset{\overset{CH_3}{|}}{CH}-\overset{\overset{O}{\|}}{C}-OCH_2CH_3 \rightleftharpoons$$

Unchanged ester

Anion of the ester

$$\left[CH_3CH_2\underset{\underset{OCH_2CH_3}{|}}{\overset{\overset{:\ddot{O}:^{-}}{|}}{C}}-\overset{\overset{CH_3}{|}}{CH}-\overset{\overset{O}{\|}}{C}OCH_2CH_3 \right]$$

New anion

The new anion might then stabilize itself by ejecting the ethoxide ion:

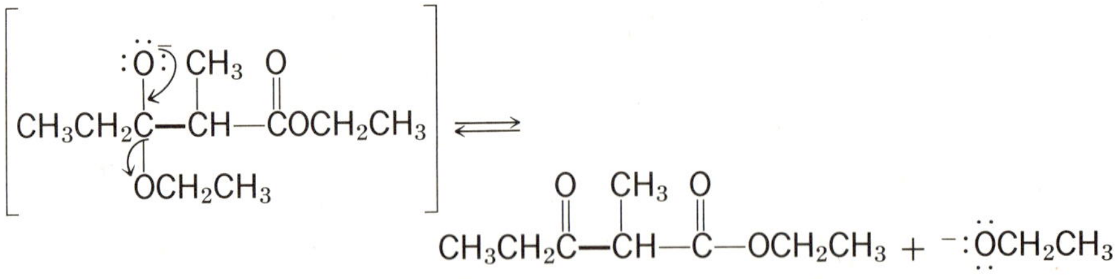

$$\left[CH_3CH_2\underset{\underset{OCH_2CH_3}{|}}{\overset{\overset{:\ddot{O}:^{-}}{|}}{C}}-\overset{\overset{CH_3}{|}}{CH}-\overset{\overset{O}{\|}}{C}OCH_2CH_3 \right] \rightleftharpoons$$

$$CH_3CH_2\overset{\overset{O}{\|}}{C}\mathbf{-}\overset{\overset{CH_3}{|}}{CH}-\overset{\overset{O}{\|}}{C}-OCH_2CH_3 + {}^{-}:\ddot{O}CH_2CH_3$$

A β-keto ester
(The new carbon-carbon bond is shown in boldface.)

[3] The reader may ask why the ethoxide ion does not attack the partially positively charged carbonyl carbon. Very likely it does, but no *new* product results. If such an attack causes ejection of the ethoxy group on the ester, the products are identical with the reactants. Such events undoubtedly occur but go nowhere.

The new carbon-carbon bond is shown in boldface. This type of reaction wherein two esters, in the presence of a strong base, react to form a β-keto ester is known as the Claisen ester condensation. Our interest in it stems from very similar reactions in biochemistry in which enzymes act as the catalysts.

Exercise 15.5 Write the structures of the products that would be expected to occur in each of the following situations. If no reaction is to be expected, write "No reaction."

(*a*) $CH_3CH_2\overset{O}{\overset{\|}{C}}OCH_3 + H_2O \text{ (excess)} \xrightarrow[\text{heat}]{H^+}$

(*b*) $CH_3CH_2O\overset{O}{\overset{\|}{C}}CH_3 + NaOH \xrightarrow[\text{heat}]{H_2O}$

(*c*) $CH_3{-}\underset{}{\overset{CH_3}{\overset{|}{C}}}HCH_2\overset{O}{\overset{\|}{C}}{-}O{-}CH_2CH_3 + H_2O \xrightarrow[\text{heat}]{H^+}$

(*d*) $CH_3CH_2CH_2O{-}\overset{O}{\overset{\|}{C}}{-}\overset{CH_3}{\overset{|}{C}}H{-}CH_3 + H_2O \xrightarrow[\text{heat}]{H^+}$

(*e*) $CH_3\overset{O}{\overset{\|}{C}}{-}O{-}CH_2$ | $CH_3\overset{O}{\overset{\|}{C}}{-}O{-}CH_2$ $+ \text{NaOH (excess)} \xrightarrow[\text{heat}]{H_2O}$

(*f*) $2CH_3CO_2CH_2CH_3 \xrightarrow[CH_3CH_2OH]{NaOCH_2CH_3}$

(*g*) $2C_6H_5CH_2\overset{O}{\overset{\|}{C}}OCH_2CH_3 \xrightarrow[CH_3CH_2OH]{NaOCH_2CH_3}$

Properties of β-Keto Esters and β-Keto Acids. Hydrolysis. Decarboxylation. Acid Cleavage. β-Keto esters are both ketones and esters, and they exhibit typical reactions of these functional groups. They can be hydrolyzed, for example, to β-keto acids:

$$\underset{\text{Ethyl acetoacetate}}{CH_3\overset{O}{\overset{\|}{C}}CH_2\overset{O}{\overset{\|}{C}}OCH_2CH_3} + H_2O \xrightarrow{H^+} \underset{\text{Acetoacetic acid}}{CH_3\overset{O}{\overset{\|}{C}}CH_2\overset{O}{\overset{\|}{C}}OH} + CH_3CH_2OH$$

Such acids are subject to rather easy decarboxylation, that is, loss of carbon dioxide:

$$\underset{\text{Acetoacetic acid}}{CH_3\overset{O}{\overset{\|}{C}}CH_2{-}\overset{O}{\overset{\|}{C}}{-}O{-}H} \xrightarrow[\text{heat}]{H^+} \underset{\text{Acetone}}{CH_3\overset{O}{\overset{\|}{C}}CH_3} + CO_2$$

The reaction shown occurs in the body, and we shall encounter it again in connection with lipid metabolism. β-Keto esters undergo another reaction important in lipid metabolism, cleavage of the α,β-carbon-carbon bond. *In vitro,* the reaction requires strong alkali.

$$CH_3\overset{\overset{O}{\|}}{C}{-}CH_2\overset{\overset{O}{\|}}{C}{-}OC_2H_5 \xrightarrow[\text{heat}]{\text{concd alkali}} 2CH_3\overset{\overset{O}{\|}}{C}{-}O^- + C_2H_5OH$$

(attacked at each carbonyl carbon by ^-OH)

$$2CH_3\overset{\overset{O}{\|}}{C}{-}O^- \xrightarrow{2H^+} 2CH_3CO_2H$$

Acetic acid

In vivo, an important example of this reaction is the enzyme-catalyzed cleavage of a thio analog of a β-keto ester. In the following example the symbol CoA—SH represents coenzyme A, which is a thio alcohol.

$$RCH_2\overset{\overset{O}{\|}}{C}{-}CH_2\overset{\overset{O}{\|}}{C}{-}S{-}CoA + H{-}S{-}CoA \longrightarrow RCH_2\overset{\overset{O}{\|}}{C}{-}SCoA + CH_3\overset{\overset{O}{\|}}{C}{-}SCoA$$

(CoAS—H adds across the cleaved bond)

to fatty acid cycle; to citric acid cycle

IMPORTANT INDIVIDUAL ESTERS

The Salicylates. Salicyclic acid can function either as an acid or as a phenol, for it possesses both groups. Salicyclic acid, its esters, and its salts, taken internally, have both an analgesic effect (depressing sensitivity to pain) and an antipyretic action (reducing fever). As analgesics they raise the threshold of pain by depressing pain centers in the thalamus region of the brain. As antipyretics they increase perspiration as well as circulation of the blood in capillaries near the surface of the skin. Both mechanisms have cooling effects. Salicyclic acid itself irritates the moist membranes lining the mouth, gullet, and the stomach because it is too acidic. The sodium or calcium salts, however, are less irritating. In the early history of aspirin[4] the salts had to be administered in aqueous solutions which had a very disagreeable sweetish taste. The search for a more palatable form led to the discovery of acetyl salicyclic acid, or aspirin, first sold by the German company Bayer in 1899 (which still owns the name Aspirin as a trademark in Germany). Aspirin is the most widely used drug in the world. During the 1960s the annual United States production exceeded 27 million pounds from which, annually, over 15 billion aspirin tablets were compounded.

[4] H. O. J. Collier, "Aspirin," *Scientific American,* November 1963, p. 97.

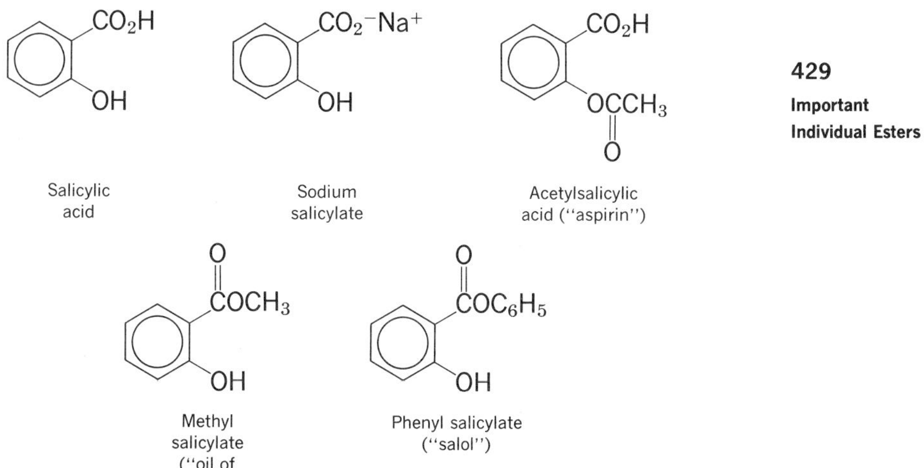

Salicylic acid — Sodium salicylate — Acetylsalicylic acid ("aspirin")

Methyl salicylate ("oil of wintergreen") — Phenyl salicylate ("salol")

Methyl salicylate is used in liniments, for it is readily absorbed through the skin. Phenyl salicylate is sometimes a constituent of ointments that protect the skin against ultraviolet rays.

Dacron. Dacron shares most of the desirable properties of nylon. Especially distinctive is its ability to be set into permanent creases and pleats. It does not tend to become grayish or yellowish with age, as does some white nylon. Chemically, Dacron is a polyester, polyethylene terephthalate, made by ester interchange (trans-esterification) between ethylene glycol and dimethyl terephthalate.

$$CH_3O\text{-}\overset{O}{\overset{\|}{C}}\text{-}C_6H_4\text{-}\overset{O}{\overset{\|}{C}}\text{-}OCH_3 + HOCH_2CH_2OH + CH_3O\text{-}\overset{O}{\overset{\|}{C}}\text{-}C_6H_4\text{-}\overset{O}{\overset{\|}{C}}\text{-}OCH_3 + HOCH_2CH_2OH$$

Dimethyl terephthalate — Ethylene glycol

$$\downarrow -CH_3OH$$

$$\text{etc.}\text{—}O\overset{O}{\overset{\|}{C}}\text{-}C_6H_4\text{-}\overset{O}{\overset{\|}{C}}\left(\text{-}OCH_2CH_2O\overset{O}{\overset{\|}{C}}\text{-}C_6H_4\text{-}\overset{O}{\overset{\|}{C}}\right)_n\text{-}OCH_2CH_2O\text{—etc.}$$

Repeating unit in Dacron

As is characteristic of all chemicals that can be formed into useful fibers, molecules of Dacron are extremely long and very narrow and symmetrical, properties that resemble those of a natural fiber. A significant use of knitted Dacron is illustrated in Figure 15.1.

Dacron is the trade name for the fiber made from polyethylene terephthalate. To increase the strenth of the fibers, the freshly spun filaments are slowly stretched, allowing the long molecules to become better aligned parallel with the fiber axis. Forces of attraction between neighboring molecules become more effective and the fiber gains considerable strength.

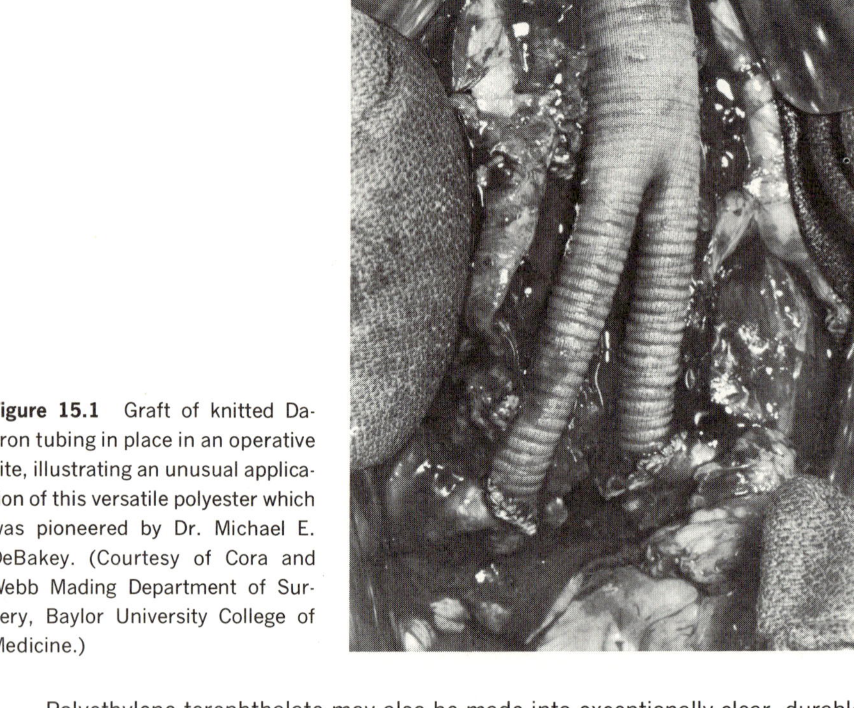

Figure 15.1 Graft of knitted Dacron tubing in place in an operative site, illustrating an unusual application of this versatile polyester which was pioneered by Dr. Michael E. DeBakey. (Courtesy of Cora and Webb Mading Department of Surgery, Baylor University College of Medicine.)

Polyethylene terephthalate may also be made into exceptionally clear, durable films, and in this form it is known as Mylar. Other trade names for polyethylene terephthalate are Terylene and Cronar.

Glyptal Resins. Ethylene glycol has but two hydroxyl groups and can form only a linear polymer with terephthalic acid. Glycerol, however, has three hydroxyl groups, the third group making the formation of cross-linked polymers possible. The glyptal resin, prized as a tough, durable surface coating in automobile and other finishes, forms when phthalic anhydride and glycerol interact in a mole ratio of 3:2 (Figure 15.2).

Primary alcohols are more reactive in esterification than secondary alcohols. Consequently, in the early stage of making glyptal resin, mixtures of mono- and diphthalate esters of glycerol are present. Actual cross-linking must bring into play glycerol's 2° alcohol function. When the resin is to be used as a surface coating, material that is not for the most part cross-linked is applied to the surface. Baking then completes resinification as the material becomes essentially one gigantic molecule. Understandably, this coating is exceptionally durable.

Acrylates. Esters of acrylic acid and methacrylic acid can be made to polymerize in a fashion exactly analogous to vinyl polymerization. The products are well-known plastics.

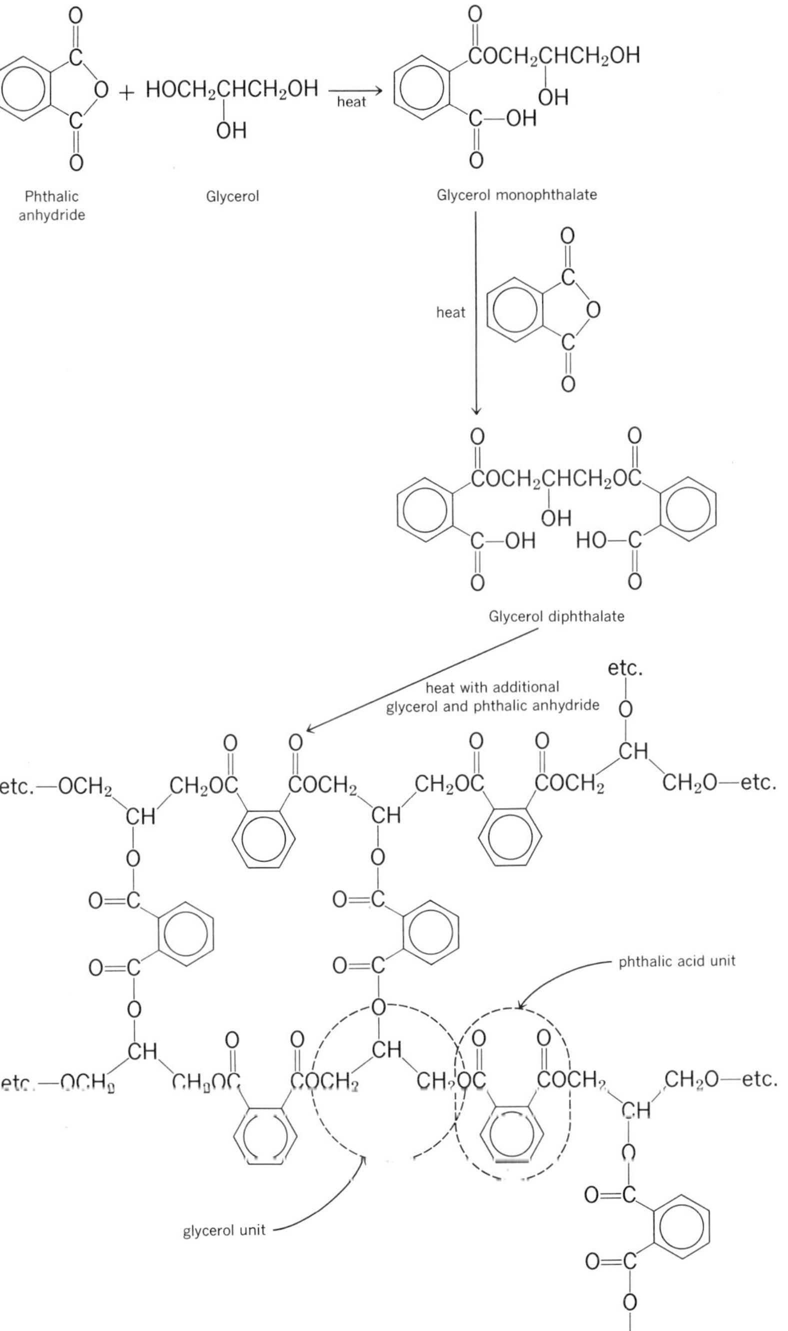
Phthalic anhydride
Glycerol
heat
Glycerol monophthalate
heat
Glycerol diphthalate
heat with additional glycerol and phthalic anhydride
phthalic acid unit
glycerol unit
etc.

Figure 15.2 Glyptal resin, basic structural features. The cross-linking shown here is for illustrative purposes only. In the actual resin these links do not occur at every possible position.

$$nCH_2{=}CH\overset{O}{\overset{\|}{C}}OCH_3 \longrightarrow \text{(}CH_2{-}\underset{COOCH_3}{\underset{|}{CH}}\text{)}_n$$

Methyl acrylate → Acryloid

$$nCH_2{=}\overset{CH_3}{\overset{|}{C}}{-}\overset{O}{\overset{\|}{C}}OCH_3 \longrightarrow \text{(}CH_2{-}\underset{COOCH_3}{\underset{|}{\overset{CH_3}{\overset{|}{C}}}}\text{)}_n$$

Methyl methacrylate → Plexiglas, Lucite

AMIDES

Physical Properties. With the exception of formamide, all the simple amides, those derived from ammonia, melt at fairly high temperatures. Their molecules are apparently quite polar and can serve as both hydrogen bond acceptors and donors. Several are listed in Table 15.8.

Chemical Properties. Hydrolysis. One chemical property of amides that we shall need to know for later work is simple hydrolysis:

$$R{-}\overset{O}{\overset{\|}{C}}{-}NH_2 + H_2O \xrightarrow[\text{heat}]{\text{acids or bases}} R{-}\overset{O}{\overset{\|}{C}}{-}OH + NH_3$$

$$R{-}\overset{O}{\overset{\|}{C}}{-}NHR' + H_2O \longrightarrow R{-}\overset{O}{\overset{\|}{C}}{-}OH + R'NH_2$$

$$R{-}\overset{O}{\overset{\|}{C}}{-}NR'_2 + H_2O \longrightarrow R{-}\overset{O}{\overset{\|}{C}}{-}OH + R'_2NH$$

Table 15.8 Amides of Carboxylic acids

Name	Structure	Mp (°C)	Bp (°C)	Solubility in Water
Formamide	$HCONH_2$	2.5	210 (decomposes)	
N-Methylformamide	$HCONHCH_3$	−5	131 (at 90 mm pressure)	
N,N-Dimethylformamide	$HCON(CH_3)_2$	−61	153	
Acetamide	CH_3CONH_2	82	222	very soluble
N-Methylacetamide	$CH_3CONHCH_3$	28	206	very soluble
N,N-Dimethylacetamide	$CH_3CON(CH_3)_2$	−20	166	
Propionamide	$CH_3CH_2CONH_2$	79	213	soluble
Butyramide	$CH_3(CH_2)_2CONH_2$	115	216	soluble
Valeramide	$CH_3(CH_2)_3CONH_2$	106	sublimes	soluble
Caproamide	$CH_3(CH_2)_4CONH_2$	100	sublimes	slightly soluble
Benzamide	$C_6H_5{-}CONH_2$	133	290	slightly soluble

It is the carbonyl-to-nitrogen bond that breaks. The reaction usually requires prolonged refluxing of the amide with either acid or base. *In vivo,* however, enzymes are available for catalyzing the reaction. The digestion of proteins is nothing more than hydrolysis of the amide linkages of proteins.

Exercise 15.6 When acetamide is allowed to react with sodium hydroxide on a 1:1 mole basis, the products are sodium acetate and ammonia. If hydrochloric acid is used, however, again on a 1:1 mole basis, the products are acetic acid and ammonium chloride. Explain.

Exercise 15.7 Write the structures of the products that would be expected to form in each of the following. If no reaction is to be expected, write "No reaction."

(*a*) $CH_3C(=O)NH_2 + H_2O \xrightarrow[\text{heat}]{}$

(*b*) $CH_3NHC(=O)CH_2CH_3 + H_2O \xrightarrow[\text{heat}]{}$

(*c*) $C_6H_5NHC(=O)CH_3 + H_2O \xrightarrow[\text{heat}]{}$

(*d*) $CH_3CH_2C(=O)N(CH_3)_2 + H_2O \xrightarrow[\text{heat}]{}$

(*e*) $NH_2CH_2C(=O)NHCH(CH_3)C(=O)OH + H_2O \xrightarrow[\text{heat}]{}$

Nylon. Nylon is the name of a family of synthetic materials consisting of long-chain, fiber-forming polymer molecules having repeating amide linkages. One member of this family is called nylon 66, suggestive of the six carbons in the diamine and the six carbons in the dicarboxylic acid.

$$NH_2(CH_2)_6NH_2 + \underset{\text{Adipic acid}}{HOC(=O)(CH_2)_4C(=O)OH} + \underset{\text{Hexamethylene diamine}}{NH_2(CH_2)_6NH_2} + HOC(=O)(CH_2)_4C(=O)OH + $$

$$\downarrow$$

$$\text{etc.}{-}NH(CH_2)_6NH{-}\underbrace{\left(C(=O)(CH_2)_4C(=O){-}NH(CH_2)_6NH\right)_n}_{\text{Repeating unit in nylon 66}}C(=O)(CH_2)_4C(=O){-}\text{etc.}$$

To be useful as a fiber, each molecule in a batch of nylon 66 should contain from 50 to 90 of each of the monomer units. Shorter molecules lead to weak or brittle fibers. The long polymer molecules have geometries that permit side-by-side align-

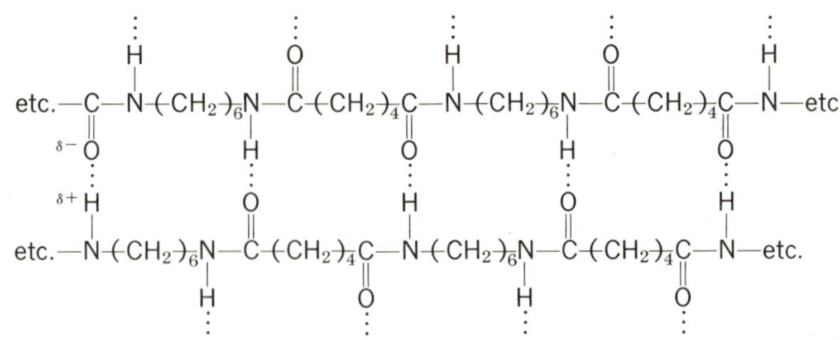

Figure 15.3 In this somewhat idealized representation of nylon 66, neighboring molecules adher to each other via hydrogen bonds (dotted lines).

ment and overlapping. Opportunities for hydrogen bonds (Figure 15.3) result in forces of attraction that are the main source of strength of the fiber.

The unusual resistance of nylon 66 to breakage by stretching, mechanical abrasion, moisture, action of mild acids and alkalis, light, dry-cleaning solvents, mildew, bacteria, and rotting conditions make it one of the most desirable of all fibers. It is more resistant to burning than wool, rayon, cotton, or silk and is almost as immune to insect attack as Fiberglas. No known insects metabolize nylon molecules, although some will cut their way through a nylon fabric if they are entrapped by it! Figure 15.4 pictures only a few of the many industrial applications made of nylon.

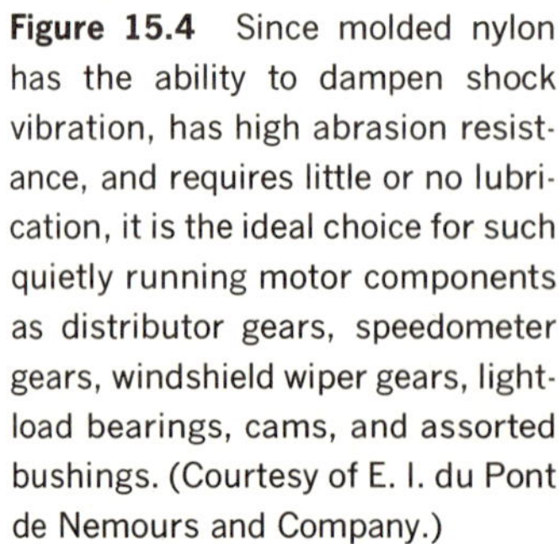

Figure 15.4 Since molded nylon has the ability to dampen shock vibration, has high abrasion resistance, and requires little or no lubrication, it is the ideal choice for such quietly running motor components as distributor gears, speedometer gears, windshield wiper gears, light-load bearings, cams, and assorted bushings. (Courtesy of E. I. du Pont de Nemours and Company.)

REFERENCES AND ANNOTATED READING LIST

BOOKS

R. T. Morrison and R. N. Boyd. *Organic Chemistry,* second edition. Allyn and Bacon, Boston, 1966. Chapter 18 ("Carboxylic Acids") and Chapter 20 ("Functional Derivatives of Carboxylic Acids") are particularly good references in a standard organic textbook.

L. O. Smith, Jr. and S. J. Cristol. *Organic Chemistry.* Reinhold Publishing Corporation, New York, 1966. Chapter 10 ("Acidity and Basicity in Organic Compounds") is a good survey of organic acids and bases and the structural factors which affect their strengths.

C. A. Vanderwerf. *Acids, Bases, and the Chemistry of the Covalent Bond.* Reinhold Publishing Corporation, New York, 1961. This book has been cited before as an excellent reference to the topics of the title. (Paperback.)

ARTICLES

L. E. Alexander. "Nylon—From Test Tube to Counter." *Chemistry,* September 1964, page 8.

H. O. Collier. "Aspirin." *Scientific American,* November 1963, page 97. The history of the salicylates and how they work are discussed.

PROBLEMS AND EXERCISES

1. Write the structure of each of the following compounds.
   (*a*) acetamide (*b*) methyl propionate (*c*) acetyl chloride
   (*d*) acetic anhydride (*e*) *N*-methylformamide (*f*) isopropyl *n*-butyrate
   (*g*) glycerol triacetate (*h*) ethyl acetoacetate (*i*) malonic acid
   (*j*) nylon-66 (Show enough of its structure to illustrate its features.)
   (*k*) Dacron (Give enough of its structure to illustrate its features.)

2. Write common names for each of the following compounds.
   (*a*) $CH_3CH_2CH_2CH_2{-}O{-}\overset{O}{\overset{\|}{C}}CH_2CH_3$ (*b*) $CH_3CH_2CH_2\overset{O}{\overset{\|}{C}}{-}NH_2$
   (*c*) $CH_3{-}\overset{O}{\overset{\|}{C}}{-}CH_2{-}\overset{O}{\overset{\|}{C}}{-}OH$ (*d*) $CH_3CH_2\overset{O}{\overset{\|}{C}}{-}Cl$
   (*e*) $C_6H_5\overset{O}{\overset{\|}{C}}{-}O{-}CH_2CH_3$ (*f*) $CH_3CH_2OCH_2CH_3$
   (*g*) $CH_3CH_2CH_2CH_2NH_2$ (*h*) $CH_3CH_2CH_2CH_2CH_2\overset{O}{\overset{\|}{C}}{-}OH$
   (*i*) $(CH_3)_2CH{-}O{-}\overset{O}{\overset{\|}{C}}{-}H$ (*j*) $CH_3CH_2\overset{O}{\overset{\|}{C}}{-}NH$

3. Go back through the chapters on organic chemistry and make a list of all the reactions we have studied, writing a specific example of each, for introducing in one step from some starting compound each of the following functional groups.
   (*a*) —OH (as in an alcohol) (*b*) —Cl (as in an alkyl halide)
   (*c*) C=C (as in an alkene) (*d*) —CHO (as in an aldehyde)

(*e*) $-\overset{\overset{\displaystyle O}{\|}}{C}-$ (as in a ketone) (*f*) $-\overset{\overset{\displaystyle O}{\|}}{C}-O-C$ (as in an ester)

(*g*) $-\overset{\overset{\displaystyle O}{\|}}{C}-N$ (as in an amide)

4. List all the reactions we have studied for making a carbon-carbon single bond.
5. List all the reactions we have studied for breaking a carbon-carbon single bond.
6. List all the reactions we have studied for making a carbon-hydrogen bond.
7. List all the reactions we have studied for breaking a carbon-hydrogen bond.
8. List all the reactions we have studied for making a carbon-oxygen single bond.
9. List all the reactions we have studied for breaking a carbon-oxygen single bond.
10. What functional groups have we studied that are susceptible to oxidation, and what products form when they are oxidized?
11. What functional groups have we studied that are susceptible to reduction by any means, and what are the products they form?
12. Give specific examples of all the *types* of reactions we have studied in which water is a reactant rather than just a solvent. Assume that whatever catalyst is normally used—acid, base, enzyme—is available.
13. Complete the following equations by writing the structure(s) of the principal organic product(s) that are expected to form. If no reaction is expected, write "No reaction."

(*a*) $CH_3CH_2\overset{\overset{\displaystyle O}{\|}}{C}H \xrightarrow{K_2Cr_2O_7}$

(*b*) $CH_3OH + CH_3\overset{\overset{\displaystyle O}{\|}}{C}OH \xrightarrow[\text{heat}]{H^+}$

(*c*) $CH_3CH_2CH_3 + H_2SO_4 \longrightarrow$

(*d*) $CH_3\overset{\overset{\displaystyle OH}{|}}{C}HCH_3 \xrightarrow[\text{heat}]{KMnO_4}$

(*e*) $CH_3\overset{\overset{\displaystyle O}{\|}}{C}NH_2 + H_2O \xrightarrow[\text{heat}]{H^+}$

(*f*) $CH_3\overset{\overset{\displaystyle OH}{|}}{C}HCH_3 \xrightarrow[\text{heat}]{H_2SO_4}$

(*g*) $CH_3O\overset{\overset{\displaystyle O}{\|}}{C}CH_3 + H_2O \xrightarrow[\text{heat}]{H^+}$

(*h*) $CH_3CH{=}CH_2 + HCl \longrightarrow$

(*i*) $CH_3CH(OCH_3)_2 + H_2O \xrightarrow[\text{heat}]{H^+}$ (two OCH_3 groups on the same carbon)

(*j*) $CH_3\overset{\overset{\displaystyle O}{\|}}{C}Cl + CH_3CH_2OH \longrightarrow$

(*k*) $CH_3CH_2OCH_2CH_3 + H_2O \xrightarrow{^-OH}$

(*l*) $CH_3CH_2\overset{\overset{\displaystyle O}{\|}}{C}H + 2CH_3OH \xrightarrow{HCl}$

(*m*) $CH_3CH{=}CHCH_3 + H_2 \xrightarrow[\text{heat, pressure}]{Ni}$

(*n*) $CH_3\overset{\overset{\displaystyle OH}{|}}{\underset{\underset{\displaystyle CH_3}{|}}{C}}CH_2CH_3 \xrightarrow{K_2Cr_2O_7}$

(*o*) $(CH_3)_2CH\overset{\overset{\displaystyle O}{\|}}{C}OH + CH_3OH \xrightarrow{H^+}$

(*p*) $CH_3Br + NH_3 \longrightarrow$

(*q*) $CH_3CH_2SH \xrightarrow{\text{oxidizing agent}}$

(*r*) $CH_3\underset{\underset{\displaystyle Br}{|}}{C}HCH_3 + KOH \xrightarrow{\text{alcohol}}$

(*s*) $C_6H_6 + Br_2 \xrightarrow{Fe}$

(*t*) $CH_3CH_2NH\overset{\overset{\displaystyle O}{\|}}{C}CH_3 + H_2O \xrightarrow{\text{heat}}$

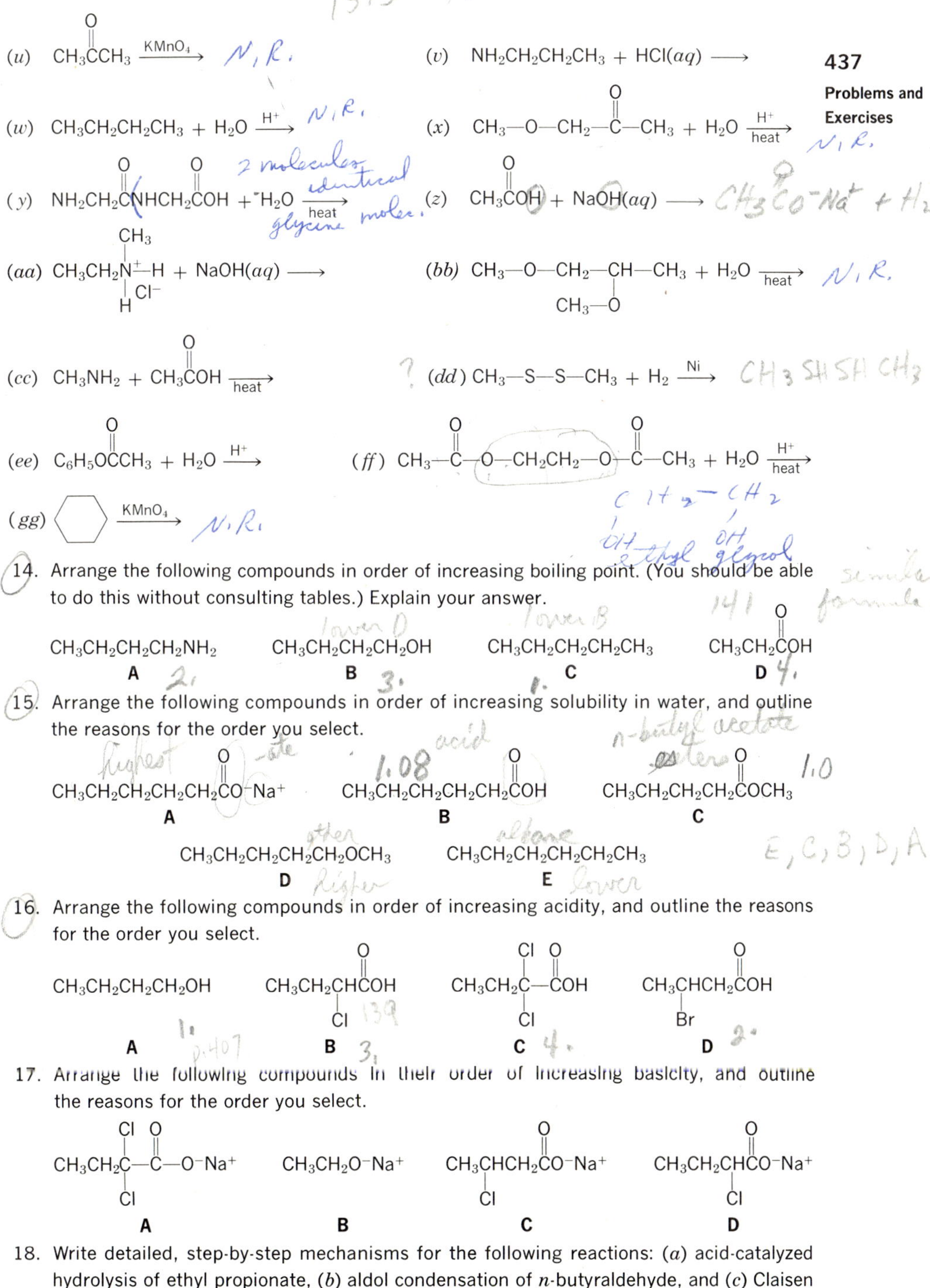

(*u*) $CH_3\overset{O}{\overset{\|}{C}}CH_3 \xrightarrow{KMnO_4}$

(*v*) $NH_2CH_2CH_2CH_3 + HCl(aq) \longrightarrow$

(*w*) $CH_3CH_2CH_2CH_3 + H_2O \xrightarrow{H^+}$

(*x*) $CH_3{-}O{-}CH_2{-}\overset{O}{\overset{\|}{C}}{-}CH_3 + H_2O \xrightarrow[heat]{H^+}$

(*y*) $NH_2CH_2\overset{O}{\overset{\|}{C}}NHCH_2\overset{O}{\overset{\|}{C}}OH + H_2O \xrightarrow{heat}$

(*z*) $CH_3\overset{O}{\overset{\|}{C}}OH + NaOH(aq) \longrightarrow$

(*aa*) $CH_3CH_2\overset{+}{N}(CH_3)(H){-}H\ Cl^- + NaOH(aq) \longrightarrow$

(*bb*) $CH_3{-}O{-}CH_2{-}\underset{CH_3{-}O}{CH}{-}CH_3 + H_2O \xrightarrow{heat}$

(*cc*) $CH_3NH_2 + CH_3\overset{O}{\overset{\|}{C}}OH \xrightarrow{heat}$

(*dd*) $CH_3{-}S{-}S{-}CH_3 + H_2 \xrightarrow{Ni}$

(*ee*) $C_6H_5O\overset{O}{\overset{\|}{C}}CH_3 + H_2O \xrightarrow{H^+}$

(*ff*) $CH_3{-}\overset{O}{\overset{\|}{C}}{-}O{-}CH_2CH_2{-}O{-}\overset{O}{\overset{\|}{C}}{-}CH_3 + H_2O \xrightarrow[heat]{H^+}$

(*gg*) cyclohexane $\xrightarrow{KMnO_4}$

14. Arrange the following compounds in order of increasing boiling point. (You should be able to do this without consulting tables.) Explain your answer.

A	B	C	D
$CH_3CH_2CH_2CH_2NH_2$	$CH_3CH_2CH_2CH_2OH$	$CH_3CH_2CH_2CH_2CH_3$	$CH_3CH_2\overset{O}{\overset{\|}{C}}OH$

15. Arrange the following compounds in order of increasing solubility in water, and outline the reasons for the order you select.

A: $CH_3CH_2CH_2CH_2CH_2\overset{O}{\overset{\|}{C}}O^-Na^+$

B: $CH_3CH_2CH_2CH_2CH_2\overset{O}{\overset{\|}{C}}OH$

C: $CH_3CH_2CH_2CH_2\overset{O}{\overset{\|}{C}}OCH_3$

D: $CH_3CH_2CH_2CH_2CH_2CH_2OCH_3$

E: $CH_3CH_2CH_2CH_2CH_2CH_3$

16. Arrange the following compounds in order of increasing acidity, and outline the reasons for the order you select.

A: $CH_3CH_2CH_2CH_2OH$

B: $CH_3CH_2\underset{Cl}{CH}\overset{O}{\overset{\|}{C}}OH$

C: $CH_3CH_2\underset{Cl}{\overset{Cl}{C}}{-}\overset{O}{\overset{\|}{C}}OH$

D: $CH_3\underset{Br}{CH}CH_2\overset{O}{\overset{\|}{C}}OH$

17. Arrange the following compounds in their order of increasing basicity, and outline the reasons for the order you select.

A: $CH_3CH_2\underset{Cl}{\overset{Cl}{C}}{-}\overset{O}{\overset{\|}{C}}{-}O^-Na^+$

B: $CH_3CH_2O^-Na^+$

C: $CH_3\underset{Cl}{CH}CH_2\overset{O}{\overset{\|}{C}}O^-Na^+$

D: $CH_3CH_2\underset{Cl}{CH}\overset{O}{\overset{\|}{C}}O^-Na^+$

18. Write detailed, step-by-step mechanisms for the following reactions: (*a*) acid-catalyzed hydrolysis of ethyl propionate, (*b*) aldol condensation of *n*-butyraldehyde, and (*c*) Claisen ester condensation of ethyl *n*-butyrate.

CHAPTER SIXTEEN

Carbohydrates

PHOTOSYNTHESIS, AN INTRODUCTION

In the sweeping panorama of biological life, man and the animals are parasites, for they ultimately depend on the plant kingdom for food. Plants can take simple inorganic compounds such as carbon dioxide, water, and certain nitrogenous minerals and fashion the molecules that make up the three great classes of foods, carbohydrates, lipids, and proteins. They do this without any direct help from the animal kingdom, depending only on a constant flow of energy from the sun. Plants, not animals, use solar energy to transform the energy-poor and structurally simple substances, carbon dioxide, water, and minerals, into energy-rich and structurally complex materials, including the foods. This overall process is known as photosynthesis, and it is introduced here because its primary products are carbohydrates.

When we use the term *energy-poor compound,* we mean that the particular arrangement of electrons and nuclei that is the structure of the molecules of the compound is quite stable. There is no other structure to which its molecules can spontaneously and quickly rearrange, even when we give the compound an initial boost, as in striking a match to it, and even when it is in the presence of many types of other chemicals, for example, air, water, soil components, etc. An *energy-rich* compound, on the other hand, possesses an arrangement of electrons and nuclei that is relatively unstable. Given the correct initial conditions—activation and other potential reactants—the molecules of an energy-rich compound and those of the other reactants will rearrange in the direction of increasing stability and increasing probability. Released during this chemical event is the energy that is the difference between the energy-rich and the energy-poor arrangements. Often this energy appears as heat, which means simply that the average kinetic energy of the molecules of the products is greater than that of the molecules of the reactants. In an exothermic process molecules of the products move more energetically than those of the reactants. Billions and billions of molecules in violent motion, which give us the sensation of heat, can be made to drive pistons, blast ore deposits, and push jet planes and space rockets.

The energy released when, for example, a piece of wood is burned ultimately came from the sun. A piece of wood, however, is certainly not just a "bottle of sunlight"; it is a complex and highly organized mixture of compounds, mostly organic. The solar energy needed to make these compounds is temporarily stored in the form of distinctive electron-nuclei arrangements of energy-rich molecules that only plants can make from the fundamental materials, carbon dioxide, water, and minerals.

Man has not yet been able to duplicate the plant's remarkable feat of converting the energy of sunlight into the internal energy of chemicals. In fact, the mechanism by which plants perform this conversion, photosynthesis, is not fully understood. The simplest statement of the photosynthesis phenomenon is the "equation"

$$CO_2 + H_2O + \text{solar energy} \xrightarrow[\text{plant enzymes}]{\text{chlorophyll}} \underset{\substack{\text{Basic unit in}\\ \text{carbohydrates}}}{(CH_2O)} + O_2$$

This equation is a compromise between clarity and accuracy, but it is useful as a reference when the balance of nature is considered.

The process of photosynthesis has been estimated yearly to fix about 200 billion tons of carbon from carbon dioxide. Every three centuries all the carbon dioxide found in the atmosphere and in the dissolved state in the waters of the earth goes through the cycle photosynthesis, decay, photosynthesis. All the oxygen in our atmosphere is renewed by this cycle once in approximately twenty centuries. It has been estimated that close to 400 billion tons of oxygen are set free by photosynthesis each year. About 10 to 20% of this activity is carried out on land by familiar land plants. The remaining 80 to 90% is conducted below the surface of the oceans by algae. (The lowly oceanic algae constantly resupply us with oxygen, and yet very few of the thousands and thousands of substances man flushes into the oceans—residues of pesticides and herbicides, detergents, radioactive wastes—have been tested for their effects on algae.) Photosynthesis can of course occur only during daylight hours and, over land, only during growing seasons. The great atmospheric patterns of air movements ensure steady supplies of oxygen for people and land animals both at night and during the winter months.

The importance of green plants and the photosynthesis they accomplish is emphasized by the variety of substances they make, as outlined in Figure 16.1. Among the important primary products of photosynthesis stand the energy-rich carbohydrates. Before we can consider how the body uses these compounds or any of the foodstuffs, we must first find out more precisely what they are.

CLASSES OF CARBOHYDRATES

Molecules that comprise the carbohydrates are polyhydroxy aldehydes, polyhydroxy ketones, or substances that by simple hydrolysis yield these. Carbohydrates that cannot be hydrolyzed to simpler molecules are called *monosaccharides* or sometimes simple sugars. They are classified according to

1. How many carbons there are in one molecule:

If the carbon number is	the monosaccharide is a
3	triose
4	tetrose
5	pentose
6	hexose, etc.

2. Whether there is an aldehyde or a keto group present:

   If an aldehyde is present, the monosaccharide is an aldose.
   If a keto group is present, the monosaccharide is a ketose.

These terms may be combined. For example, a hexose that has an aldehyde group is called an aldohexose. A ketohexose possesses a keto group and a total of six carbons.

Carbohydrates that can be hydrolyzed to two monosaccharides are called *disaccharides.* Sucrose, maltose, and lactose are common examples. Starch and cellulose are common *polysaccharides;* their molecules yield many monosaccharide units when they are hydrolyzed.

			The More Important Secondary Products	
Raw Materials	Catalysts	Primary Products	Formula Weight Less than 1000	Formula Weight Very High
			Polyene pigments	
		Oxygen	Organic acids	
CO_2		Sugars	Alcohols	Cellulose
H_2O		Amino acids	Terpenes	Hemicellulose
Light energy	chlorophyll → plant enzymes	↓	Sterols	Gums
Inorganic nitrogen compounds			Waxes	Pectins
		Reserve materials:	Phosphatides (e.g., lecithin)	Resins
			Inositol	Rubber
		Proteins	Aromatic oils and fragrances	Tannins
		Fats and oils		
		Polysaccharides	Alkaloids	Lignins
			Pyrrole pigments (e.g., antho-cyanine pigments)	
			Nucleic acids	

Figure 16.1 The synthesis of organic compounds in green plants. (Adapted from J. B. Conant, *Chemistry of Organic Compounds,* revised edition, The Macmillan Company, New York, 1939, page 597.)

OPTICAL ISOMERISM OF MONOSACCHARIDES

D- and L-Families. Glyceraldehyde. The simplest polyhydroxy aldehyde is glyceraldehyde; each molecule has one asymmetric carbon. It can therefore exist in two configurations related as object and a mirror image that cannot be superimposed (Figure 16.2). Until the advent of spectroscopic techniques, scientists could not determine which of two enantiomers was the dextrorotatory form and which the levorotatory; that is, they did not know what the *absolute configuration* of the molecules in samples of each of the two glyceraldehydes is. They noted with considerable interest, however, that a sample of glyceraldehyde that is dextrorotatory can be oxidized to a glyceric acid that is levorotatory. Or if the sample of glyceraldehyde is levorotatory, oxidation produces the dextrorotatory glyceric acid. As we can see from the equation

$$\underset{\text{Glyceraldehyde}}{HOCH_2{-}CH(OH){-}CHO} \xrightarrow{(O)} \underset{\text{Glyceric acid}}{HOCH_2{-}CH(OH){-}COOH}$$

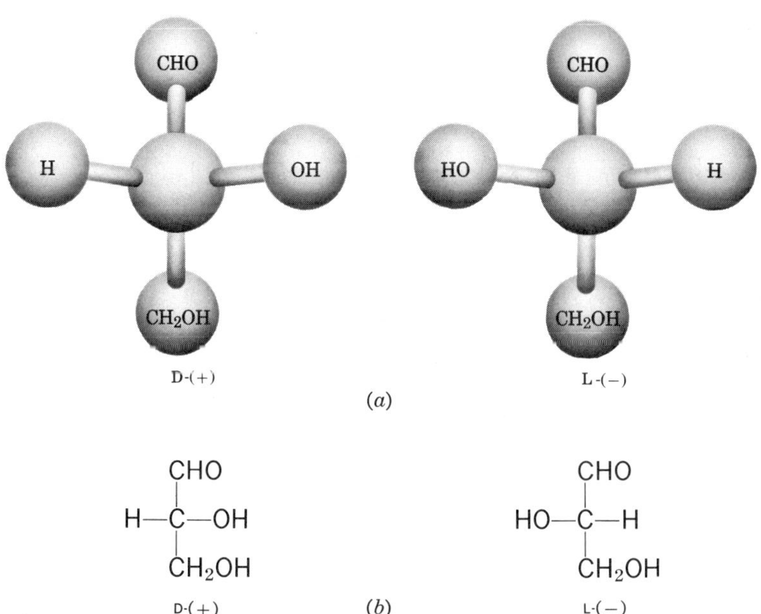
CHO
H
OH
CH₂OH
D-(+)
CHO
HO
H
CH₂OH
L-(−)
(a)
CHO
H—C—OH
CH₂OH
D-(+)
(b)
CHO
HO—C—H
CH₂OH
L-(−)

Figure 16.2 D- and L-Glyceraldehyde, absolute configurations, represented (*a*) as ball and stick models, (*b*) in plane projection diagrams.

this oxidation does not touch any of the bonds holding the four groups to the asymmetric carbon. Whatever the absolute configuration of the starting glyceraldehyde, it is retained during the oxidation. The relative configurations of the glyceraldehyde and its oxidation product, glyceric acid, are identical. There is no special correlation between the *sign* of rotation that the *substance* shows and the absolute configuration that its *molecules* have. Whether or not a reaction affects the *configuration* of an asymmetric carbon is of greater interest and importance than any (unpredictable) changes in the direction of optical rotation.

Early in the history of organic chemistry an arbitrary decision with a 50:50 chance of being correct was made. The molecules of the samples of glyceraldehyde that tested dextrorotatory were said to have the absolute configuration indicated in Figure 16.2. It did not matter whether the choice turned out to be wrong, because the only reason for the decision was to have a standard for assigning *relative* configurations to compounds related to glyceraldehyde, such as glyceric acid or the simple sugars. Thus *if* (+)-glyceraldehyde has the absolute configuration shown, then (−)-glyceric acid will have the same relative configuration. The next step was to devise a simple way to avoid the confusion the signs of rotation might cause. Thus it was agreed that all substances related to (+)-glyceraldehyde, regardless of their specific signs of rotation, would be said to be in the D-*family*. All those related to (−)-glyceraldehyde would be in the L-*family*. The symbols D and L do not stand for any words, nor do they necessarily say anything about the sign of rotation. They are merely small capital letters that give simple names to families of configurations.

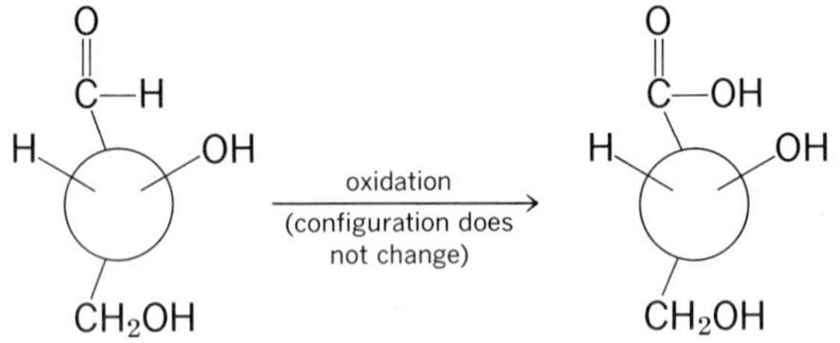

O ‖ C—H, H, OH, CH_2OH —oxidation (configuration does not change)→ O ‖ C—OH, H, OH, CH_2OH

D-(+)-Glyceraldehyde D-(−)-Glyceric acid

For the designations D and L to have clear meanings, the route from the standard, D-(+)-glyceraldehyde, to the compound in question must be specified. It is conceivable that a series of steps could convert D-(+)-glyceraldehyde to the dextrorotatory glyceric acid. Instead of simply oxidizing the aldehyde group, we reduce it to a 1° alcohol function; the $HOCH_2$— group at the other end might be the one oxidized to the —COOH group. Conventions must therefore be spelled out to preserve clarity. We say that (−)-glyceric acid is in the D-family because we arbitrarily choose to match the —CHO of the D-(+)-glyceraldehyde and the —COOH of the glyceric acid.

Plane Projection Formulas. In writing simple molecules with one asymmetric center, even untalented artists can sketch a perspective stick and ball picture. Drawing complicated systems with perspective, especially those having two or more asymmetric centers, becomes too unwieldy and frustrating. Chemists long ago began to use *plane projection* representations instead of attempting three-dimensional, perspective sketches. Again, certain conventions must be heeded. This method attempts a strictly two-dimensional diagram, creating traps for the unwary. By convention the carbon skeleton that includes the asymmetric carbon is positioned *vertically*. The carbons immediately above and below the asymmetric carbon are imagined to project to the *rear*, as in Figure 16.2. The carbon normally given the lowest number for naming purposes is at the *top*. The two groups positioned horizontally are imagined to project *forward*. The asymmetric carbon is imagined as occupying the intersection of the two perpendicular lines. When more than one asymmetric carbon is present (i.e., two or more such intersections), the vertical bonds *at each intersection* must be imagined to project rearward, the horizontal bonds forward. D- and L-Glyceraldehydes are shown in both their perspective and plane projection formulas in Figure 16.2.

With this brief introduction to D- and L-families, we are in a position to understand what it means that all the nutritionally important sugars are in the D-family. The human animal cannot use members of the L-family of sugars. We can assimilate D-glucose, but we are unable to metabolize L-glucose. Among the simple sugars the configuration of the asymmetric carbon farthest from the carbonyl group determines family membership. If in a plane projection diagram, constructed by the conventions, the hydroxyl at this carbon projects to the right (as in D-glyceraldehyde), the molecule is in the D-family. Figure 16.3 shows the family tree for the D-family of aldoses —from the triose, D-glyceraldehyde, through all the D-aldohexoses, including the all-important glucose. The diagram may be called a family tree because it is possible to start with D-glyceraldehyde and by a series of several reactions make all the compounds shown. In all these plane projection diagrams the asymmetric carbon farthest from the carbonyl has a hydroxyl projecting to the right. Another family tree could be constructed from L-glyceraldehyde.

CYCLIC FORMS OF GLUCOSE

Mutarotation. When ordinary crystalline glucose is freshly dissolved in water, it shows a specific rotation $[\alpha]_D$ $+113°$. (Here the subscript D refers to the D-line of sodium vapor light.) While the solution remains at room temperature, this value gradually changes, stopping at $[\alpha]_D$ $+52°$. This aged solution can be made to yield the original glucose by suitable conditions of crystallization. Apparently, no deep-seated chemical change has occurred to produce this gradual shift in specific rotation. The change in rotatory power of a solution of an optically active compound is called *mutarotation*.

If glucose is recovered from a solution by evaporating it at a temperature above 98°C, another form is obtained. Its specific rotation, in a freshly prepared solution, is $[\alpha]_D$ $+19°$, but this solution also mutarotates until a final value of $[\alpha]_D$ $+52°$, is obtained, the same final value cited for ordinary glucose. In fact, glucose with a specific rotation, in a fresh solution, of $[\alpha]_D$ $+113°$ can be separated from this second aged solution. The structure of D-(+)-glucose that we have used thus far does not provide any obvious clue to explain this behavior. Moreover, other chemical properties of glucose that do not fit too well with this structure are known.

Glucosides. Another peculiarity of glucose is that conversion of its aldehyde group to an acetal requires only one mole of alcohol, instead of two, per mole of glucose (cf. pp. 394).

$$\underset{\text{Glucose}}{C_6H_{12}O_6} + \underset{\substack{\text{Methyl}\\\text{alcohol}}}{HOCH_3} \xrightarrow{H^+} \underset{\text{Methyl glucoside}}{C_6H_{11}O_5\text{—}OCH_3} + H_2O$$

The product, called in the example methyl glucoside, is stable in a basic solution. It does not react positively to Tollens' test. In other words, it does not behave as

D-Glyceraldehyde
O=C—H
H—C—OH
CH₂OH

D-Erythrose	D-Threose
O=C—H	O=C—H
H—C—OH	HO—C—H
H—C—OH	H—C—OH
CH₂OH	CH₂OH

D-Ribose	D-Arabinose	D-Xylose	D-Lyxose
O=C—H	O=C—H	O=C—H	O=C—H
H—C—OH	HO—C—H	H—C—OH	HO—C—H
H—C—OH	H—C—OH	HO—C—H	HO—C—H
H—C—OH	H—C—OH	H—C—OH	H—C—OH
CH₂OH	CH₂OH	CH₂OH	CH₂OH

D-Allose	D-Altrose	D-Glucose	D-Mannose	D-Gulose	D-Idose	D-Galactose	D-Talose
O=C—H	O=C—H	O=C—H	O=C—H	O=C—H	O=C—H	O=C—H	O=C—H
H—C—OH	HO—C—H	H—C—OH	HO—C—H	H—C—OH	HO—C—H	H—C—OH	HO—C—H
H—C—OH	H—C—OH	HO—C—H	HO—C—H	H—C—OH	H—C—OH	HO—C—H	HO—C—H
H—C—OH	H—C—OH	H—C—OH	H—C—OH	HO—C—H	HO—C—H	HO—C—H	HO—C—H
H—C—OH	H—C—OH	H—C—OH	H—C—OH	H—C—OH	H—C—OH	H—C—OH	H—C—OH
CH₂OH	CH₂OH	CH₂OH	CH₂OH	CH₂OH	CH₂OH	CH₂OH	CH₂OH

Figure 16.3 The D-family of aldoses through the aldohexoses. In carbohydrate chemistry, if the —OH on the asymmetric carbon farthest from C-1 projects to the right in a plane projection diagram, the substance is in the D-family. There are eight other aldohexoses, all in the L-family, each being a mirror image of a member of the D-family. The aldoses most important in our study are three, D-ribose, D-glucose, and D-galactose.

though it were a potential aldehyde, that is, a hemiacetal. It behaves as though it were an acetal. The reaction shown can in fact be reversed. Like any acetal, methyl glucoside can be easily hydrolyzed. It does not mutarotate, however.

Just as glucose can be prepared in two forms—called α-glucose and β-glucose—methyl glucoside can also be made in two forms—α-methyl glucoside and β-methyl glucoside.

Cyclic Hemiacetal Forms for Glucose. These observations make sense in terms of structural formulas if glucose can exist in a cyclic hemiacetal form (Figure 16.4). If glucose is already a hemiacetal, then while still being a potential aldehyde capable of showing virtually all the reactions of an aldehyde, it would require only one additional molecule of an alcohol to become an acetal. The two forms of glucose that mutarotate are possible because the hemiacetal —OH group may be on either one side of the ring or the other.

All three forms are present in equilibrium in aged solutions of either α-glucose or β-glucose. The proportions at room temperature are 36% alpha, 64% beta, and 0.02% open form. Thus the phenomenon of mutarotation is explained as resulting from the slow establishment of a dynamic equilibrium between all three forms in solution, regardless of which form in the solid state, the alpha or beta, is used to make the glucose solution. During ring closure of the open form, the newly generated —OH group at C-1 may come out either above the ring or below it, in relation to the $—CH_2OH$ group which is our reference. In the closed forms C-1 is now asymmetric. The alpha and beta forms, of course, are not related as object to mirror image. They are not enantiomers but diastereomers. They differ only in the configuration at C-1.

What happens during mutarotation might be more easily understood if we try to isolate only the reacting sites. The original carbonyl carbon is shown in boldface.

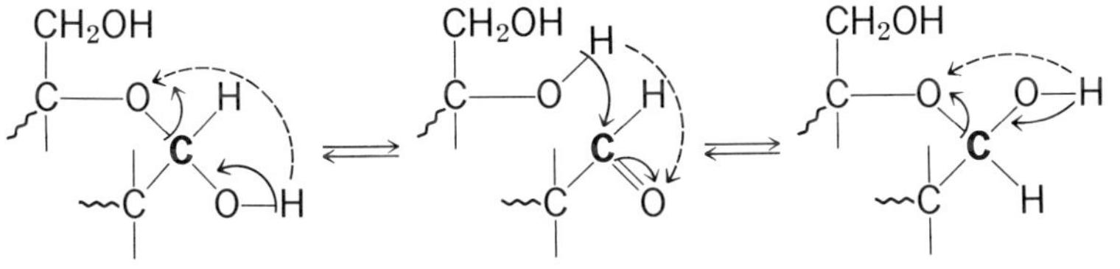

CH_2OH
C—O H
C
C O—H
CH_2OH H
C—O H
C
C O
CH_2OH
C—O O—H
C
C H

α-Glucose ⇌ ⇌ β-Glucose

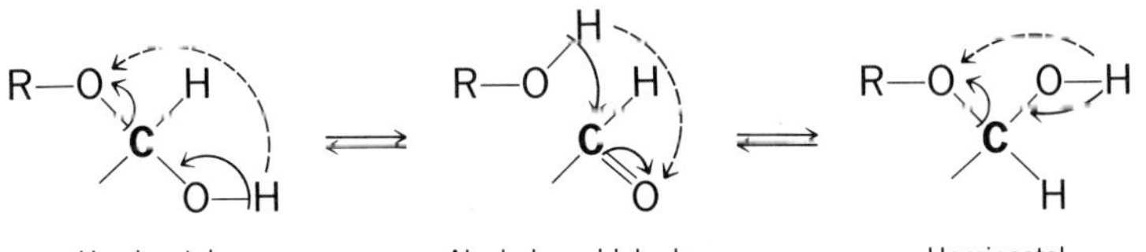

R—O H
C
O—H
H
R—O H
C
O
R—O O—H
C
H

Hemiacetal ⇌ Alcohol + aldehyde (or ketone) ⇌ Hemiacetal

A hemiacetal forms from the simple addition of the elements of an alcohol group across the double bond of a carbonyl group. If the same molecule contains both the —OH and the C=O, the reaction is intramolecular and the hemiacetal will necessarily be cyclic.

The glucosides are C-1 ethers (actually acetals) of the closed forms of the aldoses. Two forms, alpha and beta, are possible because alpha and beta forms for starting materials are possible.

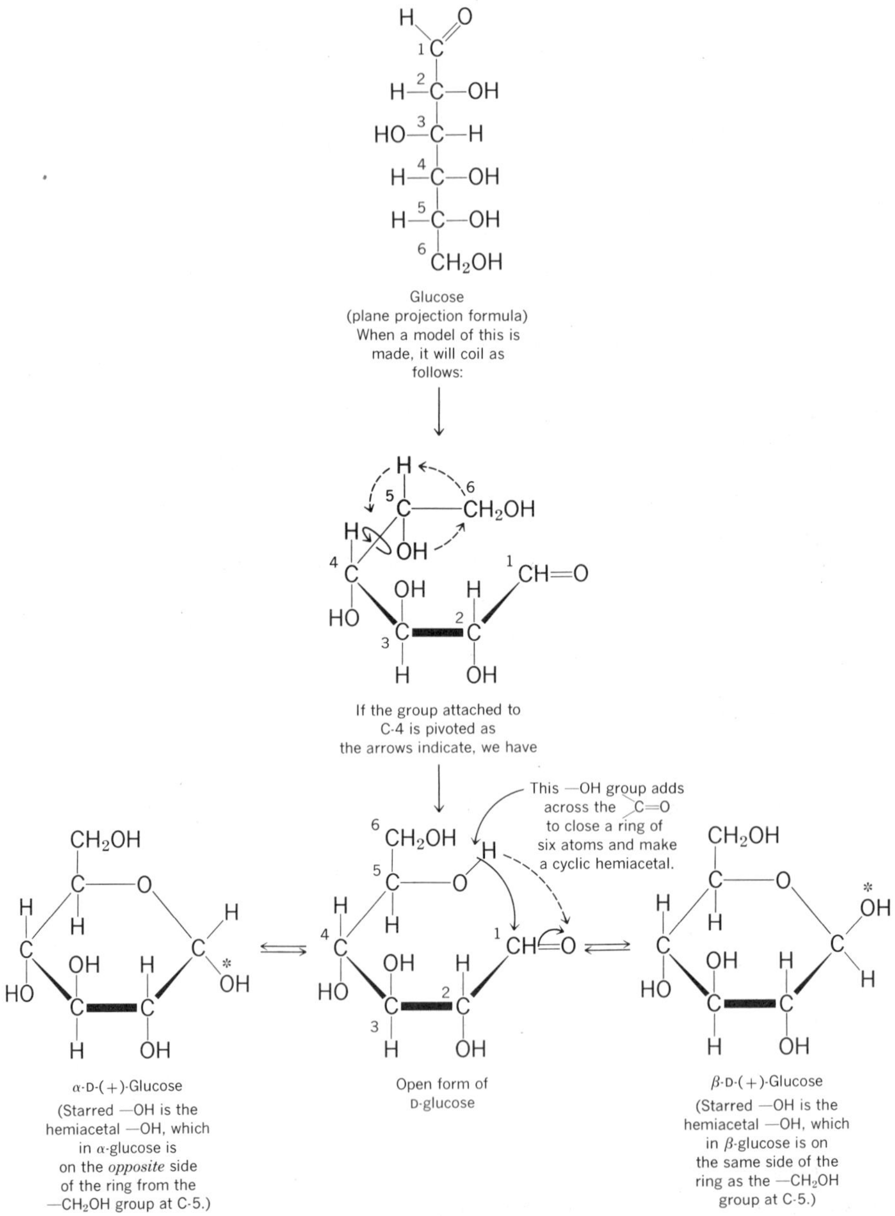

Glucose
(plane projection formula)
When a model of this is made, it will coil as follows:

If the group attached to C-4 is pivoted as the arrows indicate, we have

This —OH group adds across the C=O to close a ring of six atoms and make a cyclic hemiacetal.

α-D-(+)-Glucose
(Starred —OH is the hemiacetal —OH, which in α-glucose is on the *opposite* side of the ring from the —CH_2OH group at C-5.)

Open form of D-glucose

β-D-(+)-Glucose
(Starred —OH is the hemiacetal —OH, which in β-glucose is on the same side of the ring as the —CH_2OH group at C-5.)

Figure 16.4 The cyclic hemiacetal forms of D-glucose and their relation to its open-chain polyhydroxy aldehyde structure.

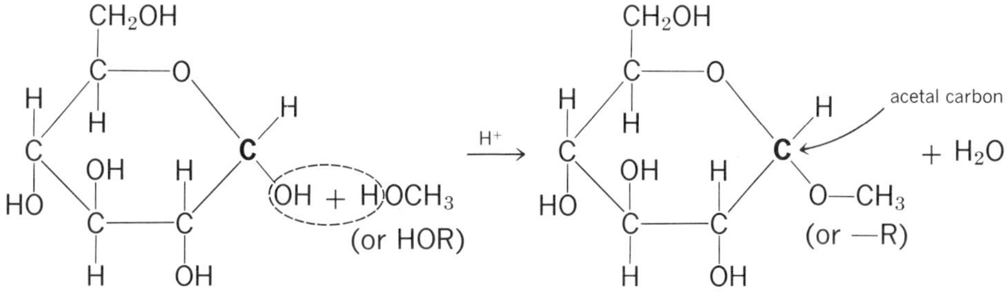

$\xrightarrow{H^+}$ + H_2O

acetal carbon

(or HOR)

(or —R)

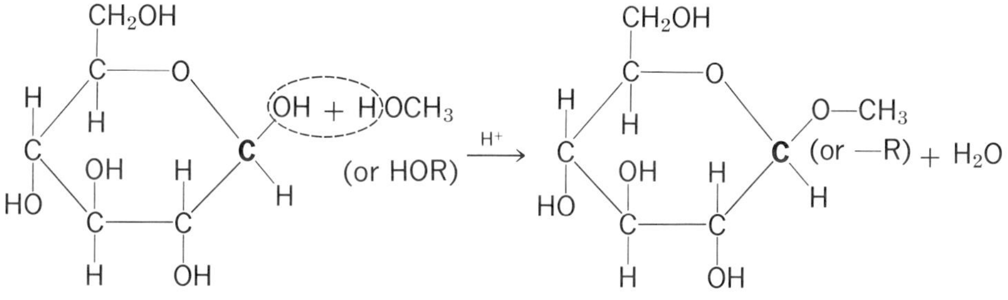

α-Glucose

α-Methyl glucoside

(or HOR)

$\xrightarrow{H^+}$ (or —R) + H_2O

β-Glucose

β-Methyl glucoside

Since glucose is a hemiacetal, it is a potential aldehyde. The fact that only a trace amount (0.02%) of the open-chain aldehyde is present at equilibrium in an aqueous solution does not reduce the availability of the aldehyde group to any reagent that is capable of demanding it (e.g., oxidizing agents such as Benedict's or Tollens' reagents). The hemiacetal linkage apparently opens easily enough so that when the trace amount of aldehyde form is consumed, the equilibrium shifts to provide a fresh supply. For this reason it is still logical to define the monosaccharides as we did, as polyhydroxy aldehydes or ketones. The central ideas that we shall need to carry forward to our study of the metabolism of glucose are these:

1. Glucose easily forms glucosides at its hemiacetal —OH of the closed forms.
2. Glucose is easily oxidized, especially at its aldehyde group, open form.
3. Glucose has other —OHs that might (*a*) be oxidized, (*b*) form esters (especially with phosphoric acid, to anticipate later needs), or (*c*) form acetals.

The structure of glucose in its open-chain and cyclic forms should be learned. Convenient condensations of the cyclic structures are the following.

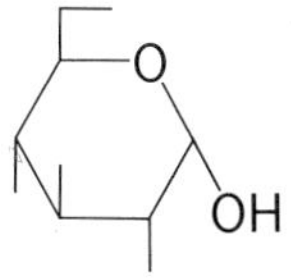

α-Glucose

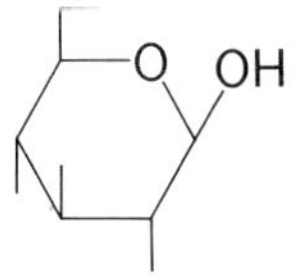

β-Glucose

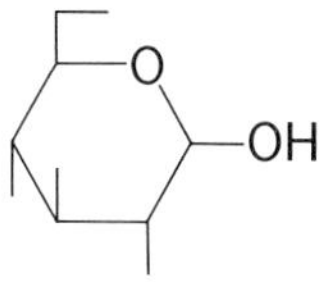

May be used when alpha and beta designations are unimportant or when the equilibrium mixture is intended.

Where a vertical line appears there is a hydroxyl group, and the inverted ell, ⌈, designates a $\underset{|}{CH_2}$—OH group. The direction of a vertical line in relation to the inverted ell indicates the direction in which the hydroxyl group projects.

Occurrence of Glucose. Glucose, the most important hexose, is found in most sweet fruits, especially in ripe grapes. It is therefore sometimes called grape sugar. A normal constituent of the bloodstream, it is also often called blood sugar. Because glucose is dextrorotatory, a third nickname is dextrose. The energy-rich molecules of glucose constitute one of the chief sources of chemical energy for animals. It is the building block for several other carbohydrates such as maltose, starch, glycogen, the dextrins, and cellulose. It is also a structural unit in lactose (milk sugar) and sucrose (table sugar, cane sugar, beet sugar).

GALACTOSE

Most galactose occurs naturally in combined forms, especially in the disaccharide, lactose (milk sugar). Galactose, a diastereomer of glucose, differs from it only in the orientation of the C-4 hydroxyl. Like glucose, it is a reducing sugar, it mutarotates, and it exists in three forms, alpha, beta, and open.

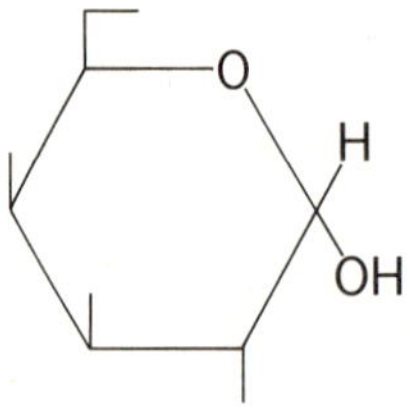

α-Galactose

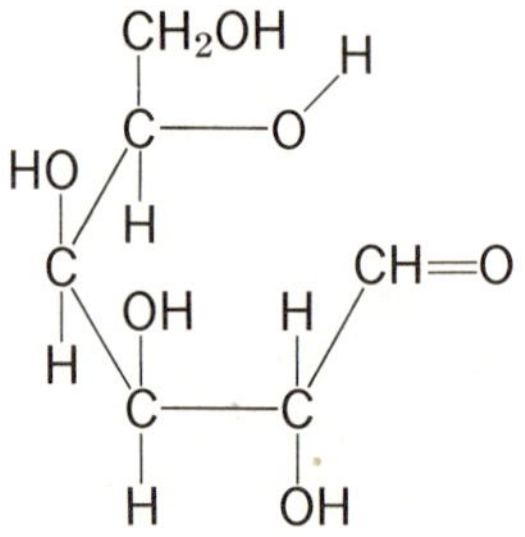

Galactose, open form

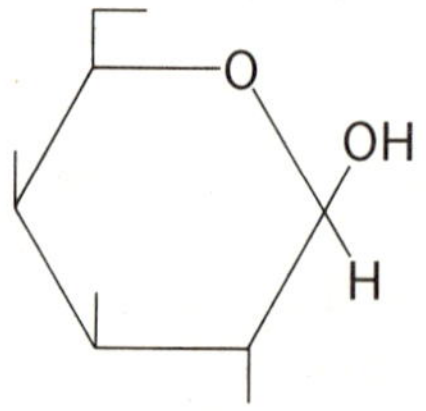

β-Galactose

KETOHEXOSES. FRUCTOSE

Fructose, the only important ketohexose, is found together with glucose and sucrose in honey and fruit juices. It can exist in more than one form, and as a building unit in sucrose it exists in a cyclic, five-membered hemiketal form.[1]

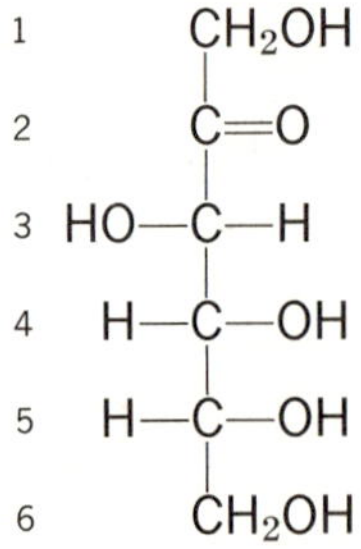

D-(−)-Fructose

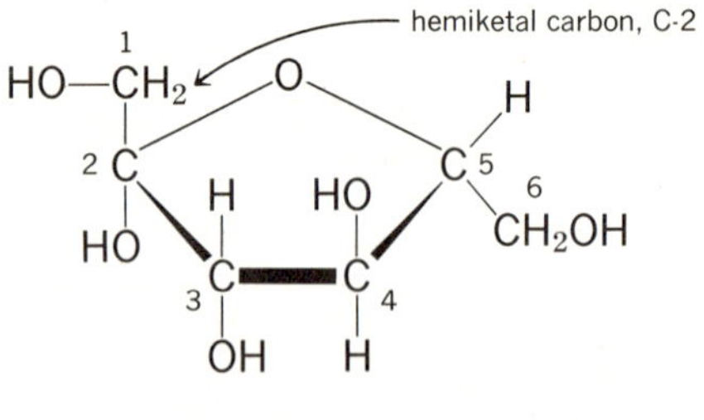

One of the cyclic forms of fructose

[1] When the hemiacetal type of compound is made from a ketone instead of an aldehyde, it is called a hemiketal.

Because fructose is strongly levorotatory, we sometimes call it levulose; the specific rotation is $[\alpha]_D$ $-92.4°$.

DISACCHARIDES

Nutritionally, the three important disaccharides are maltose, lactose, and sucrose. A fourth will be mentioned, cellobiose, which is obtained by partial hydrolysis of cellulose. How these are related to the monosaccharides just discussed may be seen in the following word equations.

$$\begin{aligned}\text{Maltose} + H_2O &\longrightarrow \text{glucose} + \text{glucose}\\ \text{Lactose} + H_2O &\longrightarrow \text{glucose} + \text{galactose}\\ \text{Sucrose} + H_2O &\longrightarrow \text{glucose} + \text{fructose}\\ \text{Cellobiose} + H_2O &\longrightarrow \text{glucose} + \text{glucose}\end{aligned}$$

All these disaccharides are glycosides and are formed by one monosaccharide acting as a hemiacetal and the other as an alcohol to make an acetal.

MALTOSE

Although present in germinating grain, maltose or malt sugar does not occur widely in the free state in nature. It can be prepared by partial hydrolysis of starch. As with glucose, two forms of maltose are known and each mutarotates. Maltose is a reducing sugar and hydrolyzes to yield two molecules of glucose. The enzyme *maltase* catalyzes this reaction, and it is known to act only on α-glucosides, not on β-glucosides. We could imagine maltose being formed by the splitting out of water between the hemiacetal —OH of one glucose unit and the C-4 —OH of the other.

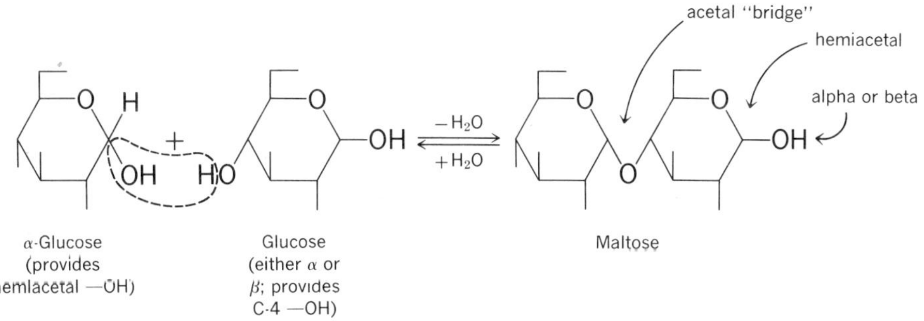

acetal "bridge"
hemiacetal
alpha or beta

O H + OH HO O —OH $\underset{+H_2O}{\overset{-H_2O}{\rightleftharpoons}}$ O O —OH O

α-Glucose (provides hemiacetal —OH)

Glucose (either α or β; provides C-4 —OH)

Maltose

LACTOSE

Lactose or milk sugar occurs in the milk of mammals, 4 to 6% in cow's milk, 5 to 8% in human milk. It is obtained commercially as a by-product in the manufacture of cheese. Like maltose, it exists in alpha and beta forms, is a reducing sugar, and mutarotates. Its hydrolysis yields galactose and glucose, and its formation may be visualized as the splitting out of water between β-galactose, acting as the hemiacetal, and glucose, furnishing the C-4 —OH group.

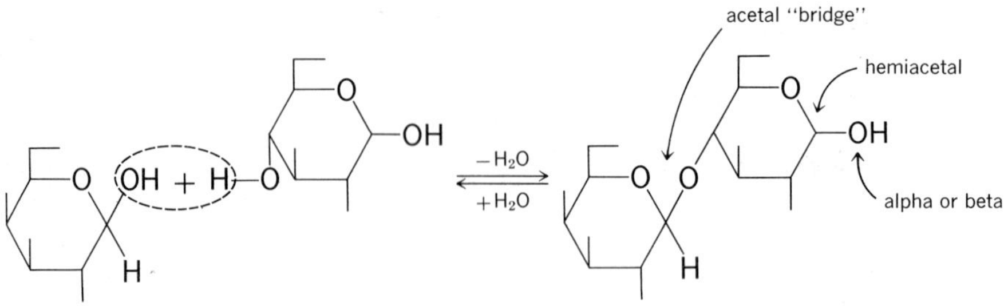

acetal "bridge"
hemiacetal
O
OH
O
OH + H
O
−H2O
+H2O
O
O
OH
alpha or beta
H
H

β-Galactose   Glucose   Lactose

SUCROSE

The juice of sugarcane, which contains about 14% sucrose, is obtained by crushing and pressing the canes. It is then freed of protein-like substances by the precipitating action of lime. Evaporation of the clear liquid leaves a semisolid mass from which raw sugar is isolated by centrifugation. (The liquid that is removed is *blackstrap molasses.*) Raw sugar (95% sucrose) is processed to remove odoriferous and colored contaminants. The resulting white sucrose, our table sugar, is probably the largest volume *pure* organic chemical produced. Much of our supply of sucrose now comes from sugar beets, and so-called beet sugar and cane sugar are chemically identical. Structurally, a molecule of sucrose is derived from one glucose unit and one fructose unit:

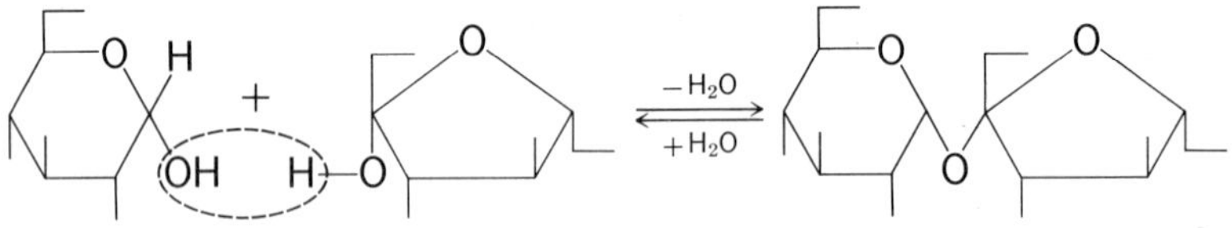

O H
+
O
OH H
O
−H2O
+H2O
O
O
O

α-Glucose   β-Fructose   Sucrose

An acetal oxygen bridge links the two units. No hemiacetal group is present in the sucrose molecule. Sucrose is not a reducing sugar.

The 50:50 mixture of glucose and fructose that forms when sucrose is hydrolyzed is often called *invert sugar.* Honey, for example, consists of this sugar for the most part. The term "invert" comes from the change or inversion in sign of optical rotation that occurs when sucrose, $[\alpha]_D$ +66.5°, is converted into the 50:50 mixture, $[\alpha]_D$ −19.9°.

CELLOBIOSE

Cellobiose ("cello-," cellulose; "bi-," two; "-ose," sugar) is the disaccharide unit in cellulose (e.g., cotton fiber). Like maltose, it delivers two glucose molecules when it is hydrolyzed. It differs structurally from maltose only in the orientation of the acetal oxygen bridge.

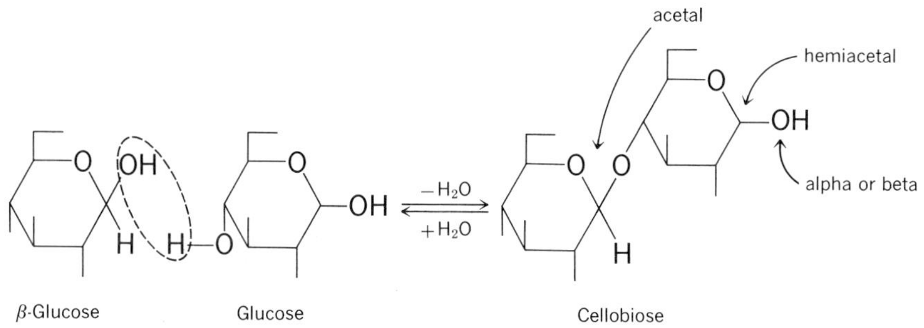

acetal
hemiacetal
OH
alpha or beta

β-Glucose Glucose $\underset{+H_2O}{\overset{-H_2O}{\rightleftharpoons}}$ Cellobiose

POLYSACCHARIDES

Much of the glucose produced in a plant by photosynthesis is used to make its cell walls and its rigid fibers. Much is also stored for the food needs of the plant. But instead of being stored as glucose molecules, which are too soluble in water, it is converted to a much less soluble form, starch. This polymer of glucose is particularly abundant in plant seeds. We use these two polymers of glucose in about the same way that plants do. Cellulose and its more woody relatives are made into enclosures ranging from frame houses to shirts and blouses. Starch is used for food. During digestion we break starch down eventually to glucose, and what we do not need right away for energy we store. We do not normally excrete excess glucose but convert it to a starch-like polymer, glycogen, or to fat. In this section we concentrate on the three types of glucose polymers, starch, glycogen, and cellulose.

STARCH

The skeletal outline of what is believed to constitute the basic structure of starch is shown in Figure 16.5. It is actually a mixture of polyglucose molecules, some linear and some branched. One type, *amylose,* consists of long, unbranched polymers of α-glucose. The other, *amylopectin,* is the branched polymer of α-glucose. Natural starches are about 10 to 20% amylose and 80 to 90% amylopectin.

Physical Properties. The huge molecules found in starch, with their many hydroxyl groups, stick to each other quite strongly, but in a disorderly way. These hydrogen bond networks are not so effectively broken up by water molecules that starch is soluble in water. It is only slightly so, and starch "solutions," prepared by grinding starch in a small amount of water and then pouring the slurry into a larger volume of boiling water, are really colloidal dispersions.

Chemical Properties. The acetal oxygen bridges linking glucose units together in starch are easily hydrolyzed, especially in the presence of acids or certain enzymes. We have learned to expect this behavior of acetals and, when we digest starch, digestive juices do nothing more than hydrolyze it.

The partial breakdown products of amylopectin are still large molecules called the *dextrins*. They are used to prepare mucilages, pastes, and fabric sizes. The step-by-step hydrolysis of amylopectin to glucose through the dextrins and maltose is described in Figure 16.6.

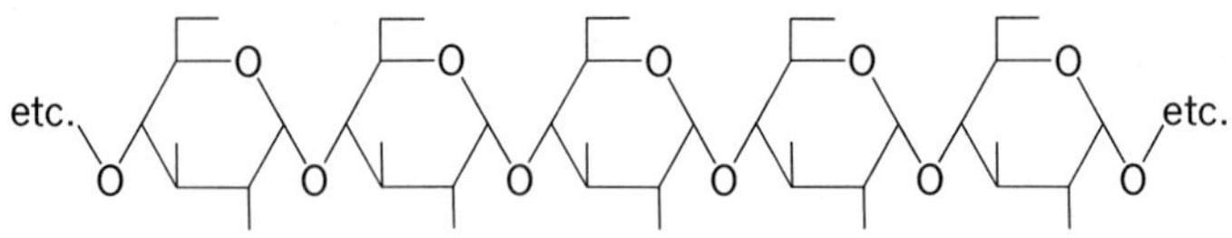

Amylose

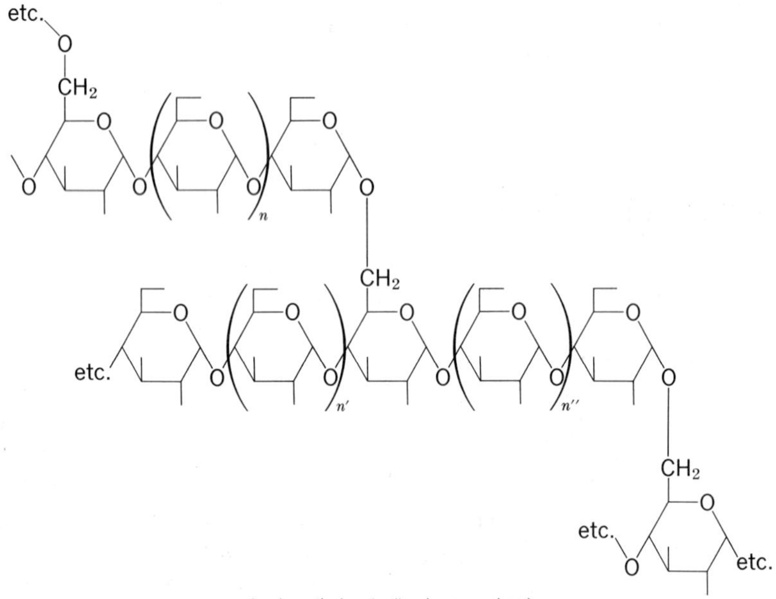

Amylopectin (n, n', n'' = large numbers)

Figure 16.5 Basic structural features of the types of glucose polymers found in starches. Molecular weight measurements on various preparations of starches have given results varying from 50,000 to several million. The higher values probably more truly represent the sizes of natural starch molecules, for during the isolation of this substance some partial hydrolysis very likely occurs. An amylose molecule with a molecular weight of 1 million would contain slightly over 6000 glucose units.

Starch is not a reducing carbohydrate. The potential aldehyde groups would be at the ends of chains only in percentages too low for detection by Benedict's reagent. Starch does, however, give an intense, brilliant blue-black color with iodine.[2] This *iodine test* for starch can detect extremely minute traces of starch in solution. The chemistry of the test is not definitely known, but iodine molecules are believed to become trapped within the vast network of starch molecules. Should this network disintegrate, as it does during the hydrolysis of starch, the test will fail. In fact, the course of the hydrolysis of starch in a test tube can be followed by this test, for as the reaction proceeds the color produced by iodine gradually changes from blue-black to purple to red to no color.

[2] The starch-iodine reagent is iodine, I_2, dissolved in an aqueous solution of sodium iodide, NaI. Iodine, by itself, is only slightly soluble in water. When iodide ion is present, however, it combines with iodine molecules to form a triiodide ion, I_3^-, which liberates iodine easily on demand.

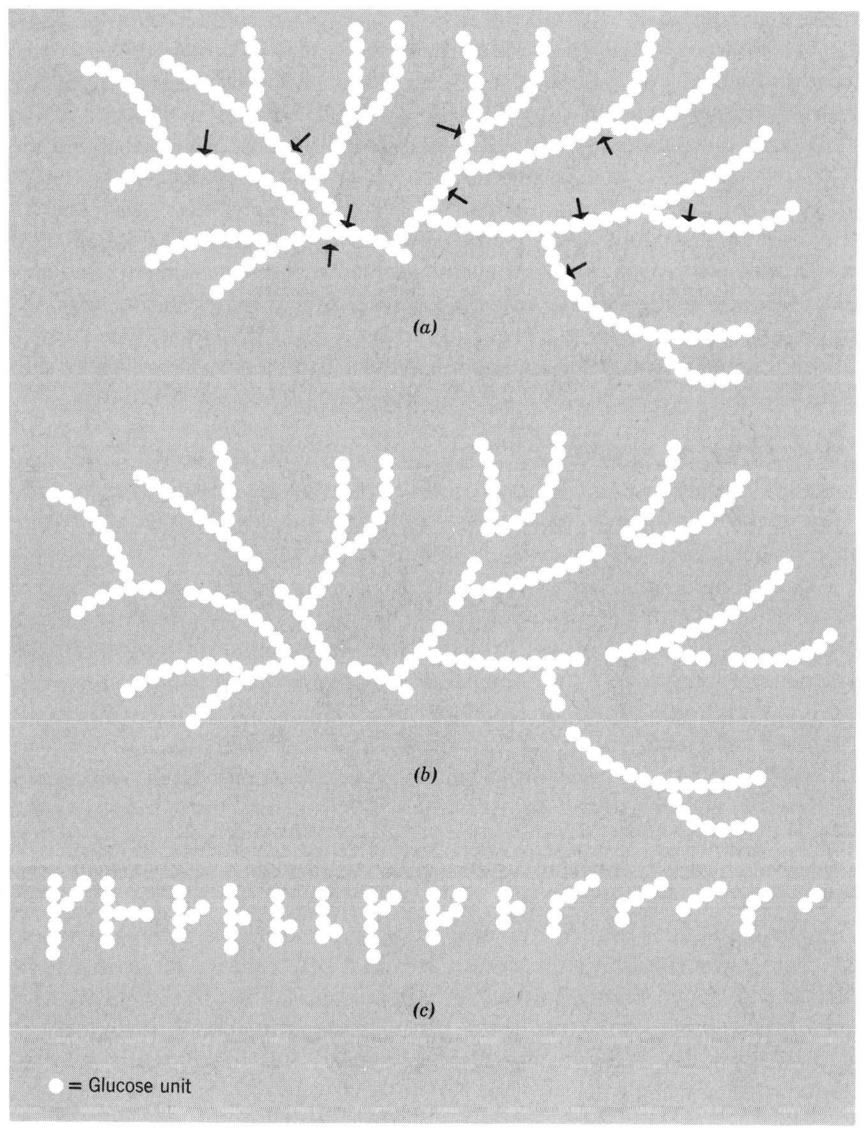
(a)
(b)
(c)
= Glucose unit

Figure 16.6 The hydrolysis of amylopectin according to P. Bernfeld. (*a*) Model of amylopectin. (*b*) Dextrins of medium molecular weight that give purple to red colors with iodine. These dextrins result from random hydrolysis of about 4% of the oxygen bridges of amylopectin. See arrows in part *a*. (c) Low-formula-weight dextrins (limit dextrins) that are actually hepta-, hexa-, and pentasaccharides. These give a brown color with iodine. (From P. Bernfeld, *Advances in Enzymology,* Vol. 12, 1951, page 390.)

GLYCOGEN

Liver and muscle tissue are the main sites of glycogen storage in the body. Under the control of enzymes, some of which are in turn controlled by hormones, glucose molecules can be mobilized from these glycogen reserves to supply chemical energy for the body.

Glycogen differs from starch by the apparent absence of any molecules of the unbranched, amylose type. It is branched very much like amylopectin, perhaps even more so. Molecular weights of various glycogen preparations have been reported over a range of 300,000 to 100,000,000, corresponding roughly to 1800 to 60,000 glucose units. During the digestion and absorption of a meal containing carbohydrates, the body builds up its glycogen deposits. Between meals, during fasts, the deposits are made to deliver glucose.

Dispersions of glycogen in water are opalescent, turning a violet-red color with iodine.

CELLULOSE

Starch and glycogen are polymers of the alpha form of glucose. Cellulose is a polymer of the beta form. Its molecules are unbranched and resemble amylose. A portion of the structure of cellulose is shown in Figure 16.7.

The structural difference between amylose and cellulose is one of the tremendous trifles in nature. In humans amylose is digestible but cellulose is not. Yet the only difference between them structurally is the orientation of the oxygen bridges. Human beings and carnivorous animals do not have the enzyme needed to make the hydrolysis of cellulose rapid enough to be of any use. Many microorganisms, snails, and ruminants (cud-chewing animals) can use cellulose, however. The ruminants have suitable microorganisms in their alimentary tracts whose own enzyme systems catalyze the conversion of cellulose into small molecules the ruminants can use. It is fortunate that *some* systems contain enzymes capable of catalyzing the hydrolysis of cellulose. Otherwise the land would soon be covered by the dead debris of grasses, leaves, and trees that have fallen, and the remains of the annual plants. Soil organisms, fortunately, go to work on these remains and break them down. (Civilization is indeed lucky that DDT does not kill soil bacteria. Its application since World War II has been so widespread that traces of DDT have been found in arctic snows where no one has deliberately brought it. The effect that DDT might have on soil microorganisms, however, was not thoroughly tested before we started using it!)

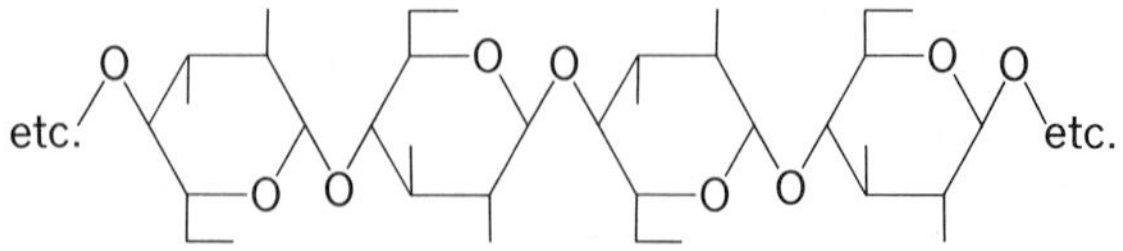

etc. O O O O O O O O etc.

Figure 16.7 Cellulose, a linear polymer of β-glucose. Note that, in order to indicate the linearity of the structure, every other glucose unit is "flipped over" from the normal way of writing it.

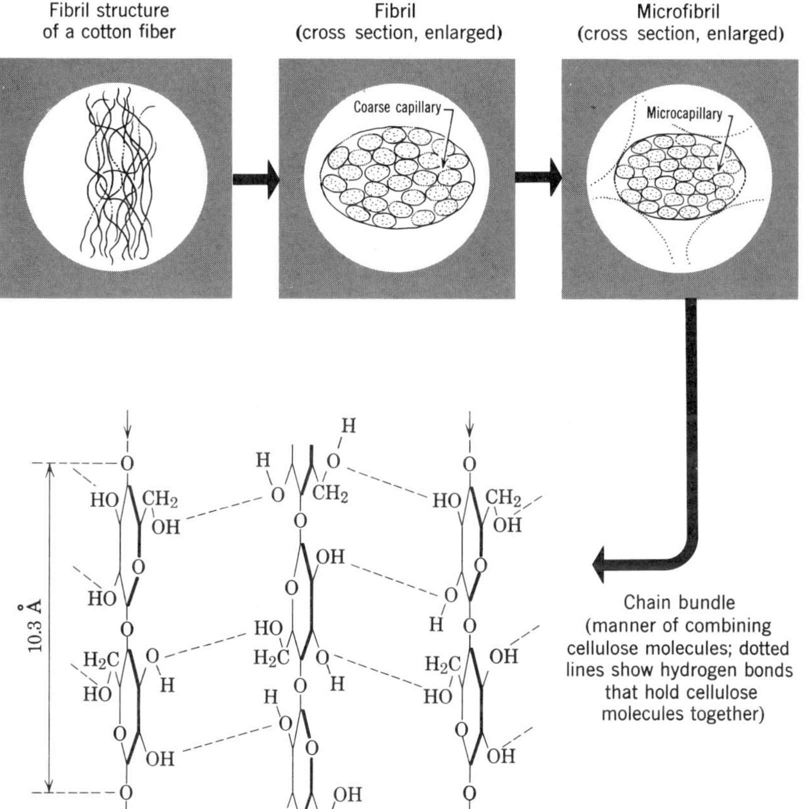
Fibril structure of a cotton fiber
Fibril (cross section, enlarged)
Coarse capillary
Microfibril (cross section, enlarged)
Microcapillary
Chain bundle (manner of combining cellulose molecules; dotted lines show hydrogen bonds that hold cellulose molecules together)
10.3 Å

Figure 16.8 Details of the cotton fiber. In this series of diagrams successively smaller portions of a cotton fiber are enlarged and depicted. (Adapted from illustrations appearing in H. R. Mauersberger, editor, *Matthews' Textile Fibers,* sixth edition, John Wiley and Sons, New York, 1954, pages 73 and 77.)

Cellulose makes up the cell membranes of the higher plants and gives them their rigidity. It is easily the most abundant organic compound in the world. Cotton fiber is almost 98% cellulose; each molecule has from 2000 to 9000 β-glucose units, arranged roughly parallel to one another and to the axis of the fiber. Structural details are illustrated in Figure 16.8. Several cellulose molecules twist and overlap to form the next to smallest unit in a cotton fiber, the *chain bundle.* A vast interlacing network of hydrogen bonds provides the necessary forces of attraction. When several chain bundles overlap, they form a microscopic unit of the visible fiber, a *microfibril.* Within it, running between the chain bundles, is a fine capillary network. A collection of intertwined, overlapping microfibrils makes up a visible *fibril,* and within it runs a coarse capillary network. A single cotton fiber is made up of a twisting, intertwining, overlapping aggregation of fibrils.

REFERENCES AND ANNOTATED READING LIST

BOOKS

V. M. Ingram. *The Biosynthesis of Macromolecules.* W. A. Benjamin, New York, 1966.

M. Stacey and S. A. Barker. *Carbohydrates of Living Tissues.* Van Nostrand Company, Princeton, N.J., 1962.

C. R. Noller. *Chemistry of Organic Compounds,* third edition. W. B. Saunders Company, Philadelphia, 1965.

ARTICLES

E. I. Rabinowitch. "Photosynthesis." *Scientific American,* August 1948, page 34. Although this is an old article, its impressive illustrations of the carbon, hydrogen, and oxygen cycles of nature on this planet make it a fine general introduction to photosynthesis.

R. D. Preston. "Cellulose." *Scientific American,* September 1957, page 156. The structure, technology, and industrial uses of cellulose are discussed.

I. A. Pearl. "Lignin Chemistry, Century-Old Puzzle." *Chemical and Engineering News,* July 6, 1964, page 81. Cellulose in wood is embedded in the natural plastic, lignin. Research departments of the paper industry have toiled long and hard to make something useful from it. This article describes part of this work and its fruits.

PROBLEMS AND EXERCISES

1. Write structures for α-maltose, β-maltose, and the open form of maltose. (*Hint.* Only one ring is open.)
2. Write an equation for the hydrolysis of maltose.
3. Define each of the following terms. Illustrative structures may be used.
   (*a*) ketohexose (*b*) disaccharide (*c*) mutarotation
   (*d*) photosynthesis (*e*) absolute configuration (*f*) D-family
   (*g*) plane projection formula (*h*) a reducing sugar
4. Write equations that illustrate how lactose undergoes mutarotation.
5. If two glucose units are linked together by the reaction

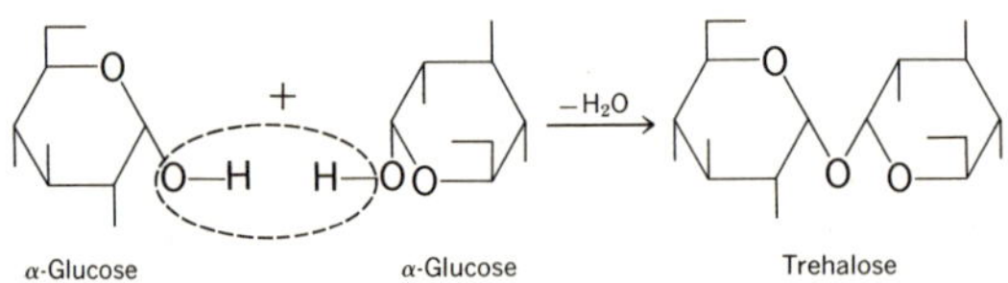
O
+
O—H
H—O
O
$-H_2O$
O
O
O
α-Glucose
α-Glucose
Trehalose

   the disaccharide formed is trehalose, a disaccharide found in young mushrooms and yeasts and the chief carbohydrate in the blood (hemolymph) of certain insects. On the basis of its structural features, answer the following questions about trehalose. If a reaction is predicted, write it.
   (*a*) Is trehalose a reducing sugar? Why?
   (*b*) Can it mutarotate? Why?
   (*c*) Can it be hydrolyzed?
6. Benedict's test, or some variation of it, is commonly used to detect glucose in urine. What are the structural features in glucose that make it give a positive reaction?
7. In some rare instances galactose appears in the urine. How might an analyst erroneously report the presence of glucose?

CHAPTER SEVENTEEN

Lipids

We defined carbohydrates in terms of their structures, but lipids have an operational definition, one stated in terms of the operation used to isolate them. When plant or animal material is crushed and ground with nonpolar solvents such as benzene, chloroform, or carbon tetrachloride (fat solvents), the portion that dissolves in the solvent is classified as lipid. Carbohydrates and proteins are insoluble. Depending on the plant or animal origin, lipid material may include such a wide variety of compounds that it is difficult to make a precise structural definition. Included in the group could be neutral fats containing only carbon, hydrogen, and oxygen; phosphorus-containing compounds called phospholipids, which usually also contain nitrogen; aliphatic alcohols; waxes; steroids; terpenes; and "derived lipids," those substances resulting from partial or complete hydrolysis of some of the foregoing. Table 17.1 provides a summary of the important lipids.

Some workers and writers in the field have arbitrarily restricted the term lipid or fat to esters of long-chain carboxylic acids and alcohols or closely related substances. This practice will be followed in this chapter, and we shall study primarily the neutral fats, the simple lipids.

SIMPLE LIPIDS. THE TRIGLYCERIDES

The most abundant group of lipids in plants and animals are the simple lipids or triglycerides. Included in this group are such common substances as lard, tallow, butterfat, olive oil, cottonseed oil, corn oil, peanut oil, linseed oil, coconut oil, and soybean oil. Their molecules consist of esters of the trihydroxy alcohol, glycerol, with various long-chain fatty acids. It is believed that a typical triglyceride molecule contains one unit from each of three different fatty acids rather than three units from a single fatty acid.

Fats and oils of whatever source are mixtures of different glyceride molecules; the differences are in specific fatty acids incorporated, not in the general structural

Table 17.1 Lipid Classification

General Class	Subclasses	Generic Class Structure	Nature of Hydrolysis Products	Examples and Occurrences
Simple Lipids	Waxes	$R{-}O{-}C({=}O){-}R'$	Long-chain carboxylic acids Long-chain alcohols	Beeswax, cuticle waxes (on flower petals and fruit skins)
	Triglycerides	$CH_2{-}O{-}C({=}O){-}R$ $CH{-}O{-}C({=}O){-}R'$ $CH_2{-}O{-}C({=}O){-}R''$	Mixture of long-chain acids Glycerol	Animal fats (solids): lard, tallow Vegetable oils (liquids): olive oil, corn oil, peanut oil, cottonseed oil, linseed oil
Phospholipids (Phosphoglycerides)	Lecithins	$CH_2{-}O{-}C({=}O){-}R$ $CH{-}O{-}C({=}O){-}R'$ $CH_2{-}O{-}P({=}O)(O^-){-}O{-}CH_2{-}CH_2{-}\overset{+}{N}(CH_3)_3$	Mixture of long-chain acids Glycerol Phosphoric acid Choline (an amino alcohol): $HOCH_2CH_2\overset{+}{N}(CH_3)_3\ X^-$	Found in nerve tissue
	Phosphatidylethanolamines	$CH_2{-}O{-}C({=}O){-}R$ $CH{-}O{-}C({=}O){-}R'$ $CH_2{-}O{-}P({=}O)(O^-){-}O{-}CH_2{-}CH_2{-}\overset{+}{N}H_3$	Mixture of long-chain acids Glycerol Phosphoric acid Aminoethanol: $HOCH_2CH_2NH_2$	Found in nerve tissue
	Phosphatidylserines	$CH_2{-}O{-}C({=}O){-}R$ $CH{-}O{-}C({=}O){-}R'$ $CH_2{-}O{-}P({=}O)(O^-){-}O{-}CH_2{-}CH(COOH){-}\overset{+}{N}H_3$	Mixture of long-chain acids Glycerol Phosphoric acid Serine: $HOCH_2CH(CO_2H)NH_2$	Found in nerve tissue
Sphingolipids	Sphingomyelins	$CH_3{-}(CH_2)_{12}{-}CH{=}CH{-}CH(OH){-}CH(NH{-}C({=}O){-}R){-}CH_2{-}O{-}P({=}O)(O^-){-}OCH_2CH_2{-}\overset{+}{N}(CH_3)_3$	Sphingosine (an unsaturated amino alcohol): $CH_3(CH_2)_{12}CH{=}CH{-}CHOH{-}CH(NH_2){-}CH_2OH$ A long-chain acid Phosphoric acid Choline	Found in brain tissue
	Cerebrosides	$CH_3{-}(CH_2)_{12}{-}CH{=}CH{-}CH(OH){-}CH(NH{-}C({=}O){-}R){-}CH_2{-}O{-}$(galactose ring)	Sphingosine A long-chain acid Galactose	Found in brain tissue (e.g., kerasin, phrenosin, nervon, and oxynervon)

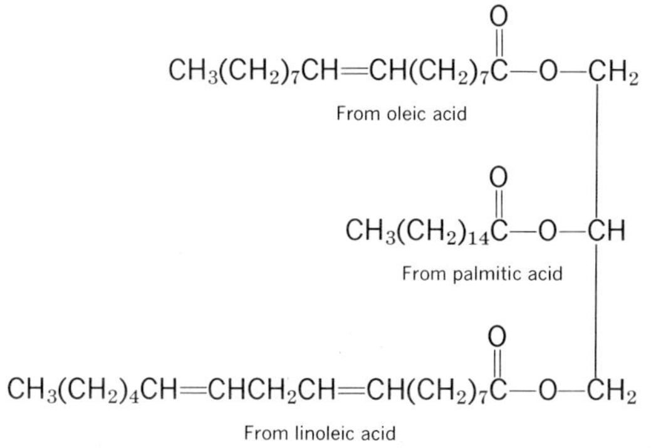

$$\begin{array}{l} CH_3(CH_2)_7CH{=}CH(CH_2)_7\overset{\overset{\displaystyle O}{\|}}{C}{-}O{-}CH_2 \\ \text{From oleic acid} \\ CH_3(CH_2)_{14}\overset{\overset{\displaystyle O}{\|}}{C}{-}O{-}CH \\ \text{From palmitic acid} \\ CH_3(CH_2)_4CH{=}CHCH_2CH{=}CH(CH_2)_7\overset{\overset{\displaystyle O}{\|}}{C}{-}O{-}CH_2 \\ \text{From linoleic acid} \end{array}$$

Figure 17.1 A typical mixed glyceride showing oleic acid, palmitic acid, and linoleic acid bound by ester linkages to one molecule of glycerol.

features. Thus it is not possible to write the structure of, say, cottonseed oil except to describe what a typical molecule is like. One such molecule is represented in Figure 17.1. In a particular fat or oil, certain fatty acids tend to predominate, certain others are either absent or are present in trace amounts, and virtually all the molecules are triglycerides. Data for several fats and oils are summarized in Table 17.2.

Table 17.2 Composition of the Fatty Acids Obtained by Hydrolysis of Common Neutral Fats and Oils

	Average Composition of Fatty Acids (%)						
Fat or Oil	Iodine Number	Myristic Acid	Palmitic Acid	Stearic Acid	Oleic Acid	Linoleic Acid	Others
Animal fats							
Butter	25–40	8–15	25–29	9–12	18–33	2–4	a
Lard	45–70	1–2	25–30	12–18	48–60	6–12	b
Beef tallow	30–45	2–5	24–34	15–30	35–45	1–3	b
Vegetable Oils							
Olive	75–95	0–1	5–15	1–4	67–84	8–12	
Peanut	85–100	—	7–12	2–6	30–60	20–38	
Corn	115–130	1–2	7–11	3–4	25–35	50–60	
Cottonseed	100–117	1–2	18–25	1–2	17–38	45–55	
Soybean	125–140	1–2	6–10	2–4	20–30	50–58	c
Linseed	175–205	—	4–7	2–4	14–30	14–25	d
Marine Oils							
Whale	110–150	5–10	10–20	2–5	33–40		e
Fish	120–180	6–8	10–25	1–3			e

[a] Three to four percent butyric acid, 1 to 2% caprylic acid, 2 to 3% capric acid, 2 to 5% lauric acid.
[b] One percent linolenic acid.
[c] Five to ten percent linolenic acid.
[d] Forty-five to sixty percent linolenic acid.
[e] Large amounts of other highly unsaturated fatty acids.

Fatty Acids. Fatty acids obtained from the lipids of most plants and animals tend to share the following characteristics.

1. They are usually monocarboxylic acids, R—CO_2H.
2. The R— group is usually an unbranched chain.
3. The number of carbon atoms is almost always even.
4. The R— group may be saturated, or it may have one, two, or three double bonds (sometimes four).

The most abundant saturated fatty acids are palmitic acid, $CH_3(CH_2)_{14}CO_2H$, and stearic acid, $CH_3(CH_2)_{16}CO_2H$, having sixteen and eighteen carbons, respectively. Others are included in Table 15.1 (p. 406), the acids above acetic with an even number of carbons, but they are present in only small amounts.

The most frequently occurring unsaturated fatty acids are listed in Table 17.3. The eighteen-carbon skeleton of stearic acid is duplicated in oleic, linoleic, and linolenic acids. Oleic acid is the most abundant and most widely distributed fatty acid in nature.

Among the eighteen-carbon fatty acids, the greater the degree of unsaturation, the lower the melting point:

stearic acid (saturated)	mp 70°C
oleic acid (one double bond)	4°C
linoleic acid (two double bonds)	−5°C
linolenic acid (three double bonds)	−11°C

A double bond produces a stiffened kinkiness in regions of the otherwise flexible chains, thus reducing the number of ways in which neighboring molecules can get close enough to each other that forces of attraction between them operate. With net forces between molecules weakened, they are freer to move about and the material is liquid at lower temperatures than the saturated relatives. For this reason most animal fats are solids and most vegetable oils are liquids at room temperature. The oils are more unsaturated than the fats. There are more double bonds per molecule in the oils and, like the unsaturated acids, they are liquids.

Table 17.3 Common Unsaturated Fatty Acids

Number of Double Bonds	Total Number of Carbons	Name	Structure	Mp (°C)
1	16	palmitoleic acid	$CH_3(CH_2)_5CH{=}CH(CH_2)_7CO_2H$	32
1	18	oleic acid	$CH_3(CH_2)_7CH{=}CH(CH_2)_7CO_2H$	4
2	18	linoleic acid	$CH_3(CH_2)_4CH{=}CHCH_2CH{=}CH(CH_2)_7CO_2H$	−5
3	18	linolenic acid	$CH_3CH_2CH{=}CHCH_2CH{=}CHCH_2CH{=}CH(CH_2)_7CO_2H$	−11
4	20	arachidonic acid	$CH_3(CH_2)_4CH{=}CHCH_2CH{=}CHCH_2CH{=}CHCH_2CH{=}CH(CH_2)_3CO_2H$	−50

The properties of the fatty acids are those to be expected of compounds having a carboxyl group, double bonds (in some), and long hydrocarbon chains. They are insoluble in water and soluble in nonpolar solvents. They can form salts, be esterified, and be reduced to the corresponding long-chain alcohols. Where present, alkene linkages react with bromine and take up hydrogen in the presence of a catalyst.

Iodine Number. The degree of unsaturation in a lipid is measured by its iodine number, defined as the number of grams of iodine that will add to the double bonds present in 100 g of the lipid. Saturated fatty acids, having no alkene linkages, have zero iodine numbers. Oleic acid has an iodine number of 90, linoleic acid 181, and linolenic acid 274. Animal fats have low iodine numbers, vegetable oils (the polyunsaturated oils of countless advertisements) higher values, as the data of Table 17.2 show. The iodine number of the mixed triglyceride of Figure 17.1 is calculated to be 89, in the range of the iodine number of olive oil or peanut oil.

CHEMICAL PROPERTIES OF TRIGLYCERIDES

Hydrolysis in the Presence of Enzymes. Enzymes in the digestive tracts of human beings and animals act as efficient catalysts for the hydrolysis of the ester links of triglycerides. In general:

$$\begin{array}{l} RC(=O)-OCH_2 \\ R'C(=O)-OCH \\ R''C(=O)-OCH_2 \end{array} + 3H_2O \xrightarrow{\text{enzyme}} \begin{array}{l} RC(=O)-OH \\ + \\ R'C(=O)-OH \\ + \\ R''C(=O)-OH \end{array} + \begin{array}{l} HOCH_2 \\ HOCH \\ HOCH_2 \end{array}$$

Triglyceride — Fatty acids — Glycerol

Example (using the triglyceride of Figure 17.1):

$$\begin{array}{r} CH_3(CH_2)_7CH{=}CH(CH_2)_7C(=O)-OCH_2 \\ CH_3(CH_2)_{14}C(=O)-OCH \\ CH_3(CH_2)_4CH{=}CHCH_2CH{=}CH(CH_2)_7C(=O)-OCH_2 \end{array} + 3H_2O \xrightarrow{\text{enzyme}} \begin{array}{r} CH_3(CH_2)_7CH{=}CH(CH_2)_7C(=O)OH \\ \text{Oleic acid} \\ + \\ \text{glycerol} + CH_3(CH_2)_{14}C(=O)OH \\ \text{Palmitic acid} \\ + \\ CH_3(CH_2)_4CH{=}CHCH_2CH{=}CH(CH_2)_7C(=O)OH \\ \text{Linoleic acid} \end{array}$$

Dilute acids are not particularly effective as catalysts for this reaction because it is not possible to put much of the lipid into the water phase where the acid is. This problem of solubility exists, of course, in the digestive tract, for the digestive juices are mostly water. One of them, bile, contains detergents, however, the *bile salts,* which emulsify the lipid and thereby facilitate attack by water and enzymes. This enzyme-catalyzed hydrolysis of triglycerides is the only reaction they undergo during digestion.

Saponification. When the ester links in triglycerides are saponified by the action of a strong base (e.g., NaOH, KOH), glycerol plus the salts of the fatty acids are produced. These salts are soaps, and how they exert detergent action will be described later in this chapter. In general:

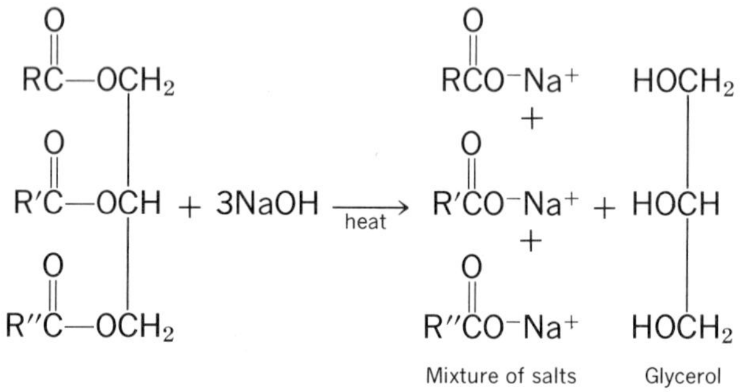

$$\begin{array}{l} RC(=O){-}OCH_2 \\ \quad | \\ R'C(=O){-}OCH \\ \quad | \\ R''C(=O){-}OCH_2 \end{array} + 3NaOH \xrightarrow[\text{heat}]{} \begin{array}{l} RC(=O)O^-Na^+ \\ \quad + \\ R'C(=O)O^-Na^+ \\ \quad + \\ R''C(=O)O^-Na^+ \end{array} + \begin{array}{l} HOCH_2 \\ \ \ | \\ HOCH \\ \ \ | \\ HOCH_2 \end{array}$$

Mixture of salts Glycerol

Example (using the triglyceride of Figure 17.1):

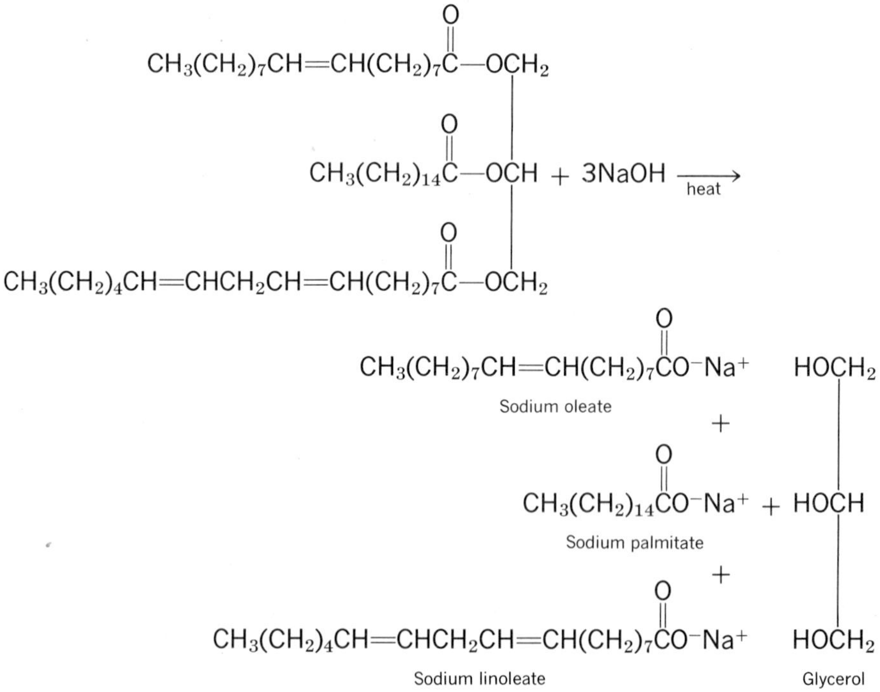

$$\begin{array}{l} CH_3(CH_2)_7CH{=}CH(CH_2)_7C(=O){-}OCH_2 \\ \qquad | \\ CH_3(CH_2)_{14}C(=O){-}OCH \\ \qquad | \\ CH_3(CH_2)_4CH{=}CHCH_2CH{=}CH(CH_2)_7C(=O){-}OCH_2 \end{array} + 3NaOH \xrightarrow[\text{heat}]{}$$

$$\begin{array}{l} CH_3(CH_2)_7CH{=}CH(CH_2)_7C(=O)O^-Na^+ \quad \text{Sodium oleate} \\ \qquad + \\ CH_3(CH_2)_{14}C(=O)O^-Na^+ \quad \text{Sodium palmitate} \\ \qquad + \\ CH_3(CH_2)_4CH{=}CHCH_2CH{=}CH(CH_2)_7C(=O)O^-Na^+ \quad \text{Sodium linoleate} \end{array} + \begin{array}{l} HOCH_2 \\ \ \ | \\ HOCH \\ \ \ | \\ HOCH_2 \end{array}$$

Glycerol

Ordinary soap is a mixture of the sodium salts of long-chain fatty acids. The potassium salts are more soluble in water and are marketed as soft soaps in some shaving creams and in liquid soap. Soaps that float on water have simply had air whipped into them. Perfumes and germicides are usually added. Scouring soaps have had fine sand, pumice, sodium carbonate, or other materials added.

Hydrogenation. If some of the double bonds in vegetable oils were hydrogenated, the oils would become like animal fats. They would be solids at room temperature, for example. Complete hydrogenation, to an iodine value of zero, is not desirable, for the product would be as brittle and unpalatable as tallow. Manufacturers of commercial hydrogenated vegetable oils such as Crisco, Fluffo, Mixo, Spry, etc., limit the degree of hydrogenation. Some double bonds are left. The product has a lower iodine number (e.g., about 50 to 60), a higher degree of saturation, and a melting point that makes it a creamy solid at room temperature, similar to lard or butterfat. The oils of soybean and cottonseed, abundant and inexpensive, are common raw materials for hydrogenated products. If just one molecule of hydrogen were added to the molecule of the mixed triglyceride of Figure 17.1, the iodine number would drop from 89 to 59, from the olive or peanut oil range to that of lard. The peanut oil in popular brands of peanut butter has been partially hydrogenated.

Oleomargarine is made by hydrogenating carefully selected and highly refined oils and fats. One goal is to produce a final product that will melt readily on the tongue, a feature that makes butter so desirable. The product of this hydrogenation is emulsified with about 17% (by weight) of milk that has been cultured with a suitable microorganism to add flavor. Vitamins A and D plus a yellow vegetable dye are usually added as well as biacetyl and acetoin, compounds that give butter its characteristic and highly prized flavor. In some oleomargarines liquid vegetable oils are whipped into the material to make them "more unsaturated," to quote the advertisements. Considerable artifice is expended to provide a cheaper spread that still tastes like butter. The commercial production of oleomargarine exceeds that of butter.

$$CH_3{-}\overset{\overset{\displaystyle O}{\|}}{C}{-}\overset{\overset{\displaystyle O}{\|}}{C}{-}CH_3$$

Diacetyl

$$CH_3{-}\underset{\underset{\displaystyle OH}{|}}{CH}{-}\overset{\overset{\displaystyle O}{\|}}{C}{-}CH_3$$

Acetoin

Rancidity. When fats and oils are left exposed to warm, moist air for any length of time, changes produce disagreeable flavors and odors. That is, the material becomes *rancid.* Two kinds of reactions are chiefly responsible, hydrolysis of ester links and oxidation of double bonds.

The hydrolysis of butterfat would produce a variety of relatively volatile and odorous fatty acids, as the data in the first footnote of Table 17.2 indicate. Water for such hydrolysis is, of course, present in butter, and airborne bacteria furnish enzymes. If the butter is not kept cold (heat accelerates reactions), it will turn rancid. If the mechanism producing rancidity is oxidation, attack by atmospheric oxygen occurs at unsaturated side chains in mixed triglycerides. Eventually, short-chain and volatile carboxylic acids and aldehydes are formed. Both types of substances have extremely disagreeable odors and flavors.

People who exercise much and bathe little develop surface films containing, among many other substances, triglycerides. If these films are not washed off before too long, they turn rancid. Hence the "essence of stale locker room."

Hardening of Oils. "Drying." Oxygen attacks unsaturated glycerides not only at their double bonds but also at carbons attached to them, at allylic positions (cf. p. 278). Linolenic acid has four such activated sites, marked by asterisks in the following structure:

$$CH_3\overset{*}{C}H_2CH{=}CH\overset{*}{C}H_2CH{=}CH\overset{*}{C}H_2CH{=}CH\overset{*}{C}H_2(CH_2)_6CO_2H$$

Linolenic acid (asterisks mark allylic sites)

Oxygen attacks these positions to form, first, hydroperoxides (—OOH):

$$-CH{=}CH-\overset{*}{C}H_2- + O_2 \longrightarrow -CH{=}CH-\underset{\displaystyle O-O-H}{\underset{|}{C}H}-$$

Portion of a side chain in an unsaturated triglyceride

Hydroperoxy group introduced at an allylic site

Once these hydroperoxy groups are randomly introduced, they begin to interact with unchanged allylic sites to form peroxide bridges, building a vast interlacing network as illustrated in Figure 17.2. As these chemical changes occur, important modifications take place in the material. If it has been spread on a surface, a film that is dry, tough, and durable forms, constituting a painted surface. The most highly unsaturated vegetable oils work best, for they have the most allylic sites. Linseed oil, which is quite rich in linolenic acid, is the most widely used drying oil. Fresh (raw) linseed oil still drys too slowly, however, and is made faster drying by heating it with driers, such as lead, manganese, or cobalt salts of certain organic acids. Pigments and dyes are added to give the desired color to the paint. Nearly a billion pounds of drying oils, mostly linseed oil, are used each year in the United States.

Before the widespread introduction of plastics, materials such as oilcloth and linoleum were more widely used than today. Oilcloth is made by coating woven canvas with several layers of a linseed oil paint. To make linoleum, thickened linseed oil is mixed with rosin and cork particles, and the mixture is allowed to harden.

How Detergents Work. Triglycerides and other types of "greases," such as hydrocarbon oils, are the binding agents that hold dirt to surfaces. The problem of cleansing reduces itself to finding a way to loosen or to dissolve this "glue." If it can no longer stick to the surface, neither can the dirt particles. Water alone is a poor cleansing agent, for its molecules are so polar that they stick to each other through hydrogen bonds rather than penetrate into a nonpolar region such as a surface film of grease.

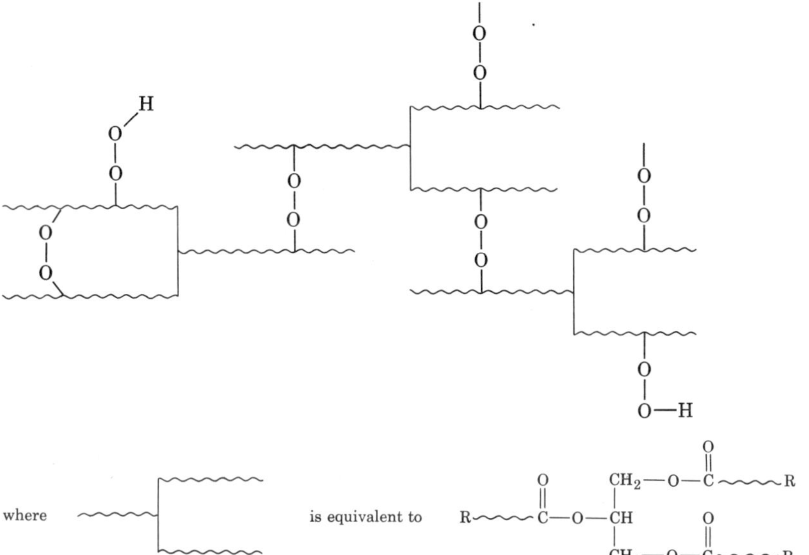
where
is equivalent to

Figure 17.2 When unsaturated glycerides harden, as in the drying of a paint, peroxide cross-links ultimately develop.

Molecules of a typical soap and a typical synthetic detergent are shown in Figure 17.3. Each has a long, nonpolar hydrocarbon "tail" and a very polar, water-insoluble "head." The tail should be easily soluble in nonpolar materials (triglycerides and greases), and the head should be quite soluble in water.

When a detergent, either a soap or a *syndet,* a synthetic detergent, is added to water and the solution is poured onto a surface coated with grease, the tails of the detergent molecules tend to dissolve in the grease layer. The polar head remains in the polar atmosphere of the solution. A little mechanical agitation—stirring, tumbling, or boiling—will help loosen the film (Figure 17.4). As it breaks up into tiny globules, each is "pin-cushioned" by detergent molecules (Figure 17.5). The grease is now in the form of a stable emulsion, because each globule has, in effect, a surface studded with electrical charges, all the same sign. Hence the globules will repel each other. They cannot stick and coalesce.

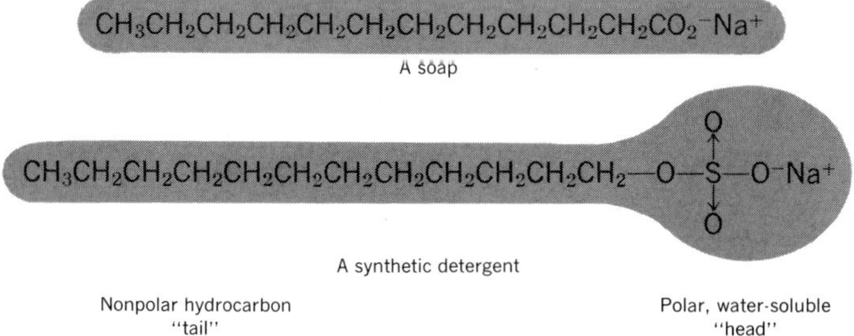
$CH_3CH_2CH_2CH_2CH_2CH_2CH_2CH_2CH_2CH_2CH_2CO_2^-Na^+$
A soap
$CH_3CH_2CH_2CH_2CH_2CH_2CH_2CH_2CH_2CH_2CH_2CH_2—O—S(O)(O)—O^-Na^+$
A synthetic detergent
Nonpolar hydrocarbon "tail"
Polar, water-soluble "head"

Figure 17.3 Two organic salts that have detergent properties.

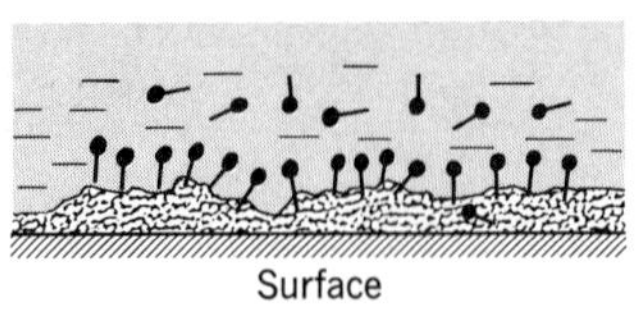
Surface
(a)

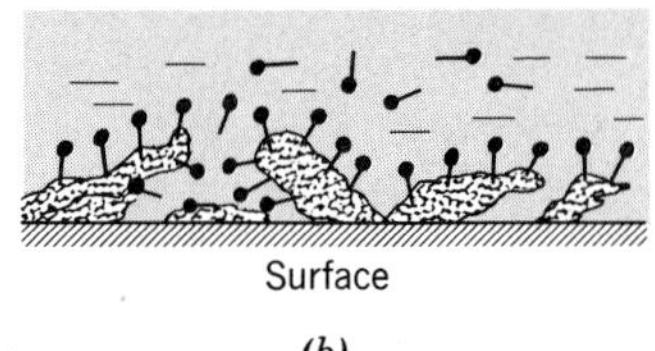
Surface
(b)

Figure 17.4 Detergent action. (*a*) Nonpolar "tails" of detergent molecules become embedded in the grease layer. (*b*) Polar "heads" of detergent molecules tend to urge the grease layer away from the surface.

One of the advantages of the syndets is that they are not precipitated by the heavier metal ions found in hard water, Ca^{2+}, Mg^{2+}, and Fe^{2+} or Fe^{3+} or both. The salts of these ions and the fatty acids are insoluble. If ordinary soap is added to hard water, the familiar scum that forms is a mixture of these salts.

Biodegradability. Although ordinary soaps have their disadvantages in hard water, they cause virtually no problem when waste water containing them is discharged into sewage disposal systems and eventually into the ground. They are *biodegradable,* that is, microorganisms metabolize them. The search for economical detergents that would not precipitate in hard water led to a variety of types as several companies worked on unique products that they could patent. A few such types that were placed on the market are shown in Table 17.4. Whether these syndets were biodegradable was not considered a serious question until the sewage disposal systems of major metropolitan areas slowly became glutted with the foam and suds of undegraded detergents and homeowners with private wells reported that their tap water foamed.

As tetrapropylene-based alkylbenzene sulfonates (Table 17.4) were more and more widely used, the problem increased because these detergents are the least biodegradable. The enzyme systems of microorganisms are not equipped to degrade the highly branched side chains. But by 1966 the major American syndet makers had

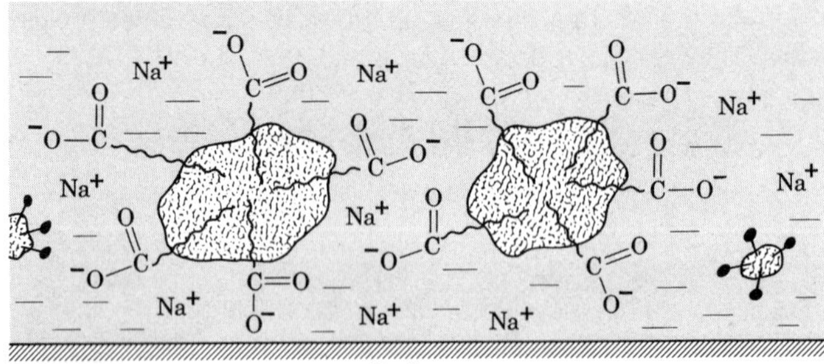
Surface

Figure 17.5 These two grease globules, pin-cusioned by negatively charged groups, cannot coalesce. Both carry negative charges, and like charges repel. They are now solubilized and are easily washed away.

Table 17.4 Biodegradability of Representative Types of Synthetic Detergents

Types of Synthetic Detergents	Examples to Illustrate Structural Features	Biodegradable?
Sodium alkyl sulfate	$CH_3(CH_2)_{10}CH_2OSO_2O^-Na^+$	yes
Sodium alkylbenzene sulfonate, tetrapropylene-based	$CH_3CH(CH_3)CH_2CH(CH_3)CH_2CH(CH_3)CH_2—CH(CH_3)—C_6H_4—SO_2—O^-Na^+$ (S with two $\rightarrow$O)	no
Sodium alkylbenzene sulfonate, n-paraffin or α-olefin-based	$CH_3(CH_2)_nCH(CH_3)—C_6H_4—SO_2—O^-Na^+$ ($n = 7 - 11$)	largely so
Sodium alkane sulfonate	$C_nH_{2n+1}SO_3^-Na^+$ (chain is largely straight, n is 15 to 18)	almost completely

switched to n-paraffin-based alkylbenzene sulfonates or similar materials. Although these are not completely degradable in groundwater, microorganisms manage to reduce their concentrations to acceptable levels.

STEROIDS

The operation of extracting solvents, which brings out triglycerides and other complex lipids from plant and animal material, also extracts a group of nonsaponifiable compounds, the *steroids*. Their structures are generally characterized by a polycyclic carbon skeleton (Figure 17.6). The alcohol group, the keto group, and the double bond are common, but the ester linkage is seldom present. Hence steroids are not saponifiable.

The physiological effects of the steroids vary greatly from compound to compound, ranging from vitamin activity to the action of sex hormones. Some steroids are fat emulsifiers found in bile; others are important hormones; one stimulates the heart; another has been implicated in hardening of the arteries; still another ruptures red blood cells. The structures, names, and chief physiological properties of several are given in the following list.

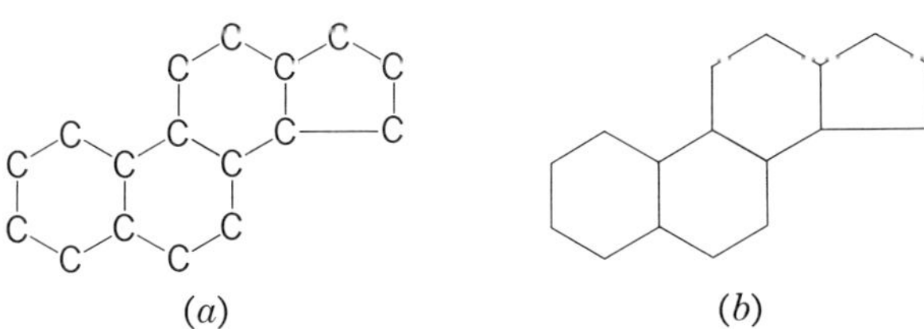
(a)
(b)

Figure 17.6 (*a*) Carbon skeleton characteristic of steroids. (*b*) Most common condensation of the steroid "nucleus."

This is one human estrogenic hormone.

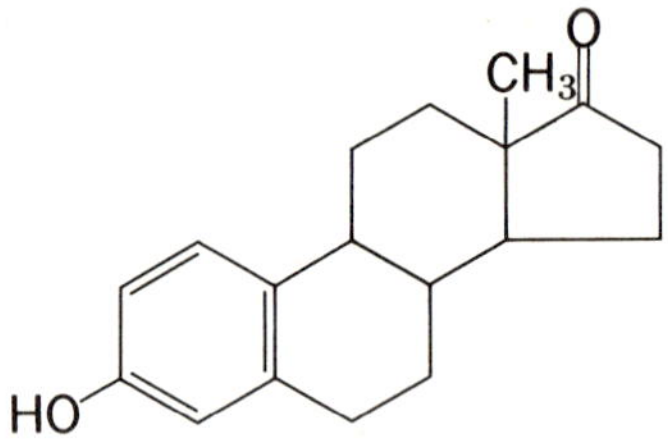

Estrone

This human pregnancy hormone is secreted by the corpus luteum.

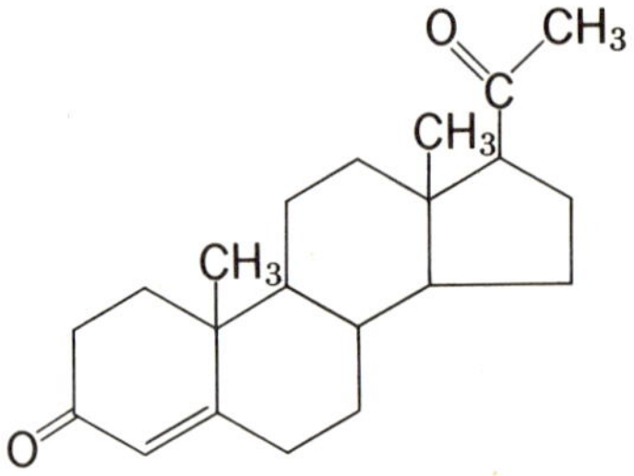

Progesterone

This male sex hormone regulates the development of reproductive organs and secondary sex characteristics.

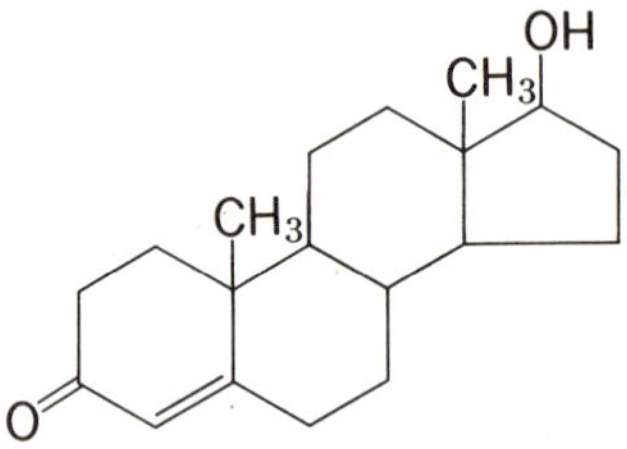

Testosterone

Bile acid

Cholic acid is found in bile in the form of its sodium salt. This and closely related salts are the bile salts that act as powerful emulsifiers of lipid material awaiting digestion in the upper intestinal tract. The sodium salt of cholic acid is soaplike because it has a very polar head and a large hydrocarbon tail.

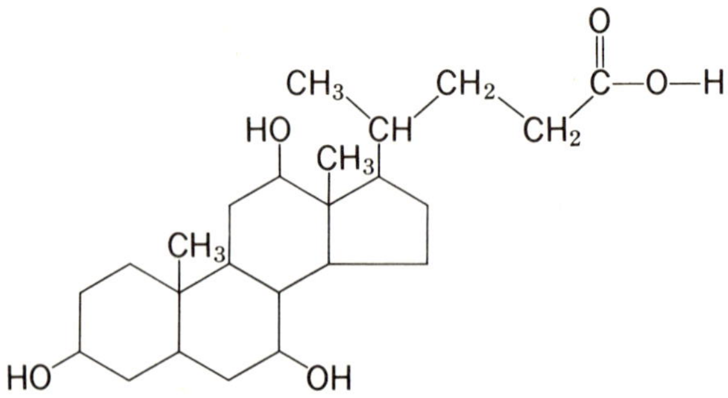

Cholic acid

Antiarthritic compound

One of the twenty-eight adrenal cortical hormones, cortisone is not only important in the control of carbohydrate metabolism but also effective in relieving the symptoms of rheumatoid arthritis. Hench and Kendall of the Mayo Clinic earned Nobel prizes for demonstrating the antiarthritic properties of cortisone. Their supply of this drug was made available by a research program at Merck and Company, from which came a thirty-two-step synthesis of cortisone from material isolated from ox bile. Cortisone is now made on a fairly large scale from diosgenin, a steroid found in tubers of a Mexican yam, a material more abundant than ox bile.

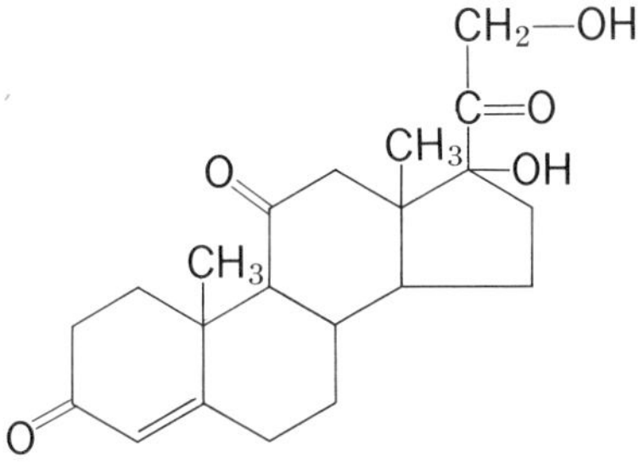

CH_2—OH, C=O, CH_3, OH, O, CH_3, O

Cortisone

Cholesterol. Cholesterol is a steroid alcohol or *sterol.* It is found in nearly all tissues of vertebrates, particularly in the brain and spinal cord, and is the main constituent of gallstones. In recent years it has been associated with circulatory problems such as hardening of the arteries. Its physiological function is still not understood, but we generally assume that the body needs it as a raw material for building other usable steroids.

Human beings and the higher animals can synthesize cholesterol. Most recent evidence indicates that the body can make it from acetate units in about thirty-six steps and in a matter of seconds.

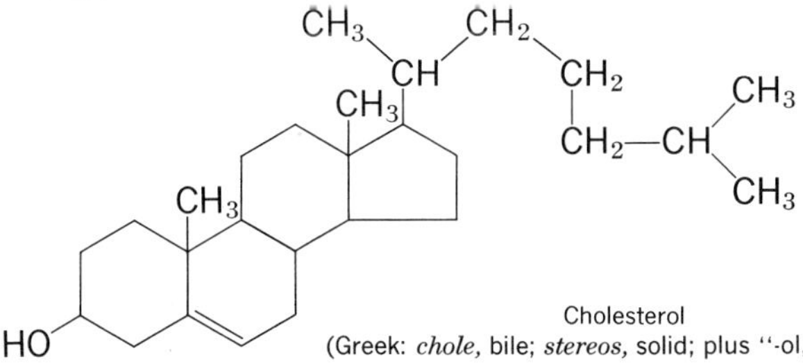

CH_3, CH, CH_2, CH_2, CH_2—CH, CH_3, CH_3, CH_3, CH_3, HO

Cholesterol
(Greek: *chole,* bile; *stereos,* solid; plus "-ol," alcohol)

REFERENCES

BOOKS

C. R. Noller. *Chemistry of Organic Compounds,* third edition. W. B. Saunders Company, Philadelphia, 1965. Chapter 12, "Waxes, Fats, and Oils."

D. J. Hanahan, F. R. N. Gurd, and I. Zabin. *Lipide Chemistry.* John Wiley and Sons, New York, 1960.

L. H. Meyer. *Food Chemistry.* Reinhold Publishing Corporation, New York, 1960. Chapter 2, "Fats and Other Lipids."

ARTICLE

V. P. Dole. "Body Fat." *Scientific American,* December 1959, page 71. This article is a discussion of obesity.

PROBLEMS AND EXERCISES

1. Write the structure of a mixed glyceride between glycerol and each of the following three acids: palmitic acid, oleic acid, and linolenic acid.
   (*a*) Write an equation for the saponification of this triglyceride.
   (*b*) Write an equation for its digestion.
   (*c*) Write an equation for a reaction that would reduce the iodine number to zero.
   (*d*) Calculate the iodine number of this triglyceride. Is it likely to be a solid or a liquid at room temperature?
2. When soap is made, the excess alkali and the glycerol may be washed out by stirring and mixing the crude product with salt water. Explain how salt water will act to inhibit the soap from dissolving.
3. When hydrochloric acid is added to a solution of sodium palmitate, an organic compound precipitates. What is its structure and name?
4. Write structures for a monoglyceride, a diglyceride, and a triglyceride. Use lauric acid, to simplify matters, in all cases. Arrange the structures in order of increasing solubility in water, and explain your reasons for selecting this order.
5. Explain *how* a monoglyceride acts as an emulsifying agent to help disperse a mixture of oil in water.
6. Define or illustrate the following.
   (*a*) iodine number (*b*) steroid (*c*) phospholipid
   (*d*) saponification (*e*) oleomargarine (*f*) syndet
   (*g*) biodegradability
7. Explain how a drying oil "dries."
8. Explain how a detergent emulsifies oily substances.
9. Explain how butter becomes rancid.

CHAPTER EIGHTEEN

Proteins

Proteins are found in all cells and in virtually all parts of cells; of the body's dry weight proteins constitute about half. Proteins hold a living organism together and run it. As skin they give it a shell; as muscles and tendons they provide levers; and as one substance inside bones they, like steel to reinforced concrete, constitute a reinforcing network. Some proteins, in buffers, antibodies, and hemoglobins, serve as policemen and long-distance haulers. Others form the communications network or nerves. Certain proteins, as enzymes, hormones, and gene regulators, direct and control all forms of body repair and construction. No other class of compounds is involved in such a variety of functions, all essential to life.

Diversity of functions might suggest diversity of molecular types. Yet the huge variety of proteins have enough molecular features in common that we can lump them together in one chemical family. One purpose of this chapter is to indicate how a few subtle variations on a theme of *similar structures* might be responsible for the many *different functions* proteins have.

Proteins have one characteristic in common: they are all polymers. The monomer units are classified as α-amino acids; of all the proteins ever analyzed, only about twenty different α-amino acids occur commonly. A few others occur here and there in unusual instances.

We sought to understand certain characteristics of starch, glycogen, and cellulose by studying their monomer unit, glucose. In a similar approach we begin our study of proteins by examining their monomers.

AMINO ACIDS AS BUILDING BLOCKS FOR PROTEINS

The system common to all the α-amino acids is given by the general structure

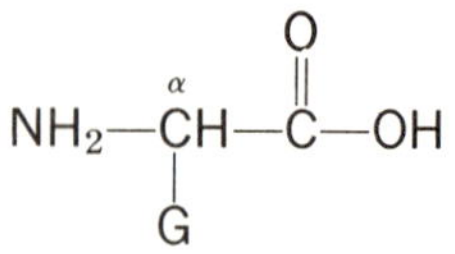

$$NH_2-\overset{\alpha}{C}H(G)-\overset{O}{\overset{\|}{C}}-OH$$

The G— stands for an organic group or system that is different in each of the various amino acids. The list of twenty-four amino acids in Table 18.1 indicates the nature of these side chains.

Optical Activity among the Amino Acids and Proteins. In all the naturally occurring amino acids except the simplest, glycine, the α-carbon is asymmetric, and the substances are optically active. Serine is used as the reference for assigning amino acids to configurational families. Naturally occurring serine is in the L-family. Its relation to L-glyceraldehyde shows in the plane projection formulas.

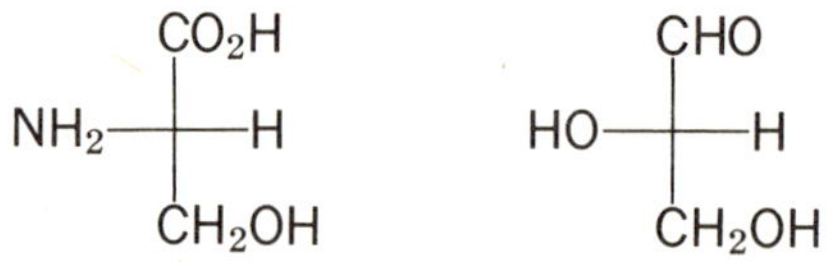

$$\begin{array}{c} CO_2H \\ NH_2-\!\!\!+\!\!\!-H \\ CH_2OH \end{array} \qquad \begin{array}{c} CHO \\ HO-\!\!\!+\!\!\!-H \\ CH_2OH \end{array}$$

L-Serine (naturally occurring enantiomer) L-Glyceraldehyde

When the other naturally occurring amino acids are compared with L-serine, they are all found to be in the same configurational family. All the optically active, naturally occurring amino acids have the L-configuration. Except in extremely rare instances nature does not use the D-forms. The proteins in our bodies are made only from L-amino acids. If we were fed the D-enantiomers, we could not use them. Our enzymes, which themselves are made from the L-forms of amino acids, work with L-forms only.

How Amino Acids Are Incorporated into Proteins. Primary Structural Feature of Proteins. If the α-amino group of one amino acid is acted upon by the carboxyl group of another amino acid, an amide link forms between the two. In general:

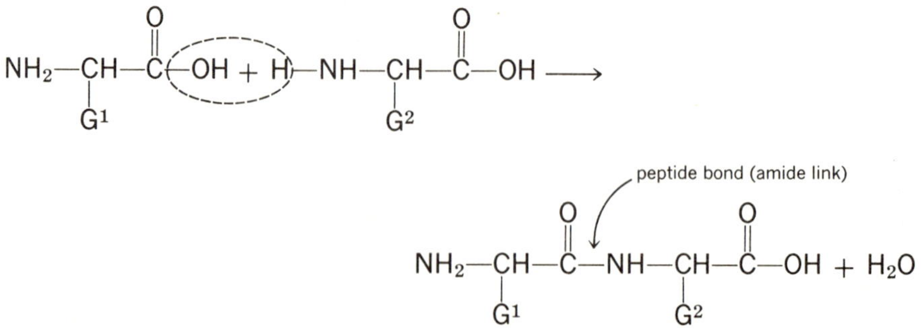

$$NH_2-CH(G^1)-\overset{O}{\overset{\|}{C}}-OH + H-NH-CH(G^2)-\overset{O}{\overset{\|}{C}}-OH \longrightarrow$$

peptide bond (amide link)

$$NH_2-CH(G^1)-\overset{O}{\overset{\|}{C}}-NH-CH(G^2)-\overset{O}{\overset{\|}{C}}-OH + H_2O$$

A dipeptide

Specific example:

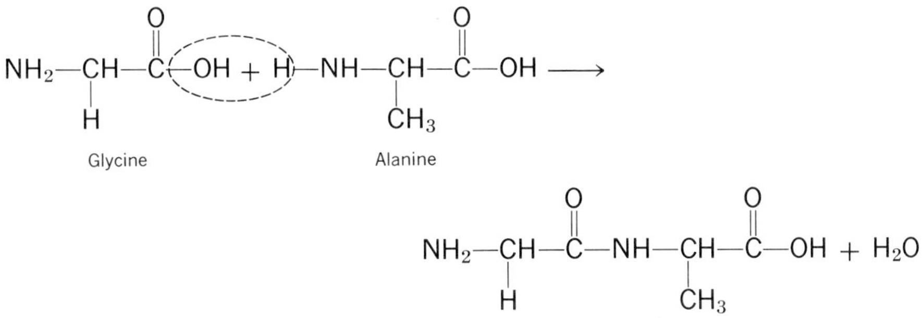

$$NH_2—\underset{\underset{\displaystyle H}{|}}{CH}—\overset{\overset{\displaystyle O}{\|}}{C}—OH + H—NH—\underset{\underset{\displaystyle CH_3}{|}}{CH}—\overset{\overset{\displaystyle O}{\|}}{C}—OH \longrightarrow$$

Glycine Alanine

$$NH_2—\underset{\underset{\displaystyle H}{|}}{CH}—\overset{\overset{\displaystyle O}{\|}}{C}—NH—\underset{\underset{\displaystyle CH_3}{|}}{CH}—\overset{\overset{\displaystyle O}{\|}}{C}—OH + H_2O$$

Glycylalanine
Gly-Ala[1]

In protein chemistry the amide link is called the *peptide bond.* The product in the example, glycylalanine, is a *dipeptide,* a molecule that can be hydrolyzed to two molecules of amino acid.

The other sequence of the two amino acids could just as well have been used. In general:

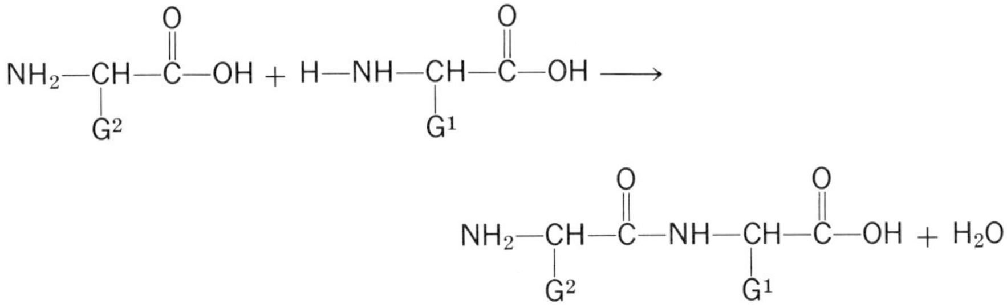

$$NH_2—\underset{\underset{\displaystyle G^2}{|}}{CH}—\overset{\overset{\displaystyle O}{\|}}{C}—OH + H—NH—\underset{\underset{\displaystyle G^1}{|}}{CH}—\overset{\overset{\displaystyle O}{\|}}{C}—OH \longrightarrow$$

$$NH_2—\underset{\underset{\displaystyle G^2}{|}}{CH}—\overset{\overset{\displaystyle O}{\|}}{C}—NH—\underset{\underset{\displaystyle G^1}{|}}{CH}—\overset{\overset{\displaystyle O}{\|}}{C}—OH + H_2O$$

Specific example:

$$NH_2—\underset{\underset{\displaystyle CH_3}{|}}{CH}—\overset{\overset{\displaystyle O}{\|}}{C}—OH + H—NH—\underset{\underset{\displaystyle H}{|}}{CH}—\overset{\overset{\displaystyle O}{\|}}{C}—OH \longrightarrow$$

Alanine Glycine

$$NH_2—\underset{\underset{\displaystyle CH_3}{|}}{CH}—\overset{\overset{\displaystyle O}{\|}}{C}—NH—\underset{\underset{\displaystyle H}{|}}{CH}—\overset{\overset{\displaystyle O}{\|}}{C}—OH + H_2O$$

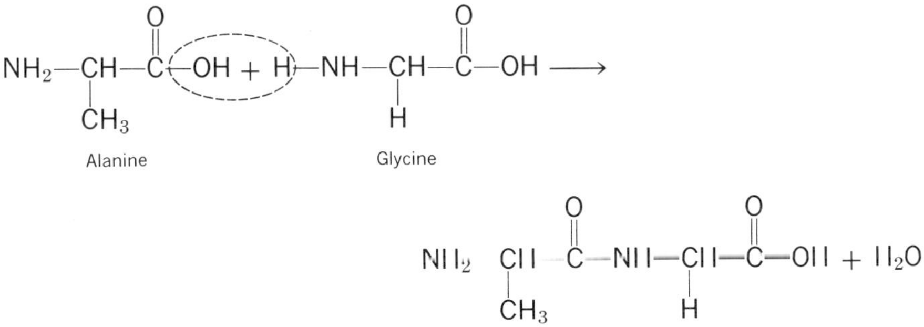

Alanylglycine
Ala-Gly

[1] By convention the symbol Gly-Ala means that the first designated amino acid unit, Gly for glycine, has a free amino group and that the last designated unit, Ala for alanine, has the free carboxyl group. In accordance with this convention amino acids are often written with the $—NH_2$ to the left and the $—CO_2H$ to the right.

Table 18.1 Amino Acids $HO-\overset{O}{\overset{\|}{C}}-\underset{NH_2}{\underset{|}{CH}}-G$

G (Side Chain)	Name	Symbols Used in Representing Amino Acids in Protein Structures: Common*	Wellner-Meister Symbol for Side Chain
Side chain is nonpolar			
$-H$	glycine	Gly	
$-CH_3$	alanine	Ala	
$-CH(CH_3)_2$	valine	Val	
$-CH_2CH(CH_3)_2$	leucine	Leu	
$-\underset{CH_3}{\underset{\vert}{C}}HCH_2CH_3$	isoleucine	Ile	
$-CH_2-C_6H_5$	phenylalanine	Phe	
$-CH_2$ (indole ring, N—H)	tryptophan	Trp	
$HO\overset{O}{\overset{\|}{C}}-CH-CH_2$, NH, CH_2, CH_2 (complete structure)	proline	Pro	
Side chain has a hydroxyl group			
$-CH_2OH$	serine	Ser	
$-\underset{CH_3}{\underset{\vert}{C}}HOH$	threonine	Thr	
$-CH_2-C_6H_4-OH$	tyrosine	Tyr	
$-CH_2-C_6H_2I_2-OH$	diiodotyrosine	—	—
$-CH_2-C_6H_2I_2-O-C_6H_2I_2-OH$	thyroxine	—	—
$HO\overset{O}{\overset{\|}{C}}-CH-CH_2$, NH, $CH-OH$, CH_2 (complete structure)	hydroxyproline	Hyp	

Table 18.1 Amino Acids $HO-\overset{O}{\overset{\|}{C}}-\underset{NH_2}{\underset{|}{CH}}-G$ **(continued)**

G (Side Chain)	Name	Symbols Used in Representing Amino Acids in Protein Structures: Common*	Wellner-Meister Symbol for Side Chain
Side chain has a carboxyl group (or the corresponding amide)			
$-CH_2CO_2H$	aspartic acid	Asp	
$-CH_2CH_2CO_2H$	glutamic acid	Glu	
$-CH_2CONH_2$	asparagine	Asn	
$-CH_2CH_2CONH_2$	glutamine	Gln	
Side chain has a basic amino group			
$-CH_2CH_2CH_2CH_2NH_2$	lysine	Lys	
$-CH_2CH_2CH_2NH-\overset{NH}{\overset{\|}{C}}-NH_2$	arginine	Arg	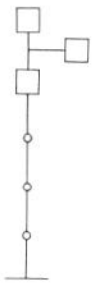
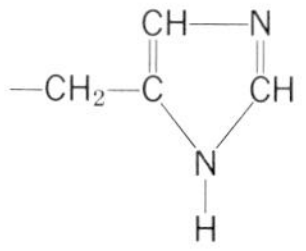 $-CH_2-C$ (ring: $C=CH-N=CH-NH$)	histidine	His	
Side chain contains sulfur			
$-CH_2S-H$	cysteine	Cys	
$-CH_2-S$ CH_2-S (S–S bonded)	cystine	Cys or Cys†	
$-CH_2CH_2SCH_3$	methionine	Met	

* These three-letter symbols are recommended by the Joint Commission on Biochemical Nomenclature of IUPAC-IUB.

† For half the cystine unit.

Just as the two letters N and O can be arranged to give two different words, NO and ON, so two different amino acids can be joined to give two isomeric dipeptides that have identical "backbones" and differ only in the sequences of side chains.

Dipeptides, of course, are still amino acids, although they are no longer α-amino acids. A third α-amino acid can react at one end or the other. In general:

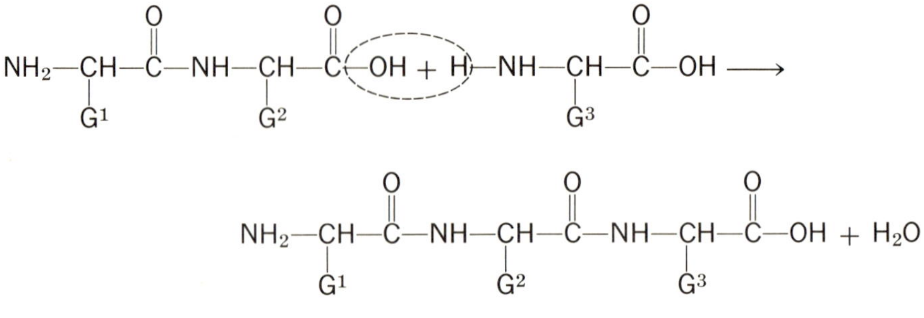

$$NH_2-\underset{\underset{G^1}{|}}{CH}-\overset{\overset{O}{\|}}{C}-NH-\underset{\underset{G^2}{|}}{CH}-\overset{\overset{O}{\|}}{C}-OH + H-NH-\underset{\underset{G^3}{|}}{CH}-\overset{\overset{O}{\|}}{C}-OH \longrightarrow$$

$$NH_2-\underset{\underset{G^1}{|}}{CH}-\overset{\overset{O}{\|}}{C}-NH-\underset{\underset{G^2}{|}}{CH}-\overset{\overset{O}{\|}}{C}-NH-\underset{\underset{G^3}{|}}{CH}-\overset{\overset{O}{\|}}{C}-OH + H_2O$$

A tripeptide

Specific example:

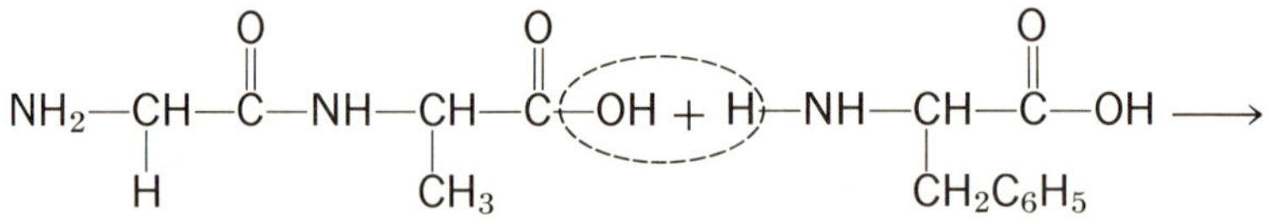

$$NH_2-\underset{\underset{H}{|}}{CH}-\overset{\overset{O}{\|}}{C}-NH-\underset{\underset{CH_3}{|}}{CH}-\overset{\overset{O}{\|}}{C}-OH + H-NH-\underset{\underset{CH_2C_6H_5}{|}}{CH}-\overset{\overset{O}{\|}}{C}-OH \longrightarrow$$

Glycylalanine
Gly-Ala

Phenylalanine
Phe

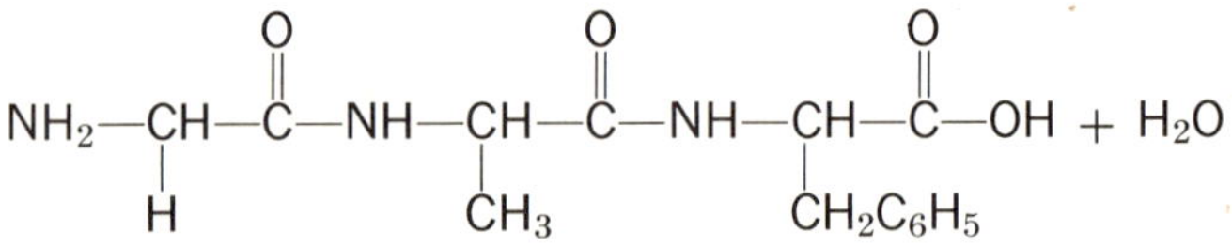

$$NH_2-\underset{\underset{H}{|}}{CH}-\overset{\overset{O}{\|}}{C}-NH-\underset{\underset{CH_3}{|}}{CH}-\overset{\overset{O}{\|}}{C}-NH-\underset{\underset{CH_2C_6H_5}{|}}{CH}-\overset{\overset{O}{\|}}{C}-OH + H_2O$$

Glycylalanylphenylalanine
Gly-Ala-Phe

Or, in Wellner-Meister symbols:

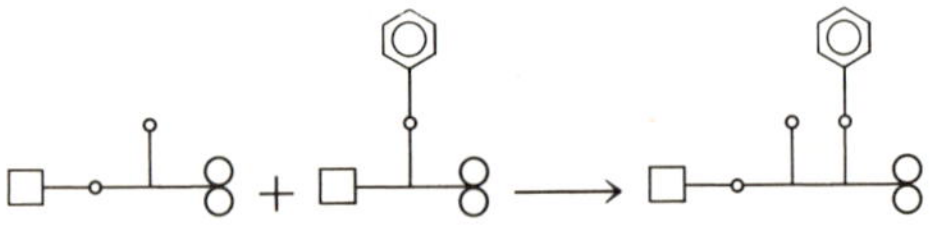

Gly-Ala + Phe $\longrightarrow$ Gly-Ala-Phe

These Wellner-Meister symbols represent protein structures. For the beginner, to whom memorizing twenty or so common names for the amino acids may seem a formidable task, these symbols convey meaning almost immediately, once the very simple system is learned. This chemical shorthand permits rapid recognition of the kind of amino acid units present, their frequency, and their sequences in a polypeptide molecule.

a square	□	means nitrogen, NH_2, or NH
a large circle	○	means oxygen, —OH, or C=O (the context together with rules of valence making it clear which is represented)
a small circle	∘	represents a carbon atom (and whatever —Hs are needed within the rules of valence)
two large circles	8	means a carboxyl group
a large circle and a square	○□	means an amide group, $—CONH_2$

Sulfur atoms are designated by S and aromatic rings by the appropriate, easily recognized geometric figures we have always used for them. Peptide bonds are "understood" to be present.[2]

The tripeptide shown in the last example is only one of the possible isomers involving three different amino acids. If we think in terms of three different letters, A, E, and T, there are six different ways of ordering them:

AET	EAT	TAE
ATE	ETA	TEA

These are not all words in the English language, but then neither are all possible polypeptides found in a particular species. The uniqueness of any tripeptide—or polypeptide, for that matter—lies in the sequence of its side chains. The set of all possible sequences for a tripeptide made from glycine, alanine, and phenylalanine is as follows.

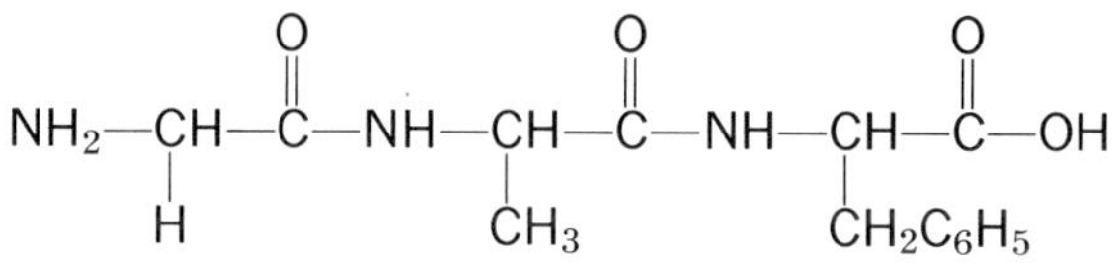

$$NH_2—\underset{H}{CH}—\overset{O}{\overset{\|}{C}}—NH—\underset{CH_3}{CH}—\overset{O}{\overset{\|}{C}}—NH—\underset{CH_2C_6H_5}{CH}—\overset{O}{\overset{\|}{C}}—OH$$

Gly-Ala-Phe

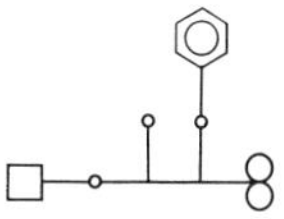

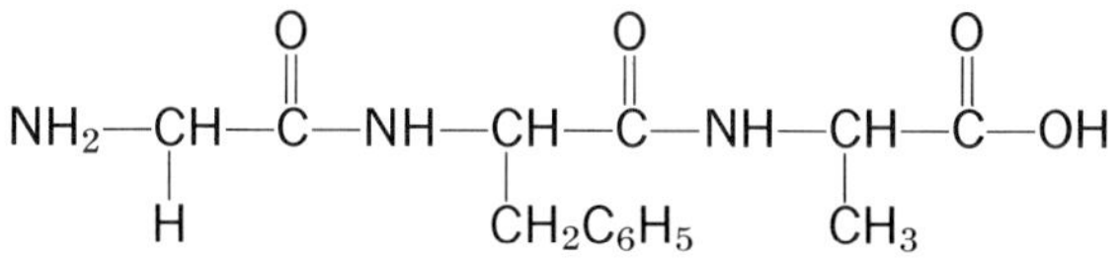

$$NH_2—\underset{H}{CH}—\overset{O}{\overset{\|}{C}}—NH—\underset{CH_2C_6H_5}{CH}—\overset{O}{\overset{\|}{C}}—NH—\underset{CH_3}{CH}—\overset{O}{\overset{\|}{C}}—OH$$

Gly-Phe-Ala

$$NH_2—\underset{CH_3}{CH}—\overset{O}{\overset{\|}{C}}—NH—\underset{H}{CH}—\overset{O}{\overset{\|}{C}}—NH—\underset{CH_2C_6H_5}{CH}—\overset{O}{\overset{\|}{C}}—OH$$

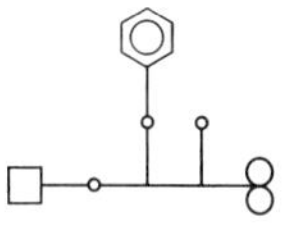

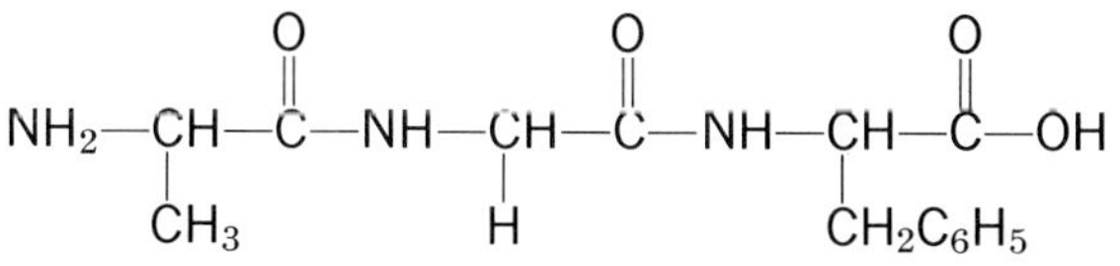

Ala-Gly-Phe

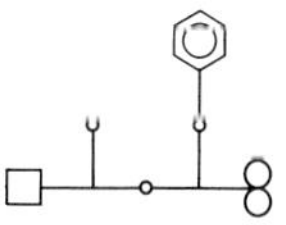

[2] The Wellner-Meister system, not the first of its kind, was introduced in 1966. Whether or not it will be widely accepted remains to be seen. It has so many advantages for an introductory treatment that it seems appropriate to use it here, whatever its future.

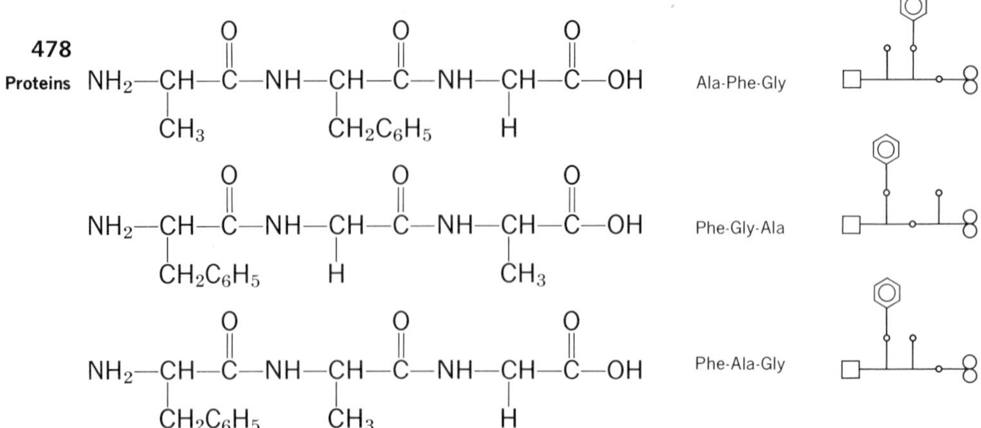
Ala-Phe-Gly
Phe-Gly-Ala
Phe-Ala-Gly

Each of these structures represents a different compound with its own unique set of physical properties. Their chemical properties will be quite similar, however, for the same functional groups are present in all. Each tripeptide is of course an amino acid, although none is an α-amino acid. Each can combine with another amino acid at one end or the other, and with a repetition of this process we can envision how a protein molecule is structured. All protein molecules have the repeating sequence,

$$-\overset{|}{N}-\underset{|}{\overset{|}{C}}-\overset{O}{\overset{\|}{C}}-$$

in the backbone, which together with the sequence of side chains constitutes the *primary structural feature*.

$$H-\overset{H}{\overset{|}{N}}-\underset{|}{C}-\overset{O}{\overset{\|}{C}}\left(\overset{|}{N}-\underset{|}{\overset{|}{C}}-\overset{O}{\overset{\|}{C}}\right)_n\overset{|}{N}-\underset{|}{\overset{|}{C}}-\overset{O}{\overset{\|}{C}}-OH$$

N-terminal unit   *C*-terminal unit   (n may vary from dozens to thousands)

In practice, the words *polypeptide* and *protein* are often used interchangeably. We follow this practice, but *polypeptide* will generally mean a smaller protein, one with lower than normal formula weight, perhaps 50 amino acid units or fewer. Many proteins contain upward of 5000 amino acid units. In such systems many amino acids have obviously been used several times. Table 18.2 contains data on the amino acid content of a few representative proteins to illustrate this point.

By long, laborious processes the structures of a few proteins have been determined. Both chemical methods and X-ray diffraction techniques have been used. Among the proteins whose structures are completely known are the pituitary hormones, oxytocin and vasopressin (Figure 18.1); the adrenocorticotropic hormone (ACTH) (Figure 18.2); the hormone of the pancreas, insulin (Figure 18.3); and an enzyme, ribonuclease (Figure 18.4). The structures of hemoglobin and of the protein associated with the tobacco mosaic virus are also known.

Table 18.2 Amino Acid Composition of Proteins

Expressed as number of amino acid residues per molecule when the formula weight is known, otherwise as the moles of amino acid residues per 100,000 g protein.

Amino Acid	Formula weight of protein:	Human Insulin	Collagen	Horse Hemoglobin	Egg Albumin	Silk Fibroin	Wool Keratin
		6000	—	68,000	45,000	—	—
Glycine		4	363	48	19	581	87
Alanine		1	107	54	35	334	46
Valine		4	29	50	28	31	40
Leucine		6	28	75	32	7	86
Isoleucine		2	15	0	25	8	—
Phenylalanine		3	15	30	21	20	22
Tryptophan		0	0	5	3	0	9
Proline		1	131	22	14	6	83
Serine		3	32	35	36	154	95
Threonine		3	19	24	16	13	54
Tyrosine		4	5	11	9	71	26
Hydroxyproline		0	107	0	0	0	0
Aspartic acid (or asparagine)		3	47	51	32	21	54
Glutamic acid (or glutamine)		7	77	38	52	15	96
Lysine		1	31	38	20	5	19
Arginine		1	49	14	15	6	60
Histidine		2	5	36	7	2	7
Cysteine		0	0	4	5	—	—
Cystine		3	0	0	1	—	49
Methionine		0	5	4	16	—	7
Percent hydrocarbon side chains		41	72	47	45	77	—

Data from G. H. Haggis, D. Michie, A. R. Muir, K. B. Roberts, and P. B. M. Walter. *Introduction to Molecular Biology*, John Wiley and Sons, New York, 1964, pages 40–41.

Besides illustrating what protein structures are like, Figures 18.1 through 18.4 teach us other lessons. Among the proteins studied thus far, only proteins performing like functions in different species (e.g., insulins from various animals) have similar amino acid sequences. What seem to be small changes in other sequences result in considerable differences in function. Oxytocin and vasopressin (Figure 18.1) illustrate this. The first is a hormone that stimulates contraction of uterine muscles. The second, in mammals, is able to cause contraction of peripheral blood vessels and a rise in blood pressure. Yet where oxytocin uses leucine (Leu), vasopressin has arginine (Arg); where oxytocin has isoleucine (Ile), vasopressin has phenylalanine (Phe). *Protein function* is highly sensitive to *protein structure* in ways we are just beginning to understand. One of the most dramatic and well-authenticated examples of this sensitivity is the hemoglobin change of sickle-cell anemia.

Sickle-Cell Anemia and Altered Hemoglobin. The hemoglobin molecule, as it occurs in adult human beings, has a molecular weight of about 65,000. It consists of four long polypeptide chains, intricately folded, two designated as alpha and two as beta. Each enfolds a large, flat, nonprotein molecule called heme (see Figure 18.5)

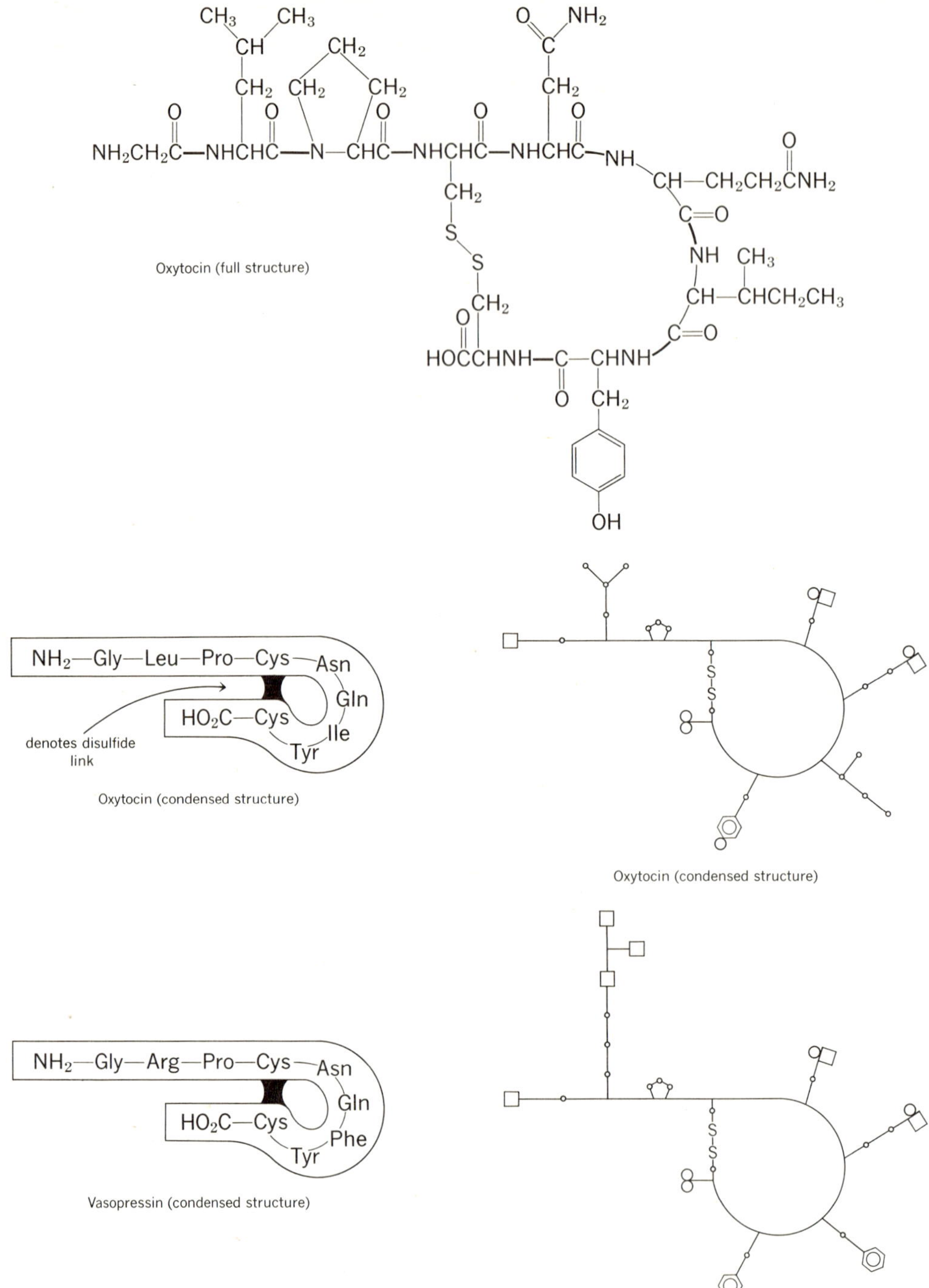
Oxytocin (full structure)
NH₂—Gly—Leu—Pro—Cys—Asn
Gln
HO₂C—Cys
Ile
Tyr
denotes disulfide link
Oxytocin (condensed structure)
Oxytocin (condensed structure)
NH₂—Gly—Arg—Pro—Cys—Asn
Gln
HO₂C—Cys
Phe
Tyr
Vasopressin (condensed structure)
Vasopressin (condensed structure)

Figure 18.1 Oxytocin and vasopressin, two low-formula-weight proteins, symbolized in several ways.

in which a ferrous ion is held. Each heme unit, in cooperation with its associated globin chain, can bind one oxygen molecule. Hence a complete hemoglobin molecule will carry four oxygen molecules from the lungs. (See also page 488.)

The polypeptide chains of normal hemoglobin contain about 300 amino acid units. In the hemoglobin of those suffering from sickle-cell anemia, *only one of these is changed.* A glutamic acid unit sixth in from the N-terminus of the β-chain is replaced by a valine unit. Glutamic acid has a carboxyl group, —COOH, on

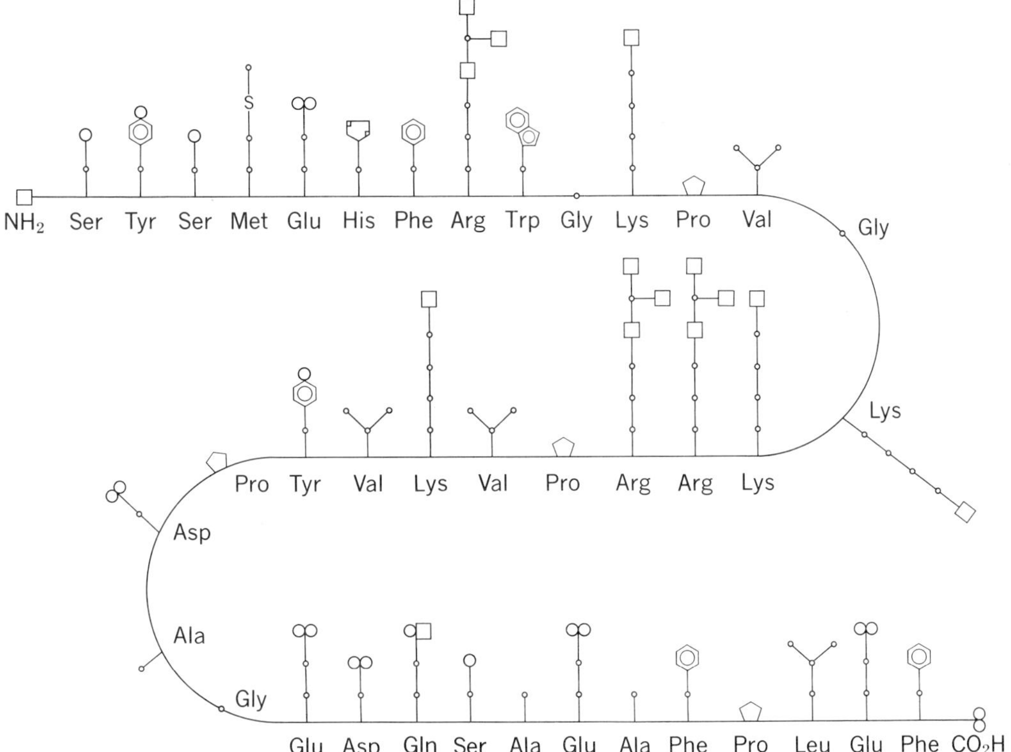
S
NH_2 Ser Tyr Ser Met Glu His Phe Arg Trp Gly Lys Pro Val
Gly
Lys
Pro Tyr Val Lys Val Pro Arg Arg Lys
Asp
Ala
Gly
Glu Asp Gln Ser Ala Glu Ala Phe Pro Leu Glu Phe CO_2H

Figure 18.2 Amino acid sequence in human adrenocorticotropic hormone (ACTH, adrenocorticotropin, corticotropin). This hormone, made from thirty nine amino acid units, stimulates the adrenal gland to produce and secrete its steroid hormones. It has been used to treat rheumatoid arthritis, rheumatic fever, and other inflammatory diseases. P. H. Bell and R. G. Shepherd (American Cyanamid) worked out the amino acid sequence of ACTH obtained from pork adrenals. The ACTH molecules from various animal sources are identical up through the first twenty-four amino acid units (counting from the free amino end, upper left). K. Hofmann (University of Pittsburgh) synthesized a polypeptide with a sequence identical to that of the first twenty-three amino acid units of ACTH; its biological activity is also similar to that of the natural hormone. Apparently the remaining residues from twenty-four through thirty-nine are not essential to biological activity.

The two folds shown in the structure are made only to fit it onto the page. The actual gross shape of the molecule is not indicated by this figure.

its side chain; the side chain of valine is the isopropyl group. The —COOH can ionize, but isopropyl cannot. Normal and sickle-cell hemoglobins are therefore capable of having different electric charges which affect their relative solubilities as well as other properties.

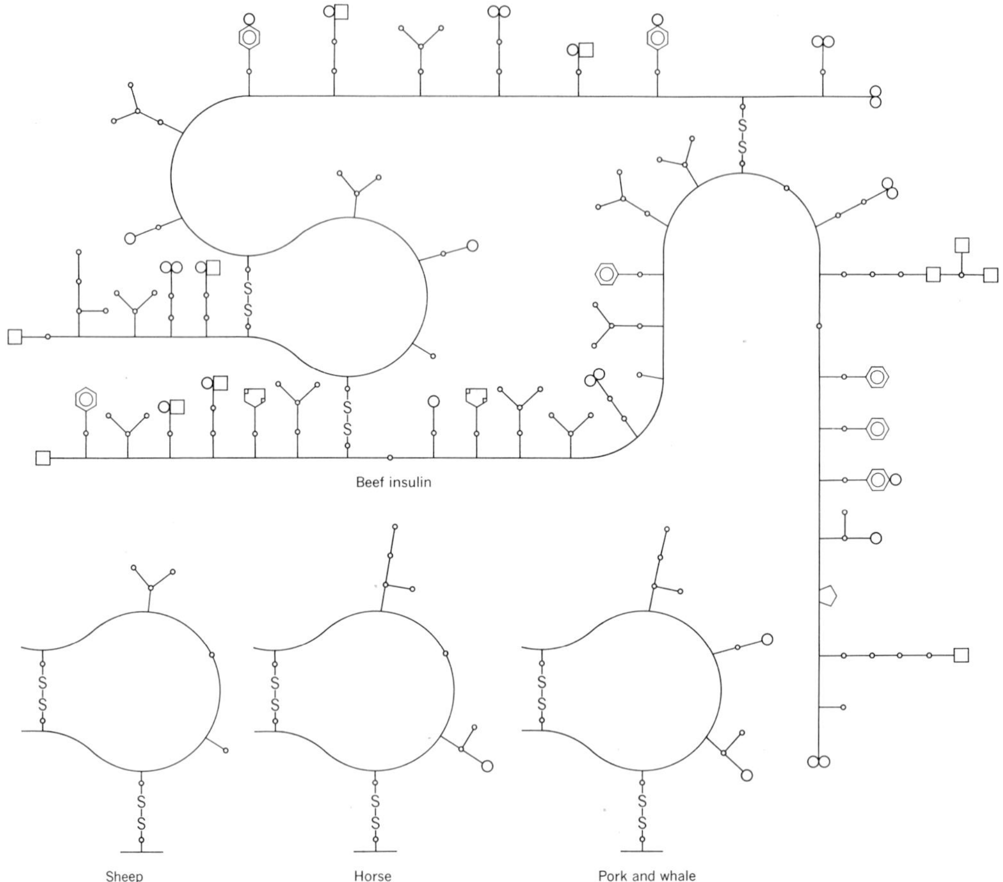
S
S
Beef insulin
Sheep
Horse
Pork and whale

Figure 18.3 The primary structure of insulin. F. Sanger, British biochemist, and a team of co-workers at Cambridge worked from 1945 to 1955 on the determination of this structure. He received the 1958 Nobel prize in chemistry for this work and the development of the techniques. Insulin from different species have the slight differences noted. Fortunately, these are not serious for diabetics on insulin therapy. (Usually, when an alien protein gets into circulation, the body makes antibodies to combine with it. The rejection of skin grafts from one person to another and the extraordinary problems of transplanting organs are related to this formation of antibodies.) Sheep insulin, however, can be given to human diabetics. Some antibodies form, but the insulin activity is reduced only slightly. The synthesis of human insulin *in vitro* was reported in 1966 at the Brookhaven National Laboratory by a team of biochemists headed by P. G. Katsoyannis. At about the same time Western scientists learned that a team of biochemists in mainland China had succeeded in synthesizing insulin in what was acknowledged to be a much purer state.

When the oxygen supply is high, as in arterial blood, both normal hemoglobin, HHb, and sickle-cell hemoglobin, HbS, have about the same solubility in the bloodstream. When the oxygen supply is lower, as in venous blood, the solubility of HbS is less than that of HHb. Consequently, when circulating blood has delivered its oxygen, the altered hemoglobin tends to precipitate in red cells containing it. This precipitate distorts the shapes of the red cells, the blood becomes more

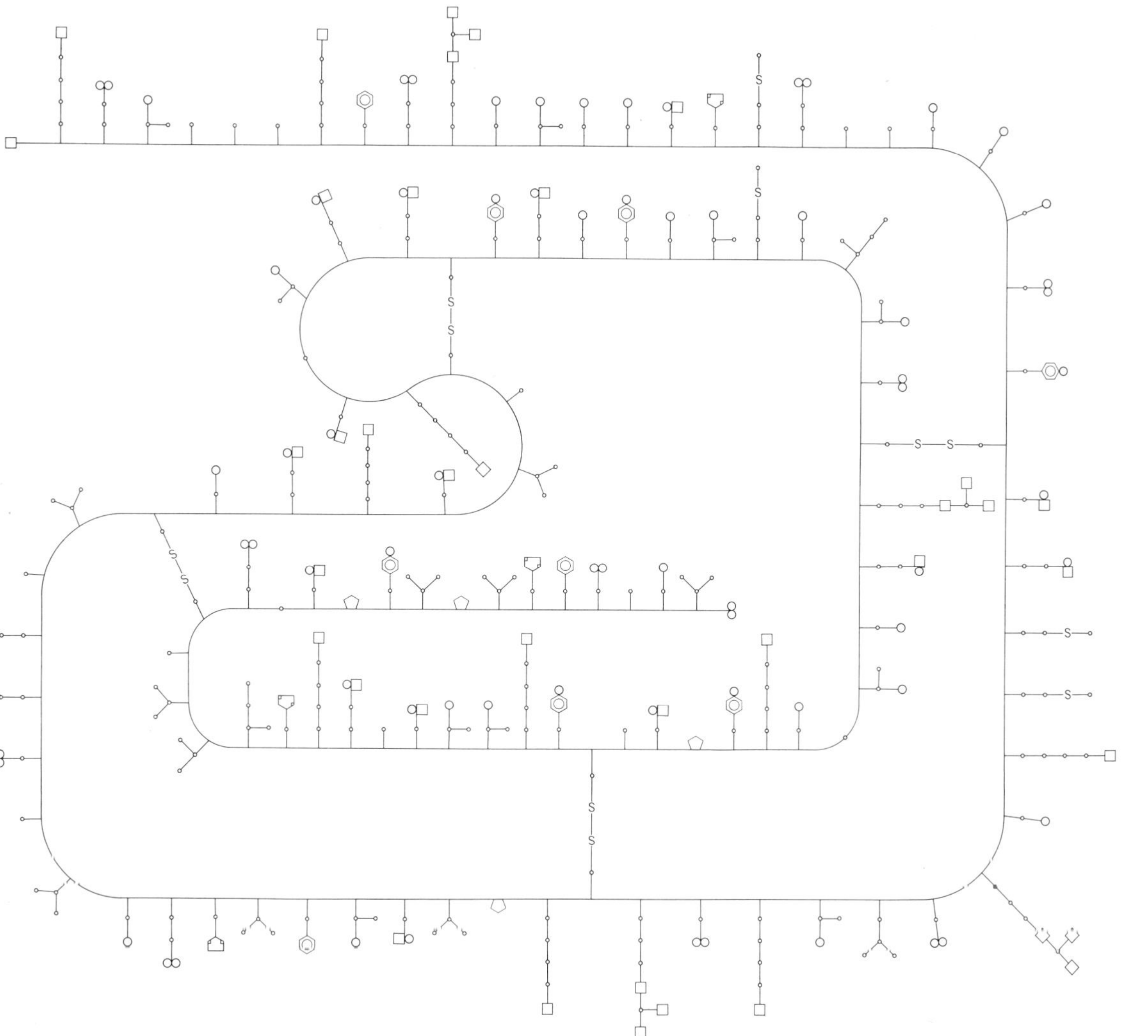

Figure 18.4 The structure of ribonuclease (RNase), which catalyzes the hydrolysis of ribonucleic acid (RNA) in biological systems. M. Kunitz (Rockefeller Institute) first isolated and crystallized this enzyme in 1940. Other scientists at the Rockefeller Institute, Hirs, Spackman, Smythe, Stein, and Moore, worked out its amino acid sequence. There are 124 amino acid units in the whole molecule.

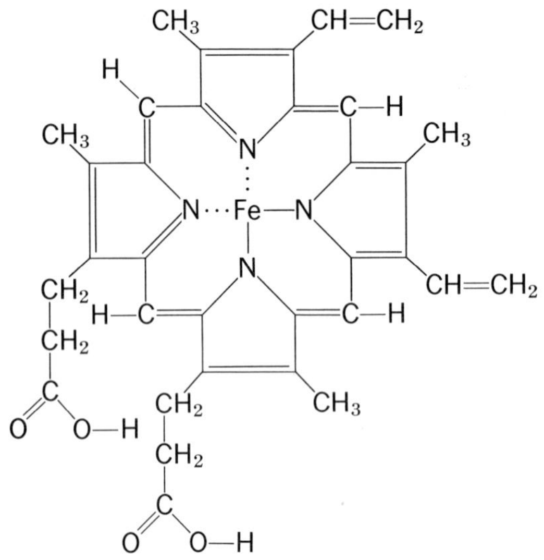

CH3
CH=CH2
H
C
C—H
CH3
CH3
N
N···Fe—N
N
CH=CH2
CH2
H—C
C—H
CH2
C
CH2
CH3
O
O—H
CH2
C
O
O—H

Figure 18.5 The heme molecule. The iron in heme holds its Fe^{2+} state in *oxygeneated* hemoglobin, that is, hemoglobin which has picked up oxygen molecules. In *oxidized* hemoglobin, however, iron is in the Fe^{3+} state, and the substance is brownish in color. When meat is cooked, the color changes from red to brown largely because Fe^{2+} oxidizes to Fe^{3+}.

difficult to pump, and a greater strain is placed on the heart. The red cells may also clump together enough to plug a capillary here and there. Sometimes the red cells split open. The molecular difference responsible for these changes, a matter of life or death for the victims, is startlingly small. One amino acid in 300 is substituted for another. Children with severe forms of the disease, those to whom both parents transmitted the genetic trait, usually die before age two. The disease is especially widespread in central and western Africa.

The physiological properties of proteins are related not only to the configurations of their amino acid units but also to the kinds of side chains present. We shall therefore examine briefly some generalizations about these side chains that will be useful to our study of proteins.

Amino Acids with Nonpolar Side Chains. The first group of amino acids in Table 18.1 are those with essentially nonpolar side chains.[3] They are said to be *hydrophobic* groups (''hydro-,'' water; ''-phobic,'' hating); when a huge protein molecule folds into its distinctive shape (cf. tertiary structural features, p. 488), these hydrophobic groups tend to be folded next to each other rather than next to highly polar groups.

[3] What looks like a 2° amino group in tryptophan is unable to coordinate with a proton. The nitrogen's unshared pair is delocalized into the aromatic system. Tryptophan is therefore put in this group because for all practical purposes its side chain is nonbasic and only slightly polar.

Amino Acids with Hydroxyl-Containing Side Chains. The second set of amino acids in Table 18.1 consists of those whose side chains carry alcohol or phenol groups. In cellular environments they are neither basic nor acidic, but they are polar and *hydrophilic* (water loving). They can be either hydrogen bond donors or acceptors, and they help bind neighboring groups into ordered structures as the long protein chain folds into its final shape. Amino acids rich in these groups are more soluble in water than otherwise.

Amino Acids with Acidic Side Chains. Aspartic acid and glutamic acid have an extra carboxyl group each. When these amino acids are included in a protein, their side chains make the protein more acidic than otherwise. They are proton donors and can serve as both hydrogen bond acceptors and donors. Their presence in a protein tends to make it more soluble in water, provided the pH of the medium is right (see p. 493).

In proteins these acids frequently occur in the form of their corresponding amides, arparagine and glutamine. Although still very polar, in this form they can no longer function as acids.

Amino Acids with Basic Side Chains. Lysine has an extra primary amino group that makes its side chain basic, hydrophilic, and a hydrogen bond donor or acceptor. The side chain in arginine has the guanidinium group, $—NH—\overset{\overset{\displaystyle NH}{\|}}{C}—NH_2$, represented as —□—┴—□ (□ above) in the Wellner-Meister symbolism. One of the most powerful proton-accepting groups found in organisms, it exists almost exclusively in its protonated form, $—NH—\overset{\overset{\displaystyle NH_2^+}{\|}}{C}—NH_2$.

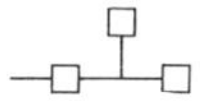

Amino Acids with Sulfur-Containing Side Chains. Cysteine is a particularly important amino acid because its mercapto group makes it especially reactive toward mild oxidizing agents. Cysteine and cystine, in fact, are interconvertible, a property of far-reaching importance in some proteins.

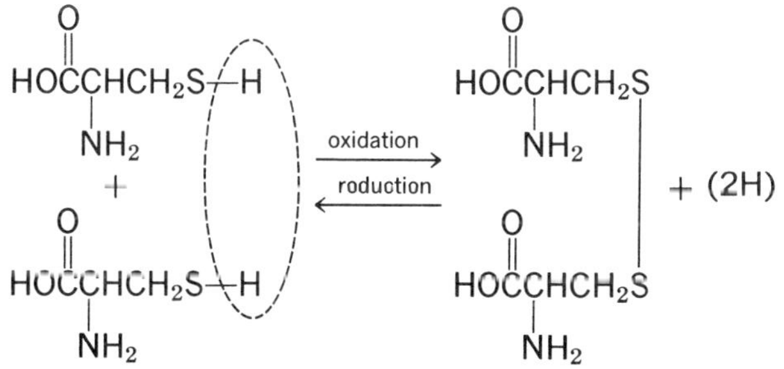

$$\begin{array}{c} HO\overset{\overset{\displaystyle O}{\|}}{C}\underset{\underset{\displaystyle NH_2}{|}}{C}HCH_2S—H \\ + \\ HO\overset{\overset{\displaystyle O}{\|}}{C}\underset{\underset{\displaystyle NH_2}{|}}{C}HCH_2S—H \end{array} \underset{\text{reduction}}{\overset{\text{oxidation}}{\rightleftharpoons}} \begin{array}{c} HO\overset{\overset{\displaystyle O}{\|}}{C}\underset{\underset{\displaystyle NH_2}{|}}{C}HCH_2S \\ | \\ HO\overset{\overset{\displaystyle O}{\|}}{C}\underset{\underset{\displaystyle NH_2}{|}}{C}HCH_2S \end{array} + (2H)$$

Two molecules of cysteine — One molecule of cystine

The disulfide linkage in cystine is common to several proteins, as we have already seen (Figures 18.1, 18.3, and 18.4). It is especially prevalent in proteins having a protective function, such as those forming hair, fingernails, and shells.

SECONDARY STRUCTURAL FEATURES OF PROTEINS

If protein molecules were simply long polymeric chains, they might be expected to behave something like pieces of string, clustering together in no particular order and forming randomly tangled bunches similar to sweepings from a sewing room floor. But they do not. Their molecules are usually found in more orderly groups, folded in various ways to make compact and ordered units.

X-ray diffraction analysis is the tool that has revealed this ordering in proteins. Although an explanation of how this technique works is beyond the scope of this book, we can with little trouble accept the results. Highly crystalline materials such as inorganic salt crystals, in which there is a considerable degree of order, give distinctive X-ray diffraction pictures. Solids in which there is little order, *amorphous* substances like waxes, show no particular diffraction pattern. If proteins were clumped together as a tangled brush heap, they would be amorphous. Many proteins, however, give X-ray diffraction patterns indicating the presence of considerable order. Many regions are crystalline, although some parts are amorphous.

The so-called *secondary structural features* of proteins make order in their aggregations possible. (The primary structural feature is the amino acid sequence along the backbone.) We shall discuss these structural features as we examine one of the most characteristic protein shapes, the α-helix.

α-Helix. In many proteins studied thus far, the chains are coiled into a spiral called an α-helix. Pauling and Corey, in 1951, were the first to publish extensive evidence for the helix (Figure 18.6). The opportunity for carbonyl oxygens to form hydrogen bonds to amide hydrogens provides the forces that cause the coiling. Thus hydrogen bonds of the type

$$\gt C{=}\overset{\delta-}{O}\cdots\overset{\delta+}{H}{-}\overset{|}{N}{-}$$

make up one of the most important of the secondary structural features.

The nature of the side chains affects intrachain hydrogen bonding and determines whether or not an α-helix can form. Polylysine, a synthetic polypeptide, does not coil in an acidic medium in which all its side chain groups are in the protonated form $-(CH_2)_4NH_3^+$. Repulsions between the rather closely spaced, like-charged groups inhibit any tendency to coil, and in acid the molecule is in a randomly flexing, open form. If now the acid is neutralized and all the side chains become uncharged, $-(CH_2)_4NH_2$, polylysine takes up an α-helix configuration. In contrast, polyglutamic acid, another synthetic polypeptide, is coiled in an acid but open in a base. Here the side chains are $-CH_2CH_2COOH$. In a base they are in the ionized form, $-CH_2CH_2COO^-$, and repulsions between like-charged side chains prevent the polymer from assuming an α-helix configuration. But in an acidic medium all the side chains are electrically neutral, $-CH_2CH_2COOH$, and coiling is not inhibited.

Salt Bridges. When lysine and glutamic acid (or others with similar side chains) are in the same protein, side chains of unlike charge are possible. Depending on the pH of the medium, the $-NH_3^+$ and $-CO_2^-$ groups may both be present. When sites of opposite full charge are close enough neighbors, forces of attraction called *salt*

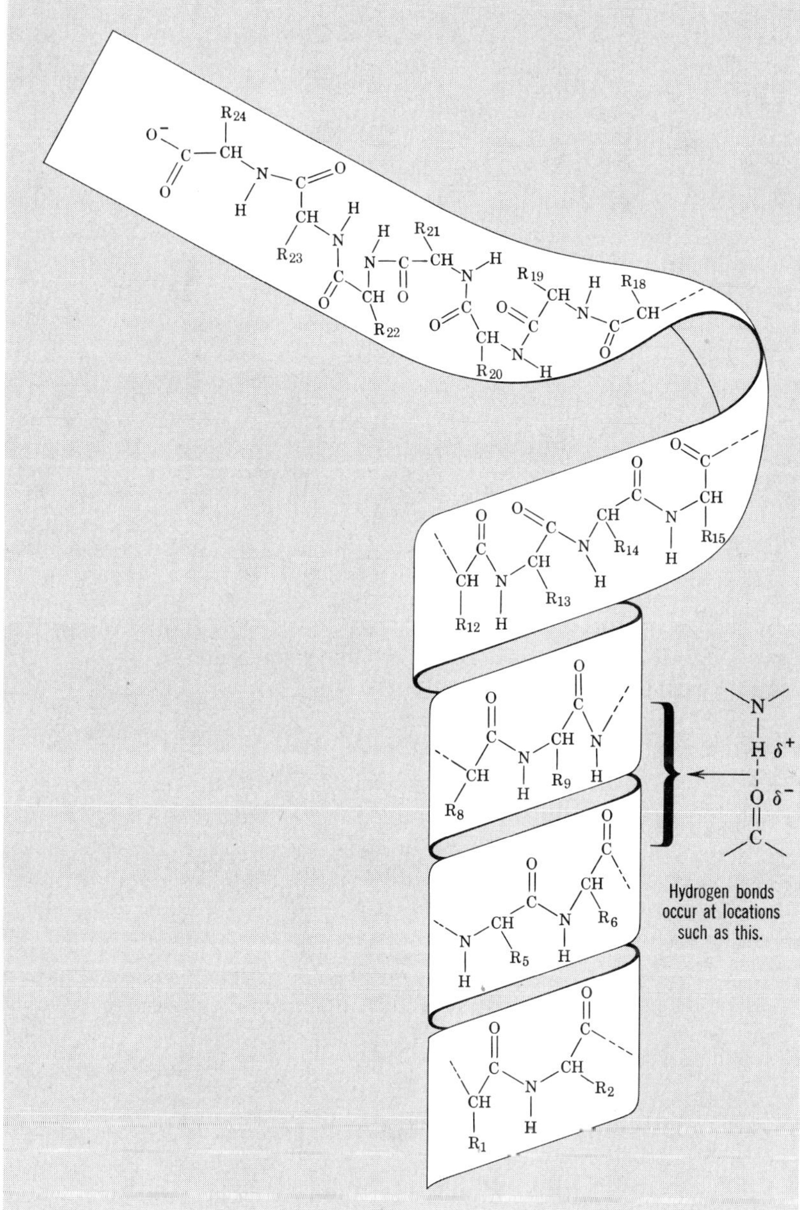
Hydrogen bonds occur at locations such as this.

Figure 18.6 This representation of a section of a protein chain as a ribbon shows the right-handed coiling of an α-helix. Hydrogen bonds exist between carbonyl oxygens and hydrogens on the amide nitrogens. Although each represents a weak force of attraction between the turns of the coil, there are many of them, and the total force stabilizing the helix is more than enough to counteract the natural tendency for the chain to adopt a randomly flexing form. It is understandable that the repeating series of hydrogen bonds up and down the helix is called the "zipper" of the molecule. (From G. H. Haggis, D. Michie, A. R. Muir, K. B. Roberts, and P. M. B. Walker, *Introduction to Molecular Biology,* John Wiley and Sons, New York, 1964, page 51.)

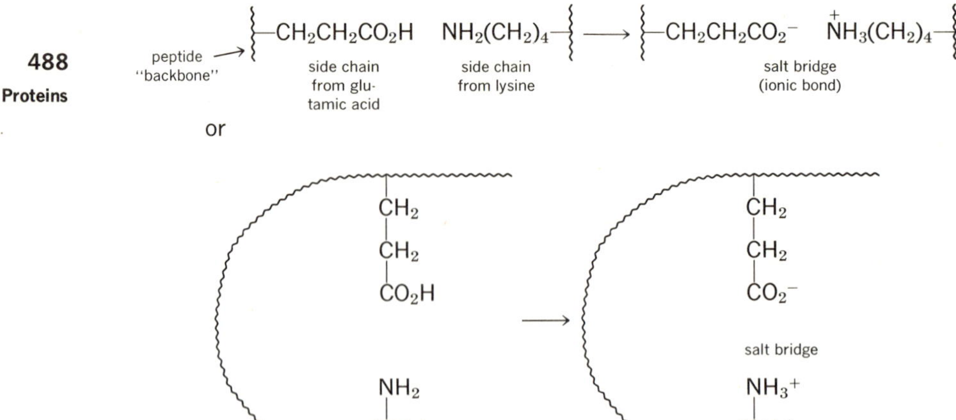
peptide
"backbone"
$-CH_2CH_2CO_2H$
side chain
from glu-
tamic acid
$NH_2(CH_2)_4-$
side chain
from lysine
$-CH_2CH_2CO_2^-$
$\overset{+}{N}H_3(CH_2)_4-$
salt bridge
(ionic bond)
or
CH_2
CH_2
CO_2H
NH_2
$(CH_2)_4$
CH_2
CH_2
CO_2^-
salt bridge
NH_3^+
$(CH_2)_4$

Figure 18.7 The salt bridge. By acid-base interaction two peptide strands can acquire opposite charges which act to hold them together (top) or stabilize a coil within a strand (bottom).

bridges exist (Figure 18.7). The salt bridge is another important secondary structural feature of proteins.

TERTIARY STRUCTURAL FEATURES

Helix formation is not always the final shaping, for helices may in turn fold, twist, or assume some other final configuration of minimum free energy for the particular environment. The data on the bottom line of Table 18.2 show that the percent of hydrocarbon side chains in proteins is quite high, seldom less than 40%. These groups are not soluble in water. The whole system will be more stable if contact between hydrocarbon side chains and the aqueous medium (if that is what the protein is in) is minimized. Further folding of the molecule, especially common among the globular proteins occurring in organisms in an aqueous environment, minimizes the contact. Once the organism has put together the particular amino acid sequence that makes a protein, folding and coiling apparently follow automatically without any further need for enzymes and other chemicals.

Myoglobin and Hemoglobin—Tertiary Structural Features. These two proteins have been extensively examined for their tertiary structural features. John Kendrew and Max Perutz of England (Cambridge) shared the 1962 Nobel prize in chemistry for their X-ray diffraction analysis of myoglobin and hemoglobin.

Kendrew concentrated on myoglobin, a protein with 153 amino acid residues. He found that its molecules are folded as illustrated in Figure 18.8. Perutz worked on the structure of hemoglobin and found that its β-chains (146 amino acid residues) are folded very much like the myoglobin system of Figure 18.8. The α-chains are folded in a different way. In the whole hemoglobin molecule two α-chains and two β-chains, each folded as shown in Figure 18.9*a*, aggregate as the model in Figure 18.9*b* indicates.

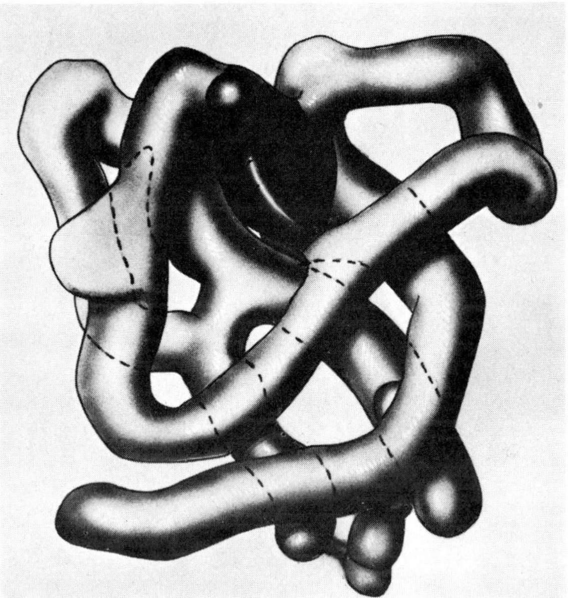

Figure 18.8 Tertiary structural features of the myoglobin molecule. The sausagelike portion contains the protein chain; where it is relatively straight, the chain is coiled in an α-helix. It is estimated that 70% of the molecule has this secondary feature. The darker, disklike section represents a heme unit. Myoglobin stores and transports oxygen at muscles. (Courtesy of John C. Kendrew, Cambridge University.)

A Word of Caution and a Disclaimer. Protein structures are studied in materials isolated from a living system. The secondary or tertiary structure that a protein has in the isolated crystalline form is not necessarily the same as the structure it has *in vivo.* (The fundamental sequence of the amino acids will be the same, assuming that techniques for isolation are gentle enough.) When they are in an aqueous medium, proteins may have up to 30% by weight of bound water—water molecules attracted to and held by the protein molecule, roughly like water of hydration. Water has an important effect on the activity of proteins in organisms. If it has affected activity, water has very likely affected structure, at least at the secondary and tertiary levels of organization.

This complication serves to remind us, if any reminder is needed, that knowledge of the isolated items in the inventory of a cell is one thing. Translating such information into an understanding of what goes on in a cell is quite another. If the reader ever thought that we shall soon synthesize a living cell *in vitro* from its scrambled molecules without the aid of other living cells, he should by now be disabused of this idea. Such synthesis could happen someday, but not soon. The complexities of dynamic cellular architecture are so vast that biology is a very long way from being reduced to chemistry. The services of all the specialists in biology—anatomists, physiologists, geneticists, microbiologists, pathologists, taxonomists, etc.—will be in demand for a long, long time. Many insights are provided by a

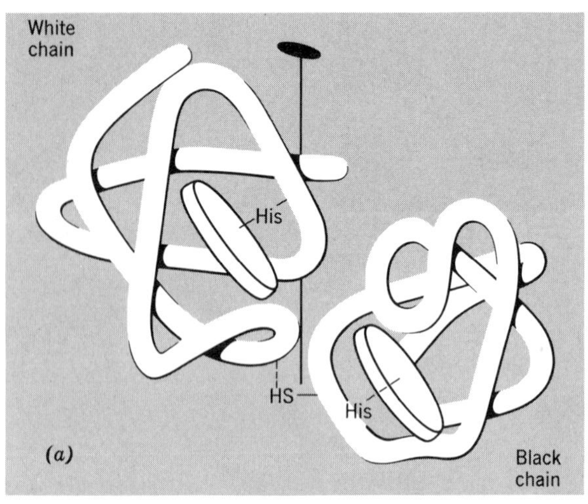
White chain
His
HS
His
(a)
Black chain

HS
O_2
(b)

Figure 18.9 Hemoglobin. (*a*) Tertiary structures of the α- and β-subunits of hemoglobin. The disks represent heme molecules. As with myoglobin (Figure 18.8), the straight sections consist of material in an α-helix form. (From G. H. Haggis, D. Michie, A. R. Muir, K. B. Roberts, and P. B. M. Walter, *Introduction to Molecular Biology,* John Wiley and Sons, New York, 1964, page 62.)

(*b*) This model, derived from X-ray diffraction data, shows how the most electron-dense regions of hemoglobin are arranged. The white section represents the two β-subunits; the black has two α-subunits. The heme molecules (disks) are for the most part folded inside the system with enough exposure to permit oxygen molecules to attach themselves for transport from lungs to regions requiring them. (Photograph by Edward Leigh, courtesy of *Chemistry in Britain.*)

study of the molecular basis of life, the principal subject of this book, but the foregoing disclaimer is stated for the sake of a balanced view.

AMINO ACIDS AND PROTEINS AS BUFFERS

Dipolar Ionic Character of Amino Acids. For convenience as well as to conform to common usage, structures of amino acids have been written as having —NH_2 and —COOH groups. Neither their chemical properties nor their physical properties agree too well with this picture, however.

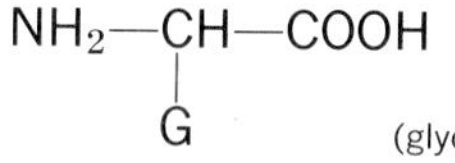

$$NH_2\text{—}\underset{\displaystyle G}{\underset{|}{CH}}\text{—}COOH \qquad \text{(glycine, if G = H—)}$$

Conventional way of writing the structure of an amino acid

Most carboxylic acids, for example, have acid dissociation constants K_a on the order of 10^{-5}, but for glycine K_a is 1.6×10^{-10}. This value is low enough to be in the range of phenols. Similarly, most primary aliphatic amines have basic dissociation constants K_b on the order of 10^{-5}, but for glycine K_b is 2.5×10^{-12}, making it so weak a proton acceptor that it is virtually neutral in this respect.

Amino acids have other peculiarities. Even the smallest ones are nonvolatile, crystalline solids without true melting points. When heated, they decompose and char before melting can occur. Furthermore, the amino acids tend to be insoluble in nonpolar solvents and soluble in water. These facts do not correlate well with the structure we have been using. The *dipolar ionic structure* is more consistent:

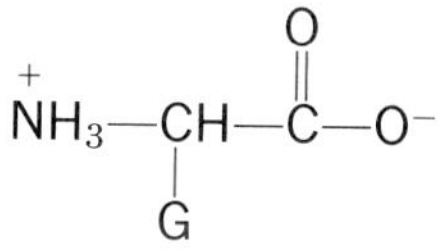

$$\overset{+}{N}H_3\text{—}\underset{\displaystyle G}{\underset{|}{CH}}\text{—}\overset{\displaystyle O}{\overset{\|}{C}}\text{—}O^-$$

Dipolar ionic form of an amino acid

In a crystalline solid made up of such particles, forces of attraction between them will be strong, as in any ionic crystal (cf. p. 113). Before heating supplies enough energy to melt such crystals, covalent bonds are disrupted within the dipolar ions, and decomposition occurs. Ionic substances, of course, are known for their insolubility in nonpolar solvents. As for their low K_a and K_b values, the dipolar ionic structure offers a ready explanation. In an amino acid we are not dealing with a free carboxyl group, —COOH, as the proton donor, but rather with a substituted ammonium ion, —NH_3^+. The proton is more strongly bound within this group than it would be on the —COOH group. Similarly, the basic group, the proton-accepting group, in an amino acid is not the free —NH_2 group. Rather it is the carboxylate group, COO^-. The latter is a weaker proton acceptor. The net effect is that amino acids are both very weak acids and very weak bases.

Amino Acids and Proteins as Buffers. If a strong base such as hydroxide ion, OH^-, is added to a solution of an amino acid, the following reaction will occur:

$$HO^- + H-\overset{+}{N}H_2-CH_2\overset{O}{\overset{\|}{C}}-O^- \longrightarrow HO-H + NH_2CH_2\overset{O}{\overset{\|}{C}}-O^-$$

Glycine

Stronger base — Stronger acid — Weaker acid — Weaker base

If a strong acid (e.g., H_3O^+) is added to a solution of an amino acid, it will be neutralized as follows:

$$\overset{+}{N}H_3CH_2\overset{O}{\overset{\|}{C}}-O^- + H-\overset{+}{O}(H)-H \longrightarrow \overset{+}{N}H_3CH_2\overset{O}{\overset{\|}{C}}-O-H + H-OH$$

Stronger base — Stronger acid — Weaker acid — Weaker base

Amino acids therefore have the ability to neutralize both stronger acids and stronger bases and can thus serve as buffers; they can hold fairly constant the pH of their solutions in water.

To the extent that protein molecules have groups such as those present in dipolar ionic amino acids, they will also act as buffers. Some of the proteins in the bloodstream serve chiefly as buffers, constituting the third buffer system of blood. (The other two are the H_2CO_3/HCO_3^- and the $H_2PO_4^-/HPO_4^{2-}$ pairs.)

Isoelectric Points of Amino Acids and Proteins. When an amino acid is dissolved in water, proton transfers occur, as illustrated with glycine:

I $$\underset{\mathbf{1}}{\overset{+}{N}H_3-CH_2\overset{O}{\overset{\|}{C}}-O^-} + H-OH \rightleftharpoons \underset{\mathbf{2}}{\overset{+}{N}H_3CH_2\overset{O}{\overset{\|}{C}}-O-H} + {}^-OH$$

II $$H_2O + \underset{\mathbf{1}}{H-\overset{+}{N}H_2-CH_2\overset{O}{\overset{\|}{C}}-O^-} \rightleftharpoons H_3O^+ + \underset{\mathbf{3}}{NH_2CH_2\overset{O}{\overset{\|}{C}}-O^-}$$

In its dipolar ionic form glycine is a slightly better proton donor (equilibrium II) than proton acceptor (equilibrium I). All three forms, **1, 2,** and **3,** coexist at equilibrium, although **2** and **3** are present only in trace amounts. Form **3** is in slight excess of **2.** If a trace of mineral acid were added, equilibrium II would be shifted to the left, and the concentration of **3** could be made equal to that of **2.** At this new pH glycine would exist almost exclusively as **1.** If an electric current were passed through the solution, there would be essentially no migration of glycine units in either direction, toward either electrode. *The pH at which an amino acid exhibits no migration in an electric field is called the isoelectric pH or the isoelectric point.*

Molecules of proteins, because they usually have several amino and carboxyl groups distributed among side chains, have ionization properties like those of simple amino acids. Protein molecules are normally capable of possessing several separate negatively charged and positively charged sites. The *net* charge may be negative, positive, or zero, depending on the specific protein, the pH of the solution, and the presence (or absence) of metallic ions that may complex with the protein and affect the net electric charge.

Protein molecules with no net electric charge are said to be in an isoelectric state.[4] Each soluble protein will have its own unique isoelectric point, the pH at which its net charge is zero, and no migration will occur in an electric field.

The purpose of this brief study of isoelectric points is to show that proteins are least soluble at their isoelectric pHs. At other pH values their molecules will each bear some net charge. They must therefore repel each other, and they cannot aggregate and precipitate. But if the molecules are all isoelectric, they are able to clump together (Figure 18.10) until the mass becomes too large to remain in solution and the protein precipitates.

The curdling of milk illustrates this effect. The isoelectric point of casein, the chief protein of cow's milk, is 4.7. It is only slightly soluble in water at this pH. The pH of cow's milk is normally in the range 6.3 to 6.6. When milk sours, bacterial growth is usually taking place and lactic acid is produced. The pH of the milk drops, approaching 4.7. As this happens, the casein molecules become more and more isoelectronic and they precipitate—or in ordinary terms the milk curdles. As long as the pH is something other than the isoelectric pH, however, the casein remains dispersed in the medium.

The importance of maintaining a constant pH of the blood is made clear by these considerations. Many enzymes and hormones circulate in the blood. If the pH of the blood varied too much up or down, presumably the isoelectric points for some of these proteins would be approached. If they were to precipitate, not only would they no longer be able to function but they would tend to plug blood vessels, thereby placing what might be too great a strain on the heart. The ability of hemo-

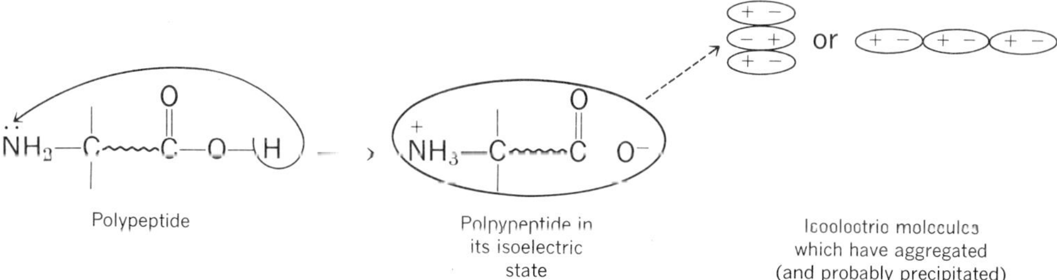
or
Polypeptide
Polypeptide in its isoelectric state
Isoelectric molecules which have aggregated (and probably precipitated)

Figure 18.10 Isoelectric protein molecules can be expected to aggregate in a manner reminiscent of oppositely charged ions collecting together to form ionic crystals. The effect is to convert already large protein molecules into even larger aggregates, reducing their solubility in water.

[4] If metallic ions, by complexing with the protein, make a contribution to the isoelectric state, it is more proper, strictly speaking, to refer to this condition as the isoionic state. Under these circumstances the pH at which no net migration occurs is called the *isoionic point.*

globin to transport oxygen is also closely related to the pH of the blood, a subject we shall study in Chapter 20. The importance of the buffers in the blood cannot be overemphasized.

DIGESTION OF PROTEINS. HYDROLYSIS

The digestion of proteins is equivalent to their hydrolysis, which is nothing more than the hydrolysis of amide linkages (Figure 18.11). Several digestive enzymes act as catalysts, about which more will be said in Chapter 20.

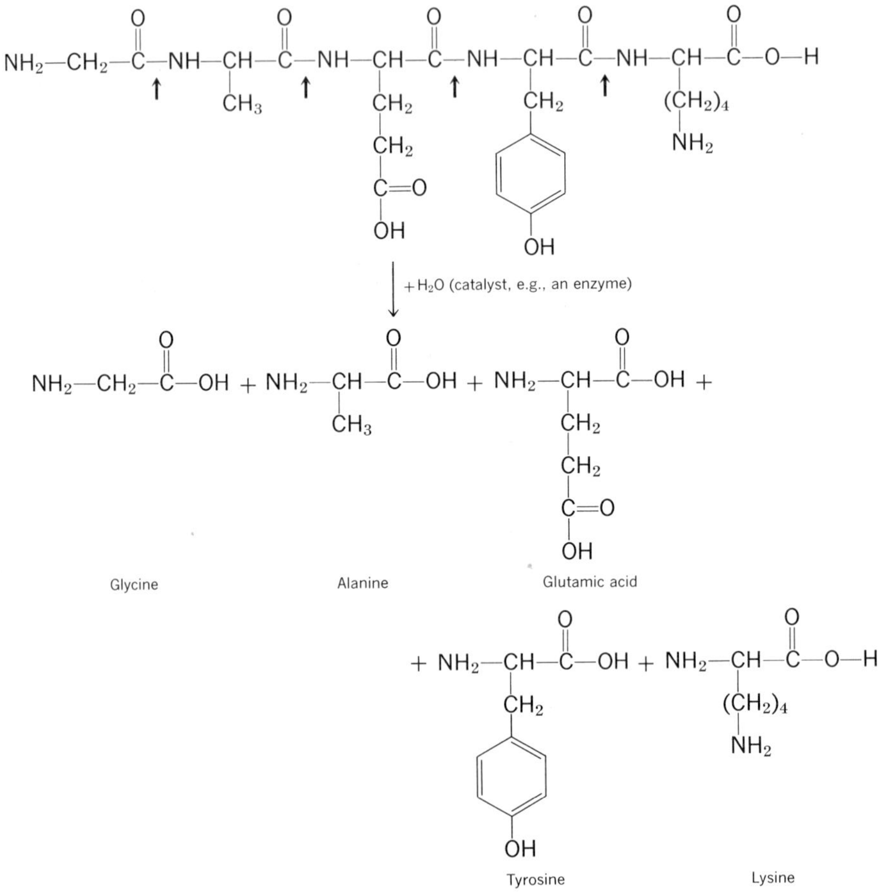

$$NH_2{-}CH_2{-}\overset{O}{\overset{\|}{C}}{-}NH{-}\underset{CH_3}{\underset{|}{CH}}{-}\overset{O}{\overset{\|}{C}}{-}NH{-}\underset{CH_2CH_2C(=O)OH}{\underset{|}{CH}}{-}\overset{O}{\overset{\|}{C}}{-}NH{-}\underset{CH_2C_6H_4OH}{\underset{|}{CH}}{-}\overset{O}{\overset{\|}{C}}{-}NH{-}\underset{(CH_2)_4NH_2}{\underset{|}{CH}}{-}\overset{O}{\overset{\|}{C}}{-}O{-}H$$

$$\Big\downarrow\ +H_2O\ \text{(catalyst, e.g., an enzyme)}$$

$$NH_2{-}CH_2{-}\overset{O}{\overset{\|}{C}}{-}OH + NH_2{-}\underset{CH_3}{\underset{|}{CH}}{-}\overset{O}{\overset{\|}{C}}{-}OH + NH_2{-}\underset{CH_2CH_2C(=O)OH}{\underset{|}{CH}}{-}\overset{O}{\overset{\|}{C}}{-}OH +$$

Glycine Alanine Glutamic acid

$$+ NH_2{-}\underset{CH_2C_6H_4OH}{\underset{|}{CH}}{-}\overset{O}{\overset{\|}{C}}{-}OH + NH_2{-}\underset{(CH_2)_4NH_2}{\underset{|}{CH}}{-}\overset{O}{\overset{\|}{C}}{-}O{-}H$$

Tyrosine Lysine

Figure 18.11 Protein hydrolysis (digestion). In the complete hydrolysis of this hypothetical polypeptide, only carbonyl-to-nitrogen bonds are ruptured (see arrows). A hydroxyl group, from water, becomes attached to each carbonyl carbon; a hydrogen atom, also from water, becomes attached to each nitrogen:

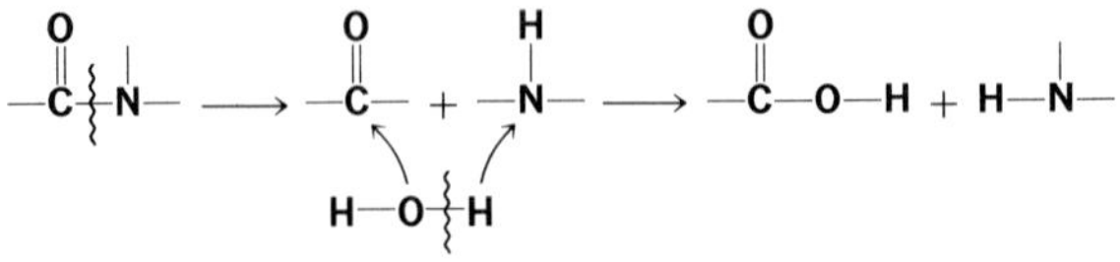

$$-\overset{O}{\overset{\|}{C}}\wr\overset{|}{N}- \longrightarrow -\overset{O}{\overset{\|}{C}}- + -\overset{H}{\overset{|}{N}}- \ (\text{from } H{-}O\wr H) \longrightarrow -\overset{O}{\overset{\|}{C}}{-}O{-}H + H{-}\overset{|}{N}-$$

DENATURATION

A wide variety of reagents and conditions that do not hydrolyze peptide bonds will destroy the biological nature and activity of the protein. When this happens, the protein is said to have been *denatured.* After denaturation the protein usually coagulates. Several of the more common chemicals and conditions that denature proteins are listed in Table 18.3.

At the molecular level denaturation is a disorganization of the shape of a protein. It can occur as an unfolding or uncoiling of a pleated or coiled structure or as the separation of the protein into subunits which then unfold or uncoil (Figure 18.12).

Some denaturations have positive medical advantages. The fact that certain heavy-metal salts denature and coagulate proteins is the basis of a method for

Table 18.3 Chemicals and Conditions That Cause Denaturation

Denaturing Agent	How the Agent May Operate
Heat	Disrupts hydrogen bonds and salt bridges by making molecules vibrate too violently. Produces coagulation as in the frying of an egg.
Solutions of urea $(NH_2-\overset{O}{\overset{\Vert}{C}}-NH_2)$	Disrupt hydrogen bonds. Being amide-like, urea can form hydrogen bonds of its own.
Ultraviolet radiation	Appears to operate the same way that heat operates (e.g., sunburning).
Organic solvents (e.g., ethyl alcohol, acetone, isopropyl alcohol)	May interfere with hydrogen bonds in protein, since alcohol molecules are themselves capable of hydrogen bonding. Quickly denatures the proteins of bacteria, thus killing them (e.g., disinfectant action of ethyl alcohol, 70% solution).
Strong acids or bases	Can disrupt hydrogen bonds and salt bridges: 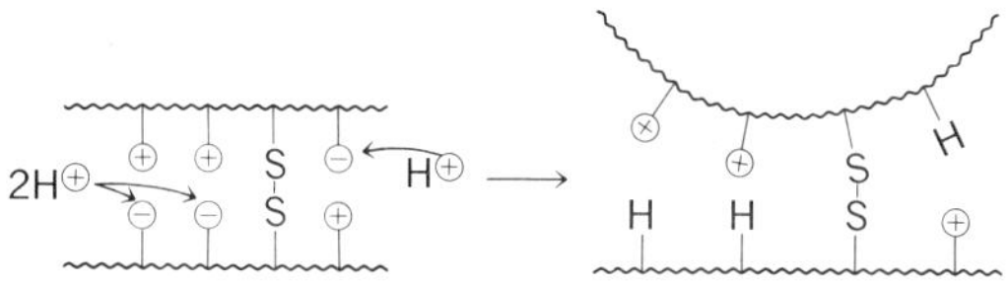 Prolonged action of aqueous acids or bases leads to actual hydrolysis of proteins.
Detergents	May affect salt bridges (by forming new salt bridges of their own), or may affect hydrogen bonds.
Salts of heavy metals (e.g., salts of the ions Hg^{2+}, Ag^{+}, Pb^{2+})	May disrupt salt bridges (by forming new salt bridges to themselves). These ions usually precipitate proteins (coagulation).
Alkaloidal reagents (e.g., tannic acid, picric acid, phosphomolybdic acid)	May affect both salt bridges and hydrogen bonds. These reagents precipitate proteins.
Violent whipping or shaking	May form surface films of denatured proteins from protein solutions (e.g., beating egg white into meringue).

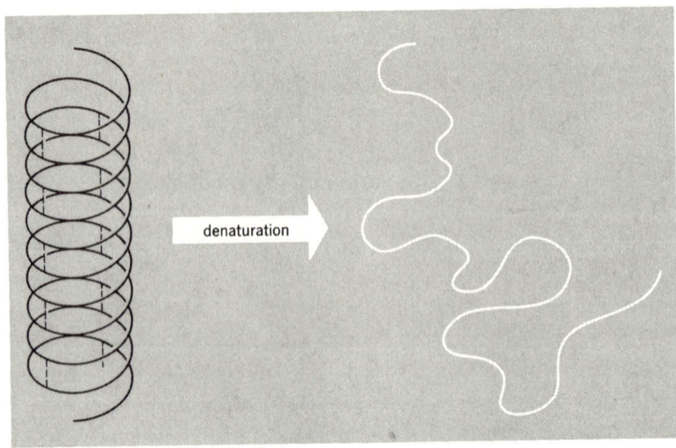
denaturation

Figure 18.12 Denaturation of a protein is fundamentally a disorganization of its molecular configuration without necessarily breaking any peptide bonds. It happens whenever the secondary structural features such as hydrogen bonds and salt bridges (represented here as dotted lines) are interfered with.

treating poisons made of these salts. Mercuric, silver, and lead salts are dangerous primarily because these metallic ions wreak havoc among important proteins of the body, particularly enzymes. When the salts are accidentally taken orally, their ions will be kept from reaching general circulation if they can somehow be precipitated in the stomach. The handiest protein available, raw egg white (albumin), should be swallowed. As the heavy-metal ions denature it, they become tangled in the coagulated mass. Next, *an emetic must be given.* The individual must be made to vomit, thus removing this material from the stomach. Otherwise the digestive juices will go to work on it and eventually release the poisonous ions.

Another medically important denaturation is the treatment of burns. When a large area of the skin is burned, one of the many serious problems is the loss of water. If the surface proteins are denatured and coagulated, the crust acts to inhibit this loss. In an emergency, if it is simply not possible to get the patient to a doctor, strong tea may be sprayed or dripped onto the burned area. Tea contains tannic acid, a denaturing agent. We must emphasize, however, that this is not a preferred treatment for a burn. A severe burn requires the prompt attention of a doctor. If it is at all possible to get this attention, *nothing* should be placed on the burn that might complicate the doctor's work.

Denaturing agents differ widely in their action, depending largely on the protein. Some proteins (e.g., skin, hide, hair) strongly resist most denaturing actions, as they must.

HOW PROTEINS ARE CLASSIFIED

Having studied the general principles of protein structure and behavior, let us turn our attention to the myriad types of proteins. They may be classified in the various ways listed.

I. Gross Structure and Solubility

A. *Fibrous proteins.* As the name implies, these proteins consist of fibers. At the molecular level both the α-helix and the pleated sheet (cf. p. 498) are found. These proteins can be stretched, and they contract when the tension is released. They perform important structural, supporting, and protective functions and are insoluble in aqueous media.

1. Collagens. These are the proteins of connective tissue, making up about one-third the body's protein. When acted upon by boiling water, they are converted into more soluble *gelatins.* In contrast to collagen, gelatin is readily digestible. Hence the collagen-to-gelatin conversion that takes place when meat is cooked is an important preliminary to digestion. At the molecular level the change is thought to be simply an unfolding or uncoiling of collagen molecules to expose the peptide bonds to the hydrolytic action of water and digestive enzymes.
2. Elastins. Elastic tissues such as tendons and arteries are elastins. They are similar to collagens, but they cannot be changed into gelatin.
3. Keratins. These proteins make up such substances as wool, hair, hoofs, nails, and porcupine quills. They are exceptionally rich in cystine, which has a disulfide link.
4. Myosins. Muscle tissue is rich in myosin, a protein directly involved in the extension and contraction of muscle.
5. Fibrin. Fibrin is the protein that forms from fibrinogen, its soluble precursor, when a blood vessel breaks. The long fibrin molecules tangle together to form a clot.

B. *Globular proteins.* Members of this broad class are soluble in aqueous media, some in pure water, others in solutions of certain electrolytes. In contrast to the fibrous proteins, the globular proteins are easily denatured. Examples include the following.

1. Albumins. Egg albumin is the most familiar member of this class. Albumins are soluble in pure water (forming, actually, colloidal dispersions), and they are easily coagulated by heat. In the bloodstream albumins contribute to osmotic pressure relations and to the pool of buffers.
2. Globulins. These are soluble in solutions of electrolytes, and they are also coagulated by heat. The γ-globulins in blood are very important elements in the body's defensive mechanisms against infectious diseases.

II. Function

Another way of classifying proteins, including those already discussed, is by the biological functions they serve.

A. *Structural proteins.* Fibrous proteins are examples.

B. *Enzymes.* These are the catalysts of a living organism without which it could not live.

C. *Hormones.* Many, but not all, hormones are proteins.

D. *Toxins.* These proteins produced by bacteria in the living organism act as poisons to that organism.

E. *Antibodies.* The body makes these proteins to destroy foreign proteins that invade it during an attack by an infectious agent.

F. *Oxygen-transporting protein.* Hemoglobin is the name of this important protein.

Many other special functions could be listed. With this as a survey of the main classes of proteins, we shall single out one of them, the fibrous proteins, for detailed study. In succeeding chapters many of the other kinds of proteins will be discussed.

FIBROUS PROTEINS

In our earlier studies of cellulose (p. 454), nylon (p. 433), and Dacron (p. 429) we learned that long molecules are characteristic of fiber-forming materials. It is therefore not surprising that many fibers are proteins. Silk is an example. Wool and hair belong to a family of proteins called the *keratins;* feathers, claws, porcupine quills, horns, hooves, and nails are also in this family. *Collagen,* the reinforcing "rod" material of bones, is also a fibrous protein. So are *myosin* of muscles and *fibrin* of a blood clot.

Silk Fibroin. The Pleated Sheet as a Tertiary Structural Feature. α-Helices are found among fibrous proteins, but another distinctive structural feature is the *pleated sheet.* Pauling, Corey, and Marsh recognized its existence in silk fibroin through X-ray diffraction analysis. Silk fibroin is unusual in that the two simplest amino acids, glycine and alanine, make up most of the molecule, as the data in Table 18.2 show. The polypeptide strand is coiled to some extent, but the groups that can form hydrogen bonds, C=O and H—N, protrude from the coil at almost right angles to its long axis. This orientation makes it possible for hydrogen bonds to form *between* coils. The coils line up more or less side by side, and this succession of neighboring helices gives an overall sheetlike appearance (Figure 18.13).

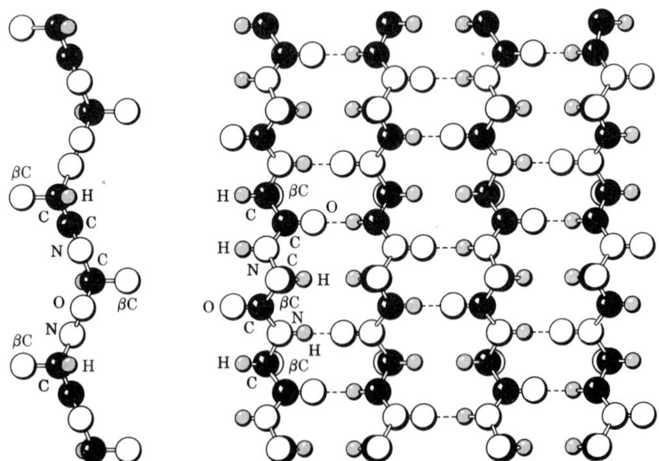
βC
H
C
C
N
C
O
βC
N
βC
H
C
H
βC
C
O
H
C
C
N
H
O
βC
N
C
H
H
βC
C

Figure 18.13 A drawing of a short segment of the antiparallel-chain, pleated-sheet network of protein molecules in silk fibroin. The folding of the "pleats" is horizontal. (From R. E. Marsh, R. B. Corey, and L. Pauling, *Biochimica et Biophysica Acta,* Vol. 16, 1955, page 13.)

The individual chains, because they are not fully extended, make the sheet pleated rather than flat. The greatest number of interchain hydrogen bonds can form if adjacent chains run in opposite directions. Details of the packing of protein molecules in silk fibroin are shown in Figure 18.14.

Keratins. In hair and wool the keratin molecules appear to have the α-helix configuration. If they are soaked in water, they can be stretched to almost twice their lengths, and they then apparently adopt the arrangement found in silk fibroin. Adding considerable strength to wool fibers are interchain disulfide links, for the

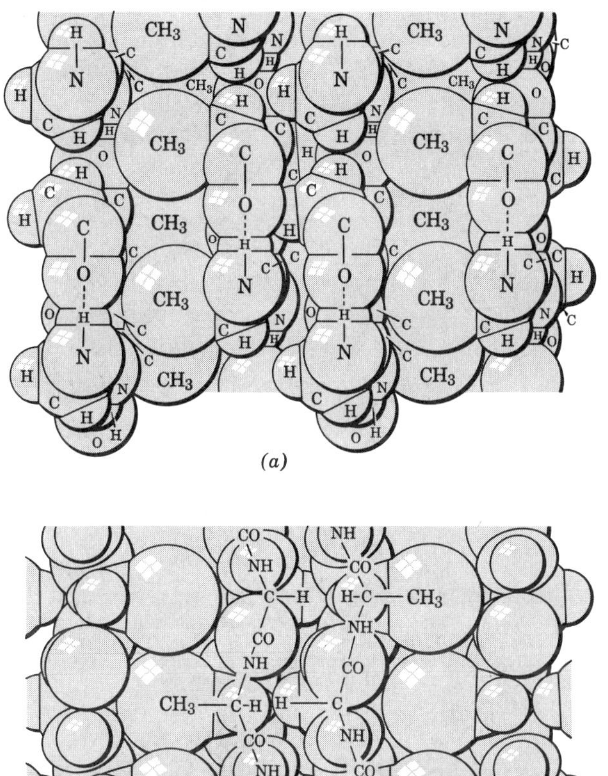
(a)
(b)
CO
NH
C—H
H-C—CH3
CH3—C-H
H—C
NH
CO

Figure 18.14 Packing drawings of silk fibroin sections: (*a*) looking down along the axis of the fiber, (*b*) viewing the fiber from an angle perpendicular to its axis and parallel to the plane of the pleated sheets. (From R. E. Marsh, R. B. Corey, and L. Pauling, *Biochimica et Biophysica Acta,* Vol. 16, 1955, page 18.)

keratins are especially rich in cystine. In some keratins several α-helices are believed to twist together into giant helices much like the several orders of twisting in rope. Salt bridges, hydrogen bonds, and disulfide links are the vulnerable sites in wool at which various reagents and conditions attack and weaken the fibers. Acids and bases, for example, interfere with salt bridges and catalyze the hydrolysis of peptide bonds. Even if only 10 to 15% of these are broken, the loss of fiber strength is severe. Oxidizing and reducing agents weaken wool fibers by attacking disulfide bridges.

Oxidizing agents such as hydrogen peroxide, a common hair bleach, or the combined action of ultraviolet light and oxygen at the beach or on the ski slope generate trace amounts of sulfonic acids and sulfuric acid from the sulfur in hair, causing a certain amount of "frizziness." Reducing agents, on the other hand, convert disulfide links into separate mercaptan groups, weakening the fiber. This reaction is the basis for the first step in some home permanents. The hair is first wetted and then treated with a reducing agent, such as ammonium thioglycollate:

$$\text{\{}\!-CH_2S-SCH_2-\!\text{\}} + 2H-SCH_2CO_2^-NH_4^+ \longrightarrow$$

Disulfide link in keratin of hair

Ammonium thioglycollate

$$\text{\{}\!-CH_2S-H + H-SCH_2-\!\text{\}} + \begin{array}{l} SCH_2CO_2^-NH_4^+ \\ | \\ SCH_2CO_2^-NH_4^+ \end{array}$$

Two mercaptan groups, now on separated peptide strands

Ammonium dithioglycollate

The hair is now curled and clamped in some desired fashion. Next a mild oxidizing agent, such as potassium bromate, is used in the expectation that as disulfide links re-form, they will do so in different patterns that stabilize the "wave":

$$3\text{\{}\!-CH_2S-H + H-SCH_2-\!\text{\}} + KBrO_3 \longrightarrow 3\text{\{}\!-CH_2S-SCH_2-\!\text{\}} + KBr + 3H_2O$$

Collagen. Probably the most abundant protein in the animal kingdom, collagen is the principal fibrous component in cartilage and bone, in tendons and ligaments, and in skin and corneas. In man it accounts for about one-third of body protein. The high content of glycine, proline, and hydroxyproline (cf. Table 18.2) and the fact that the hydroxyl group of hydroxyproline is put in place *after* the peptide has been strung together make collagen different from other proteins. As the structures of proline and hydroxyproline reveal, when either is incorporated into a protein chain, no hydrogen is left attached to the nitrogen of the amide group. The possibilities for interchain hydrogen bonding are reduced, and the means for stabilizing an α-helix are removed. For this reason collagen molecules are incapable of forming

α-helices, and they adopt a different and unusual arrangement, a three-stranded, twisting system (Figure 18.15) called *tropocollagen*. Collagen fibers are made from the systematic aggregation of these super helices.

In bone tissue collagen fibrils form a matrix within which the mineral hydroxyapatite, $3Ca_3(PO_4)_2 \cdot Ca(OH)_2$, crystallizes. Salts of citric acid and carbonic acid are present, and any remaining spaces are filled by a semiliquid material which provides a means for transporting substances between the bone and the circulatory system.

Proteins in Skeletal Muscles. Actin, Myosin, and the Molecular Basis of Muscular Work. Several proteins have been isolated from skeletal muscles, but two are the most directly involved in muscle contraction and mechanical work, *actin* and *myosin*. Actin is known in two forms. Globular actin, or *G-actin*, is a protein with a formula weight of about 45,000. In dilute salt solutions it aggregates into extended linear filaments seemingly without forming any covalent bonds, for the aggregation is easy to break up *in vitro*. If two of these linear filaments intertwine to form an open double helix, the protein is called *F-actin*, the type believed to be present *in vivo*.

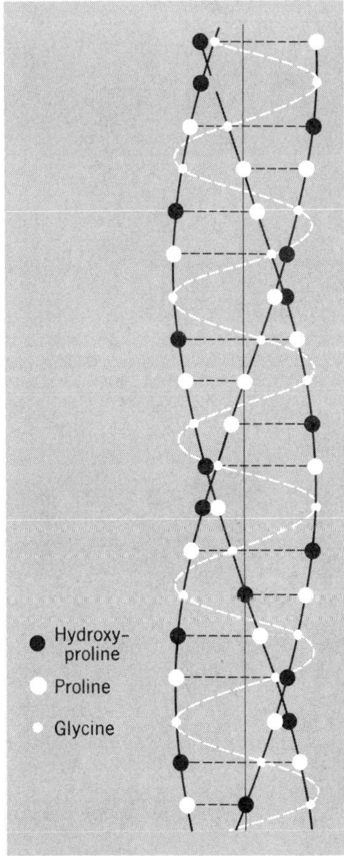
Hydroxy-
proline
Proline
Glycine

Figure 18.15 An idealized picture of the helical intertwining of three protein strands in collagen. Each chain consists of about a thousand amino acid units and has an average formula weight of roughly 120,000. This superhelix formed by these three chains is called tropocollagen. Hydrogen bonds between subunits are believed to stabilize it. The three amino acid units symbolized are not the only ones present, but glycine, which occurs at about every fourth position, is followed immediately by proline and hydroxyproline. (From G. H. Haggis, D. Michie, A. R. Muir, K. B. Roberts, and P. B. M. Walter, *Introduction to Molecular Biology,* John Wiley and Sons, New York, 1964, page 93.)

Myosin has a very high formula weight, about 500,000, and is largely α-helical. Its molecules, rodlike with globular ends, appear to be made of two identical long-chain polypeptide subunits and three much smaller subunits. The two chains are intertwined for most of the length of the myosin molecule, resembling a two-stranded rope.

In mammalian skeletal muscle tissue, myosin and F-actin are arranged as illustrated in Figure 18.16, forming myofibrils. Figure 18.17 is a diagram of some of the structural features of a typical striated muscle fiber such as might occur in a leg or arm muscle of a vertebrate. The many mitochondria that are present furnish sources of energy for contraction in ways to be described in Chapters 21 and 22. Interlacing the myofibrils and the mitochondria is the sarcoplasmic reticulum, a fluid-filled space which provides for movement of substances between parts of the cell.

Returning now to the myosin-actin units of Figure 18.16, we note that along the main axis of the myosin unit are projections or side chains. They are distributed around the myosin is a spiral manner and point toward actin units. Just what these side chains are, how they are involved in muscle contraction, and how they interact with the actin units are questions still being actively investigated. And what happens at the molecular level when contraction occurs? Do the individual protein molecules all coil up simultaneously? This would surely cause a shortening of the muscle. Or do they slide past each other in the manner of telescope segments? A British scientist, H. E. Huxley of Cambridge, using the electron microscope, furnished con-

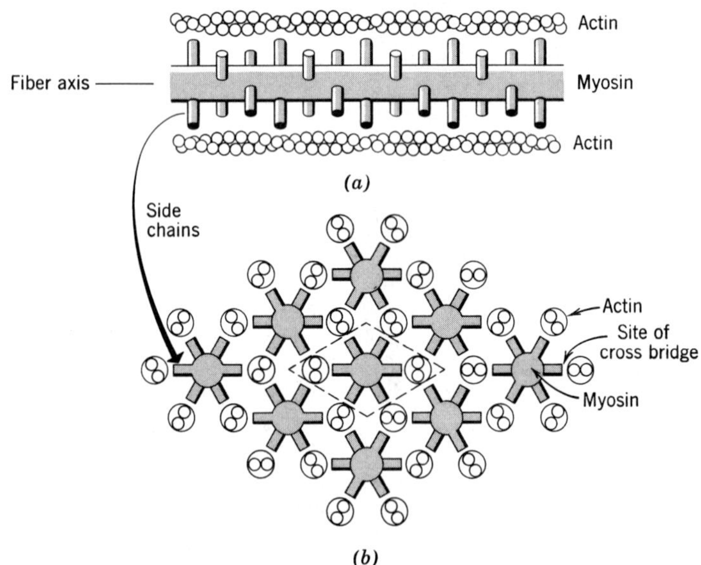
Actin
Fiber axis
Myosin
Actin
(a)
Side chains
Actin
Site of cross bridge
Myosin
(b)

Figure 18.16 Arrangement of filaments of actin and myosin in fixed striated muscle: (*a*) looking perpendicular to the fiber axis, (*b*) looking down the fiber axis. (From R. E. Davies, *Nature,* Vol. 199, 1963, page 1068, as modified by information in H. E. Huxley and W. Brown, *Journal of Molecular Biology,* Vol. 30, 1967, page 394.)

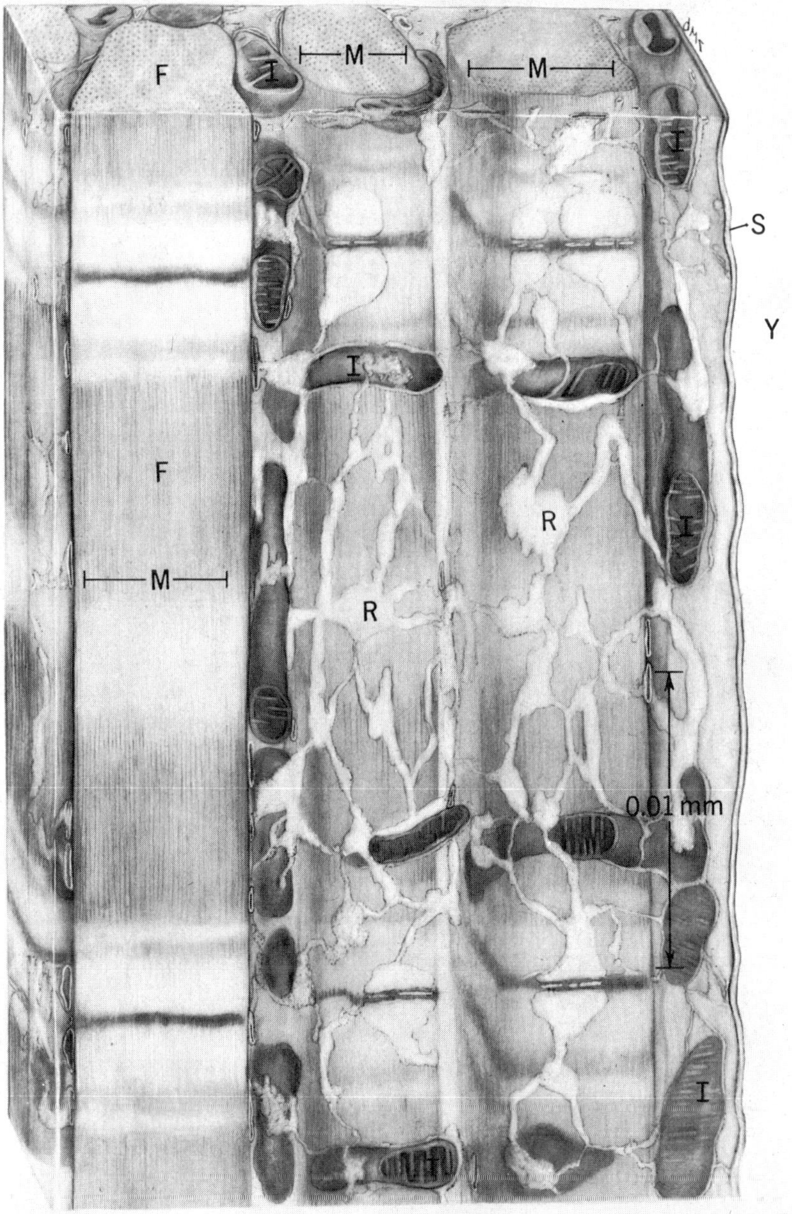

Figure 18.17 Structural features of striated muscle. This sketch represents a semitransparent block cut out of the muscle: Y, extracellular space; S, the plasma membrane or sarcolemma of the muscle cell; M, myofibrils; F, myofilaments of molecular dimensions that make up a myofibril; I, mitochondria scattered among the myofibrils; R, sarcoplasmic reticulum (or sarcotubules) interlacing the myofibrils. (From H. Stanley Bennett, "Structure of Muscle Cells," in *Biophysical Science—A Study Program,* edited by J. L. Oncley, John Wiley and Sons, New York, 1959, page 396.)

vincing evidence that muscles contract by filaments sliding past each other (Figure 18.18). Just how this sliding is made to happen, in chemical terms, has been the subject of intensive study still in progress. One theory of muscle contraction proposed by R. E. Davies (University of Pennsylvania) is outlined here with the realization that such conjectures are subject to both minor and major overhaul. The reader should keep in mind the fact that what follows, although an excellent theory well founded on experimental data, is a theory nonetheless, one that is still being tested.

The Davies Theory of Muscle Contraction. The side chains on the myosin filaments, described in greater detail in Figure 18.19, are important in this theory. According to Davies, each of these side chains terminates in a protein-bound molecule of adenosine triphosphate, universally called ATP. (ATP will be discussed in greater detail in Chapters 21 and 22. For the present all that interests us is its triphosphate system, which is simultaneously an ester, an anhydride, and an acid, with the acid groups probably being ionized as shown in Figure 18.19. The anhydride system, $-\overset{\overset{\displaystyle O}{\|}}{\underset{|}{P}}-O-\overset{\overset{\displaystyle O}{\|}}{\underset{|}{P}}-$, was discussed on page 422.) Near the base of the side chain is a fixed negative charge, probably $-COO^-$ of the side group of glutamic or aspartic acid. Repulsion between this fixed charge and the net negative charge at the ATP tip is thought to keep the side chain extended in the resting molecule. Otherwise the side chain polypeptide would tend to coil up. Should it do this, according to the theory, the phosphoric anhydride system would be drawn into a region of the myosin that actually provides a local catalyst for hydrolyzing such

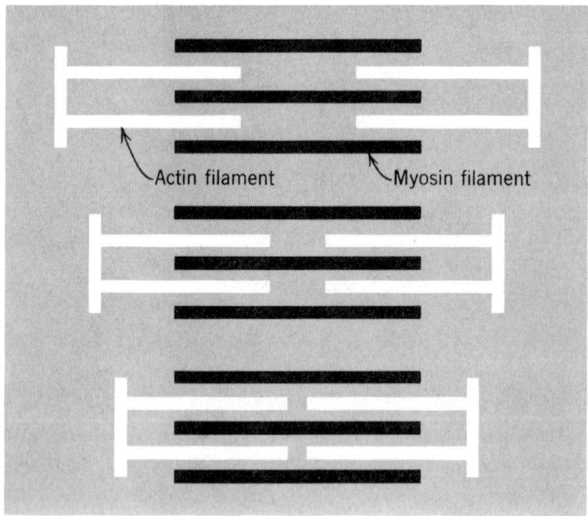
Actin filament
Myosin filament

Figure 18.18 The relations between filaments of actin and myosin: at the top, the extended state; in the middle, the resting state; at the bottom, a partly contracted state. By examining many electron microscope pictures of muscles, H. E. Huxley (*Endeavour,* Vol. 15, 1956, page 177) deduced this sliding filament model for protein contraction.

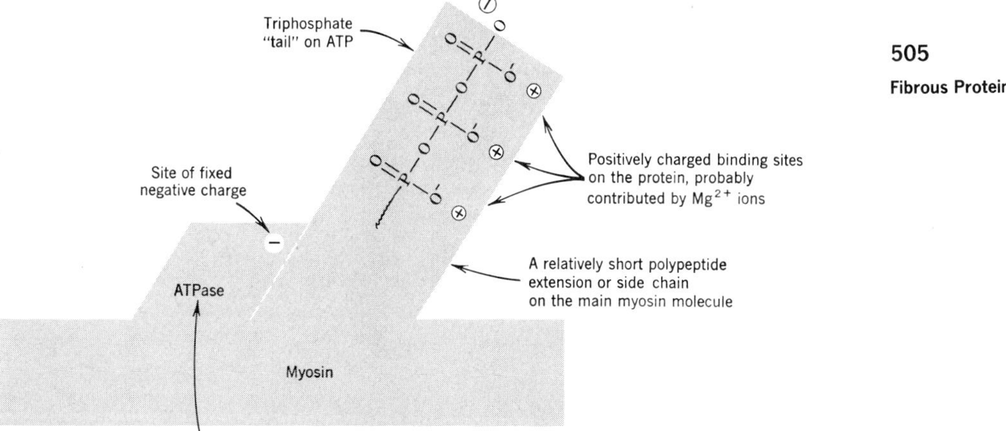
Triphosphate "tail" on ATP
Site of fixed negative charge
Positively charged binding sites on the protein, probably contributed by Mg^{2+} ions
A relatively short polypeptide extension or side chain on the main myosin molecule
ATPase
Myosin
Region of molecule having catalytic activity for hydrolyzing phosphate units in ATP.

Figure 18.19 The cross-bridge in muscle myosin is depicted here in its resting or extended state. The net charge at the ATP tip is −1. Repulsion between it and the fixed charge keeps the side chain extended. The side chain is called H-meromyosin. (Adapted, by permission, from R. E. Davies, *Nature,* Vol. 199, 1963, page 1068.)

anhydrides. In Figure 18.19 this region is labeled ATPase. (The names of enzymes end in -ase, and ATPase is an enzyme that acts on ATP.)

The side chain on myosin is called a cross-bridge because it can be part of an ionic bond or bridge to a neighboring actin molecule. Relative arrangements of actin and myosin in the resting state are shown in greater detail in Figure 18.20.

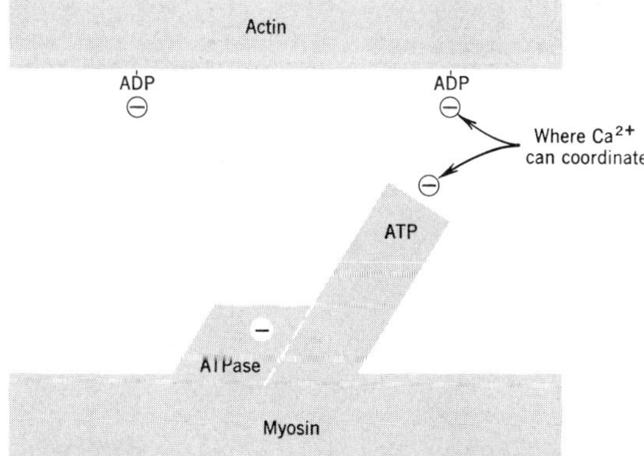
Actin
ADP
ADP
Where Ca^{2+} can coordinate
ATP
ATPase
Myosin

Figure 18.20 Detail of how actin and myosin are arranged in the resting state. When the muscle is activated, calcium ions are released from the sarcoplasmic reticulum, and they move to the negatively charged sites, as shown. Then a cyclic series of events (Figure 18.21) makes the actin and myosin filaments slide past each other. (ADP stands for adenosine diphosphate, which is identical with ATP except that it has one less phosphate unit.)(Adapted, by permission, from R. E. Davies, *Nature,* Vol. 199, 1963, page 1068.)

This bridge is completed in the moment of activation when calcium ions are released from a region of the cell called the sarcoplasmic reticulum. This ion with two plus charges can fit between the two sites with minus charges, one on myosin and one on actin. Once this happens, the minus charge at the ATP tip is neutralized. The force of repulsion between this tip and the fixed negative charge at the base of the side chain is removed. The side chain at once coils into a helix (Figure 18.21). It cannot do this, of course, without pulling on the actin molecule, for as long as the calcium ion is there the actin and myosin are for all practical purposes bound together. Referring back to Figure 18.16, which shows actin units and myosin neighbors surrounded on all sides, we see that this "pull" can only cause the actin and myosin filaments to slide by each other, as Huxley said they must. This sequence of events is described further in Figure 18.22, which carries the story through the relaxation stage.

When the side chain is in its coiled phase, the ATP tip is exposed to the action of the ATPase (Figure 18.23) and a phosphate unit is clipped off by hydrolysis.

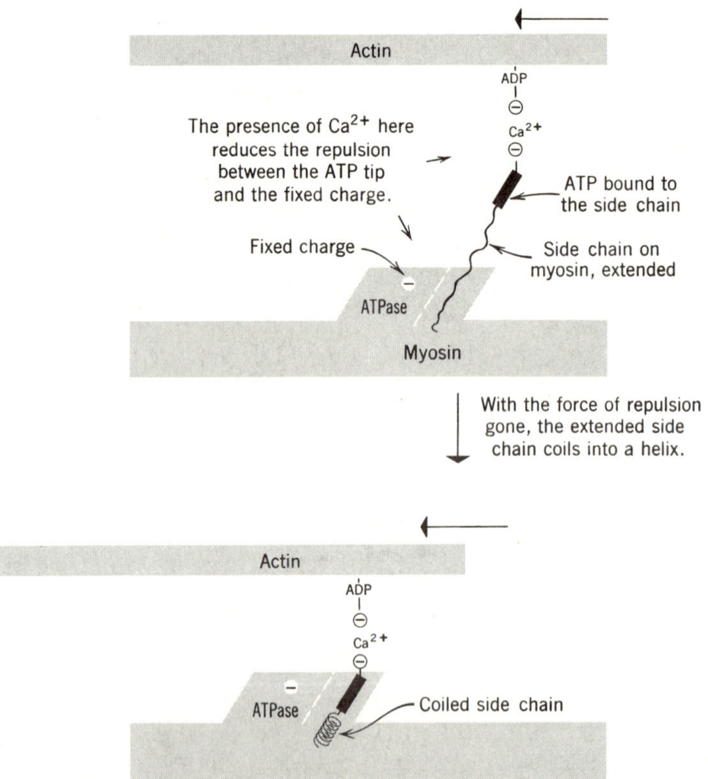
Actin
ADP
Ca2+
The presence of Ca2+ here reduces the repulsion between the ATP tip and the fixed charge.
ATP bound to the side chain
Fixed charge
Side chain on myosin, extended
ATPase
Myosin
With the force of repulsion gone, the extended side chain coils into a helix.
Actin
ADP
Ca2+
ATPase
Coiled side chain

Figure 18.21 The calcium ion removes the force that keeps the side chain extended, and it coils back toward the main chain of myosin. As it does so, the actin filament is pulled in the direction of the arrows, causing it to slip along the myosin chain. (Adapted, by permission, from R. E. Davies, *Nature,* Vol. 199, 1963, page 1068.)

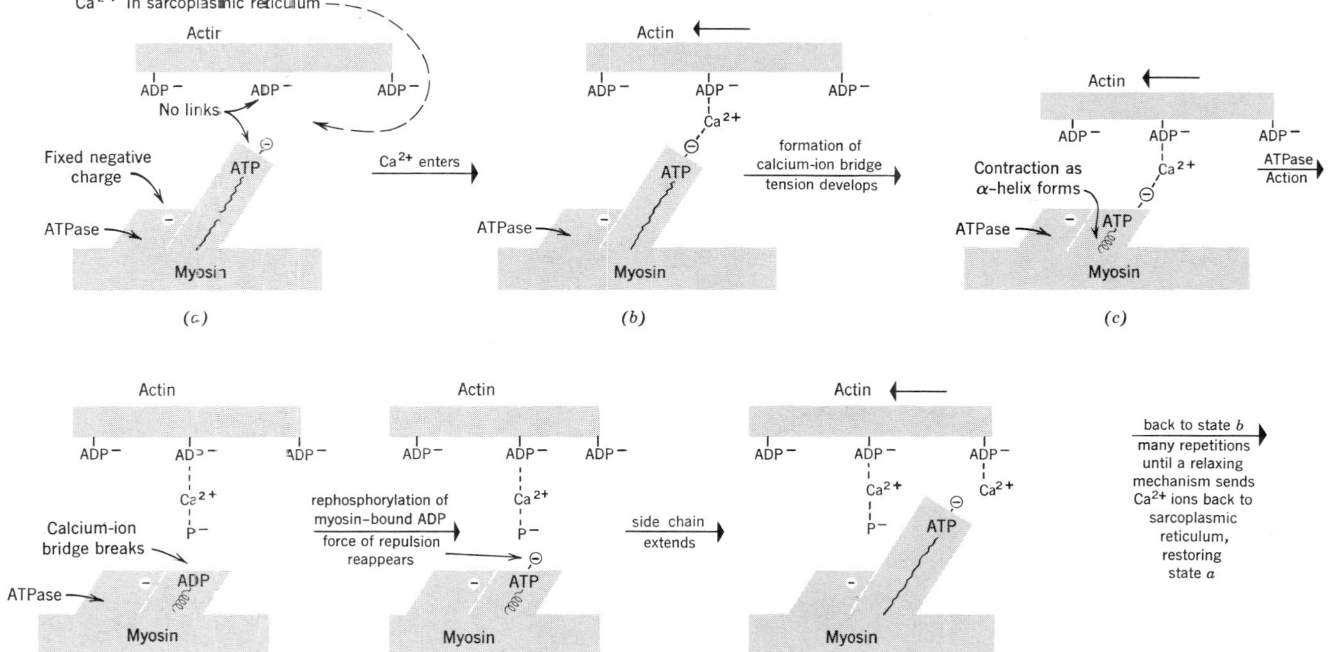
Ca 2+ In sarcoplasmic reticulum
Actin
ADP −
ADP −
ADP −
No links
Fixed negative charge
ATPase
ATP
Myosin
(a)
Ca2+ enters
Actin
ADP −
ADP −
Ca 2+
ADP −
ATPase
ATP
Myosin
(b)
formation of calcium-ion bridge
tension develops
Actin
ADP −
ADP −
Ca 2+
ADP −
Contraction as α-helix forms
ATPase
ATP
Myosin
(c)
ATPase
Action
Actin
ADP −
ADP −
Ca 2+
P −
ADP −
Calcium-ion bridge breaks
ATPase
ADP
Myosin
(d)
rephosphorylation of myosin–bound ADP
force of repulsion reappears
Actin
ADP −
ADP −
Ca 2+
P −
ADP −
ATP
Myosin
(e)
side chain
extends
Actin
ADP −
ADP −
Ca2+
P −
ADP −
Ca2+
ATP
Myosin
(f)
back to state b
many repetitions until a relaxing mechanism sends Ca2+ ions back to sarcoplasmic reticulum, restoring state a

Figure 18.22 The contraction of a skeletal muscle according to the theory of R. E. Davies. (*a*) Resting state (cf. Figure 18.19). (*b*) When Ca^{2+} enters, the negative charge at the ATP tip on the myosin side chain is neutralized (cf. Figure 18.21). (*c*) The side chain rapidly and automatically coils into a helix. Because of the calcium-ion bridge to actin, the actin filament is pulled along.

(*d*) When the helix forms, the ATP tip is dragged into the part of the myosin molecule, the ATPase region, that has catalytic activity for hydrolyzing a phosphate unit. For details see Figure 18.23. The phosphate (represented here simply as P^-) stays with the calcium ion, and the bridge between actin and myosin is broken at this point. Until now the actin has been given one tiny jerk. For continued work the system must be recharged, which translates at the molecular level into the need to put a phosphate unit back on the ATP tip. (*e*) The negative charge at the ATP tip reappears, and since it is quite close to the fixed negative charge, the charges repel each other and the side chain becomes reextended. (*f*) A phosphate ion is displaced from the vicinity of a calcium ion, and the calcium-ion bridge is reestablished. This action of course again removes the negative charge at the ATP tip, the side chain coils again, and the whole cycle starts over. Many repetitions all up and down the actin and myosin systems result in muscular work. (From R. E. Davies, *Nature*, Vol. 199, 1963, page 1068. Used by permission.)

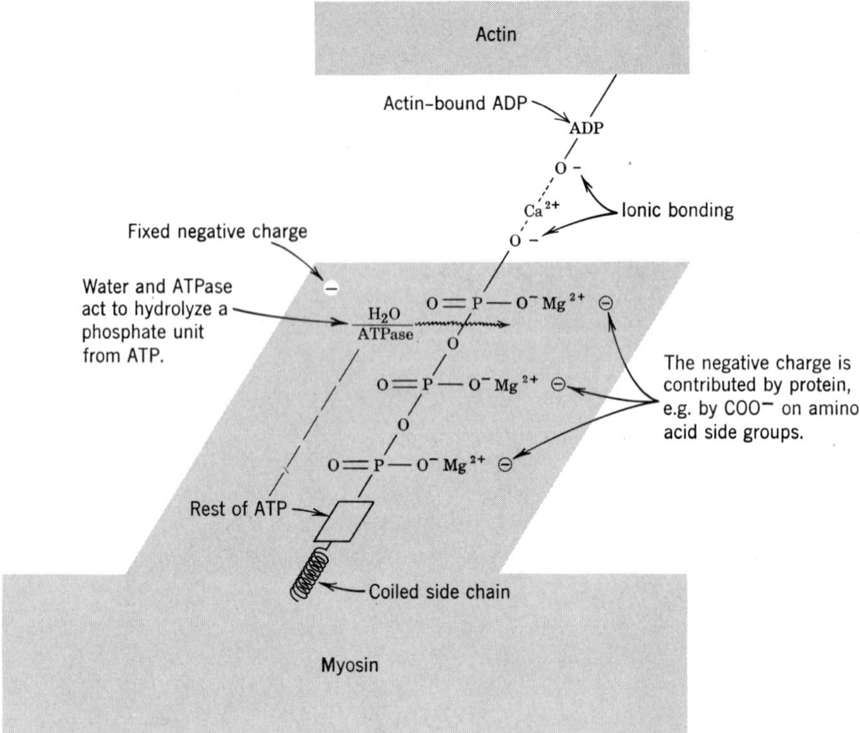

Actin
Actin–bound ADP
ADP
O –
Ca 2+
O –
Ionic bonding
Fixed negative charge
Water and ATPase act to hydrolyze a phosphate unit from ATP.
H2O
ATPase
O=P—O⁻ Mg 2+
O
O=P—O⁻ Mg 2+
O
O=P—O⁻ Mg 2+
The negative charge is contributed by protein, e.g. by COO⁻ on amino acid side groups.
Rest of ATP
Coiled side chain
Myosin

Figure 18.23 A site on the myosin molecule acts as an enzyme to catalyze the hydrolysis of a phosphate unit from the myosin-bound ATP, breaking the calcium-ion bridge to the actin at this point. It will be re-formed to a different actin-bound ADP unit in the next step (see Figure 18.22). (Adapted, by permission, from R. E. Davies, *Nature,* Vol. 199, 1963, page 1068.)

Now comes a key step, putting a phosphate unit back on. The resynthesis of ATP is the major metabolic function of glucose and fatty acids. We shall devote three chapters to this story. It is enough for the present simply to note that ATP is resynthesized and the ATP tip, now regenerated with its net negative charge, has this charge repelled by the fixed negative charge. The side chain uncoils and shoots back out ready for another cycle. Many rapid repetitions of these stages produce the overall effect of contraction. The presence of calcium in the sarcoplasm activates a relaxing mechanism which pumps the calcium ions back into the sarcoplasmic reticulum. When the calcium has been removed, the muscle can no longer contract.

The Huxley model of muscle contraction (Figure 18.18) is thus given a molecular interpretation by the theory of R. E. Davies. The modified and amplified forms are shown in Figure 18.24.

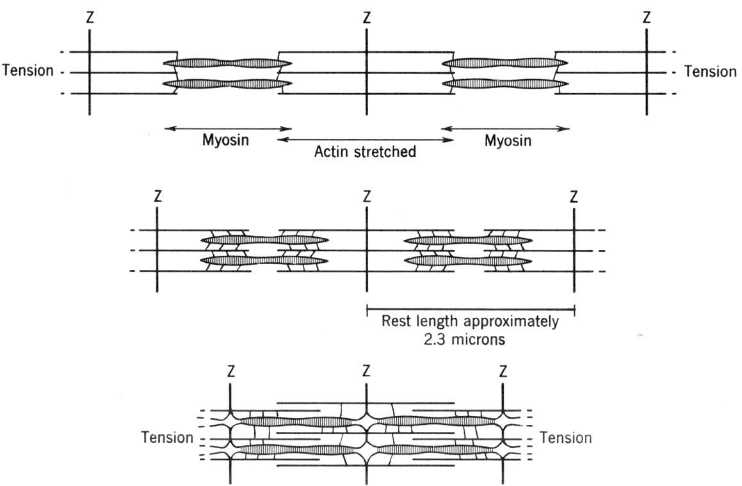
Z
Z
Z
Tension
Tension
Myosin
Actin stretched
Myosin
Z
Z
Z
Rest length approximately
2.3 microns
Z
Z
Z
Tension
Tension
Shortened

Figure 18.24 The telescoping of actin and myosin during the contraction of skeletal muscle. The Z-lines are believed to be continuous membranes bisecting bundles of filaments, and individual filaments probably terminate here. The cross-bridges are indicated by thin lines between myosin and actin units. The successive formation and breaking of these cross-bridges produce the telescoping. (Hanson-Huxley model as amplified by R. E. Davies, *Nature,* Vol. 199, 1963, page 1068. Used by permission.)

REFERENCES AND ANNOTATED READING LIST

BOOKS

R. E. Davies. "On the Mechanism of Muscular Contraction," in *Essays in Biochemistry,* edited by P. N. Campbell and G. D. Greville, Vol. 1. Academic Press, New York, 1965, page 29.

S. W. Fox and J. F. Foster. *Introduction to Protein Chemistry.* John Wiley and Sons, New York, 1957.

G. H. Haggis, D. Michie, A. R. Muir, K. B. Roberts, and P. B. M. Walker. *Introduction to Molecular Biology.* John Wiley and Sons, New York, 1964.

J. L. Oncley, editor in chief. *Biophysical Science—A Study Program.* John Wiley and Sons, New York, 1959.

A. White, P. Handler, and E. L. Smith. *Principles of Biochemistry,* third edition. McGraw-Hill Book Company, New York, 1964.

K. D. Kopple. *Peptides and Amino Acids.* W. A. Benjamin, New York, 1966.

ARTICLES

D. Wellner and A. Meister. "New Symbols for the Amino Acid Residues of Peptides and Proteins." *Science,* Vol. 151, 1966, page 77.

M. F. Perutz. "The Hemoglobin Molecule." *Scientific American,* November 1964, page 64. In an easy-to-understand style, the 1962 Nobel prize winner in chemistry discusses the work that earned this award.

E. Zuckerkandl. "The Evolution of Hemoglobin." *Scientific American,* May 1965, page 110. In this interesting discussion of how evolution might proceed at the molecular level, the evolution of a specific chemical, human hemoglobin, is reconstructed by comparing its amino acid sequences with those in hemoglobin from other animals.

Choh Hao Li. "The ACTH Molecule." *Scientific American,* July 1963, page 46. One of the leaders in the successful isolation and purification of ACTH writes about this work and about how the function of this compound is related to its structure.

J. C. Kendrew. "The Three-Dimensional Structure of a Protein Molecule." *Scientific American,* December 1961, page 96. The work on myoglobin described in this article earned for Kendrew a share (with Perutz) of the 1962 Nobel prize in chemistry.

K. R. Porter and C. Franzini-Armstrong. "The Sarcoplasmic Reticulum." *Scientific American,* March 1965, page 73. Electron micrographs and elegant drawings clarify the function of the complex network of tubules and sacs in a striated muscle fiber, the sarcoplasmic reticulum.

J. Gross. "Collagen." *Scientific American,* May 1961, page 121. Studies of the molecular nature of this important protein are described. The article contains some beautiful electron photomicrographs.

F. C. McLean. "Bone." *Scientific American,* February 1955, page 84. Bone is an active tissue. This article describes its structure and functions.

H. E. Huxley. "The Mechanism of Muscular Contraction." *Scientific American,* December 1965, page 18. The electron microscope techniques upon which were based the sliding filament model of muscle contraction are described by the chief architect of this theory.

R. E. Davies. "A Molecular Theory of Muscle Contraction: Calcium-Dependent Contractions with Hydrogen Bond Formation Plus ATP-Dependent Extensions of Part of the Myosin-Actin Cross-Bridges." *Nature,* Vol. 199, 1963, page 1068.

PROBLEMS AND EXERCISES

1. Describe in structural terms and symbols what all proteins have in common.
2. What are the structural factors that are the basis of differences among proteins?
3. Write the structures of the isomeric dipeptides that could be hydrolyzed to a mixture of serine and methionine.
4. Draw the structures of the following pentapeptides: (*a*) Ala-Gly-Phe-Val-Leu, (*b*) Glu-Lys-Tyr-Thr-Ser. Make the best judgment about the placement of protons and the location of charged sites. (*c*) Which of the two compounds would tend to be more soluble in a nonaqueous medium? Why? (*d*) Which would tend to be more soluble in water? Why? (*e*) Which would be better able to participate in salt bridges? Why? (*f*) If in the pentapeptide of part *b* the serine residue (Ser) were replaced by arginine, would the new compound be more or less basic?
5. Write equations to show how valine in its dipolar ionic form can act as a buffer.
6. Describe what would happen, chemically, to the following pentapeptide under these conditions: (*a*) digestion, (*b*) action of a reducing agent, (*c*) addition of hydroxide ion (not intending hydrolysis of peptide bonds).

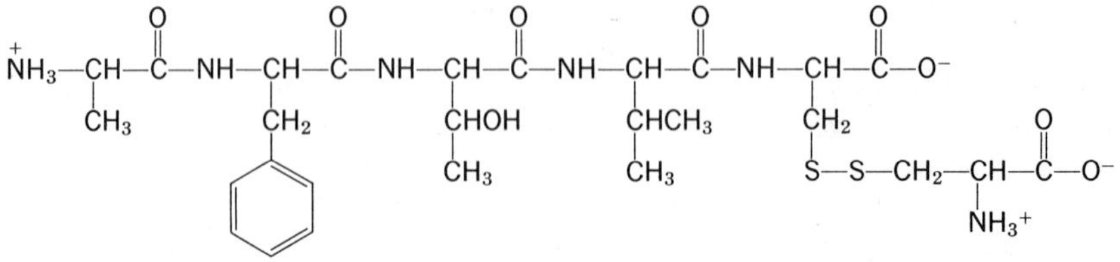

$$\overset{+}{N}H_3-CH(CH_3)-\overset{O}{\overset{\|}{C}}-NH-CH(CH_2C_6H_5)-\overset{O}{\overset{\|}{C}}-NH-CH(CHOH\,CH_3)-\overset{O}{\overset{\|}{C}}-NH-CH(CHCH_3\,CH_3)-\overset{O}{\overset{\|}{C}}-NH-CH(CH_2-S-S-CH_2-CH(NH_3^+)-\overset{O}{\overset{\|}{C}}-O^-)-\overset{O}{\overset{\|}{C}}-O^-$$

7. Discuss the role of the hydrogen bond in protein structure.
8. Explain why there is no net migration of amino acid molecules in an electric field when they are in a medium at their isoelectric point.
9. Why is a protein least soluble in a medium at its isoelectric point?
10. What is an isoionic point?
11. In terms of events at the molecular level, what happens during denaturation?
12. What is the relation between collagen and gelatin?
13. Discuss functions for the following: (*a*) elastins, (*b*) albumins, (*c*) keratins, (*d*) fibrin, (*e*) collagen, (*f*) keratin.
14. Discuss in molecular terms what might be responsible for the fact that collagen molecules do not form α-helices.
15. What is the difference between F-actin and G-actin?
16. In your own words and by means of your own drawings, discuss the Davies theory of muscle contraction and how it fits into the Huxley model.

CHAPTER NINETEEN

Biochemical Regulation and Defense

Within both the plant and animal kingdoms much is shared in common from species to species. The higher organisms in the animal kingdom have, for example, carbohydrates, lipids, and proteins as part of their composition and their diets. Yet any thoughtful person who has walked anywhere on the face of the earth is sooner or later struck by the bewildering variety of living things. Because a mere handful of the chemical elements are involved, both the similarities among species as well as their differences must be accounted for in terms of these elements. One of the outstanding triumphs of studies in the molecular basis of life and health has been the success in understanding both sameness and variety in common terms. The practice of medicine would hardly be possible without a great deal of sameness in people, but neither Olympic games nor great literature would exist without their differences.

In a sense this chapter is Part I of the "molecular basis of human uniqueness." It is about enzymes and hormones and vitamins. At one level of understanding the uniqueness of a species is related to the uniqueness of its set of enzymes and hormones. The survival of the species requires that each new offspring somehow gain possession of substantially the same set of enzymes and hormones that the parents have. The description of how this happens molecularly could be called Part II of the molecular basis of human uniqueness and is the subject of Chapter 25.

Survival also requires defensive mechanisms against a host of enemies, the most dangerous of which attack the body's biochemical regulatory systems. This topic is also considered in this chapter.

ENZYMES

CHEMICAL NATURE OF ENZYMES

Life as we know it would be unthinkable if there were no catalysts to aid and control the processes of metabolism. These catalysts are the enzymes.

All the enzymes that have thus far been studied consist of large polypeptide or protein units plus some other type of substance called a *cofactor*. The protein portion is designated the *apoenzyme* (Greek *apo,* away from). Without its cofactor it is catalytically inactive. The cofactor is sometimes a simple, divalent metallic ion (e.g., Mg^{2+}, Ca^{2+}, Zn^{2+}, Co^{2+}, or Mn^{2+}), sometimes a nonprotein organic compound. Some enzymes require both kinds of cofactors. If the cofactor is firmly bound to the apoenzyme, it is called a *prosthetic group* (Greek *prosthesis,* an addition). If, instead of being more or less permanently bound to the apoenzyme, the organic cofactor is brought into play during the act of catalysis, it is called a *coenzyme.* The fully intact enzyme is sometimes referred to as the *holoenzyme.* The relation expressed in a word equation is

$$\text{Cofactor} + \text{apoenzyme} \longrightarrow \text{holoenzyme} \quad \text{(or enzyme, for short)}$$

Classifying and Naming Enzymes. Before continuing our examination of how enzymes work, it will be helpful to note how names for enzymes are derived. The names of most end in "-ase," for example, maltase, oxidase, esterase, peptidase, transferase, reductase. Prefixes to this ending indicate either the substance acted upon or the kind of reaction performed. For example, maltase is an enzyme that acts as a catalyst for a reaction, the hydrolysis, of maltose; an esterase acts on an ester; a transferase assists the transfer of a group from one substrate to another. The names of some very common enzymes, such as several in the digestive system —pepsin, trypsin, and chymotrypsin, for example—stand as exceptions to the rules.

The following classification of enzymes includes only the major types together with specific examples.

A. *Enzymes that catalyze hydrolysis. Hydrolases*

1. Esterases and lipases. These act on esters in general or lipids in particular.

$$R-\overset{\displaystyle O}{\overset{\|}{C}}-O-R' + H_2O \overset{1}{\rightleftharpoons} R-\overset{\displaystyle O}{\overset{\|}{C}}-OH + HO-R'$$

2. Carbohydrases (or glycosidases). There are several of these, and each has

[1] Double arrows indicate equilibria in all these examples because a catalyst does not affect the *direction* of a reaction. It affects *how rapidly an equilibrium between reactants and products will be established.* A catalyst acts to lower the energy of activation for a change, and the lowering of this energy barrier for the forward reaction simultaneously means a lowering for the reverse reaction. If the forward reaction is to be favored, some other event (e.g., removal of a product as it forms, perhaps by a second reaction) must occur, and in cells it does. Therefore the discussions are written with the left-to-right reaction as the forward reaction.

the responsibility for catalyzing the hydrolysis of a specific kind of acetal oxygen bridge (glycosidic link) in a di- or polysaccharide. Examples are maltase, lactase, sucrase, and amylase.

$$H_2O + \text{sugar}'\text{—O—sugar}'' \rightleftharpoons \text{sugar}' + \text{sugar}''$$

3. Proteolytic enzymes. These assist hydrolysis of peptide linkages in proteins. Many specialized proteolytic enzymes are known, each being very effective for catalyzing hydrolysis of a peptide bond between two particular amino acids or between two types of amino acids. Examples include pepsin, trypsin, and chymotrypsin which participate in the digestion of proteins.

$$R\text{—}\overset{\overset{\displaystyle O}{\|}}{C}\text{—NH—R}' + H_2O \rightleftharpoons R\text{—}\overset{\overset{\displaystyle O}{\|}}{C}\text{—OH} + NH_2\text{—R}'$$

4. Phosphatases. Many metabolic intermediates are esters or anhydrides of phosphoric acid (or its anhydrides), and several reactions consist of the hydrolysis of such groups.

$$R\text{—O—}\underset{\underset{\displaystyle OH}{|}}{\overset{\overset{\displaystyle O}{\|}}{P}}\text{—OH} + H_2O \rightleftharpoons R\text{—OH} + HO\text{—}\underset{\underset{\displaystyle OH}{|}}{\overset{\overset{\displaystyle O}{\|}}{P}}\text{—OH}$$

B. *Respiratory enzymes.* Respiration in this usage is not confined to events in the lungs. Tissue cells are the main sites for reactions with oxygen, and the body can oxidize a wide variety of chemicals that are stable in air. Such reactions are usually quite productive of energy, and any enzyme participating in them may be classified as a respiratory enzyme.

1. Oxidases. An enzyme that catalyzes the transfer of hydrogen from a molecule directly to molecular oxygen is classified as an oxidase.
2. Hydroperoxidases. Hydrogen peroxide is a (temporary) product in certain metabolic pathways, and hydroperoxidases are enzymes that catalyze its further breakdown.

Catalase: $2H_2O_2 \rightleftharpoons 2H_2O + O_2$

Peroxidase: $H_2O_2 + \text{some acceptor molecule} \rightleftharpoons H_2O + \text{oxidized form of acceptor molecule}$

3. Dehydrogenases. A dehydrogenase catalyzes the transfer of hydrogen from some donor molecule to an organic acceptor (in contrast to oxidases, which transfer hydrogen directly to oxygen). Although numerous examples have been studied, only general types of hydrogen transfer processes will be given.

Alcohol dehydrogenation:

$$\underset{R'}{\overset{R}{>}}C(H)\text{—O(H)} + \text{acceptor} \rightleftharpoons \underset{R'}{\overset{R}{>}}C\text{=O} + \text{acceptor-}H_2$$

Aldehyde oxidation:

$$R{-}\overset{\displaystyle O}{\overset{\|}{C}}{-}H + H{-}OH \xrightleftharpoons{\text{acceptor}} R{-}\overset{\displaystyle O}{\overset{\|}{C}}{-}OH + \text{acceptor-}H_2$$

The most important dehydrogenases are enzymes whose coenzymes are pyridine nucleotides, NAD and NADP (cf. p. 518).

C. *Transferases.* Many metabolic events consist of the transfer of a group from one molecular species to another with the preservation of the energies of the bonds associated with the reactants. (The energy is not liberated as heat, for example.) The principal transferases are

—Transphosphorylases:	catalyzing the transfer of a phosphate unit from one molecule to another.
—Transglycosidases:	catalyzing the transfer of a saccharide unit as, for example, when a glucose unit is transferred to fructose in the formation of sucrose.
—Transaminases:	catalyzing the transfer of an amino group, making it possible for an organism to make some of its needed amino acids internally from molecular fragments of nonprotein origin.
—Transacylases:	catalyzing the transfer of an acyl group, for example, acetyl.

D. *Isomerases.* Several metabolic reactions result only in an isomerization of a particular molecule, either structural isomerization or racemization. Enzymes catalyzing such changes are classified as isomerases or sometimes as mutases.

E. *Carboxylases.* Enzymes responsible for stripping carbon dioxide from an organic reactant or for inserting it are called carboxylases.

Apoenzymes. Much of the information we need to understand the circumstances under which enzymes do or do not work can be deduced from the fact that they are substantially protein in nature. They can be expected, for example, to be sensitive to any or all of the denaturing agents. They retain maximum catalytic activity over very narrow ranges of temperature and pH. In the previous chapter the influence of these conditions on secondary and tertiary structural features of proteins was described. Apparently, if a particular protein is to function as an apoenzyme, it must have both the correct sequence of amino acids and a particular final gross shape that leaves just the right groups and polar sites exposed in just the right pattern. It is in terms of this pattern that we seek to understand one of the most remarkable properties of enzymes, their specificity. As far as test tube experiments are concerned, we know that both acids and bases catalyze a wide variety of reactions. In contrast, an enzyme that catalyzes the hydrolysis of an ester may be (and usually is) ineffective in catalyzing the hydrolysis of an amide. Various digestive juices have different enzymes for hydrolyzing terminal amino acid groups, internal amino acid groups, ester linkages in lipids, and acetal linkages in amylose and separate enzymes for hydrolyzing maltose, sucrose, and lactose. None of these enzymes may be at all efficient in catalyzing the chemical breakdown of the tissues lining the digestive tract.

Coenzymes. Vitamins. Not all diseases are caused by germs. If the diet lacks vitamins or if the system cannot use the vitamins it receives, some very debilitating maladies result, for example, scurvy, beriberi, and pernicious anemia. Science is a long way from a complete knowledge of how vitamins are involved in health. In fact, all the vitamins may not have been discovered yet. But certain of the vitamins are known to function as parts of enzyme systems. The term vitamin is applied to an organic compound if

1. It cannot be synthesized by the host, and it must therefore be provided by the diet.
2. Its absence causes a specific disease, a "vitamin deficiency disease."
3. Its presence is required for normal growth and health.
4. It is present in ordinary foods in very small concentrations and is not a carbohydrate, a lipid, or a protein.

The list of vitamins needed by man includes between sixteen and twenty independent types, with some variations within a type. Each species apparently has its own vitamin requirements; the list for man is not the same as the list for, say, microorganisms,[2] and some vitamins may appear on more than one list.

Since vitamins are required in very small daily amounts, we have generally thought that they are somehow involved in the catalysis of biochemical reactions. This assumption has not been shown to be true for all vitamins, but a large number of coenzymes either are identical with certain of the vitamins or are simple derivatives of them. Table 19.1 contains a partial list of vitamins that have been implicated in enzymic activity. Names and structures of one or more coenzymes derived from the vitamins are usually given. In some structures parts have been shaded to draw attention to the way the vitamin unit is fitted into the coenzyme. Most of the coenzymes are esters of phosphoric acid.

ENZYME ACTION

One of the most rapidly advancing fields of research is concerned with the question how enzymes work. Before examining the general features of current theory, we must look at some of the facts with which the theory makers must work.

1. *Enzyme molecules not only are largely protein in nature but are very often much, much larger than the molecules of the chemical or chemicals whose reactions they catalyze.*
2. *The catalysis produced by enzymes is usually selective.* Some enzymes, for example, possess *absolute specificity.* They catalyze one specific reaction of one specific compound. The enzymes urease and fumarase are examples. Urease catalyzes the hydrolysis of urea—only the hydrolysis and only that for urea:

[2] Vitamin designates those compounds required by the higher animals. The term nutrilite is applied to substances required only by microorganisms.

Table 19.1 Vitamins in Coenzyme Systems

Vitamin	Structure	Coenzyme System	General Area of Need
Thiamine		Thiamine pyrophosphate (cocarboxylase)	Removing CO_2 from acids
Riboflavin		Riboflavin phosphate (flavin mononucleotide "yellow coenzyme," FMN)	Oxidation-reduction reactions
Pyridoxal		Pyridoxal phosphate	Utilization of amino acids

(continued on next page)

Table 19.1 Vitamins in Coenzyme Systems (continued)

Vitamin	Structure	Coenzyme System	General Area of Need
Nicotinic acid Nicotinamide	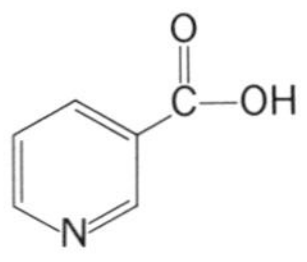 Nicotinic acid (niacin) 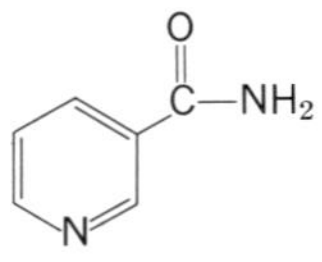 Nicotinamide (niacinamide)	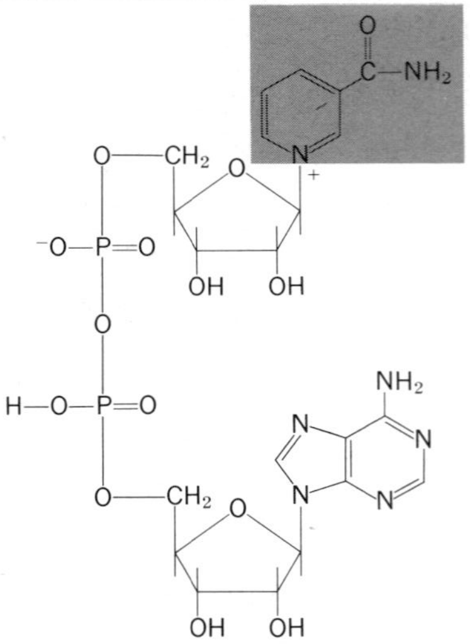 Nicotinamide adenine dinucleotide (NAD), older usage: diphosphopyridine nucleotide (DPN)	Oxidation-reduction reactions; hydrogen transfers; occurs in several important enzymes

Table 19.1 Vitamins in Coenzyme Systems (continued)

Vitamin	Structure	Coenzyme System	General Area of Need
		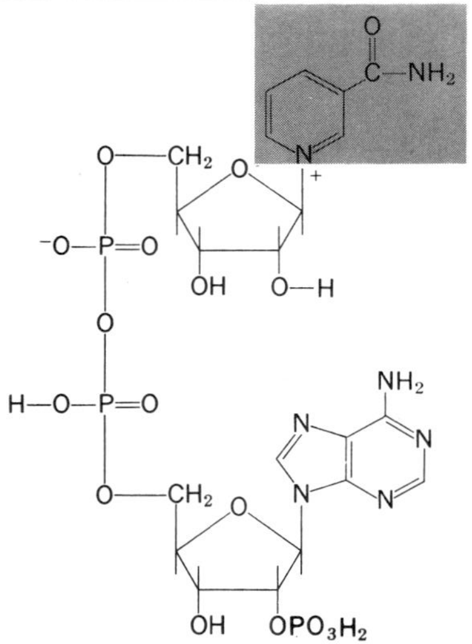 Nicotinamide adenine dinucleotide phosphate (NADP), older usage: triphosphopyridine nucleotide (TPN)	Oxidation-reduction reactions; hydrogen transfers; occurs in many key enzyme systems

(continued on next page)

Table 19.1 Vitamins in Coenzyme Systems (continued)

Vitamin	Structure	Coenzyme System	General Area of Need
Pantothenic acid	H—O—CH_2—C(CH_3)(CH_3)—CH(OH)—C(=O)—N(H)—CH_2—CH_2—C(=O)—O—H	Coenzyme A (CoA—SH)	Oxidation of fatty acids; resynthesis of fatty acids from acetic acid
p-Aminobenzoic acid (a nutrilite)	H—N(H)—C_6H_4—C(=O)—O—H	Folic acid (pteroylglutamic acid; also classed as a B vitamin by some)	Utilization of amino acids

Table 19.1 Vitamins in Coenzyme Systems (continued)

Vitamin	Structure	Coenzyme System	General Area of Need
Lipoic acid (thioctic acid)	Five-membered ring CH_2–CH_2–CH–S–S with side chain $(CH_2)_4C(=O)$—O—H on the CH	Lipoic acid itself appears to be a coenzyme.	Biological oxidation-reductions (note the easily reduced disulfide linkage)
Biotin	Fused rings: H–N–C(=O)–N–H; H–C—C–H; CH_2–S–CH with side chain $(CH_2)_4C(=O)$—O—H	This vitamin is a coenzyme for at least two enzyme systems.	May be needed to make other enzymes; may be needed in fatty acid synthesis; may be involved in the metabolism of carbohydrates and proteins

(continued on next page)

Table 19.1 Vitamins in Coenzyme Systems (continued)

Vitamin	Structure	Coenzyme System	General Area of Need
Vitamin B_{12} (cyanocobalamin)	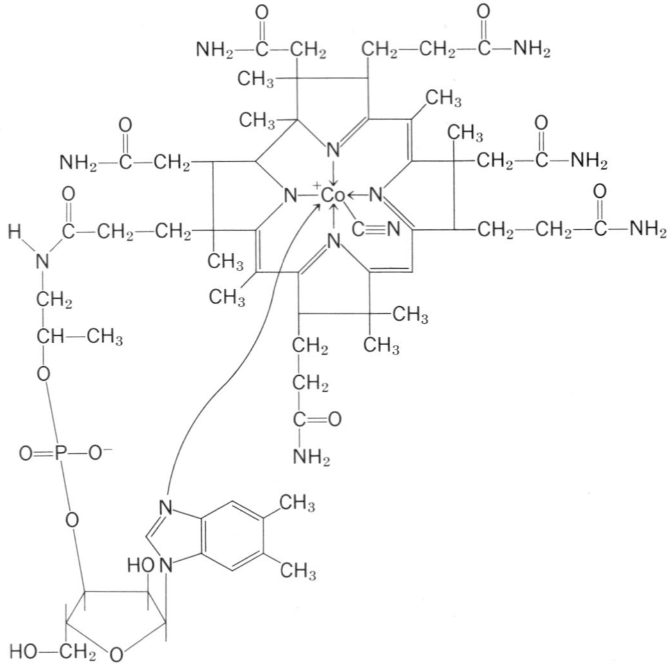 (The 1964 Nobel prize in chemistry went to British scientist Dorothy Crowfoot Hodgkin for determining this structure, and others, with X-ray diffraction techniques.)	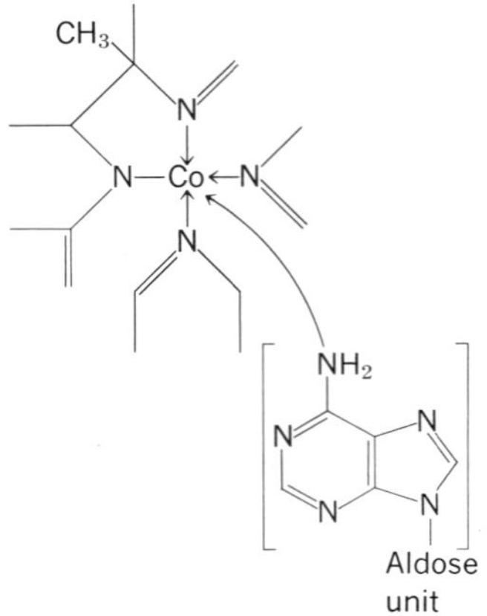 Aldose unit. This unit (structure uncertain) replaces the —C≡N group of the vitamin in the coenzyme.	Anti-pernicious anemia factor

$$NH_2{-}\overset{\displaystyle O}{\overset{\|}{C}}{-}NH_2 + H_2O \xrightarrow{\text{urease}} 2NH_3 + CO_2$$

Although it is essentially nothing more than the hydrolytic cleavage of two amide linkages, urease has no catalytic effect on the hydrolysis of biuret, closely related to urea, or on that of any other amide. This specificity is in striking contrast with

$$NH_2{-}\overset{\displaystyle O}{\overset{\|}{C}}{-}NH{-}\overset{\displaystyle O}{\overset{\|}{C}}{-}NH_2$$

Biuret

the fact that mineral acids will catalyze the hydrolysis of virtually any amide. Similarly, the enzyme fumarase catalyzes the addition of water to the double bond of fumaric acid but has no catalytic activity for this kind of reaction on even such a closely related compound as maleic acid (the cis isomer of fumaric acid).

$$\begin{array}{c} H{-}C{-}CO_2H \\ \| \\ HO_2C{-}C{-}H \end{array} + H_2O \xrightarrow[\text{fumarase}]{} \begin{array}{c} HO{-}CH{-}CO_2H \\ | \\ HO_2C{-}CH_2 \end{array}$$

Fumaric acid — Malic acid

$$\begin{array}{c} H{-}C{-}CO_2H \\ \| \\ H{-}C{-}CO_2H \end{array} + H_2O \underset{\text{fumarase}}{\not\longrightarrow} \text{no reaction}$$

Maleic acid
(cis isomer of fumaric acid)

Most enzymes are characterized by *group specificity*. They catalyze reactions involving only a certain functional group, for example, an ester linkage. Such group specificity may be absolute or relative. An enzyme with absolute group specificity acts on only one kind of functional group. An enzyme with relative group specificity acts predominately on one type of linkage, such as an amide, but it could act on another, for example, an ester. The digestive enzyme trypsin appears to be an example of the second type.

3. *The rates of enzyme-catalyzed reactions are extraordinarily more rapid than the same or similar reactions subject to nonenzymic catalysis.* This fact about enzyme-catalyzed reactions has stubbornly resisted explanation and has been one of the major subjects of research in this field. Reactions catalyzed by enzymes frequently proceed from 100,000 to 10,000,000 times as rapidly as they do under ordinary catalysis. The digestion of amylose (cf. p. 451) is catalyzed by the enzyme β-amylase, and it involves the hydrolysis of the acetal-oxygen bridges between glucose units. β-Amylase works at a rate equivalent to the hydrolysis of 4000 oxygen bridges per *second* per *molecule* of enzyme! If the β-amylase, in a separate experiment, is hydrolyzed to a mixture of its constituent amino acids, this mixture has no detectable catalytic effect on the hydrolysis of starch.

4. *Enzymes promote reactions under relatively mild temperatures.* The healthy existence of a human being requires that his body temperature be maintained fairly constantly at about 98.6°F, the temperature of a hot but not intoler-

able summer day. Enzymes must therefore do their work at this temperature. Normally we can accelerate reactions by simply applying heat, but too much heat applied to an enzyme-catalyzed reaction, even if it occurs in a test tube, may do nothing more than denature the enzyme.

5, *Enzymes promote reactions at nearly neutral pHs.* The variations in the pHs of body fluids are not large. Neither strongly acidic conditions nor strongly basic conditions prevail in the fluids of the body,[3] but enzymes will catalyze reactions that might, in a test tube in the absence of the enzyme, require either a fairly high pH (strongly basic) or a fairly low one (strongly acidic). Of course, both acids and bases can denature some proteins. Furthermore, at its isoelectric pH an enzyme is likely to have its lowest solubility and quite likely its lowest activity. Not all enzymes, however, have to be in solution. Many are incorporated into the molecular matrix making up the wall of, for example, a mitochondrion, a type of small granule present in cells (see p. 589).

THEORY OF ENZYME ACTION

We know almost intuitively that enzyme action must involve a coming together of the enzyme molecule and the molecule(s) of reactant(s). With the name *substrate* (or substrates) given to what the enzyme acts upon, the enzyme (E) and the substrate (S) are thought to form an unstable intermediate called the *enzyme-substrate complex* (E-S). During the very brief existence of this complex, the enzyme activates the substrate for further reaction (E-S*),[4] and the end product (P) soon starts to form (E-P) and disengages, freeing the enzyme for more work.

$$E + S \rightleftharpoons E\text{-}S \rightleftharpoons E\text{-}S^* \rightleftharpoons E\text{-}P \rightleftharpoons E + P$$

(The shifting of the equilibrium from left to right could be mediated by a variety of factors we need not consider here.) In at least one example, reported in 1964, the enzyme-substrate complex for a particular reaction was actually isolated.

The first fact cited in the previous section, concerning the relative sizes of the enzyme and the substrate molecules, strongly indicates that only a small segment of the apoenzyme is directly involved. It also implies that the substrate does not take up just any position on the enzyme molecule but rather selects some one segment. These sites on the enzyme molecule that act as substrate binders or activators or both are called *active sites.*

The second fact in the earlier section, relating to specificity, requires that active substrate-binding sites be able to discriminate between very similar substrate molecules. The high specificity makes almost irresistible the analogy between enzyme-substrate interaction and the fit of a key to a tumbler lock. In fact, the theory is often called the *lock and key theory* (Figure 19.1). An enzyme-substrate complex of reasonable stability cannot be expected to form unless the two can fit very closely and comfortably together. A left foot can be forced into a right shoe, a right hand into a left-hand glove, although the results are uncomfortable and not long endured. But a right hand will slip easily into a right-hand glove. Moreover,

[3] The stomach is something of an exception to this rule. The pH of gastric juice is about 2, and the enzyme pepsin works in this medium.

[4] The asterisk means that the substrate is activated.

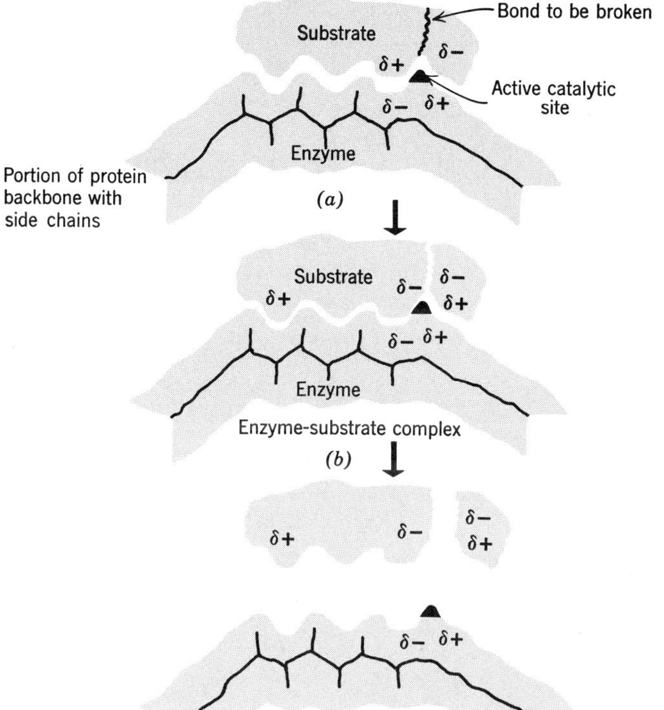
Bond to be broken
Substrate
δ+
δ−
Active catalytic site
δ−
δ+
Enzyme
Portion of protein backbone with side chains
(a)
Substrate
δ+
δ−
δ−
δ+
δ−
δ+
Enzyme
Enzyme-substrate complex
(b)
δ+
δ−
δ−
δ+
δ−
δ+
(c)

Figure 19.1 The "lock and key" theory of enzymic action. (*a*) Enzyme and substrate have complementary shapes, and electric forces assist their association. (*b*) While locked together in the enzyme-substrate complex, the reaction proceeds. (c) A change in relative polarities accompanying the reaction causes the pieces of the substrate to leave the enzyme.

the process of fitting involves some flexing of the glove, it is believed that the act of a substrate fitting to an active binding site of an enzyme may sometimes involve some flexing of the enzyme molecule. Figure 19.2 illustrates how the process of binding between enzyme and substrate may induce a flexing in the enzyme to bring its active catalytic sites into proper alignment. (Some scientists have suggested that a few hormones may exert their action by causing protein chains of enzymes to flex into configurations capable of entering into otherwise unlikely processes.) Thus the specificity of an enzyme is rationalized in terms of fitting together enzyme and substrate. If anythng happens to alter the geometry of an active binding site of an enzyme, it cannot function. The binding site can be altered by incorrect synthesis of the enzyme by the organism, for example, by mutation of a gene (discussed in Chapter 25), or by some denaturing action that changes the shape of the protein or blocks its active site, the effect of some poisons.

The enormously fast rates of enzyme-catalyzed reactions are more difficult to explain. Suppose, for example, that an enzyme is to catalyze some reaction between

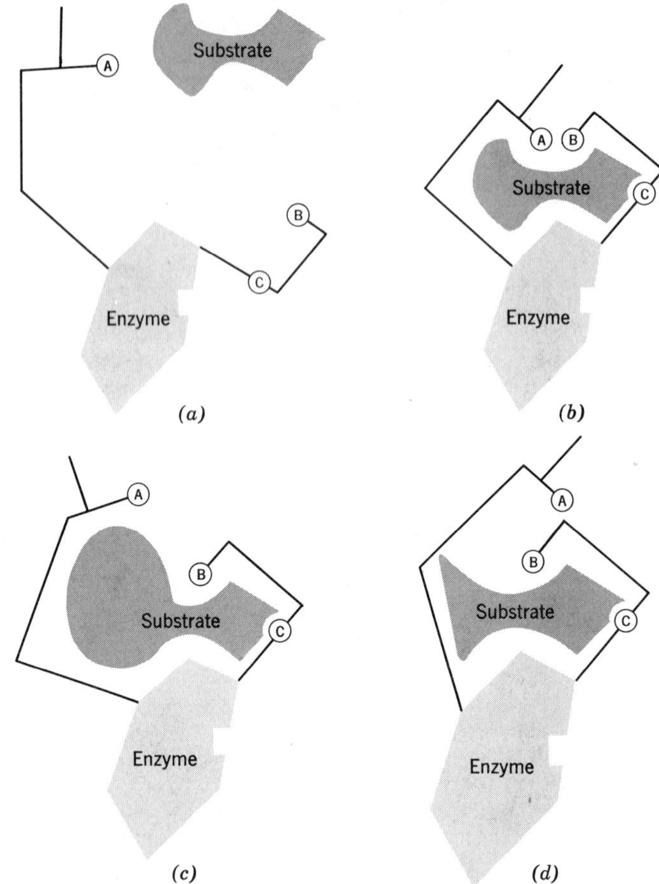
Substrate
A
B
C
Enzyme
(a)
Substrate
A
B
C
Enzyme
(b)
A
B
Substrate
C
Enzyme
(c)
A
B
Substrate
C
Enzyme
(d)

Figure 19.2 The concept of a flexible active site. As a substrate of proper shape and size nestles to the binding site of the enzyme (*a*), it causes other portions of the enzyme to flex (*b*) and bring about a proper alignment of catalytic groups A and B. A substrate that is either too large (*c*) or too small (*d*) may be bound to the enzyme, but neither can induce the proper flexing, and the reaction therefore fails. (From D. E. Koshland, Jr., *Science,* Vol. 142, 1963, page 1539. Copyright 1963 by the American Association for the Advancement of Science. Used by permission.)

two substrates, A and B, as in Figure 19.3. Suppose, furthermore, that three active sites, R, S, and T, are needed on the enzyme. As drawn in the figure, the reaction will require a carefully oriented coming together of five pieces. If five particles were to come together by random collision in solution, with R, T, and S on physically separated particles, the probability of their colliding in exactly one orientation (as in Figure 19.3) would be vanishingly remote. Just to make A and B collide as indicated and not in some other way (e.g., (←A)(←B) or 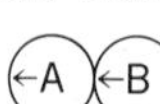 or (A↑)(B↓),

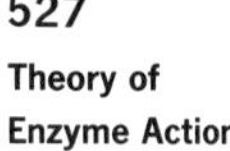

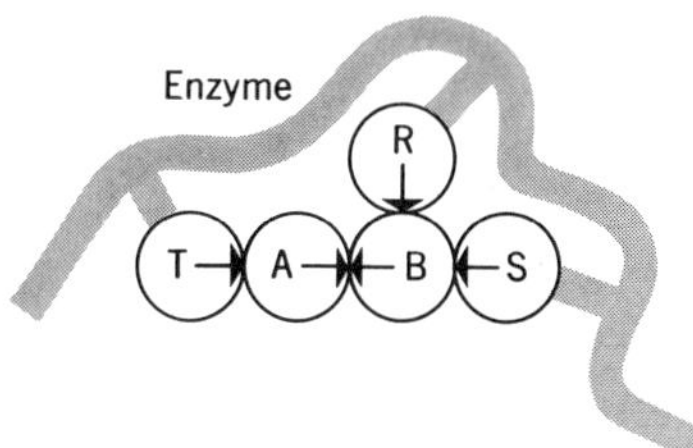
Enzyme
R
T
A
B
S

Figure 19.3 An active site in an enzyme may involve two or more side chains on the protein, or one may be part of the coenzyme. Shown here are three catalytic side chains, R, S, and T. If they have the correct shapes and sites of partial charge, they will force the reactants A and B to become oriented in just one way, and a reaction between them will occur. (From D. E. Koshland, Jr., *Science,* Vol. 142, 1963, page 1538. Copyright 1963 by the American Association for the Advancement of Science. Used by permission.)

etc.) is of low enough probability. A three-body collision of precise orientation is even more remote; and a five-body collision by a chance coming together of all five in exactly the right orientation and right collision energy is statistically almost impossible. But if the three active sites are fixed and are from the start properly oriented, and if A and B not only can fit as shown but may also be attracted to these sites by electric forces, the probabilities for the reaction are vastly improved. In fact, Koshland has calculated that the rate of a reaction involving five bodies like those in Figure 19.3 would be accelerated by a factor of 10^{17} to 10^{23} compared to a random five-body collision in solution. Incredible as it may seem, even this acceleration is not enough to explain some known velocities of enzyme-catalyzed reactions.

When molecules of substrate and enzyme approach each other, electric forces of attraction are believed sometimes to assist the swift formation of the enzyme-substrate complex. These forces are quite likely those of electrically polar regions, partially charged groupings, or ionic sites—all of which are possible in the side chains made available by the amino acid units in proteins. The formation of the complex is illustrated schematically in Figure 19.1. The reaction that the substrate undergoes may then introduce new functional groups with radically changed polarities. After the reaction the enzyme and the chemically changed substrate, which still adheres momentarily to the enzyme's surface, may repel each other. The enzyme therefore "peels" away quite naturally and rapidly and is freed for further action.

Thus far the theory has not explained how coenzymes and, indirectly, vitamins are involved. In many complexes studied, the coenzyme contributes the active catalytic site, illustrated schematically in Figure 19.4. There is good evidence, for example, that the coenzyme *cocarboxylase* (made from thiamine; cf. Table 19.1) catalyzes decarboxylations of α-keto acids by temporarily transforming them to resemble a β-dicarboxylic acid. This type of acid, we have learned (p. 427), undergoes decarboxylation rather readily. Figure 19.5 illustrates how cocarboxylase might accomplish this.

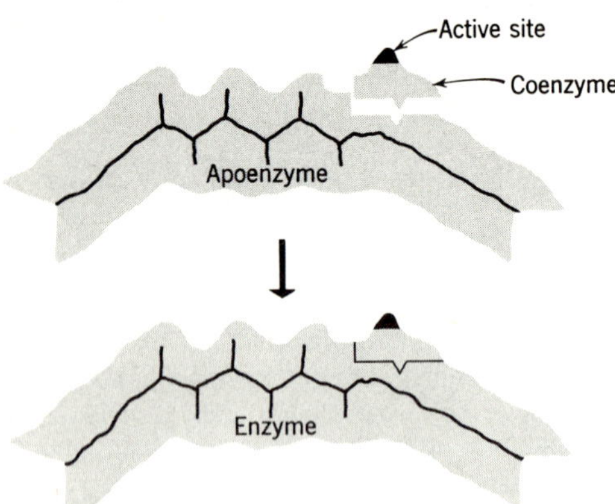
Active site
Coenzyme
Apoenzyme
Enzyme

Figure 19.4 The formation of a complete enzyme by the union of an apoenzyme and a coenzyme.

Nicotinamide (niacinamide) is a vitamin used by the body in making the coenzyme nicotinamide adenine dinucleotide (NAD).[5] Its structure is given in Table 19.1, but we need pay direct attention only to one small portion of it, the nicotinamide unit. Figure 19.6 illustrates how NAD acts to catalyze a biological oxidation by removing the elements of hydrogen. This is one coenzyme for which the associated apoenzyme is known to impart *substrate binding* specificity and the cofactor (NAD) is known to provide the active *catalytic* site.

Numerous other examples illustrating how coenzymes and their vitamin units participate in biological reactions could be discussed. Our purpose, however, is to be illustrative rather than comprehensive. These rather complicated examples make one point that should not be overlooked: an understanding of life at the molecular level does not appear to require general physical principles any different from those used successfully to account for other natural processes. We have not always believed this to be true. In spite of the commonplaceness of the idea now, it can still inspire awe in the minds of those who, as students or scientists, have examined it in depth. The power of this idea for inspiring further research in all areas pertaining to the processes of life is immense. The field of chemotherapy (p. 532) is but one of many living monuments to the idea. Endocrinology has also profited, in spite of the fact that molecular explanations for the activities of hormones have stubbornly resisted discovery.

HORMONES

In higher forms of life, as in complex societies, communication between highly specialized sections is essential. Although specialization creates its own unique problems, without it we would have difficulty imagining what we call "higher living forms," with their specialized organs and tissues. In a living human body, as one example, communication between tissues and organs is accomplished in two

[5] Still called in many sources diphosphopyridine nucleotide (DPN).

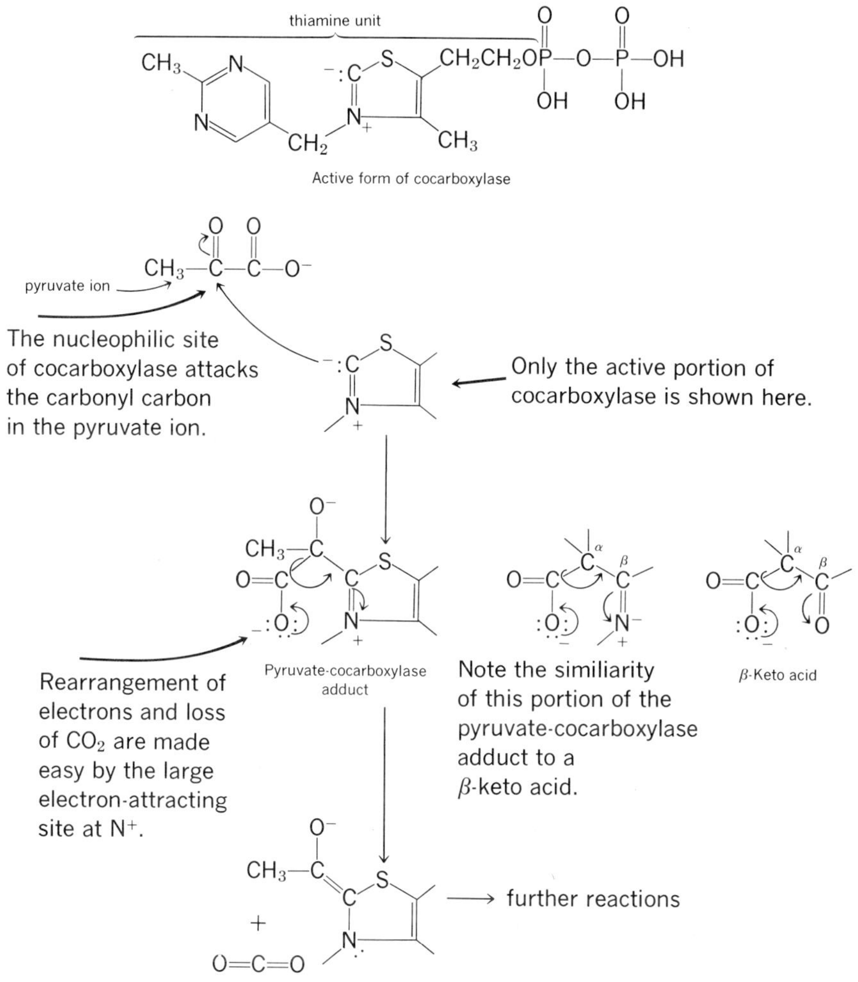

thiamine unit

Active form of cocarboxylase

pyruvate ion

The nucleophilic site of cocarboxylase attacks the carbonyl carbon in the pyruvate ion.

Only the active portion of cocarboxylase is shown here.

Rearrangement of electrons and loss of CO_2 are made easy by the large electron-attracting site at N^+.

Pyruvate-cocarboxylase adduct

Note the similiarity of this portion of the pyruvate-cocarboxylase adduct to a β-keto acid.

β-Keto acid

further reactions

Figure 19.5 Cocarboxylase catalyzes the decarboxylation of pyruvic acid. When the pyruvate ion becomes attached to the cocarboxylase as shown in this figure, the new system is analogous to a β-keto acid. These easily lose the elements of carbon dioxide (cf. p. 427).

ways, neural and humoral. Neural transmission occurs in the nervous system. Humoral communication takes place via the circulating fluids of the body, notably the bloodstream. Certain organs of the body are so specialized toward sending "chemical signals" via humoral circulation that they are considered a specific group, the *endocrine glands,* the glands of internal secretion. (Endocrine comes from the Greek: *endon,* within; *krinein,* to separate.) These glands specialize in the synthesis and secretion of compounds called *hormones* (Greek *hormon,* arousing, exciting). For each hormone there is a certain organ or a certain kind of cell whose biochemical reactions are most affected by it—a *target organ* or a *target cell.* The overabundance or the lack of various hormones has been related to such dramatic gross

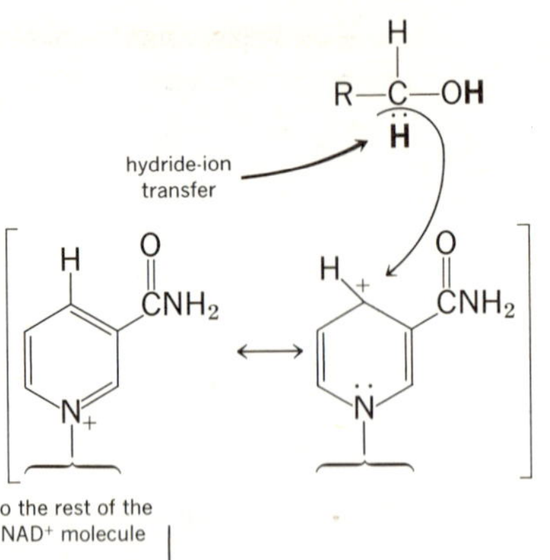
hydride-ion transfer
to the rest of the NAD+ molecule

An alcohol is shown as a substrate, and it is in the process of being dehydrogenated to an aldehyde.

The nicotinamide portion of NAD^+ is best represented as a hybrid of the two contributing structures shown here. The position opposite the ring's nitrogen bears a partial postive charge, and it is a hydride-ion acceptor.

to the rest of the NADH molecule, the reduced form of NAD+
Aldehyde

Figure 19.6 Removal of the elements of hydrogen ($H:^-$ and H^+) from an alcohol is catalyzed by an enzyme whose coenzyme is NAD^+ (nicotinamide adenine dinucleotide). Only the nicotinamide unit is shown here. The first step is the transfer of a hydride ion from the alcohol to the coenzyme, which is structured to be a good hydride-ion acceptor.

effects as cretinism, goiter, diabetes, dwarfism, and giantism. Sexual development is especially dependent on normal hormonal action.

In spite of decades of research and hundreds of reported investigations by hundreds of scientists, a detailed biochemical picture of *how* hormones act is still in its early stages of development. We shall limit our discussion to the theories that have been proposed to explain the effects of hormones and to serve as the basis for devising further research. Then in succeeding chapters the action of specific hormones will be mentioned from time to time, reflecting the intimate relation between hormonal action and a wide variety of metabolic reactions.

Consider a very generalized metabolic event. Compound A reacts with compound B to establish an equilibrium with compound C; a specific enzyme E is required:

$$A + B \overset{E}{\rightleftarrows} C$$

The rate at which this equilibrium is established will depend on (1) the availability of A, (2) the availability of B, (3) the availability of E, (4) the specific activity of E, and (5) the concentration of C already present. Hormones do not appear to *initiate* reactions between substances that would not otherwise react at all, without an enzyme. Instead, hormones affect the *rates* of metabolic events. In the example a specific hormone could therefore control the overall rate at which the equilibrium is established by altering any one (or more) of the factors. Not all hormones alter factors in the same way.

Since 1950 the evidence has increasingly pointed to hormonal control of the availability of reactant A or B or both, at least in a wide variety of (but not all) instances. If the reaction occurs inside a cell, to inhibit it the hormone need merely deny reactant A or B or both access to the cell's interior. Even when enzyme E is already present inside the cell in a highly active form, if A or B or both cannot get through the cell wall, the reaction obviously will not occur. The same argument would apply to a reaction occurring inside a subcellular particle such as a mitochondrion; these have their own specialized membranes.

Suppose a so-called "growth hormone" were needed for normal development. Cells grow by dividing. Before they divide a great deal of chemical synthesis must occur to provide for the new daughter cells. Amino acids, for example, must be allowed into any cell before it can divide in a normal way; amino acids are required for protein synthesis. But suppose the absence of the growth hormone means that amino acids cannot be transported through the cell membrane (or at least not in sufficient amounts). Then growth will be retarded. But if the growth hormone is available, it can exert an effect on the cell membrane that will allow amino acids through. The molecules of the hormone itself need not penetrate the cell wall to have this effect. Many of the hormones are proteins or polypeptides, that is, large molecules. This theory of hormonal action, as affecting transport properties of cell membranes, is attractive because it does not require large hormone molecules to move inside the cells of the target organ.

One protein hormone, insulin, has been studied particularly well; evidence indicates that its principal but not exclusive action is altering the permeability of the walls of its target cells for glucose molecules. The five most important hormones involved in regulating secretion of digestive juices all appear to exert their effects by modifying the transport properties of cellular membranes. (We shall return to these hormones in Chapter 20.) The hormones that regulate secretion and reabsorption in the tubules of the kidneys also appear to work by affecting permeabilities of cell walls.

This theory of hormonal action has another intriguing aspect, its *possible* relation to embryonic cell differentiation. The fertilized egg is one cell. After one division there are two cells. When these divide there are four, and the embryo is on its way to development. Sooner or later, however, some of the new cells must begin to take on specialized characteristics. Some must eventually become brain tissue, others bones, still others various internal organs, etc. Such cell differentiation must imply *selective* admission or exclusion of some nutrients rather than others from the outside medium. In fact, in the fully developed organism, each kind of cell must have a way of picking and choosing from the fluid in which it is bathed just those nutrients it needs. Without a mechanism for this selection, there can be no differentiation, development, or even long-term maintenance. It appears that in many instances hormones must be considered vital factors in this mechanism. We do not mean to imply that hormones exert their influence in this one way only. We shall briefly examine another theory that probably applies to other hormonal actions.

Most individual reactions in the body are but single steps in a long *metabolic pathway,* a sequence of chemical events wherein the product of one step is a reactant for the next. Each step requires an enzyme. The speed of the overall sequence cannot be any faster than the speed of the slowest step in the pathway, the "bottleneck step." If the pathway is to be accelerated, the hormone may exert its action by increasing the activity, the catalytic "ability," of the enzyme for the slowest step. Alternatively, if the hormone functions to slow down a particular pathway, it may do so by inhibiting one of the enzymes in the sequence to the point that this step becomes the slowest of all.

The hormone may exert its effect on the enzyme by interacting with the apoenzyme and changing its molecular shape, the nature of the folding or coiling of its protein chain. The hormone may also affect the concentration of the enzyme, perhaps by influencing the gene that directs its synthesis (cf. p. 683). Or the hormone and the cofactor may interact in some way. Conceivably the hormone is itself a cofactor, or perhaps it competes with a cofactor for binding to the apoenzyme, inhibiting the enzyme.

Thus there are several possibilities, several ways in which hormonal action may be explained. The theory postulating that hormones affect transport of nutrients through cellular membranes appears to apply to the majority of hormones, but in no instance do we yet understand the detailed chemistry. Some of the most sophisticated research is being directed to this problem. The profound overall effects that hormones can have make this research particularly important. Control of cancer could conceivably require complete information about what regulates the transport of nutrients through cell membranes. Rapidly growing cancer cells are obviously able to let in more than their share of nutrients.

CHEMOTHERAPY

Chemotherapy is the use of chemicals, or drugs, to destroy infectious organisms without seriously harming human protoplasm. Chemicals that simply inhibit the growth of a microbe are called *antimetabolites* and are classified as chemotherapeutic agents. Excluded from this class of chemotherapeutic substances are such general bactericidal agents as iodine, phenol (carbolic acid), and silver compounds which, although good germicides outside the body, destroy most kinds of tissue and cannot be administered internally.

SULFA DRUGS

Although the parent sulfa drug is sulfanilamide, thousands of variations have been prepared and tested. Figure 19.7 shows a few that have been found to be the best.

Sulfa drugs are antimetabolites and work by producing deficiency diseases in bacteria. According to D. D. Woods, bacteria sensitive to sulfa drugs cannot discriminate between a molecule of sulfanilamide and a molecule of *p*-aminobenzoic

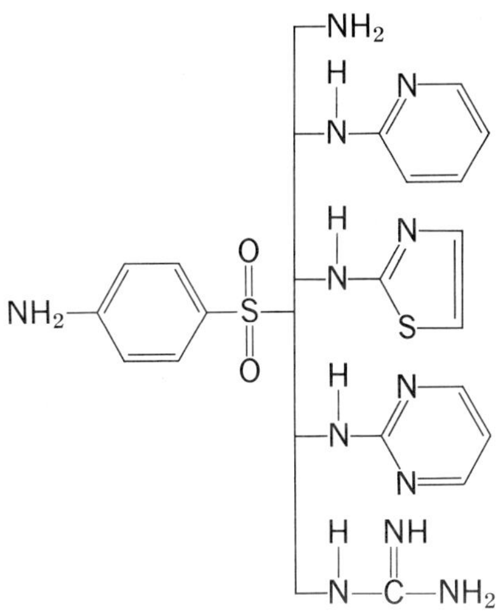

Sulfanilamide
Sulfapyridine
Sulfathiazole
Sulfadiazine
Sulfaguanidine

Figure 19.7 Some important sulfa drugs. Sulfapyridine was shown to be effective against pneumonia by Whitby, an English scientist, in 1938. A scourge which had brought sudden, premature death to countless people was removed. Sulfadiazine is credited with saving the lives of thousands of battlefront casualties in World War II and later wars.

acid, which it needs to complete one of its enzyme systems, folic acid, a coenzyme. A molecule of a sulfa drug resembles this *p*-aminobenzoic acid enough to permit its incorporation into the coenzyme (Figure 19.8). The altered coenzyme fails to function as required. An important metabolic process in the microbe is not initiated, its growth is inhibited, and it may die. In the meantime the normal defensive mechanisms of the body are able to cope with the infection before it spreads too far.

ANTIBIOTICS. PITTING MICROBE AGAINST MICROBE

In 1928 the British bacteriologist Sir Alexander Fleming, experimenting with cultures of *Micrococcus pyogenes* var. *aureus,* noticed that one culture had become contaminated by a mold, *Penicillium notatum,* similar to common bread mold. Spores of the mold, borne by the breeze, had evidently alighted on the petri dish in which the micrococci were multiplying. Both the mold spores and the micrococci used the nutrient medium in the dish as their food and competed for it. The mold spores, however, secreted a chemical that was fatal to the micrococci. Micrococcal colonies failed to grow in the vicinity of the mold, or those that had started there showed signs of dissolution. One living thing, the mold, had apparently killed the other living organism. Fleming named the active chemical produced by the mold *penicillin.* Although Fleming's chance observation was made in 1928, he could not find anyone who would or could isolate and purify the mold's active substance. Not until 1940 did Florey and Chain succeed in isolating it.

The term *antibiotic* means "against life." Antibiotics may be defined as chemicals that are produced by microorganisms and that can, in small concentrations, inhibit the growth of or even destroy other microorganisms. Microbes fruitful

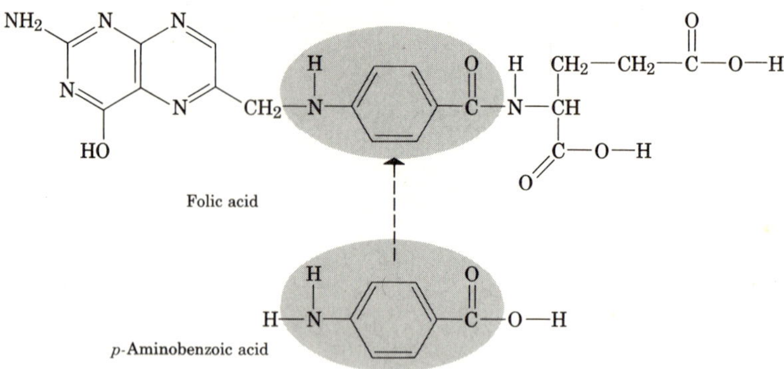
Folic acid
p-Aminobenzoic acid

Altered folic acid in which a portion of a sulfa drug molecule has been incorporated

Sulfa drug (G = a group attached to nitrogen)

Figure 19.8 Sulfa drugs as antimetabolites. When a microbe selects a sulfa drug to synthesize folic acid, it fashions an altered folic acid that is ineffective as a coenzyme. Reactions requiring it do not occur, and the growth of the microbe is inhibited. It suffers, in effect, from a vitamin deficiency disease.

in producing antibiotics have been found chiefly in the soil. Different soil samples contain different antibiotic-producing microbes. Tens of thousands of soil samples from all over the world have been examined.

In 1944, in a handful of New Jersey soil, Dr. Selman Waksman found a strain of actinomycetes that produces the antibiotic *streptomycin,* effective against many microbes not attacked by penicillin, including the tough tubercle bacillus. In 1947 Dr. Paul Burkholder discovered another strain of actinomycetes in soil sent to him from Caracas, Venezuela. It produces *chloramphenicol* (Chloromycetin), effective against typhus, spotted fever, typhoid fever, and certain types of dysenteries.

Dr. Benjamin Duggar found a strain of actinomycetes in Missouri River mud that produces *Aureomycin,* useful against a broad spectrum of infectious agents, especially Asiatic cholera and brucellosis. In Indiana soil was found still another strain that produces Terramycin, an extremely efficient, broad-spectrum antibiotic.

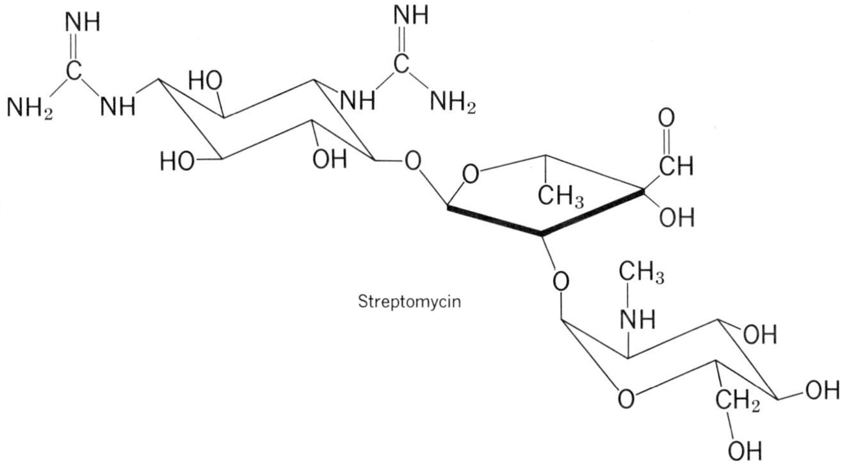

Streptomycin

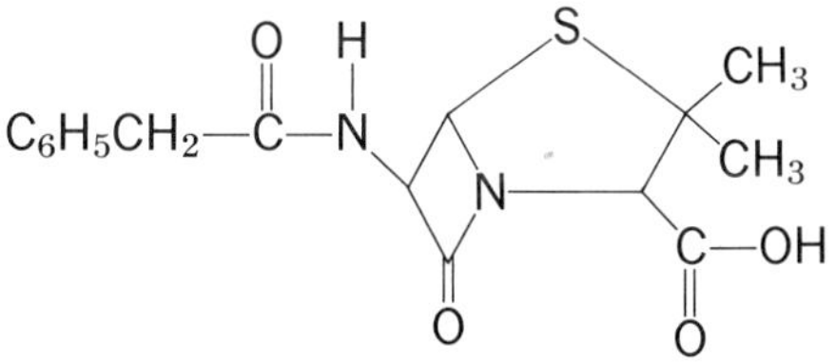

Penicillin G

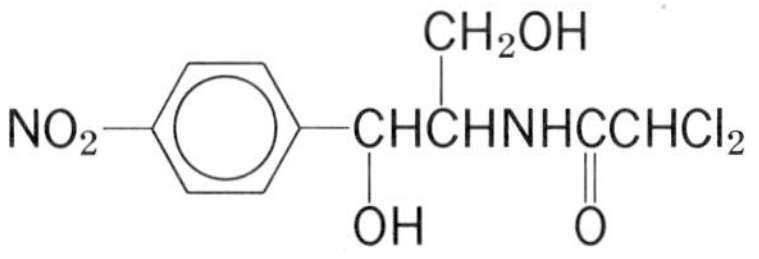

Chloramphenicol

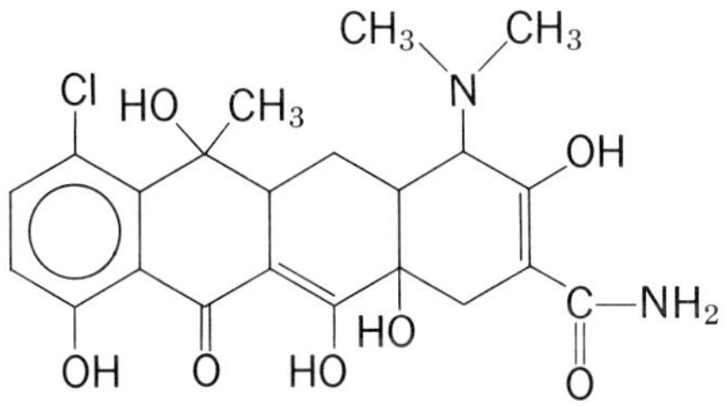

Aureomycin

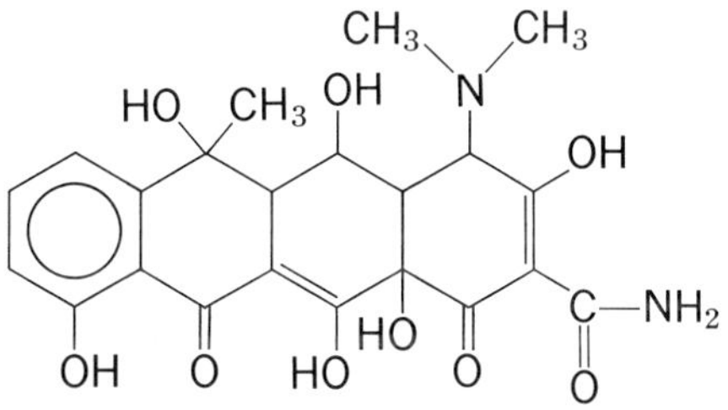

Terramycin

How do antibiotics work? No single answer can be given, for each works differently. Many act as antimetabolites. Penicillin, for example, inhibits the passage of an important amino acid across the cell wall of bacteria. Deprived of this compound, they cannot properly maintain their cell walls, and the walls deteriorate, spilling the contents of the cell. Chloromycetin inhibits protein synthesis in bacteria, quite likely by interfering with important enzymes or even by interfering with the synthesis of apoenzymes.

The chemical aspects of chemotherapy just studied illustrate how *struggles between the invading microbe and the host are, at root, interactions between chemicals.* These interactions must involve the ordinary laws of chemical behavior, even though they occur among uncommonly complicated molecules. This last statement amounts to a declaration of faith. Belief in it impels scientists to seek *chemical* answers and explanations to the problems of infections. The more that is learned about the chemistry of diseases, the better they can be controlled.

INHIBITION OF ENZYME ACTIVITY. POISONS

Anything that prevents formation of an enzyme from its apoenzyme and cofactor(s) will clearly inhibit or prevent the reaction for which the enzyme is needed. Anything that inhibits an enzyme and its substrate from forming their complex will give the same negative results. In these inhibiting effects the connection between poisons and enzymes can be seen. *The most powerful poisons known wreak their devastation by inhibiting key enzymes.* All are effective in very small doses, for only catalytic amounts of poisons, after all, are needed to inhibit catalytic amounts of enzyme.

Cyanides. The cyanide ion reacts with many metal-ion activators of enzymes and makes these ions unavailable to an apoenzyme. Cells that consume oxygen require the presence of the enzyme cytochrome oxidase, which contains iron. Cyanide ion inhibits this enzyme, so that cellular respiration ceases instantly.

Arsenic Compounds. Arsenic poisons such as sodium arsenate, Na_3AsO_4, block enzymes that contain phosphate units. Such units are present in many of the coenzymes listed in Table 19.1. Arsenate ions, AsO_4^{3-}, closely resemble phosphate ions, PO_4^{3-}, and can take their place. If that happens, however, the coenzyme or the enzyme does not function properly.

Nerve Poisons. A variety of deadly poisons act by attacking enzymes of the nervous system. In certain parts of the system, the so-called cholinergic nerves, arrival of a nerve impulse at one nerve cell stimulates formation of a trace amount of acetylcholine all along its length. This compound then reacts with a *receptor protein* on the next cell, making it possible for that cell to send the signal along. Once the signal has been sent, the acetylcholine must be hydrolyzed to leave the cell in readiness for the next signal. Otherwise excess acetylcholine forms, resulting in overstimulation of muscles, glands, and nerves; choking, convulsions, and paralysis cause death. The hydrolysis of acetylcholine must occur within a few millionths of a second after it appears. The equation for this reaction, the hydrolysis of an ester, is

$$\underset{\text{Acetylcholine}}{\underset{HO^-\quad\quad}{CH_3-\overset{CH_3}{\underset{CH_3}{\overset{|+}{\underset{|}{N}}}}-CH_2-CH_2-O-\overset{O}{\overset{\|}{C}}-CH_3}} + H_2O \xrightarrow{\text{cholinesterase}}$$

$$\underset{\text{Choline}}{\underset{^-OH\quad\quad}{CH_3-\overset{CH_3}{\underset{CH_3}{\overset{|+}{\underset{|}{N}}}}-CH_2-CH_2-OH}} + \underset{\text{Acetic acid}}{CH_3-\overset{O}{\overset{\|}{C}}-O-H}$$

As summarized in Figure 19.9, some very deadly substances interfere either with the synthesis of acetylcholine or with its hydrolysis, causing paralysis. The

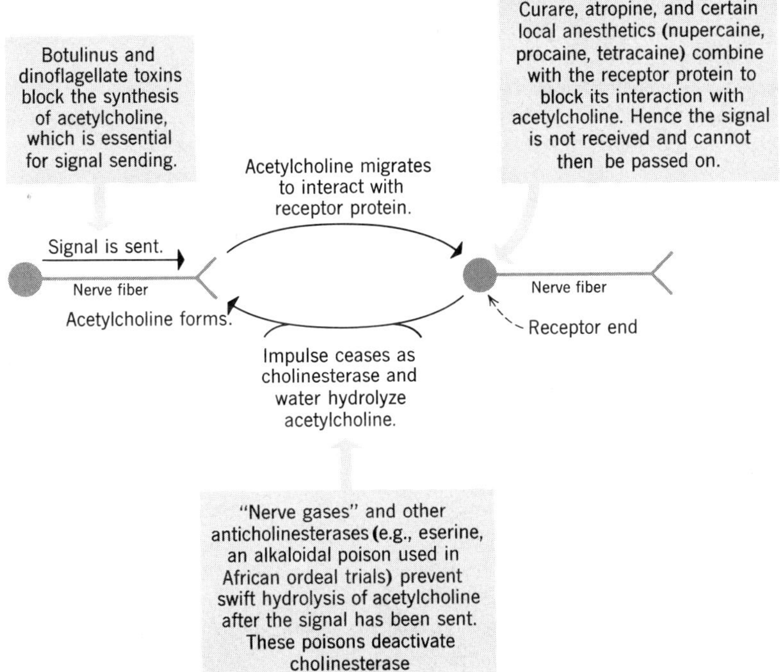
Botulinus and dinoflagellate toxins block the synthesis of acetylcholine, which is essential for signal sending.
Curare, atropine, and certain local anesthetics (nupercaine, procaine, tetracaine) combine with the receptor protein to block its interaction with acetylcholine. Hence the signal is not received and cannot then be passed on.
Acetylcholine migrates to interact with receptor protein.
Signal is sent.
Nerve fiber
Nerve fiber
Acetylcholine forms.
Receptor end
Impulse ceases as cholinesterase and water hydrolyze acetylcholine.
"Nerve gases" and other anticholinesterases (e.g., eserine, an alkaloidal poison used in African ordeal trials) prevent swift hydrolysis of acetylcholine after the signal has been sent. These poisons deactivate cholinesterase

Figure 19.9 The acetylcholine cycle and the action of nerve poisons.

toxic protein produced by the botulinus bacillus, *Clostridium botulinum,* a microorganism that multiplies in spoiled food, is lethal to mice in a dose averaging only eight molecules per nerve cell. It is estimated that about half a pound of this protein would be enough to destroy all the human beings on earth.

REFERENCES AND ANNOTATED LIST

BOOKS

F. R. Jevons. *The Biochemical Approach to Life.* Basic Books, New York, 1964. Chapter 6 introduces vitamins and coenzymes with an historical treatment.

G. Litwack and D. Kritchevsky. *Actions of Hormones on Molecular Processes.* John Wiley and Sons, New York, 1964. Chapter 1, written by T. R. Riggs, is a discussion of the influence of hormones on the transportation of nutrients across cell membranes.

ARTICLES

D. E. Koshland, Jr. "Correlation of Structure and Function in Enzyme Action." *Science,* Vol. 142, 1963, page 1533. This paper was an important reference for the writing of this chapter.

I. E. Frieden. "The Enzyme-Substrate Complex." *Scientific American,* August 1959, page 119. "Life is essentially a system of cooperating enzyme reactions." Thus the author closes this article, which traces the idea of the enzyme-substrate complex from the first insights of Jons Jakob Berzelius (1835), to the first isolation of an enzyme in crystalline form (urease, by J. B. Summer of Cornell University, in 1926), through a succession of experiments by several teams of scientists who have fixed the concept firmly in science.

Sir Solly Zuckerman. "Hormones." *Scientific American,* March 1957, page 77. For a general survey of the field of hormones, this article is both well written and well illustrated.

E. Adams. "Poisons." *Scientific American,* November 1959, page 76. Some of the most lethal poisons are also important drugs. Adams discusses this relationship as part of a general description of poisons and the way they work. Nerve poisons receive special treatment.

F. Wroblewski. "Enzymes in Medical Diagnosis." *Scientific American,* August 1961, page 99. The detection of cancer before it is a malignancy may be possible by assaying extracellular fluids for enzymes. Wroblewski's article explains how.

PROBLEMS AND EXERCISES

1. Define the following terms.
   (*a*) apoenzyme (*b*) cofactor (*c*) prosthetic group
   (*d*) coenzyme (*e*) holoenzyme (*f*) esterase
   (*g*) lipase (*h*) proteolytic enzyme (*i*) dehydrogenase
   (*j*) vitamin (*k*) absolute enzyme specificity (*l*) substrate
   (*m*) enzyme-substrate complex (*n*) hormone (*o*) endocrine glands
   (*p*) target organ (*q*) chemotherapy (*r*) antimetabolite
2. Outline the relations of apoenzymes, coenzymes, vitamins, minerals, and enzymes.
3. Explain the specificity of enzymes.
4. Explain why enymes have optimum pHs and temperatures.
5. Discuss the factors that help to account for the high rates of enzyme-catalyzed reactions.
6. How might certain hormones influence chemical events inside cells, even though their own molecules are not able to get inside?
7. How do the most dangerous poisons usually work?

CHAPTER TWENTY

Important Fluids of the Body

THE INTERNAL ENVIRONMENT

To live and move and have our being in this world, we must be able to cope with wide fluctuations in outside temperature, humidity, wind velocity, amount of sunshine, and other factors. We also have an internal environment, a *milieu interieur*. Claude Bernard, the great French physiologist of the nineteenth century, was the first to call attention to its importance and to the fact that higher animals are able to exert nearly perfect control over their composition, regardless of the external conditions. The internal environment consists of the extracellular fluids, all the fluids that are not inside cells. They amount to about 20% of the weight of the body. Three-fourths of this is the *interstitial fluid* within which most cells are constantly bathed. Virtually all the remaining one-fourth is blood plasma. Its circulation provides the main link for chemical communication between the internal and the external environments. (Other forms of communication are provided by eyes, ears, nose, and the nerves that are sensitive to heat and touch.) The digestive juices, cerebrospinal fluid, synovial fluid, aqueous and vitreous humor, and the lymph are also "compartments" of the extracellular fluid.

In this chapter we study only certain selected aspects of the internal environment, concentrating for the most part on the digestive juices and the bloodstream. The processes of digestion dismantle the large molecules of the diet into small molecules which are distributed by the bloodstream wherever they are needed. Since we do not eat constantly, traffic from the digestive tract into general circulation has its rush hour periods. Through feast and famine, however, the composition of the bloodstream fluctuates narrowly, for means of storing excess nutrient molecules are available. The concentration of electrolytes and the pH of the blood are kept remarkably constant. For this service we are very much dependent on the work of the kidneys, which we shall study; the formation of urine (diuresis) is an important part of the chemistry of blood. Then we shall study how other chemicals are exchanged between cells and the bloodstream; here we must look at the lymphatic system, which is also important to blood chemistry. Because oxygen is one of the important chemicals delivered to cells, we study oxygen–carbon dioxide transport. Finally, we examine the clotting mechanism that stands guard over tissues enclosing the bloodstream.

DIGESTION

Food is something we seldom take for granted, even when we have plenty. The digestion of food, however, we always take for granted, until something goes wrong with the process. No small amount of contentment depends on the efficiency of chemical systems to digest almost any combination of food and drink. In this section we study the molecular basis of this life process.

THE DIGESTIVE TRACT

The digestive tract is essentially a tube running through the body; its principal parts and organs are located and named in Figure 20.1. The major subdivisions are the mouth, throat, and alimentary canal, which consists of the esophagus, stomach, and intestines. Connected to these at various places are the salivary glands, liver, and pancreas. These manufacture juices that are released into the digestive tract and are needed for the chemical processes of digestion. Other digestive fluids are made in and released from cells in the walls of the stomach or intestines.

When food is taken into the mouth, chewing shreds it to small particles with greater total surface area. Digestive fluids, are therefore given larger areas on which to make their chemical attacks. During chewing saliva lubricates the food while launching a preliminary digestive attack on one substance in food, starch. The tongue then pushes the food back toward the throat where powerful muscles in the throat and esophagus send it downward to the stomach. Protein digestion starts here.The well-churned contents of the stomach are eventually released, portion by portion, through the pyloric valve (Greek *pylorus,* a gate guard) into the duodenum,[1] the first section of the small intestine. The digestion of lipids, carbohydrates, and proteins is completed in the small intestine. As the undigested material moves on, most of the water is reabsorbed into the body through the walls of the first section of the large intestine. This part of the alimentary tract contains a multitude of beneficial bacteria which act on residues passing through and produce some vitamin K and some B-complex vitamins. They also synthesize or modify compounds that give the characteristic odor to the final residue of undigested food, the feces.

The overall processes of digestion are principally those of breaking large molecules into small ones that can be readily absorbed through the intestinal walls into the bloodstream or the lymph system. The chemical changes are quite simple.

[1] Duodenum is from the Latin for "twelve." Herophilus measured this section as the width of twelve fingers. Actually it is about 10 inches long; the total length of the small intestine is about 21 feet.

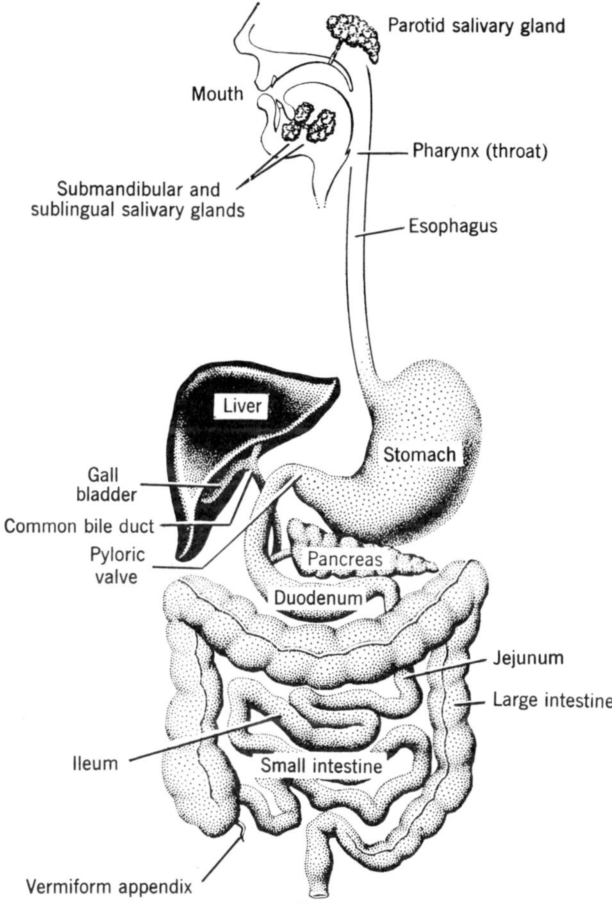
Parotid salivary gland
Mouth
Pharynx (throat)
Submandibular and sublingual salivary glands
Esophagus
Liver
Stomach
Gall bladder
Common bile duct
Pyloric valve
Pancreas
Duodenum
Jejunum
Large intestine
Ileum
Small intestine
Vermiform appendix
Anus

Figure 20.1 Organs of the digestive tract. (Adapted from Eva D. Wilson, Katherine H. Fisher, and Mary E. Fuqua, *Principles of Nutrition,* second edition, John Wiley and Sons, New York, 1965.)

Ester linkages in lipids are hydrolyzed, and so too are amide (peptide) bonds in proteins and acetal-oxygen bridges in di- and polysaccharides. Each of these reactions requires specific enzymes which it is the business of the digestive juices to provide. With the exception of bile, all the digestive juices are dilute solutions of enzymes together with some simple inorganic ions.

DIGESTIVE PROCESSES. DIGESTIVE JUICES AND ENZYMES

We now study the various chemical events of digestion by examining each digestive juice, its components, and its action on foodstuff molecules.

Saliva. The flow of saliva, controlled by nerves, is stimulated by the sight, smell, taste, or even the thought of food. Its flow, on the average, is 1 to 1.5 liters

Table 20.1 Components of Saliva

Nonenzymic substances:	1. Water (99.5%) 2. Mucin, a glycoprotein that gives saliva its characteristic consistency 3. Inorganic ions: Ca^{2+}, Na^+, K^+, Mg^{2+}, $H_2PO_4^-$, Cl^-, and HCO_3^-; average pH of saliva, 6.8 4. Miscellaneous molecules found also in blood and urine: urea, ammonia, cholesterol, amino acids, uric acid
Enzyme:	α-Amylase (ptyalin)

per person per day. Table 20.1 shows its principal components. Mucin functions to lubricate the food. α-Amylase, the principal enzyme in saliva, begins the hydrolysis or digestion of amylose.

Gastric Juice. The chief components of gastric juice, secreted into the stomach, are listed in Table 20.2. It is a mixture of several secretions originating in various glands or areas of the stomach wall. Its hydrochloric acid comes from one group, the parietal cells, its enzymes (or enzyme precursors) from another, the chief cells. An average adult's stomach receives about 2 to 3 liters of gastric juice per day from the ducts of several million gastric glands located in the stomach walls.

The secretion from the parietal cells of gastric glands is a solution that contains $0.16M$ HCl and $0.007M$ KCl plus trace amounts of other inorganic ions and virtually no organic substances. Its flow is stimulated in part by the action of a hormone, *gastrin*, which apparently consists of two active peptides, gastrin I and gastrin II. Its mechanism of action is unknown, but the hormone is released into circulation by mechanical distension of the stomach as well as by the appearance of chemicals (of the foods). When this hormone reaches its target cells (parietal cells), they are made to release their particular contribution to gastric juice. The ability of parietal cells to make hydrochloric acid is quite a remarkable feat, since the final hydrogen-ion concentration of gastric juice is about 10^6 times that of

Table 20.2 Components of Gastric Juice

Nonenzymic substances:	1. Water (99%) 2. Mucin 3. Inorganic ions: Na^+, K^+, Cl^-, $H_2PO_4^-$ 4. HCl (0.5%); average pH, 1.6–1.8
Enzymes:	1. Pepsin, a proteolytic enzyme 2. Gastric lipase*

* A lipase is known to be present in the stomach contents. Whether this enzyme is produced as part of gastric secretions is an unanswered question. Lipases are secreted into the duodenum, and the lipase in the stomach may result from regurgitation of lipase-containing intestinal fluids back through the pylorus. In any event, the optimum pH for this lipase is near neutrality, and it would not be capable of much activity in the relatively high acidity of the stomach. (It may be functional in infants, however, for they have a lower gastric acidity and their important lipid is the highly emulsified fat in milk, a form readily attacked.)

blood plasma. This acid is apparently made by an energy-demanding process such as

$$\text{NaCl} + \text{H}_2\text{CO}_3 \rightleftharpoons \text{HCl (to gastric juice)} + \text{NaHCO}_3 \text{ (to blood plasma)}$$

In vitro the right-to-left reaction is spontaneous. Parietal cells appear to be able to force the left-to-right change.

The chief cells of gastric glands synthesize a polypeptide, *pepsinogen,* in a slightly alkaline medium. This polypeptide is the enzymically inactive precursor for the enzyme *pepsin.* Pepsinogen is converted to pepsin when it encounters the hydrochloric acid from the parietal cells. Once some pepsin forms, it acts to catalyze a more rapid conversion of unchanged pepsinogen to more pepsin. The reaction is believed to consist of a simple hydrolytic splitting of a small segment of the polypeptide chain. The remainder automatically takes up a new secondary or tertiary shape (cf. p. 488) in which the active site of the enzyme is now exposed. No cofactor is involved; therefore pepsinogen is not really an apoenzyme. Rather, it is classified as a *zymogen,* a type we shall encounter elsewhere.

Pepsin, a proteolytic enzyme, catalyzes the hydrolysis of peptide bonds in proteins, the digestion of which commences in the stomach. The optimum pH for its activity is in the neighborhood of 2, and at pHs higher than 6 it is readily denatured. Thus when pepsin together with the contents of the stomach moves into the duodenum (pH = 6.0 to 6.5 during digestive activity), its enzymic work ceases and other proteolytic enzymes take over. Pepsin has its best specificity for amide linkages when an amino acid with an aromatic side group furnishes the amino group for the bond. In contrast with some of the proteolytic enzymes, pepsin can help cleave peptide bonds in the interior of a protein chain. Others, to be studied shortly, favor terminal amino acid groups only. The action of pepsin therefore generates intermediate breakdown products of proteins with varying chain lengths; their names are given in the following word equation:

$$\text{Proteins} \rightarrow \text{proteoses} \rightarrow \text{peptones} \rightarrow \text{polypeptides} \rightarrow \text{amino acids}$$

Fragments become progressively smaller ⟶

Proteoses, peptones, and polypeptides are the principal products of peptic digestion. A theory proposing to explain how pepsin works is described in Figure 20.2.

Because the membranes constituting the digestive tract are themselves made up of proteins, we may wonder why they are not digested. In most people digestion of the stomach or intestinal walls obviously does not occur. But for a few it does, and they have ulcers. The defensive mechanism has several features. In addition to the chief and parietal cells, the gastric glands possess a third type, the *mucous cells.* These make and secrete a thick viscous fluid containing a mucoprotein, a complex between a polysaccharide and a protein. Other cells in the inner surface layer of the stomach also secrete a "mucus," which acts to coat the stomach walls. Since the mucus is constantly replenished, it is only slowly digested and has both acid-binding and peptic-inhibitory abilities. It therefore helps to keep enzymes away from the walls of the stomach. (A similar mucus is found in the intestines.) Should the concentration of hydrochloric acid in the stomach become and remain too high

Step 1. Two neighboring carboxyl groups on side chains of the pepsin molecule interact to form the more reactive anhydride system (page 420).

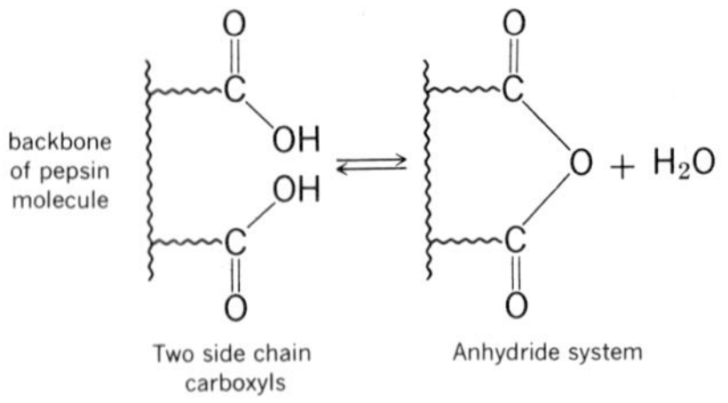

backbone of pepsin molecule

Two side chain carboxyls

Anhydride system

Step 2. A molecule of protein substrate, assisted by the geometrical and electrical factors described in the lock and key theory of enzymic action, collides with the anhydride unit of the enzyme.

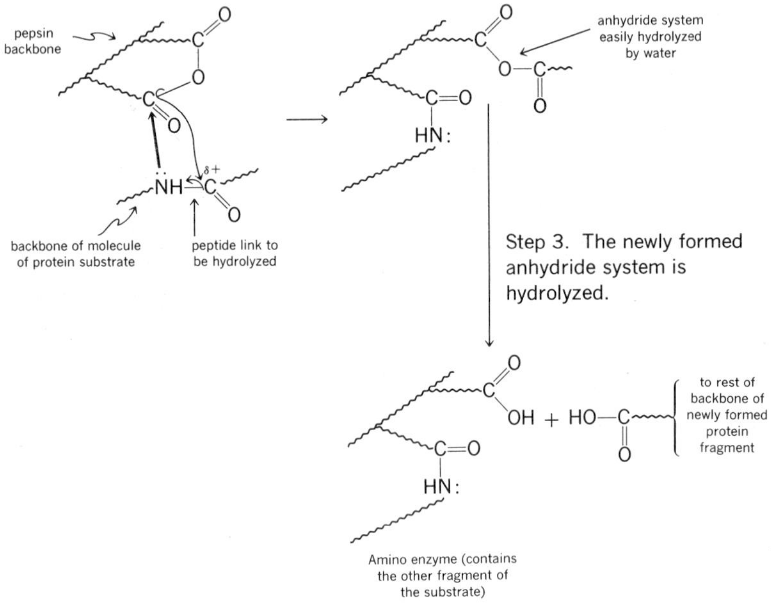

pepsin backbone

backbone of molecule of protein substrate

peptide link to be hydrolyzed

anhydride system easily hydrolyzed by water

Step 3. The newly formed anhydride system is hydrolyzed.

to rest of backbone of newly formed protein fragment

Amino enzyme (contains the other fragment of the substrate)

Step 4. The "amino enzyme" unit has a free carboxyl group as a close neighbor. Cleavage of the amino group is relatively easy in such systems, and the mechanism might be something like the following.

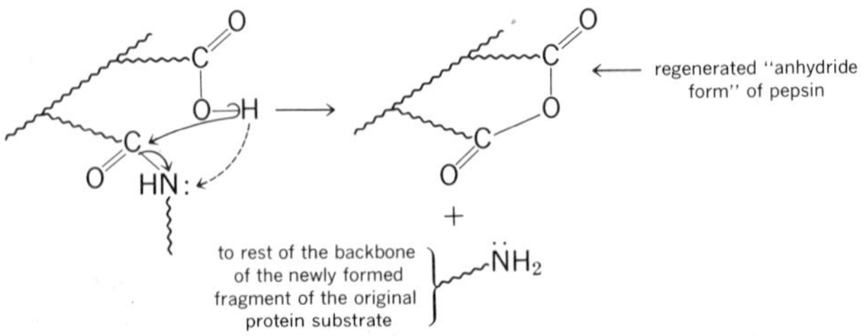

regenerated "anhydride form" of pepsin

to rest of the backbone of the newly formed fragment of the original protein substrate

Figure 20.2 Proposed mechanism for the action of pepsin. (From M. L. Bender and F. J. Kezdy, *Annual Review of Biochemistry,* Vol. 34, 1965, page 63.)

(*hyperacidity*), the mucus itself is more rapidly digested, and the stomach walls are exposed to digestive action. If this goes too far, the wall is perforated. Perforation may happen to occur at a blood vessel, producing serious hemorrhaging which can be fatal.

The secretion of hydrochloric acid is controlled in part by the action of another hormone, *enterogastrone*. When food leaves the stomach and enters the duodenum, cells in its walls are stimulated to release this substance (which may consist of two polypeptides) into the bloodstream. When circulation has brought enterogastrone to the gastric glands, it acts to shut down their secretion of hydrochloric acid. Thus a regulatory mechanism controls the secretion of gastric juice. When it is needed, it is produced. When the need diminishes, its flow is reduced.

The churning, semidigested mixture of food and fluids present in the digestive tract is called *chyme* (Latin *chymus*, juicelike). Over a two- to five-hour period the gastric chyme is released in portions through the pyloric valve into the duodenum where several enzyme-rich secretions complete the digestion of food.

Clinical Importance of Gastric Analysis. After stimulating the secretion of gastric juice by feeding the subject a meal or alcohol or by injecting histamine, samples of gastric juice are withdrawn at about fifteen-minute intervals and analyzed for both free and combined acid. The results may reveal disorders that are very remote from the stomach. A few are listed.

1. Achlorhydria: the absence of free acid. Although this condition is rare, if it is discovered and if pepsin is also absent, pernicious anemia or stomach cancer is indicated.

2. Hypoacidity (hypochlorhydria): too low a concentration of acid. Only infrequently is this condition serious, but it can indicate stomach cancer, chronic constipation, or inflammation of the stomach (chronic gastritis).

3. Hyperacidity (hyperchlorhydria): an excess of acid. Relatively more common, this condition is associated with chronic "heartburn," chronic "indigestion," gastric or duodenal ulcers, or even inflammation of the gallbladder (cholecystitis).

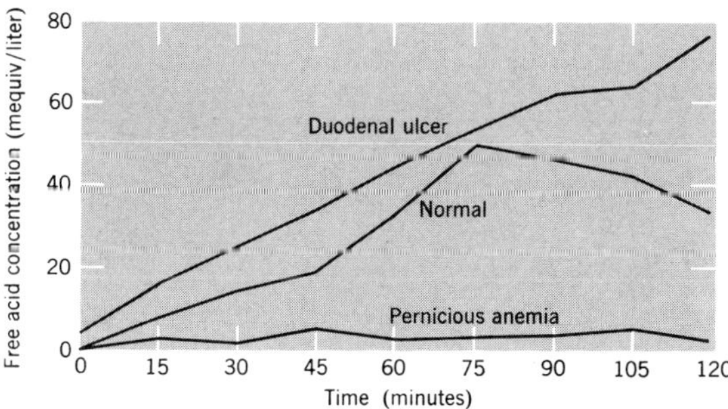
Free acid concentration (mequiv/liter)
80
60
40
20
0
Duodenal ulcer
Normal
Pernicious anemia
0
15
30
45
60
75
90
105
120
Time (minutes)

Figure 20.3 Gastric analysis. The graph shows for three situations how the concentration of free acid varies with the time lag between the initial stimulation of gastric secretion and the withdrawal of a sample for analysis. (From A. White, P. Handler, and E. L. Smith, *Principles of Biochemistry,* third edition, McGraw-Hill Book Company, New York, 1964, page 712. Used by permission.)

Figure 20.3 shows how the concentration of free acid varies with time in the normal individual, in the patient with pernicious anemia, and in the patient who has a duodenal ulcer.

Pancreatic Juice. The pancreas produces two secretions. The internal secretion which contains insulin is discharged into the bloodstream. The external secretion, pancreatic juice, empties into the duodenum. Its principal components are listed in Table 20.3.

The flow of pancreatic juice and its composition are regulated by two hormones, *secretin* and *pancreozymin.* Both are released into the bloodstream from cells of the upper intestinal mucosa whenever it is exposed to certain chemicals in the chyme. Release of secretin is prompted primarily by the presence of an acid that comes into the duodenum from the stomach. When molecules of this hormone reach the pancreas, they stimulate an increase in the volume flow and the electrolyte content (particulary HCO_3^-) of pancreatic juice. Secretin does not itself stimulate the release of either digestive enzymes or zymogens. Their release requires the cooperation of the second hormone, pancreozymin, which is secreted when chemicals originating from ingested food appear in the duodenum.

Part of the enzymic activity of pancreatic juice requires the preliminary action of substances in bile and intestinal juice. Our discussion of the enzyme-catalyzed reactions in the upper intestinal tract will therefore be continued after these other digestive juices have been described.

Intestinal Juice (*Succus entericus*). Mucosal glands of the duodenum secrete this digestive juice whenever chyme enters from the stomach. The regulation of its flow is not very well understood, for it appears to be stimulated in more than one way, including hormonal action. The principal components of intestinal juice are listed in Table 20.4.

The enzyme-rich intestinal juice is important to the digestion of all foods. Various peptidases complete the hydrolysis of proteins to amino acids. Sucrase, maltase, and lactase handle conversion of the important disaccharides to glucose, fructose, and galactose. Simple lipids are acted upon by a lipase; more complex

Table 20.3 Components of Pancreatic Juice

Nonenzymic substances:	1. Water; average daily volume, 500–800 cc 2. Inorganic ions: Na^+, K^+, Ca^{2+}, Cl^-, HCO_3^-, HPO_4^{2-}; average pH, 7–8*
Enzymes:	1. Pancreatic lipase ("steapsin") 2. α-Amylase 3. Maltase 4. Ribonuclease
Zymogens:	1. Trypsinogen 2. Chymotrypsinogen 3. Procarboxypeptidase

* Under physiological conditions the contents of pancreatic juice will not function in a slightly alkaline fluid. The mixing of acidic chyme from the stomach with this juice plus bile and intestinal juice produces a neutral or slightly acidic material.

Table 20.4 Components of Intestinal Juice

Nonenzymic substances:	1. Water 2. Inorganic ions: Na^+, K^+, Ca^{2+}, Cl^-, HCO_3^-, HPO_4^{2-}
Enzymes:	1. Aminopeptidase 2. Dipeptidase 3. Nucleases, nucleotidase, nucleosidase 4. Maltase 5. Sucrase 6. Lactase 7. Intestinal lipase 8. Lecithinase 9. Phosphatase 10. Enterokinase

lipids are hydrolyzed by the catalytic action of lecithinase and phosphatase. Nucleic acids (cf. Chapter 25) respond to enzymes peculiar to their hydrolytic needs.

One of the most important enzymes found in intestinal juice is *enterokinase,* whose major function is to convert trypsinogen to trypsin. Once trypsin is formed, it serves to activate the other zymogens in pancreatic juice. Figure 20.4 presents

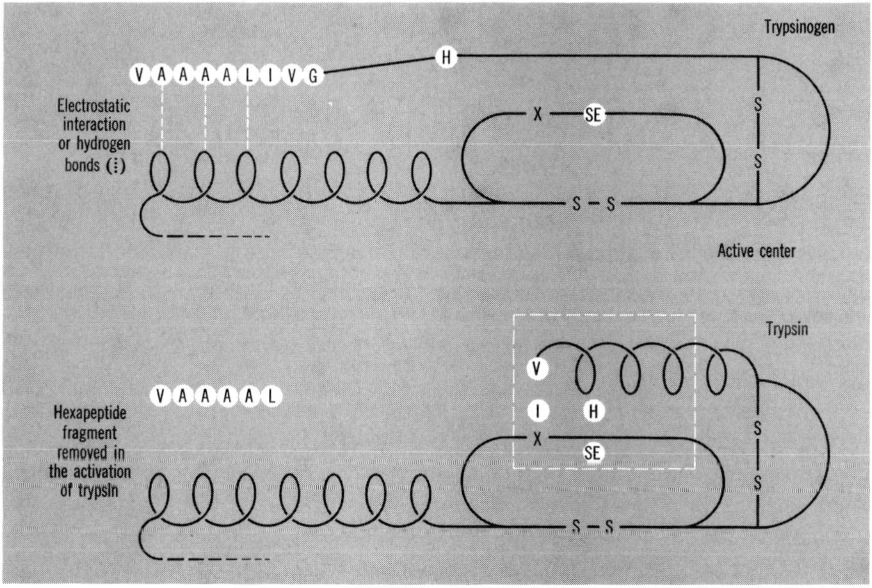
Trypsinogen
V A A A A L I V G
H
Electrostatic interaction or hydrogen bonds (⁝)
X
SE
S
S
S – S
Active center
Trypsin
V A A A A L
V
I
H
X
SE
Hexapeptide fragment removed in the activation of trypsin
S
S
S – S

Figure 20.4 The structural changes involved in the conversion of trypsinogen to trypsin. When the small hexapeptide is split off, the remainder of the chain at this end of trypsinogen automatically adopts a more helical configuration, permitting the histidine unit (H) to come close to the serine unit (SE), a juxtaposition required for enzymic activity. A, aspartic acid unit; G, glycine unit; H, histidine unit; I, isoleucine unit; SE, serine unit; V, valine unit; X, active site. (From H. Neurath, "Protein Structure and Enzyme Action," in *Biophysical Science—A Study Program,* edited by J. L. Oncley, John Wiley and Sons, New York, 1959, page 189.)

a theory to explain the activation of trypsin. Thus for pancreatic juice to be effective in the duodenum, the participation of intestinal juice is needed.

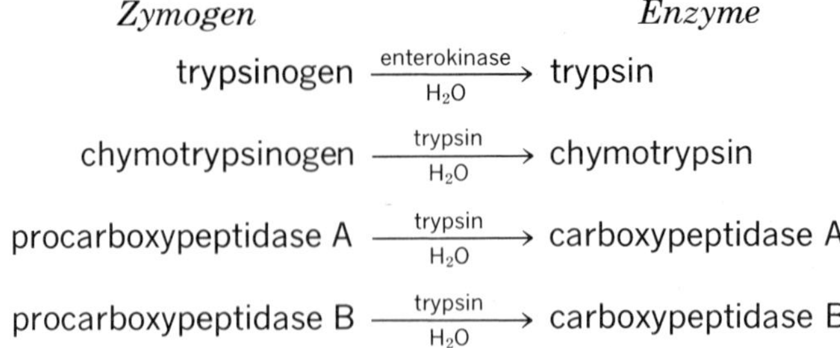
Zymogen
Enzyme
trypsinogen
enterokinase
H2O
trypsin
chymotrypsinogen
trypsin
H2O
chymotrypsin
procarboxypeptidase A
trypsin
H2O
carboxypeptidase A
procarboxypeptidase B
trypsin
H2O
carboxypeptidase B

Some of the enzymic influences of the intestinal juice are exerted within the cells of the duodenal mucosa rather than in the duodenum itself. If nutrient molecules that are not completely hydrolyzed begin to dialyze through the intestinal walls, their hydrolysis is completed during the journey by the action of the intestinal juice enzymes.

Bile. Bile, the third juice that empties into the duodenum, contains no digestive enzymes. Nevertheless, it is vitally important to digestion, for it contains *bile salts* whose fat-emulsfying action is essential to rapid digestion of lipids. The chief problem in lipid digestion is that lipids are notoriously water-insoluble. They tend to form large globules, and any digestive attack on their molecules can occur only at the surface of the globule. The bile salts, structures for two of which are shown in Figure 20.5, have all the structural features of soaps, that is, large hydrocarbon-like portions joined to highly polar (ionic) ends (cf. p. 465). Emulsification of an oil, of course, means that large globules are broken into many times their number of smaller globules. Although the total amount of oil is obviously the same, the total surface area of all the globules is enormously increased, and lipases plus water can more readily hydrolyze the lipid material. Some scientists believe that the bile soaps act primarily to emulsify fats, although this subject is a matter of controversy in a few circles. Bile soaps may also be needed for washing fatty coatings from partly digested particles of other foods.

Bile also contains *bile pigments* (from the breakdown of hemoglobin) and cholesterol. If the concentrations of cholesterol and bile salts in bile become too high while they are in the gallbladder, they may precipitate to form gallstones. When these become large enough, they can plug the duct from the gallbladder to the duodenum. Usually the only remedy is to remove the gallbladder, and those who have had such an operation must henceforth exercise careful control over the lipid content of their diet.

The flow of bile is regulated by a hormone, *cholecystokinin,* which is synthesized in cells of the upper intestinal mucosa. Whenever chyme that contains fats or fatty acids enters the duodenum, this hormone is released into general circulation. When its molecules arrive at the gallbladder, they cause a powerful contraction of its muscles, forcing bile out of it. Details of the mechanism are not known.

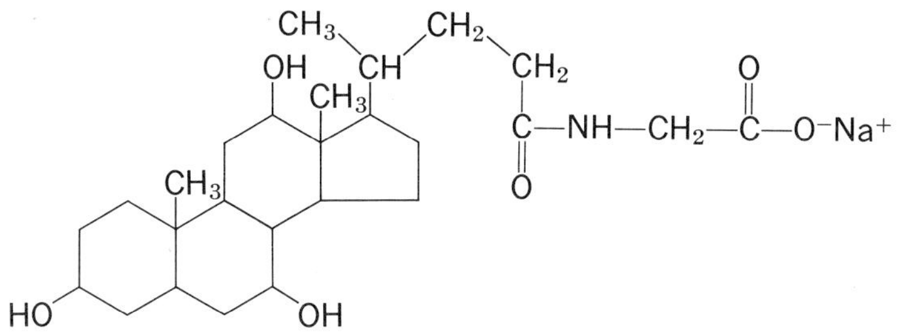

Sodium salt of glycocholic acid

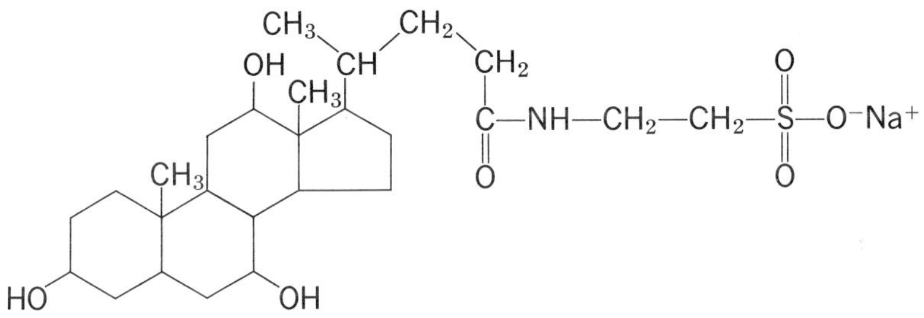

Sodium salt of taurocholic acid

Figure 20.5 Two of the more common bile salts that act as lipid-emulsifying soaps in the duodenum.

Digestion in the Upper Intestinal Tract

1. **Proteolytic digestion.** The digestion of proteins begins in the stomach, but gastric chyme that enters the duodenum can still contain an abundance of large protein molecules. Trypsin and chymotrypsin are the two enzymes chiefly responsible for handling such materials. The smaller fragments of proteins that they produce are further degraded by several specialized proteolytic enzymes, carboxypeptidases, tripeptidase, and dipeptidase. Table 20.5 contains a summary of the specificity of these enzymes. With a variety of proteolytic enzymes available, it is apparent that almost any protein of whatever unique or unusual amino acid sequence can be digested, provided the amino acids belong to the L-family. The principal lack in human beings is an enzyme for digesting keratins. These are rich in disulfide linkages for which man has no digestive enzyme.

2. **Carbohydrate digestion.** Pancreatic juice contains an amylase that acts on acetal linkages in amylose and amylopectin molecules that have escaped hydrolysis in the mouth. The end product is maltose, which is hydrolyzed to glucose by the action of maltase. (Mentioned earlier, p. 546, were the actions of lactase and sucrase.)

3. **Lipid digestion.** The pancreatic lipase, steapsin, is the most important fat-splitting enzyme in the digestive tract. It catalyzes the hydrolysis of the ester linkages in triglycerides formed originally from the primary hydroxyls of glycerol.

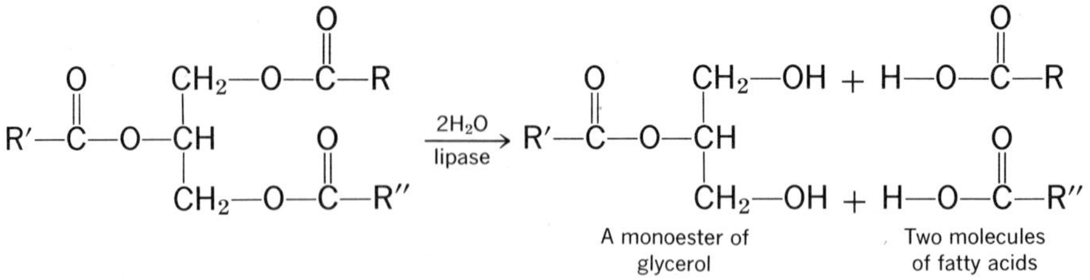

$$\underset{\text{triglyceride}}{\mathrm{R'{-}\overset{O}{\overset{\|}{C}}{-}O{-}CH(CH_2{-}O{-}\overset{O}{\overset{\|}{C}}{-}R)(CH_2{-}O{-}\overset{O}{\overset{\|}{C}}{-}R'')}} \xrightarrow[\text{lipase}]{2H_2O} \underset{\text{A monoester of glycerol}}{\mathrm{R'{-}\overset{O}{\overset{\|}{C}}{-}O{-}CH(CH_2{-}OH)(CH_2{-}OH)}} + \underset{\text{Two molecules of fatty acids}}{\mathrm{H{-}O{-}\overset{O}{\overset{\|}{C}}{-}R + H{-}O{-}\overset{O}{\overset{\|}{C}}{-}R''}}$$

One of the ester linkages survives, and there is evidence that the successful mobilization of lipid material out of the intestinal tract does not require further digestion

Table 20.5 Specificity of Proteolytic Enzymes

Enzyme	Preferred Site of Action*	
Endopeptidases (act on nonterminal sites)		
Pepsin	$-\overset{O}{\overset{\|}{C}}-NH-\overset{Ar'}{\overset{\vert}{CH}}-\overset{O}{\overset{\|}{C}}\wr NH-\overset{Ar}{\overset{\vert}{CH}}-$	Ar′ or Ar = side chain from a tyrosine or phenylalanine unit
Trypsin	$-\overset{O}{\overset{\|}{C}}-NH-\overset{NH_3^+-G}{\overset{\vert}{CH}}-\overset{O}{\overset{\|}{C}}\wr NH-$	Positively charged side chain from arginine or lysine
Chymotrypsin	$-\overset{O}{\overset{\|}{C}}-NH-\overset{Ar}{\overset{\vert}{CH}}-\overset{O}{\overset{\|}{C}}\wr NH-$	Ar = an aromatic side chain
Exopeptidases (act on terminal amino acid units)		
Carboxypeptidase A	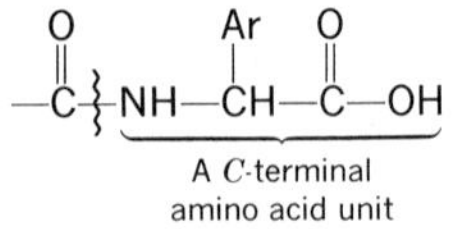 $-\overset{O}{\overset{\|}{C}}\wr \underbrace{NH-\overset{Ar}{\overset{\vert}{CH}}-\overset{O}{\overset{\|}{C}}-OH}_{\text{A } C\text{-terminal amino acid unit}}$	Ar = an aromatic side chain
Carboxypeptidase B (aminopeptidase)	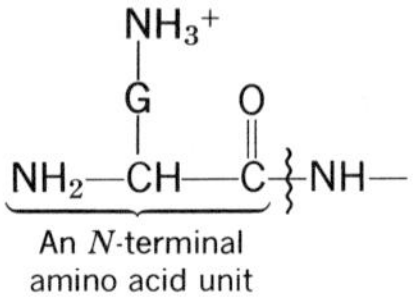 $\underbrace{NH_2-\overset{NH_3^+-G}{\overset{\vert}{CH}}-\overset{O}{\overset{\|}{C}}}_{\text{An } N\text{-terminal amino acid unit}}\wr NH-$	Positively charged side chain from arginine or lysine
Prolidase	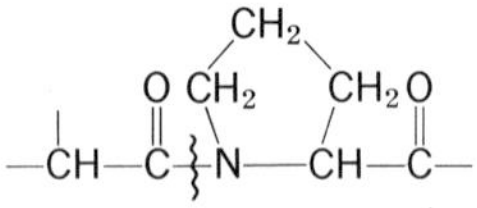 $-\overset{\vert}{CH}-\overset{O}{\overset{\|}{C}}\wr N-CH-\overset{O}{\overset{\|}{C}}-$ (N and CH joined in a ring through $CH_2-CH_2-CH_2$)	
Tripeptidase	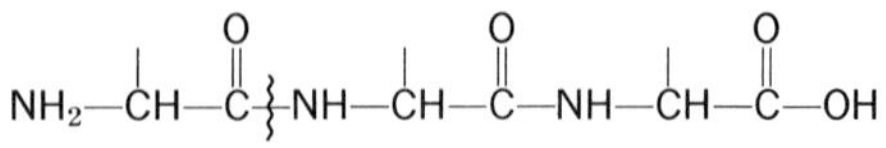 $NH_2-\overset{\vert}{CH}-\overset{O}{\overset{\|}{C}}\wr NH-\overset{\vert}{CH}-\overset{O}{\overset{\|}{C}}-NH-\overset{\vert}{CH}-\overset{O}{\overset{\|}{C}}-OH$	
Dipeptidase	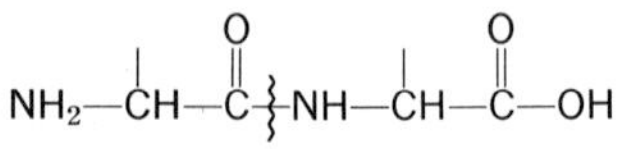 $NH_2-\overset{\vert}{CH}-\overset{O}{\overset{\|}{C}}\wr NH-\overset{\vert}{CH}-\overset{O}{\overset{\|}{C}}-OH$	

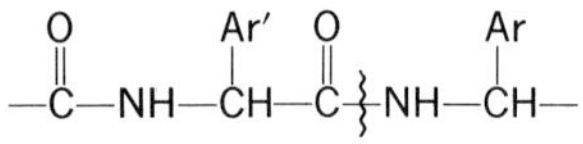

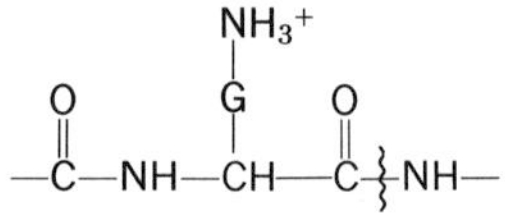

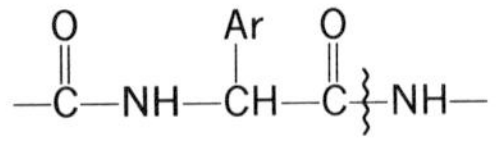

* A vertical wavy line shows where the hydrolysis occurs.

of this monoester. The fatty acids plus the monoester may be absorbed into cells of the intestinal mucosa where another lipase acts rapidly to finish the hydrolysis of the last ester linkage, provided it is of a short-chain fatty acid. If it is not, lipid metabolism in these cells takes what might seem to be a surprising turn. Reesterification occurs. The newly formed glycerides enter circulation via the lymph system, and when they or fatty acids reach the bloodstream, they are believed to be picked up by circulating protein molecules and are transported as lipoprotein complexes.

BRIEF SUMMARY OF THE DIGESTION OF THE THREE BASIC FOODS

Carbohydrates. Starch and the three disaccharides, maltose, lactose, and sucrose, are the chief carbohydrates in nutrition. Digestion of cooked starch begins in the mouth, where α-amylase is present, and continues for a time in the stomach until the hydrochloric acid in the gastric juice inactivates the α-amylase that has accompanied swallowed food. α-Amylase is not considered active toward uncooked starch. Dextrins and maltose are the end products of the work of this enzyme.

The disaccharides probably undergo some nonenzymic hydrolysis in the stomach, where the action of hydrochloric acid hydrolyzes acetal linkages in these molecules. This hydrolysis is not of great importance, however.

The amylase in pancreatic juice completes the hydrolysis to maltose of both raw and cooked starch. The three carbohydrases, maltase, lactase, and sucrase, that appear in the duodenum complete the digestion of the disaccharides. *The end products of carbohydrate digestion then are fructose, galactose, and glucose.*

Cellulose in the diet (e.g., in leafy vegetables) passes through essentially unchanged, furnishing much of the bulk needed for efficient operation of the excretory system in the lower intestinal tract.

Lipids. For all practical purposes lipids undergo no digestive action until they reach the duodenum where bile salts rapidly emulsify them. Once emulsified, pancreatic lipase smoothly hydrolyzes lipids to *glycerol, fatty acids, and monoglycerides, the end products of the digestion of the neutral fats and oils.* The lipase of the intestinal juice acts on lipids that escape hydrolysis catalyzed by steapsin. Reesterification of glycerol begins during absorption.

Proteins. Simple proteins are not acted on in the mouth. In the stomach pepsin catalyzes the conversion of proteins to proteoses and peptones. Trypsin and chymotrypsin, in the duodenum, act rapidly on proteoses and peptones (and, to a certain extent, on native proteins), converting them to simple peptides and amino acids. Peptidases in intestinal juice and pancreatic juice complete the digestion of proteins. *The end products of protein digestion are amino acids and, perhaps, some simple soluble di- and tripeptides.*

BLOOD

FUNCTIONS

A schematic diagram of the circulatory system is given in Figure 20.6. The functions of the bloodstream are quite well known. It transports oxygen, other nutrients, and waste products, thereby establishing communication between the external and the internal environments. It also transports substances from one site to another within the body, for example, moving fatty acids from adipose tissue to

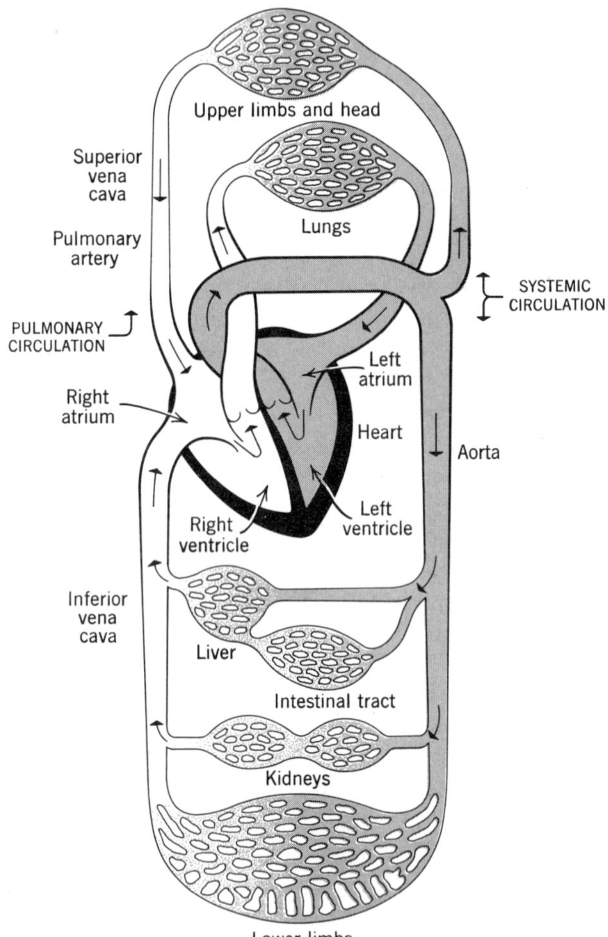
Upper limbs and head
Superior vena cava
Lungs
Pulmonary artery
PULMONARY CIRCULATION
SYSTEMIC CIRCULATION
Left atrium
Right atrium
Heart
Aorta
Right ventricle
Left ventricle
Inferior vena cava
Liver
Intestinal tract
Kidneys
Lower limbs

Figure 20.6 The circulatory system in man. Shaded areas designate oxygenated blood. Blood received from the veins at the right atrium of the heart passes into the right ventricle, and from there it is forced through the spongy capillary networks of the lungs. In the capillaries carbon dioxide is removed and oxygen is taken up. The blood then goes to the left atrium for delivery to the left ventricle and on to the rest of the body.

the liver, and moving freshly made glucose and ketone bodies, from the liver to tissues that use them. The bloodstream is important in heat exchange. By bringing some of the flowage through capillaries near the skin, heat can be pumped out of the system. The body's defensive mechanisms, which guard against infectious diseases as well as physical damage to the tissues that enclose the moving blood, depend greatly on substances carried by it. To meet all these needs, the blood contains a host of specialized chemicals and cells.

COMPOSITION

Blood consists of a solution called *plasma* in which are dispersed *formed elements* or cellular bodies: erythrocytes (red blood cells), leukocytes (white blood cells), and thrombocytes (blood platelets). *Blood serum* is what is left of the plasma after the formed elements and fibrinogen (the clotting protein) have been separated. So-called *defibrinated blood* is whole blood from which this clotting protein has been removed; the cellular bodies are still present.

Formed Elements. Oxygen transport is the most important function of erythrocytes. The blood of a normal adult contains about five million of these red cells per cubic millimeter of blood. (A drop of blood has a volume of about 20 to 30 cubic millimeters.) Variations in the concentration of red cells can be measured ("red cell count") and have diagnostic value. In anemia, for example, the red cell count is abnormally low.

Leukocytes, or white blood cells, number from 5 to 10,000 per cubic millimeter of blood. Several varieties are recognized. One, the polymorphonuclear leukocyte, attacks invading bacteria and other foreign bodies. Another, the lymphocyte, is involved in the synthesis of antibodies and the fixation of toxins. The fatal disease of the blood-forming organs, leukemia, is accompanied by huge increases in the number of leukocytes in the blood. Thus the white cell count is also an important diagnostic tool.

Thrombocytes or platelets are important in blood clotting.

Blood Plasma. The average adult has about 5 liters of blood, 55% of which is plasma. The erythrocytes make up another 42 to 43%. The white cells and the platelets are each present in trace percentages. The other principal components of blood are various proteins; upward of seventy have been identified, and more are believed to be present. The proteins have been classified according to certain solubility relations, certain properties that make rough separations of these proteins possible. When plasma, for example, is mixed with a saturated solution of ammonium sulfate, some of the proteins precipitate, and they are called the *globulins*. After these have been removed, addition of solid ammonium sulfate causes another batch of proteins to precipitate, the *albumins*. It may seem altogether arbitrary to base a system for classifying proteins on such a procedure, but the proteins in each class do share common functional purposes.

One of the important functions of albumins, which account for about 55% of the total protein in blood, is the regulation of osmotic pressure. Another general function is to act as transporting agents for other relatively insoluble chemicals.

Globulins constitute about 40% of the total protein in blood. Three general types, α-globulin, β-globulin, and γ-globulin, are recognized; each type consists of several different proteins. The γ-globulins are important in the body's defense against infectious disease. The other globulins appear to be necessary for the transport of metal ions such as Fe^{2+} and Cu^{2+}, which would otherwise be insoluble in the slightly basic medium of blood.

Fibrinogen, the third major protein found in plasma, makes up about 5 to 6% of the total protein. It is the immediate precursor of fibrin, a plasma-insoluble protein that precipitates as a tangled brush heap of molecules in a developing blood clot.

Electrolytes constitute another general type of plasma solute. Inorganic ions are of course found in all body fluids; Figure 20.7 shows the kinds and concentrations present in plasma, interstitial fluid, and cell fluid. The distribution of ions in cell fluid is by no means uniform, however. Localized pockets of higher or lower concentrations are found in the cell. The most significant difference to be noted in the data of Figure 20.7 is that inside the cell the ions K^+, Mg^{2+}, and $SO_4{}^{2-}$ and various phosphate ions predominate. Outside the cell in plasma or interstitial fluid the ions Na^+, Cl^-, and $HCO_3{}^-$ predominate. Furthermore, the concentration of large molecules, proteins, is much greater inside a cell than outside.

Osmotic pressure, we have learned, is related to particle concentration without substantial regard to the sizes of these particles or the charges on them. Since cells have walls that are freely permeable to water molecules, and since cells normally neither burst nor shrink, it must be true that the osmotic pressure inside a cell is the same as that outside. Therefore any water that migrates out of a cell is replaced by water that moves in. The effective particle concentration is the same inside a cell as ouside, but the kinds of particles (sodium or potassium ions, etc.) are different. If the fluid is locally diluted either inside or outside a cell (the second being the more common problem), ions are moved from one side to another to minimize the effects of the imbalance. Which ions are moved is kept under control. Normally, for example, sodium ions would not move from the inside of a cell (low concentration) to the outside (high concentration). We should expect the reverse. A similar statement could be made about potassium ions. To make the separation of these ions possible, as indicated by the data in Figure 20.7, the cell must do work. Energy-demanding chemical processes take place, in effect to pump sodium ions in one direction. If the sodium ions are *actively* transported by chemical processes that use energy to pass them to the outside, some positive ions must take their place. Potassium ions do, and they need not be actively transported. Passive diffusion will accomplish it. For example, if red cells, at the temperature of an ice bath, are kept in contact with saline solution containing glucose, they lose their potassium ions and sodium ions move in until a general distribution of these ions has occurred. If the system is warmed to body temperature (37°C), sodium ions are pumped out of the red cells and potassium ions diffuse back until the two are for the most part separated. If the glucose is omitted, when the system is warmed to body temperature, the *active transport* of sodium ions outside and potassium ions inside does not occur. Then if the glucose is added, sodium ions move out and potassium ions diffuse back. Glucose is apparently necessary to the

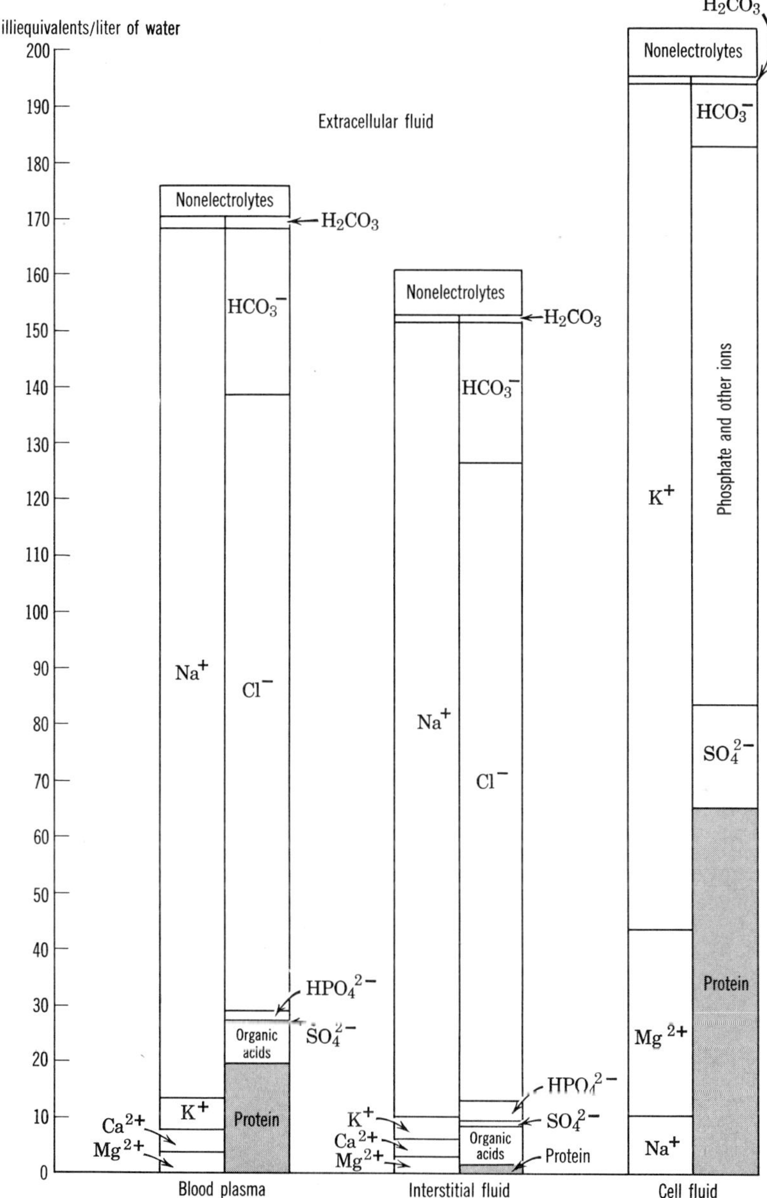

Milliequivalents/liter of water
200
190
180
170
160
150
140
130
120
110
100
90
80
70
60
50
40
30
20
10
0
Extracellular fluid
Nonelectrolytes
H2CO3
HCO3−
Na+
Cl−
HPO42−
SO42−
Organic acids
K+
Protein
Ca2+
Mg2+
Blood plasma
Interstitial fluid
Cell fluid
Phosphate and other ions

Figure 20.7 Electrolyte composition of body fluids. (Adapted, by permission, from J. L. Gamble, *Chemical Anatomy, Physiology and Pathology of Extracellular Fluid,* sixth edition, Harvard University Press, Cambridge, Mass., 1954.)

process, probably as a source of ATP or some other high-energy compound needed to supply the energy for moving and keeping the system away from the ionic equilibrium. What we have described about red cells is true for most types of cells. Significant portions of their energy needs are for active transport of certain substances against an *electrochemical gradient*—from a region of low to high concentration of charged particles. It is apparent that cells do a great deal of work not directly connected with such obvious functions as muscular exercise, thinking, and the like. It is also obvious that the delivery of a fresh supply of chemicals to the bloodstream from digested food could alter the sensitive balance of the concentrations of electrolytes. The cells of one organ, the kidneys, therefore perform specialized tasks in regulating the volume of water left in the bloodstream and its concentrations of ions and other small molecules. Whether or not the bloodstream does its job depends in large measure on the work of the kidneys, which we shall describe next.

URINE

DIURESIS

The whole process of urine formation is called diuresis (Greek *ouresis,* to make water or urine). The kidneys, in cooperation with certain hormones, exert control over the concentrations of electrolytes, including those affecting the acid-base balance of the blood. It will be helpful to our understanding of what kidneys do to know something about what they are.

The kidneys are located slightly above the waist, behind the other organs on either side of the spinal column. Into each a single large artery and a single large vein are connected. While in the kidneys the blood undergoes a filtering process of immense proportions. Liquid leaves the bloodstream and returns at a rate of 130 ml per minute or about 190 liters per day. Only about 1 to 2 liters is converted into urine. The blood fluid that temporarily leaves the bloodstream carries with it sodium chloride (at the rate of over 1 kg per day), sodium bicarbonate (about 0.5 kg per day), glucose (150 g per day), amino acids, other ions (notably phosphates), and other small molecules. Virtually all this material reenters the bloodstream. Very little remains in the urine being formed. To accomplish such feats of filtration, enormous reabsorption surface areas are required, and these are provided by the successive branching of the circulatory system into smaller and smaller tubes until the individual filtration and reabsorption units, the *nephrons,* are reached. Each kidney has about a million of them, and their principal parts are described in Figure 20.8. Arterial blood finally reaches a tuft of parallel capillaries called the *glomerulus;* here the blood water together with dissolved ions and small molecules leaves the bloodstream and enters the *renal capsule.* This liquid is now called the *glomerular filtrate.* The cellular elements, plasma proteins, and lipids remain behind in the blood. The glomerular filtrate passes first through the *proximal tubule* where about seven-eighths is reabsorbed by the capillary beds interlacing the nephron. In this manner nearly all the electrolytes, all the blood sugar, and most of the water

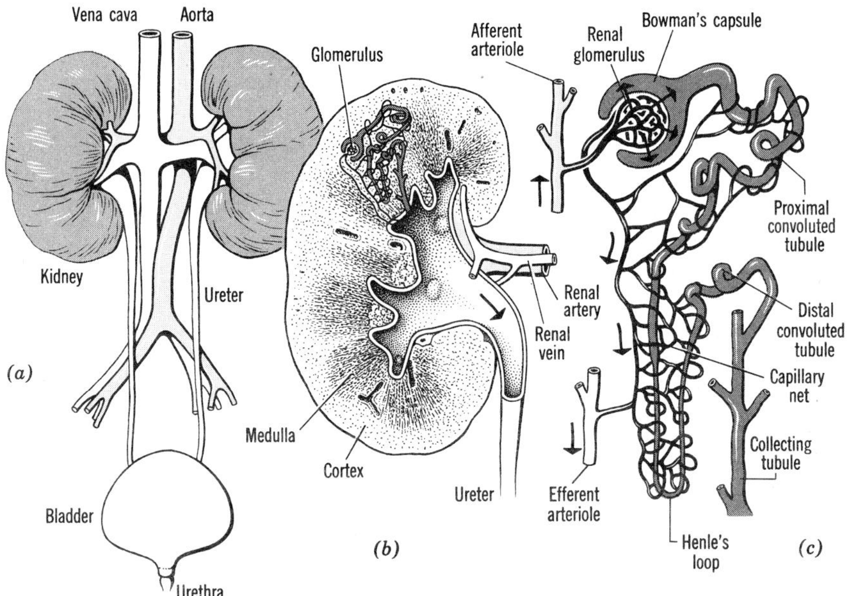
Vena cava
Aorta
Kidney
Ureter
(a)
Bladder
Urethra
Glomerulus
Medulla
Cortex
Ureter
(b)
Afferent arteriole
Renal glomerulus
Bowman's capsule
Renal artery
Renal vein
Proximal convoluted tubule
Distal convoluted tubule
Capillary net
Collecting tubule
Efferent arteriole
Henle's loop
(c)

Figure 20.8 Operating units of the kidneys: (*a*) principal parts, (*b*) cross section of the kidney, (*c*) nephron. (Adapted, with permission, from G. E. Nelson, G. G. Robinson, and R. A. Boolootian, *Fundamental Concepts of Biology,* John Wiley and Sons, New York, 1967, page 131.)

are soon returned to circulation. In the region of the proximal tubule *obligatory reabsorption* is said to occur. Urea, uric acid, and creatinine tend to remain in the urine being formed. The remaining one-eighth of the glomerular filtrate may or may not be reabsorbed farther along the line, in the *distal tubule.* In other words, distal reabsorption is *optional reabsorption.* The exercise of this option is regulated by an antidiuretic hormone called *vasopressin.*

When vasopressin is present, optional absorption occurs, as much as 100%. The volume of urine made is therefore small. When vasopressin is absent, optional absorption does not occur and relatively larger amounts of urine are made. Vasopressin is secreted by cells of the posterior, or neural, region of the hypophysis, a small endocrine gland lying under the brain. These cells respond to the osmotic pressure of the blood in the following ways.[2]

If the osmotic pressure of the blood goes up, the hypophysis secretes vasopressin, which means that optional reabsorption occurs and the water initially in the blood is kept there. If the osmotic pressure of the blood is high, the system acts to keep it from going higher, as it would if water were removed. The thirst mechanism is now also stimulated to encourage drinking, thus bringing in more water.

[2] Since the concept of osmotic pressure will be used frequently in this chapter, it would be well to review it; see pages 157–160. In general, a concentrated solution has a high osmotic pressure, a dilute solution a low osmotic pressure.

If the osmotic pressure of the blood goes down, little or no vasopressin is secreted. Hence optional reabsorption tends not to occur, and at least some of the water taken out of the bloodstream in the glomerulus is kept out and voided as urine. By removing this water from the blood, the concentration of the remaining blood rises. These changes in osmotic pressure of the blood need be no more than 2% up or down for the hypophysis to act to restore the osmotic pressure to normal. If the pressure goes up, the kidneys do not make as much urine, water is conserved, and the thirst mechanism is stimulated to bring in more water. If the osmotic pressure goes down, water is removed. The hypophysis acts as the "thermostat" for this regulation, relaying signals not by electrical means but by the chemical messenger vasopressin.

In a rare disease called diabetes insipidus, the secretion of vasopressin is blocked. Virtually no optional reabsorption occurs, and the volume of the urine increases rather dramatically. Upward of 5 to 12 liters of urine may be voided per day. In contrast, with average intake of fluids a normal person eliminates about 1.5 liters of urine each day.

Another hormone, *aldosterone,* participates in diuresis by regulating the reabsorption of specific ions. As noted earlier, major portions of Na^+, Cl^-, K^+, and HCO_3^- leave the bloodstream and return in the obligatory reabsorption of the proximal tubule. In the distal tubule final adjustments in the concentrations of these ions are made, with aldosterone acting in a regulatory manner.

Regulation of the Acid-Base Balance of the Blood. As measured at room temperature, the pH of the blood must be kept within the narrow range of 7.0 to 7.9; for venous plasma it is normally in the extremely narrow range of 7.38 to 7.41. A drop in pH, that is, a shift toward the acid side, is called *acidosis. Alkalosis* is a rise in pH, a shift to greater basicity. Acidosis is the more common tendency because ordinary metabolism produces acids. Two agents, the buffers of the blood and the kidneys, remove this acid and maintain a steady pH. The kidneys "pump" hydrogen ions from the blood to the urine being formed; in their place (to preserve electrical neutrality) sodium ions are returned via optional reabsorption.

The buffers of the blood were discussed on pages 235 and 492. The most important is the system HCO_3^-/H_2CO_3. In neutralizing acids, bicarbonate ions in blood combine with hydrogen ions:

$$\underset{\text{Hydrogen ions of developing acidosis}}{H^+} + \underset{\text{Bicarbonate ions in the blood}}{HCO_3^-} \rightleftharpoons \underset{\text{Carbonic acid, a weak acid with little tendency to dissociate}}{H_2CO_3}$$

Too much carbonic acid can be handled by decomposing it to water and carbon dioxide:

$$H_2CO_3 \xrightarrow{\text{carbonic anhydrase}} H_2O + CO_2 \quad \text{(excreted at the lungs)}$$

Removal of carbon dioxide in the lungs means a net replacement of a hydrogen ion by a water molecule at the expense of a bicarbonate ion, thus depleting part of the buffer system. The kidneys come to the rescue, however, for they are able to generate a fresh supply of bicarbonate ions for the bloodstream and at the same time produce a more acidic urine.

Acidification of the Urine. R. F. Pitts proposed the following mechanism to explain how cells of the distal tubules increase the supply of bicarbonate ion in blood and feed hydrogen ions to the urine. In this way they restore the acid-base balance and prevent acidosis from developing further. Pitts's scheme may best be studied by referring to Figure 20.9, with steps indicated by encircled numbers.

The oxidation of some nutrient molecule in a distal tubule cell produces carbon dioxide ①. Action of the enzyme carbonic anhydrase causes it to combine with water to form carbonic acid ②, which ionizes to produce bicarbonate ions and hydrogen ions ③. In a sense the ionization is forced, for sodium ions from tubular urine are made to exchange themselves for hydrogen ions ④. Aldosterone appears essential to the promotion of this exchange. The urine thereby becomes more acidic. Bicarbonate ions leave the distal cell and enter the bloodstream ⑤, replacing bicarbonate ions used earlier to neutralize hydrogen ions. Freshly voided urine normally has a pH of about 6. In severe acidosis the kidneys, as they battle to control it, may produce urine with a pH as low as 4.6. Such urine has a hydrogen-ion concentration several hundred times that of normal blood.

When the kidneys fail to function and the formation of urine falls off or stops, a condition called *uremia* exists. (Uremia means urine in the blood, in the sense that the substances normally removed from the bloodstream by the kidneys stay instead in the blood. The term uremic poisoning is sometimes used in this connection.) Clinical evidence for uremia is obtained by measuring the rise in nonprotein nitrogen

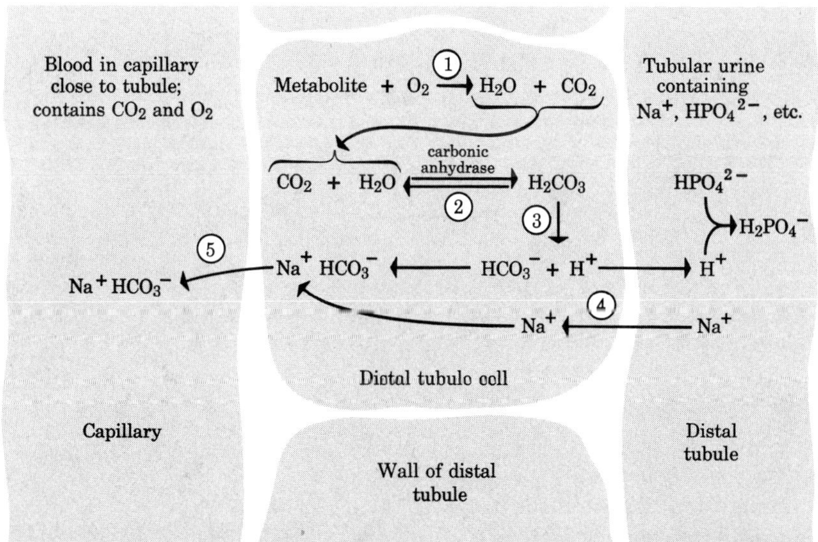
Blood in capillary close to tubule; contains CO_2 and O_2
Metabolite + O_2 ① → H_2O + CO_2
Tubular urine containing Na^+, HPO_4^{2-}, etc.
carbonic anhydrase
CO_2 + H_2O ⇌ H_2CO_3
②
③
HPO_4^{2-}
$H_2PO_4^-$
⑤
$Na^+ HCO_3^-$
$Na^+ HCO_3^-$ ← HCO_3^- + H^+ → H^+
④
Na^+ ← Na^+
Distal tubule cell
Capillary
Distal tubule
Wall of distal tubule

Figure 20.9 The acidification of the urine, according to R. F. Pitts. Circled numbers are explained in text. (From *American Journal of Medicine,* Vol. 9, 1950, page 356.)

(NPN)—chiefly urea with smaller quantities of uric acid and creatinine. The blood urea nitrogen (BUN) can also be measured, as can the pH and the concentrations of several ions. Symptoms of uremic poisoning include nausea, vomiting, drowsiness, a tendency to bleed, and a developing coma. In some instances of renal failure the kidneys have not been permanently injured, and the body can recover. Treatment varies, but artificial kidneys have been developed to clean the blood until normal recuperative processes have regenerated kidney function. In principle, the device by dialysis selectively removes dissolved substances from the blood shunted through it. Cellophane has served as a dialyzing membrane in these instruments, which have saved many lives. Because their cost and scarcity have limited their availability, hospitals providing them have had to set up committees of doctors, clergymen, professional men, and others to make the agonizing decisions of who will have a place in the time schedule of the artificial kidney and who will not.

Proteinuria (Albuminuria). Protein molecules do not normally leave the bloodstream in the glomerulus. When protein is detected in urine, it usually indicates some injury or disease in the kidneys. Pores in the glomerulus are believed to become enlarged to permit the passage of the larger protein molecules. Since the albumins are the major proteins in blood, the condition is frequently called albuminuria, but other proteins also escape. Loss of the blood proteins upsets osmotic pressure relations, affecting functions throughout the body.

LYMPH

THE LYMPHATIC SYSTEM

The bloodstream is one of the circulatory systems in the body. The second one, the lymphatic, performs the following functions: (1) recirculates interstitial fluid to the bloodstream, (2) transports some chemicals from where they are made to general circulation and conveys freshly absorbed, resynthesized lipid material from the intestinal tract to the bloodstream, and (3) participates in the defensive mechanisms of the body.

The lymphatic system (Figures 20.10 and 20.11) is made up of ducts that branch successively to terminate in a spongy mesh of tiny, thin-walled capillaries, all with closed ends, bedded within most of the soft tissue of the body. The larger lymph vessels, which have valves, receive lymph fluid from the branches and convey it to large blood vessels near the neck. The thoracic duct drains the lower limbs and the tissues of all the organs except the lungs, the heart, and the upper part of the diaphragm. The *right lymphatic duct* takes care of these organs, and the *cervical ducts* take lymph from the head and neck. Spaced along the larger lymphatic vessels are *lymph nodes,* specialized capillary beds which filter out solid matter in lymph before it reaches the bloodstream. More importantly, these nodes contain both white cells that destroy bacteria and other foreign substances and special cells that make antibodies. The lymph nodes are very important to the defenses of the body. Conversely, however, when cancer strikes the lymph ducts are avenues for its spread from one tissue to another, a phenomenon known as metastasis.

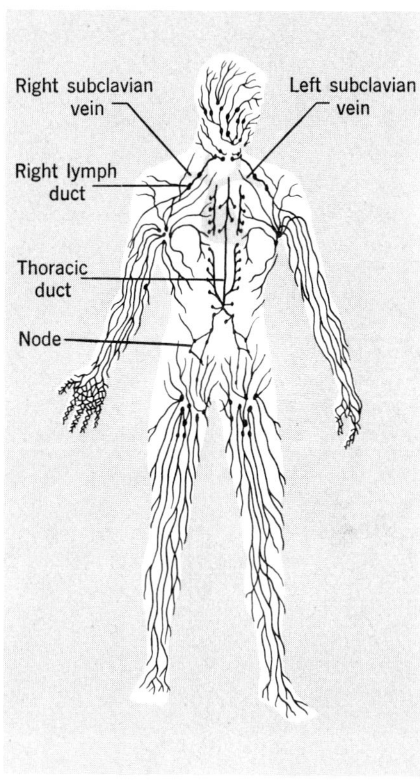
Right subclavian
vein
Left subclavian
vein
Right lymph
duct
Thoracic
duct
Node

Figure 20.10 The relations of the larger vessels of the lymphatic system to one another and the bloodstream. (From G. E. Nelson, G. G. Robinson, and R. A. Boolootian, *Fundamental Concepts of Biology,* John Wiley and Sons, New York, 1967, page 126.)

TRANSPORT OF CHEMICALS

Having learned something about the bloodstream and some of the ways in which its composition is controlled, and having looked briefly at the second circulatory system, the lymphatics, we are now in a position to discuss how these two systems bring nutrients to cells and carry waste products away.

EXCHANGE OF CHEMICALS AT TISSUE CELLS

The walls of capillaries of the bloodstream may be treated as selectively permeable membranes or filters. They allow water with dissolved nutrient molecules, including oxygen, to pass in one direction and water with dissolved waste chemicals to return in the other. Fluids that leave the bloodstream should return in equal volume; two mechanisms cooperate to make this possible while ensuring that the direction of the flow is correct. The pumping action of the heart creates a pressure which is higher on the arterial side of the capillary than on the venous side. Since the fluids of the blood, the cells, and the interstitial spaces have dissolved materials, osmosis and dialysis also occur. The cooperation of these two factors plus the involvement of the lymph system are illustrated in Figure 20.12.

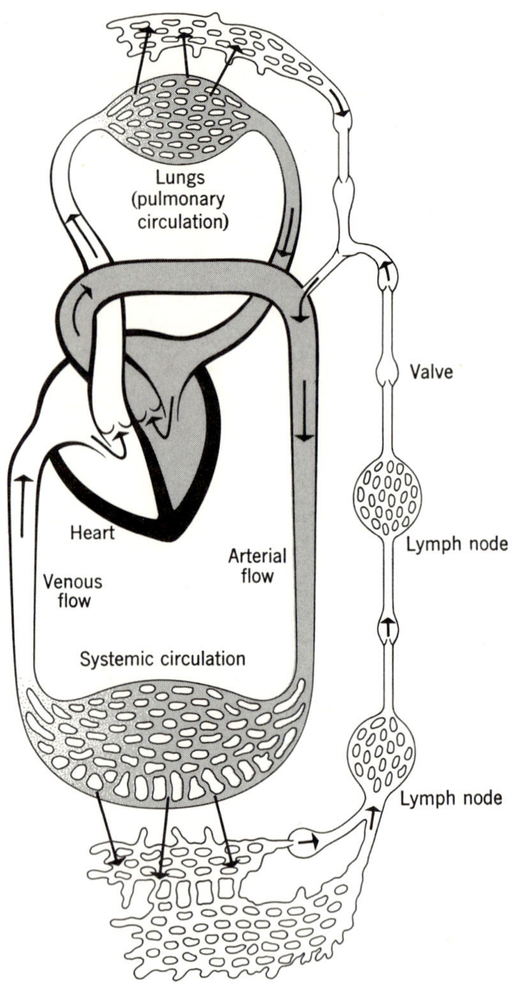
Lungs
(pulmonary
circulation)
Valve
Heart
Arterial
flow
Venous
flow
Lymph node
Systemic circulation
Lymph node

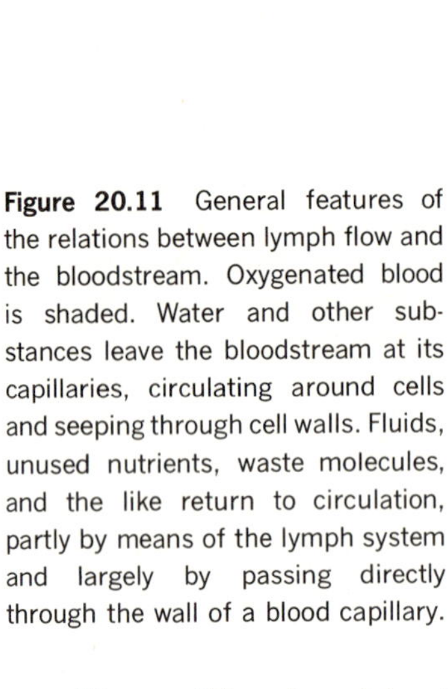

Figure 20.11 General features of the relations between lymph flow and the bloodstream. Oxygenated blood is shaded. Water and other substances leave the bloodstream at its capillaries, circulating around cells and seeping through cell walls. Fluids, unused nutrients, waste molecules, and the like return to circulation, partly by means of the lymph system and largely by passing directly through the wall of a blood capillary.

Plasma differs from interstitial fluid primarily in being more concentrated in proteins and should therefore more effectively induce osmosis and dialysis. The natural *net* direction for dialysis would be from the interstitial compartment into the bloodstream, on both the arterial and the venous ends of the capillary constriction. But on the arterial end the higher blood pressure reverses this tendency, making the net direction of fluid flow be from the capillary into the interstitial space. Nutrients are brought into contact with the tissue cells which selectively abstract those they need and send out waste chemicals such as carbon dioxide and excess water. The *net concentration* of dissolved solutes does not change, but the solutes themselves vary as nutrients are consumed and by-products are produced. The foregoing description constitutes a hypothesis formulated by E. H. Starling (1866–1927), one of the great British physiologists. He did not allow for "leakage" of proteins and lipids from the capillaries, but later work using radioactively labeled

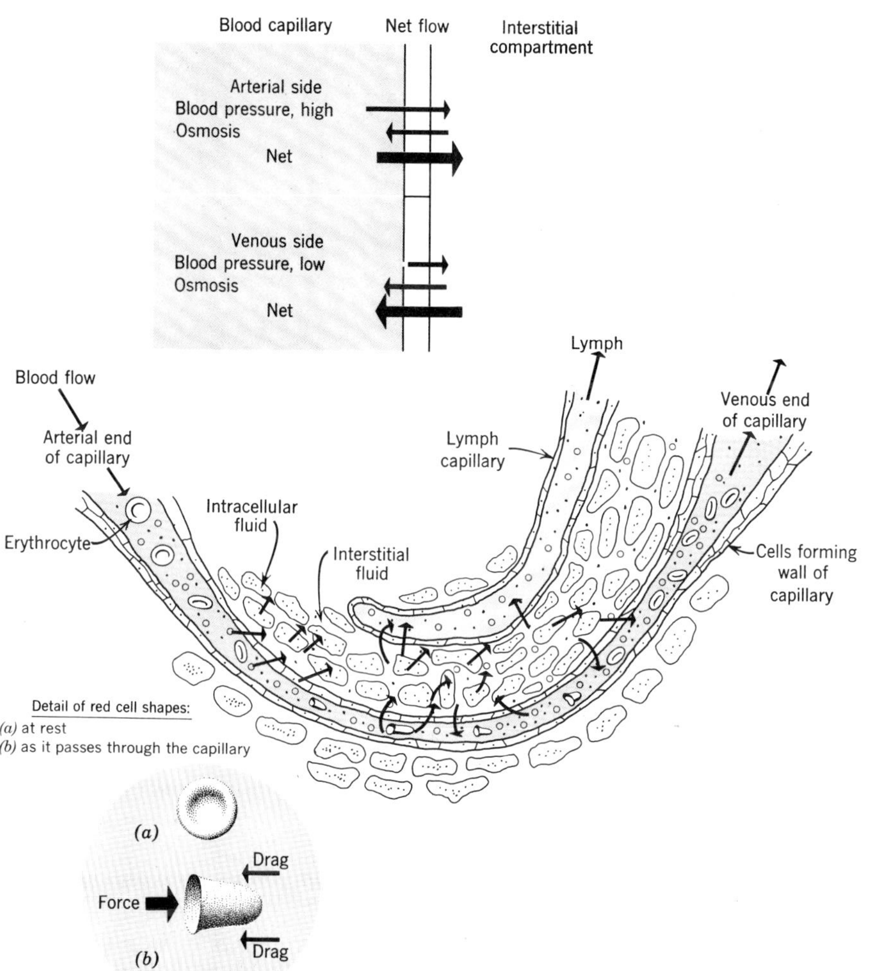
Blood capillary
Net flow
Interstitial compartment
Arterial side
Blood pressure, high
Osmosis
Net
Venous side
Blood pressure, low
Osmosis
Net
Lymph
Blood flow
Arterial end of capillary
Lymph capillary
Venous end of capillary
Intracellular fluid
Erythrocyte
Interstitial fluid
Cells forming wall of capillary
Detail of red cell shapes:
(a) at rest
(b) as it passes through the capillary
(a)
Drag
Force
(b)
Drag

Figure 20.12 Fluid exchange at capillaries. Dots represent proteins, circles lipids. Erythrocytes (red cells) have the shape of a disk dented on opposite sides, but during their passage through the most constricted region of a capillary, their shapes are distorted as shown. This distortion is believed to aid somehow in the release of oxygen. As much as half the total protein in the blood leaves the bloodstream each day and returns via the lymph.

The schematic diagram at the top summarizes Starling's hypothesis, explaining how fluids leave the capillaries at their arterial sides and reenter at their venous sides. On the arterial side blood pressure counterbalances osmosis and dialysis, and the net movement is from capillary to interstitial space. Blood pressure drops as it goes through the constricted capillary; on the venous side it no longer overpowers osmosis and dialysis, and the net movement is reversed.

proteins and other techniques (notably by the Americans H. S. Mayerson, K. Wasserman, and S. J. LeBrie, who built on hypotheses by C. K. Drinker) has shown that these larger molecules do, in fact, routinely leave the bloodstream. Because it is very difficult for them to reenter the bloodstream directly, they return via lymph fluid.

Enormous amounts of fluid diffuse in the various capillary beds of the body within a short period of time. It has been estimated that in the entire capillary network of a 160-pound man, fluid diffusion proceeds at a *rate* of about 400 gallons per minute. Delicate balances in fluid exchange at cells may be upset by important breakdowns in function.

Shock. The amount of proteins leaked through capillaries appears to increase when an individual has been subjected to sudden, severe injury, extensive burns, or even major surgery. Loss of protein from the blood upsets osmotic-pressure relations. Fluids forced out on the arterial side of a capillary loop do not return as they should. The blood volume drops, and the circulatory system no longer brings enough oxygen and nutrients to the cells. Those in the central nervous system are the most seriously affected, and a condition known as shock ensues. Shock may also result from extensive hemorrhage, for any failure of the circulatory system will cause it. Restoration of blood volume is mandatory for recovery from shock.

Edema. Fluids accumulating in the interstitial and cellular regions in abnormally large amounts produce a condition called edema. It may have one of several causes, only a few of which can be mentioned here.

If proteins leak through the kidneys into the urine being formed (albuminuria, p. 560), osmotic pressure relations may be upset and enough fluid retained in the interstitial compartment to produce noticeable swelling or edema in some tissues, notably those of the lower limbs. The swelling that usually accompanies malnutrition has similar causes. Prolonged malnutrition amounting to starvation will eventually result in less concentrated blood fluid, with the same result, edema.

Localized edema or swelling occurs around an injured area. The injury may have damaged either blood vessels or lymph capillaries, preventing normal drainage, and the fluid that accumulates produces the swelling. In a particularly ugly tropical disease, elephantiasis, small worms (filari) enter the body and live in the lymphatics, thereby reducing drainage through them. The swelling assumes grotesque proportions.

Obstructions in veins, as in certain cancers or in varicose veins, may cause a back pressure on the venous side of the capillary loops, hindering dialysis of interstitial fluids into veins. Again not quite as much fluid returns as enters the interstitial compartment, and swelling occurs.

RESPIRATORY FUNCTIONS OF THE BLOOD

Moving oxygen from air to tissue cells is a task of enormous magnitude. If an adult generates 3000 kcal per day from his food, he requires 600 liters of oxygen, and he generates about 480 liters of carbon dioxide. The movement of these gases to and from air and cells is explained in Figure 20.13. The rate of this movement requires large surface areas across which contact between the blood

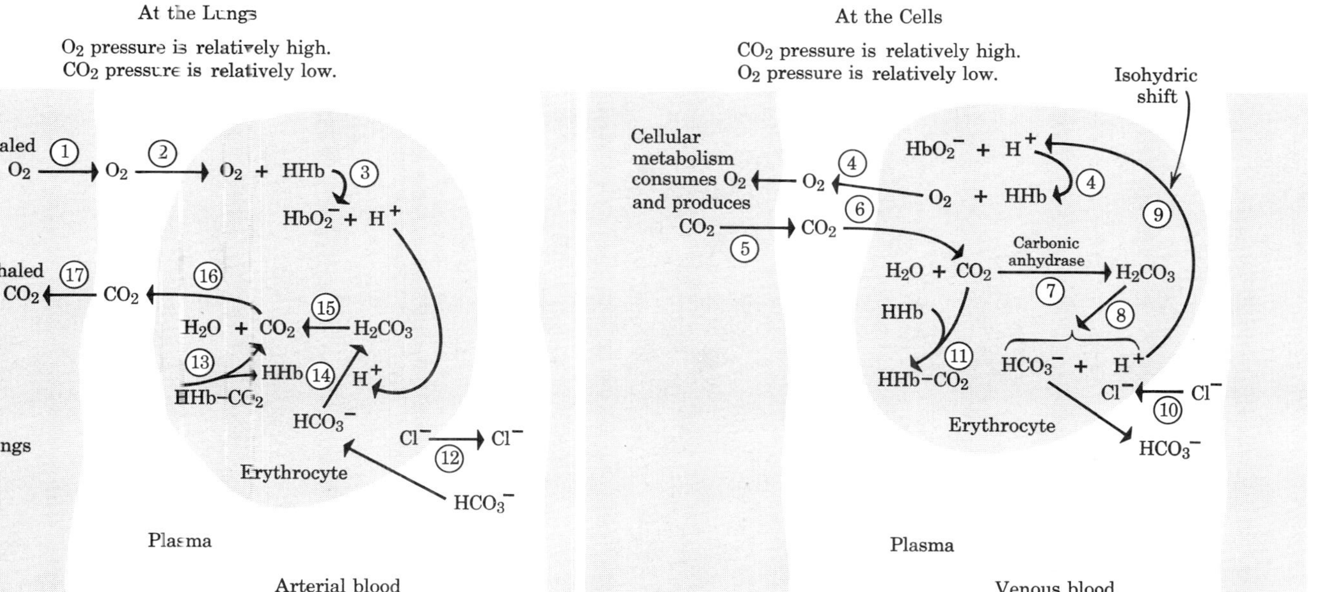

At the Lungs
O2 pressure is relatively high.
CO2 pressure is relatively low.
Inhaled
O2
O2
O2 + HHb
HbO2⁻ + H+
Exhaled
CO2
CO2
H2O + CO2
H2CO3
HHb
H+
HHb–CO2
HCO3⁻
Cl⁻
Cl⁻
HCO3⁻
Lungs
Erythrocyte
Plasma
Arterial blood
At the Cells
CO2 pressure is relatively high.
O2 pressure is relatively low.
Isohydric shift
Cellular metabolism consumes O2 and produces CO2
O2
CO2
HbO2⁻ + H+
O2 + HHb
Carbonic anhydrase
H2O + CO2
H2CO3
HHb
HHb–CO2
HCO3⁻ + H+
Cl⁻
Cl⁻
HCO3⁻
Erythrocyte
Plasma
Venous blood

Figure 20.13 Oxygen–carbon dioxide transport by the blood. Starting at the lungs with oxygen transport, the mechanics of breathing make oxygen pressure relatively high in the lungs. Oxygen is forced to diffuse into the bloodstream (1). Although blood may carry some dissolved gaseous oxygen, virtually all is drawn into erythrocytes (2), where it combines with hemoglobin, HHb, to form oxyhemoglobin, HbO_2^-, and H^+ (3):

$$HHb + O_2 \rightleftharpoons HbO_2^- + H^+$$

Two factors help shift this equilibrium from left to right in the lungs. The oxygen pressure is relatively high, and the hydrogen ion is neutralized by bicarbonate ion (14), something that must happen anyway if carbon dioxide is to be released. Thus arrival of bicarbonate ion helps draw oxygen into the cell. In cellular areas needing oxygen, two factors help in its release from HbO_2^-. Oxygen pressure is now relatively low, and carbon dioxide is poised to enter at (5) and (6). To release oxygen from HbO_2^- (4), a hydrogen ion is needed (9); the influx of carbon dioxide makes it available in (7) and (8). If carbon dioxide is not available, the cell presumably does not need oxygen anyway. This shift of hydrogen ion (9) is called the *isohydric shift*. Bicarbonate ions leave the red cell during transport, and to replace their charge chloride ions move in (10). (This switch is called the *chloride shift*.) Most carbon dioxide is taken to the lungs as bicarbonate ion, but some combines with hemoglobin to be carried as carbamino hemoglobin (11), written as $HHb–CO_2$ for short. Very little moves simply as the gas dissolved in the plasma. When the erythrocyte returns to the lungs, a reversal of events, (12) to (17), sends carbon dioxide out into exhaled air and the cyclical process repeats itself.

flow and the inhaled air can be rapidly made. To provide this surface, the trachea ("wind pipe") undergoes a succession of branchings until the final, smallest functional unit, the individual *alveolus,* is reached. Incoming air is distributed to about 300 million alveoli, each one of which is little more than a bundle of fine capillaries suspended in air. An estimated 1500 miles of such exposed capillaries provide about 800 square feet of diffusing surface. Through these capillaries blood flows at a rate of about 5 liters per minute during resting periods; during severe exercise the rate is from 20 to 30 liters per minute.

Oxygen Toxicity. If an individual breathes pure oxygen, enough simply dissolves in plasma to supply nearly all the needs of tissues. Consequently, oxyhemoglobin has difficulty in dissociating at cells (④ in Figure 20.13). The isohydric shift ⑨ cannot therefore be used, and the hydrogen ions produced at ⑧, instead of being buffered by a hemoglobin molecule, tend to force the pH of the blood down. The resulting acidosis may be severe and very dangerous.

Oxygen Deficiency. Anoxia. Up to an altitude of about 18,000 feet, air pressure is high enough to provide the air needed, when inhaled into the lungs, to saturate erythrocytes with oxygen, provided little exercise is engaged in. At higher altitudes the oxygen pressure is not enough, and symptoms of oxygen deficiency—restlessness, confused thinking, and eventually unconsciousness ("blackout")—become evident. Portable oxygen or air tanks and pressurized cabins are familiar solutions to the problem.

Oxygen deficiency occurs in pneumonia, when the alveoli fill with fluids that hinder rapid exchange of gases. *Fibrosis* in alveoli, a condition which causes tissue to harden and become fibrous, also interferes with this exchange. Miners who breathe ore dusts and farmers who breathe organic dusts are taking risks that may eventually lead to lung failure. Certain organic substances in cigarette smoke have been implicated in lung cancer, the only known cure for which is surgical removal of the affected area. Only about 10% of such patients survive more than five years, unless the cancer is caught in its earliest stages. The lungs have some capacity for repair, and medical experts insist that those who smoke excessively should cut down or quit, however long they have been smoking.

Carbon monoxide is a poison that affects oxygen transport by combining with hemoglobin about 200 times more firmly than oxygen. In such a situation nearly pure oxygen must be administered to displace the molecules of carbon monoxide and to make up for oxygen deficiency.

BLOOD CLOTTING

Blood must clot when it is shed, and it must not clot when it is not shed. These simply stated rules are a matter of life and death, and a complicated apparatus controls whether or not clotting occurs. Because the details are imperfectly understood, we confine our study to a broader view.

When blood clots, the soluble protein *fibrinogen* is converted to an insoluble protein *fibrin,* whose fibers aggregate in a matlike way to seal off the sites where the blood vessel has been broken or cut. The formation of a clot can be studied *in vitro.* In fact, special precautions must be taken to prevent freshly drawn blood

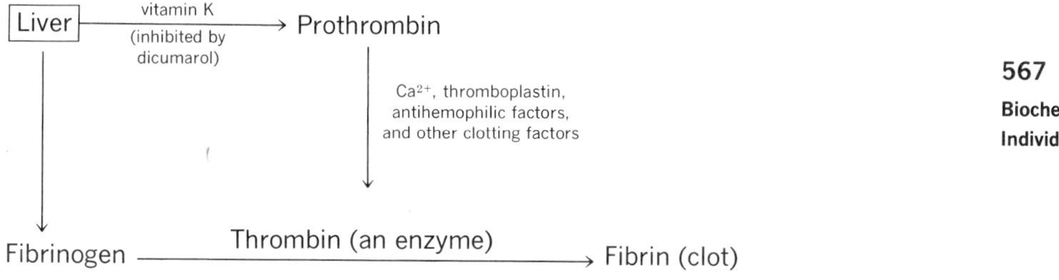
Liver
vitamin K
(inhibited by dicumarol)
Prothrombin
Ca²⁺, thromboplastin, antihemophilic factors, and other clotting factors
Fibrinogen
Thrombin (an enzyme)
Fibrin (clot)

Figure 20.14 Probable mechanism of blood clotting. Both fibrinogen and prothrombin are made in the liver. Some clotting factors occur in the blood, some in platelets, and some in cells that release them when cut or injured.

from clotting. To change fibrinogen to fibrin, an enzyme is needed, but it can hardly be expected to be present in blood at all times. The key to the control of blood clotting is restricting the formation of this enzyme to the times when it is needed.

The clotting enzyme is called *thrombin,* and it exists in the bloodstream in an inactive form called *prothrombin.* To convert prothrombin to the active enzyme requires at least ten factors—calcium ion, antihemophilic factors, a coenzyme, and a group of molecules collectively called thromboplastin. Some are present in blood, and others are released when tissues are cut. Thus the clotting mechanism is not thrown into operation unless a cut or injury occurs. When blood is removed from the body, it is not easy to prevent any and all contact with tissues slightly injured by the act of inserting a needle. Clotting under these circumstances can be prevented simply by tying up the calcium ions through the administration of sodium oxalate or sodium citrate. With the first sodium salt calcium precipitates as calcium oxalate; with the second salt calcium ions complex with citrate ions and are effectively removed.

Figure 20.14 illustrates the principal features of a probable mechanism for clot formation. Vitamin K deficiency, as in some hemorrhagic diseases, reduces the formation of the inactive prothrombin in the liver. The same effect can be achieved with dicumarol, a chemical formed in decaying sweet clover and sometimes given to prevent thrombosis, the unwanted formation of a clot within a blood vessel. Heparin, an extremely potent anticoagulant, is found in cells located near the walls of the tiniest capillaries. It acts by slowing the conversion of prothrombin to thrombin.

BIOCHEMICAL INDIVIDUALITY

Many facets of nutrition have been studied in this chapter, but we shall leave to other references such matters as specific nutritional requirements and the best sources of nutrients. One topic, however, is inadequately covered in the literature. Roger J. Williams (University of Texas) deserves most of the credit for calling our attention to it, biochemical individuality.

Understanding of the molecular basis of life and health or of illness and disease is not the same thing as applying this knowledge to the control and cure of a metabolic disorder. One major emphasis of this book has been and will be the likenesses and similarities at the molecular level of the members of any given species. But obvious differences exist among human beings. External differences are completely apparent, but few people are aware of internal differences.

In an effort to simplify the study of any subject, we try to generalize. Such expressions as "a typical cell," "a normal adult," "a normal child," and an "average appetite" are concepts of some usefulness in the beginning stages of a serious study. Generalizations too often become dogmas, however, and when practical application is attempted to an *individual* case, any deviation from the normal is greeted with surprise at best or outrage at worst. Each human being has a very distinctive body chemistry. There are many likenesses, to be sure, but Williams emphasizes that differences exist and are ignored only to our peril. Every individual organism, because of its distinctive genetic background, has distinctive nutritional needs that must be satisfied for maximum well-being.

Heredity cannot be ignored; we simply would not *be* without it. Environment cannot be ignored; we simply could not *become* without it. A great deal of research in nutrition has focused on what *all* people require. Such research is clearly important, and more is needed. Williams' work should alert us to the fact that research in nutrition must also be directed to identifying individual needs.

REFERENCES AND ANNOTATED READING LIST

BOOKS

I. N. Kugelmass. *Biochemistry of Blood in Health and Disease.* Charles C Thomas, Springfield, Ill., 1959.

J. L. Gamble. *Chemical Anatomy, Physiology and Extracellular Fluid,* sixth edition. Harvard University Press, Cambridge, Mass., 1954.

A. White, P. Handler, and E. L. Smith. *Principles of Biochemistry,* third edition. McGraw-Hill Book Company, New York, 1964.

R. J. Williams. *Biochemical Individuality: The Basis for the Genetotrophic Concept.* John Wiley and Sons, New York, 1956.

ARTICLES

H. Neurath. "Protein-Digesting Enzymes." *Scientific American,* December 1964, page 68. By studying the molecular structures of certain digestive enzymes, Neurath and his group at the University of Washington hope to discover clues to the actions of all enzymes.

D. M. Surgenor. "Blood." *Scientific American,* February 1954, page 54. The author describes the functions of the blood, discusses its preservation and handling, and states the problems attendant on separating its components.

J. E. Wood. "The Venous System." *Scientific American,* January 1968, page 86. Veins are more than tubes. They are flexible reservoirs for blood.

H. S. Mayerson. "The Lymphatic System." *Scientific American,* June 1963, page 80. After discussing the anatomy of the lymphatic system and its general functions, Mayerson describes experiments he and others have performed to prove that it is normal for capillaries to "leak" proteins and lipids.

J. H. Comroe, Jr. "The Lung." *Scientific American,* February 1966, page 57. The anatomy of the respiratory system is described, and pulmonary circulation and the mechanics of breathing are discussed.

H. W. Smith. "The Kidney." *Scientific American,* January 1953, page 40. After a well-illustrated discussion of the structure and function of the kidney, the author speculates on its evolution.

J. P. Merrill. "The Artificial Kidney." *Scientific American,* July 1961, page 132. While discussing the problems of devising an artificial kidney, the author explains normal kidney function.

K. Laki. "The Clotting of Fibrinogen." *Scientific American,* March 1962, page 60. How the enzyme thrombin works is discussed in detail.

L. Pauling. "Orthomolecular Psychiatry." *Science,* Vol. 159, 1968, page 265. Biochemical individuality is probably nowhere more significant than in the brain. As Pauling marshals evidence for this supposition, he proposes that the methods used in treating patients with mental disease—psychotherapy, chemotherapy, shock therapy—be augmented by another general method, orthomolecular psychiatric therapy. He defines this as treatment that provides the optimum molecular environment for the mind, especially the optimum concentrations of chemicals normally present in the human body.

PROBLEMS AND EXERCISES

1. Define each of the following terms.

(*a*) internal environment (*b*) extracellular fluids (*c*) interstitial fluids
(*d*) plasma (*e*) serum (*f*) defibrinated blood
(*g*) red cell count (*h*) leukocytes (*i*) lymphocytes
(*j*) thrombocytes (*k*) isohydric shift (*l*) thrombin
(*m*) prothrombin (*n*) anoxia (*o*) α-amylase
(*p*) parietal cells (*q*) chief cells (*r*) mucous cells
(*s*) gastrin (*t*) pepsinogen (*u*) zymogen
(*v*) pepsin (*w*) proteolytic enzymes (*x*) proteoses
(*y*) peptones (*z*) mucoprotein

(*aa*) gastric hyperacidity (*bb*) enterogastrone (*cc*) chyme
(*dd*) pyloric valve (*ee*) duodenum (*ff*) gastric achlorhydria
(*gg*) gastric hypoacidity (*hh*) secretin (*ii*) pancreozymin
(*jj*) enterokinase (*kk*) cholecystokinin (*ll*) steapsin
(*mm*) active transport (*nn*) diuresis (*oo*) glomerular filtrate
(*pp*) diabetes insipidus (*qq*) acidosis (*rr*) alkalosis
(*ss*) BUN (*tt*) NPN (*uu*) uremia
(*vv*) albuminuria (*ww*) edema (*xx*) alveolus
(*yy*) thrombin (*zz*) prothrombin

2. Explain how proteins of the stomach wall are protected from being digested.
3. Describe the function of each of the following.

(*a*) trypsin (*b*) bile salts (*c*) serum globulins
(*d*) serum albumins (*e*) fibrinogen (*f*) vasopressin
(*g*) aldosterone (*h*) acidification of the urine (*i*) lymph
(*j*) lymph nodes

4. Explain how the flow of each of the following is controlled.
   (*a*) saliva (*b*) gastric juice (*c*) pancreatic juice
   (*d*) bile (*e*) vasopressin
5. How does carbon monoxide act as a poison?
6. Alcohol in blood suppresses the secretion of vasopressin. How does this affect diuresis?
7. In anemia the supply of hemoglobin is low. How does this affect respiration?
8. Referring to this chapter and to previous ones, what are the three main buffers in blood? Show equations that illustrate *how* they work. Explain *why* they *must* work.
9. Explain how malnutrition will upset osmotic pressure relations between blood and the interstitial fluids.
10. Explain how the kidneys act to reduce developing acidosis.
11. Explain how fluid exchange occurs between plasma and cells? (How does arterial blood become venous blood?)
12. Describe ways in which edema may originate.
13. Explain how the H^+ needed to help release O_2 from HbO_2^- is produced precisely where it is needed.
14. Explain how H^+ is tied up at precisely the right point in order to assist the uptake of O_2 as HbO_2^-.
15. Explain in your own words how a blood clot forms.
16. Why is it unwise to breathe pure oxygen if your respiratory organs are functioning properly?
17. How does the administration of vitamin K assist the clotting mechanism?

CHAPTER TWENTY-ONE

Energy for Living

"In the sweat of your face you shall eat bread until you return to the ground" said an ancient writer, "for out of it you were taken; you are dust, and to dust you shall return" (Genesis 3:19).

This gloomy commentary on the human predicament recognized our kinship with nature and its laws. All our ancestors, including our parents, dined on plants and on animals which fed on plants. Indirectly we are indeed made of dust, born out of stable, randomly dispersed chemicals such as carbon dioxide, water, and a few mineral substances which plants can convert into foods with the indispensable help of energy from the sun.

Sooner or later the stuff of our bodies will become these common things again, but there is a marvelous interlude between dust and dust, and we call it life. However long or briefly it endures, it is possible only as long as the organism can maintain itself in a condition quite remote from the ultimate equilibrium. The unrelenting tendency in nature to achieve universal equilibrium is easily life's most implacable foe. Universal equilibrium is the impossibility, or the extreme improbability, of any further *net* change—the existence of maximum disorder and maximum chaos, the universal "sea level" of existence, dust and ashes scattered to the four winds. A living organism, by comparison, is a highly organized, a finely ordered, and a relatively high-energy state of affairs. To keep it that way, even temporarily, requires information, energy, and "spare parts." We obtain energy and spare parts from the chemicals, from the foods in our diet and the air that we breathe. We obtain information genetically, as discussed in Chapter 25. The emphasis in this chapter will be on the *energy* we receive, how we obtain it, how we store it, how we transform it, and how we use it.

By now we should be somewhat used to the notion that we can obtain energy from chemicals. We know that certain combinations or mixtures undergo changes that are accompanied by the evolution of energy. Sometimes the change is little more than an explosion, but if it is made to occur in certain ways, objects are usefully pushed—ore in a mine, a cylinder in a car engine, a rocket engine. We are also accustomed to the idea that chemical changes are rearrangements or redistributions of electrons in relation to atomic nuclei. With a little clever engineering we can sometimes make this electron "flow" occur through circuits, through motors or heaters or light bulbs, and useful work is done.

Because chemicals can be made to do work for us, whatever in chemicals makes this work possible is important enough to have a name. We call it chemical energy. We do not say that the chemicals have *work,* because we use this word when we wish to talk about energy actually being expended *in a purposeful way.* The chemicals simply have *a potential for doing work,* and thus chemical energy is a special kind of potential energy.

In many instances we convert heat energy derived from chemical changes directly into useful work. In steam engines, for example, wood or coal is burned to give heat to make steam. The steam expands against a piston, and when the piston is pushed as far as it will go, the steam is either vented to the air or internally cooled, condensed to liquid water, and used over again. Meanwhile the piston returns to its original position for another push. In the gasoline engine of a car, fuel is burned and the gases so produced expand against a piston before they are vented to the air, and again the piston returns for another push. The process is *cyclic.* For practical reasons efficient and continuous conversion of heat into work requires a *cyclic process* with different parts of the device operating at different temperatures. Work is tapped from heat as heat "flows" from a higher temperature to a lower one.

In contrast, a living organism, for all practical purposes, has its different parts at the *same* temperature. It cannot by any cyclic process convert heat into the work of moving muscles, of sending electrical signals along the nervous system, or of moving ions and molecules across barriers. A living organism is healthy, by and large, only as long as it can maintain a fairly constant temperature—in humans about 98.6°F. Experience has taught scientists (as well as would-be inventors of perpetual motion machines) that *it is just not possible to convert heat into work in a cyclic process at constant temperature.*[1] This rule is part of the givenness of nature.

The human body is not analogous to a heat engine, in spite of all the references to the "machinery of the body" we see in popular print. Although the body is in some ways like a machine, it is not powered by heat. To be sure, metabolism produces heat, and we need it to maintain a steady working temperature. This heat, however, is not the energy that is converted into work. The body has other ways of deriving work from its ingested or stored chemicals. In a steam engine the heat stored in the steam is the immediate reservoir of energy for work; the energy is stored in the form of violently and rapidly moving molecules constantly pushing at the walls of their container, one of which, the surface of the piston, can move. In the body the immediate storehouses of energy are not steam but other chemicals. Energy is

[1] The second law of thermodynamics is so stated for isothermal processes.

stored in the relatively unstable bonds of these chemicals, not in their rapid motion. The organophosphates are among the most important of these substances.

ORGANOPHOSPHATES IN FERMENTATION AND IN HUMAN BIOCHEMISTRY

Fermentation. Some of the major advances in understanding human biochemistry can be traced to successes in discovering what happens during fermentation, for we now know that many of its chemical steps are duplicated in the body. The discovery of fermentation cannot be dated. It goes far back into prehistory. Anthropologists have never discovered a group of people who did not take advantage of it in some way. Although making alcohol or alcoholic beverages may be the fermentation that first comes to mind, the word applies to any process in which microorganisms (bacteria or fungi, molds and yeasts) act on agricultural materials to produce definite products and definite changes. The brewing of ales, meads, beers, and other "spirits," wine making, aging of meat, and making sauerkraut, pickles, cheeses, butter, yogurt, soy sauce, and leavened bread—all these and other processes involve various fermentations. Antibiotics, vitamins, and many important industrial chemicals are also made by the quiet action of unseen microorganisms, and for such work they have been bred as carefully as prize race horses.

The word fermentation comes from a Latin term *fervere,* meaning "to boil." It is descriptive of the frothing and foaming action that often accompanies the process and is now known to be caused by the evolution of carbon dioxide. The work of Lavoisier, Gay-Lussac, and Thenard in the early years of the nineteenth century established fermentation as a *chemical* change rather than as magic or some spiritous process. Alcohol fermentation was summarized by Gay-Lussac as follows:

$$\underset{\text{Sugar}}{C_6H_{12}O_6} \xrightarrow[\text{yeast}]{} \underset{\substack{\text{Carbon}\\\text{dioxide}}}{2CO_2} + \underset{\text{Ethyl alcohol}}{2CH_3CH_2OH}$$

Just what the connection is between the yeast present in fermentation and the fermentation itself was not known at the beginning of the nineteenth century. In fact, no one was sure just how yeast, as matter, ought to be classified. In 1803 Thenard announced that his microscopic studies had shown that the yeasts present in and during wine making actually caused the fermentation. Furthermore, the yeasts ought to be classified as animal matter. The implication of this statement for the mechanism of fermentation was obscure, and it remained so for years.

Louis Jacques Thenard (1777–1857). French chemist and discoverer of hydrogen peroxide. His greatest contribution to the advancement of science was his four-volume *Traité élémentaire de chimie theorique et pratique* ("Elementary Treatise of Theoretical and Practical Chemistry").

Conceding that yeasts were present, most workers insisted that fermentation was a purely chemical reaction, not a "life process." The most widely accepted explanation, proposed the most strongly by Justus von Liebig, was that molecular vibrations produced by dead, decomposing yeast sparked the chemical changes that converted sugar to alcohol and carbon dioxide. Controversies over the nature of fermentation raged all during the nineteenth century, embroiling some of the greatest men of science.

The idea of catalysis was emerging, and by 1820 a few complex organic substances (undoubtedly mixtures) engaging in catalytic activity were known. These catalysts of biological origin were initially called *ferments*. Later, in 1878, the German physiologist Wilhelm Kühne renamed them *enzymes* (meaning "in yeast"). *Diastase* was one of them. Isolated in 1833 from malt (sprouted barley), it was found to split starch to maltose units. But this was not fermentation. *Invertase*, prepared from yeast in 1860, was another enzyme; it split sucrose to fructose and glucose. Again, this reaction was not fermentation. These "ferments" were unquestionably dead chemicals, even though they had been isolated from plant and animal juices; and none could induce fermentation.

By 1838 virtually all scientists interested in the problem agreed that yeasts were living organisms, although they were now classified as plants. However, such distinguished chemists as Berzelius, Wöhler, and von Liebig (especially von Liebig) vigorously opposed any notion that *living* yeast cells were necessary for fermentation—dead and decaying cells, perhaps, but not living cells. These men, after all, had just succeeded in ousting the theory of vitalism (cf. p. 242), and they were not about to allow it to sneak back into science via the biochemical door. They had yet to reckon with one of the greatest and most versatile scientists of the nineteenth century, Louis Pasteur of France, who entered the lists of the fermentation jousts in or about the year 1855.

Louis Pasteur (1822–1895). Born in Dole, France, the son of a tanner, Pasteur's interest in science was aroused at an early age by his father's business and his mother's encouragement of intellectual pursuits. In his first chemistry class he persisted in asking his professor difficult questions in the presence of the other students. For his pains he was rebuked with the admonition that it was the right of the teacher to ask questions of the students, not the other way around. At the famous École Normale in Paris the intellectual climate was different. Pasteur first became interested in molecular physics and the properties of crystals. In scientific circles about this time there was great interest in optical activity. Pasteur studied the optical activity of isomeric tartaric acids, by-products of the wine industry. His conclusions concerning asymmetric molecules were of such importance that he is credited with being the "father of modern stereochemistry." In 1854 he went to Lille to organize the new faculty of science. Responding to wishes of local businessmen who wanted something practical from their new science facilities, Pasteur decided to study fermentation. (Lille's chief industry was the manufacture of alcohol by fermentation of grain and sugar beets.) His lifelong study led to *pasteurization* of wines and beers as a means of preventing them from spoiling. Pasteurization of milk (heating it for 30 minutes at 145°F or for 15 seconds at 161°F), directed against microbes that might harm man, has been of inestimable lifesaving importance in the modern world. His interests in the "infectious diseases" of wines and beers developed into a general interest in all microbial infections, and he discovered inoculations against anthrax in animals and rabies in people bitten by rabid dogs. He saved the silk industry of France by his investigations of the diseases plaguing silkworms. He performed extraordinarily careful experiments to demonstrate that the microorganisms causing fermentation did not spontaneously arise in the medium but came from the air, which is virtually impregnated with such creatures. The theory of the spontaneous generation of life was dealt a fatal blow. His series of outstanding discoveries in a variety of fields, many of very obvious lifesaving value, made Pasteur a popular hero in France. A national subscription for funds from a grateful public made possible the Pasteur Institute. Biographers justifiably call him one of the greatest benefactors the human race has ever produced.

What might literally be called diseases of wine and of beer plagued brewers and winemakers at the time. Wines would turn sour or bitter or become cloudy without apparent explanation, and Pasteur was asked to see what might be done. By careful work he succeeded in growing brewer's yeast in a medium composed of simple substances. During its growth fermentation also occurred. There was no question about it now; the two processes, the growth of yeast and fermentation, were related. Von Liebig and the other chemists had been proved wrong, or so it seemed. But could the one, fermentation, occur without the *life* of the other, the yeast? Many insisted that the ferments inside the yeast cells and not the life process of yeast caused fermentation. Pasteur and others tried to remove from yeast the catalysts, that is, the ferments, that might be responsible. All such efforts failed, and Pasteur was forced to conclude that the *living* yeast cell was essential. Dead and decaying cells would not do. Pasteur's great fame and authority served to strengthen the position, reminiscent of the way the authority of Sir Isaac Newton dominated physics during the same period. It is one of the ironies of the history of science that, not long after Pasteur's death, others succeeded where he had failed. A substance that could catalyze fermentation was isolated from yeast. Both sides of the long conflict were partly right. Living yeast could cause fermentation, and so could "dead" chemicals taken from yeast. An accidental discovery by an open mind prepared the way.

Eduard Buchner was interested in the proteins in yeast and in their possible pharmaceutical properties. He had discovered a way of rupturing yeast cells by mechanically grinding them with quartz sand and kieselguhr. After water was added, the mixture was squeezed in a hydraulic press through filter cloths. A clear, yellowish, cell-free press juice could be obtained, but it did not keep well. To preserve it, Buchner introduced the substance commonly used to protect food material. He added large amounts of raw sugar. To his surprise, after a brief time some foaming which lasted a day was observed. It was Buchner's genius to recognize this as true fermentation, even though no yeast cells were present in either a living or a decayed state—that is, in spite of Pasteur's widely held dictum that no fermentation could occur without life. For the first time cell-free alcoholic fermentation had been observed. Apparently some chemical in the yeast, apart from intact, living cells, can bring about fermentation; Buchner gave the name *zymase* to the catalyst.

Buchner's discovery attracted the attention of many scientists, among them Arthur Harden of England. Freshly pressed yeast juice did not indefinitely retain its ability to cause fermentation. It could be completely inactivated by boiling. (We know now, of course, that heat denatures enzymes.) Yet when boiled yeast juice was added to "tired" yeast juice, the fermenting ability of the later was restored. Blood serum also seemed to reactivate "worn out" yeast juice. In 1905 Harden, together with W. J. Young, reported that yeast juice must be supplied with simple inorganic phosphate ion to maintain its fermenting ability outside the cell. Plasma or boiled (but otherwise fresh) yeast juice happens to have this ion. Apparently, as fermentation progresses, catalyzed by pressed yeast juice, the initial supply of phosphate ion is used up. Boiled yeast juice can rejuvenate "spent" yeast juice because phosphate ions are stable to heat. A mechanism for regenerating phosphate ion must also be present when fermentation is catalyzed by *living* yeast cells. Harden found that before glucose can be fermented, it is converted into an ester of phosphoric acid. He found that other intermediates between glucose and ethyl alcohol are also esters of phosphoric acid. One of them, fructose 1,6-diphosphate, was for years called the Harden-Young ester.

Eduard Buchner (1860–1917). German biochemist awarded the Nobel prize in chemistry in 1907 for his research in fermentation and enzymes.

Arthur Harden (1865–1940). This English biochemist, as head of the Department of Biochemistry of the Lister Institute (London), devoted thirty years of study to the steps in fermentation. He found that Buchner's zymase was actually a mixture of several enzymes and several substances later to be called coenzymes. This and his discovery of the existence of phosphate esters as intermediates in alcohol fermentation were his principal contributions. In 1929 he shared the Nobel prize in chemistry with Hans von Euler-Chelpin. He was knighted in 1936.

Gustav Embden (1874–1933). German physiological chemist, professor at the university in Frankfort on the Main.

Otto Meyerhof (1884–1951). This German biochemist was awarded a Nobel prize in physiology and medicine in 1922 (jointly with A. V. Hill) for determining the fixed relation between oxygen uptake by muscle tissue and its metabolism of lactic acid. He was forced out of Germany in 1938 by Nazi authorities, and in 1940 he fled Paris just before the German army overran it. He became a research professor in physiological chemistry at the University of Pennsylvania.

In 1914 Gustav Embden discovered that enzymes present in muscles transform the Harden-Young ester into lactic acid. This discovery of a common intermediate was an early indication that alcohol fermentation by yeast may resemble the breakdown of glucose in muscle cells. At least, as a rising tide of research increasingly demonstrated, the similarities are much more numerous than the differences.

A major pathway of glucose catabolism in the body, the Embden-Meyerhof pathway, is named after Embden and another important contributor in the field, Otto Meyerhof.

To prepare ourselves for the study of virtually any metabolic pathway, we must become more familiar with the organic compounds of phosphoric acid. Some of them are the means by which the body stores chemical energy just before its final use for muscular work or for any of the many other energy-demanding events at the molecular level of life.

Low-Energy and High-Energy Phosphates and Free Energy Relations. Some organophosphates are relatively unreactive to water; others are not. Although the border line between them is not sharp, it does give us two groups. In the one are simple esters of phosphoric acid, some of which are so unreactive to water that they require prolonged digestion with hot acid to be hydrolyzed. Two relatively unreactive organophosphates are α-glycerophosphate and glucose 6-phosphate.

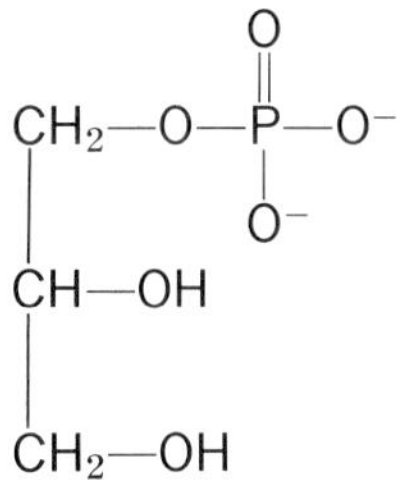

α-Glycerophosphate
($\Delta G^\circ_{\text{hydr}} = -1000$ cal/mole)

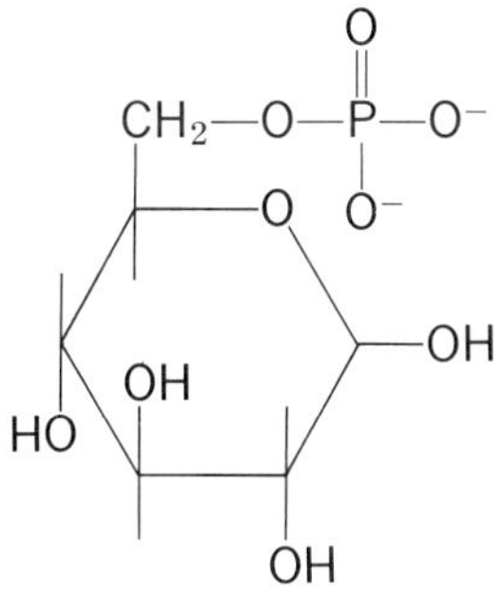

Glucose 6-phosphate
($\Delta G^\circ_{\text{hydr}} = -4000$ cal/mole)

The Gibbs free energy data for the hydrolysis of each compound are shown to indicate the extent to which they react with water as well how potentially useful this reaction might be as a source of energy. We recall that negative ΔG° values indicate that the equilibrium constant for the reaction will be greater than one. If we start with one mole each of reactants, at equilibrium the products will be favored over the starting materials. One worked example, the hydrolysis of α-glycerophosphate, illustrates the reaction.

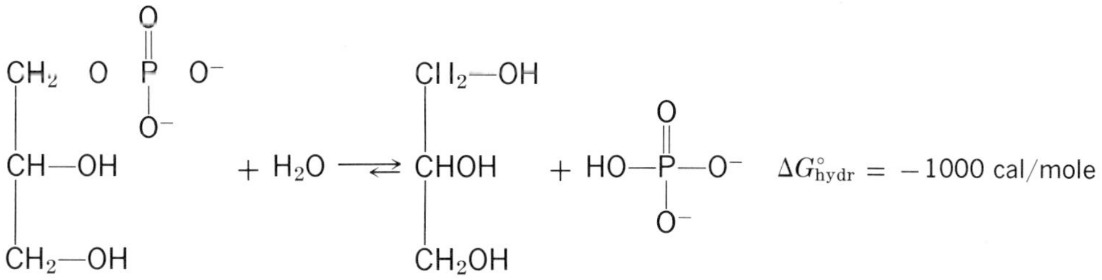

$$\alpha\text{-Glycerophosphate} + H_2O \rightleftharpoons \text{Glycerol} + HPO_4^{2-} \qquad \Delta G^\circ_{\text{hydr}} = -1000 \text{ cal/mole}$$

α-Glycerophosphate

Glycerol

Monohydrogen phosphate ion $(HPO_4)^{2-}$

Since

$$\Delta G^\circ = -2.303RT \log K$$

then

$$-1000 \text{ cal/mole} = -2.303 \times 1.987 \text{ cal/(°K)(mole)} \times 310\text{°K} \times \log K$$

where 310°K = 37°C, body temperature; or

$$\frac{1000}{2.303 \times 1.987 \times 310} = \log K \qquad \text{(the units having been canceled)}$$

$$0.7049 = \log K$$

$$K = 5.1$$

(Note that °K refers to degrees Kelvin whereas K refers to equilibrium constant.) In other words,

$$K = 5.1 = \frac{[\text{glycerol}][\text{HPO}_4{}^{2-}]}{[\alpha\text{-glycerophosphate}][\text{H}_2\text{O}]}$$

The calculation provides some generalizations we wish to carry forward.

—Negative ΔG° values mean K values greater than one.
—K values greater than one mean that at equilibrium the products tend to be favored.
—The larger the ΔG° value is in a *negative* sense, the larger the equilibrium constant and the more favored the products at equilibrium.

Thus, for the hydrolysis of glucose 6-phosphate for which $\Delta G^\circ = -4000$, the calculated equilibrium constant, at 37°C, is $K = 660$. This may be compared with $K = 5.1$ when $\Delta G^\circ = -1000$.

In connection with these generalizations, three major warnings must be stated. First, the expression "tend to be favored" can be seriously misleading. Second, we must keep in mind the difference between an equilibrium constant and a rate. Third, conditions of equilibrium seldom prevail in the body.

How "tend to be favored" is misleading can be illustrated in the reaction

$$\text{A} + \text{B} \rightleftharpoons \text{C} + \text{D} \qquad \Delta G^\circ = -1000 \text{ cal/mole}$$

$$K = 5.1 = \frac{[\text{C}][\text{D}]}{[\text{A}][\text{B}]}$$

What if, for example, A and B are allowed to react in a solution *that already has a large concentration of* C? The equation defining the equilibrium constant will still be correct, *but not much additional* C *will be made* and hence not much new D can be made either. To say then that the products "tend to be favored" is not very meaningful. This situation prevails throughout body chemistry. The consumption of one set of reactants is tied to the consumption of the products that they form or their physical removal by migration to another area. And whether the products are used up depends often on whether *their* products in turn have accumulated or are in demand. Thus a whole series of successive reactions can be controlled by just one or two factors, one or two concentrations. And this control relates to the second warning, the difference between a rate and an equilibrium constant or better, between a large negative free energy for a reaction and *how fast* it will proceed.

We use $\Delta G°$ values only as indications of *tendencies* to undergo specified reactions. Thus we can compare tendencies for α-glycerophosphate and glucose 6-phosphate to undergo hydrolysis by examining the values of $\Delta G°$ or of equilibrium constants for these reactions. We must remember, however, that these comparisons say nothing about *rates* of the reactions. They indicate only the relative completeness of each reaction *if it does proceed.* The presence or absence of a catalyst, for example, may influence the rate in a powerful way (but it will not itself affect the *position* of equilibrium). Of course, values of $\Delta G°$ also indicate the potentiality of the reaction for supplying useful energy.

Exercise 21.1 Perform calculations to show that the reaction of one mole of water with one mole of glucose 6-phosphate at 25°C will be roughly 96% complete when equilibrium is established. Thus, *given time,* glucose 6-phosphate will be almost completely hydrolyzed by water.

Our third warning cautions that very few individual reactions proceed to equilibrium in living cells. In fact, if they all did, we would be dead! Thermodynamic equilibrium for all the chemicals in the body is the very thing that must not occur if the organism is to live. From one point of view, life is the involved effort to prevent such equilibrium. For example, when a protein is in contact with water, equilibrium is represented by the completely hydrolyzed form, with only the constituent amino acids left. Because proteins inside the body are constantly in contact with water, their tendency to hydrolyze must somehow be counterbalanced; work in the form of chemical work or free energy must be expended. Protein repair and synthesis are but two energy-demanding processes that must go on for life and health. In a very general way the free energy made available while some nutrients are converted into waste products is required to overcome the tendencies toward a fatal equilibrium characteristic of all cellular constituents essential to the life of the organism. Free energy values for individual reactions as they occur in a test tube may seem to help little in understanding dynamic processes in living cells, but they are about the only quantitative data we have that indicate the size of the job of maintaining life against the tendencies toward equilibrium. In particular, they help us understand how processes with negative free energies can drive others with positive free energies. One very commonplace example occurring innumerable times daily, but outside the body, is the powering of an automobile. It is not natural for a car to move up a hill, but it is natural (spontaneous) for gasoline to burn. The second process, a free energy producer, powers the first, a free energy consumer. The engineering mechanics of coupling these two so that the one can power the other is familiar to all boys and very likely to most girls. The expanding gases from the combustion drive a piston connected in a clever way to a drive shaft which is geared to wheels that go around, carrying the car with them.

In cellular metabolism the coupling of free-energy-producing events with free-energy-consuming ones is obviously much less familiar but is really no more difficult to understand. To study it we must look at the second group of organophosphates, the high-energy phosphates.

Somewhat by definition, if $\Delta G°$ for the hydrolysis of a phosphate is no more negative than about -4000 cal/mole, it is called a low-energy phosphate. If $\Delta G°$ lies between about -5000 and $-11{,}000$ cal/mole, it is a high-energy phosphate. The structures of four such compounds are shown.

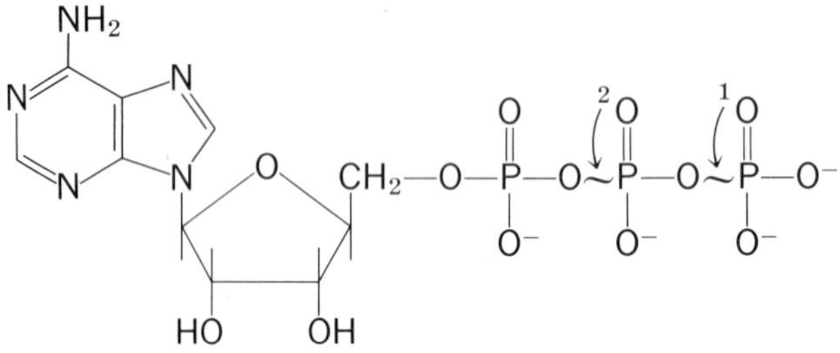

Adenosine triphosphate (ATP)
$\Delta G°_{hydr} = -7000$ cal/mole at bond 1
$= -6000$ cal/mole at bond 2

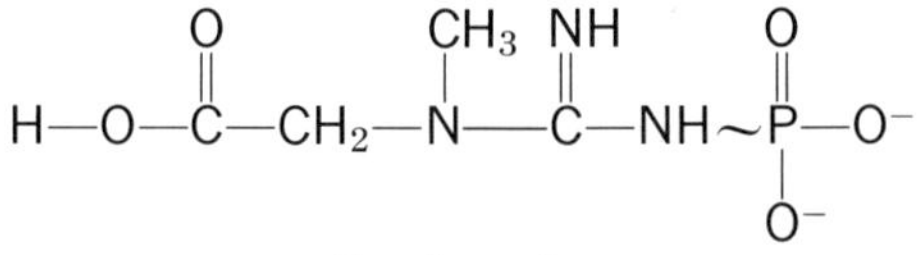

$$H{-}O{-}\overset{\overset{O}{\|}}{C}{-}CH_2{-}\underset{}{\overset{\overset{CH_3}{|}}{N}}{-}\overset{\overset{NH}{\|}}{C}{-}NH{\sim}\overset{\overset{O}{\|}}{\underset{\underset{O^-}{|}}{P}}{-}O^-$$

Phosphocreatine
$\Delta G°_{hydr} = -10{,}500$ cal/mole

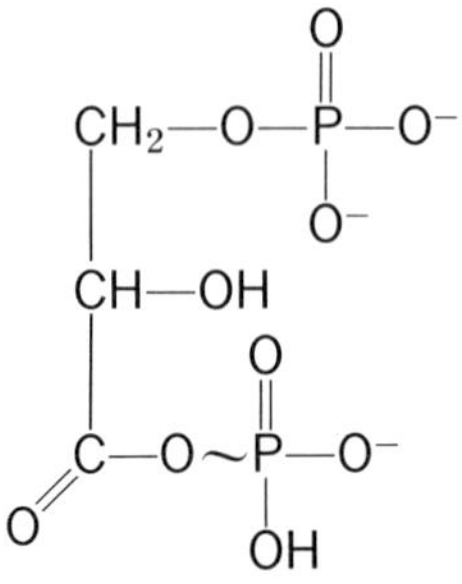

1,3-Diphosphoglyceric acid
$\Delta G°_{hydr} = -11{,}800$ cal/mole

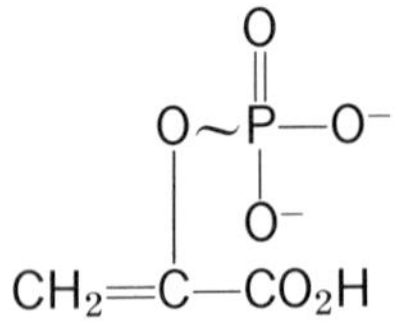

Phosphoenolpyruvic acid[2]
$\Delta G°_{hydr} = -12{,}800$ cal/mole

[2] Phŏs′phô·ē′nŏl·pĭ·ro͞ov′ĭc acid. Pyruvic acid is $CH_3{-}\overset{\overset{O}{\|}}{C}{-}\overset{\overset{O}{\|}}{C}{-}OH$. Its enol form (cf. p. 311) is $CH_2{=}\overset{\overset{OH}{|}}{C}{-}\overset{\overset{O}{\|}}{C}{-}OH$. Thus phosphoenolpyruvic acid is the phosphate ester of the enol form of pyruvic acid. The equation for its hydrolysis is

$$CH_2{=}\overset{\overset{OPO_3^{2-}}{|}}{C}{-}CO_2H + H_2O \longrightarrow HPO_4^{2-} + \left[CH_2{=}\overset{\overset{OH}{|}}{C}{-}CO_2H\right] \longrightarrow CH_3{-}\overset{\overset{O}{\|}}{C}{-}CO_2H$$

The enol of pyruvic acid spontaneously rearranges to the more stable keto form.

The squiggle, ~, indicates a bond that can be broken by the action of water or some other chemical. The $\Delta G°$ values given for each compound are for their hydrolysis and are only approximate. They refer to reaction with water at pHs that in each case make the ionic forms shown necessarily the ones present.[3] The experimental determination of these free energies is difficult, and account has not always been taken of the exact influence of pH on $\Delta G°$. The values cited represent orders of magnitude. Phosphoenolpyruvic acid is believed to have the highest free energy of hydrolysis of any metabolite. When $\Delta G°_{\text{hydr}}$ is $-11{,}000$ cal/mole, then K is 60,000,000. Such an equilibrium constant is so high that for all practical purposes if and when phosphoenolpyruvic acid reacts with water, it does so quantitatively. (Pyruvic acid and P_i are the products.)

Because of the large negative free energy changes that occur with hydrolysis of these high-energy phosphates, it has become general practice to refer to the bonds represented by squiggles as "energy-rich phosphate bonds." The free energy, it must be emphasized, is actually a property of the whole molecule, and the free energy released does not come from just one particular bond. Moreover, it is not the reaction with *water* in which we are the most interested. We are merely using the data for this reaction for comparison purposes.

Adenosine triphosphate, by far the most widely distributed of the high-energy phosphates, is found in the lowest and highest forms of life of both plants and animals. For the processes that are believed to occur in evolutionary change, nature long ago selected ATP as a universal medium of energy transactions. It has a prominence not shared by other high-energy compounds. Ordinary carboxylic acid anhydrides (as well as acid chlorides) also have large negative $\Delta G°_{\text{hydr}}$ values. Fritz Lipmann, one of the pioneers in focusing the attention of scientists on free energy factors in biochemical changes, suggested that the phosphate system, rather than anhydrides of carboxylic acids (or acid chlorides), is selected by living organisms not because it is more (or less) energy-rich. It is not. Rather the phosphate system reacts less rapidly with water at physiological pH conditions. (Again we are reminded that a large negative value of $\Delta G°_{\text{hydr}}$ does not necessarily mean that the system

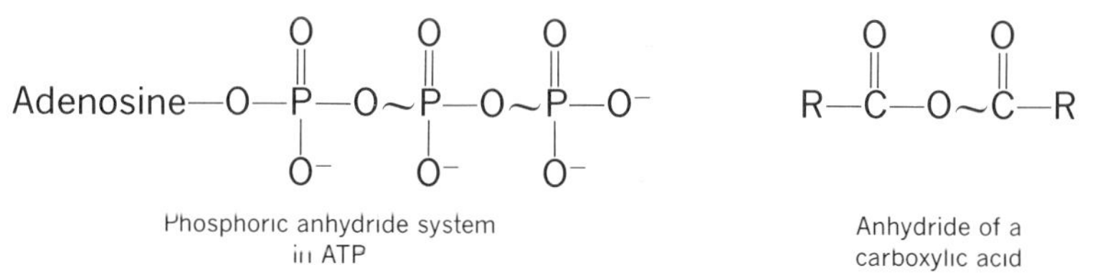
Adenosine—O—P—O~P—O~P—O⁻
R—C—O~C—R

Phosphoric anhydride system in ATP

Anhydride of a carboxylic acid

[3] Since these organophosphates are derivatives of phosphoric *acid*, they have ionizable hydrogens. The exact state of their ionization is determined by the pH of the medium. For example, at room temperature and in a solution of pH 5, ordinary phosphoric acid, H_3PO_4, exists almost entirely in the form of the dihydrogen phosphate ion, $H_2PO_4^-$. At the slightly alkaline pH of 8, it is in the form of the monohydrogen phosphate ion, HPO_4^{2-}. Between pH 5 and 8 there are essentially no phosphate ions, PO_4^{3-}, and no un-ionized phosphoric acid, H_3PO_4. In this range inorganic phosphate exists as a mixture of $H_2PO_4^-$ and HPO_4^{2-}. Having noted all this for the sake of accuracy, we shall now for the sake of simplicity ignore it and not try to be exact about the state of ionization of organophosphates or phosphoric acid. In fact, we shall frequently use the symbol P_i for inorganic phosphate *of whatever state of ionization.*

reacts *rapidly*.) The reaction of ATP with water at physiological pH and temperature is not so rapid that it causes a serious leakage of free energy.

Fritz Lipmann (1899–   ). United States biochemist. In 1953 he shared the Nobel prize in physiology and medicine with Hans Krebs for his discovery of coenzyme A. He was born in Germany, but he came to the United States and eventually became a citizen in 1944.

Bernard and Alberte Pullman of France have provided insights into the source of the energy wealth of the triphosphate system. By quantum mechanical calculations they have been able to assign numerical values to the partial charges in this network.[4]

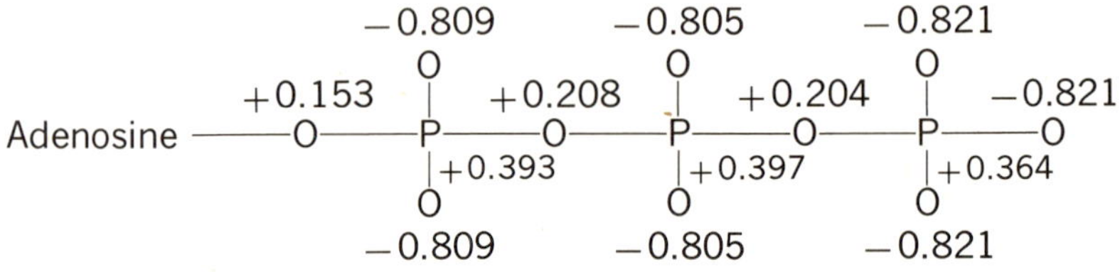

```
                    −0.809          −0.805          −0.821
                      O               O               O
          +0.153      |    +0.208     |    +0.204     |     −0.821
Adenosine ──────O─────P───────O───────P───────O───────P───────O
                      |+0.393         |+0.397         |+0.364
                      O               O               O
                    −0.809          −0.805          −0.821
```

ATP

The net charge, the algebraic sum of the partial charges, is −3.972 or, roughly, −4. The structure shown therefore refers to the quadruply ionized or maximally ionized ATP molecule. The significant feature of these results is that like charges are nearest neighbors. All along the chain O—P—O—P—O—P there is a succession of partial positive charges, and like charges repel. Around the oxygen-populated outskirts is a series of partial negative charges. As long as this molecule hangs together, there is an internal tension. Some relief can be expected by shortening that chain. Even by losing one phosphate unit, some gain in stability will be expected—and not just for the reason already suggested. Total internal energy will be further reduced by the increased resonance stabilization possible in the products (ADP and phosphate) and by increased opportunities for solvating the ions. Yet, as we have said, ATP does not react rapidly with water, especially when compared with other high-energy systems such as acid chlorides and anhydrides. To account for this lower rate we might seek a steric explanation. As depicted in Figure 21.1, the "least-resistance" or lowest-energy approach by a nucleophile such as water to the carbonyl carbon of a carboxylic acid anhydride or chloride is along a line somewhat perpendicular to the plane in which the carbonyl carbon and its three attached groups lie. The four groups attached to the phosphorus of the terminal phosphate unit in ATP, however, do not lie in a plane. The arrangement is tetrahedral as

[4] A. Pullman and B. Pullman, "π Molecular Orbitals and the Process of Life," in *Molecular Orbitals in Chemistry, Physics, and Biology,* edited by P. Löwdin and B. Pullman, Academic Press, New York, 1964, page 561.

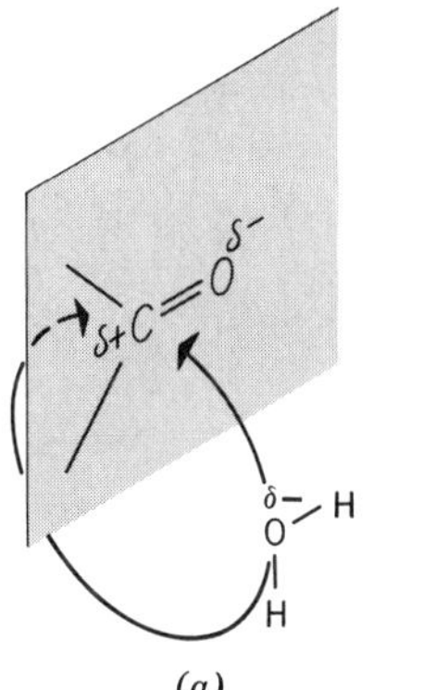
δ−
δ+
δ−
H
O
H

(a)

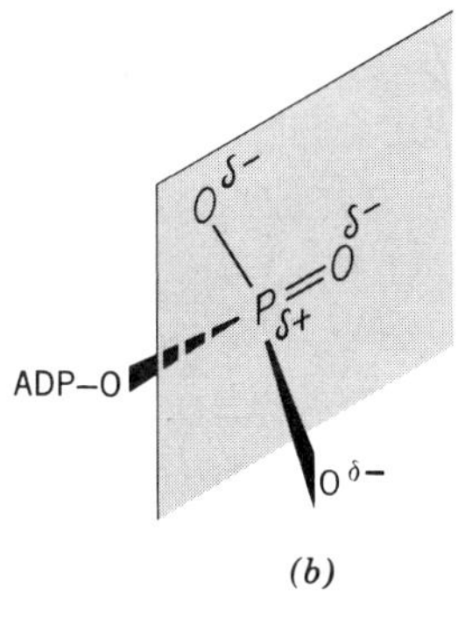
δ−
O
δ−
O
P
δ+
ADP—O
O δ−

(b)

Figure 21.1 Speculation about the lower reactivity of ATP toward water when compared with those of anhydrides and acid chlorides derived from carboxylic acids. Even though these three are "high-energy" systems, the phosphoric acid anhydride network is less reactive. (*a*) Attack by a nucleophile such as water at the carbonyl carbon of an anhydride or acid chloride derived from a carboxylic acid is made relatively easy by the openness of the approach from either side and by the partial positive change at this site. (*b*) Attack by a nucleophile at the phosphorus of a terminal phosphate unit of ATP is hindered because less space is available through which to make an approach, and the neighboring oxygens have partial negative charges that tend to repel an electron-rich nucleophile.

in methane and is therefore more crowded. Moreover, for the nucleophile to approach the phosphorus from almost any angle, it must pass through a region of negative charge density created by the oxygens.

We are faced with the obvious conclusion that the *energy of activation* for hydrolysis of ATP must be *high* if the *rate* for it is *low*. And this conclusion may be interpreted as meaning that of all the collisions or potential collisions between water molecules and the terminal phosphorus in ATP, only a very small fraction, those with potentially the highest collision energy, are successful. This fraction of fruitful collisions is dependent on obstacles in the path of the two collision centers, the terminal phosphorus in ATP and the oxygen in water. There are more obstacles in the ATP system, but the energy payoff for a successful collision is still high. Thus ATP is apparently uniquely adapted to be the coinage for a large number of energy exchanges in the cellular realm. Foreign molecules bearing free energy (e.g., glucose, fatty acids) must have it converted into this form. We are now in a better position to study how this conversion takes place.

Nonscientific Interlude. "Science means most not to the person who thinks it means everything but to the one who is most keenly aware also of man's search for beauty, his capacity for enjoyment of life and his desire for personal fulfillment."[5]

Man's search for truth has sometimes been inspired by an awareness of

[5] The passage, from a publication of the Society of Friends, was approvingly quoted in an autobiographical sketch by Max Rudolf Lemberg, "Chemist, Biochemist, and Seeker in Three Countries," *Annual Review of Biochemistry*, Vol. 34, 1965, page 1.

beauty. The beauty of a living organism prompts in some the query "I wonder how it works?" and the plunge is made into laboratories and test tubes, into chemicals, instruments, blackboards, and computors, and into frustrations. The leap may carry the seeker so deep that the awareness of beauty is buried in the profusion of bare fact. But then the pieces begin to fit, the picture becomes clearer, and the sense of standing before what is beautiful is greatly heightened by the deeper insights into how the organism works. From its rim the Grand Canyon of the Colorado River in northern Arizona is beautiful and awe-inspiring. Idle chattering dies away as one stands in silence before it, as before a great work of art. Those who plunge sweating and toiling into its depths temporarily give up this view. When they return, however, the canyon is for most people even more moving than before, and their lives are richer for the experience, even though they enjoy it only once in a lifetime.

We are about to study some of the most fundamental aspects of human life, the rock-bottom conditions without which life in any fuller meaning would not be possible. Of course, there is great beauty even in rocks, especially if they stand in some relation to one another, fitted together in an intricate pattern which is colorful, harmonious, or simply rugged. Analogies now fail us, however. Rocks are commonplace. Canyons are familiar to anyone beyond grade school, even if only from pictures. We have already been traveling in strange country, and what we are about to enter is even stranger. The beauty of it cannot be anticipated. It can only be discovered, and this requires both effort and an open mind. No guarantee can be made that the trip will be worthwhile, but even hardened cynics have returned from this land with a recovered sense of awe and wonder.

We shall be looking at some structural formulas of enormous complexity. It should ease the reader's way to know that he is not expected to memorize them, for they can always be referred to again. They should be treated like maps, for which we have learned a number of important map signs in studying the organic functional groups and their properties. Sometimes, by shading key portions of complicated structures, we can draw attention to particularly important parts and make it easier to follow reaction pathways. Once we attain some visual familiarity with the structure, we can simplify many matters by using highly condensed abbreviations. (We have already done this; ATP is an example.) We might ask why these structures must be so complicated if only certain parts react. The answer probably has something to do with lock and key requirements of enzyme-controlled events.

We shall be examining some sequences of reactions, some metabolic pathways that are both long and intricate. The reason why the synthesis of ATP occurs in so many steps probably has something to do with the problem of delivering an enormous amount of free energy, the difference between that contained in the final products and that of the initial metabolites (glucose or fatty acids and oxygen). If the starting materials went directly to carbon dioxide and water, and if all the energy were delivered at once, much more of it would be wasted as heat. By delivering this energy in small packages in several successive steps, it is more easily managed and controlled. Less of the available energy is lost as heat and more is retained in storage as ATP.

BIOENERGETICS

RESPIRATION

In everyday usage respiration means breathing, but scientists have found it convenient to give the word other meanings. In this section the word describes the chemical events most closely connected with the reduction of molecular oxygen to water. Reference will be made, for example, to "respiratory enzymes" and to the "respiratory chain," a series of chemical events leading to oxygen consumption. We intend to illustrate in some depth the connection between the synthesis of ATP and the uptake of atmospheric oxygen.

Our study of the molecular basis of muscular work began in Chapter 18, and proceeded as far as connecting mechanical work with the conversion of ATP to ADP and P_i. The rest of the story is the use of oxygen and some nutrient (e.g., glucose or fatty acids) to restore the ATP. The reduction of oxygen to water requires for each oxygen atom a pair of hydrogen ions and a pair of electrons, or in its barest statement,[6]

$$:\ddot{\underset{\cdot\cdot}{O}} + : + 2H^+ \longrightarrow H:\ddot{\underset{\cdot\cdot}{O}}:H$$

Viewed this way, the problem is threefold: (1) how to obtain a pair of electrons from an organic molecule and transport them to oxygen, (2) how to obtain the hydrogens, and (3) how to deliver the large free energy associated with the reduction of oxygen at least partly into storage as ATP.

MAJOR OXIDATIVE PATHWAYS. AN OVERVIEW

Two of our foods, the carbohydrates and lipids, provide us with virtually all our energy for living. Each follows its own pathway for several steps, and then they converge into two of the most important metabolic sequences in the body, the *citric acid cycle* and the *respiratory chain* (Figure 21.2). Before studying each in detail, let us first describe them in a general way.

Glycolysis. The word may be thought of as derived from "glycogen loosening" (Greek *lysis,* act of loosening) or from "glycose loosening" (glycose is a generic name for any of the isomeric aldohexoses, including glucose). Glycolysis is a series of reactions in the body by which glucose is broken down to lactic acid. Because oxygen is not required, the reactions are sometimes called the *anaerobic sequence.* The sequence accomplishes two things: it powers the synthesis of a small amount of ATP, and it yields a chemical that can be passed eventually to the citric acid cycle (leading to more ATP). Leaving the details for a later chapter, we find that the over-

[6] We do not believe of course that electrons are really available in the way indicated by this purely hypothetical equation. The equation merely serves to trim the problem to its minimum statement.

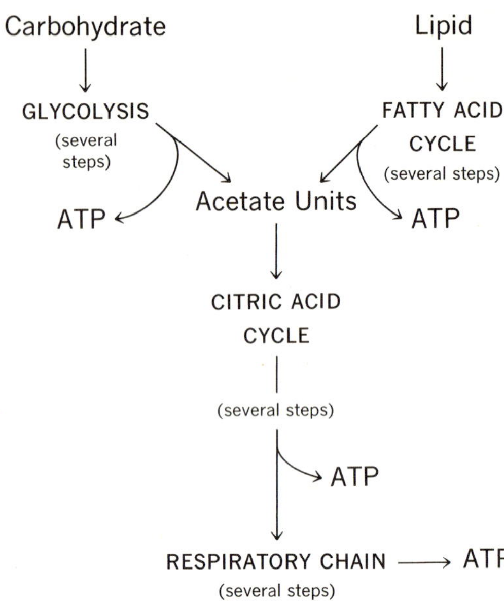
Carbohydrate
Lipid
GLYCOLYSIS
(several steps)
FATTY ACID CYCLE
(several steps)
ATP
Acetate Units
ATP
CITRIC ACID CYCLE
(several steps)
ATP
RESPIRATORY CHAIN
ATP
(several steps)

Figure 21.2 Major oxidative pathways in the body.

all change balances as follows:

$$\underset{\text{Glucose}}{C_6H_{12}O_6} + 2ADP + 2P_i \xrightarrow{\text{glycolysis}} \underset{\text{Lactic acid}}{2CH_3\underset{\displaystyle OH}{\underset{|}{C}}HCO_2H} + 2ATP$$

Fatty Acid Cycle. In this series of reactions a fatty acid molecule has its chain shortened two carbons at a time. The two-carbon fragments, acetate units, are fed into the citric acid cycle leading to ATP production. The shortened fatty acid goes around the fatty acid cycle again and again until it has no more two-carbon units to lose. The cycle itself produces considerable ATP in its many turns, but it does so indirectly through the participation of the respiratory chain. The details will be given in Chapter 23. The provision of a raw material, acetate units, for the citric acid cycle interests us here.

Citric Acid Cycle. In this sequence a four-carbon carrier molecule picks up a two-carbon acetate unit from either glycolysis or the fatty acid cycle. The resulting six-carbon molecule, citric acid, is successively degraded until two carbons have been broken off as carbon dioxide, hydrogen and electrons have been fed to the respiratory chain for ATP production, and the four-carbon carrier molecule has been regenerated. The cycle is then ready for another "turn."

Respiratory Chain. The chain may be regarded as a series of reactions, as a series of carrier molecules, or as both, depending on the context of the discussion. It is the mechanism for delivering the elements of hydrogen ($H:^-$ and H^+) to molecular oxygen taken in by breathing. Water, of course, is the end product. The chief purpose of the chain is to tap some of the energy potentially available from reducing oxygen to water and to use this energy in powering the production of ATP.

ATP Production from the Major Oxidative Pathways. A Summary. The overall equation for the oxidation of glucose in the body is

$$C_6H_{12}O_6 + 6O_2 + 36ADP + 36P_i \longrightarrow 6CO_2 + 42H_2O + 36ATP$$

This may be regarded as the sum of the following reactions:

$$\begin{array}{lll} C_6H_{12}O_6 + 6O_2 \longrightarrow 6CO_2 + 6H_2O & & \Delta G = -680{,}000 \text{ cal} \\ 36ADP + 36P_i \longrightarrow 36ATP + 36H_2O & & \Delta G = +324{,}000 \text{ cal} \\ & \text{Net} & \Delta G = -356{,}000 \text{ cal} \end{array}$$

Considering the amount of free energy available and the amount actually used, the efficiency of the operation, in terms of conserving free energy, is

$$\frac{324{,}000}{680{,}000} \times 100 = 48\%$$

Thus when the body uses glucose to obtain energy (as ATP), roughly 48% of what is potentially available is actually conserved.

The complete oxidation of a typical fatty acid, palmitic acid, and the simultaneous production of ATP proceeds by the equation

$$CH_3(CH_2)_{14}CO_2H + 23O_2 + 130ADP + 130P_i \longrightarrow 16CO_2 + 146H_2O + 130ATP$$

This breaks down as follows:

$$\begin{array}{lll} CH_3(CH_2)_{14}CO_2H + 23O_2 \longrightarrow 16CO_2 + 16H_2O & & \Delta G = -2{,}340{,}000 \text{ cal} \\ 130ADP + 130P_i \longrightarrow 130H_2O + 130ATP & & \Delta G = +1{,}170{,}000 \text{ cal} \\ & \text{Net} & \Delta G = -1{,}170{,}000 \text{ cal} \end{array}$$

The efficiency is

$$\frac{1{,}170{,}000}{2{,}340{,}000} \times 100 = 50\%$$

That is, of all the free energy potentially available from the oxidation of palmitic acid, the body taps 50% of it to make ATP.

Regulation of ATP Production. Homeostasis. Many factors must be present to make ATP production possible. Organic nutrients are needed to provide electrons and hydrogens for the reduction of oxygen; oxygen, of course, is needed; ADP and P_i must be available; all the enzymes must be functioning. Let us therefore assume that the subject is healthy, his diet has been adequate, nutrients are available, and he is breathing normally. Suppose that he has been resting and is completely relaxed. It is at least probable that his ATP production will not be as rapid as that of an identical subject who happens to be exercising, for the synthesis of ATP is tied to the demand for it.

For the reaction

$$ADP + P_i \longrightarrow ATP$$

we must think in terms of the ratio

$$\frac{[ATP]}{[ADP][P_i]}$$

When this ratio is high, there is probably not much tendency to make it even higher, that is, not much tendency to make more ATP. Conversely, when the ratio is small, the time is ripe to make fresh ATP. When this ratio drops, because ATP has been used in some energy-demanding process, reactions start spontaneously to bring it back to normal. The response works like a thermostat, a device for translating information (room temperature has dropped) into action (the furnace is turned on) until new information (the temperature comes back up) shuts the system off again. In engineering this reponse is called an inverse feedback mechanism, in molecular biology *homeostasis*. Walter Cannon coined this word for the behavior of an organism toward stimuli (e.g., an energy-demanding event uses up ATP and the ratio changes) that commences a series of metabolic events (e.g., respiratory chain) to restore the system to normal (e.g., produce ATP to replace what was used). Several conditions in the human body are so constant that diseases can be diagnosed if they change measurably. Body temperature is perhaps the most obvious. The pH of the blood, its white cell count and red cell count, and concentrations of various dissolved substances are others. Mechanisms that operate to keep these constant are called homeostatic.

Walter Bradford Cannon (1871–1945). Professor of physiology at Harvard from 1906 to 1942, Cannon was one of the greatest physiologists of the United States.

Cellular respiration is stimulated not so much by the presence of oxygen as by some activity (e.g., muscular work) that uses up ATP and generates ADP plus P_i. A rise in concentration of ADP and P_i triggers reactions that use them, reactions that lead to ATP. We therefore begin our more detailed study of bioenergetics with the sequence that stands closest to actual oxygen consumption, the respiratory chain.

THE RESPIRATORY CHAIN

Of the many major metabolic sequences in the body, our knowledge of the respiratory chain is still the most sketchy. The following discussion is a reasonable approximation of the state of our knowledge in this area at the present time. The respiratory enzymes are quite well known and understood. The sequence of electron transport from enzyme to enzyme is also believed to be fairly well known, although our discussion will not include all the details. (There will be, in fact, considerable simplification, and it is acknowledged that what is simplification for one is oversimplification for another or vexatious difficulty for still another.) The coupling of the flow of electrons down the respiratory chain of enzymes to the conversion of ADP and P_i into ATP is the most speculative part of the explanation. We shall be talking about coupling enzymes and transfer enzymes, but thus far they are as much idea as hard fact. Evidence exists, to be sure, but a goodly portion of what follows is just one plausible theory that explains many aspects of *oxidative phosphorylation*—that is, the making of ATP by using energy from oxidation. We begin with the best-understood aspect of the respiratory chain, its enzymes.

RESPIRATORY ENZYMES AND COENZYMES

Mitochondria. The enzymes whose coenzymes are to be described are organized primarily in membranes of mitochondria (Greek: *mitos,* a thread; *chondros,* a grain). Frequently called the "powerhouses of cells," they are the primary sites for oxidative phosphorylation. Individual cells can have hundreds, sometimes thousands, of these cigar- or sausage-shaped mitochondria (Figure 21.3). The inside surface of

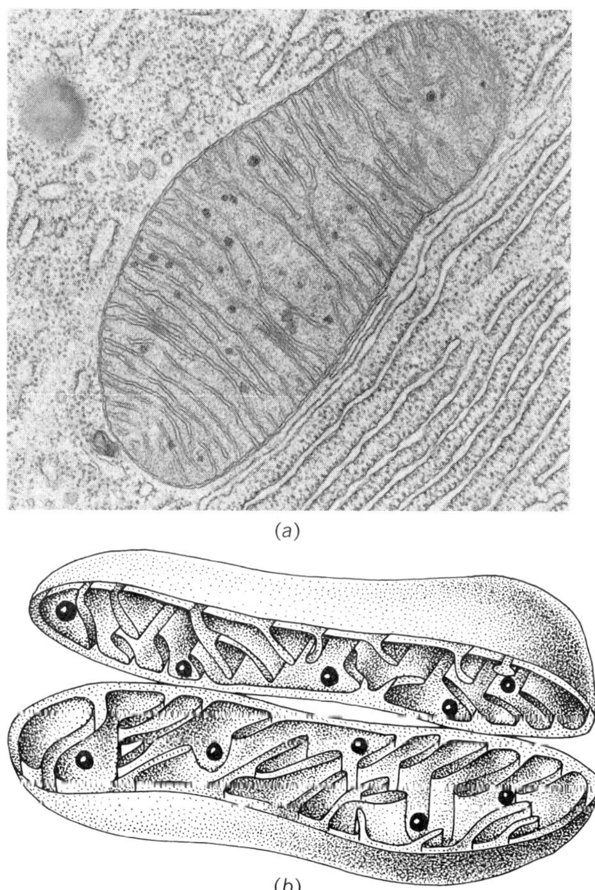
(a)
(b)

Figure 21.3 A mitochondrion. (*a*) An electron micrograph of a mitochondrion in a pancreas cell of a bat (×53,000). (Courtesy of Dr. Keith R. Porter.) (*b*) Perspective drawing of an "opened" mitochondrion. The dots represent units of mitochondrial activity. Larger granules of uncertain composition are also present. (Courtesy of G. E. Nelson, G. G. Robinson, and R. A. Boolootian, *Fundamental Concepts of Biology,* John Wiley and Sons, New York, 1967, page 40.)

the inner membrane and the outside surface of the outer membrane of the mitochondrion are covered with thousands of tiny particles, each one an elementary unit of mitochondrial activity. A single mitochondrion in the flight muscle of a wasp, for example, has as many as 100,000 such units—a reflection of the exceedingly intense energy demands within this particular tissue. The particles on the outside membrane of the mitochondrion accept electrons and hydrogens from suitable donors which we generally call metabolites.[7] These particles also catalyze reactions powered by ATP. The inner particles convey the electrons along the chain of respiratory enzymes until ATP is made and atmospheric oxygen is reduced to water. These enzymes make up about 15% of the proteins of the membranes. They are therefore probably important not only to cell *function* but also to mitochondrial structure.[8] The enzymes of the citric acid cycle are also associated with mitochondrial structure. Thus mitochondria can receive metabolites, ADP, P_i, and oxygen, and they can discharge carbon dioxide, water, and ATP.

Coenzymes of the Respiratory Chain. Only four types of coenzymes or prothestic groups—pyridine nucleotides, flavin nucleotides, ubiquinones, and cytochromes—participate intimately in the enzymes of the respiratory chain. We may ignore the protein portions of these enzymes, the apoenzymes that give binding site uniqueness to the individual catalysts, and study only the respiratory coenzymes.

Pyridine Nucleotides. NAD^+ and $NADP^+$. A *nucleotide* is a compound made from a molecule of phosphoric acid, plus a five-carbon sugar, plus one of about half a dozen heterocyclic amines. In addition to being building blocks for genes (cf. p. 664), they are parts of several coenzymes. Adenosine monophosphate (AMP) is a typical nucleotide.

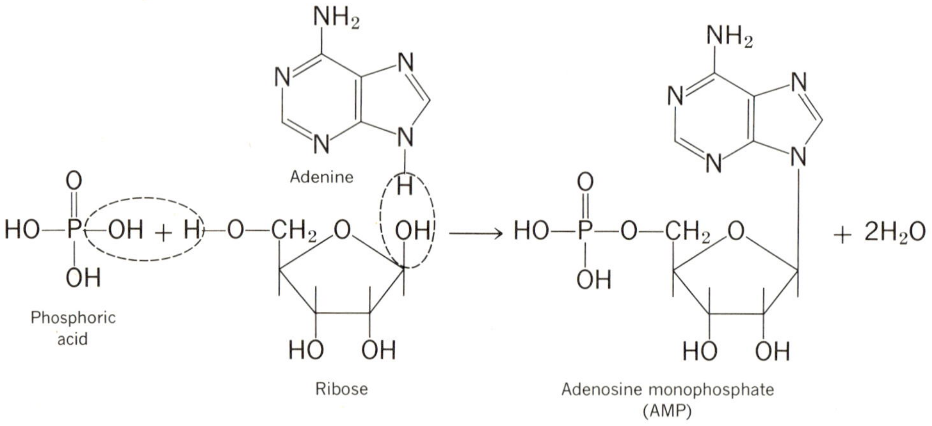

NH2
N
N
N
N
Adenine
H
O
HO—P—OH + H—O—CH2 O OH ⟶ HO—P—O—CH2 O + 2H2O
OH
Phosphoric acid
HO OH
Ribose
NH2
N
N
N
N
OH
HO OH
Adenosine monophosphate (AMP)

[7] We use this word to mean an intermediate that is produced by chemical action on some *nutrient* (any product of digestion) and will eventually be used for another chemial change. The word metabolite therefore arbitrarily excludes enzymes and final products such as urea, water, and carbon dioxide.

[8] This fact accounts for the major difficulty in unraveling the workings of the respiratory chain. Experiments designed to isolate its enzymes and coenzymes disrupt their cooperative existence. When separated from each other, they sometimes fail to give the reactions of the intact mitochondrion. Furthermore, deducing what happens *in vivo* from activity *in vitro* is always open to question.

One of the important coenzymes of the respiratory chain is a derivative of AMP, *nicotinamide adenine dinucleotide* (NAD^+). Its structure consists of two nucleotides joined together at the phosphate groups. One nucleotide has adenine as its side chain amine, the other nicotinamide, one of the vitamins. Only the shaded portion of the complicated structure, where the catalytic activity occurs, concerns us now, although the rest is essential in ways not well understood.

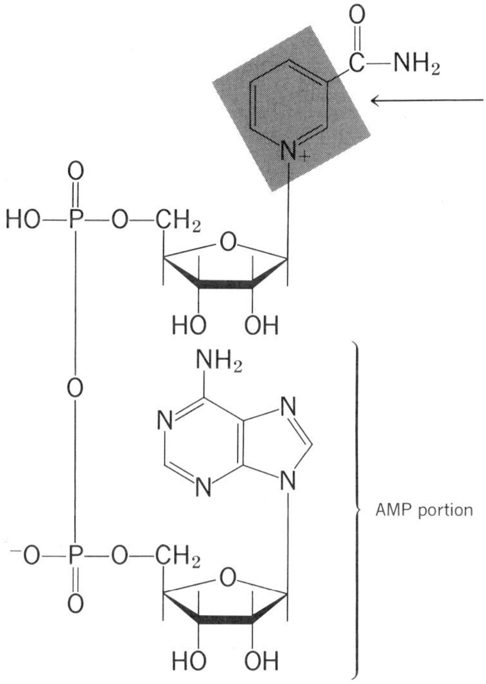

Pyridine unit (shaded)—the hydride acceptor in NAD^+

AMP portion

Nicotinamide adenine dinucleotide (NAD^+)

At its pyridine unit NAD^+ is a hydride acceptor. The following equation shows how it accepts the hydride:

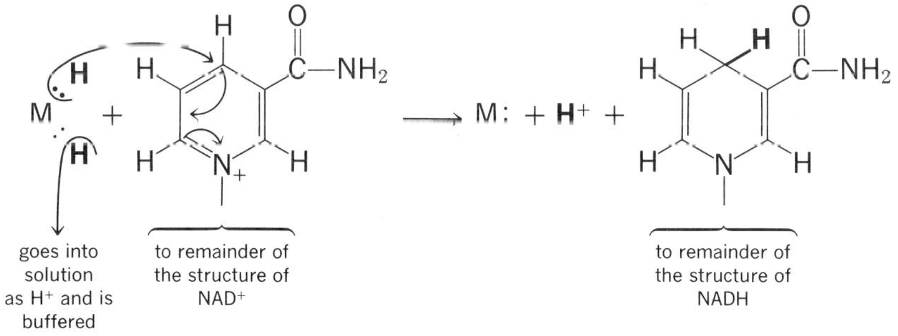

$$M\mathbf{H}_2 + NAD^+ \longrightarrow M: + \mathbf{H}^+ + NADH$$

goes into solution as H^+ and is buffered

to remainder of the structure of NAD^+

to remainder of the structure of NADH

Boldface indicates the pair of electrons and the two hydrogens that are about to depart from the metabolite, MH_2. Stated in more simplified form, the equation may be written as

$$MH_2 + NAD^+ \longrightarrow M\!: + H^+ + NADH$$

Or, to use a method of organizing biochemical events that we shall frequently employ,

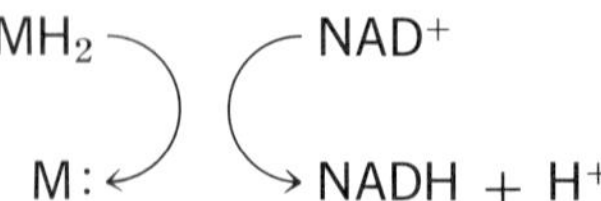
MH_2
NAD^+
M:
$NADH + H^+$

where MH_2 is shorthand for any intermediate or metabolite that can lose the elements of hydrogen (i.e., $H:^-$ and H^+)

NAD^+ will henceforth be our only symbol for the *oxidized* form of nicotinamide adenine dinucleotide (read as "en–ay–de–plus")[9]

M: represents what is left of the metabolite after it has lost the elements of hydrogen

NADH will henceforth be our only symbol for the *reduced* form of nicotinamide adenine dinucleotide.

For many metabolites, the equation given represents the first step in a journey that a pair of electrons makes from their molecules down the respiratory chain for final delivery to molecular oxygen. The hydrogens trail along, and water is the final product. Our bodies do not have little metallic wires to conduct electrons. Electrons must be passed from enzyme to enzyme, but we still consider the process electron flow. We find NAD^+ not only in enzymes bound to the surfaces of mitochondria but also in soluble enzymes.

A close relative of NAD^+, *nicotinamide adenine dinucleotide phosphate* ($NADP^+$), another of the pyridine nucleotides,[10] is encountered less often. It functions in the same way as NAD^+; at least the equation for its action is very similar:

$$NADP^+ + MH_2 \longrightarrow M\!: + H^+ + NADPH$$

When the body converts excess sugar to fat, a considerable amount of chemical reduction is necessary. A molecule of a fatty acid has many more carbon-hydrogen bonds than does a molecule of glucose. The principal supplier of hydrogen for this reduction is NADPH.

[9] In some contexts it is simply written NAD. The plus sign is useful in writing balanced equations for hydride transfers, but the actual *net* electric charge on the molecule will be a function of the pH of the immediate environment. There is, after all, a diphosphate system present; in a footnote on page 581 the policy followed in this text for such systems is stated.

[10] The symbols introduced in this section for the pyridine nucleotides conform to the recommendations of the Commission on Enzymes of the Joint Commission on Biochemical Nomenclature of the International Union of Pure and Applied Chemistry and the International Union of Biochemistry, 1965. These symbols have been recommended as replacements for older ones which will undoubtedly continue to be used for some time in many texts and references. Thus NAD^+ was formerly DPN^+ (for diphosphosphopyridine nucleotide) and $NADP^+$ was formerly TPN^+ (for triphosphopyridine nucleotide). The older terms are considered objectionable because in neither compound is a "phospho-" group attached to a pyridine ring.

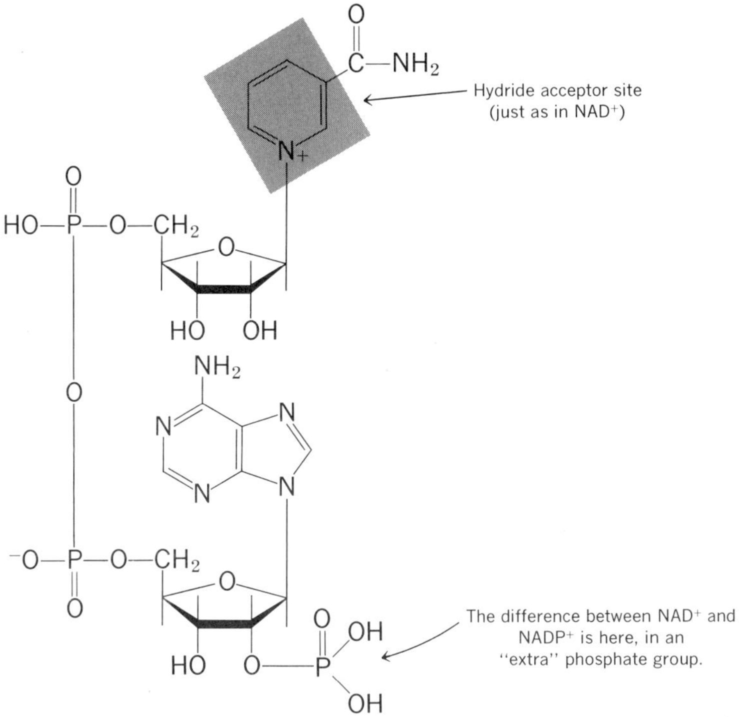

Hydride acceptor site
(just as in NAD^+)
The difference between NAD^+ and $NADP^+$ is here, in an "extra" phosphate group.

Nicotinamide adenine dinculeotide phosphate ($NADP^+$)

Flavin Nucleotides. Flavoproteins. A large group of respiratory enzymes use as their cofactor one of two derivatives of riboflavin, one of the vitamins. They are

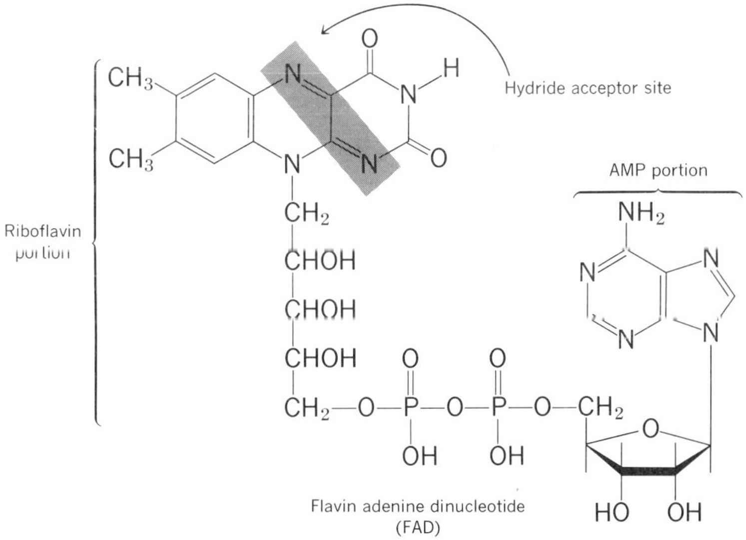

Hydride acceptor site
AMP portion
Riboflavin portion

Flavin adenine dinucleotide (FAD)

flavin mononucleotide (FMN) and *flavin adenine dinucleotide* (FAD). We shall be concerned for the most part with FAD, particularly the shaded portion of its structure. The combination of either FMN or FAD with an apoenzyme is called a flavoprotein (FP). The flavoproteins catalyze the removal of hydride ion, H:⁻, and hydrogen ion, H^+, from a metabolite:

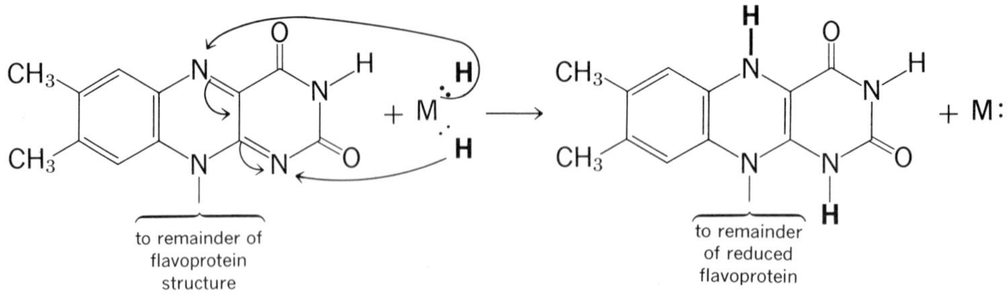

to remainder of flavoprotein structure

to remainder of reduced flavoprotein

or

$$FAD + MH_2 \longrightarrow FADH_2 + M:$$

or

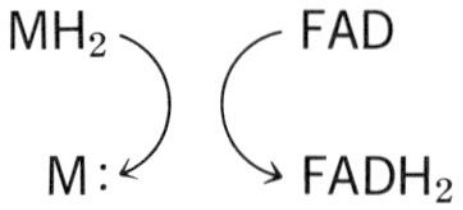

$$MH_2 \rightarrow M: \quad FAD \rightarrow FADH_2$$

An important source of hydrogen for this reaction is a reduced pyridine nucleotide:

$$H^+ + NADH + FAD \longrightarrow NAD^+ + FADH_2$$

This reaction is a key part of the respiratory chain and the second stage in the trip of a pair of electrons and a pair of protons from a metabolite to oxygen. Putting the first two stages together, we have

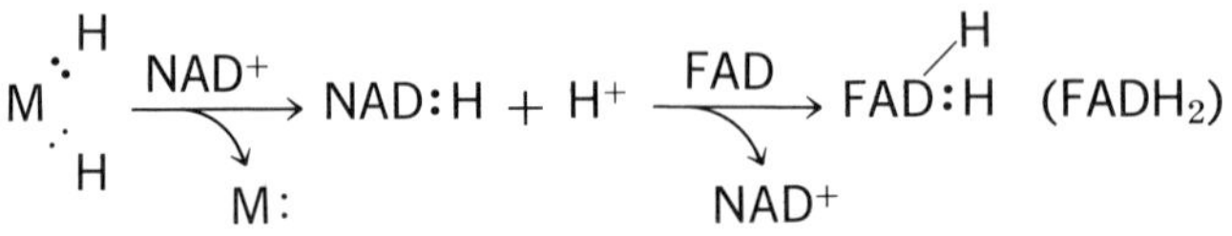

$$M{:}H_2 \xrightarrow[M:]{NAD^+} NAD{:}H + H^+ \xrightarrow[NAD^+]{FAD} FAD{:}H \text{ (}FADH_2\text{)}$$

or

$$MH_2 \rightarrow M: \quad NAD^+ \rightarrow NADH + H^+ \quad NADH + H^+ \rightarrow NAD^+ \quad FAD \rightarrow FADH_2$$

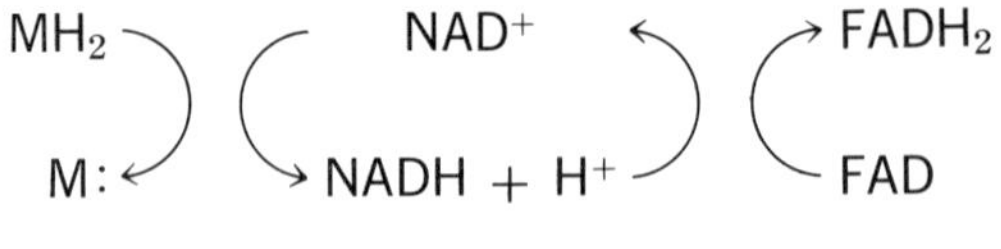

Ubiquinones. Coenzyme Q. The ubiquinones are a group of coenzymes having the general structure shown. The name ubiquinone reflects both structure and occurrence. The coenzymes are quinones, and they are found not just in mitochondria but also in cell nuclei, microsomes, and elsewhere. Hence the contraction of "ubiquitous quinones" to ubiquinones. We use the synonym coenzyme Q (CoQ).

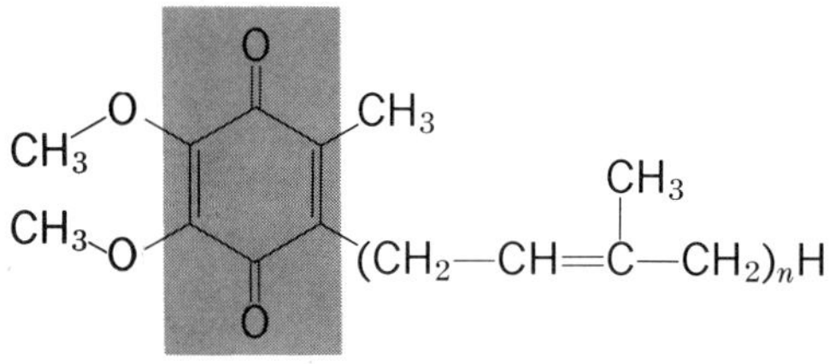

Ubiquinone, Q_n (coenzyme Q, CoQ) $(n = 6, 7, \ldots, 10)$

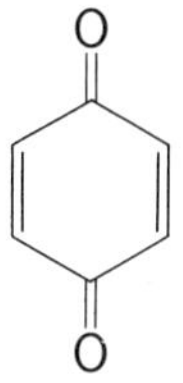

p-Quinone (mp 116°C)

The shaded portion of the structure, the quinone system itself, is of interest here. *One* of the ways by which coenzyme Q *appears* to be involved is in hydrogen transfer from $FADH_2$, a reaction which may be represented as

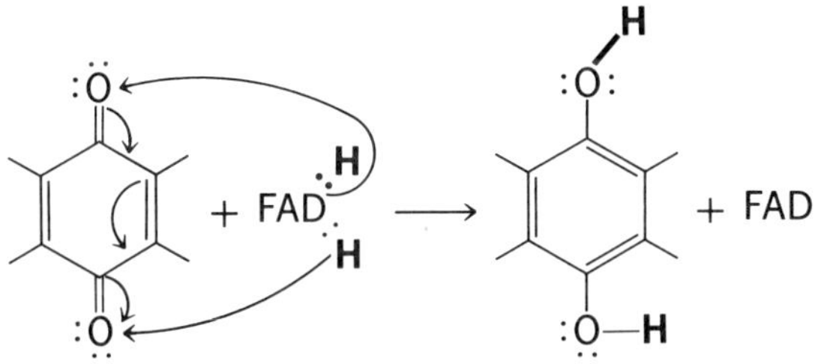

Quinone system + FAD:H₂ ⟶ Hydroquinone system (or reduced quinone) + FAD

Quinone system

Hydroquinone system (or reduced quinone)

or $$CoQ + FADH_2 \longrightarrow CoQH_2 + FAD$$

To keep current the log of the electron and proton travels, this is the third stage in those sequences involving the pyridine nucleotides. Putting the three stages together, we have

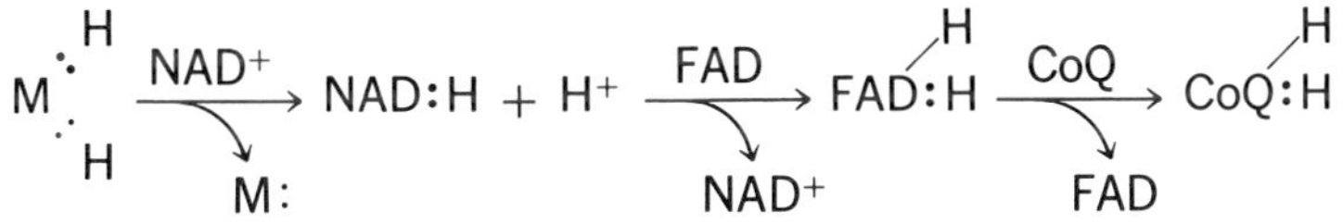

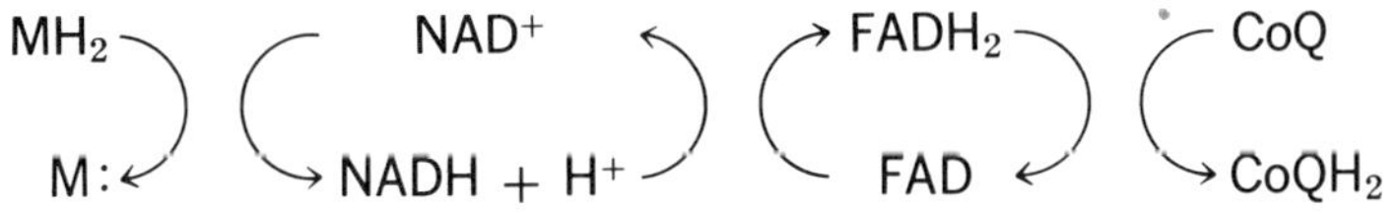

$$M{:}H_2 \xrightarrow{NAD^+} NAD{:}H + H^+ \xrightarrow{FAD} FAD{:}H_2 \xrightarrow{CoQ} CoQ{:}H_2$$
(with M:, NAD^+ and FAD released at the successive steps)

$$MH_2 \rightarrow M: \quad NAD^+ \rightarrow NADH + H^+ \quad FAD \rightarrow FADH_2 \quad CoQ \rightarrow CoQH_2$$

The pair of electrons has been passed from MH_2 to NAD^+ and then to FAD and then to CoQ. Hydrogens (more correctly, protons) have gone along, but the trip is only about half complete. From here on the electrons are handed successively through a series of cytochromes, the last group in the respiratory chain.

Cytochromes. Several members of this family are known, but only general characteristics concern us. Cytochromes belong to the family of *hemoproteins,* which includes hemoglobin (cf. p. 479). As their name implies (Greek: *kyto-,* cell;

chroma, pigment), they are colored. The molecular unit responsible for the color is the *porphin* skeleton, made of four pyrrole units linked by —CH— bridges (Figure 21.4).

Substituted porphins, called *porphyrins,* are capable of forming organometallic complexes with several metal ions, notably Fe^{2+} and Fe^{3+}. The structure of one of the cytochromes, *cytochrome c,* is shown in Figure 21.5. The iron-porphyrin unit is partially wrapped by a protein coat. Just how electrons can shuttle in and out of these units is not understood, and the structures are so complicated that some simplification is obviously necessary. We therefore refer only to the iron ions. One Fe^{3+} ion is able to accept one electron and become Fe^{2+}, rather than becoming Fe^{+} or Fe^{0} by taking more. Hence, if a cytochrome is to accept a *pair* of electrons, we must visualize two Fe^{3+} ions being present. Not enough is known about the cytochromes, however, to say categorically how many metallic ions are part of a single enzyme. To strike a chemical balance, we imagine the involvement of *two* iron ions in one enzyme. No effort will be made to analyze current speculations

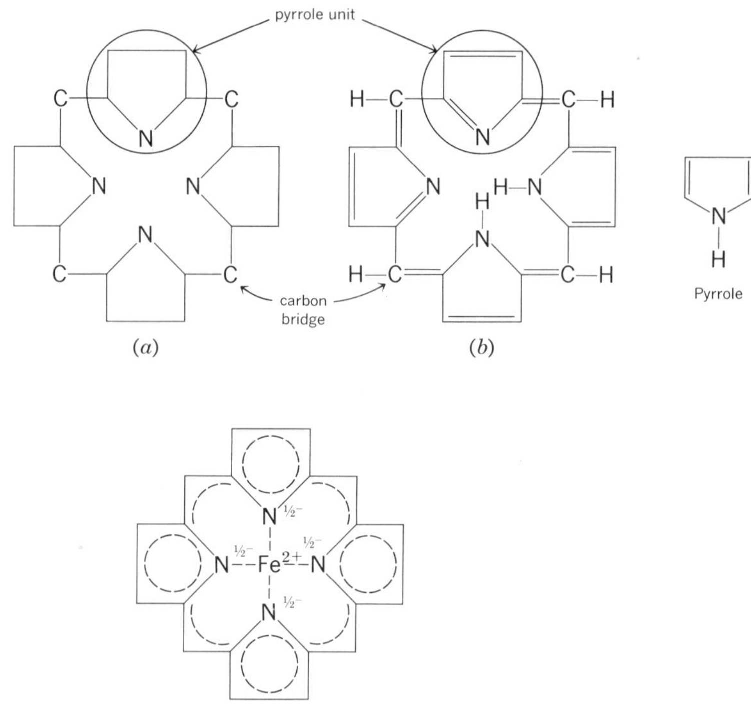
pyrrole unit
carbon bridge
(a)
(b)
Pyrrole
(c)

Figure 21.4 Cytochrome structure, general features. (*a*) Tetrapyrrole skeleton of prophin; (*b*) porphin, showing the extensive conjugated system; (*c*) the Fe^{2+}–porphin complex as it occurs in heme and in at least some of the cytochromes. By the loss of an electron Fe^{2+} can become Fe^{3+} and still remain complexed within the porphin skeleton. Not all the complexing potential of the iron ion is used in what is shown here. In cytochrome c (Figure 21.5) two more groups, provided by the protein portion, are complexed to the central iron by means of their nitrogens.

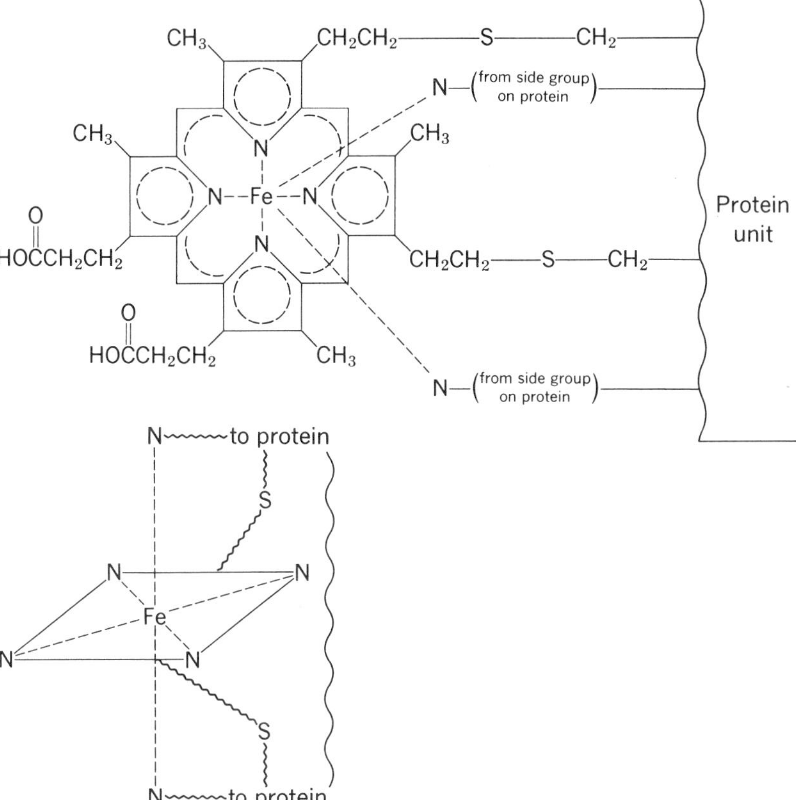
CH3
CH2CH2—S—CH2
N—(from side group on protein)
CH3
CH3
N--Fe--N
N
N
O
HOCCH2CH2
CH2CH2—S—CH2
Protein unit
O
HOCCH2CH2
CH3
N—(from side group on protein)
N~~~to protein
S
N
N
Fe
N
N
S
N~~~to protein

Porphryn unit, edge-on view

Figure 21.5 Cytochrome c. The complete amino acid sequence of this compound, as isolated from heart cells of several species, is known but is not shown here. The porphyrin portion is attached to the protein chain by two sufide links (via side chains of the amino acid cysteine) and by complexing with nitrogens from the side chains of two histidine units.

and the evidence for them. To complicate matters further, *cytochrome a* also contains Cu^{+} and Cu^{2+} ions; we henceforth ignore them at some sacrifice of accuracy.

Cytochrome b is believed to be capable of accepting electrons from the reduced form of coenzyme Q, $CoQH_2$:

$$CoQH_2 + 2Fe^{3+} \longrightarrow CoQ + 2H^+ + 2Fe^{2+}$$

To represent this reaction in terms of our running log on the travels of a pair of electrons from some metabolite to oxygen, we have

$$\underset{\text{(reduced form)}}{\underset{\text{Coenzyme Q}}{CoQ\begin{matrix}:\mathbf{H}\\ \cdot\\ \cdot\,\mathbf{H}\end{matrix}}} + \underset{\text{(oxidized form)}}{\underset{\text{Cytochrome b}}{\begin{matrix}Fe^{3+}\\ \\ Fe^{3+}\end{matrix}}} \longrightarrow \underset{\text{(oxidized form)}}{\underset{\text{Coenzyme Q}}{CoQ}} + 2\mathbf{H}^+ + \underset{\text{(reduced form)}}{\underset{\text{Cytochrome b}}{\begin{matrix}Fe\cdot^{2+}\\ \\ Fe\cdot^{2+}\end{matrix}}}$$

Once the pair of electrons has entered the cytochrome system, it is passed from cytochrome to cytochrome until it finally reaches one that can catalyze its transfer to oxygen, while hydrogen ions come in and water forms. We can now put together the sequence of electron flow in the respiratory chain. Then we shall backtrack and discuss how this electron flow is made to do work, putting ADP and P_i back together to form ATP.

ELECTRON TRANSPORT IN THE RESPIRATORY CHAIN

Piecing together material from the previous section, we summarize the electron transport chain in Figure 21.6. Some metabolites can deliver the elements of hydrogen directly to FAD. Others pass them first to NAD^+, which accounts for the "branching" in the figure. Only one coenzyme Q is indicated, but probably two or more are involved. The circled numbers in Figure 21.6 identify reactions that may be written as follows:

① $MH_2 + NAD^+ \longrightarrow M\!: + H^+ + NADH$

② $NADH + FAD + H^+ \longrightarrow NAD^+ + FADH_2$

③ $FADH_2 + CoQ \longrightarrow FAD + CoQH_2$

④ $CoQH_2 + 2Fe_b^{3+} \longrightarrow CoQ + 2H^+ + 2Fe_b^{2+}$

⑤ $2Fe_b^{2+} + 2Fe_{c_1}^{3+} \longrightarrow 2Fe_b^{3+} + 2Fe_{c_1}^{2+}$

⑥ $2Fe_{c_1}^{2+} + 2Fe_c^{3+} \longrightarrow 2Fe_{c_1}^{3+} + 2Fe_c^{2+}$

⑦ $2Fe_c^{2+} + 2Fe_a^{3+} \longrightarrow 2Fe_c^{3+} + 2Fe_a^{2+}$

⑧ $2Fe_a^{2+} + \frac{1}{2}O_2 + 2H^+ \longrightarrow 2Fe_a^{3+} + H_2O$

Sum: $MH_2 + \frac{1}{2}O_2 \longrightarrow M\!: + H_2O$

①′ $Succinate + FAD \longrightarrow fumarate + FADH_2$

②′ $FADH_2 + CoQ \longrightarrow FAD + CoQH_2$ (then to step ④)

$Succinate + \frac{1}{2}O_2 \longrightarrow fumarate + H_2O$

The net effect, or sum, represents a considerable fall down the free energy hill. The exact change in free energy depends on the metabolite involved and the concentrations, but whatever its amount, it powers the phosphorylation of ADP to form ATP.

OXIDATIVE PHOSPHORYLATION

What follows is an outline of one theory attempting to explain the most speculative part of ATP synthesis. One of the problems in oxidative phosphorylation is the large *positive* free energy needed for the job:

$$ADP + P_i \longrightarrow ATP + H_2O \qquad \Delta G = +9000 \text{ cal/mole (under } in\ vivo \text{ conditions)}$$

The free energy fall in the respiratory chain is ample enough to provide for ATP

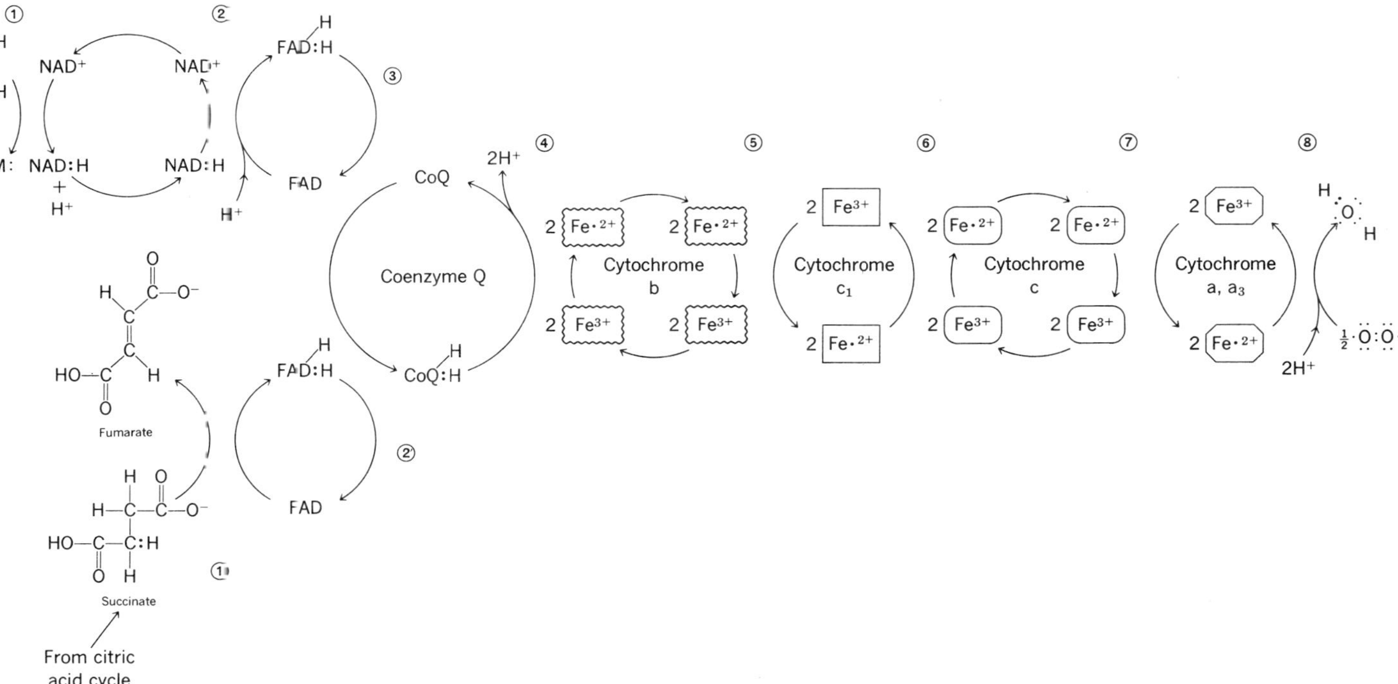

NAD+
NAD:H
FAD:H
FAD
CoQ
2H+
Coenzyme Q
CoQ:H
Cytochrome b
Cytochrome c1
Cytochrome c
Cytochrome a, a3
Fumarate
Succinate
From citric acid cycle

Figure 21.6 The respiratory chain, tentative pathway for electron transfer. In this schematic outline not all postulated cytochromes are indicated, and the positioning of coenzyme Q is for convenience. The succinate ion, for example, may feed to one CoQ system (later tied to cytochrome b), and NADH may feed to another CoQ unit. The NAD+/NADH system is not often bypassed. The circled numbers identify reactions summarized in another way in the textual discussion. (References: D. R. Sanadi, *Annual Review of Biochemistry,* Vol. 34, 1965, page 22; A. L. Lehninger, *The Mitochondrion,* W. A. Benjamin, New York, 1964 pages 60 and 115; and D. A. Green, "The Mitochondrion," *Scientific American,* January 1964, page 63.)

synthesis, but to use the free energy made available by one reaction to drive another, the two reactions must be coupled mechanistically. It is not enough that they happen in the same solution. There must be a common intermediate, and its identity is our next problem. How do we coordinate enzymic actions in electron transport with enzymic actions in ATP synthesis? How do we divert energy released by electron transport to do the chemical work of making ATP?

At least two *types* of enzymes are believed to intervene between ADP and ATP. One is called a *coupling enzyme* (CE), the other a *phosphate transfer enzyme* (TE). These enzymes are believed to be located physically adjacent to the group of respiratory enzymes, and all of them together constitute one unit of respiration, a *respiratory assembly*. These are the units that spangle the membrane surfaces of a mitochondrion.

The coupling of electron transport to ATP synthesis is believed to occur in at least three places along the respiratory chain. We illustrate what has been postulated to happen at one such place, the region in which NADH interacts with FAD (step 2 of Figure 21.6). Analogous reactions for other sites are summarized as part of Figure 21.7.

According to the theory, then, for ATP synthesis at the NADH-FAD zone of the respiratory assembly:

1. A coupling enzyme attaches itself temporarily to NADH.
$$\text{NADH} + \text{CE} \longrightarrow \text{NADH—CE}$$

2. When FAD pulls H:^- from this complex, a high-energy bond is left behind.
$$\text{NADH—CE} + \text{FAD} + \text{H}^+ \longrightarrow \text{NAD}^+\sim\text{CE} + \text{FADH}_2$$

3. The coupling enzyme is now activated to the point that it can be transferred to phosphate ion.
$$\text{NAD}^+\sim\text{CE} + \text{P}_i \longrightarrow \text{NAD} + \text{CE}\sim\text{P}$$

4. CE~P is a high-energy phosphate and as such transfers phosphate to the transfer enzyme.
$$\text{CE}\sim\text{P} + \text{TE} \longrightarrow \text{CE} + \text{P}\sim\text{TE}$$

5. P~TE is a higher-energy phosphate than ATP, and it can therefore transfer its phosphate to ADP.
$$\text{P}\sim\text{TE} + \text{ADP} \longrightarrow \text{TE} + \text{ATP}$$

Net effect:

$$\text{NADH} + \text{FAD} + \text{H}^+ + \text{ADP} + \text{P}_i \longrightarrow \text{NAD}^+ + \text{FADH}_2 + \text{ATP}$$

Step 2 is the key step. Instead of a simple transfer of H:^- from NADH to FAD (which releases free energy), *the transfer results in the formation of a new high-energy bond* in what is left behind. Free energy is therefore conserved in that bond,[11] and it powers the remaining sequences. Thus in step 2 energy available from one

[11] In a manner of speaking only. Free energy, of course, is a property of the entire molecule, not just the bond.

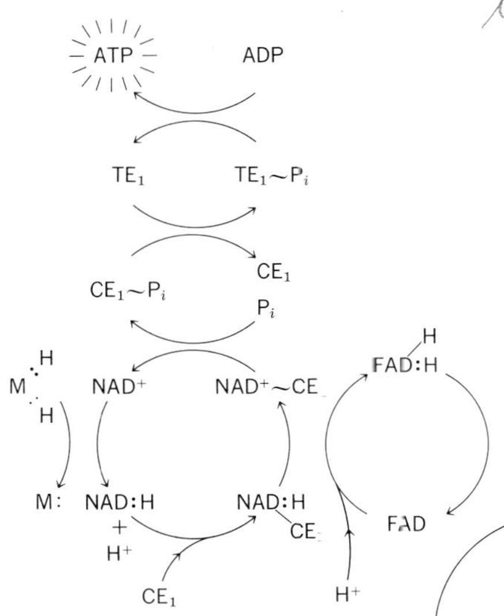

ATP
ADP
TE_1
$TE_1 \sim P_i$
$CE_1 \sim P_i$
CE_1
P_i
M:H:H
NAD^+
$NAD^+ \sim CE$
FAD:H:H
M:
NAD:H
+
H^+
NAD:H CE
FAD
CE_1
H^+
CoQ
$2H^+$
Coenzyme Q
CoQ:H:H
CE_2
2 $Fe \cdot {}^{2+}$
2 $Fe \cdot {}^{2+}$
Cytochrome b
2 Fe^{3+}
2 Fe^{3+}
CE_2
$CE_2 \sim P_i$
P_i
CE_2
TE_2
$TE_2 \sim P_i$
ATP
ADP
2 Fe^{3+}
Cytochrome c_1
2 $Fe \cdot {}^{2+}$
CE_3
2 $Fe \cdot {}^{2+}$
2 $Fe \cdot {}^{2+}$
Cytochrome c
2 Fe^{3+}
2 Fe^{3+}
CE_3
$CE_3 \sim P_i$
P_i
CE_3
TE_3
$TE_3 \sim P_i$
ATP
ADP
2 Fe^{3+}
Cytochrome a, a_3
2 $Fe \cdot {}^{2+}$
H:O:H
$2H^+$
$\frac{1}{2} \cdot O:O \cdot$

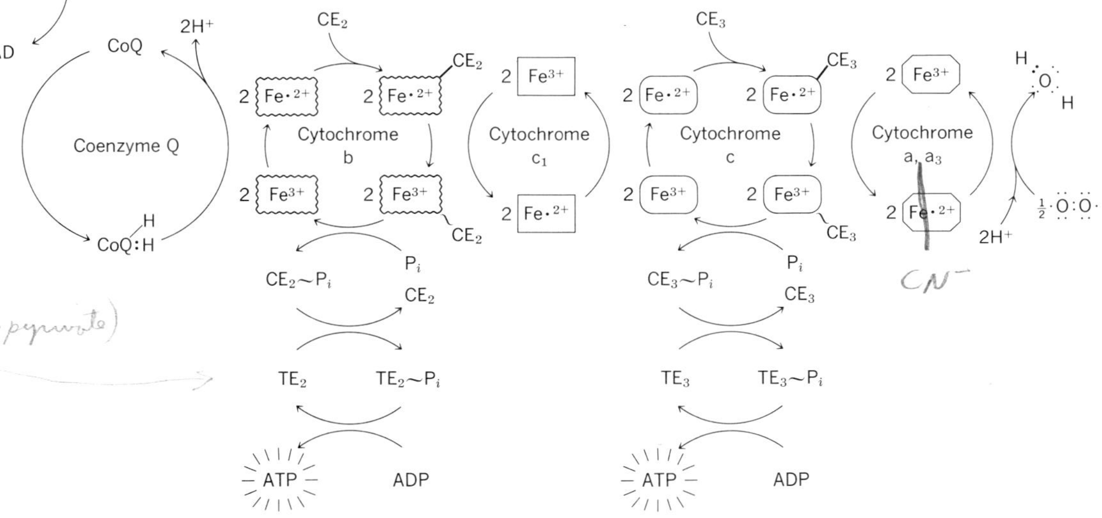

Figure 21.7 Oxidative phosphorylation, tentative pathway for coupling oxidation (i.e., electron transport) with the phosphorylation of ADP. This highly schematic diagram is meant to convey only the general idea of one theory. Details are subject to change, and alternative but similar proposals exist. Coupling enzymes are identified as CE_1, CE_2, CE_3, phosphate transfer enzymes as TE_1, TE_2, and TE_3. (See references cited for Figure 21.6 plus the following: A. L. Lehninger, "The Transfer of Energy Within Cells," in *Ideas in Modern Biology,* edited by J. A. Moore, Natural History Press, Garden City, N.Y., 1965, page 195; and A. F. Brodie and J. Adelson, "Respiratory Chains and Sites of Coupled Phosphorylation," *Science,* Vol. 149, 1965, page 265.)

sequence, namely electron transport, is tapped to drive another sequence, oxidative phosphorylation. The events are organized in a more schematic way in Figure 21.7.

The respiratory chain cannot function, of course, unless many conditions are met, one of which is obviously an adequate supply of the metabolites that can furnish the elements of hydrogen. The citric acid cycle is the principal source of this supply.

CITRIC ACID CYCLE

A GENERAL VIEW

The series of reactions sometimes referred to as the *Krebs cycle,* sometimes as the tricarboxylic acid cycle, but more usually as the citric acid cycle is outlined in Figure 21.8. Only the major intermediates are shown, and all details concerning enzymes and possible mechanisms are omitted. Strictly speaking, the cycle itself does not include pyruvic acid, but by showing it we indicate the connection between this cycle and glycolysis. From pyruvic acid the overall balanced equation is

$$2CH_3-\overset{\overset{\displaystyle O}{\|}}{C}-\overset{\overset{\displaystyle O}{\|}}{C}-OH + 5O_2 \longrightarrow$$

$$6CO_2 + 4H_2O \qquad \Delta G_{est} = -560{,}000 \text{ cal} \quad \text{(for two moles pyruvic acid)}$$

Hans Krebs (1900–), German-born British biochemist who shared the 1953 Nobel prize in medicine and physiology with Fritz Lipmann (p. 582). Forced out of Germany by Nazi pressures in the early 1930s, Krebs eventually joined the faculty of Oxford University. Using evidence he had discovered himself and piecing together the work of others, Krebs together with W. A. Johnson postulated much of the citric acid cycle in 1937. Since then he has been spontaneously honored by scientists the world over who very often refer to the cycle as the Krebs cycle. So many people have contributed to it that citric acid cycle, the name Krebs used, will be employed in this text.

Oxygen, shown as a reactant even though it does not react directly with any intermediate in the citric acid cycle, is included because this cycle does not occur unless the respiratory chain operates. Intermediates in the cycle deliver hydrogens and electrons to the chain, and eventually these combine with oxygen to form water. The respiratory chain is not called into action unless ATP has for some reason been used up. If we were to ignore the respiratory chain, the equation just given would read as follows:

$$2CH_3-\overset{\overset{\displaystyle O}{\|}}{C}-\overset{\overset{\displaystyle O}{\|}}{C}-OH + 6H_2O \longrightarrow 6CO_2 + 10(H{:}^- + H^+)$$

delivered to the respiratory chain ⟶

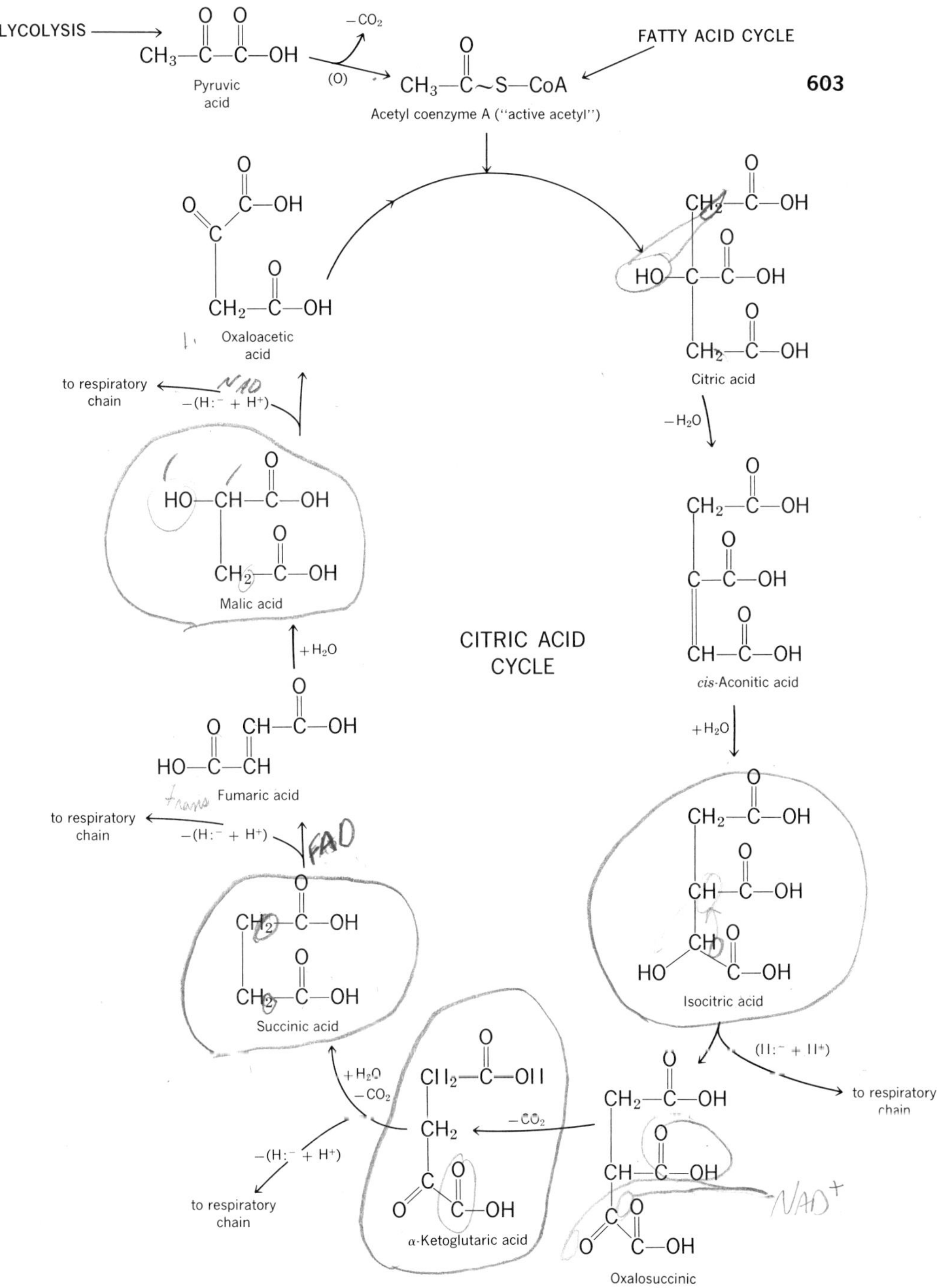
GLYCOLYSIS
Pyruvic acid
$-CO_2$
(O)
FATTY ACID CYCLE
Acetyl coenzyme A ("active acetyl")
Oxaloacetic acid
Citric acid
$-H_2O$
cis-Aconitic acid
$+H_2O$
Isocitric acid
(H:⁻ + H⁺)
to respiratory chain
Oxalosuccinic acid
$-CO_2$
α-Ketoglutaric acid
$+H_2O$
$-CO_2$
−(H:⁻ + H⁺)
to respiratory chain
Succinic acid
to respiratory chain
−(H:⁻ + H⁺)
Fumaric acid
$+H_2O$
Malic acid
to respiratory chain
−(H:⁻ + H⁺)
CITRIC ACID CYCLE

Figure 21.8 The citric acid cycle. Acetyl groups from acetyl coenzyme A are fed into the cycle, and by a succession of steps two carbons are broken off as carbon dioxide and four units of the elements of hydrogen ($H:^- + H^+$) are sent on to the respiratory chain.

The large free energy drop for the oxidation of pyruvic acid powers the formation of ATP, and for every two molecules[12] of pyruvic acid ATP production is

$$30\text{ADP} + 30\text{P}_i \longrightarrow 30\text{ATP} + 30\text{H}_2\text{O} \qquad \Delta G_{\text{est}} = +270{,}000 \text{ cal}^{13}$$

Figure 21.9 shows how the citric acid cycle couples to the respiratory chain and at what points the elements of hydrogen are transferred.

Many of the individual reactions in the citric acid cycle are very much like the ordinary organic reactions we have already studied. The principal difference is that enzymes rather than acids or bases intervene as catalysts. The first step, when acetyl coenzyme A enters the citric acid cycle, is like a Claisen condensation (p. 425). Succeeding steps involve relatively minor molecular surgery: a molecule of water is taken out or put in; the elements of hydrogen are removed; carbon dioxide is broken off.

Sources of Acetyl Coenzyme A. As indicated in Figures 21.8 and 21.9, the citric acid cycle must start with a supply of acetyl coenzyme A. Several substances serve as sources of the acetyl group, and acetyl coenzyme A is at one of the major "crossroads" of metabolic pathways in the body.

Proteins are one source. Several amino acids can be degraded by enzyme systems of various cells to produce acetyl coenzyme A. Some of them do so by way of pyruvic acid, which is also a source of acetyl groups.

Fatty acids are a rich and abundant source of acetyl groups. By steps to be described in Chapter 23, a fatty acid having n carbons (n is an even number) can be broken into $n/2$ molecules of acetyl coenzyme A.

Carbohydrates are another major supply of acetyl groups, for the monosaccharides available from the diet, fructose, galactose, and glucose, can each be converted to two molecules of pyruvic acid (p. 609). The metabolic pathways of fructose and galactose converge on that of glucose.

To change pyruvic acid to acetyl coenzyme A requires both a decarboxylation and an oxidation:

$$\underset{\text{Pyruvic acid}}{\text{CH}_3\text{—}\overset{\overset{\text{O}}{\|}}{\text{C}}\text{—}\overset{\overset{\text{O}}{\|}}{\text{C}}\text{—OH}} + \underset{\text{Coenzyme A}}{\text{CoA—S—H}} + \text{NAD}^+ \longrightarrow$$

$$\underset{\substack{\text{Acetyl}\\ \text{coenzyme A}\\ (\text{"active acetyl"})}}{\text{CH}_3\text{—}\overset{\overset{\text{O}}{\|}}{\text{C}}\text{—S—CoA}} + \text{CO}_2 + \text{NADH} + \text{H}^+$$

There are several intermediate steps which are not fully understood and are still being investigated.

[12] The data are for *two* molecules of pyruvic acid because that many are obtainable from one molecule of glucose via glycolysis.
[13] Based on an estimated +9000 cal for each mole of ATP when it is formed under physiological conditions rather than standard conditions.

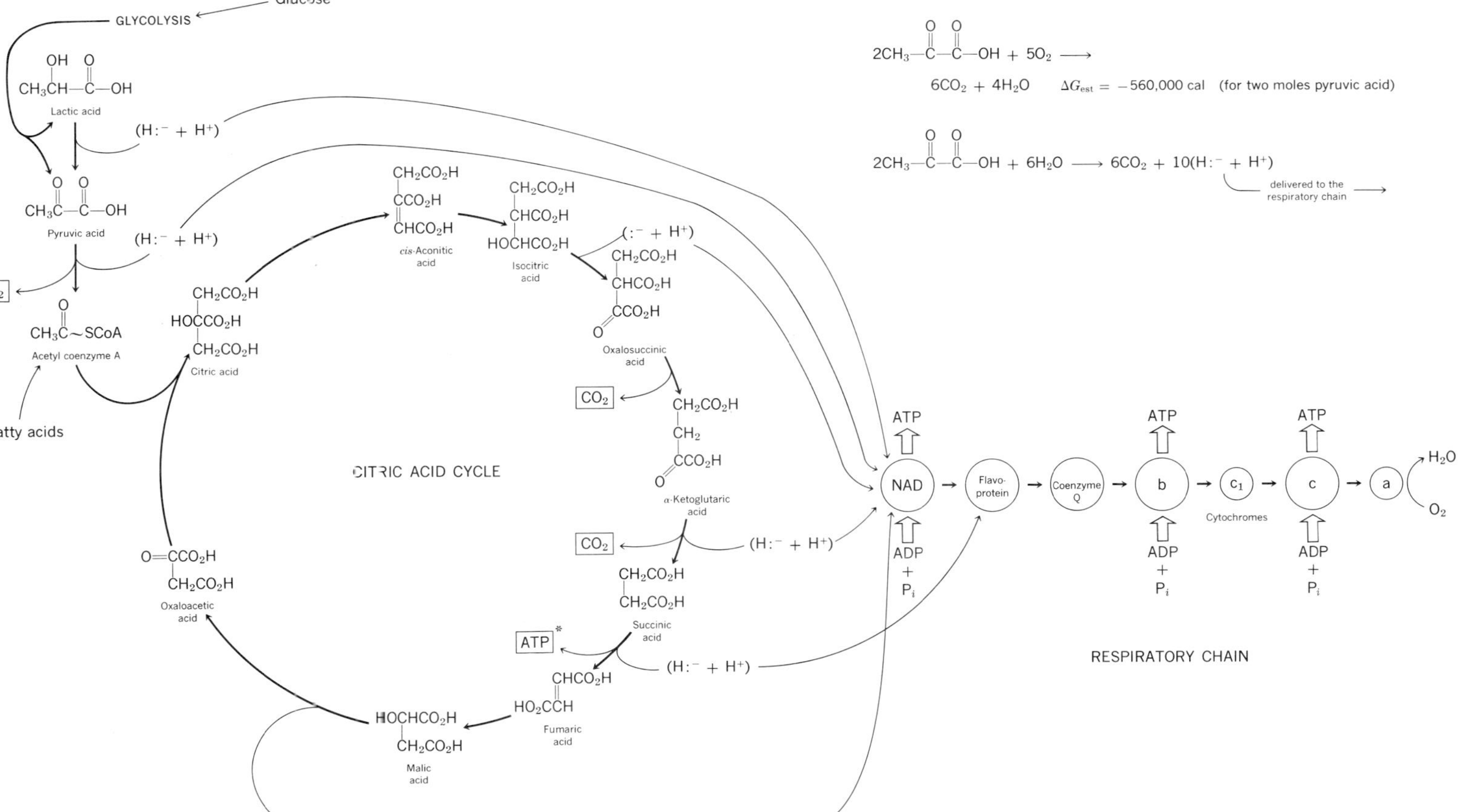
Glucose
GLYCOLYSIS
Lactic acid
Pyruvic acid
Acetyl coenzyme A
Fatty acids
Citric acid
cis-Aconitic acid
Isocitric acid
Oxalosuccinic acid
α-Ketoglutaric acid
Succinic acid
Fumaric acid
Malic acid
Oxaloacetic acid
CITRIC ACID CYCLE
NAD
Flavo-protein
Coenzyme Q
Cytochromes
RESPIRATORY CHAIN
ATP
ADP + P_i
delivered to the respiratory chain

$$2CH_3-\overset{O}{\overset{\|}{C}}-\overset{O}{\overset{\|}{C}}-OH + 5O_2 \longrightarrow 6CO_2 + 4H_2O \qquad \Delta G_{est} = -560{,}000 \text{ cal} \quad \text{(for two moles pyruvic acid)}$$

$$2CH_3-\overset{O}{\overset{\|}{C}}-\overset{O}{\overset{\|}{C}}-OH + 6H_2O \longrightarrow 6CO_2 + 10(H:^- + H^+)$$

Figure 21.9 Coupling of the citric acid cycle with the respiratory chain. The ATP production marked with an asterisk occurs by a mechanism not discussed in this book. (See references cited for Figures 21.6 and 21.7.)

Table 21.1 Yield of ATP from Pyruvic Acid in the Citric Acid Cycle and the Respiratory Chain

Steps			Receiver of (H:$^-$ + H$^+$) in the Respiratory Chain	Molecules ATP Formed
Pyruvic acid	$\longrightarrow$	acetyl CoA	NAD^+	3
Isocitric acid	$\longrightarrow$	α-ketoglutaric acid	NAD^+	3
α-Ketoglutaric acid	$\longrightarrow$	succinyl-CoA*	NAD^+	3
Succinyl-CoA	$\xrightarrow{GDP}$	succinic acid*	—	1
Succinic acid	$\longrightarrow$	fumaric acid	FAD	2
Malic acid	$\longrightarrow$	Oxaloacetic acid	NAD^+	3
Total ATP per molecule pyruvic acid:				15 molecules

* Details of how ATP forms in this step are omitted.

Table 21.1 summarizes the yield of ATP obtained from pyruvic acid via the citric acid cycle and the respiratory chain. For every hydrogen unit (H:$^-$ + H$^+$) delivered to NAD^+, three ATPs are produced. If NAD^+ is bypassed, as at the succinic acid–to–fumaric acid stage, and delivery is to FAD, only two ATPs form.

REFERENCES AND ANNOTATED READING LIST

BOOKS

A. L. Lehninger. *Bioenergetics.* W. A. Benjamin, New York, 1965. This outstanding book will serve as a reference for much of the rest of this text. Written for college undergraduates, it begins with a survey of elementary thermodynamic principles and discusses the flow of energy in the biological world, the importance of ATP, and most of the important sources and uses of ATP. (Paperback.)

A. L. Lehninger. *The Mitochondrion.* W. A. Benjamin, New York, 1964. Although the physiological aspects of bioenergetics receive more emphasis in this book, the principal topic is still energy for living.

D. E. Griffiths. "Oxidative Phosphorylation," in *Essays in Biochemistry,* edited by P. N. Campbell and G. D. Greville, Vol. 1. Academic Press, New York, 1965. This chapter is a summary discussion of theories (as of about 1964) pertaining to the coupling of electron flow in the respiratory chain to the synthesis of ATP from ADP and P_i.

Annual Review of Biochemistry (published by Annual Reviews, Palo Alto, Calif.) has been consulted in the preparation of this chapter.

A. White, P. Handler, and E. L. Smith. *Principles of Biochemistry,* third edition. McGraw-Hill Book Company, New York, 1964.

ARTICLES

A. L. Lehninger. "Energy Transformation in the Cell." *Scientific American,* May 1960, page 102. Lehninger writes for a more general audience in this article, and much work has appeared since 1960.

E. Racker. "The Membrane of the Mitochondrion." *Scientific American,* February 1968, page 32. This article describes studies on how chemicals making up the folded inner membrane of mitochondria contribute to ATP synthesis.

D. E. Green. "Biological Oxidation." *Scientific American,* July 1958, page 56. The role of mitochondria is emphasized, and fewer biochemical details are given.

D. E. Green. "The Mitochondrion." *Scientific American,* January 1964, page 63. Particular attention is given to the molecular structure of a mitochondrion and its serviceability in performing its functions. Green has taken vigorous exception to many aspects of the most widely held views on oxidative phosphorylation. For a report see *Scientific Research,* May 13, 1968, page 33.

M. A. Amerine. "Wine." *Scientific American,* August 1964, page 46. This interesting, relaxing article on the history and technology of wine making includes a brief summary of the chemistry of fermentation.

PROBLEMS AND EXERCISES

1. Define the following terms. (Neither equations nor full structures, if they are especially complicated, are called for.)

(*a*) fermentation	(*b*) zymase	(*c*) respiration
(*d*) glycolysis	(*e*) anaerobic sequence	(*f*) respiratory chain
(*g*) homeostasis	(*h*) oxidative phosphorylation	(*i*) mitochondrion
(*j*) metabolite	(*k*) NAD	(*l*) NADP
(*m*) DPN	(*n*) FAD	(*o*) CoQ
(*p*) cytochromes	(*q*) coupling enzyme	(*r*) respiratory assembly
(*s*) Krebs cycle	(*t*) "active acetyl"	

2. Briefly describe the contribution each of the following made to our understanding of fermentation or of biochemical energetics.

(*a*) Gay-Lussac	(*b*) Thenard	(*c*) Pasteur
(*d*) Buchner	(*e*) Harden	(*f*) Embden
(*g*) Meyerhof	(*h*) Cannon	(*i*) Krebs
(*j*) Lipmann		

3. Explain how the body is not a "heat engine."
4. Look up "fermentation" in some good encyclopedias and list substances made via fermentation processes.
5. Why did von Liebig oppose the notion that the life of the yeast cell was necessary for fermentation?
6. Explain how spent yeast juice can be rejuvenated by boiled yeast juice if boiling denatures proteins.
7. What was the significance of Embden's discovery?
8. What is the difference between high-energy and low-energy phosphates?
9. Discuss how the sentence "K values greater than one mean that at equilibrium the products tend to be favored" can be misleading when applied to biochemical events *in vivo*.
10. If a mixture of glucose and oxygen is less stable than a mixture of carbon dioxide and water (the combustion products of glucose), explain how glucose can be stored almost indefinitely in air.
11. In the reduction of oxygen to water in the body, what are some sources of H^+ and electrons? List the compounds (their structures) that supply these.
12. Discuss how ATP production is geared to ATP need.
13. Write equations to show how NAD^+ abstracts the elements of hydrogen from (*a*) lactic

acid, (*b*) isocitric acid, and (*c*) malic acid. Use as a symbol for NAD^+ the following partial structure:

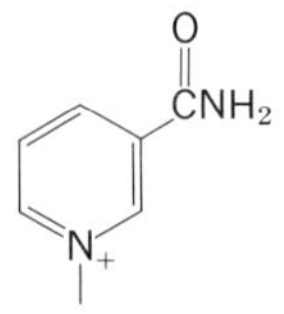

(pyridinium ring bearing $\overset{O}{\overset{\|}{C}}NH_2$, with N^+ bonded below)

14. Write the following in equation form:

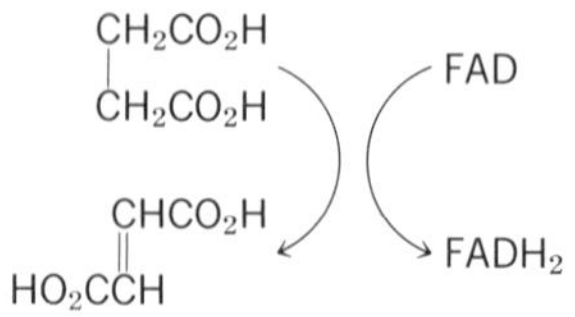

$$\begin{matrix} CH_2CO_2H \\ | \\ CH_2CO_2H \end{matrix} \quad \xrightarrow[\quad FAD \;\rightarrow\; FADH_2 \quad]{} \quad \begin{matrix} CHCO_2H \\ \| \\ HO_2CCH \end{matrix}$$

15. In this chapter the expression "homeostatic mechanism" was introduced. Even though this name was not used, we encountered an example of homeostasis in Chapter 20. It had to do with vasopressin. What was it? Describe how it illustrates homeostasis.

CHAPTER TWENTY-TWO

Metabolism of Carbohydrates

No one eats continually. Even the most compulsive nibbler goes without while he sleeps. The body has remarkable resources for storing nutrients in various forms and for releasing them as needed. These needs of course vary considerably from hour to hour. Everyone has known periods of severe exercise when breathing becomes increasingly vigorous, and sooner or later he must stop to "catch his breath." At times the system simply cannot take in and distribute oxygen as fast as it is needed, and yet the person keeps going, at least for awhile. At other times ATP demand in certain tissues is higher than the supply available from their respiratory chains. If oxygen does not arrive at respiratory units of mitochondria rapidly enough, ATP production by the respiratory chain slows down and its rate adjusts to the rate of oxygen supply. Assuming adequate supplies of stored fat and glycogen, and assuming a body in shape, the energy output of a long-distance runner is limited primarily by the rate at which oxygen can enter at the lungs and be distributed.

Sometimes, when emergencies arise, the body needs a brief, very intense burst of energy, as in a short sprint or a hundred-yard dash. The body then relies on a mechanism for generating ATP in the absence of oxygen—an *anaerobic sequence* for making ATP known as *glycolysis*. In glycolysis glucose units are broken down to lactic acid while a small amount of ATP is made.

Glucose is not the only carbohydrate that can undergo glycolysis. We concentrate our attention on it because the major metabolic pathways of all the dietary carbohydrates—starch, sucrose, maltose, lactose, fructose, galactose, and glucose—converge (Figure 22.1).

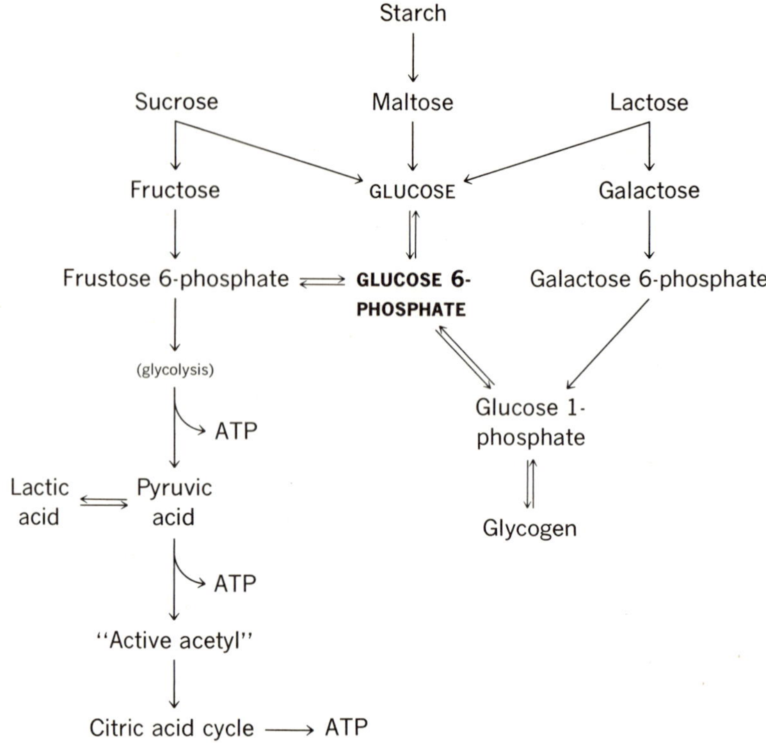
Starch
Sucrose
Maltose
Lactose
Fructose
GLUCOSE
Galactose
Frustose 6-phosphate
GLUCOSE 6-PHOSPHATE
Galactose 6-phosphate
(glycolysis)
ATP
Glucose 1-phosphate
Lactic acid
Pyruvic acid
Glycogen
ATP
"Active acetyl"
Citric acid cycle
ATP

Figure 22.1 Convergence of pathways of dietary carbohydrates.

GLYCOLYSIS

GENERAL VIEW

ATP Production. The amount of ATP made by glycolysis depends on the starting point. If it is glycogen, three ATPs per glucose unit are made; if it is glucose, only two ATPs per glucose molecule are produced. Glucose 6-phosphate can be made from glycogen without sacrificing any ATP, but when it is made from glucose one ATP has to be used, lowering the net yield. The following equations represent the overall effects of glycolysis.

Glycolysis from glycogen:

[terminal glucose unit in a branch of a glycogen molecule: CH_2OH, O, H, OH, HO, O, OH] } to rest of glycogen structure $+ H_2O + 3P_i + 3ADP \longrightarrow$

HO–[ ] } rest of glycogen $+ 2CH_3\underset{\displaystyle OH}{CH}\overset{\displaystyle O}{\overset{\|}{C}}OH + 3ATP$

Lactic acid

Glycolysis from glucose:

$$\text{Glucose} + 2P_i + 2ADP \longrightarrow 2CH_3CH(OH)COOH + 2ATP$$

Glucose   Lactic acid

Between the left and right sides of these equations, several steps occur (Figure 22.2). From beginning to end, two ATPs are consumed while four are produced for a net of two ATPs per initial glucose molecule (Table 22.1). The yield of ATP may not seem much, but it can be produced quickly without oxygen. Moreover, lactic acid can be catabolized further when oxygen becomes available, or it may migrate to other tissues and cells where the supply of oxygen is favorable for running the respiratory chain. As indicated in Figure 21.9, we obtain eighteen ATPs from each lactic acid molecule when it is processed by the citric acid cycle and the respiratory chain. Therefore glycolysis does not waste the free energy available from glucose; it merely taps some of it in an anaerobic way while leaving its end product, lactic acid, available for further ATP production.

The Role of Lactic Acid in Glycolysis. Even though glycolysis does not require oxygen, an oxidation does occur, the conversion of glyceraldehyde 3-phosphate to 1,3-diphosphoglyceric acid. In Figure 22.2 the conversion is shown as a dehydrogenation with hydrogen transferred to NAD^+.

$$(HO)_2P(=O)OCH_2CH(OH)C(=O){-}H + H_2O + NAD^+ + P_i \longrightarrow$$

Glyceraldehyde 3-phosphate

$$(HO)_2P(=O)OCH_2CH(OH)C(=O){-}O{\sim}P(=O)(OH)_2 + NADH + H^+$$

1,3-Diphosphoglyceric acid

If the hydrogen stays there (i.e., as $NADH + H^+$), the next molecule of glucose to be processed this far can go no further. The enzyme (having NAD^+ as its coenzyme) is out of service; glycolysis and production of ATP will stop, for this is an anaerobic sequence. The respiratory chain is not working, meaning that the elements of hydrogen in $NADH + H^+$ cannot be passed to it, thereby regenerating the NAD^+-dependent enzyme. The problem here is the regeneration of this NAD^+. In animals the problem is solved by using something produced later in glycolysis, pyruvic acid, as the acceptor for the "H_2" in $NADH + H^+$. Thus the last step in glycolysis is regeneration of this NAD^+-dependent enzyme, and the lactic acid that forms becomes at least a temporary reservoir for hydrogen. This key aspect of glycolysis is discussed again in Figure 22.3.

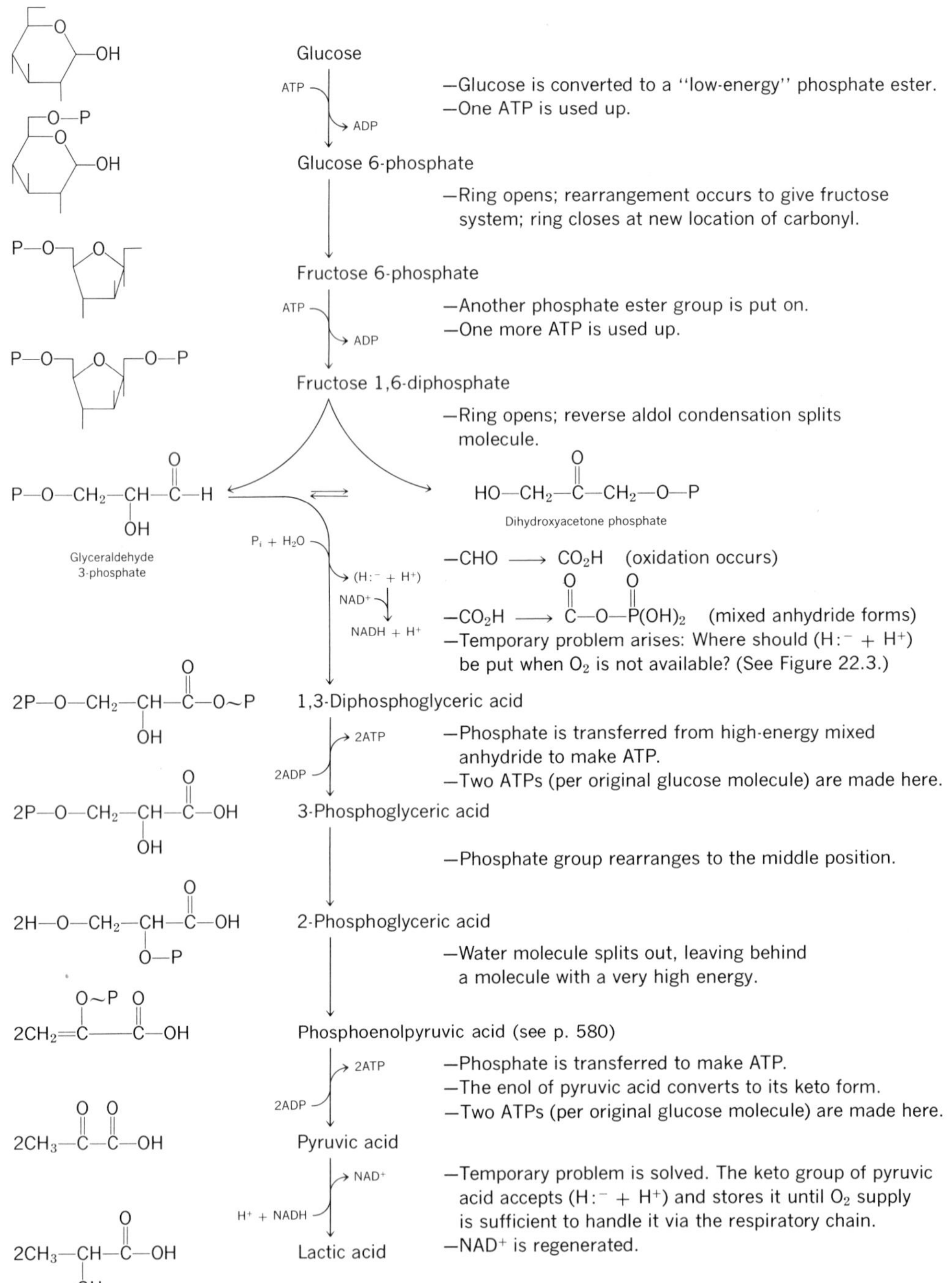
Glucose
ATP
ADP
—Glucose is converted to a "low-energy" phosphate ester.
—One ATP is used up.
Glucose 6-phosphate
—Ring opens; rearrangement occurs to give fructose system; ring closes at new location of carbonyl.
Fructose 6-phosphate
ATP
ADP
—Another phosphate ester group is put on.
—One more ATP is used up.
Fructose 1,6-diphosphate
—Ring opens; reverse aldol condensation splits molecule.
P—O—CH2—CH—C(=O)—H
Glyceraldehyde 3-phosphate
HO—CH2—C(=O)—CH2—O—P
Dihydroxyacetone phosphate
Pi + H2O
(H:⁻ + H⁺)
NAD⁺
NADH + H⁺
—CHO ⟶ CO2H (oxidation occurs)
—CO2H ⟶ C(=O)—O—P(=O)(OH)2 (mixed anhydride forms)
—Temporary problem arises: Where should (H:⁻ + H⁺) be put when O2 is not available? (See Figure 22.3.)
2P—O—CH2—CH(OH)—C(=O)—O~P
1,3-Diphosphoglyceric acid
2ATP
2ADP
—Phosphate is transferred from high-energy mixed anhydride to make ATP.
—Two ATPs (per original glucose molecule) are made here.
2P—O—CH2—CH(OH)—C(=O)—OH
3-Phosphoglyceric acid
—Phosphate group rearranges to the middle position.
2H—O—CH2—CH(O—P)—C(=O)—OH
2-Phosphoglyceric acid
—Water molecule splits out, leaving behind a molecule with a very high energy.
2CH2=C(O~P)—C(=O)—OH
Phosphoenolpyruvic acid (see p. 580)
2ATP
2ADP
—Phosphate is transferred to make ATP.
—The enol of pyruvic acid converts to its keto form.
—Two ATPs (per original glucose molecule) are made here.
2CH3—C(=O)—C(=O)—OH
Pyruvic acid
NAD⁺
H⁺ + NADH
—Temporary problem is solved. The keto group of pyruvic acid accepts (H:⁻ + H⁺) and stores it until O2 supply is sufficient to handle it via the respiratory chain.
—NAD⁺ is regenerated.
2CH3—CH(OH)—C(=O)—OH
Lactic acid

Figure 22.2 Glycolysis. Anaerobic catabolism, glucose to lactic acid. In the formulas P represents a phosphate ester group.

Table 22.1 Yield of ATP from Glucose in Glycolysis—Anaerobic Catabolism, Glucose to Lactic Acid

Steps			Molecules of ATP Gained (+) or Lost (−) per Initial Glucose Unit
Glucose + ATP	→	glucose 6-phosphate + ADP	−1*
Fructose 6-phosphate + ATP	→	fructose 1,6-diphosphate + ADP	−1
1,3-Diphosphoglyceric acid (2 molecules) + 2ADP	→	3-phosphoglyceric acid (2 molecules) + 2ATP	+2
Phosphoenolpyruvic acid (2 molecules) + 2ADP	→	pyruvic acid (2 molecules) + 2ATP	+2
Net gain in ATP via glycolysis			+2

* If the starting point is glycogen instead of glucose, this consumption of ATP is unnecessary, and the net ATP production is three per glucose unit.

Reversal of Glycolysis. When lactic acid forms during anaerobic muscular exercise, it migrates from muscle cells. Circulation carries it to other tissues, many of which can use it as a metabolite by changing it back to pyruvic acid and from that to acetyl coenzyme A. At this point it can be oxidized by the citric acid cycle and the respiratory chain for ATP synthesis. At the liver (or the kidneys), however, lactic acid will encounter enzymes that could change it back to glycogen instead of oxidizing it. In other words, the effect of glycolysis can be reversed. In fact, the steps for accomplishing this are, for the most part, the exact reverse of steps in glycolysis. There are, however, two important differences. At least two steps in glycolysis are not readily reversible, and methods for bypassing them are available, making the control of glycolysis easier. By exerting special control over the enzymes of the bypasses, the system can determine whether or not glycolysis will be reversed.[1]

Since glycolysis *produces* ATP, its reverse must *require* some. If glycolysis is going downhill, the reverse is going uphill. To provide energy for this, some of the lactic acid is channeled into the citric acid cycle to generate the ATP needed to carry the remaining back to glycogen. Only a small fraction of lactic acid (one molecule out of six) need be oxidized further, for the aerobic sequence (citric acid cycle and respiratory chain) is particularly productive of ATP. Starting from one molecule of *lactic acid,* this pathway yields eighteen molecules of ATP. One-sixth of this is three molecules. Glycolysis *from glycogen* (rather than from glucose) yields a net three molecules of ATP. Its reverse will therefore need at least this much.

Glycolysis in Microorganisms. Alcohol Fermentation. When certain yeasts act to convert sugars to alcohol, they employ essentially the same reactions of glycolysis that we have already studied. The major difference is the way in which they handle the anaerobic storing of "H_2," a problem considered in Figure 22.2 and particularly

[1] Referring to Figure 22.2, although not explicitly shown there, these bypasses occur at the stages producing phosphoenolpyruvic acid from pyruvic acid and fructose 6-phosphate from fructose 1,6-diphosphate.

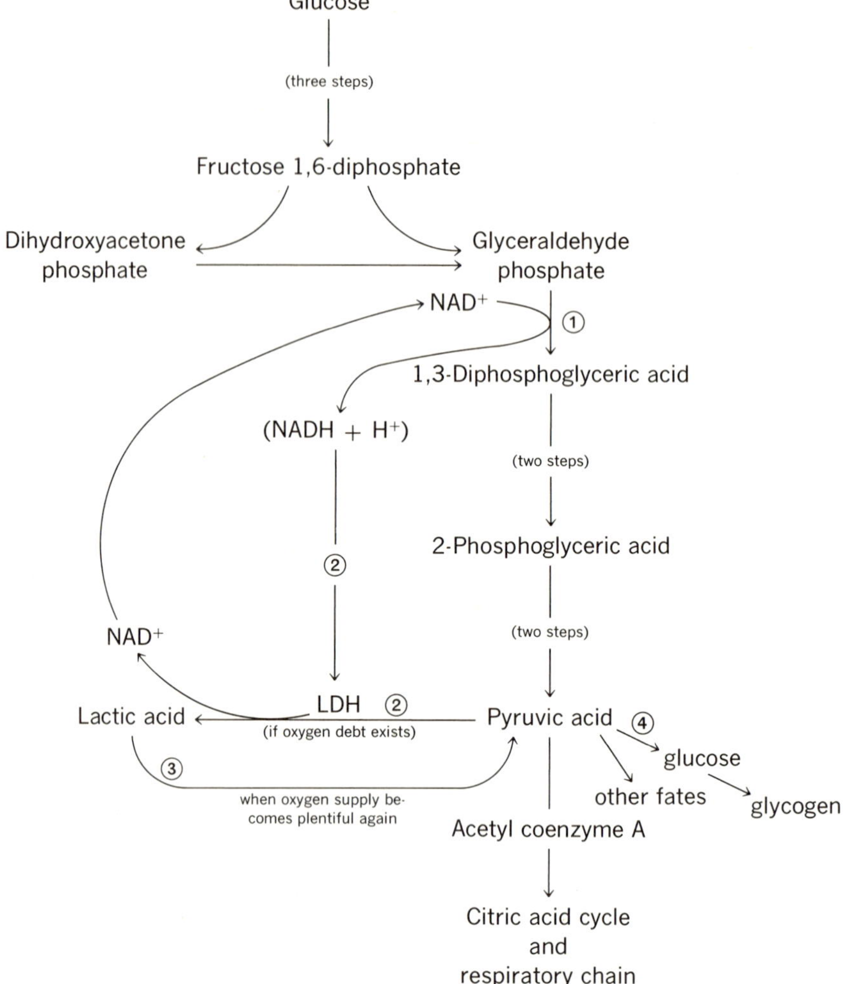
Glucose
(three steps)
Fructose 1,6-diphosphate
Dihydroxyacetone phosphate
Glyceraldehyde phosphate
NAD+
①
1,3-Diphosphoglyceric acid
(NADH + H+)
(two steps)
2-Phosphoglyceric acid
②
NAD+
(two steps)
LDH ②
Lactic acid
(if oxygen debt exists)
Pyruvic acid
④
glucose
③
other fates
glycogen
when oxygen supply becomes plentiful again
Acetyl coenzyme A
Citric acid cycle
and
respiratory chain

Figure 22.3 Lactic acid as a reservoir of hydrogen. Principal fates of pyruvic acid. During glycolysis the NAD^+-dependent enzyme at ① is reduced and, unless it is regenerated, glycolysis will stop. The enzyme can be regenerated in two ways. If the oxygen supply is sufficient, the elements of hydrogen removed at ① by NAD^+ can be passed on down the respiratory chain. If the oxygen supply is insufficient (a state of oxygen debt), the enzyme lactic acid dehydrogenase (LDH) catalyzes the transfer ② of ($H:^-$ + H^+) to pyruvic acid, and lactic acid becomes the storage depot. The enzyme for ① is restored. Once the oxygen supply is rebuilt, the hydrogen stored in lactic acid is removed ③ and delivered to the respiratory chain. The pyruvic acid that is made simultaneously may go into the citric acid cycle, or it may be converted ④ to glucose. (Reference: C. L. Markert, "Mechanisms of Cellular Differentiation," in *Ideas in Modern Biology,* edited by J. A. Moore, Natural History Press, Garden City, N.Y., 1965, page 234.)

Figure 22.3. In muscle cell glycolysis this "H_2" is transferred to the keto group of pyruvic acid to make lactic acid, the temporary reservoir. In alcohol fermentation, once yeasts have made pyruvic acid, they decarboxylate it to form a new "H_2" acceptor, acetaldehyde:

$$CH_3-\overset{\overset{O}{\|}}{C}-\overset{\overset{O}{\|}}{C}-O-H \xrightarrow{\text{in certain yeasts}} CH_3-\overset{\overset{O}{\|}}{C}-H + O{=}C{=}O\uparrow$$

Pyruvic acid — Acetaldehyde — Carbon dioxide

When it accepts "H_2" from reduced NAD^+, ethyl alcohol is the final product:

$$CH_3-\overset{\overset{O}{\|}}{C}-H + NADH + H^+ \xrightarrow{\text{in certain yeasts}} CH_3-CH_2-O-H + NAD^+$$

Acetaldehyde — Ethyl alcohol

The liberated carbon dioxide causes the frothing or foaming action that accompanies fermentation.

PHOSPHOGLUCONATE OXIDATIVE PATHWAY OF GLUCOSE CATABOLISM

A second way of catabolizing glucose in the absence of oxygen is by the phosphogluconate pathway or the "hexose monophosphate shunt." Our discussion of this pathway is brief, for even indirectly it is not important for ATP production in skeletal muscle. It is, however, a way of converting glucose ultimately to carbon dioxide and water without going through the citric acid cycle. *Its most important function is generating NADPH* (the reduced form of NADP, nicotinamide adenine dinucleotide phosphate—cf. p. 592), which is needed to furnish hydrogen for reduction steps whenever the body synthesizes fatty acids and steroids (referred to again in Chapter 23).

Starting from glucose 6-phosphate, the balanced equation representing the net effect of this pathway is

$$6\text{ glucose 6-phosphate} + 12NADP^+ \xrightarrow{\text{hexose monophosphate shunt}} 5\text{ glucose 6 phosphate} + 6CO_2 + 12NADPH + 12H^+ + P_i$$

There are of course several intermediate steps, some of them using enzymes of the glycolysis pathway. Among the intermediates are certain five-carbon sugars (e.g., ribose) needed by the body to make nucleotides and nucleic acids. For our future studies, however, we should remember that this shunt is an important way to make NADPH.

GLUCOSE DELIVERY AND TRANSPORT

ABSORPTION

Glucose, fructose, and galactose, the principal end products of the digestion of carbohydrates, readily move across the intestinal barrier, partly by simple diffusion and partly by active transport. In migration by active transport enzyme-catalyzed reactions occur, consuming energy (ATP) and converting the substances temporarily into forms that themselves diffuse. Interlacing the walls of the various parts of the gastrointestinal tract are capillaries that are tributaries of larger and larger blood vessels which finally join the hepatic portal vein[2] (Figure 22.4). This network provides intimate access between the digestive tract and the circulatory system. Nutrient molecules formed during digestion are therefore carried first by portal circulation through the liver. The blood passes through innumerable capillary networks (hepatic sinusoids in Figure 22.4) before it emerges into the inferior vena cava and travels from there to the right atrium of the heart. While the blood percolates through the liver, nutrients may or may not be removed, depending on several factors. This largest organ of the body has an extraordinary number of important functions involving all classes of foods. We are concerned here about what it does with glucose and how it participates in regulating glucose metabolism elsewhere. In describing these activities, we shall find certain terms useful. Since they are compounded from reasonably familiar word parts, they are almost self-defining.

Glucogenesis (glucose genesis or glucose making). When the body synthesizes glucose from any product of glycolysis (e.g., lactic acid or pyruvic acid) or from other carbohydrate sources, glucogenesis is said to have occurred. The liver is the principal site for it (Figure 22.5).

Gluconeogenesis (gluco-neo-genesis; "neo-" is a word part meaning "new"). Noncarbohydrate materials such as proteins and lipids can be converted into glucose by certain cells, notably in the liver. Glucose can also be made from the skeletons of *glucogenic* amino acids (sometimes trimming may be necessary). Fatty acids must normally participate to provide hydrogen. When glucose is made from noncarbohydrates, which constitute a new source, gluco*neo*genesis is said to take place.

Glycogenesis (glycogen genesis or glycogen making). Whatever the source of glucose molecules, some of them may be taken out of circulation and converted into glycogen, a process called glycogenesis which occurs principally in the liver and in muscles. Such glycogen is in a dynamic state, meaning that its glucose units enter and leave constantly. Figure 22.6 outlines the principal steps in glycogenesis from glucose.

[2] *Hepatic* is from the Greek: *hepat-*, liver; *-ikos,* akin to—that is, of, relating to, or affecting the liver. *Portal* is from the Latin *porta,* gate—that is, relating to a place of entry, a communicating part or region of an organ. The portal region of the liver, the *porta hepatis,* is the transverse fissure on its underside, where a number of vessels enter.

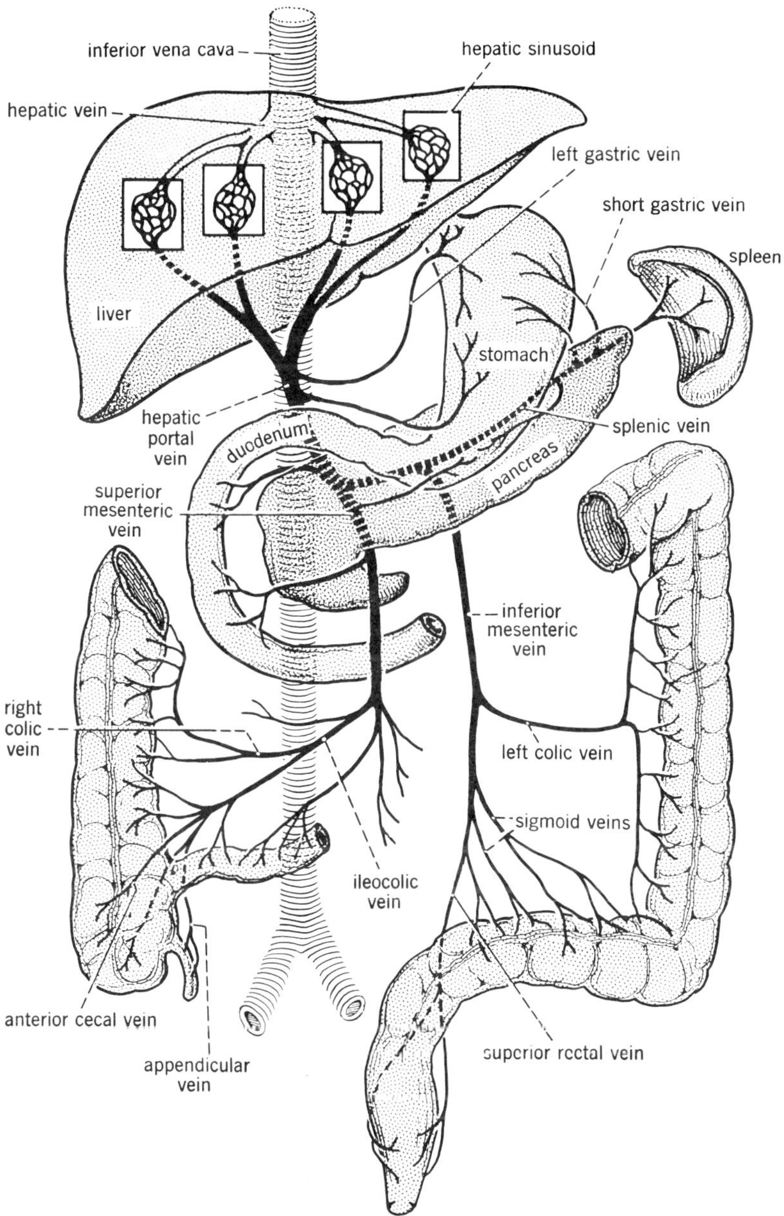
inferior vena cava
hepatic sinusoid
hepatic vein
left gastric vein
short gastric vein
spleen
liver
stomach
hepatic
portal
vein
duodenum
splenic vein
pancreas
superior
mesenteric
vein
inferior
mesenteric
vein
right
colic
vein
left colic vein
sigmoid veins
ileocolic
vein
anterior cecal vein
appendicular
vein
superior rectal vein

Figure 22.4 The major tributaries of the hepatic portal vein. (From A. J. Berger, *Elementary Human Anatomy,* John Wiley and Sons, New York, 1964, page 372.)

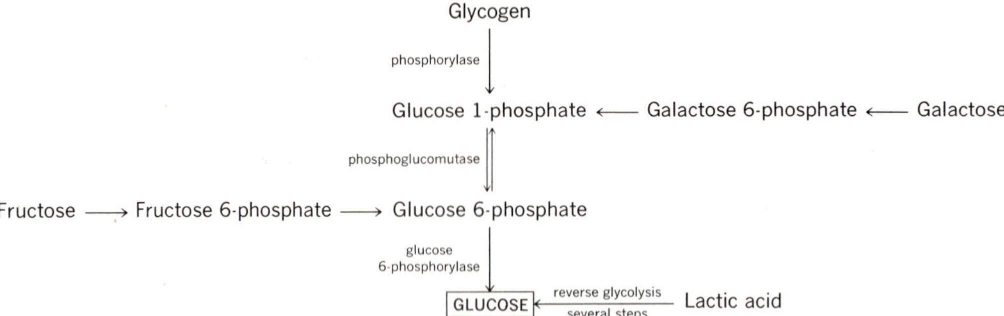

Glycogen
phosphorylase
Glucose 1-phosphate
Galactose 6-phosphate
Galactose
phosphoglucomutase
Fructose
Fructose 6-phosphate
Glucose 6-phosphate
glucose 6-phosphorylase
GLUCOSE
reverse glycolysis
several steps
Lactic acid

Figure 22.5 Glucogenesis. Sources of glucose other than direct absorption of dietary glucose. Phosphorylase is activated by epinephrine. The activity of glucose 6-phosphorylase appears to be enhanced in diabetes mellitus.

Glycogenolysis (glycogen-lysis; i.e., glycogen breakdown or hydrolysis). When glycogen reserves are depleted because glycogen breaks down or is hydrolyzed faster than glucose units are added, the process is called glycogenolysis. The glycogen content of muscles is rapidly used up in severe exercise, and fresh glucose molecules must be taken in from the bloodstream. To replenish this supply, liver glycogen can break down. Several factors initiate it: physical exercise, fasting, epinephrine (a hormone released during periods of fright), failure of the insulin supply, and other hormonal activities. Glucose released from glycogen reserves in the liver reaches muscles by the circulatory system.

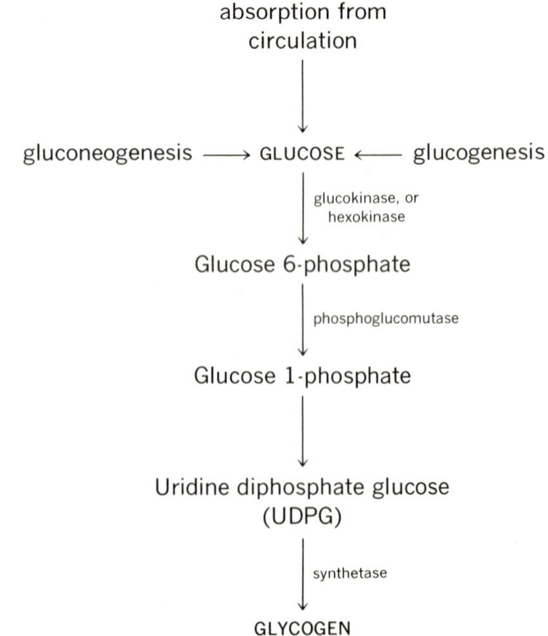

absorption from circulation
gluconeogenesis
GLUCOSE
glucogenesis
glucokinase, or hexokinase
Glucose 6-phosphate
phosphoglucomutase
Glucose 1-phosphate
Uridine diphosphate glucose (UDPG)
synthetase
GLYCOGEN

Figure 22.6 Glycogenesis. The activity of the enzyme *synthetase* is normally stimulated by an excess of glucose 6-phosphate, which means that the extra glucose 6-phosphate will be removed and made into glycogen. The activity of the enzyme *glucokinase* drops in diabetes. (UDPG is a complex derivative of glucose whose structure we shall not examine.)

Glucose found in the bloodstream is called *blood sugar*. Like the circulation of money in a healthy economy, the circulation of blood sugar in a healthy body is controlled by routine supply and demand and must be able to respond to emergencies. The connections between liver glycogen, muscle glycogen, blood sugar, and lactic acid are displayed by means of a modified Cori cycle in Figure 22.7.

Carl and Gerty Cori, who shared the Nobel prize in physiology and medicine with B. Houssay in 1947, made their greatest contributions to the chemistry of glycogenesis and glycogenolysis. They succeeded in synthesizing glycogen *in vitro* from glucose, ATP, and certain enzymes.

BLOOD SUGAR LEVEL

The word *level* in this term means concentration, and it is expressed in milligrams of glucose per 100 ml of blood. After eight to twelve hours of fasting, the blood sugar level in the venous blood of an average adult is usually in the range of 60 to 100 mg per 100 ml, called the *normal fasting level.*

Special words, describing levels outside this range, are constructed from word parts that make the terms almost self-defining.[3] When concentrations of blood sugar are above the normal fasting level, the condition is called *hyperglycemia. Hypoglycemia* is a condition with blood sugar below the normal fasting level. If the blood sugar level goes high enough, the kidneys remove some of the glucose and put

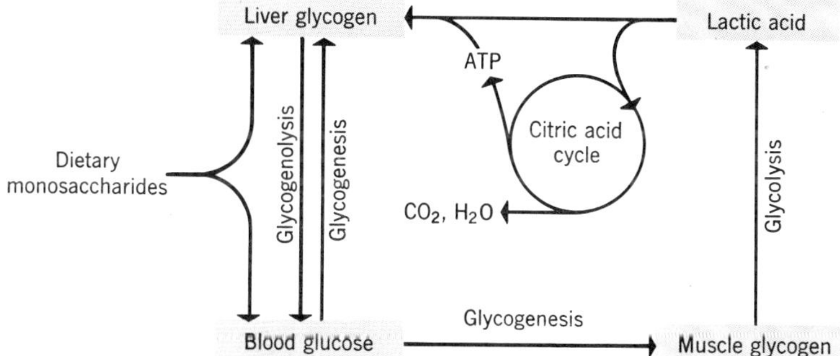
Liver glycogen
Lactic acid
ATP
Citric acid cycle
Dietary monosaccharides
Glycogenolysis
Glycogenesis
CO_2, H_2O
Glycolysis
Glycogenesis
Blood glucose
Muscle glycogen

Figure 22.7 Modified Cori cycle, showing relations of glycogenesis, glycogenolysis, glycolysis, and the citric acid cycle to one another.

[3] The word parts are as follows:

hyper-, over, above, beyond, excessive; e.g., *hyper*sensitive.

hypo-, under, beneath, down; e.g., *hypo*dermic (beneath the skin).

glyco- or *glyc-*, may refer to a "glycose" (generic name for a monosaccharide) or to glycogen; usually refers to glucose, as in hyper*glyc*emia.

gluco- or *gluc-*, refers to glucose.

-emia, refers to blood or a condition of having a specific substance in the blood.

-uria, refers to urine or a condition of having a specific substance in the urine.

it into the urine. The blood sugar level above which this happens, called the *renal threshold* (Latin *renes,* kidneys) for glucose, is normally about 140 to 160 mg of glucose per 100 ml of blood, sometimes higher. When detectable quantities of glucose appear in the urine, the condition is called *glucosuria.* These conditions are not in themselves diseases, but they are usually symptoms of a disorder. Anyone who eats a meal extremely rich in carbohydrate may temporarily become hyperglycemic, even to the point of being glucosuric. Anyone who fasts or is starving will sooner or later become hypoglycemic. People with diabetes mellitus are usually hyperglycemic and glucosuric; if they give themselves an overdose of insulin, they will rather promptly become hypoglycemic and go into "insulin shock." A very low blood sugar level is obviously serious. Hyperglycemia, on the other hand, is not too serious *in itself;* events happen in the body to reduce it (Figure 22.8).

Tissue Dependence on Blood Sugar Level. Different tissues depend to various degrees on direct access to glucose from the bloodstream as a source of energy. Heart muscle, for example, can use a variety of metabolites including lactic acid and fatty acids. Skeletal muscles use fatty acids and certain of their breakdown products to maintain themselves during resting periods, and they draw upon stored glycogen in times of stress. This glycogen must eventually be replenished, but there will be no swift muscle failure if the blood sugar level drops. The brain and the central nervous system are more vulnerable.

The glycogen content of the brain is about 0.1% by weight, a reserve supply of glucose units far too low to sustain brain function for more than a short time. Under normal conditions the brain obtains virtually all its required energy from glu-

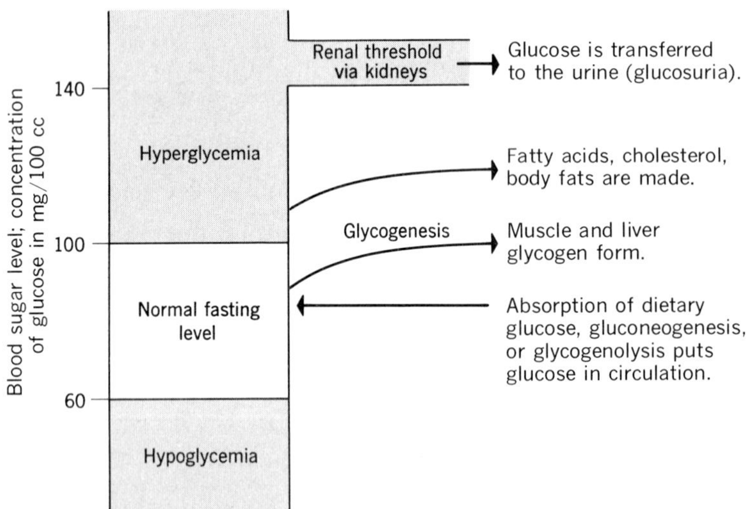
Blood sugar level; concentration of glucose in mg/100 cc
140
100
60
Hyperglycemia
Normal fasting level
Hypoglycemia
Renal threshold via kidneys
Glucose is transferred to the urine (glucosuria).
Fatty acids, cholesterol, body fats are made.
Glycogenesis
Muscle and liver glycogen form.
Absorption of dietary glucose, gluconeogenesis, or glycogenolysis puts glucose in circulation.

Figure 22.8 Reduction of hyperglycemia. Glucose that is not needed to replenish glycogen reserves in muscles or the liver or to supply the brain and central nervous system is for the most part converted to fat. Normally the renal threshold for glucose is not exceeded, and glucose is not found in the urine. In diabetes mellitus, however, glucosuria is standard.

cose taken directly from the bloodstream. Mitochondria of the brain and the central nervous system are particularly active in glycolysis. The citric acid cycle also occurs. Any interference in the smooth operation of this cycle will soon appear as malfunction of the brain and the central nervous system. Thus if the blood sugar level drops for any reason to very hypoglycemic levels, the brain loses its main source of energy. For a short while it can manage by catabolizing amino acids and lipids, but severe hypoglycemia usually means convulsions and coma. Damages to the brain may be permanent. Even a temporary, mild hypoglycemia may cause dizziness and fainting spells. Regulation of the blood sugar level is obviously a matter of considerable importance. A healthy liver and a smoothly functioning endocrine system are indispensible. Figure 22.9 shows the general means by which the endocrine system participates in controlling the blood sugar level. Proper control of the blood sugar level by the body is a measure of its *glucose tolerance,* that is, its ability to use glucose in a normal way.

THE LIVER IN GLUCOSE METABOLISM

For a detailed study of the molecular basis of a disease, diabetes mellitus, we need to know more about glucose metabolism in the liver. In the following discussion the circled numbers refer to those in Figure 22.10.

In Figure 22.10 we find glucose in circulation ① in a capillary within the liver. Some glucose molecules are leaving this capillary and migrating through a wall of a liver cell ②. Once inside the liver cell a glucose molecule will normally be phosphorylated ③, that is, converted to glucose 6-phosphate (G-6-P) at the expense of ATP. The enzyme for this is *glucokinase,* and insulin somehow affects its activity, apparently by controlling the rate of its synthesis and therefore its concentration. The presence of another enzyme, *phosphoglucomutase,* makes possible the rearrangement of G-6-P to glucose 1-phosphate (G-1-P) at ④. By steps we shall not study, G-1-P can be made into glycogen at ⑤. But if the glucose supply is low (e.g., developing starvation), glucose can be removed from storage ⑥ through the action of the enzyme *phosphorylase*—followed by reversal of ④, followed by hydrolysis of G-6-P at ⑦, followed by migration out of the cell at ⑧. Epinephrine, a hormone, makes the enzyme phosphorylase at ⑥ especially active, causing rapid depletion of the reserves of glycogen.

The enzymes at ③ and ⑦, *glucokinase* and *glucose 6-phosphorylase,* have activities that adjust to the concentrations of glucose and G-6-P. In a *fed* animal the rate at ③ will normally equal or slightly exceed that at ⑦, and a net intake of glucose occurs. In the *fasted* animal the rate at ⑦ moderately exceeds that at ③ to help take glucose out of storage and put it into circulation, especially to meet the needs of the brain and the central nervous system.

In diabetes mellitus the rate at ③ is much slower than at ⑦. In other words, this mechanism for "trapping" glucose inside the cell by converting it to a phosphate ester is no longer effective. Furthermore, the enzymes for gluconeogenesis at ⑨ become much more active in diabetes. At least the liver makes even more glucose by this method. In so doing it also makes urea and the so-called "ketone bodies," byproducts of gluconeogenesis which migrate out of the cell at ⑩. *This overproduction*

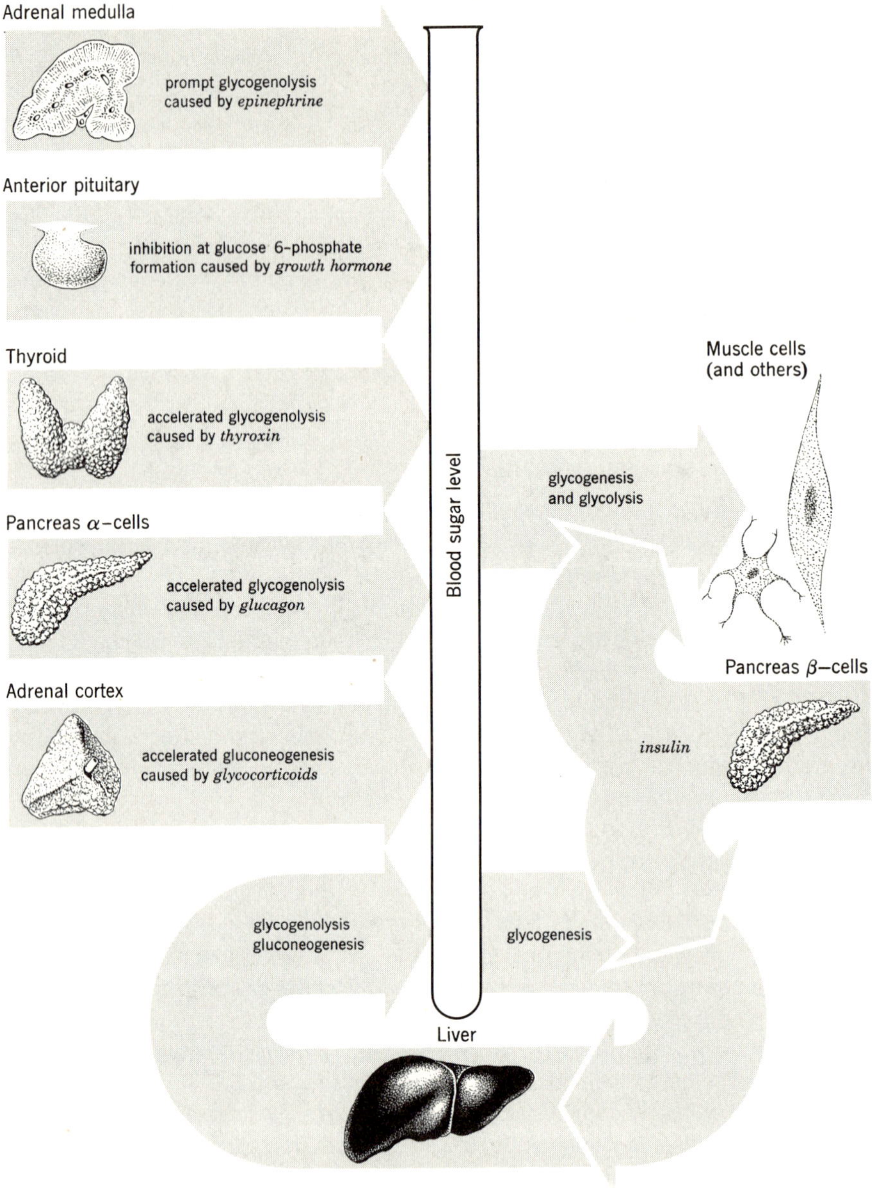

Adrenal medulla

prompt glycogenolysis caused by *epinephrine*

Anterior pituitary

inhibition at glucose 6-phosphate formation caused by *growth hormone*

Thyroid

accelerated glycogenolysis caused by *thyroxin*

Pancreas α-cells

accelerated glycogenolysis caused by *glucagon*

Adrenal cortex

accelerated gluconeogenesis caused by *glycocorticoids*

Blood sugar level

Muscle cells (and others)

glycogenesis and glycolysis

Pancreas β-cells

insulin

glycogenolysis gluconeogenesis

glycogenesis

Liver

Figure 22.9 The blood sugar level as influenced by hormones. Arrows pointing *in* signify factors that tend to make the blood sugar level rise. Arrows pointing *out* signify factors causing lower blood sugar levels. Some of the hormones exert their action within the liver, which gives this organ a central role in the regulation of the blood sugar level. (Adapted from I. N. Kugelmass, *Biochemistry of Blood in Health and Disease,* Charles C Thomas, Springfield, Ill., 1959. Courtesy of Charles C Thomas.)

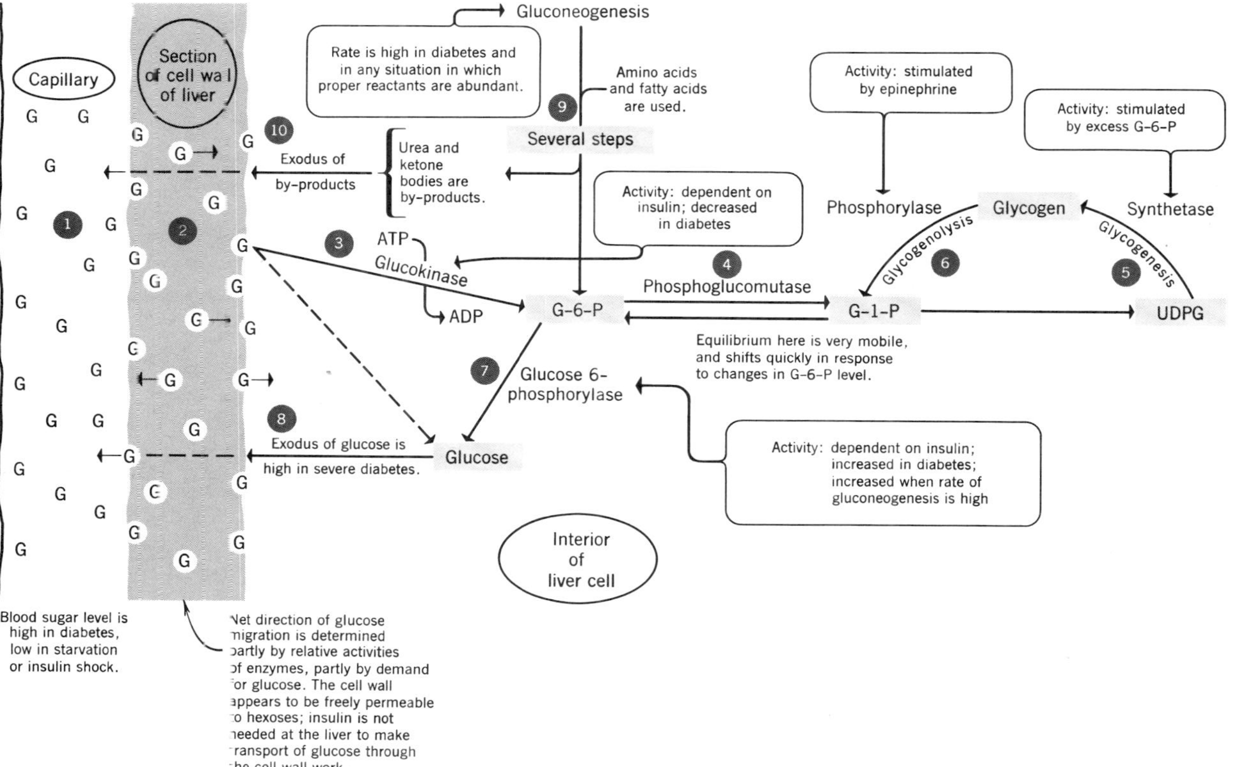
Gluconeogenesis
Capillary
Section of cell wall of liver
Rate is high in diabetes and in any situation in which proper reactants are abundant.
Amino acids and fatty acids are used.
Several steps
Activity: stimulated by epinephrine
Activity: stimulated by excess G-6-P
Exodus of by-products
Urea and ketone bodies are by-products.
Activity: dependent on insulin; decreased in diabetes
Phosphorylase
Glycogen
Synthetase
ATP
Glucokinase
ADP
Glycogenolysis
Glycogenesis
G-6-P
Phosphoglucomutase
G-1-P
UDPG
Equilibrium here is very mobile, and shifts quickly in response to changes in G-6-P level.
Glucose 6-phosphorylase
Exodus of glucose is high in severe diabetes.
Glucose
Activity: dependent on insulin; increased in diabetes; increased when rate of gluconeogenesis is high
Interior of liver cell
Blood sugar level is high in diabetes, low in starvation or insulin shock.
Net direction of glucose migration is determined partly by relative activities of enzymes, partly by demand for glucose. The cell wall appears to be freely permeable to hexoses; insulin is not needed at the liver to make transport of glucose through the cell wall work.

Figure 22.10 Some aspects of glucose metabolism in liver cells. Circled numbers are explained in text. G, glucose; G-6-P, glucose 6-phosphate; G-1-P, glucose 1-phosphate; UDPG, uridine diphosphate glucose (a derivative of glucose).

of "ketone bodies" would make unchecked diabetes mellitus fatal if some other complication did not cause death. Exactly what the ketone bodies are and why their overproduction is serious we shall discuss in Chapter 23. For the present it is sufficient to say that two of these compounds are acids, and that they lower the pH of the blood.

In summary, the liver can take glucose out of circulation, store it, catabolize it, synthesize it (glucogenesis or gluconeogenesis), or release it back into circulation. What happens at any particular moment depends on concentrations of the metabolites, including glucose, or hormones, and on the relative activities of many enzymes. As we have seen, hormones frequently act by moderating or activating enzymes. In diabetes mellitus, even though the blood sugar level is already high, liver cells make even more glucose to send the blood sugar level higher. Very harmful by-products are formed at the same time. The most important factor in regulating what the liver does to glucose is the hormone insulin.

DIABETES MELLITUS AND INSULIN

Diabetes comes from the Greek *diabetes,* to pass—here meaning to pass urine in greater than normal amounts. Mellitus signifies "honey-sweet." Diabetes mellitus therefore signifies passage of urine containing significant quantities of dissolved sugar (glucose), which is the condition glucosuria.

The primary defect in diabetes mellitus is the failure of glucose to enter into specific tissues because effective insulin is absent. Insulin is a protein manufactured and stored in the β-cells of the islets of Langerhans in the pancreas. Located as shown in Figure 22.4, the pancreas controls the release of insulin. When the blood sugar level rises, insulin is put into circulation; when the blood sugar level drops, the rate of insulin release falls off.

The primary action of insulin is to promote the transfer of glucose from circulation, across cell walls, and into the cells of several kinds of tissues. Insulin produces hundreds of changes all through the body, but most, if not all of them, are the result of its primary action, helping to take glucose out of circulation and put it into various cells. Insulin also directly enhances the activity of the enzyme glucokinase in the liver (cf. Figure 22.10). At least the activity of this enzyme is related to the presence of insulin, and there is evidence that insulin in some unknown way stimulates its synthesis.

Not all types of tissues are sensitive to insulin. Insulin is not needed, for example, to help transport glucose into cells in the brain, the kidneys, and gastrointestinal tract, and into red blood cells. It is needed to get glucose across the cell barriers in skeletal and cardiac muscle and in adipose tissue (fatty tissue), among others.

If effective insulin is absent, the blood sugar level rises much higher than normal, even in a fasting condition. In addition, liver cells are stimulated to manufacture even more glucose. The blood sugar level is eventually so high that the renal threshold for glucose is exceeded, and the kidneys transfer some of it to the urine.

Most cases of diabetes are discovered when a patient sees a doctor about some other condition (e.g., arteriosclerosis, persistent drowsiness, loss of appetite, skin infections, and many other ailments). Routine urine analysis then reveals the presence of glucose, and even if this is not observed, the doctor may use other more sensitive tests. Analysis of the glucose in the blood may reveal that the fasting blood sugar level (in blood taken from a vein, i.e., venous blood) is above 120 (but not necessarily above the renal threshold), indicating diabetes; even a level between 100 and 120 is suggestive. Further tests are made. The blood sugar level may be analyzed two hours after the patient has ingested a meal containing 100 g of glucose. By this time, in a normal person, the blood sugar level will be back below 100. If it is not, diabetes is indicated.

Glucose Tolerance Test. Perhaps the best and most useful test for diagnosing diabetes is the glucose tolerance test. For about three days before the test the patient ingests upward of 300 g of carbohydrate per day in addition to the normal components of the diet. Then after a fast of about ten hours (overnight) the subject drinks a solution of 100 g of glucose in 400 ml of water flavored with lemon. A specimen of venous blood is immediately analyzed for blood sugar level, and further analyses are made at intervals of 0.5, 1, 1.5, 2, 3, and sometimes 4 and 5 hours. Urine specimens are also analyzed. In a person without diabetes the following values are usually observed.

Fasting blood sugar level:	not above 100 mg per 100 cc of venous blood
Maximum or peak value:	less than 160
Value at two-hour interval:	less than 120

In the diabetic subject the fasting and peak levels are higher and, of special significance, the rate at which the level returns to "normal" is much slower. Typical

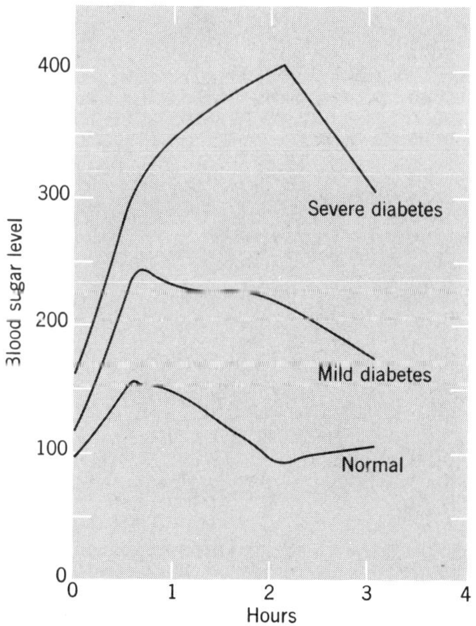
400
300
200
100
0
Blood sugar level
Severe diabetes
Mild diabetes
Normal
0
1
2
3
4
Hours

Figure 22.11 Glucose tolerance curves. (From I. N. Kugelmass, *Biochemistry of Blood in Health and Disease,* Charles C Thomas, Springfield, Ill., 1959. Courtesy of Charles C Thomas.)

glucose tolerance curves plotted from the data for diabetic and nondiabetic persons are shown in Figure 22.11.

Hyperinsulinism. "Insulin Shock." When too much insulin appears in the blood, a condition called hyperinsulinism, an excessive amount of glucose is removed from the bloodstream by any or all of the several processes of glucose utilization, and the blood sugar level drops. The ensuing hypoglycemia may cause malnutrition in the brain. Since its cells depend directly on blood sugar, hyperinsulinism may produce convulsions and "insulin shock." The remedy is immediate ingestion of some readily digestible carbohydrate such as fruit juice, candy, sugar lumps, etc.

Severe hypoglycemia can occur in diabetic people if too much insulin is inadvertently injected. Moreover, their insulin needs can vary if they become ill in other ways. If a nondiabetic individual ingests a relatively large amount of carbohydrate, the sharp rise in blood sugar level may cause a small overstimulation of the pancreas. As the normal glucose tolerance curve of Figure 22.11 indicates, about two to three hours later the blood sugar level is slightly below the normal fasting level. If it goes too low, the brain cells experience a mild scarcity of glucose, and the subject may feel irritable, bleary-eyed, and sleepy. Many whose breakfast consists almost entirely of fruit juice, a heavily sugared roll or doughnut, and sweetened coffee feel drowsy by midmorning. The coffee break becomes a welcome relief, but of course it is only another round of pastry and sugared coffee. The blood sugar level rises and the brain cells perk up, but the pancreas dutifully pours out insulin to restore the level to normal—again a bit more insulin than needed, and two or three hours later grogginess reappears. Many do not achieve any balance in their diet until the evening meal.

The most important meal of the day is breakfast, and it should be rich in proteins with some fat and carbohydrate. Proteins and fat tend to moderate the speed with which glucose leaves the digestive tract. Instead of pouring suddenly into the bloodstream, glucose enters over a longer period, and overstimulation of the pancreas is avoided.

EPINEPHRINE

Of the several hormones besides insulin that regulate glucose catabolism, epinephrine (adrenaline) is particularly important in emergencies. Relatively simple in

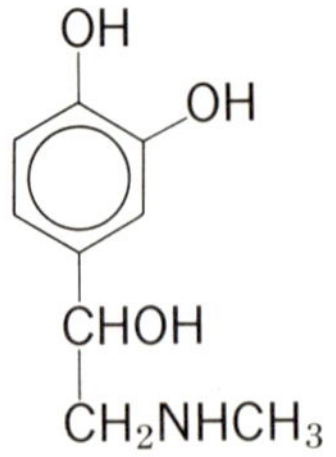

OH
OH
CHOH
CH_2NHCH_3

Epinephrine

structure, this hormone is made by cells in the adrenal medulla, and in times of danger or stress it is discharged into the bloodstream in minute amounts. It activates the enzyme phosphorylase in the liver (cf. Figure 22.10) and in muscles, leading to prompt breakdown of glycogen to glucose. Glucose thus formed in the liver migrates quickly into the bloodstream, and the blood sugar level may rise abruptly, very often to the renal threshold. This sudden appearance of blood glucose makes available to the brain its chief nutrient at the precise moment when it may be required to react the most sharply to a threatening situation. This hormone also affects lipid metabolism by stimulating the breakdown of stored triglycerides and causing the concentration of free fatty acids in the bloodstream to rise quickly.

REFERENCES

E. T. Bell. *Diabetes Mellitus.* Charles C Thomas, Springfield, Ill., 1960.

I. N. Kugelmass. *Biochemistry of Blood in Health and Disease.* Charles C Thomas, Springfield, Ill., 1959.

A. L. Lehninger. "The Transfer of Energy Within Cells," in *Ideas in Modern Biology,* edited by J. A. Moore. Natural History Press, Garden City, N.Y., 1965.

A. L. Lehninger. *Bioenergetics.* W. A. Benjamin, New York, 1965.

A. L. Lehninger. *The Mitochondrion.* W. A. Benjamin, New York, 1964.

G. Litwack and D. Kritchevsky, editors. *Actions of Hormones on Molecular Processes.* John Wiley and Sons, New York, 1964.

A. Marble and G. F. Cahill, Jr. *The Chemistry and Chemotherapy of Diabetes Mellitus.* Charles C Thomas, Springfield, Ill., 1962.

E. F. Neufeld and V. Ginsburg. "Carbohydrate Metabolism." *Annual Review of Biochemistry,* Vol. 34, 1965, page 297.

A. White, P. Handler, and E. L. Smith. *Principles of Biochemistry,* third edition. McGraw-Hill Book Company, New York, 1964.

R. H. Williams. *Disorders in Carbohydrate and Lipid Metabolism.* W. B. Saunders, Philadelphia, 1962.

H. G. Wood and M. F. Utter. "The Role of CO_2 Fixation in Metabolism," in *Essays in Biochemistry,* edited by P. N. Campbell and G. D. Greville, Vol. 1. Academic Press, New York, 1965.

PROBLEMS AND EXERCISES

1. Discuss the meaning of each of the following terms.
   (*a*) glycolysis (*b*) anaerobic sequence (*c*) oxygen debt
   (*d*) active transport (*e*) glucogenesis (*f*) gluconeogenesis
   (*g*) glycogenesis (*h*) glycogenolysis (*i*) blood sugar
   (*j*) blood sugar level (*k*) normal fasting level (*l*) portal vein
   (*m*) hyperglycemia (*n*) hypoglycemia (*o*) renal threshold
   (*p*) glucosuria (*q*) glucose tolerance (*r*) glucose tolerance test
   (*s*) hyperinsulinism
2. Describe what each of the following do. Use equations where appropriate.
   (*a*) glucokinase (*b*) glucose 6-phosphorylase (*c*) phosphoglucomutase
   (*d*) phosphorylase (*e*) insulin (*f*) epinephrine
3. What is the difference between glucogenesis and gluconeogenesis?
4. Explain how pyruvic acid serves to regenerate an enzyme needed in glycolysis.

5. Explain by equations how glycolysis by yeast cells differs from that accomplished in human muscle cells.
6. Explain why hypoglycemia is serious.
7. Discuss the relative activities of glucokinase and glucose 6-phosphorylase in (*a*) a fed animal, (*b*) a fasted animal, and (*c*) a person with diabetes mellitus.
8. Describe how the body maintains a fairly constant blood sugar level, even though it ingests glucose only two or three times a day.
9. In periods of fasting, would you expect the amount of glycogen in the liver to increase or to decrease? Why?
10. Why does severe hyperinsulinism lead quickly to "insulin shock," whereas the absence of insulin, at least for a short period, is not so serious?
11. Is glucosuria always caused by diabetes mellitus? If not, how else might it arise?

CHAPTER TWENTY-THREE

Metabolism of Lipids

People who, for whatever reason, eat much and exercise little gain weight. The body has a remarkable capacity for converting carbohydrates and proteins into fat and storing it. In this chapter we study how the body makes and uses lipids, particularly the simple triglycerides.

ABSORPTION, DISTRIBUTION, AND STORAGE OF LIPIDS

The complete digestion of simple lipids produces glycerol and a mixture of long-chain fatty acids. Such complete hydrolysis, however, is not necessary for absorption to occur, for monoesters of glycerol, the monoglycerides, can also leave the intestinal tract.

As molecules of monoglycerides, fatty acids or their anions, and glycerol migrate across the intestinal barrier, triglyceride molecules are extensively resynthesized until what is delivered for circulaton is essentially reconstituted lipids. Only a very small fraction is delivered directly into portal circulation, however. Most of the absorbed lipid emerges into the lymph system. Lymph coming from regions of the digestive system is ultimately delivered to the bloodstream via the thoracic duct; most of the lipid is present as microdroplets or chylomicra averaging one micron in diameter.

Triglycerides found in the bloodstream, whether as microdroplets or as individual molecules, are associated with plasma-soluble proteins, and thus the strong tendency for lipids to coalesce and to remain out of solution in water is defeated. Triglycerides and fatty acids are normally transported with no difficulty. The plasma is known to contain a lipase, an enzyme that can catalyze the hydrolysis of the ester linkages in triglycerides. Thus fatty acids are formed, and these too are bound to plasma-soluble proteins for ready transport. Certain tissues can remove these fatty acids and use them for ATP synthesis. Those not so used are eventually stored in the specialized connective tissue, adipose tissue.

Cells of adipose tissue can remove fatty acids from the bloodstream, and they can synthesize triglycerides not only from fatty acids but also from glucose. The fat collects as larger and larger droplets in the cytoplasm of each cell until the cell's nucleus is pushed to the edge and becomes flattened. This fat-laden tissue is found principally around abdominal organs and beneath the skin, and it constitutes the body's chief reserve of chemical energy, far exceeding the normal carbohydrate reserves. Lipid deposits around internal organs serve to cushion and protect them from mechanical shock when the jars and bumps of living are experienced. Adipose tissue found subcutaneously, that is, beneath the skin, serves also as insulation against excessive heat loss to the environment (or as some protection against cold). Once thought to be simply a passive insulation, something like an ordinary blanket, subcutaneous adipose tissue is now believed to act more like an electric blanket. Metabolic activity in the cells of this tissue generates a small amount of heat to offset heat loss when the outside temperature drops. Adipose tissue, wherever it is found, is quite active metabolically. The cells have large numbers of mitochondria, nerves extend in and among them, and the bloodstream pushes its capillary network deep into this tissue. In fact, one of the problems associated with obesity is the extra load placed on the heart not only to assist in moving the extra weight during exercise but also to pump blood throughout a larger network.

The average adult human male receiving an adequate diet has enough lipid in reserve to sustain his life for thirty to forty days, assuming he has enough water. For an individual in a modern, developed country this reserve is obviously much more than he needs, although in more primitive times it may have been essential to the survival of the race. The body need not be without food too long before the fat reserves are tapped. After a few hours the body's glycogen reserves virtually disappear. There is only enough lipid material or glucose in the bloodstream to sustain average metabolic needs for several minutes. The fat deposited in the liver provides the energy needed for slightly more than an hour. Obviously other, larger reserves are needed and the body has two. One is the unabsorbed food in the digestive tract. Although a temporary reserve, it is an important one and quite plainly much used by people who are fortunate enough to eat regularly. The other reserve is the adipose tissue. In the steady state, that is, in the normally fed individual, about 30% of the carbohydrate taken in is temporarily changed into triglyceride, and virtually all this conversion takes place within the cells of this tissue. In an area of very active research virtually no one dissents from this conclusion. (Earlier researchers believed that the liver was the main site for fatty acid synthesis, and that newly made fatty acid molecules were released into circulation, picked up by adipose cells, and converted into triglycerides.)

The advantages of storing chemical energy as triglyceride can be understood in terms of the *energy densities,* grams per calorie, of various storage materials. When glucose is in isotonic concentrations in water, it takes 5 g of this solution to yield 1 calorie. When glucose is stored as wet glycogen in tissues, it takes only about 0.6 g of such material to yield 1 calorie. Essentially no water is involved in the storage of triglycerides, however, and only about 0.13 g of the substance is needed to yield 1 calorie.

Energy density can be thought of in other terms. A molecule of a fatty acid can deliver over a hundred molecules of ATP, depending on the chain length, whereas a molecule of glucose can make available only about thirty-eight. Almost half the hydrogens in glucose are already attached to oxygen, but hydrogens attached to *carbon* are by and large the ones that fuel the respiratory chain (from which they ultimately emerge bound to oxygen). Fatty acid molecules have all but one of their hydrogens attached to carbon.

The conversion of glucose to fatty acids and thence to triglycerides is not without some cost. Energy is required, and about a quarter of the glucose supply must be sacrificed to furnish it. If 30% of carbohydrate taken in is changed into lipids, and if another 25% is used to power this change, a significant portion of all oxygen consumption and heat production in the body goes into just this operation.

If food intake (more correctly, chemical energy intake) exceeds energy expenditure, the newly synthesized triglyceride becomes part of a relatively inactive "compartment" of adipose tissue. Considerable indirect evidence indicates that there are two compartments for storing fat. Just what or precisely where they are or how they are distributed is not known, anatomically, but it is known that a portion of the deposited lipid stays put a long time, and that another portion has a more rapid turnover. Hence the *idea* of two compartments, even though the word itself may misleadingly suggest walled-off regions. The half-life of lipid in the one compartment is from 350 to 500 days, meaning that 50 g of an initial 100 g have after this length of time been replaced through the release of stored lipid and the substitution of a fresh supply. In the active "compartment" the half-life is on the order of 30 to 40 days, as determined by radioactive labeling studies. A great deal of metabolic activity occurs in this compartment. In succeeding sections we shall study in greater detail how fatty acids are made, how glycerides are put together, how fatty acids are taken from storage, and how they are oxidized to produce ATP. The events that take place in the turnover of triglyceride in adipose tissue are outlined in Figure 23.1.

SYNTHESIS OF TRIGLYCERIDES FROM GLUCOSE

General View. Glucose is in a more oxidized state than a fatty acid. To convert glucose into fatty acid therefore requires a *reducing agent,* that is, a source of the elements of hydrogen, ($H:^- + H^+$). To make triglycerides we also need a source of *glycerol,* or its equivalent. And we obviously need *carbon parts* for the fatty acids. Glucose provides essentially all these services (Figure 23.2).

Some of the glucose taken in by adipose tissue is oxidized by the phosphogluconate pathway, the hexose monophosphate shunt described briefly in Chapter 22 (p. 615) and called here simply the *shunt.* The net result of this series of reactions, which we did not study in detail, was the oxidation of glucose and the reduction of $NADP^+$, the coenzyme described on page 592 as a close relative of NAD^+. The shunt was summarized as follows:

$$\text{6 glucose 6-phosphate} + 12NADP^+ \longrightarrow \text{5 glucose 6-phosphate} + 12NADPH + 12H^+ + 6CO_2 + P_i$$

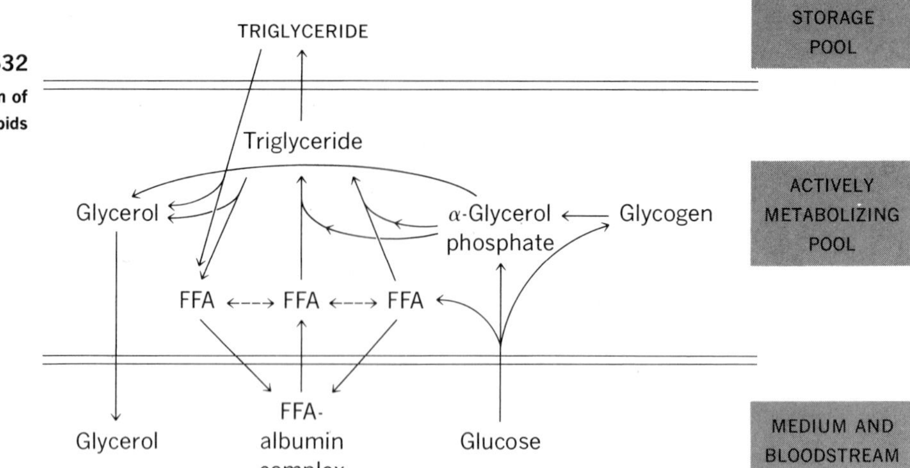
TRIGLYCERIDE
STORAGE POOL
Triglyceride
Glycerol
α-Glycerol phosphate
Glycogen
ACTIVELY METABOLIZING POOL
FFA
FFA
FFA
Glycerol
FFA-albumin complex
Glucose
MEDIUM AND BLOODSTREAM

Figure 23.1 Triglyceride turnover in adipose tissue (FFA stands for free fatty acids). In this diagram the concept of compartments for triglyceride storage represents the *idea* rather than implying that they have been identified anatomically. (From B. Shapiro, "Triglyceride Metabolism," in *Handbook of Physiology, Section 5: Adipose Tissue,* edited by A. E. Renold and G. F. Cahill, Jr., American Physiological Society, Washington, D.C., 1965, page 222. Used by permission.)

Cells of adipose tissue have the necessary enzymes for this shunt, and one of its main purposes appears to be furnishing NADPH for fatty acid synthesis.

Most of the rest of the glucose taken in by adipose tissue is catabolized by glycolysis to furnish, finally, two compounds essential to the formation of triglycerides. One is needed directly, α-glycerol phosphate, for this form of glycerol, its monoester with phosphoric acid, is required to make triglycerides. Glycerol itself will not do. It must be the α-phosphate ester. The other product of glycolysis, pyruvic acid, is needed for forming not triglyceride but acetyl coenzyme A. Fatty acids are made from acetyl coenzyme A (Figure 23.3) and in turn interact with α-glycerol phosphate to furnish triglycerides.

CATABOLISM OF TRIGLYCERIDES

To mobilize the energy reserves in adipose lipids, several successive steps must occur (Figure 23.4).

1. Triglycerides in adipose tissue (or other depots) must be hydrolyzed to give free fatty acids (FFA) and glycerol.
2. These products must be transported in the bloodstream to the sites of oxidation. Glycerol can enter the glycolysis pathway after it has been converted into glyceraldehyde 3-phosphate.
3. Fatty acids are broken down two carbons at a time to release acetyl coenzyme A units.
4. These units are fed into the citric acid cycle to furnish hydrogen donors.
5. Hydrogen donors fuel the respiratory chain that produces ATP and water.

For all these steps to occur, there must of course be a need for ATP. The principal site of fatty acid catabolism is the liver, but some of the smaller fragments produced from the fatty acids can be utilized by muscles. In fact, we have already noted that most of the energy requirements of resting muscle are met by intermediates from

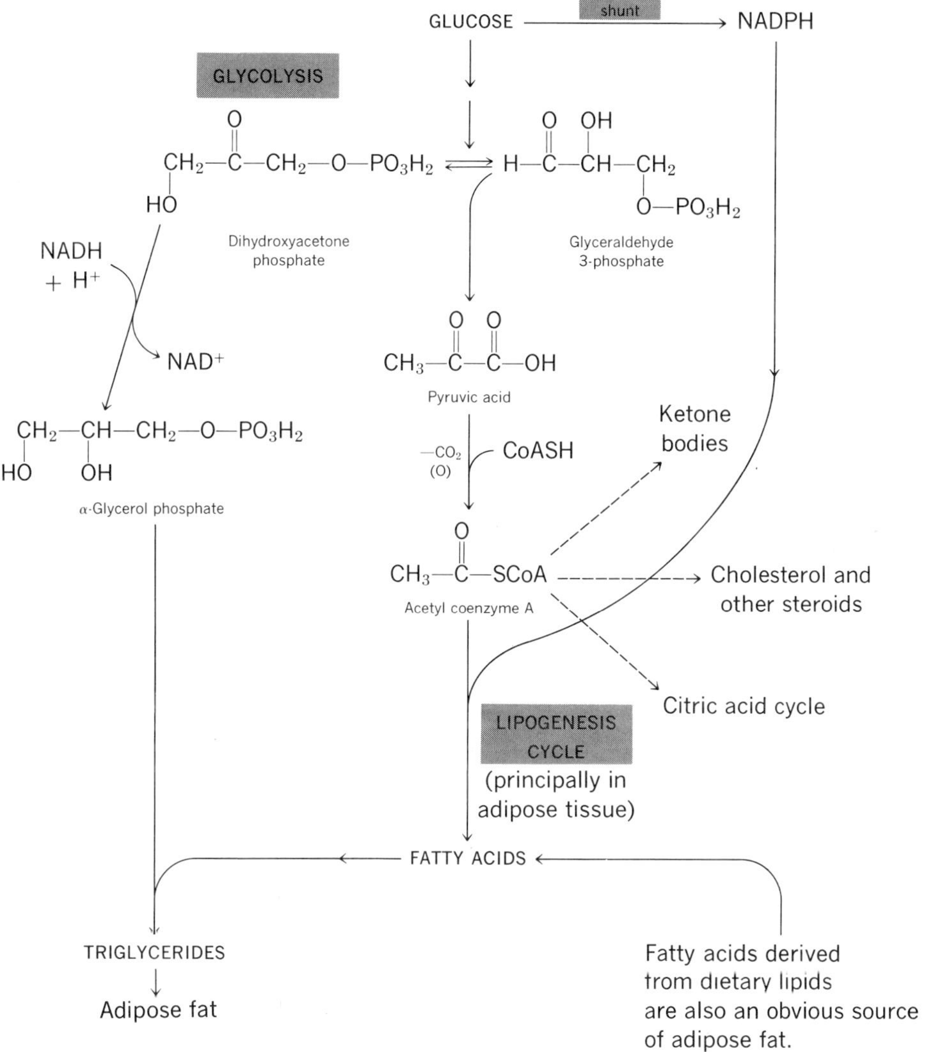
GLUCOSE
shunt
NADPH
GLYCOLYSIS
CH₂—C(=O)—CH₂—O—PO₃H₂
HO
H—C(=O)—CH(OH)—CH₂
O—PO₃H₂
NADH
+ H+
Dihydroxyacetone phosphate
Glyceraldehyde 3-phosphate
NAD+
CH₃—C(=O)—C(=O)—OH
Pyruvic acid
Ketone bodies
CH₂—CH—CH₂—O—PO₃H₂
HO OH
α-Glycerol phosphate
—CO₂
(O)
CoASH
CH₃—C(=O)—SCoA
Acetyl coenzyme A
Cholesterol and other steroids
Citric acid cycle
LIPOGENESIS CYCLE
(principally in adipose tissue)
FATTY ACIDS
TRIGLYCERIDES
Adipose fat
Fatty acids derived from dietary lipids are also an obvious source of adipose fat.

Figure 23.2 Triglyceride synthesis in adipose tissue with glucose as the principal raw material. Glucose, via the *shunt,* produces the needed reducing agent, NADPH. Via glycolysis, glucose produces acetyl units for fatty acids. Dihydroxyacetone phosphate, an intermediate in glycolysis, is also a source of α-glycerol phosphate, the form glycerol must be in if it is to be used to make triglycerides. Dotted arrows indicate pathways from acetyl coenzyme A that are alternatives to triglyceride synthesis.

fatty acid catabolism. The degradation of fatty acids occurs by a repeating series of steps known as the fatty acid cycle.

Fatty Acid Cycle. Fatty Acid Spiral. Before a molecule of a fatty acid can yield ATP, it must be activated, consuming ATP. The price is small, for the return is great.

$$R{-}\overset{\displaystyle O}{\overset{\|}{C}}{-}OH + CoASH + ATP \longrightarrow R{-}\overset{\displaystyle O}{\overset{\|}{C}}{-}SCoA + AMP + PP_i$$

In this rare reaction of ATP its molecule is cleaved to produce not simple inorganic phosphate, P_i, but the diphosphate (or pyrophosphate) ion, here symbolized simply as PP_i. Once this initial activation is achieved, a series of reactions clip two carbons at a time from the fatty acid unit. When two carbons have been removed, as an acetyl unit, the remainder of the original fatty acid goes through the cycle again to have another acetyl group removed. These reactions continue until the original fatty acid has been cut down to the size of a final acetyl group. The acetyl groups normally enter the citric acid cycle as diagramed in Figure 23.4.

The steps in the fatty acid cycle, outlined in Figure 23.5, are as follows; the circled numbers refer to this figure.

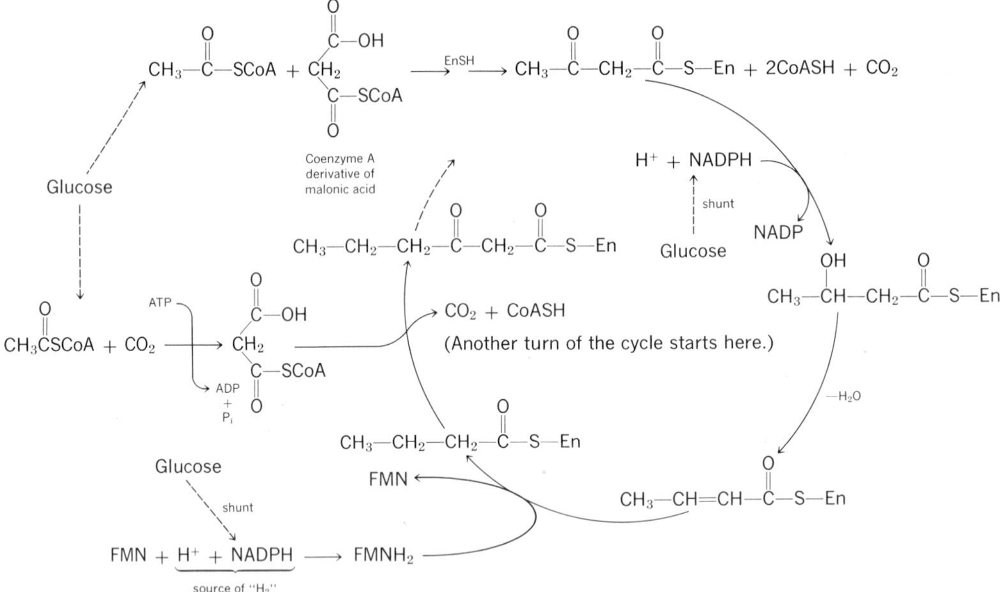
CH₃—C(=O)—SCoA + CH₂(C(=O)—OH)(C(=O)—SCoA)
EnSH
CH₃—C(=O)—CH₂—C(=O)—S—En + 2CoASH + CO₂
Coenzyme A derivative of malonic acid
Glucose
H⁺ + NADPH
shunt
NADP
Glucose
CH₃—CH₂—CH₂—C(=O)—CH₂—C(=O)—S—En
OH
CH₃—CH—CH₂—C(=O)—S—En
ATP
CH₃C(=O)SCoA + CO₂
ADP + Pᵢ
CO₂ + CoASH
(Another turn of the cycle starts here.)
—H₂O
CH₃—CH₂—CH₂—C(=O)—S—En
FMN
CH₃—CH=CH—C(=O)—S—En
Glucose
shunt
FMN + H⁺ + NADPH ⟶ FMNH₂
source of "H₂"

Figure 23.3 Lipogenesis cycle. Biosynthesis of fatty acids from acetyl units. An enzyme having an —SH group, EnSH, locks the growing chain to a multienzyme system. The growing fatty acid unit swings from specialized enzyme to specialized enzyme, pivoting about the S—En bond as the steps of the cycle take place. (FMN is flavin mononucleotide, $FMNH_2$ its reduced form.)

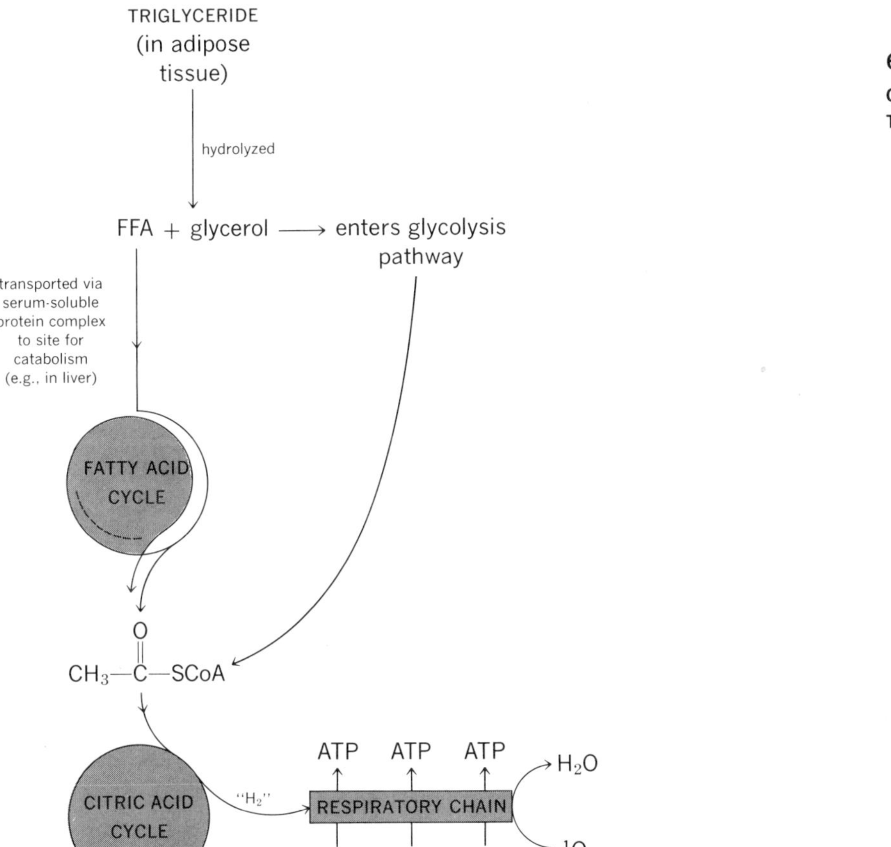
TRIGLYCERIDE
(in adipose tissue)
hydrolyzed
FFA + glycerol ⟶ enters glycolysis pathway
transported via serum-soluble protein complex to site for catabolism (e.g., in liver)
FATTY ACID CYCLE
CH3—C(=O)—SCoA
CITRIC ACID CYCLE
"H2"
RESPIRATORY CHAIN
ATP
ATP
ATP
ADP
ADP
ADP
H2O
½O2

Figure 23.4 Mobilization of the energy reserves in triglycerides.

① Dehydrogenation. FAD is the hydrogen acceptor.

$$CH_3(CH_2)_{12}CH_2—CH_2—\overset{\overset{\displaystyle O}{\|}}{C}—SCoA + FAD \longrightarrow$$

Palmitic acid CoA, from a C_{16} acid

$$CH_3(CH_2)_{12}CH{=}CH—\overset{\overset{\displaystyle O}{\|}}{C}—SCoA + FADH_2$$

An α,β-unsaturated acid derivative of CoASH

$FADH_2$ → FAD ← Respiratory chain ⟶ 2ATP

Several flavoproteins with FAD as the cofactor act as enzymes, each one differing in the size of the fatty acid unit it acts upon. Each $FADH_2$ produced presumably yields two ATPs via the respiratory chain, as implied in Figure 21.9.

② Addition of water to the double bond.

$$CH_3(CH_2)_{12}CH{=}CH{-}\overset{\overset{\displaystyle O}{\|}}{C}{-}SCoA + H_2O \longrightarrow CH_3(CH_2)_{12}\overset{\overset{\displaystyle OH}{|}}{C}H{-}CH_2{-}\overset{\overset{\displaystyle O}{\|}}{C}{-}SCoA$$

A β-hydroxy acid derivative of CoASH

③ Dehydrogenation. The 2° alcohol group is oxidized to a keto group, with NAD^+ the hydrogen acceptor.

$$CH_3(CH_2)_{12}\overset{\overset{\displaystyle OH}{|}}{C}H{-}CH_2{-}\overset{\overset{\displaystyle O}{\|}}{C}{-}SCoA + NAD^+ \longrightarrow$$

$$CH_3(CH_2)_{12}\overset{\overset{\displaystyle O}{\|}}{C}{-}CH_2{-}\overset{\overset{\displaystyle O}{\|}}{C}{-}SCoA + \underbrace{H^+ + NADH}$$

A β-keto acid derivative of CoASH

$H^+ + NADH \rightarrow$ Respiratory chain ($\rightarrow NAD^+$) $\rightarrow$ 3ATP

Each NADH produced presumably yields three ATPs via the respiratory chain (cf. again Figure 21.9).

④ Cleavage of the β-keto acid system.

$$CH_3(CH_2)_{12}\overset{\overset{\displaystyle O}{\|}}{C}{-}CH_2\overset{\overset{\displaystyle O}{\|}}{C}{-}SCoA + CoASH \longrightarrow$$

$$CH_3(CH_2)_{12}\overset{\overset{\displaystyle O}{\|}}{C}{-}SCoA + CH_3\overset{\overset{\displaystyle O}{\|}}{C}{-}SCoA$$

A C_{14} acid derivative

$CH_3COSCoA \rightarrow$ Citric acid cycle $\rightarrow$ Respiratory chain $\rightarrow$ 12ATP

A summary of the energy (ATP) yield for the complete oxidation of one molecule of palmitic acid is given in Table 23.1.

The intermediates in this catabolism do not normally accumulate to any significant extent, for as soon as they form they react further. When the acetyl coenzyme A stage is reached, several options are available. Hints of these have appeared in a few figures (e.g., Figure 23.2), and we are now at the point that we can study them. One of them has an important bearing on the synthesis of cholesterol; another is particularly ominous in diabetes.

METABOLIC FATES OF ACETYL COENZYME A

General View. As we have often seen, acetyl coenzyme A stands at one of the most important metabolic intersections in all body chemistry. Figure 23.6 serves as a reminder and also summarizes some features not yet noted. The ready interconversion of acetyl coenzyme A to acetoacetyl coenzyme A leads to another important crossroad. The formation of β-hydroxy-β-methylglutaryl coenzyme A (we call it HMG-CoA for our limited needs) makes possible the subsequent syntheses of

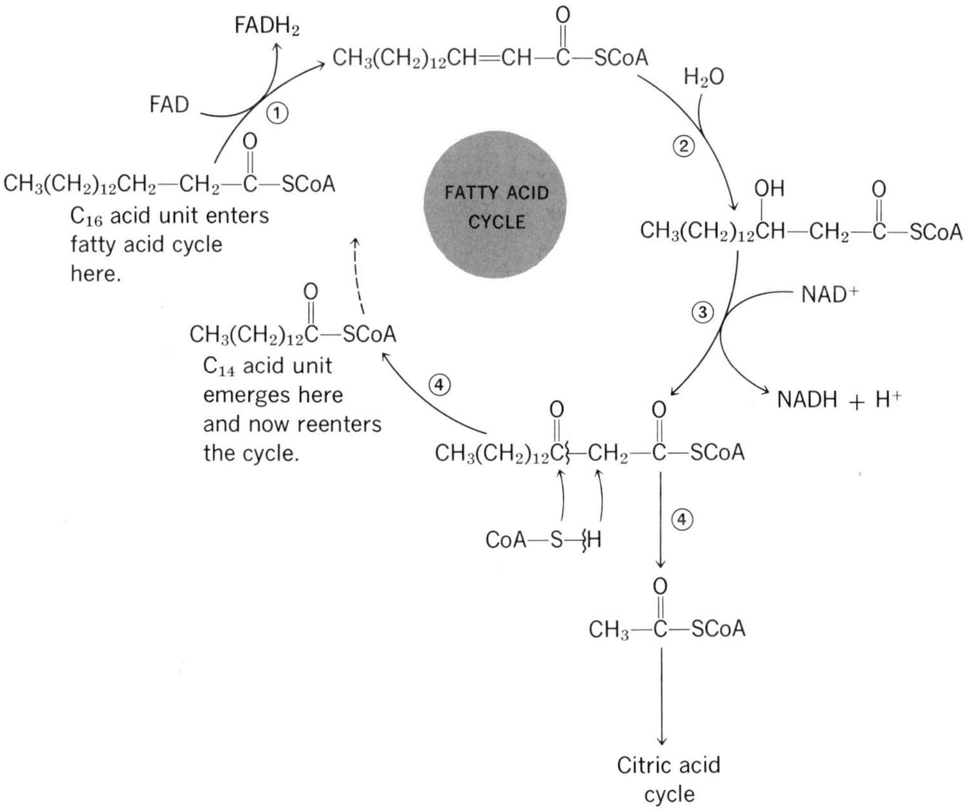

FATTY ACID CYCLE

$CH_3(CH_2)_{12}CH_2—CH_2—C(=O)—SCoA$

C_{16} acid unit enters fatty acid cycle here.

FAD → $FADH_2$ ①

$CH_3(CH_2)_{12}CH{=}CH—C(=O)—SCoA$

H_2O ②

$CH_3(CH_2)_{12}CH(OH)—CH_2—C(=O)—SCoA$

NAD^+ → $NADH + H^+$ ③

$CH_3(CH_2)_{12}C(=O)—CH_2—C(=O)—SCoA$

CoA—S—H ④

$CH_3(CH_2)_{12}C(=O)—SCoA$

C_{14} acid unit emerges here and now reenters the cycle.

$CH_3—C(=O)—SCoA$

Citric acid cycle

Figure 23.5 Fatty acid cycle. The reactions taking place at the circled numbers are discussed in the text. In this cycle the same intermediates do not recur; only types of them do, and types of reactions recur. The C_{14} acid unit produced at ④ will next be dehydrogenated to give a C_{14} unsaturated acid. This acid will be hydrated ②, and oxidized ③, and then cleaved ④ to produce a C_{12} acid unit and another acetyl coenzyme A. Seven turns of the cycle are needed to break the original palmitic acid (C_{16}) into eight acetyl coenzyme A units.

cholesterol and conceivably of the steroid hormones and the bile salts. It also makes possible the production of three compounds, somewhat inaccurately called the ketone bodies, that are important to our study of the molecular basis of diabetes mellitus.

Reactions Leading to the Ketone Bodies. Formation of acetoacetyl coenzyme A. This reaction (Figure 23.6) resembles a Claisen ester condensation (cf. p. 425). Thus

$$CH_3-\overset{O}{\overset{\|}{C}}-SCoA + H-CH_2-\overset{O}{\overset{\|}{C}}-SCoA \longrightarrow CH_3-\overset{O}{\overset{\|}{C}}-CH_2-\overset{O}{\overset{\|}{C}}-SCoA + HSCoA$$

Acetoacetyl coenzyme A

Formation of HMG-CoA. This reaction resembles the aldol condensation (cf. p. 397):

$$CoA-S-\overset{O}{\overset{\|}{C}}CH_2\text{—}H + \underset{CH_3}{\underset{|}{\overset{O}{\overset{\|}{C}}}}CH_2\overset{O}{\overset{\|}{C}}-SCoA \longrightarrow CoA-S-\overset{O}{\overset{\|}{C}}CH_2-\underset{CH_3}{\underset{|}{\overset{OH}{\overset{|}{C}}}}CH_2\overset{O}{\overset{\|}{C}}-SCoA$$

$$\xrightarrow{H_2O} HO\overset{O}{\overset{\|}{C}}CH_2-\underset{CH_3}{\underset{|}{\overset{OH}{\overset{|}{C}}}}CH_2\overset{O}{\overset{\|}{C}}-SCoA + CoASH$$

HMG–CoA
(β-hydroxy-β-methylglutaryl coenzyme A)

Table 23.1 Yield of ATP During Complete Oxidation of Palmitic Acid

Step			ATP Made or Used
$CH_3(CH_2)_{14}CO_2H$	$\longrightarrow$	$CH_3(CH_2)_{14}\overset{O}{\overset{\|}{C}}-SCoA$	-1
$CH_3(CH_2)_{14}\overset{O}{\overset{\|}{C}}-SCoA$	$\longrightarrow$	$8CH_3\overset{O}{\overset{\|}{C}}-SCoA$	
1. $7FADH_2$	$\longrightarrow$	$7FAD + 7(H:^- + H^+)$ to chain	$+14\ (7 \times 2)$
2. $7NADH + H^+$	$\longrightarrow$	$7NAD^+ + 7(H:^- + H^+)$ to chain	$+21\ (7 \times 3)$
$8CH_3\overset{O}{\overset{\|}{C}}-SCoA$	$\longrightarrow$	$16CO_2 + 8H_2O + 8CoASH$	$+96\ (8 \times 4 \times 3)$
Net ATP yield			$+130$

Formation of acetoacetic acid. If forming HMG-CoA is like an aldol condensation, this next step is like a reverse aldol condensation (cf. p. 400):

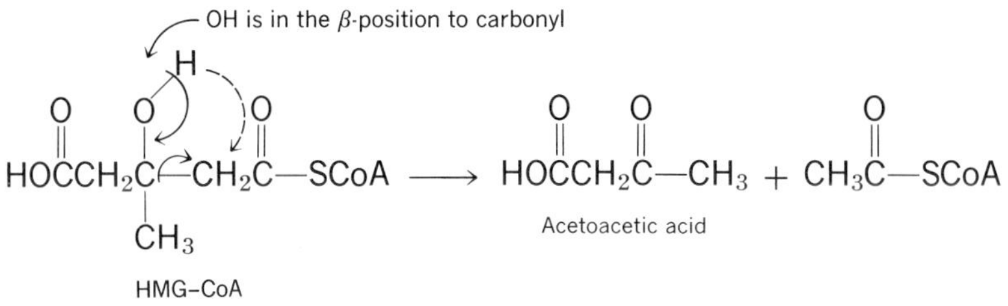

OH is in the β-position to carbonyl

$$HOCCH_2C(OH)(CH_3)CH_2C(=O)\text{—SCoA} \longrightarrow HOC(=O)CH_2C(=O)\text{—}CH_3 + CH_3C(=O)\text{—SCoA}$$

HMG–CoA → Acetoacetic acid

Formation of β-hydroxybutyric acid. The keto group in the product of the last reaction is reducible by NADH. The reaction is reversible, and both the starting acid and the product are found in circulation in a ratio that varies widely, depending on other factors.

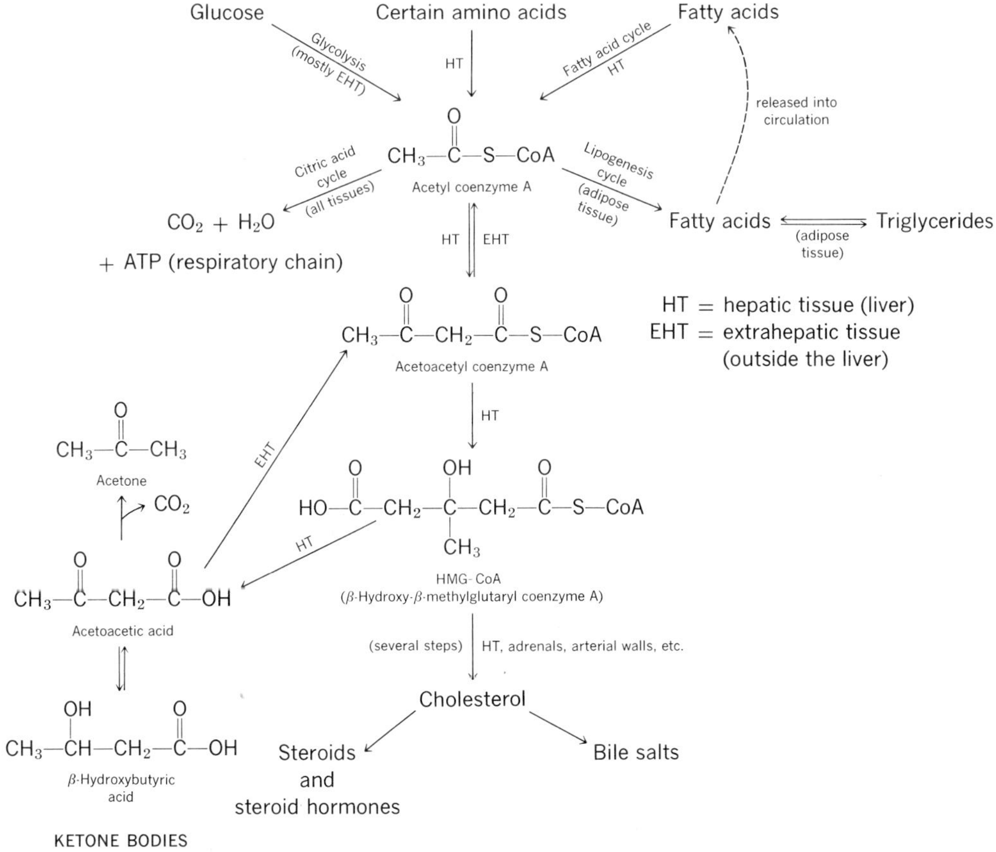

Glucose
Glycolysis (mostly EHT)
Certain amino acids
HT
Fatty acids
Fatty acid cycle HT
released into circulation
CH_3—C(=O)—S—CoA
Acetyl coenzyme A
Citric acid cycle (all tissues)
$CO_2 + H_2O$
+ ATP (respiratory chain)
Lipogenesis cycle (adipose tissue)
Fatty acids ⇌ Triglycerides
(adipose tissue)
HT
EHT
HT = hepatic tissue (liver)
EHT = extrahepatic tissue (outside the liver)
CH_3—C(=O)—CH_2—C(=O)—S—CoA
Acetoacetyl coenzyme A
HT
HO—C(=O)—CH_2—C(OH)(CH_3)—CH_2—C(=O)—S—CoA
HMG-CoA
(β-Hydroxy-β-methylglutaryl coenzyme A)
(several steps)
HT, adrenals, arterial walls, etc.
Cholesterol
Steroids and steroid hormones
Bile salts
CH_3—C(=O)—CH_3
Acetone
CO_2
EHT
HT
CH_3—C(=O)—CH_2—C(=O)—OH
Acetoacetic acid
CH_3—CH(OH)—CH_2—C(=O)—OH
β-Hydroxybutyric acid
KETONE BODIES

Figure 23.6 The HMG-CoA crossroads. The various pathways occur usually, but not exclusively, in the tissues indicated.

$$\underset{\text{Acetoacetic acid}}{CH_3\overset{O}{\overset{\|}{C}}CH_2\overset{O}{\overset{\|}{C}}OH} + NADH + H^+ \rightleftharpoons \underset{\beta\text{-Hydroxybutyric acid}}{CH_3\overset{OH}{\overset{|}{C}}HCH_2\overset{O}{\overset{\|}{C}}OH} + NAD^+$$

Formation of acetone. This reaction has its direct counterpart in *in vitro* reactions. β-Keto acids, in general, easily lose the elements of carbon dioxide (cf. p. 427). One of the few reactions in the body for which a specific enzyme catalyst has not been established, it is unimportant except in diabetes. With this condition its frequency rises, and the odor of acetone may even be detected on the breath of a person with severe diabetes.

$$CH_3-\overset{O}{\overset{\|}{C}}-CH_2-\overset{O}{\overset{\|}{C}}-O-H \longrightarrow \underset{\text{Acetone}}{CH_3-\overset{O}{\overset{\|}{C}}-CH_3} + CO_2$$

The Ketone Bodies. The three compounds acetoacetic acid, β-hydroxybutyric acid, and acetone are traditionally called the *ketone bodies*. Collective analysis of the three of them is possible, and the ketone body concentration is usually expressed in terms of β-hydroxybutyric acid. Formed slowly but relatively continuously in the liver and released into general circulation, these compounds are normally present. Yet the ketone body level in venous circulation in an adult is roughly only 1 mg per 100 ml. This low level is maintained partly because two of the three ketone bodies, the two acids, serve as sources of energy in many tissues. Muscles, for example, obtain much of their ATP for resting functions from acetoacetic acid, a supplier of acetyl groups for acetyl coenzyme A and the citric acid cycle as indicated in Figure 23.6. The kidneys also remove ketone bodies, excreting about 20 mg of these substances per day via the urine. When the ketone body level in the blood rises, a condition of *ketonemia* exists. When, as a result of high ketonemia, significant quantities of these compounds are found in the urine, the condition is called *ketonuria*. Usually "ketone breath" is by now noticeable. These three conditions—ketonemia, ketonuria, and ketone breath—are collectively called *ketosis*. The situation is dangerous. Unchecked, ever-rising ketosis means death by means we shall soon study.

KETOSIS AND ACIDOSIS

Two of the ketone bodies are carboxylic acids. Persistent overproduction of them will therefore eventually overtax the buffer systems of the blood. The kidneys can remove hydrogen ions, and the pH of urine can become as low as 4 (meaning a hydrogen-ion concentration about 1000 times that of normal blood). The kidneys can also remove the two acids of the ketone bodies as their negative ions. But if an excessive output of ketone bodies continues, the kidneys must eventually put increasing amounts of sodium ion into the urine being formed. (For every negative ion from a carboxylic acid put into the urine, a positive ion must be placed there too. Electric neutrality requires this.) Loss of sodium ion means, indirectly, loss of buffer capacity in the blood. The pH of the blood slowly starts to drop. We have learned that this drift downward in the pH of the blood, from about 7.4 to a value of 7 or lower (as measured at room temperature), is called acidosis. As the pH drops, buffers in the blood act to neutralize the acids. One of these buffers, the H_2CO_3/HCO_3^- system, is the most affected. Bicarbonate ions are neutralized, and the excess carbonic acid decomposes to water and carbon dioxide which is expelled via the lungs. Normally the bicarbonate level in blood is 22 to 30 millimoles/liter. This level drops to 16 to 20 millimoles in mild acidosis; to 10 to 16 millimoles in moderate acidosis; and below 10 millimoles in severe acidosis.

Acidosis is serious for many reasons; one disruption, that of the mechanism for transporting oxygen, was studied in Chapter 20. In moderate to severe acidosis the difficulties in taking in oxygen at the lungs have become so great that the person experiences severe "air hunger"; breathing is very painful and laborious (a condition called dyspnea). By the time the bicarbonate level has dropped to 6 to 7 millimoles/liter in the adult, he is in a coma.

In even moderate acidosis the body loses an excessive amount of fluids via the kidneys. To put the salts of the carboxylic acids (from two of the ketone bodies) into the urine being formed, large quantities of water must also be removed. The person suffers from a general dehydration. This coupled with disruption of his ability to transport oxygen will normally depress the central nervous system. Even in mild acidosis a person experiences fatigue, a desire to stay in bed, lack of appetite, nausea, headache, reduced power of concentration, an indisposition to converse, and difficulty in making even simple decisions. Acidosis has been described in some detail because it is instructive to realize how much of human well-being depends on control of the pH of the blood. And acidosis is much more common than a healthy person might realize. Everyone experiences it briefly after severe physical exercise.

Lactic acid acidosis develops during strenuous muscular work. We learned in Chapter 22 that glycolysis becomes an important source of ATP under anaerobic conditions, producing lactic acid. When it forms faster than it can be removed (be converted to glycogen at the expense of some of it), its level in the blood will rise, the level of bicarbonate ion will drop, and so will the pH. The lactic acid level in blood may rise from its resting level of 1 to 2 millimoles/liter to 10 to 12 millimoles/liter after hard work. This will cut the bicarbonate level roughly in half, to a level of 12 to 14 millimoles/liter, meaning that the moderate form of acidosis results. All athletes have experienced the violent dyspnea, the painful, gulping air hunger that accompanies maximum effort in a contest. When the lactic acid level is about 10 millimoles/liter, further work is rendered virtually impossible regardless of will power. Lactic acid acidosis is obviously a controlling factor in the improvement of athletic performance.

Lactic acid acidosis may develop during milder exercise if the supply of oxygen drops. Men living in lower altitudes who fancy themselves rugged individuals may return from even brief hikes at high altitudes with severe pains in the chest and the ego. Neither is serious. Recovery from lactic acid acidosis merely requires rest.

Causes and Prevention of Ketosis. We have seen that ketosis may lead to coma and death; we need to know now how it can be prevented or controlled. The best evidence indicates that overproduction of the ketone bodies is caused by overproduction of acetyl coenzyme A. At the acetyl coenzyme A crossroads near the top of Figure 23.6, three pathways are available for removing acetyl coenzyme A as it forms. The citric acid cycle, however, will not remove it unless there is a demand for ATP, leaving to nonactive people the production of fat and acetoacetyl coenzyme A as ways of handling extra acetyl coenzyme A. In diabetes mellitus and starvation acetoacetyl coenzyme A is the more important drain for excess acetyl coenzyme A, leading in turn to the overproduction of HMG-CoA from which the ketone bodies form. With this connection between ketone bodies and acetyl coenzyme A established, we must determine how acetyl coenzyme A may be synthesized too rapidly in the body.

The two main sources of acetyl coenzyme A are glucose and fatty acids. Glucose reserves, we have noted (p. 630), do not last long before they are depleted. After this depletion the body switches to fatty acids as the almost exclusive source of acetyl coenzyme A. Although glucose can be utilized directly out of the bloodstream by essentially all tissues, fatty acids must by and large be collected and first processed at the liver. This organ then sends out acetoacetate units which can be picked up in lieu of glucose in extrahepatic tissues and used to make acetyl coenzyme A. Glucose reserves obviously disappear in starvation. Less obvious is their disappearance in diabetes.

In starvation the supply of glucose runs out. In diabetes the glucose supply is more than adequate, but the system's ability to use it has deteriorated. We have learned that the primary defect in diabetes mellitus is the failure of the system to produce and circulate effective insulin. The pancreas may have lost its ability to make it, or the insulin made may be so tightly bound to proteins as it circulates in the blood that for all practical purposes it might as well not be there. Insulin may be present, but it is not *effective insulin.* A small amount of evidence, recently reported, indicates that the insulin produced in some individuals is structurally not quite right. A genetic defect, an inborn error issuing from a mutation, is responsible. Whatever the cause, the result is approximately the same.

The tissue of the body most sensitive to the action of insulin is adipose tissue. Its cells exhibit a response to insulin even when the concentration of this hormone drops to one-twentieth its usual fasting level. Insulin appears to exert its action by controlling the entry of glucose into the cells. When insulin is absent, glucose does not enter; when insulin is present, glucose enters. Once inside, glucose is phosphorylated and trapped. From then on roughly half of it is catabolized by the shunt to make NADPH (cf. p. 631); the other half undergoes glycolysis to furnish acetyl coenzyme A, most of which is converted, with the help of NADPH, into fatty acids and then to glycerides. All these reactions were discussed earlier in this chapter. In effect, when insulin is present, cells of adipose tissue tend to make triglycerides. A rise in blood sugar level normally signals the pancreas to release insulin to speed the removal of glucose and its conversion to lipid. In diabetes, except for the so-called *obese diabetic,* glucose is not converted to lipid.

Triglyceride is normally not only made in adipose tissue but also broken down. The *net* effect, uptake of glucose and lipogenesis *or* lipolysis and release of free fatty acids, obviously depends on the supply of effective insulin as well as other factors. *In the nonobese diabetic the release of free fatty acid predominates.* This free fatty acid is picked up by protein molecules in the bloodstream and carried to the liver which catabolizes the fatty acids as fast as it can.[1] Thus acetyl coenzyme A is produced at a rate faster than normal. Obviously acetyl coenzyme A is not needed to make fatty acid. Even if it could reach adipose cells again, it would not be taken in, converted back to fatty acids, and stored as triglyceride because to make triglyceride α-glycerol phosphate is needed. Glucose is essential to supply the α-phosphate ester, but since insulin is not present, adipose tissue is starved for glucose. The superabundance of acetyl coenzyme A cannot all be removed via the citric acid cycle. Demand for ATP simply cannot be sustained at that high level, and without ATP demand this cycle is not in action. The only pathway left for the excess acetyl coenzyme A is toward HMG-CoA and the production of ketone bodies and cholesterol (Figure 23.6).

In summary, when adipose tissue does not take in glucose, it releases fatty acids faster than it can make them. These are catabolized in the liver (primarily) at a more rapid rate than usual. The excess acetyl coenzyme A thus produced is converted into the ketone bodies faster than they can be catabolized in extrahepatic tissue. Ketosis results. If it goes unchecked long enough, acidosis ensues and leads to dehydration, disturbances in the central nervous system, and eventually coma and death. Acidosis produced by insulin deficiency is managed with insulin therapy. In some cases, when the patient is in a coma, isotonic sodium bicarbonate (1.5%) is administered intravenously to maintain life until insulin therapy becomes effective for the long-range treatment. When starvation causes acidosis, the remedy is obvious but is not always available. Starvation, defined here as total fasting whether voluntary or involuntary, is distinguished from undernutrition in which acidosis is seldom observed. Ancel Keys found during the course of the Minnesota Experiment that ketosis was essentially absent in volunteers who for six months were each given a daily diet of only 1500 calories (in contrast with an estimated daily per capita intake in the United States of over 3000 calories). During this time they lost a fourth of their weight.

The Obese Diabetic. At the time of diagnosis about 40% of all diabetics are overweight compared to a 10% incidence in nondiabetics. Quite obviously adipose tissue is overly successful in making and storing triglycerides in these individuals. In most other forms of diabetes the individual loses weight as the fats in adipose tissue are mobilized for energy. Several factors must be remembered in considering this paradox. First, diabetes is defined *clinically* in terms of symptoms and clinical measurements of impaired glucose tolerance. It is not defined in terms of any particular explanation for this impairment at the molecular level. The best, most widely accepted explanation of faulty glucose tolerance is *relative lack* of effective insulin, or at least delays and poor timing in the release of effective insulin. Conditions of glucosuria and ketosis (and subsequent acidosis) are explained at the molecular level in terms of what this primary defect does to carbohydrate-lipid-protein metabolism. Second, diabetes is not just one disease. The clinical definition

[1] When the concentration of lipid-protein complex in the blood rises, a condition of lipemia ("lipid-emia") exists, and the liver is presented with an abundance of this metabolite. The liver swells as it becomes fat-logged.

leaves open the possibilities that the symptoms and the impaired glucose tolerance may have a variety of causes, that several organs and tissues may be involved. Insulin therapy is not the complete solution to diabetes, as clinically defined, even though it has been of immeasurable help in permitting diabetics to lead fuller, more useful lives. Finally, much about diabetes must still be explained. We have been studying theories that do not necessarily account for all aspects of clinically defined diabetes.

The obese diabetic, the overweight individual with clinically defined diabetes, appears able to make insulin and release it. But either much or all of it is *atypical insulin* or it is largely *typical insulin* so tightly bound to serum proteins that it is ineffective at most tissues. At least these are two theories that have recently attracted considerable attention. Adipose tissue, we have learned, is very sensitive even to low concentrations of insulin. Thus glucose can get into cells of adipose tissue, where it becomes fat, but not into other cells. Net effects are always balances of opposing tendencies. Although the cells of adipose tissue in the obese diabetic are active in making fat, they are also active in releasing fatty acids. Fat making exceeds fatty acid release, but the rates of both are high. Weight rises (helped, no doubt, by too much food intake), and fatty acid catabolism also increases. These fatty acids compensate for the glucose that is not available to other cells for energy, and they are also used to make even more glucose.

To compensate for the difficulty with which glucose penetrates cells of other tissues, the system works to increase even further the glucose concentration of the blood. This activity is carried out primarily in the liver. We have learned that the synthesis of glucose from essentially noncarbohydrate sources, called gluconeogenesis (p. 616), requires amino acids (primarily for carbon skeletons) and fatty acids (for hydrogen). Thus fatty acids are again mobilized and catabolized faster than normal; ketosis may result. Figure 23.7 outlines these general events. Ketosis is seldom as severe in obese diabetes as it is in other forms of the disease.

Associated with diabetes, especially with relatively severe cases, is the tendency to have problems with the circulatory system. Arteriosclerosis increases in what appears to be an independent phenomenon; at least it is not well controlled by insulin. Although the course of illness is imperfectly understood, vascular diseases are now the major cause of death in diabetes. Before Banting and Best introduced insulin therapy in the 1920s, acidosis was the major cause.

Tolbutamide. In many diabetics the pancreas retains some ability to make and secrete insulin, although not enough. Frequently in such cases the oral administration of the drug tolbutamide (or chlorpropamide) is sufficient, for this compound apparently stimulates insulin secretion in the pancreas.

$$CH_3\text{-}C_6H_4\text{-}S(=O)_2NHC(=O)NHCH_2CH_2CH_2CH_3$$

Tolbutamide
(Orinase, Rastinon, Diabuton, Mobenol, Toluina, Diaben, Ipoglicone, Orabet, Oralin, Artosin, Dolipol, U 2043, D 860)

$$Cl\text{-}C_6H_4\text{-}S(=O)_2NHC(=O)NHCH_2CH_2CH_3$$

Chlorpropamide
(Diabinese, Catanil, P-607)

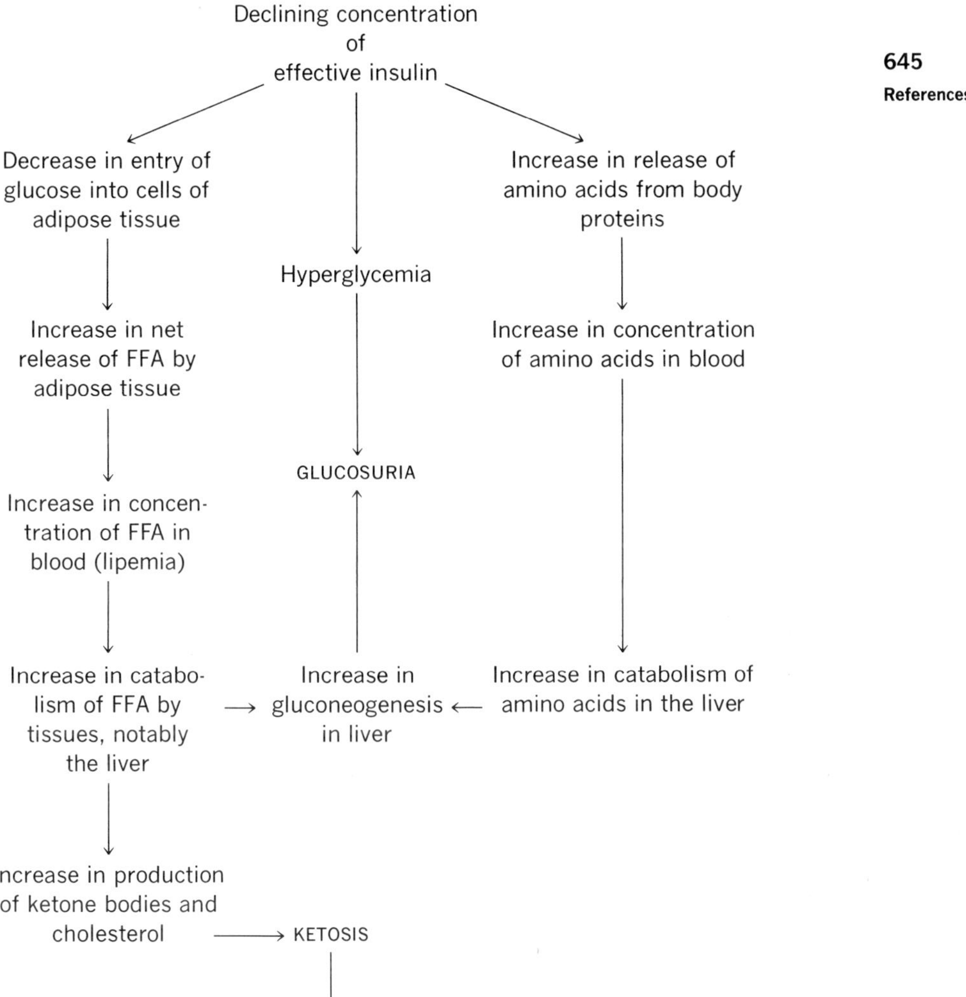
Declining concentration of effective insulin
Decrease in entry of glucose into cells of adipose tissue
Hyperglycemia
Increase in release of amino acids from body proteins
Increase in net release of FFA by adipose tissue
Increase in concentration of amino acids in blood
GLUCOSURIA
Increase in concentration of FFA in blood (lipemia)
Increase in catabolism of FFA by tissues, notably the liver
Increase in gluconeogenesis in liver
Increase in catabolism of amino acids in the liver
Increase in production of ketone bodies and cholesterol
KETOSIS
ACIDOSIS

Figure 23.7 General outline of the main sequences of events in diabetes mellitus. (From A. Marble and G. F. Cahill, Jr., *The Chemistry and Chemotherapy of Diabetes Mellitus,* Charles C Thomas, Springfield, Ill., 1962. Courtesy of Charles C Thomas.)

REFERENCES

A. White, P. Handler, and E. L. Smith. *Principles of Biochemistry,* third edition. McGraw-Hill Book Company, New York, 1964.

W. E. M. Lands. "Lipid Metabolism." *Annual Review of Biochemistry,* Vol. 34, 1965, page 313.

P. R. Vagelos. "Lipid Metabolism." *Annual Review of Biochemistry,* Vol. 33, 1964, page 139.

A. E. Renold and G. F. Cahill, Jr., section editors. *Handbook of Physiology, Section 5: Adipose Tissue.* American Physiological Society, Washington, D.C., 1965.

R. H. Williams. *Disorders in Carbohydrate and Lipid Metabolism.* W. B. Saunders Company, Philadelphia, 1962.

G. Litwack and D. Kritchevsky, editors. *Actions of Hormones on Molecular Processes.* John Wiley and Sons, New York, 1964.

H. G. Wood and M. F. Utter. "The Role of CO_2 Fixation in Metabolism," in *Essays in Biochemistry,* edited by P. N. Campbell and G. D. Greville, Vol. 1. Academic Press, New York, 1965.

A. L. Lehninger. *The Mitochondrion.* W. A. Benjamin, New York, 1964.

K. Bloch, editor. *Lipid Metabolism.* John Wiley and Sons, New York, 1960.

A. Marble and G. F. Cahill, Jr., *The Chemistry and Chemotherapy of Diabetes Mellitus.* Charles C Thomas, Springfield, Ill., 1962.

E. Kirk. *Acidosis.* William Heinemann Medical Books, Ltd., London, 1946.

A. Keys, J. Brozek, A. Henschel, O. Mickelsen, and H. L. Taylor. *The Biology of Human Starvation.* University of Minnesota Press, Minneapolis, 1950.

M. P. Cameron and M. O'Connor, editors. *Aetiology of Diabetes Mellitus and Its Complications,* Vol. 15, Ciba Foundation Colloquia on Endocrinology. Little, Brown, Boston, 1964.

R. M. C. Dawson and D. N. Rhodes, editors. *Metabolism and Physiological Significance of Lipids.* John Wiley and Sons, New York, 1964.

H. S. Mayerson. "The Lymphatic System." *Scientific American,* June 1963, page 80.

PROBLEMS AND EXERCISES

1. Define each of the following terms.
   (*a*) ketonemia (*b*) ketonuria (*c*) chylomicra
   (*d*) ketosis (*e*) energy density (*f*) lactic acid acidosis
   (*g*) lipemia (*h*) adipose tissue
2. List the functions of adipose tissue.
3. Write equations for the steps in the conversion of $CH_3CH_2CH_2\overset{O}{\overset{\|}{C}}—S—En$ to

$$CH_3CH_2CH_2CH_2CH_2\overset{O}{\overset{\|}{C}}—S—En$$

4. Write equations for the steps in the conversion of $CH_3CH_2CH_2CH_2CH_2\overset{O}{\overset{\|}{C}}—SCoA$ to

$$CH_3CH_2CH_2\overset{O}{\overset{\|}{C}}—SCoA \text{ and } CH_3\overset{O}{\overset{\|}{C}}—SCoA$$

5. How does removal of the two acidic ketone bodies as their negatively charged ions at the kidneys deplete the amount of sodium ion in the blood?
6. Referring to Figure 20.9, how does depletion of Na^+ interfere with the transfer of H^+ into the urine?
7. Referring to Figure 20.13, at what point does acidosis interfere with the uptake of O_2 and why?
8. How does a condition of being overweight place an extra burden on the heart?
9. Explain, step by step, how lack of effective insulin will lead (*a*) to overproduction of acetyl coenzyme A, (*b*) thence to overproduction of ketone bodies, (*c*) thence to acidosis, and finally (*d*) to coma and death.
10. Construct a flow chart that explains how a carbohydrate-rich diet will cause a person to become fat.

CHAPTER TWENTY-FOUR

Metabolism of Proteins

DYNAMIC STATE OF NITROGEN METABOLISM

The mixture of dietary amino acids produced by digestion of proteins is rapidly absorbed from the intestinal tract, although not by simple dialysis. The process consumes energy (ATP), and it appears to require vitamin B_6 (pyridoxal) and manganese ion, Mn^{2+}. Virtually all the dietary amino acids enter circulation via the portal vein (p. 617). They do not remain in the bloodstream long before they are removed and used in one or more processes. The following pathways are the options.

1. Synthesis of protein—the repair of tissue, the formation of new tissue, the synthesis of enzymes and some hormones.
2. Synthesis of nonprotein, nitrogenous compounds—substances needed internally such as nucleic acids, heme, creatine, some hormones, and complex lipids.
3. Synthesis of certain other amino acids that might temporarily be in short supply.
4. Conversion into nonnitrogenous compounds—eventually glycogen and/or triglyceride, with the elimination of nitrogen in the form of urea.
5. Catabolism leading to ATP, carbon dioxide, water, and urea.

When the amount of nitrogen excreted as urea equals the amount ingested in other forms in the diet, the person is in a state of *nitrogen equilibrium.* Growing infants and children as well as those recovering from a wasting disease have a *positive nitrogen balance;* more nitrogen is taken in than excreted because the body conserves nitrogen nuclei to make and repair tissue. The processes of aging, starvation, and suffering from a debilitating disease produce a *negative nitrogen balance;* what nitrogen goes out exceeds intake. We find it useful to think in terms of a nitrogen pool (Figure 24.1). Although nothing in protein metabolism resembles the lipid reserves in adipose tissue or the glycogen reserves in the liver and muscles,

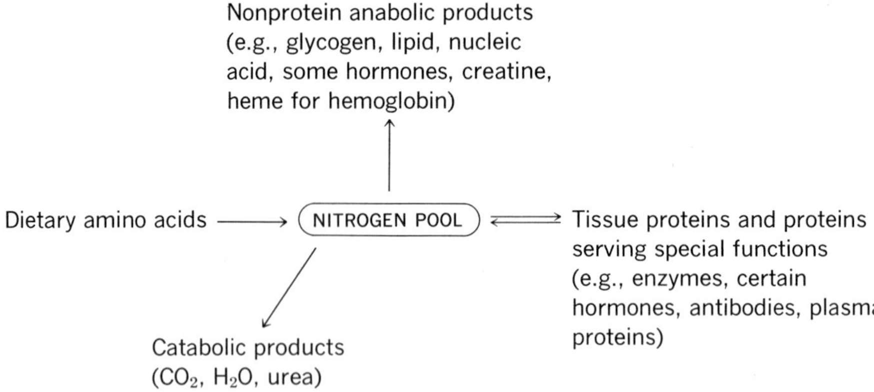
Nonprotein anabolic products (e.g., glycogen, lipid, nucleic acid, some hormones, creatine, heme for hemoglobin)
Dietary amino acids
NITROGEN POOL
Tissue proteins and proteins serving special functions (e.g., enzymes, certain hormones, antibodies, plasma proteins)
Catabolic products (CO_2, H_2O, urea)

Figure 24.1 The nitrogen pool.

we may still regard any tissue (i.e., all tissue) in which amino acids are located, whether as proteins or as dissolved amino acids, as a depot or pool for amino acids. Most tissue proteins are in a dynamic state, constantly undergoing degradation and resynthesis. This turnover is fairly rapid among proteins of the liver and blood plasma, very slow among muscle proteins. Enzyme proteins do not last indefinitely; they are broken back down to amino acids and must be replaced as needed.

The body has no mechanism for the temporary storage of a chance excess of dietary amino acids, as it has for glucose (glycogen) and lipids (depot fat). It does not lay in a supply for some emergency. Amino acids not used internally are oxidized. The carbon fragments enter one of a variety of pathways including gluconeogenesis, lipogenesis, and the citric acid cycle. Nitrogen is excreted in the form of urea.

The Synthesis of Urea. Urea, the major end product of nitrogen metabolism in human beings, is made in the liver. From this organ it is carried in the bloodstream to the kidneys which remove it and place it in the urine being formed. Urea is made from carbon dioxide and ammonia. The net equation is as follows, although a few steps are required to achieve synthesis (Figure 24.2).

$$2NH_3 + CO_2 \longrightarrow \underset{\text{Urea}}{NH_2{-}\overset{\overset{\displaystyle O}{\|}}{C}{-}NH_2} + H_2O$$

Essential Amino Acids. Although the body has the capacity to use some amino acids to make others, a few cannot be synthesized this way. Unless they are present in the proteins of the diet, nitrogen equilibrium is upset and the nitrogen balance becomes negative. Presumably human metabolism cannot form these amino acids from the carbon skeletons and the amino groups of other intermediates, and for this reason they are said to be *essential* (see Table 24.1). The *nonessential amino acids,* also listed in Table 24.1, can be made in the body. (Nonessential, in this context, means *temporarily* dispensable. Obviously all the amino acids listed are used by the body, and protein synthesis could not proceed without them.)

Table 24.1 Amino Acids, Classified as Nutritionally Essential or Nonessential in Maintaining Nitrogen Equilibrium in an Adult Man

Essential	Nonessential
Isoleucine	Alanine
Leucine	Arginine
Lysine	Aspartic acid
Methionine	Cystine
Phenylalanine	Glutamic acid
Threonine	Glycine
Tryptophan	Histidine
Valine	Hydroxyproline
	Proline
	Serine
	Tyrosine

Data from W. C. Rose, *Federation Proceedings,* Vol. 8, 1949, page 546.

Dietary proteins that contain all the essential amino acids are called *adequate proteins.* Gelatin without tryptophan and zein (protein in corn) without lysine are inadequate proteins.

Kwashiorkor. In Latin America, Asia, and Africa the death rate among children is several times that in developed, industrialized, literate societies. Children by the

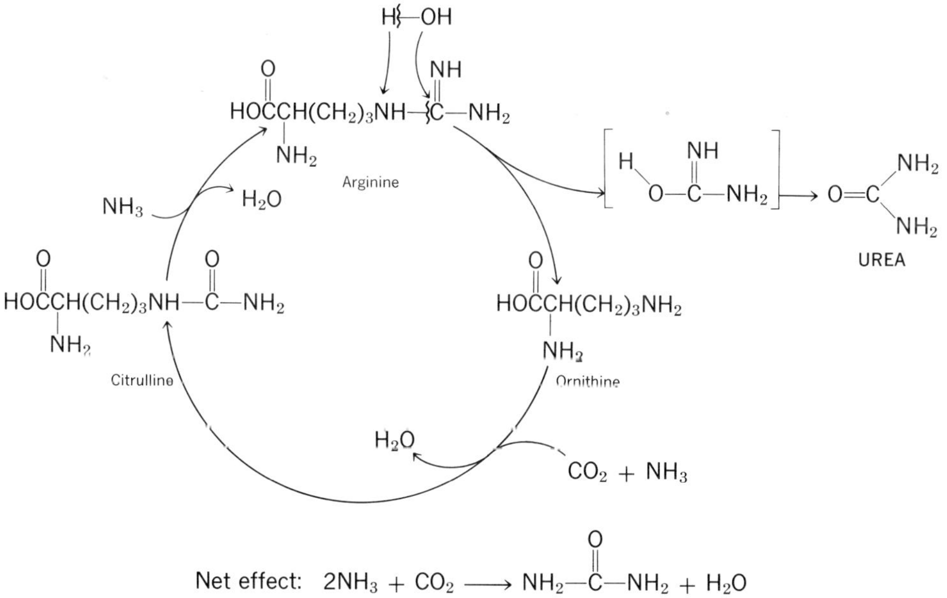

H—OH
HOCCH(CH2)3NH—C—NH2 (O, NH, NH2)
Arginine
NH3
H2O
H—O—C(=NH)—NH2
O=C(NH2)NH2
UREA
HOCCH(CH2)3NH—C—NH2 (O, O, NH2)
Citrulline
HOCCH(CH2)3NH2 (O, NH2)
Ornithine
H2O
CO2 + NH3

$$\text{Net effect:}\quad 2NH_3 + CO_2 \longrightarrow \underset{\text{Urea}}{NH_2\text{—}\overset{\displaystyle O}{\overset{\|}{C}}\text{—}NH_2} + H_2O$$

Figure 24.2 Urea synthesis by the Krebs ornithine cycle. The ornithine needed as a carrier molecule can be made from glutamic acid as shown in Figure 24.3.

thousands are doomed to short lives with bloated bellies, patchy skin, and discolored hair. As long as they are nourished at their mother's breast, they enjoy health. When the second child comes and displaces the first, the symptoms appear in the first child. The disease is called kwashiorkor, a name taken from two words of an African dialect meaning "first" and "second"—the disease that the first child contracts when the second one is born. The diet of the firstborn, instead of milk, is now starchy and contains inadequate protein. Hardly recognized until the 1940s, the ailment is now known to be a protein deficiency disease. Both undernutrition and malnutrition are responsible.[1] The initial symptoms are a loss of appetite and diarrhea—both of which lead the mother to reduce the amount of food she gives the child, thus hastening the onset of additional complications. In the weakened state the child is even more susceptible to the diseases that are a constant hazard in the tropics. Efforts to improve the quality of protein in the diet of people in these regions have been intensive. The Institute of Nutrition of Central America and Panama (INCAP) has developed a protein-rich flour from cottonseed. In areas where the major cash crop is cotton, this flour plus vitamins and minerals is added to local corn meal, which alone is an inadequate protein. Intensive research is being conducted to develop a hybrid corn that produces an adequate protein. The sea is a vast, untapped reservoir of protein-rich fish, and groups in several countries have developed fish meal flour. Plankton, seaweed, and algae are also being investigated. At the present time the population growth in Latin America is 3% per year, the growth in food production 2.5%. What may seem a small fraction of a percent actually means death from starvation (or from diseases brought on during the weakened condition of developing starvation) for millions of people. Between the 1980s and the turn of the century, this planet may well see the greatest catastrophe in recorded history, and history has recorded some monstrous famines.[2] At the present rate of population growth, by the 1980s the developed countries with high-yield agriculture will no longer have the capacity, regardless of intentions and maximum effort, to make up the food deficits of underdeveloped countries. Children are the first to suffer, for their dietary needs are greater and they must have a positive nitrogen balance to grow. Yet they are weaker than adults, and in the competition for food the battle is particularly grim.

When a body is on a starvation diet, its resources are mobilized to stave off the crisis. The glucose shortage is one of the serious consequences, for it means that gluconeogenesis must take place in an effort to supply what blood sugar is needed. Gluconeogenesis requires certain of the amino acids as well as fatty acids. Thus demand for this metabolic pathway is high at a time when the body has precious few amino acids to spare. To supply them, degradation of tissue protein must occur,

[1] Undernutrition is a general inadequacy of the diet; malnutrition is a serious imbalance in the necessary components of the diet, for example, vitamin deficiency or deficiency in the essential amino acids.

[2] For example, the Great Famine in Bengal, 1769–1770, ten million people or a third of the population perished; the Irish famine, 1846–1847, over a million people died when the potato crops failed; the famine of 1877–1878 in North China, over nine million people perished; the Russian famine of 1921–1922, three million died; World War II famines deliberately induced in the Warsaw ghetto and Nazi concentration camps; the Bengal famine of 1943, one and a half million died; and the Nigerian war famine in Biafra in 1968.

partly accounting for the wasting-away aspects of starvation. The lipid reserves also disappear. Fatty acid catabolism and gluconeogenesis generate ketone bodies which create serious problems of their own (already discussed, Chapter 23). We have mentioned that the body can make glucose and lipids from amino acids, and it can make some amino acids from other amino acids and from glucose and lipids. We shall now examine in greater detail just how these processes are carried out.

SYNTHESIS OF NONESSENTIAL AMINO ACIDS

Multiple lines of evidence indicate that the body can make some amino acids from molecular parts contributed by carbohydrates, lipids, and the amino acids that it cannot make, the essential amino acids. Much of the research work has been done with rats, but most of the results appear to apply to man. When young animals do not grow on a diet deliberately made deficient in an amino acid, that amino acid probably cannot be made internally. When the diet is made to include amino acids labeled with ^{15}N in the α-position and this radioactive form of nitrogen soon appears in all the amino acids (except lysine), it is obvious that pathways for passing the nitrogen of an amino group from amino acid to amino acid exist.

Transamination. The body contains a family of enzymes called *transaminases* that catalyze the following type of reaction:

$$R-\overset{\overset{\displaystyle O}{\|}}{C}-\overset{\overset{\displaystyle O}{\|}}{C}-OH + R'-\overset{\overset{\displaystyle NH_2}{|}}{C}H-\overset{\overset{\displaystyle O}{\|}}{C}OH \rightleftharpoons R-\underset{\underset{\displaystyle NH_2}{|}}{C}H\overset{\overset{\displaystyle O}{\|}}{C}OH + R'-\overset{\overset{\displaystyle O}{\|}}{C}-\overset{\overset{\displaystyle O}{\|}}{C}-OH$$

α-Keto acid $\qquad$ α-Amino acid $\qquad$ New α-amino acid $\qquad$ New α-keto acid

Transferring an animo group from one molecule to another is called *transamination.* Several intermediate steps occur, and vitamin B_6, pyridoxal, is apparently the essential coenzyme for all the transaminases.

Glutamic acid serves as a common provider of amino groups; most of the nonessential amino acids can be made from it as indicated in Figure 24.3.

CATABOLISM OF AMINO ACIDS

By a variety of experimental techniques (use of radioactive isotopes as labels, effects of certain diets) we now know that the majority of the amino acids can be converted into glycogen (with excretion of urea), and are therefore called the glycogenic amino acids (Table 24.2). The catabolism of a few amino acids leads to ketone bodies, called the ketogenic amino acids (Table 24.2). Since glycogen can be converted into lipids and since amino acids can be made into glycogen (gluconeogenesis), molecular parts can obviously undergo considerable shuffling in the body.

With over twenty amino acids, the discussion of their catabolism could fill books. Since our purposes are illustrative rather than documentary, only a few of the metabolic reactions of amino acids will be described.

Methods for Removing Groups. The following are two of several ways for removing the α-amino group of an amino acid.

Table 24.2 Amino Acids as Raw Materials for Glycogen and Fatty Acids

Glycogenic amino acids	
Alanine	Hydroxyproline
Arginine	Methionine
Aspartic acid and asparagine	Proline
Cysteine	Serine
Glutamic acid and glutamine	Threonine
Glycine	Tryptophan
Histidine	Valine
Ketogenic amino acids*	
Isoleucine	Phenylalanine
Leucine	Tyrosine
Lysine	

* All but leucine are also considered to be glycogenic.

1. **Transamination** (indirect oxidative deamination). The amino group is transferred to α-ketoglutaric acid which becomes glutamic acid and is oxidized.

(*a*)
$$\underset{\alpha\text{-Amino acid}}{RCH(NH_2)CO_2H} + \underset{\alpha\text{-Ketoglutaric acid}}{HO_2CCH_2CH_2C(=O)CO_2H} \rightleftarrows \underset{\alpha\text{-Keto acid}}{RC(=O)CO_2H} + \underset{\text{Glutamic acid}}{HO_2CCH_2CH_2CH(NH_2)CO_2H}$$

(*b*)
$$\underset{\text{Glutamic acid}}{HO_2CCH_2CH_2CH(NH_2)CO_2H} + NAD^+ + H_2O \longrightarrow \underset{\alpha\text{-Ketoglutaric acid}}{HO_2CCH_2CH_2C(=O)CO_2H} + NH_3 + NADH + H^+$$

Sum
$$\underset{\alpha\text{-Amino acid}}{RCH(NH_2)CO_2H} + NAD^+ + H_2O \longrightarrow \underset{\alpha\text{-Keto acid}}{RC(=O)CO_2H} + NH_3 + NADH + H^+$$

$NH_3 \rightarrow$ Urea

$NADH + H^+ \rightarrow NAD^+$ + $(H:^- + H^+)$ $\rightarrow$ Electron transport chain

2. **Direct oxidative deamination.** Step *b* in the previous method is an oxidation (NAD^+ is reduced) and a deamination. Being essentially irreversible, this step overcomes the unfavorable equilibrium in the first step. There is some evidence that flavoprotein enzymes (FAD enzymes) can catalyze this type of reaction directly on amino acids other than glutamic acid. Besides a new α-keto acid, ammonia and $FADH_2$ are produced.

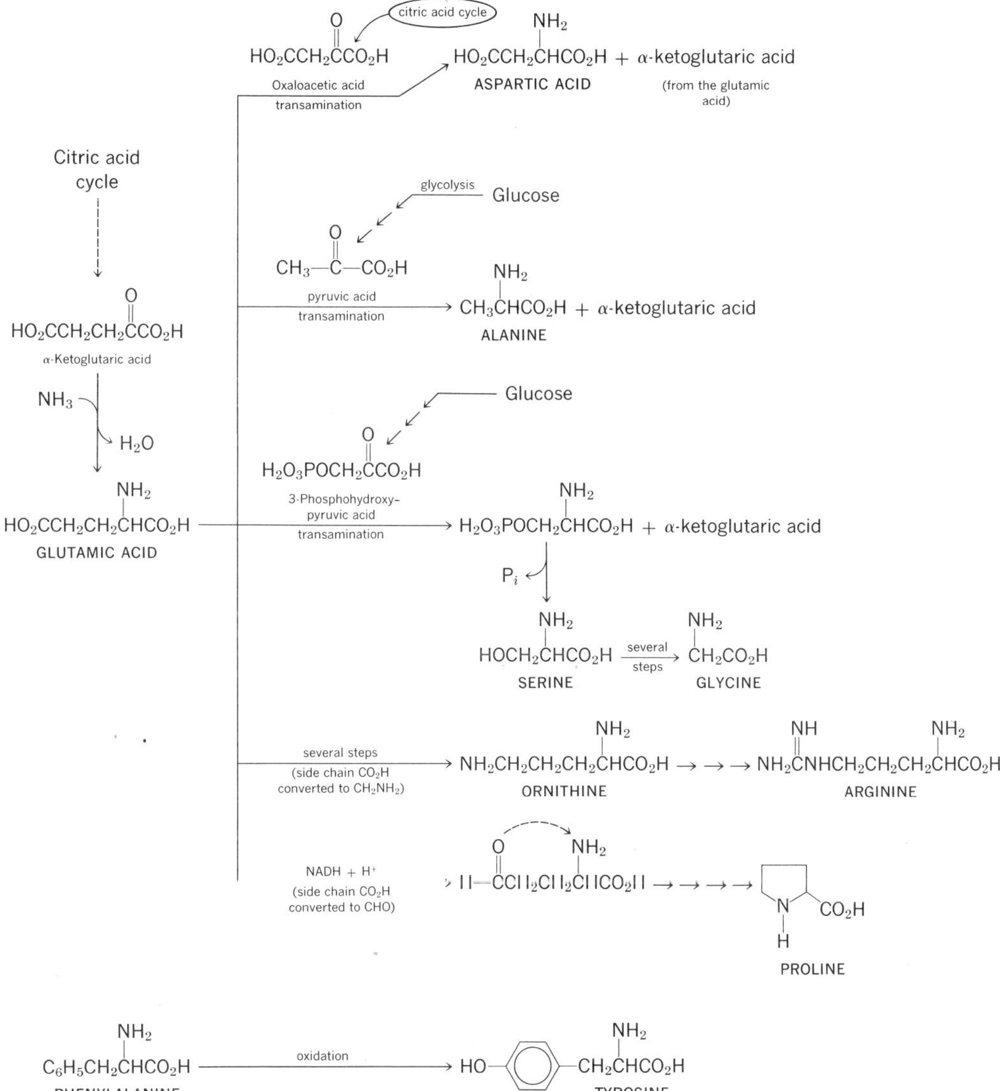
citric acid cycle
HO2CCH2CCO2H
Oxaloacetic acid
transamination
HO2CCH2CHCO2H + α-ketoglutaric acid
ASPARTIC ACID
(from the glutamic acid)
Citric acid cycle
HO2CCH2CH2CCO2H
α-Ketoglutaric acid
NH3
H2O
HO2CCH2CH2CHCO2H
GLUTAMIC ACID
glycolysis
Glucose
CH3—C—CO2H
pyruvic acid
transamination
CH3CHCO2H + α-ketoglutaric acid
ALANINE
Glucose
H2O3POCH2CCO2H
3-Phosphohydroxy-pyruvic acid
transamination
H2O3POCH2CHCO2H + α-ketoglutaric acid
Pi
HOCH2CHCO2H
SERINE
several steps
CH2CO2H
GLYCINE
several steps
(side chain CO2H converted to CH2NH2)
NH2CH2CH2CH2CHCO2H
ORNITHINE
NH2CNHCH2CH2CH2CHCO2H
ARGININE
NADH + H+
(side chain CO2H converted to CHO)
H—CCH2CH2CHCO2H
CO2H
PROLINE
C6H5CH2CHCO2H
PHENYLALANINE
oxidation
HO—CH2CHCO2H
TYROSINE

Figure 24.3 The biosynthesis of some nonessential amino acids.

$$(a)\quad R{-}\underset{\displaystyle NH_2}{\underset{|}{C}H}{-}CO_2H + H_2O + FAD \rightleftharpoons R{-}\overset{\displaystyle O}{\overset{\|}{C}}{-}CO_2H + NH_3 + FADH_2$$

$$(b)\quad FADH_2 + O_2 \longrightarrow FAD + H_2O_2$$

$$H_2O_2 \xrightarrow{\text{catalase}} H_2O + \tfrac{1}{2}O_2$$

Figure 24.4 gives a general summary of how molecular parts of some of the amino acids can be used to make glucose and lipids. The student is reminded that these are by no means the only reactions of the amino acids. Amino acids are used principally to make or to replace tissue proteins, enzymes, some hormones, and other nitrogenous substances needed for life and health. Gluconeogenesis is important in periods of inadequate diet, fasting, starvation, diabetes, and wasting diseases. Clearly if the body is to make "new" glucose, it must sacrifice some of its proteins as well as some of its fatty acids.

The following examples have been selected to illustrate the catabolism of amino acids.

Alanine. Oxidative deamination produces pyruvic acid.

$$\underset{\text{Alanine}}{CH_3\underset{}{\overset{\displaystyle NH_2}{\overset{|}{C}}}HCO_2H} \xrightarrow[\text{deamination}]{\text{oxidative}} \underset{\text{Pyruvic acid}}{CH_3\overset{\displaystyle O}{\overset{\|}{C}}\overset{\displaystyle O}{\overset{\|}{C}}OH} \begin{cases} \rightarrow \text{Gluconeogenesis} \\ \rightarrow \text{Citric acid cycle} \end{cases}$$

Aspartic acid. Again, oxidative deamination directly produces oxaloacetic acid, an intermediate in gluconeogenesis.

$$\underset{\text{Aspartic acid}}{HO\overset{\displaystyle O}{\overset{\|}{C}}CH_2\underset{\displaystyle NH_2}{\underset{|}{C}}H\overset{\displaystyle O}{\overset{\|}{C}}OH} \xrightarrow[\text{deamination}]{\text{oxidative}} \underset{\text{Oxaloacetic acid}}{HO\overset{\displaystyle O}{\overset{\|}{C}}CH_2\overset{\displaystyle O}{\overset{\|}{C}}\overset{\displaystyle O}{\overset{\|}{C}}OH} \begin{cases} \rightarrow \text{Gluconeogenesis} \\ \rightarrow \text{Citric acid cycle} \end{cases}$$

Cysteine. The mercaptan group is first oxidized to a sulfinic acid, $HO_2S{-}$, which is very much like a carboxyl group, $HO_2C{-}$. Then the amino group is lost. In the product, β-sulfinylpyruvic acid, the keto group is in the β-position to the sulfinyl group. If it were a β-keto carboxylic acid, we would expect loss of carbon dioxide (cf. p. 427). Being instead a β-keto sulfinic acid, it loses sulfur dioxide. The sulfur dioxide is oxidized to sulfate ion, but pyruvic acid is the organic product. The initial intermediate, cysteinesulfinic acid, is also used to make many biologically important compounds of sulfur. The bile acid, taurocholic acid (p. 549), is one of them. Sulfate ions produced by cysteine catabolism are excreted via the urine.

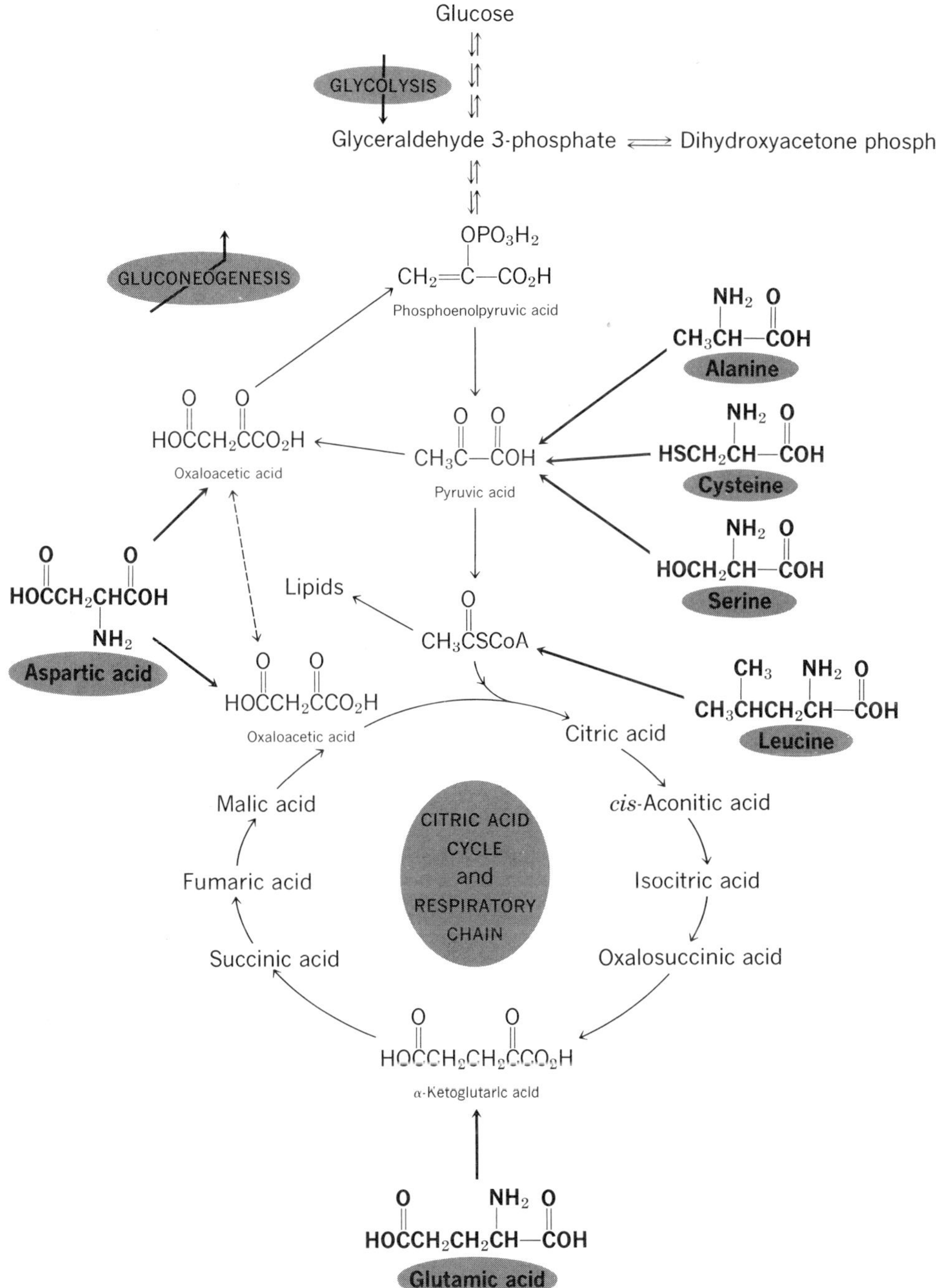

Glucose
GLYCOLYSIS
Glyceraldehyde 3-phosphate
Dihydroxyacetone phosphate
GLUCONEOGENESIS
OPO_3H_2
$CH_2{=}C{-}CO_2H$
Phosphoenolpyruvic acid
$HOCCH_2CCO_2H$
Oxaloacetic acid
$CH_3C{-}COH$
Pyruvic acid
$CH_3CH{-}COH$
Alanine
$HSCH_2CH{-}COH$
Cysteine
$HOCH_2CH{-}COH$
Serine
$HOCCH_2CHCOH$
NH_2
Aspartic acid
Lipids
CH_3CSCoA
$CH_3CHCH_2CH{-}COH$
Leucine
Citric acid
cis-Aconitic acid
Isocitric acid
Oxalosuccinic acid
CITRIC ACID CYCLE and RESPIRATORY CHAIN
Malic acid
Fumaric acid
Succinic acid
$HOCCH_2CH_2CCO_2H$
α-Ketoglutaric acid
$HOCCH_2CH_2CH{-}COH$
Glutamic acid

Figure 24.4 Catabolism of some amino acids, illustrative examples. Several amino acids can enter the gluconeogenesis pathways. A few produce acetyl coenzyme A, and from it lipids, including cholesterol and steroids, can be made. Individual steps are not shown here.

$$\underset{\text{Cysteine}}{H{-}S{-}CH_2\underset{NH_2}{\underset{|}{C}}H\overset{O}{\overset{\|}{C}}OH} \xrightarrow{(O)} \underset{\text{Cysteinesulfinic acid}}{HO\overset{O}{\overset{\|}{S}}CH_2\underset{NH_2}{\underset{|}{C}}H\overset{O}{\overset{\|}{C}}OH} \xrightarrow{\text{deamination}}$$

$$\underset{\beta\text{-Sulfinylpyruvic acid}}{H{-}O{-}\overset{O}{\overset{\|}{S}}{-}CH_2\overset{O}{\overset{\|}{C}}\overset{O}{\overset{\|}{C}}OH}$$

$$\downarrow \to [SO_2] \xrightarrow{(O)} SO_4^{2-}$$

$$\underset{\text{Pyruvic acid}}{CH_3\overset{O}{\overset{\|}{C}}\overset{O}{\overset{\|}{C}}OH} \begin{cases} \nearrow \text{Gluconeogenesis} \\ \searrow \text{Citric acid cycle} \end{cases}$$

Glutamic acid. Oxidative deamination of this amino acid gives α-ketoglutaric acid. In three more steps of the citric acid cycle (Figure 24.4) this compound is converted into oxaloacetic acid.

$$\underset{\text{Glutamic acid}}{HO\overset{O}{\overset{\|}{C}}CH_2CH_2\underset{NH_2}{\underset{|}{C}}H\overset{O}{\overset{\|}{C}}OH} \longrightarrow \underset{\alpha\text{-Ketoglutaric acid}}{HO\overset{O}{\overset{\|}{C}}CH_2CH_2\overset{O}{\overset{\|}{C}}\overset{O}{\overset{\|}{C}}OH}$$

$$\to\to\to \text{Oxaloacetic acid} \begin{cases} \nearrow \text{Glucogenesis} \\ \searrow \text{Citric acid cycle} \end{cases}$$

Threonine. Threonine does not undergo loss of its amino group by transamination. Rather, a special enzyme found principally in the liver and the kidneys catalyzes the following change.

$$\underset{\text{Threonine}}{NH_2{-}\underset{\underset{CH_3}{|}}{\underset{CH{-}OH}{\underset{|}{C}}}HCO_2H} \longrightarrow \underset{\text{Glycine}}{NH_2CH_2CO_2H} + \underset{\text{Acetaldehyde}}{CH_3\overset{O}{\overset{\|}{C}}H}$$

Acetaldehyde is oxidized to acetic acid which is changed to acetyl coenzyme A.

Serine. Dehydration of this amino acid leads to unstable intermediates which eventually become pyruvic acid.

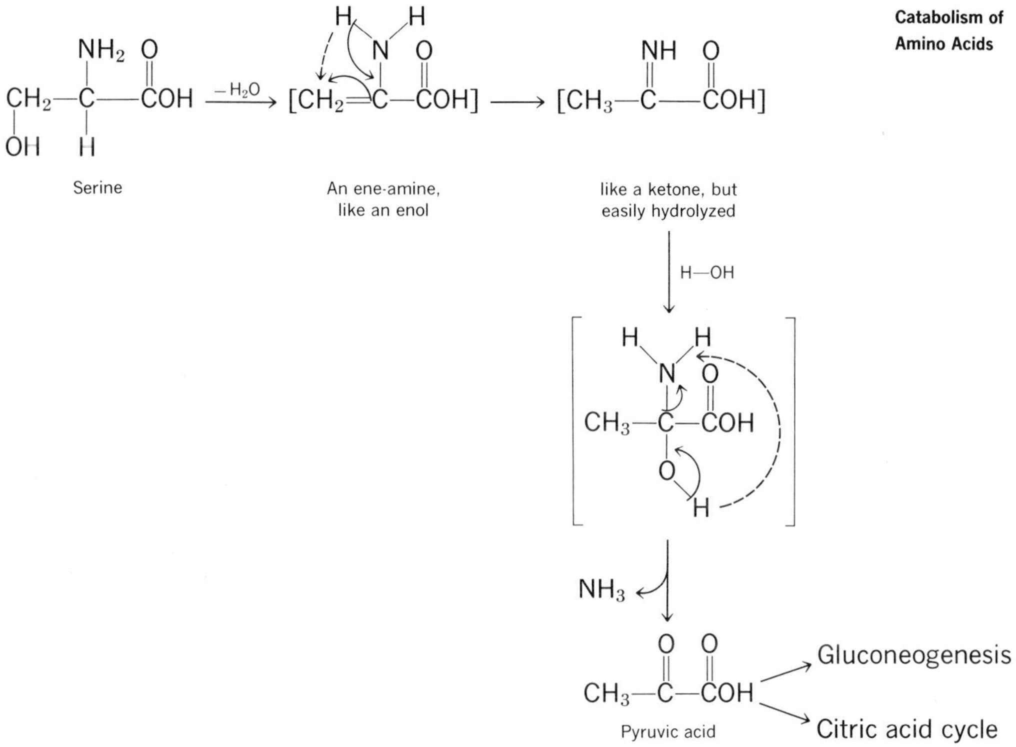

$$HOCH_2-CH(NH_2)-COOH \xrightarrow{-H_2O} [CH_2{=}C(NH_2)-COOH] \longrightarrow [CH_3-C(=NH)-COOH]$$

Serine

An ene-amine, like an enol

like a ketone, but easily hydrolyzed

$$\xrightarrow{H-OH} [CH_3-C(OH)(NH_2)-COOH] \xrightarrow{-NH_3} CH_3-CO-COOH$$

Pyruvic acid → Gluconeogenesis; → Citric acid cycle

Leucine. The catabolism of leucine illustrates the fate of a ketogenic amino acid, one that can furnish carbon fragments for the synthesis of lipids and steroids. In the reactions that follow we have

① a transamination as leucine is converted into an α-keto acid;

② an oxidative decarboxylation which is very similar to the conversion of pyruvic acid to acetyl coenzyme A (p. 604);

③ a dehydrogenation very much like step ① in the fatty acid cycle (p. 637, Figure 23.5);

④ a carboxylation resembling the addition of carbon dioxide to acetyl coenzyme A in the initial phase of the lipogenesis cycle (p. 634, Figure 23.3);

⑤ the addition of water to a double bond just as in step ② of the fatty acid cycle (p. 637); and

⑥ a reaction resembling a reverse aldol condensation or a reverse Claisen ester condensation (cf. p. 400 or p. 425).

These steps connect leucine to the HMG-CoA crossroads of metabolism as discussed in Figure 23.6 (p. 639).

Thus, by all these examples, we see that the body performs its chemical work in many small steps, each one of them amounting to relatively minor molecular surgery which is reasonable in terms of what we know about organic reactions and about energy relations.

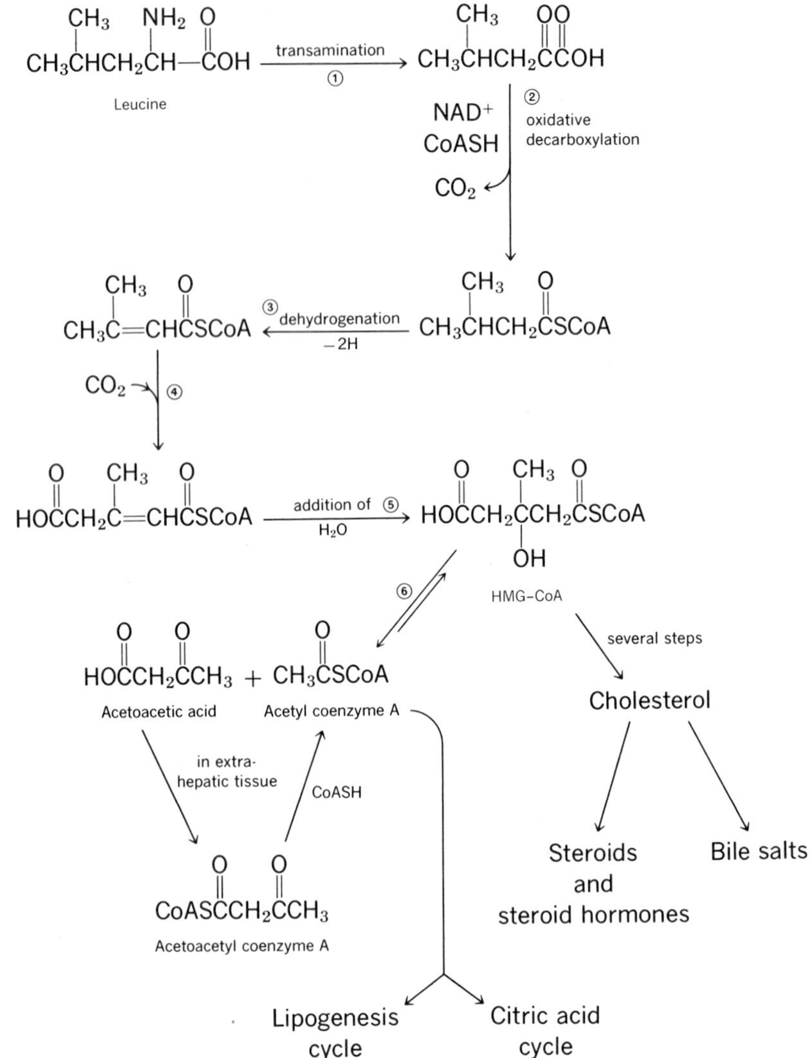
Leucine
transamination
①
②
oxidative decarboxylation
NAD+
CoASH
CO2
③ dehydrogenation
−2H
④
CO2
addition of ⑤
H2O
HMG-CoA
⑥
several steps
Acetoacetic acid
Acetyl coenzyme A
Cholesterol
in extra-hepatic tissue
CoASH
Acetoacetyl coenzyme A
Steroids and steroid hormones
Bile salts
Lipogenesis cycle
Citric acid cycle

REFERENCES

S. W. Fox and J. F. Foster. *Introduction to Protein Chemistry.* John Wiley and Sons, New York, 1957.

A. White, P. Handler, and E. L. Smith. *Principles of Biochemistry,* third edition. McGraw-Hill Book Company, New York, 1964.

E. D. Wilson, K. H. Fisher, and M. E. Fuqua. *Principles of Nutrition,* second edition. John Wiley and Sons, New York, 1965.

PROBLEMS AND EXERCISES

1. Define each of the following terms.
   (*a*) nitrogen equilibrium (*b*) positive nitrogen balance (*c*) negative nitrogen balance
   (*d*) nitrogen pool (*e*) essential amino acid (*f*) nonessential amino acid
   (*g*) kwashiorkor (*h*) adequate protein (*i*) glycogenic amino acid
   (*j*) ketogenic amino acid
2. What is the difference between undernutrition and malnutrition?
3. In general terms, how does the body try to meet its needs for glucose during developing starvation?
4. What are the principal end products of amino acid catabolism?
5. In Guatamala corn makes up about 70% of the human diet. The protein in corn is deficient in lysine. Discuss the implications of this.
6. What is the difference between gluconeogenesis and glucogenesis, and in what ways are proteins and lipids used in the former? Be specific, writing specific reactions.

CHAPTER TWENTY-FIVE

The Chemistry of Heredity

HEREDITY AND ENZYMES

The physical links between one generation and the next are the sperm cell of the male and the egg cell of the female. After the union of these two cells, a series of chemical events unfolds. All the information that is needed to ensure that the fertilized egg will develop physically into an authentic member of the parents' species must be present in that tiny and wonderful unit of matter. If the baby animal is to have fur rather than feathers, at its conception a sequence of chemical reactions is started that will lead to fur making rather than feather making. In spite of the basic similarity in diets, especially after digestion has done its work, radically different pathways are taken by various species.

We have learned that nearly every reaction occurring in a living organism requires special enzymes which act uniquely on fairly common molecules such as glucose, fatty acids and glycerol, and amino acids. Each species possesses its own peculiar set of enzymes and hormones, even though some remarkable similarities exist, especially at the coenzyme level. Fur-bearing animals have enzyme systems that catalyze the ordering of amino acid sequences to produce fur. Feathered creatures have different enzyme systems that generate the proteins of feathers from basically the same set of amino acids. Whatever the genetic message may be, whether a set of instructions to develop as a cowbird or a set to develop as a cow, it very likely concerns the generation of a distinctive enzyme system. We therefore expect any theory concerning the chemical basis of heredity to explain how a species acquires and reproduces its special set of enzymes.

GENERAL FEATURES OF THE PHYSICAL BASIS OF HEREDITY

In a typical animal, union of a sperm cell with an egg cell produces a new cell called a *zygote,* which proceeds to multiply by a process of cell division called *mitosis*. The daughter cells of the zygote themselves divide, and so on, as an embryo takes form. Early in this stage of development two fundamentally different kinds of cells can be distinguished: *germ cells,* which will give rise to either sperm or eggs, and *somatic cells,* from which will form all the myriad tissues and organs unique to the body of the species. The germ cells are cells set apart, protected from change and unaffected by the tremendous variations taking place among the somatic cells. When the somatic cells have proceeded far enough in their development that the gonads are fully elaborated and sexual maturity is reached, the germ cells become active. They develop sperm or eggs, depending on the sex of the individual. If the sperm and the egg of the parents contained the essentials to produce a unique enzyme system, the germ cells of the children must also possess these essentials, in order that they, in turn, may pass them to the next generation.

Chromosomes and Genes. The cell is the structural unit of life, and chemicals associated with living things are, by and large, organized in these units. All the material comprising the cells is called the *protoplasm.* Discrete "bodies," such as mitochondria and the cell nucleus, exist in the protoplasm. Figure 25.1 depicts a generalized animal cell, with several parts labeled. The cell nucleus contains a fluid in which twisted and intertwined filaments called chromonemata exist; these apparenty bear strings of the basic units of heredity, the *genes.* Chromonemata and their gene strings constitute individual *chromosomes.*

During mitosis division of the nucleus precedes division of the cell as a whole. In preliminary stages of nucleus division the chromonemata and the genes normally produce exact duplicates of themselves. The final cell division, then, separates

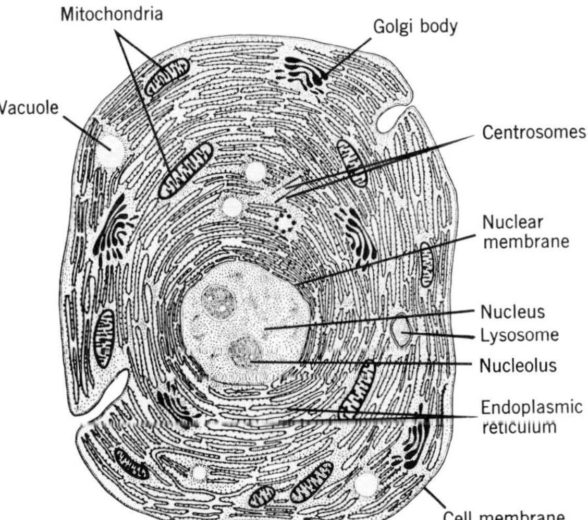
Mitochondria
Golgi body
Vacuole
Centrosomes
Nuclear membrane
Nucleus
Lysosome
Nucleolus
Endoplasmic reticulum
Cell membrane

Figure 25.1 A generalized animal cell. Cells vary greatly from tissue to tissue, but most have the common features shown here. In this chapter we are particularly concerned with the nucleus and the ribosome-studded endoplasmic reticulum. (From Jean Brachet, "The Living Cell," *Scientific American,* September 1961, page 50. Copyright © 1961 by Scientific American, Inc. All rights reserved.)

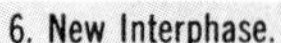

6. New Interphase.

Two new daughter cells emerge with sets of chromosomes and genes identical with each other and the "parent" cell.

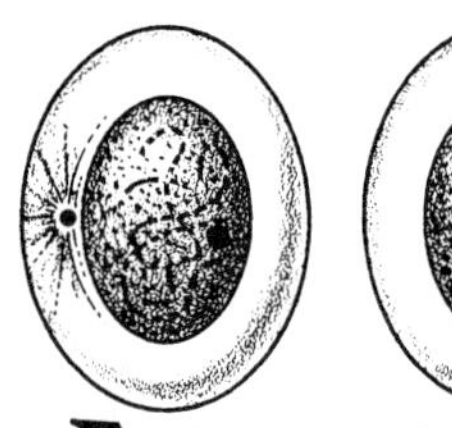

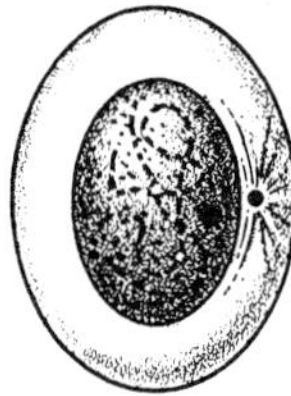

Daughter Cells

Late Telophase

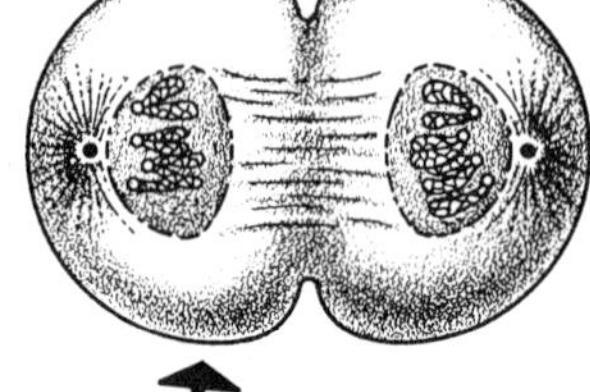

5. Telophase.

Cell begins to divide. New nuclear membranes begin to form.

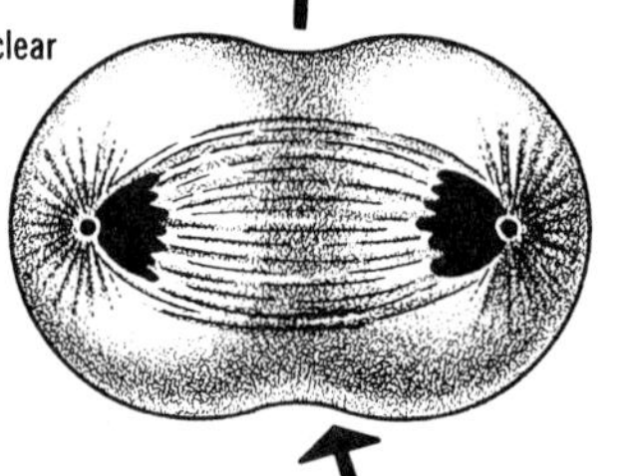

Early Telophase

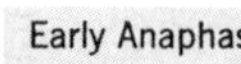

Early Anaphase

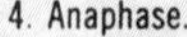

Anaphase

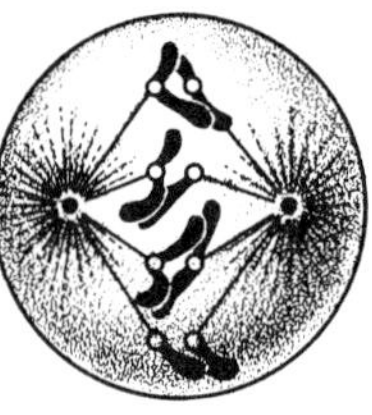

4. Anaphase.

Centromeres now divide and "daughter" chromosomes are pulled apart as the centromeres move toward opposite poles of the spindle.

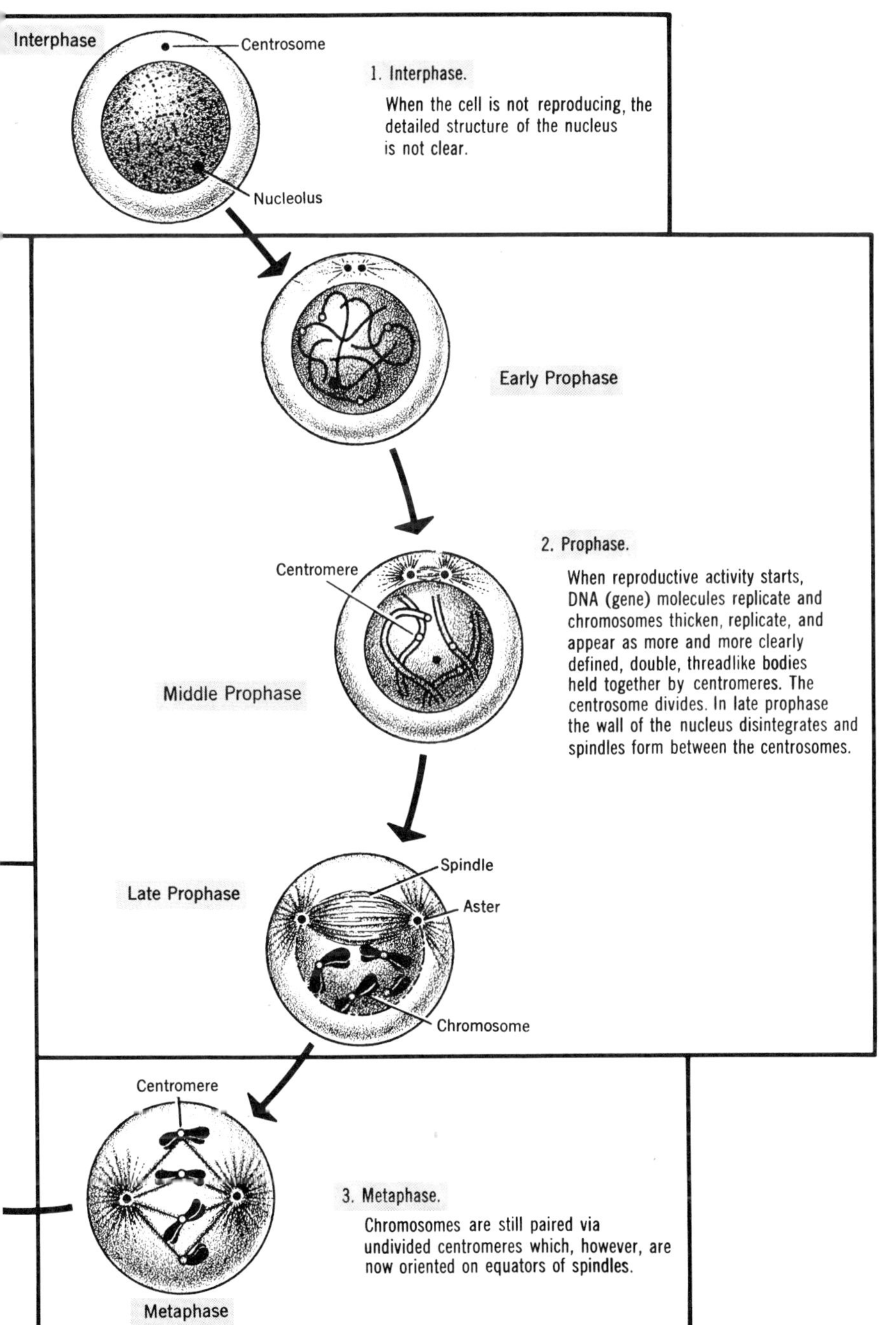
Interphase
Centrosome
1. Interphase.
When the cell is not reproducing, the detailed structure of the nucleus is not clear.
Nucleolus
Early Prophase
2. Prophase.
Centromere
When reproductive activity starts, DNA (gene) molecules replicate and chromosomes thicken, replicate, and appear as more and more clearly defined, double, threadlike bodies held together by centromeres. The centrosome divides. In late prophase the wall of the nucleus disintegrates and spindles form between the centrosomes.
Middle Prophase
Spindle
Late Prophase
Aster
Chromosome
Centromere
3. Metaphase.
Chromosomes are still paired via undivided centromeres which, however, are now oriented on equators of spindles.
Metaphase

Figure 25.2 Mitosis. This sequence applies to animal cells. Plant cells show a different mitotic sequence, especially in the telophase. (Cell structures from E. J. Gardner, *Principles of Genetics*, third edition, John Wiley and Sons, New York, 1968, page 28.)

duplicated chromonemata and gene strings into the individual nuclei of two new daughter cells. The division of one parent cell involves one duplication of each gene; this reproductive duplication is often called *replication*. Figure 25.2 shows the chief stages in mitosis. To understand how genes might be responsible for carrying heredity messages, we must study the chemistry of these heredity units.[1]

NUCLEIC ACIDS AS HEREDITY UNITS

We have long known that cell nuclei are rich in a polymeric material called deoxyribonucleic acid (DNA). All available evidence supports the conclusion that DNA is the actual chemical constituting genes. At the University of Cambridge in 1953 F. H. C. Crick and J. D. Watson proposed a structure for DNA that provides a means for correlating its physical and chemical properties with its properties as the apparent chemical of genes. For this work these two men shared with Maurice Wilkins the 1962 Nobel prize in medicine and physiology. Much about DNA was known before 1953, and good evidence suggested that this chemical was the actual constituent of genes. Data from X-ray diffraction studies even suggested that the polymeric molecules making up DNA were aligned side by side in a twisted helix. But until the work of Crick and Watson the picture was not clear enough for scientists to see just *how* the structure of DNA had something to do with genetic properties. Moreover, other polymeric substances chemically similar to DNA existed in cells, and how might these be involved?

Nucleic Acids. Deoxyribonucleic acid is a member of a family of polymers called *nucleic acids*. Their monomer units are called *nucleotides,* and these, in turn, are built from simpler parts. Figure 25.3 shows the relation of the nucleic acids to their building blocks by indicating how hydrolysis splits the polymer successively into smaller and smaller units. If the original nucleic acid is of the DNA type, one product of its hydrolysis (Figure 25.3) will be deoxyribose ("de-" means "lacking"; "deoxy-" means "lacking in an oxygen atom found in the close structural relative, ribose"). The other major type of nucleic acid, ribonucleic acid (RNA), will hydrolyze to form ribose instead of deoxyribose.

The heterocyclic bases obtained from nucleic acids are derived from either purine or pyrimidine. In Figure 25.3 all the bases but adenine are shown in a keto form (really a lactam form) and an enol form (actually lactim). In brief, the following equilibria are analogous.

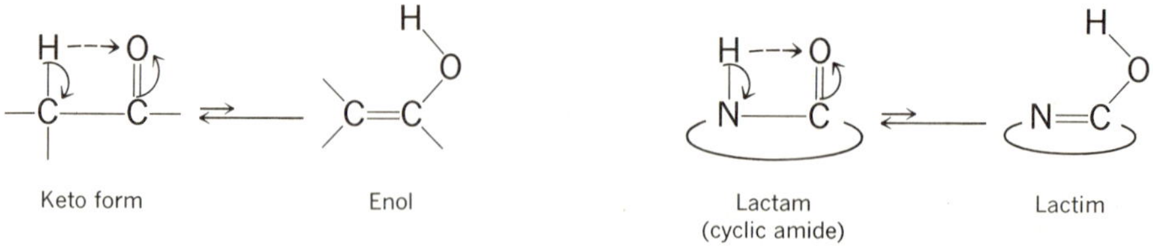

$$-\overset{H}{\underset{|}{C}}-\overset{O}{\overset{\|}{C}}- \rightleftarrows \;\; >C=C(-OH)- \qquad\qquad \overset{H}{N}-\overset{O}{\overset{\|}{C}} \rightleftarrows N=C(-OH)$$

Keto form — Enol — Lactam (cyclic amide) — Lactim

[1] It is beyond the intended scope of this book to examine the development of the gene concept. Some of the numerous outstanding references available are cited at the end of this chapter.

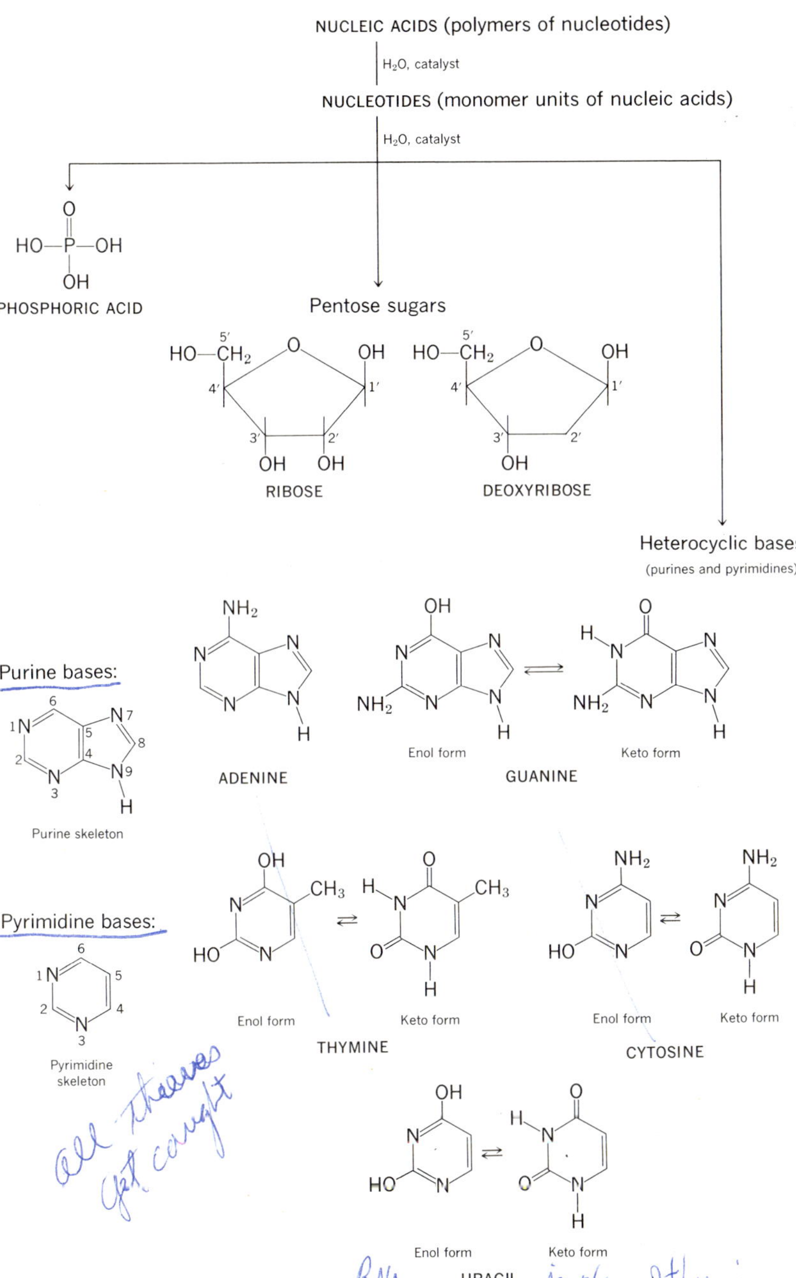
NUCLEIC ACIDS (polymers of nucleotides)
H2O, catalyst
NUCLEOTIDES (monomer units of nucleic acids)
H2O, catalyst
PHOSPHORIC ACID
Pentose sugars
RIBOSE
DEOXYRIBOSE
Heterocyclic bases
(purines and pyrimidines)
Purine bases:
Purine skeleton
ADENINE
Enol form
Keto form
GUANINE
Pyrimidine bases:
Pyrimidine skeleton
Enol form
Keto form
THYMINE
Enol form
Keto form
CYTOSINE
Enol form
Keto form
URACIL

Figure 25.3 Hydrolysis products of nucleic acids. The pentose obtained from deoxyribonucleic acid (DNA) is deoxyribose. From ribonucleic acids (RNA) ribose is obtained. The five principal heterocyclic bases are shown here, although others are known. As separate chemicals they exist largely in their enol forms. Bound into nucleic acids, they exist largely in the keto forms.

In their enol forms the systems are fully aromatic and derive stabilization via resonance. As isolated chemicals they are believed to exist largely in such forms, but when these purines and pyrimidines are bound into nucleic acids, they are in their keto forms. What might appear a loss of stabilization is recovered through the greatly increased availability of sites for possible hydrogen bonding, and stabilization is achieved in a different way.

Of the three pyrimidine bases shown, uracil is found almost exclusively in RNA. The complete hydrolysis of DNA yields the following four bases: adenine (A), guanine (G), thymine (T), and cytosine (C).

The monomer units of nucleic acids, the nucleotides, are made from one molecule each of phosphoric acid, a pentose, and a heterocyclic base. The manner of assembly of a typical nucleotide is shown in Figure 25.4. The heterocyclic amine is always attached at C-1′ of the pentose; the phosphate ester always forms at the C-5′ position.

Each nucleotide has at least two places where new ester groups may form, at hydroxyls on the pentose ring and at the phosphate terminus, making the polymerization of nucleotides and the formation of nucleic acids possible (Figure 25.5). The phosphoric acid unit on one nucleotide splits out the elements of water with an alcohol unit of the pentose of the next nucleotide. A phosphate ester linkage or bridge forms between the two nucleotides. The process is visualized as continuing until hundreds of nucleotides are incorporated into the polymer.

The backbone of any nucleic acid from any of the forms of life investigated thus far appears to be the simple alternating system phosphate-pentose-phosphate-pentose-phosphate-pentose-etc. Because each pentose unit has a heterocyclic amine attached, a more complete picture of a nucleic acid is

```
                 amine′              amine″              amine‴
                   |                   |                   |
etc.—phosphate—pentose—phosphate—pentose—phosphate—pentose—etc.
```

Adenine

Phosphoric acid

Ribose

$\xrightarrow{-2H_2O}$

Adenosine 5′-phosphate (AMP)

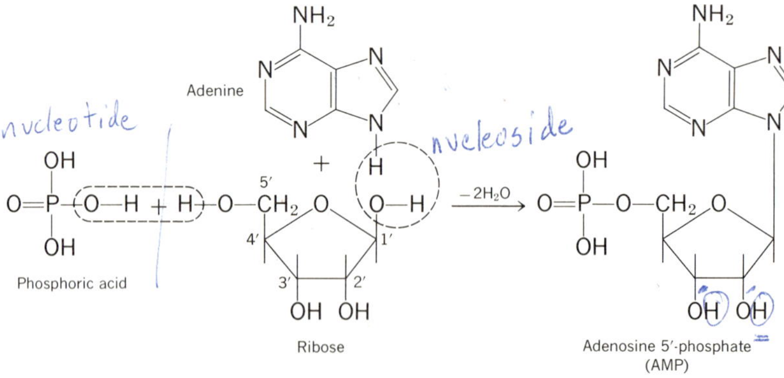

Figure 25.4 A typical nucleotide, a monomer for RNA. If the —OH group at C-2′ were replaced by —H, the nucleotide would be one of the monomers for DNA. By convention, the ring position numbers are primed for the ribose ring to distinguish them from the unprimed numbers of the heterocyclic ring.

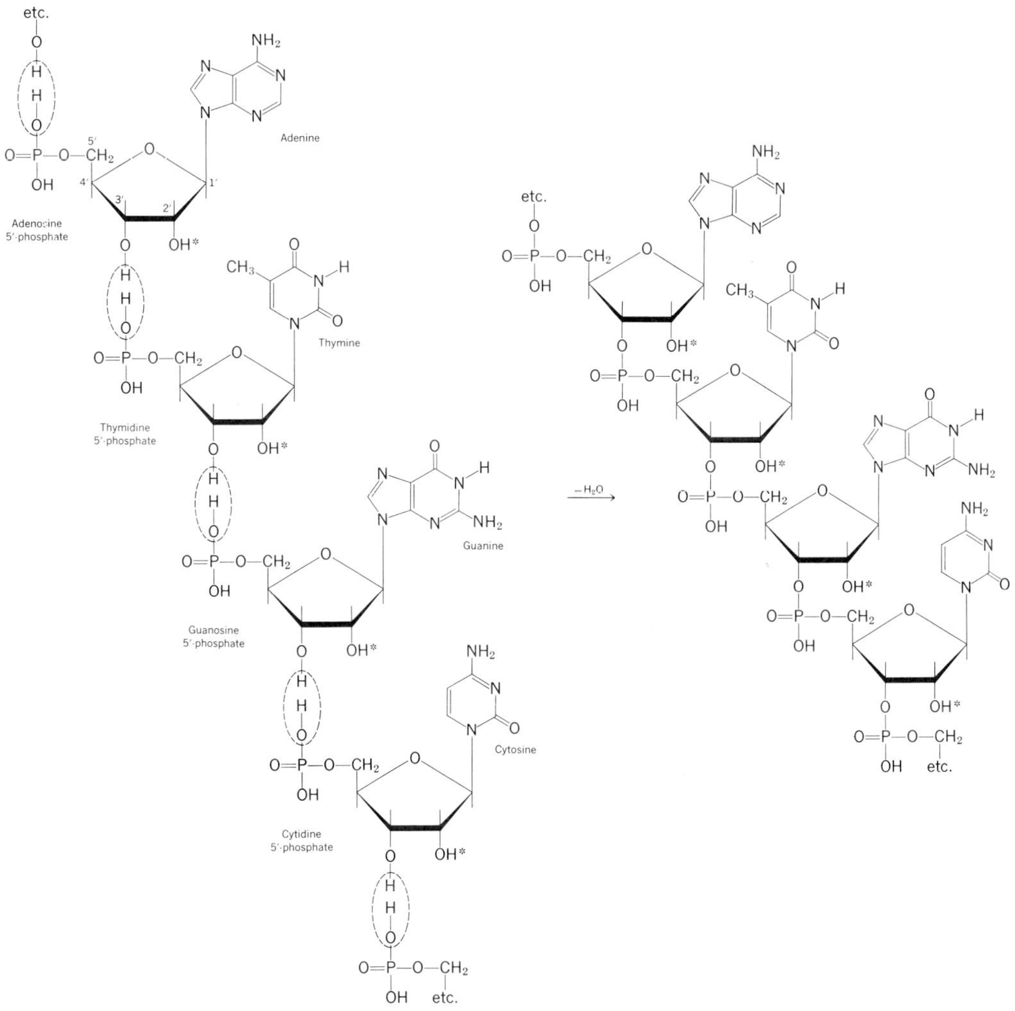

Figure 25.5 Formation of a nucleic acid chain. Shown here is a segment of an RNA chain, except that uracil replaces thymine in RNA. If the —OHs marked by asterisks were replaced by —Hs, this would be a segment of a DNA chain. The sequence of the heterocyclic amines is purely arbitrary in this drawing, but one each of the four amines common to DNA has been included. J. Cairns has found a molecular weight of 2.8×10^9 for the DNA of one species (*Escherichia coli*). If we use a value of 325 as the average formula weight of each nucleotide, this DNA will be made of 8,600,000 nucleotide units. Such a DNA molecule would probably make up a collection of genes rather than just one. Genetic studies indicate that the average gene size is 1500 nucleotide pairs (of a double helix).

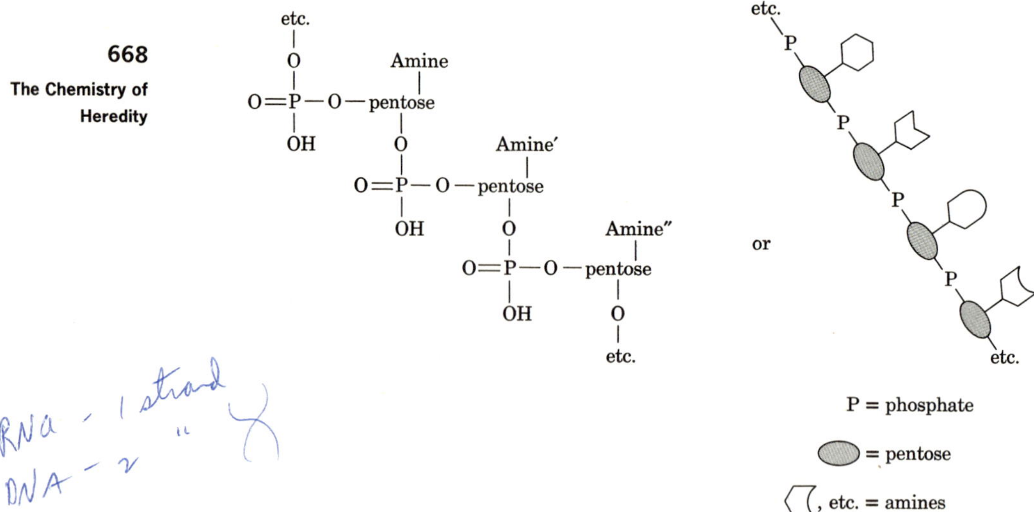
etc.
O
Amine
O=P—O—pentose
OH
O
Amine′
O=P—O—pentose
OH
O
Amine″
O=P—O—pentose
OH
O
etc.
or
etc.
P
P
P
P
etc.
P = phosphate
= pentose
, etc. = amines

Figure 25.6 Condensed structural representations of nucleic acids.

A more useful condensation of a nucleic acid structure is given in Figure 25.6.

If this structure represents what nucleic acids have in common, how they are different becomes apparent. The lengths of the backbones and the sequences in which the amines are strung along the backbones can both be different. In fact, we shall shortly identify this sequence with the genetic code itself. The facts of heredity will be correlated, in principle, with this aspect of the structure of a chemical.

Watson-Crick Theory. The DNA samples from a variety of species have been analyzed for proportions of the heterocyclic bases; Chargaff called the attention of scientists to some regularities that proved to be of great theoretical importance. To illustrate what these regularities are, some of the analytical data have been compiled in Table 25.1. The purines, we recall, are adenine (A) and guanine (G), the pyrimidines thymine (T) and cytosine (C).[2] Chargaff noted that for all the species analyzed

1. The purines equaled the pyrimidines in total percentages; that is, the purine-pyrimidine ratio was $(G + A)/(C + T) = 1$.

2. The percentage of adenine (a purine) was always very close to the percentage of thymine (a pyrimidine), that is, $A/T = 1$.

3. The percentage of guanine (a purine) was always very close to the percentage of cytosine (a pyrimidine), that is, $G/C = 1$.

4. The *characteristic composition* that was unique for a species was the ratio $(A + T)/(G + C)$.

This characteristic ratio for a particular species is not affected by factors such as the age of the species or its manner of growth. Different organs of the same species (e.g., thymus and liver of human beings, Table 25.1) exhibit substantially the same

[2] 5-Methylcytosine, one of the rarer pyrimidines, is present in some species and the percents given for cytosine include this compound.

Table 25.1 DNA Composition of Various Species

Species	Base Proportions (mole %)				$\frac{G + A}{C + T}$ (i.e., purine/pyrimidine)	$\frac{A}{T}$	$\frac{G}{C}$	$\frac{A + T}{G + C}$ (species characteristic)
	Purines		Pyrimidines					
	G	A	C	T				
Sarcina lutea	37.1	13.4	37.1	12.4	1.02	1.08	1.00	0.35
Brucella abortus	20.0	21.0	28.9	21.1	1.00	1.00	1.00	0.73
Escherichia coli K12	24.9	26.0	25.2	23.9	1.08	1.09	0.99	1.00
Wheat germ	22.7	27.3	22.8	27.1	1.00	1.01	1.00	1.19
Staphylococcus aureus	21.0	30.8	19.0	29.2	1.07	1.05	1.11	1.50
Human thymus	19.9	30.9	19.8	29.4	1.03	1.05	1.01	1.52
Human liver	19.5	30.3	19.9	30.3	0.99	1.00	0.98	1.54

Data from A. White, P. Handler, and E. L. Smith, *Principles of Biochemistry,* third edition, McGraw-Hill Book Company, New York, 1964, page 172. Used by permission.

characteristic ratios. This constancy indicates something important about the kind of chemical that was, even before Crick and Watson's theory, believed to be the stuff of genes. Whatever the genetic material may be, and however the genetic code may be structured, some things about it must be uniform, at least within a species. That there are some constant aspects was apparent from Chargaff's data. It remained for Watson and Crick to exploit these observations as well as results of X-ray analysis and the known general features of the DNA polymer. Watson and Crick's first paper on this topic, "A Structure for Deoxyribose Nucleic Acid," appeared on April 25, 1953, in the British publication *Nature.* Soon after, on May 30, 1953, a second paper, "Genetic Implications of the Structure of Deoxyribonucleic Acid," appeared in the same journal. These are two landmarks in the history of science. They were not, as many have written, documents of a revolution in science, nor were they turning points or radical departures. They were major advances. In a revolution old ideas are discarded, old institutions destroyed, and new ideas, new institutions are built on their remains. The accomplishment of Watson and Crick was built on what had gone before. They were not the first to propose that DNA might be the chemical of genes. The basic manner of hooking together the nucleotides had been established, and many proposed models for DNA gave the polymer a particular configuration in space. None of these models, however, was successful in explaining how the chemical could do what the gene was known to do—become duplicated exactly. This was the singular contribution of Watson and Crick. The model of a gene molecule that they proposed not only best fit the X-ray data—the best data appeared in a paper by M. H. F. Wilkins, A. R. Stokes, and H. R. Wilson published immediately after Watson and Crick's first paper—but could direct the essential operation required of genetic material, namely self-duplication. For this work Watson, Crick, and Wilkins shared the 1962 Nobel prize.

To begin our discussion of the Watson-Crick model for a gene, several portions of their two papers will be quoted (with permission). The first paper opens as follows:

"We wish to suggest a structure for the salt of deoxyribose nucleic acid (D.N.A.). This structure has novel features which are of considerable biological interest. [A brief discussion of structures that have been proposed by others, and the objections that Crick and Watson see in them, follows.] We wish to put forward a radically different structure for the salt of deoxyribose nucleic acid. This structure has two helical chains each coiled round the same axis (see diagram). [Their diagram is reproduced in Figure 25.7 together with their legend.] The novel feature of the structure is the manner in which the two chains are held together by the purine and pyrimidine bases. The planes of the bases are perpendicular to the fibre axis. They are joined together in pairs, a single base from one chain being hydrogen-bonded to a single base from the other chain, so that the two lie side by side with identical z coordinates. One of the pair must be a purine and the other a pyrimidine for bonding to occur . . . .

"If it is assumed that the bases only occur in the structure in the most plausible tautomer forms (that is, with the keto rather than the enol configurations), it is found that only specific pairs of bases can bond together. These pairs are: adenine (purine) with thymine (pyrimidine), and guanine (purine) with cytosine (pyrimidine).

"In other words, if an adenine forms one member of a pair, on either chain, then on these assumptions the other member must be thymine; similarly for guanine and cytosine. The sequence of bases on a single chain does not appear to be restricted in any way. However, if only specific pairs of bases can be formed, it follows that if the sequences of bases on one chain is given, then the sequence on the other chain is automatically determined. [Crick and Watson next refer to ratios of adenine to thymine and guanine to cytosine as given, for example, by Chargaff's work and in Table 25.1.]

"It has been found experimentally that the ratio of the amounts of adenine to thymine, and the ratio of guanine to cytosine, are always very close to unity for deoxyribose nucleic acid. [Their postulated structure, of course, *requires* just

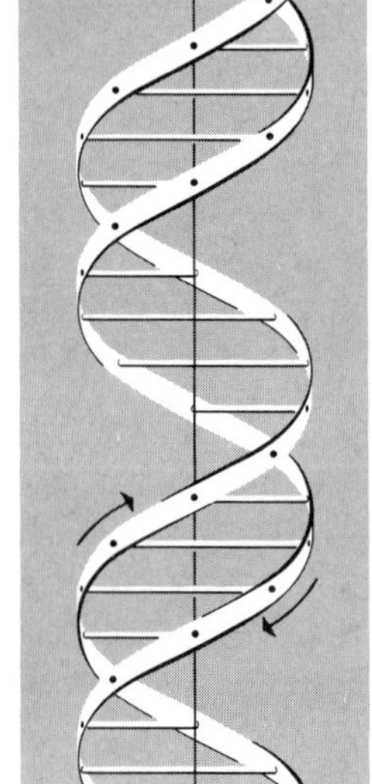

Figure 25.7 The DNA double helix. "This figure is purely diagrammatic. The two ribbons symbolize the two phosphate-sugar chains, and the horizontal rods the pairs of bases holding the chains together. The vertical line marks the fibre axis." (Figure and quotation from J. D. Watson and F. H. C. Crick, *Nature,* Vol. 171, 1953, page 737. Used by permission.)

such a ratio. They next comment on the provisional character of their postulated structure and the need for more exact X-ray studies.They are aware that Wilkins' paper, which was to give more accurate data, will follow their own in the journal, but they note that they were unaware of the details of Wilkins' results while they were devising their model. The next paragraph of their first paper opens the window.]

"It has not escaped our notice that the specific pairing we have postulated immediately suggests a possible copying mechanism for the genetic material."

In their second paper they describe the view from this window in richer detail. Before we can assimilate it ourselves, we should learn more about "base pairing" and how it is important to the structure of DNA.

The horizontal rods shown in Figure 25.7 symbolize base pairing in a very general way; details are given in Figure 25.8. Base pairing is the mutual attraction of a purine base for a pyrimidine base by means of hydrogen bonds. The geometries, the bond angles, and the availability of hydrogen-bond-donating and hydrogen-bond-accepting groups are apparently just right for this attraction when thymine and adenine are paired and cytosine and guanine are paired. (5-Methylcytosine can be substituted for cytosine.) Thus, of the four common bases in DNA, A pairs with T and C pairs with G. This peculiar specificity is described and explained by Crick and Watson in their second paper.

"The bases are joined together in pairs, a single base from one chain being hydrogen-bonded to a single base from the other. The important point is that only certain pairs of bases will fit into the structure. One member of a pair must be a purine and the other a pyrimidine in order to bridge between the two chains. If a pair consisted of two purines, for example, there would not be room for it.

"We believe that the bases will be present almost entirely in their most probable tautomeric forms. If this is true, the conditions for forming hydrogen bonds are more restrictive, and the only pairs of bases possible are:

adenine with thymine;
quanine with cytosine.

The way in which these are joined together is shown in Figures 4 and 5. [Figure 25.8 of this text corresponds to these Figures 4 and 5.] A given pair can be either way round. Adenine, for example, can occur on either chain; but when it does, its partner on the other chain must always be thymine."

If two purines paired, "there would not be room." The X-ray data revealed these limitations on the space available between the two spiral backbones (the ribbons in Figures 25.7 and 25.9). If two pyrimidines (recall that these are narrower molecules with only one ring each) paired, the space could not be filled to the point that hydrogen bonds would be possible. A purine with a pyrimidine, however, works out just right. Figure 25.9 is a fuller but still schematic representation of the Watson-Crick DNA double helix. Figure 25.10 pictures a scale model of a short section, one and a half turns, of the double helix.

REPLICATION OF DNA

Speculation about how chemicals might become duplicated in preparation for cell division has generally included the idea of a chemical mold or a template. In

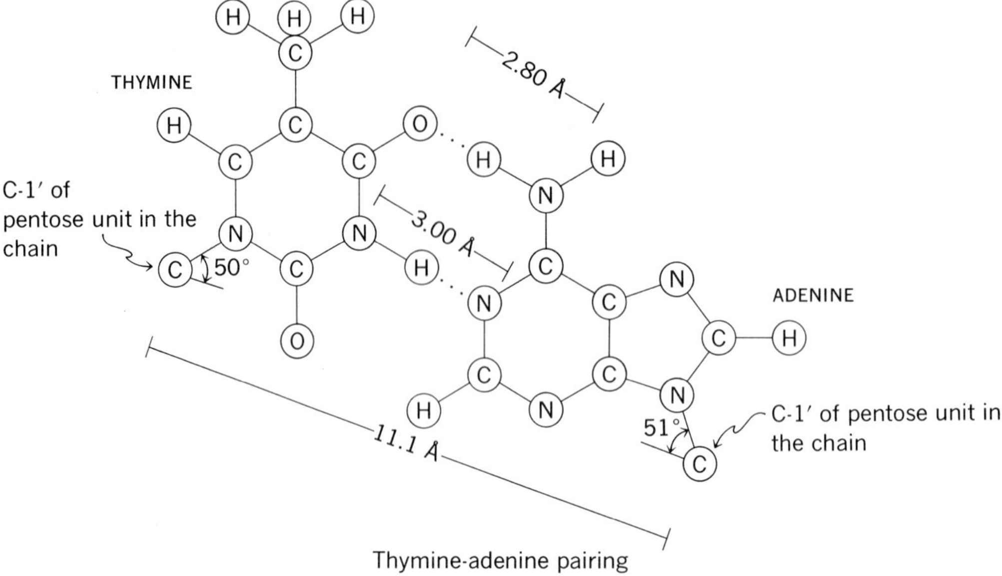
THYMINE
C-1′ of
pentose unit in the
chain
50°
2.80 Å
3.00 Å
ADENINE
51°
C-1′ of pentose unit in
the chain
11.1 Å

Thymine-adenine pairing

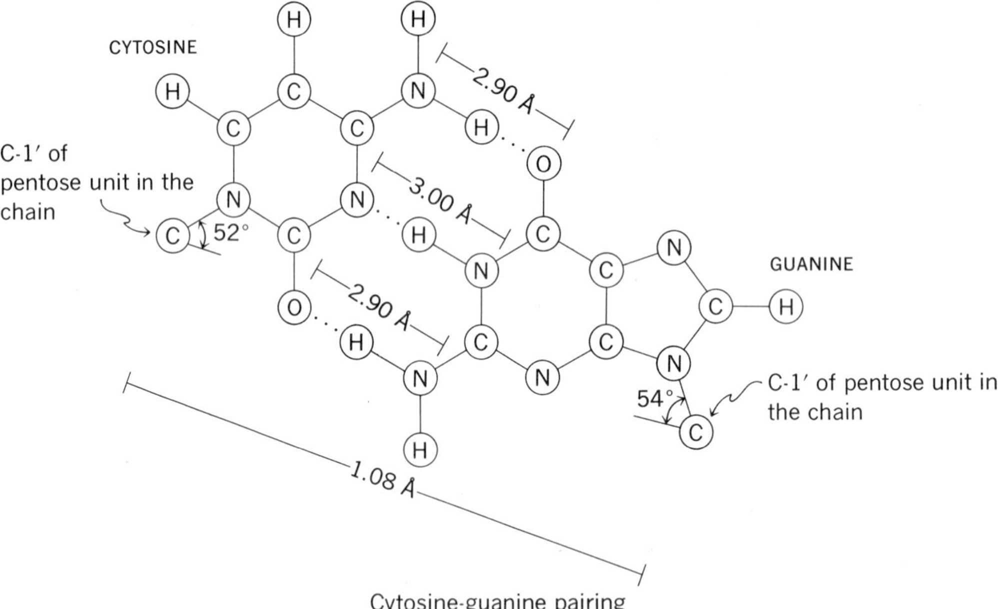
CYTOSINE
C-1′ of
pentose unit in the
chain
52°
2.90 Å
3.00 Å
2.90 Å
GUANINE
54°
C-1′ of pentose unit in
the chain
1.08 Å

Cytosine-guanine pairing

Figure 25.8 Pairing of purine-pyrimidine bases in DNA. The dimensions and geometries of these molecules are such that in their keto forms they can fit to each other via hydrogen bonds, two between thymine and adenine and three between cytosine and guanine. These pairs, which are almost entirely coplanar, form the "rungs" of the "spiral staircase" which the Watson-Crick model resembles. The perpendicular axis of Figure 25.7 would come up through the regions of the hydrogen bonds. The dimensions shown were determined by Pauling and Corey.

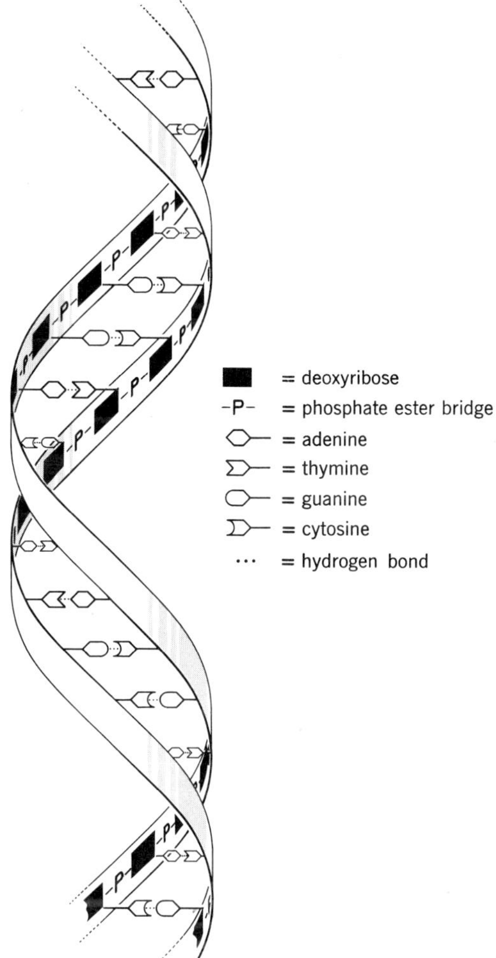
-P-
= deoxyribose
-P- = phosphate ester bridge
= adenine
= thymine
= guanine
= cytosine
··· = hydrogen bond

Figure 25.9 A DNA double helix showing base pairing.

industry a template is a pattern or guide used in laying out and scribing a workpiece. In shipbuilding it is a full-sized wooden mold or paper pattern used to make hull parts. A template can be a gauge or a pattern for checking contours and dimensions. The idea of a chemical serving as a template or a mold for making another chemical has long been considered in connection with genetics, but until the Watson-Crick model of DNA we had no detailed picture of *how* this could be done. Again, let us use the words of Watson and Crick from their second paper.

"The phosphate-sugar backbone of our model is completely regular, but any sequence of the pairs of bases can fit into the structure. It follows that in a long molecule many different permutations are possible, and it therefore seems likely that the precise sequence of the bases is the code which carries the genetical information. If the actual order of the bases on one of the pair of chains were given,

Figure 25.10 Scale model of a DNA double helix. (Courtesy of Nova Research Group, Columbus, Ohio.)

one could write down the exact order of the bases on the other one, because of the specific pairing. Thus one chain is, as it were, the complement of the other, and it is this feature which suggests how the deoxyribonucleic acid molecule might duplicate itself.

"Previous discussions of self-duplication have usually involved the concept of a template, or mould. Either the template was supposed to copy itself directly or it was to produce a 'negative,' which in its turn was to act as a template and produce the original 'positive' once again. In no case has it been explained in detail how it would do this in terms of atoms and molecules.

"Now our model for deoxyribonucleic acid is, in effect, a *pair* of templates, each of which is complementary to the other. We imagine that prior to duplication the hydrogen bonds are broken, and the two chains unwind and separate. Each chain then acts as a template for the formation on to itself of a new companion chain, so that eventually we shall have *two* pairs of chains, where we only had one before. Moreover, the sequence of the pairs of bases will have been duplicated exactly.

[The authors next discuss some of the uncertainties of their model as well as some of the evidence for it. The following quotation is from their closing paragraph.]

"Despite these uncertainties we feel that our proposed structure for deoxyribonucleic acid may help to solve one of the fundamental biological problems—the molecular basis of the template needed for genetic replication. The hypothesis we are suggesting is that the template is the pattern of bases formed by one chain of the deoxyribonucleic acid and that the gene contains a complementary pair of such templates."

It is as though genetic messages were written with a four-letter alphabet. Statistical theory enables us to calculate the number of isomers possible when four letters (four bases) are used many times. If there are, as indicated by genetic studies, about 1500 nucleotide pairs per gene, the possible number of isomers of such a gene would be 4^{1500}. J. D. Watson estimates that this number greatly exceeds the number of different genes that have ever existed in all the chromosomes since the origin of life. A four-letter alphabet used enough times, is certainly sufficient for encoding any realistic amount of genetic information.

As for the amount of DNA in the human race and its responsibility for all the individuals in all their marvelous varieties, Theodosius Dobzhansky has made some interesting calculations. According to figures by Müller, the volume of DNA in a human gamete is about 4 cubic microns and the weight is a paltry 4×10^{-12} g. The number of persons living is, very roughly, 3×10^9. They have come from twice that number of gametes, 6×10^9. The total volume of the physical carriers of genetic messages that the human species now alive has received from its ancestors is therefore about 2.4 mm^3, the total weight about 0.24 mg. This quantity, notes Dobzhansky, is about that of a raindrop.

Before cell division DNA molecules must be reproduced in duplicate; they must be *replicated.* According to the Watson-Crick theory, such replication consists of a temporary uncoiling, or partial uncoiling, of the two helices of a DNA double helix. This uncoiling occurs in a medium that contains building blocks for fresh DNA strands. From this mononucleotide pool each uncoiled DNA strand draws to itself the nucleotides that will fit with it via hydrogen bonds. Thus when an original A-T or G-C pair becomes unpaired, the old A becomes paired with a new T, the old T becomes paired with a new A, etc. Two new A-T pairs form, but on different double helices. Figure 25.11 describes this pairing in greater detail. Figure 25.12 shows in a schematic way how the growth of two new chains may occur as the parent double helix unwinds.

The Meselson-Stahl Experiment. The mechanism for DNA replication just described is called *semiconservative.* As indicated in Figure 25.13, there are three principal possibilities, conservative, semiconservative, and dispersive. In the conservative mechanism the strands of the parent double helix come back together again after serving as templates for two new strands. The dispersive mechanism implies that the original strands break up and then somehow recombine with new DNA. Meselson and Stahl in their experiment provided the most convincing evidence for the semiconservative model. They used the mass-15 isotope of nitrogen, combined in ammonia; we symbolize it as $^{15}NH_3$, the ordinary ammonia being $^{14}NH_3$. Working with the common colon bacillus, *Escherichia coli,* they grew some of this bacteria in a medium containing $^{15}NH_3$ as the only source of nitrogen. As a result, all

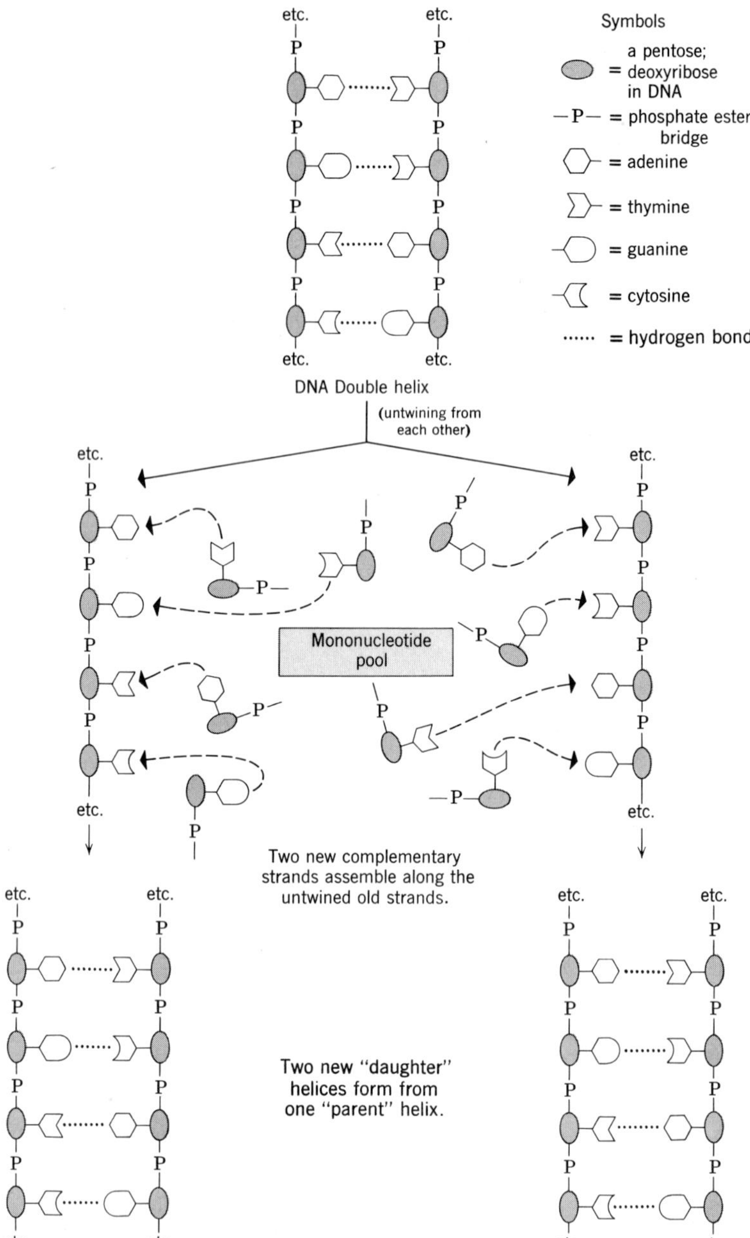

Symbols
a pentose; = deoxyribose in DNA
—P— = phosphate ester bridge
= adenine
= thymine
= guanine
= cytosine
...... = hydrogen bonds
DNA Double helix
(untwining from each other)
Mononucleotide pool
Two new complementary strands assemble along the untwined old strands.
Two new "daughter" helices form from one "parent" helix.

Figure 25.11 The replication of DNA. The complete untwining of the original double helix (top) implied here and suggested originally by Crick and Watson probably does not happen. Instead, as the long DNA double helix starts to unwind, replication may begin at the freshly exposed ends of the single strands as they appear (cf. Figure 25.12). New DNA is actually built up one nucleotide at a time, not (as implied here) all at once after the nucleotides have lined up. It is impractical to convey all relevant facts, and this figure stresses the *pairing* of bases as the most important feature in the replication of the genetic code. The actual mechanism of replication in the growing regions of DNA is unknown (1968).

Figure 25.12 According to one theory, replication of DNA occurs *while* the parent double helix unwinds rather than after it has become entirely unwound.

the nitrogens of the bases in the DNA of freshly grown bacteria were of the heavier ^{15}N. The density of this DNA was greater than that of the ordinary DNA of this species made with ^{14}N. The analytical data were obtained in this experiment by measuring densities. Some of these bacteria grown in the ^{15}N medium were then transferred to a new medium where $^{14}NH_3$ was the only source of nitrogen. By the time a new generation had developed, the density of the new DNA indicated that each DNA molecule must consist of one strand of "heavy" (^{15}N) DNA and one strand of "light" (^{14}N) DNA; we may call this "isotopic hybrid DNA." In the second generation half the DNA molecules were of the isotopic hybrid type and half were light. Figure 25.13 illustrates this experiment, reported by M. Meselson and F. W. Stahl in 1958. Referring to Figure 25.13, we see that the conservative mechanism would not yield any isotopic hybrid DNA. The dispersive model would not, at least after just two generations, yield any wholly light DNA. The semiconservative model is consistent with the results. Other details of the mechanism of the biosynthesis of DNA have been reported by Arthur Kornberg (Nobel prize, 1959). At Washington University (St. Louis) a team of scientists led by him isolated an enzyme, *DNA polymerase,* that would catalyze the formation of DNA in a test tube containing a small amount of "primer" DNA and the triphosphate forms of the four nucleotides (i.e., dATP, dTTP, dGTP and dCTP).

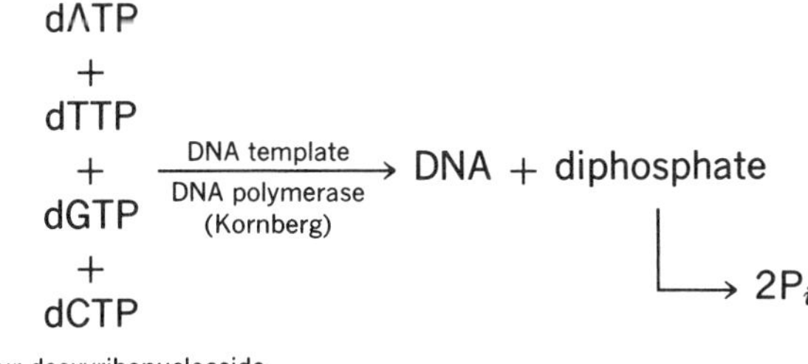
dATP
+
dTTP
+
dGTP
+
dCTP
DNA template
DNA polymerase
(Kornberg)
DNA + diphosphate
$2P_i$
The four deoxyribonucleoside triphosphates

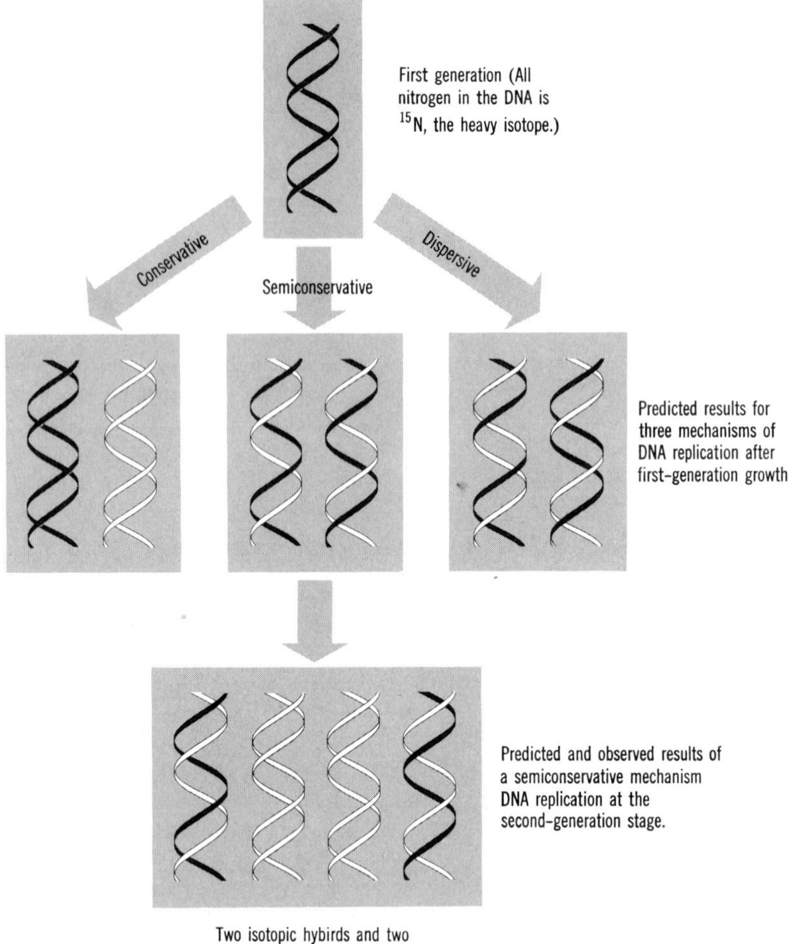
First generation (All nitrogen in the DNA is ^{15}N, the heavy isotope.)
Conservative
Semiconservative
Dispersive
Predicted results for three mechanisms of DNA replication after first-generation growth.
Predicted and observed results of a semiconservative mechanism DNA replication at the second-generation stage.
Two isotopic hybirds and two "light" DNA double helices

Figure 25.13 The Meselson-Stahl experiment. The strands drawn with the heavy line represent DNA in which all the nitrogen is in the form of the heavy isotope, ^{15}N. The lightly drawn strands stand for DNA made of ^{14}N. By measuring densities of the DNA formed after the first and the second generations, Meselson and Stahl (1958) found convincing evidence for the semiconservative mechanism.

The mechanism of DNA replication just described leaves many questions unanswered. In higher plants and animals, for example, chromosomes have as much as 50% protein. Just what does this protein have to do, if anything, with the chemistry of heredity? How are the DNA molecules incorporated together with the protein to form a chromosome? We shall not suggest any theories to answer these questions, but it is possible to say a great deal about the most important function of DNA (aside from replication), its control over the synthesis of polypeptides and thence that of enzymes.

DNA AND THE SYNTHESIS OF ENZYMES

The general scheme connecting DNA to polypeptides (enzymes, usually) is outlined in Figure 25.14; our discussion will give more details. The immediate task of chromosomal DNA is to participate in and determine the nucleotide sequence of RNA. The RNA molecules have all the features of DNA strands with two differences.

1. The —OH group at C-2′, absent in DNA, is of course present in RNA which is made from ribose rather than deoxyribose. This difference appears to be minor, but it may be a factor in determining the final configuration of the RNA strand. Most RNA exists as single strands rather than as double helices, although such formations are known for RNA.

2. There is no thymine in RNA; its place is taken by uracil (Figure 25.3), which can pair with adenine just as well as thymine. Thus we can have A-T pairing and A-U pairing.

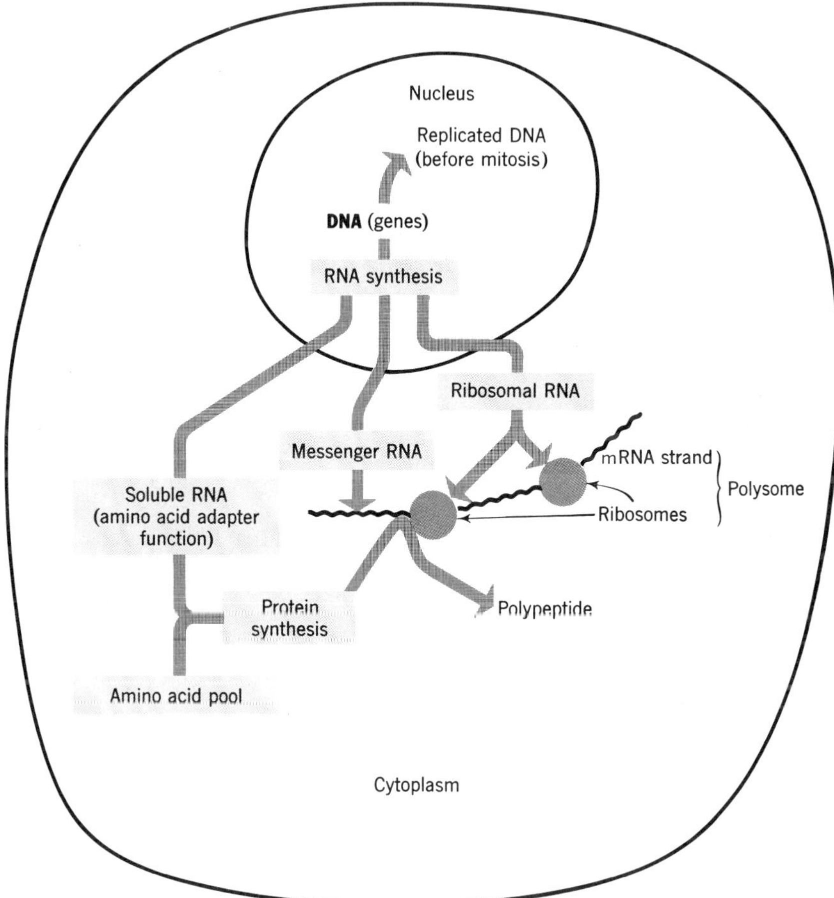
Nucleus
Replicated DNA
(before mitosis)
DNA (genes)
RNA synthesis
Ribosomal RNA
Messenger RNA
mRNA strand
Polysome
Ribosomes
Soluble RNA
(amino acid adapter
function)
Protein
synthesis
Polypeptide
Amino acid pool
Cytoplasm

Figure 25.14 The relations of DNA to various RNAs and protein synthesis.

The three types of RNA known must be distinguished. All three types are synthesized under the control of DNA. (A fourth type of RNA associated with virus particles is also known.) For the discovery of an enzyme, *RNA polymerase,* which made possible *in vitro* synthesis of RNA, Severo Ochoa shared the 1959 Nobel prize in physiology and medicine with Arthur Kornberg.

Ribosomal RNA (rRNA). The main structural framework of a cell is the *endoplasmic reticulum,* a series of canaliculi and cisternae that interconnect and permeate most of the cytoplasm (see Figure 25.1). Studding portions of the endoplasmic reticulum are granules varying from 70 to 200 Å in diameter. Sometimes found free in the cytoplasm and made of RNA and protein, these granules may bind up about 75 to 80% of all the RNA of a cell. Ribosomes are the sites of protein synthesis, and yet ribosomal RNA itself does not appear to direct this work. Messenger RNA, the second type, directs protein synthesis. When we examine the details of polypeptide synthesis later in this chapter, we shall learn that ribosomal RNA may provide some stabilizing forces to a temporary complex between messenger RNA and the third type we study, soluble RNA. Ribosomes from a common source are known to be alike. They must therefore serve some nonspecific function that is common to the synthesis of all the protein (enzymes) made by that source.

Messenger RNA (mRNA). Messenger RNA molecules vary considerably in length and account for about 5 to 10% of the total RNA in a cell. As we shall see, mRNA molecules are the bearers of the genetic code after they have been synthesized under the direct supervision of DNA (Figure 25.15). Once made, mRNAs move out of the nucleus into the cytoplasm where they attach ribosomes to themselves at intervals along their chains. Such an assembly of many ribosomes along an mRNA chain is called a *polysome* (polyribosome; see Figure 25.16). Evidence indicates that mRNA can dissociate from ribosomes and that ribosomes can associate with different mRNA molecules of different molecular weight. When protein synthesis takes place, however, it is much more difficult for mRNA to dissociate from ribosomes. Additional evidence indicates that a ribosome moves along an mRNA chain while the synthesis of a polypeptide, directed by the mRNA, takes place. For such synthesis amino acids must be brought to the site in the order in which they are to appear in the final polypeptide. The cell uses the third type of RNA for this task.

Soluble RNA (sRNA). Because soluble RNA has a much lower formula weight than rRNA or mRNA, it is more soluble in media used to isolate the higher-formula-weight forms. Thus the designation soluble refers to a physical property and says nothing about function. When its function is to be stressed, various authors refer to sRNA as *transfer RNA* (tRNA) or as *amino acid adapter RNA* (aaRNA). We shall use sRNA for most of our purposes. The different species of sRNA molecules are believed to number at least twenty, one for each of the some twenty amino acids. The function of sRNA is to attach to itself the particular amino acid for which it is coded, carry it to a protein synthesis site on a polysome (specifically an mRNA site in contact with a ribosome), and there give it up to the growing end of a polypeptide chain at the particular moment called for by the genetic code. As illustrated in Figure 25.17, which is meant to convey only the most general aspects of

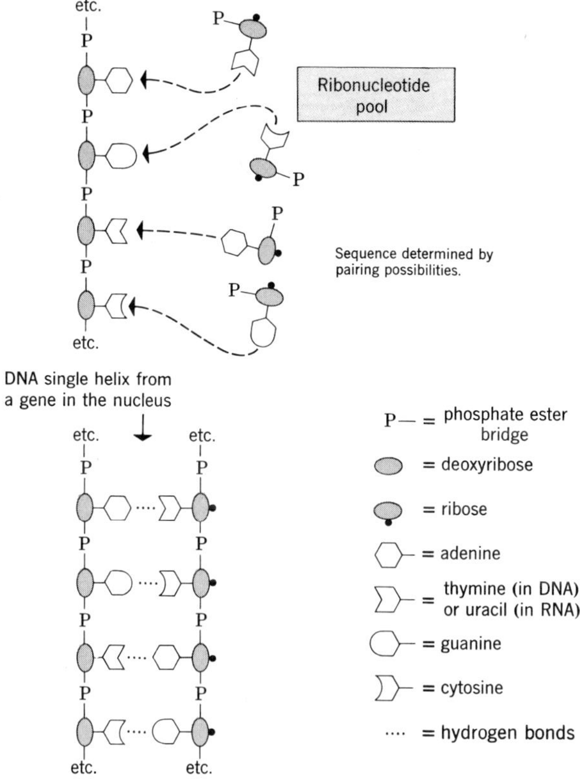
etc.
P
Ribonucleotide pool
Sequence determined by pairing possibilities.
etc.
DNA single helix from a gene in the nucleus
P— = phosphate ester bridge
= deoxyribose
= ribose
= adenine
= thymine (in DNA) or uracil (in RNA)
= guanine
= cytosine
···· = hydrogen bonds
An RNA strand, complementary in terms of the sequence of heterocyclic amines, assembles along the DNA strand in the nucleus. A hybrid complex of this type was observed (1960) by Alexander Rich of M.I.T.

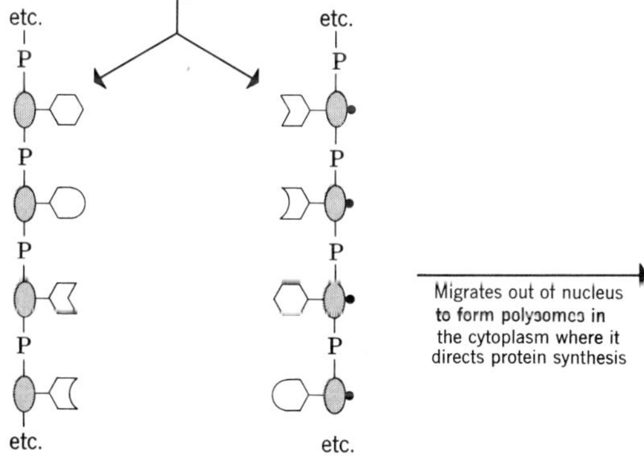
etc.
P
Migrates out of nucleus to form polysomes in the cytoplasm where it directs protein synthesis
DNA strand in readiness for synthesis of more RNA
Newly formed mRNA strand

Figure 25.15 Transmission of a genetic code from DNA to mRNA in cell nuclei. The enzyme *RNA polymerase,* itself a polypeptide that is dependent on a specific DNA molecule, controls this synthesis along with the DNA shown here (ATP is also required). The ribonucleotide "pool" consists of the triphosphate forms of the nucleotides, making ATP a member of the pool.

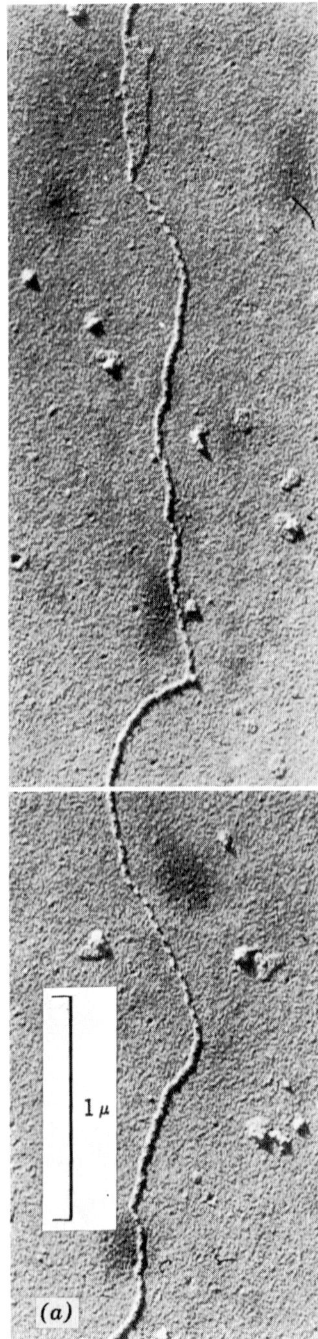
1 μ
(a)

Figure 25.16 Polysomes. (*a*) Electron micrograph of a long polysome (polyribosome) isolated from rat skeletal muscle tissue by Lederle scientists. The threadlike material is mRNA, and what appear to be rather evenly spaced bulges are ribosome units. (Courtesy of Lederle Laboratories, a division of American Cyanamid Company.) (*b*) Electron micrograph of polysomes from the rabbit reticulocyte stained with uranyl acetate. Again, the dark thread running between ribosomes (large dark shapes) is mRNA. Its diameter is very close to 15 Å. (Courtesy of Professor Alexander Rich, Massachusetts Institute of Technology.)

the synthesis, the mRNA acts as a template. Its sequence of nucleotides was determined by a sequence on a DNA molecule, or a portion thereof (Figure 25.15). Although mRNAs cannot recognize individual amino acids, they can recognize individual sRNAs, for these can assemble along an mRNA strand only in a sequence dictated by the requirements of base pairing. These requirements must inevitably determine the order in which amino acids become lined up in a growing polypeptide chain.

Thus a sequence of nucleotides on a gene (DNA) molecule and a sequence of amino acids in a protein have a linear correlation. Since a protein synthesized under this genetic control is the apoenzyme portion of an enzyme, this theory offers a molecular explanation of the one-gene, one-enzyme idea first advanced with evidence in 1941 by George Beadle and Edward Tatum, who received the 1958 Nobel prize in physiology and medicine for this work. (Joshua Lederberg also shared the prize for work done on genetic recombination.) Since the work of Beadle and Tatum many enzymes have been found to consist of more than one polypeptide or protein strand. Because more than one gene is required to direct the synthesis of such enzymes, we now have a one-gene, one-polypeptide theory instead

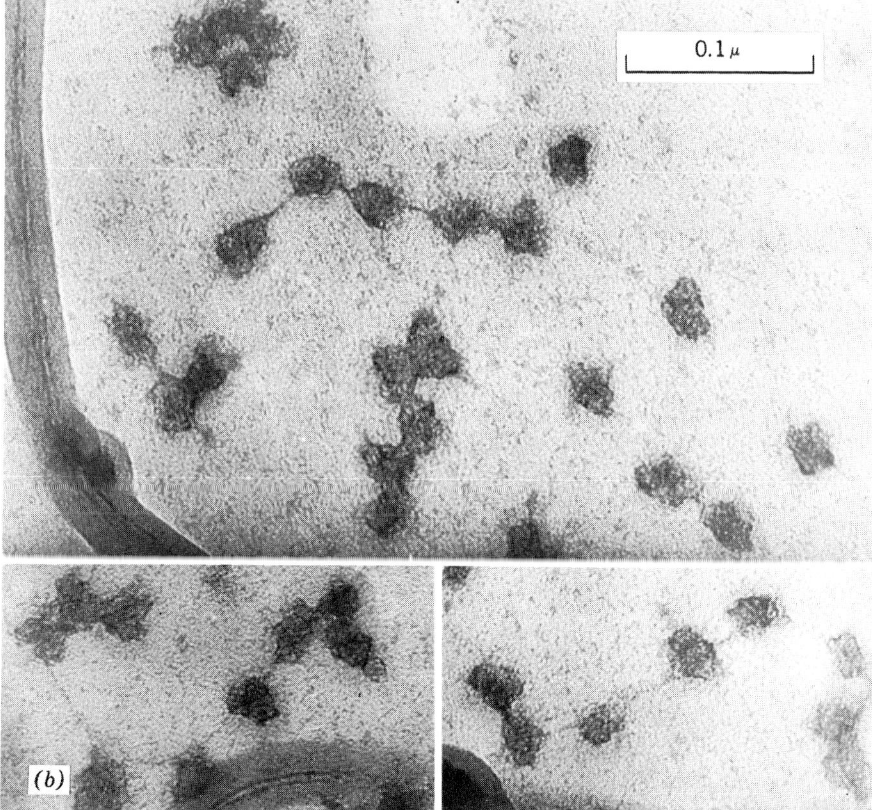
0.1 μ
(b)

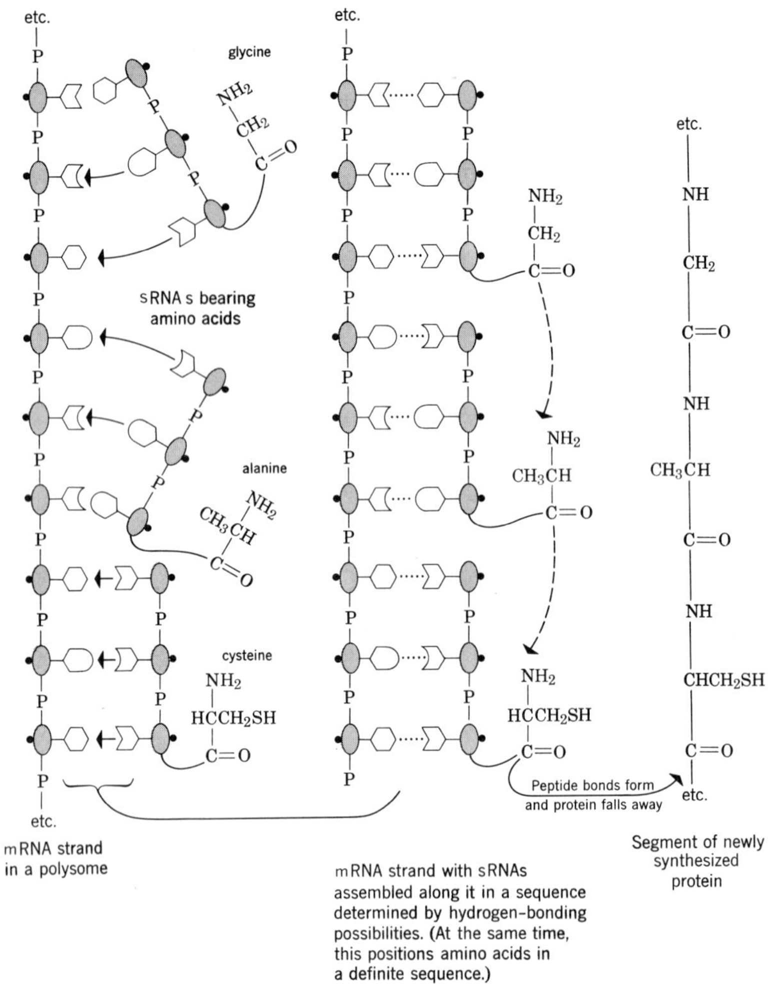
etc.
P
glycine
NH_2
CH_2
C=O
sRNA s bearing amino acids
alanine
NH_2
CH_3CH
C=O
cysteine
NH_2
$HCCH_2SH$
C=O
etc.
mRNA strand in a polysome
NH
CH_2
$CHCH_2SH$
Peptide bonds form and protein falls away
mRNA strand with sRNAs assembled along it in a sequence determined by hydrogen-bonding possibilities. (At the same time, this positions amino acids in a definite sequence.)
Segment of newly synthesized protein

Figure 25.17 A generalized view of how sRNAs participate together with gene-coded mRNA in determining the order of amino acid units in a developing polypeptide. The figure emphasizes the importance of the pairing of complementary heterocyclic amines in this phase of gene-directed protein synthesis.

of a one-gene, one-enzyme theory, with the understanding that the polypeptide is usually an enzyme or is destined to become part of an enzyme.

We now have a general idea of how the sequence of nucleotides on a gene molecule can be transcribed into a sequence on an mRNA molecule and then translated into a specific amino acid sequence. The importance of polymerizing enzymes, *DNA polymerase* and *RNA polymerase,* was indicated, and the need for triphosphates as sources of energy was noted. Many additional details are known, and those relating to how the various forms of RNA cooperate are described in the next sections. Although a few of the important experiments are mentioned, the reader should consult the reference list for further study. The names of several scientists have been and will be mentioned. Unfortunately, it is not practical to credit all the great scientists in the burgeoning field. In the opinion of some it is also unfortunate that more Nobel prizes cannot be given to reward these men for their distinguished work.

TRANSLATING THE GENETIC CODE

Codons. If we start with one set of four different symbols and another set of twenty different symbols, it is obviously impossible to find a one-to-one correspondence of one set with the other. Yet this is the task we must perform if we insist on using a four-letter alphabet (the four nucleotides in DNA) to specify unique sequences based on a twenty-letter alphabet (amino acids for proteins). We could say, of course, that the *letters* of the four-letter alphabet are not themselves the code. Rather these letters form two-letter words and the *words* make up the code. Thus from A, T, C, and G we have the following "words":

AA	AT	TC	CG
TT	TA	CT	GC
CC	AC	TG	
GG	CA	GT	
	AG		
	GA		

This list totals sixteen, still not enough, still short of twenty. By using three-letter units, however, we can form sixty-four words, more than enough. Statistical considerations strongly suggest that the genetic code is made up of "words" rather than individual letters. The genetic "alphabet" is too short. Genetic evidence not only confirms this supposition but indicates that *groups of three nucleotides are the fundamental genetic units.* When expressed as sequences on mRNA, they are called *codons.* For example, a sequence of side chain bases on some section of a DNA molecule might be cytosine-thymine-thymine, or CTT. Transcription of this into mRNA will give a sequence of guanine-adenine-adenine, or GAA. The GAA sequence of the mRNA is the codon.

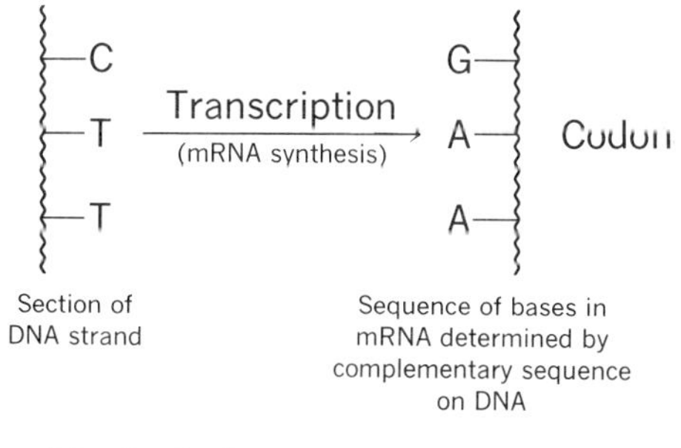

C — G
T — Transcription (mRNA synthesis) → A — Codon
T — A

Section of DNA strand

Sequence of bases in mRNA determined by complementary sequence on DNA

Sequence Recognition by Amino Acids. Few of the amino acids have side chains to make direct pairing between them and bases on the side chains of mRNA even conceivable. There is no basis for a direct recognition by amino acids of the sequence they are to take up. The sRNA molecules mediate this difficulty by providing surfaces that can recognize pairing opportunities on an mRNA chain. To put an amino acid and an sRNA molecule together, the following events must occur. Wherever an enzyme is specified, its synthesis too is controlled by a gene (DNA).

1. The amino acid must be activated for eventual attachment to an sRNA molecule (this requires an activating enzyme and ATP for energy):

$$\text{Activating enzyme} + \text{amino acid} + \text{ATP} \longrightarrow \text{activating enzyme—amino acyl} \sim \text{AMP} + \text{diphosphate ion}$$

An amino acyl group will have the form: $R{-}\underset{\underset{NH_2}{|}}{CH}{-}\overset{\overset{O}{\|}}{C}{-}$; AMP is of course adenosine monophosphate.

2. The amino acyl group is transferred to a specific sRNA molecule (the same enzyme serves):

$$\text{Activating enzyme—amino acyl} \sim \text{AMP} + \text{sRNA} \longrightarrow \text{amino acyl} \sim \text{sRNA} + \text{AMP} + \text{activating enzyme}$$

amino acyl∼sRNA	AMP	activating enzyme
↓ Will migrate to mRNA to incorporate the amino acid into a polypeptide	↓ To be "recharged" to ATP via the respiratory chain	↓ To be reused

To get the right amino acid attached to the right sRNA molecule, the activating enzyme is crucial. It must be able to bind itself not only to just one specific amino acid but also to the sRNA adapter molecule. Thus some twenty specific enzymes (polypeptides) are needed, but it is not difficult to imagine this number each with secondary and tertiary structural features that can recognize on the one hand an amino acid and on the other hand, at some other part of the enzyme, an sRNA molecule. The lock and key theory of enzyme action serves well here.

In the sRNA portion of the amino acyl∼sRNA complex we have the recognition site for a codon on an mRNA molecule. This recognition site must also be a sequence of three bases, and if the codon is GAA (example on p. 685), the recognition site must have the complementary sequence CUU, (not CTT; U replaces T in RNA). We call any sequence of three that is complementary to a codon an *anticodon*. (Some authors use the term *nodoc*.) The sRNA molecules, as noted earlier, are much longer than three nucleotides. The average molecular weight of sRNA is about 25,000, which corresponds to about eighty nucleotides linked together. The first structure determination of a specific RNA was for the sRNA molecule that attaches to itself the amino acid alanine. A team of scientists, associated with Cornell University and the U.S. Department of Agriculture and led by R. W. Holley,[3] reported in 1965 that *alanine transfer RNA* from yeast contains seventy-seven nucleotides, several of which are unusual (meaning that they are other than A, G, U, or C). In 1966 the structure of *tyrosine transfer RNA* was reported by J. Madison

[3] R. W. Holley, H. G. Khorana, and M. W. Nirenberg, who worked independently, shared the 1968 Nobel prize in physiology and medicine for their discoveries concerning the genetic code and its function in protein synthesis.

and G. Everett to contain seventy-eight nucleotides. Other sRNAs have been examined in enough detail to show that probably all of them have the CCA sequence of bases at one end of the chain. The other end (the C-5′ end) terminates in a G unit. These molecules are single-stranded, but by a hairpin fold most of the bases become paired; a double helix, as in DNA, makes up part of the configuration. Somewhere along this chain the anticodon sequence of three bases occurs, making it possible for the sRNA molecule to recognize a codon on an mRNA chain. Figure 25.18 shows how an sRNA molecule might look, both without and with an attached amino acyl group. The step-by-step growth of a polypeptide chain is pictured and discussed in Figures 25.19 and 25.20. According to a theory proposed by Alexander Rich (Massachusetts Institute of Technology), individual ribosomes move from one end of a mRNA chain to the other, making protein as they go. When the other end of the mRNA chain is reached, the ribosome drops off and the newly made polypeptide strand is released. As discussed earlier (p. 488),

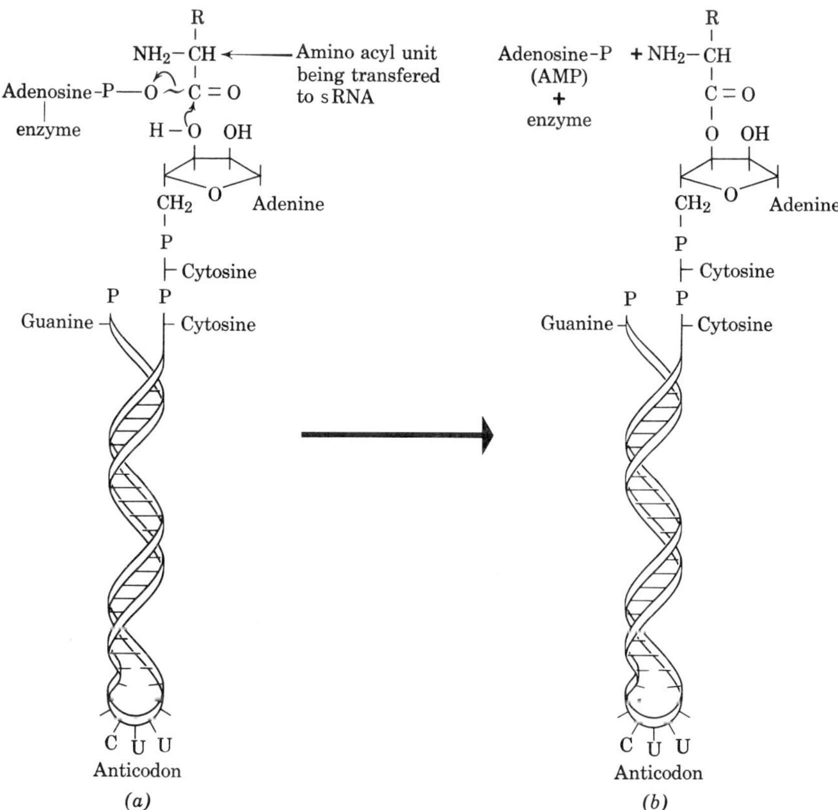
R
NH_2-CH
Amino acyl unit being transfered to sRNA
Adenosine-P—O
C = O
enzyme
H—O OH
CH_2
O
Adenine
P
Cytosine
P P
Guanine
Cytosine
C U U
Anticodon
(a)
Adenosine-P (AMP) + enzyme
+ NH_2-CH
R
C = O
O OH
CH_2
O
Adenine
P
Cytosine
P P
Guanine
Cytosine
C U U
Anticodon
(b)

Figure 25.18 (*a*) Diagrammatic view of an sRNA molecule and (*b*) the same molecule after it has been attached to an amino acyl group. The sRNA molecule is believed to be single-stranded but folded so that much of it has a double-helix configuration. Placing the triplet anticodon at the hairpin turn is for convenience only, as was the selection of the specific sequence cytosine-uracil-uracil (CUU) to illustrate an anticodon.

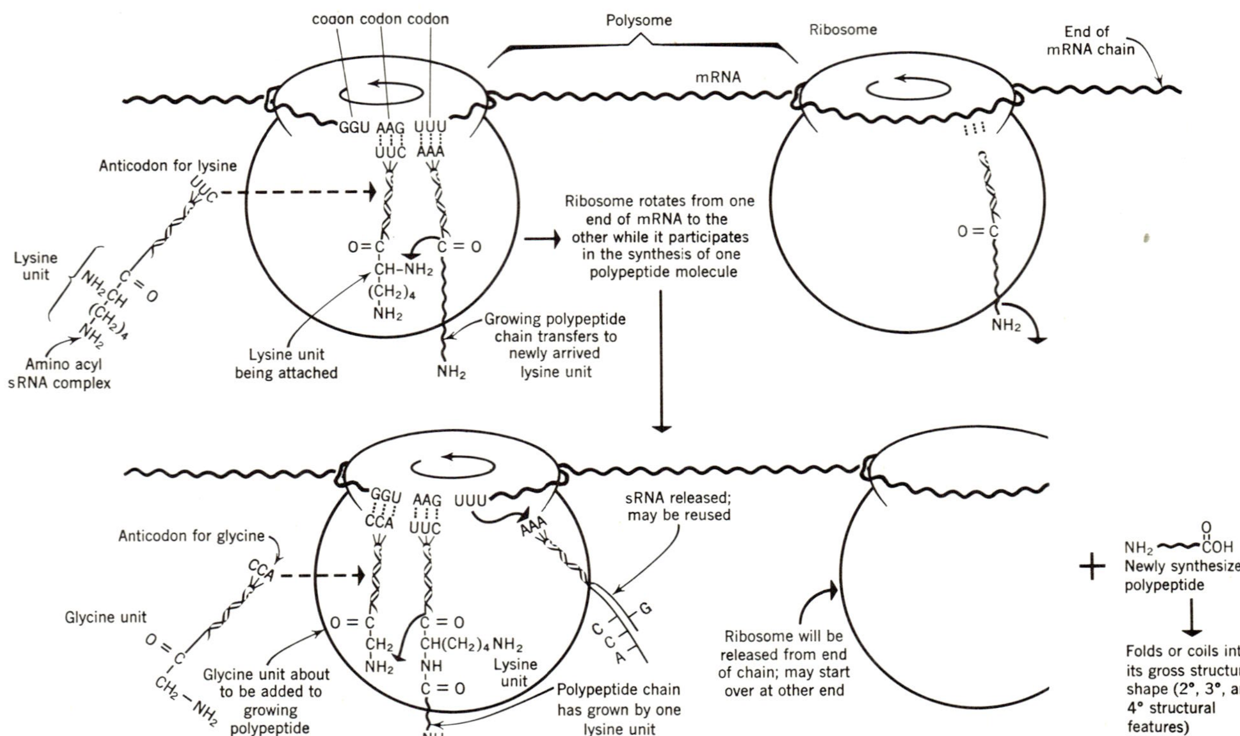

codon codon codon
Polysome
Ribosome
End of mRNA chain
mRNA
GGU AAG UUU
UUC AAA
Anticodon for lysine
UUC
Lysine unit
Amino acyl sRNA complex
Lysine unit being attached
Growing polypeptide chain transfers to newly arrived lysine unit
Ribosome rotates from one end of mRNA to the other while it participates in the synthesis of one polypeptide molecule
NH2
GGU AAG UUU
CCA UUC
AAA
Anticodon for glycine
CCA
Glycine unit
Glycine unit about to be added to growing polypeptide chain
Lysine unit
sRNA released; may be reused
Polypeptide chain has grown by one lysine unit
Ribosome will be released from end of chain; may start over at other end
Newly synthesized polypeptide
Folds or coils into its gross structural shape (2°, 3°, and 4° structural features)

Figure 25.19 Step-by-step growth of a polypeptide chain. Amino acyl sRNA complexes carry amino acyl groups to a polysome. Codon-anticodon pairing occurs between the sRNA and the mRNA at a ribosomal surface. The already partially grown polypeptide chain transfers to the amino group of the newly arrived amino acyl unit, and thus the chain grows. The ribosome moves along the mRNA strand as all this happens. At the end the ribosome drops off. This figure combines the ideas and results of many, especially those of Geoffrey Zubay (Brookhaven) and Alexander Rich (Massachusetts Institute of Technology).

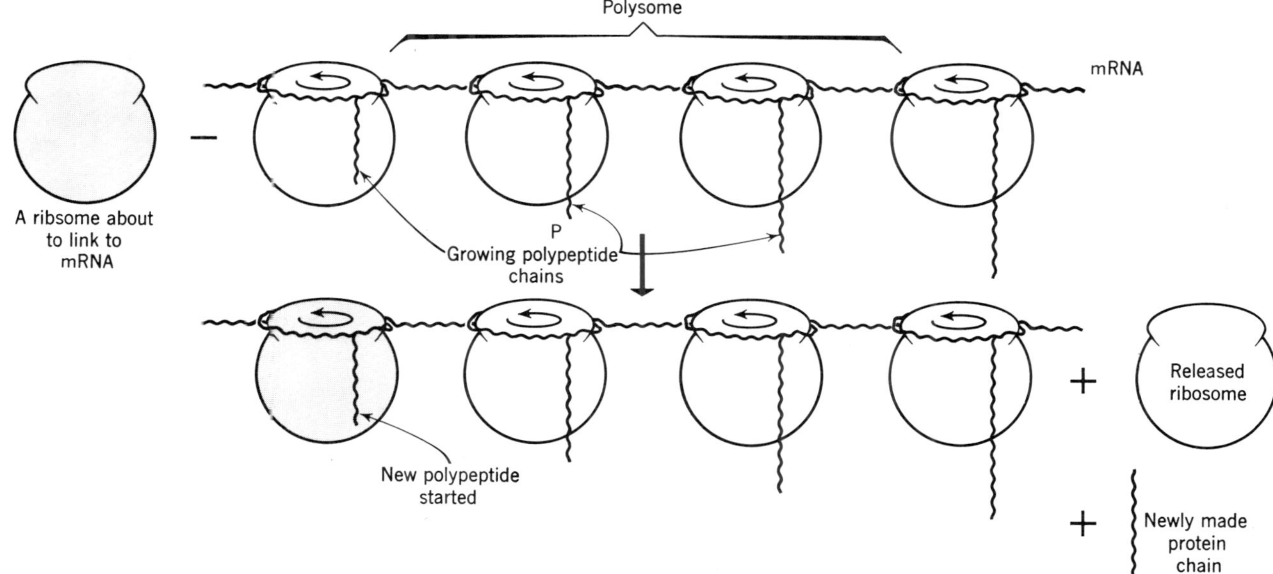
Polysome
mRNA
A ribsome about to link to mRNA
P
Growing polypeptide chains
New polypeptide started
Released ribosome
Newly made protein chain

Figure 25.20 Possible model for the dynamic function of a polysome, according to Alexander Rich. (From A. Rich, "Polyribosomes," *Scientific American*, December 1963, page 44.)

once the sequence of amino acids is put together, the protein automatically assumes its secondary or higher structural features. As one ribosome leaves the mRNA chain, a new empty one starts out at the other end according to Rich's theory. It must be borne in mind that this theory has by no means been proved. In fact, the way we have depicted the arrangement between mRNA and ribosomes in Figures 25.19 and 25.20 is still partly conjecture. The arrangement appears to be the easiest way to account for the facts, and it does help to rationalize the high rate of protein synthesis.

Rate of Formation of Proteins. The production rate of fresh hemoglobin can be estimated; on the average, one of its polypeptide chains must be made per ribosome every 90 seconds. Since there are about 150 amino acid units per polypeptide, the chain must grow at the rate of approximately two amino acid units per second.

The Nirenberg-Matthaei Experiments and the Specific Code Words. Marshall Nirenberg[4] and J. H. Matthaei of the National Institutes of Health reported on a series of experiments that gave support to the mechanisms we have just described and opened the way to assign specific three-letter code words, triplets, to specific amino acids. They worked with the colon bacillus *Escherichia coli* (not because these bacteria are more interesting than people, but because they are handier to use and because so many of their metabolic pathways are the same as those in people). By grinding and centrifuging these bacteria, Nirenberg and Matthaei were able to prepare a cell-free extract containing ribosomes, enzymes, and DNA. To this they added a mixture of the twenty amino acids that make up most proteins; they also supplied triphosphates for energy. This extract was able to synthesize protein. They then added an enzyme that catalyzes the destruction (by hydrolysis) of DNA. Yet for twenty more minutes the extract could make proteins. The implication was that during this time the synthesis of proteins already started was allowed to finish, but that to make new protein new mRNA would be needed. (It is known that mRNA is not particularly stable and must be replenished rather frequently.) But the DNA for directing the synthesis of the mRNA had been destroyed. Hence no new mRNA and no new protein. When they added RNA extracted from fresh batches of *E. coli,* protein synthesis recommenced. If at this point they added RNA not from the fresh batches of *E. coli* but rather from some entirely different organism such as yeast cells or tobacco mosaic virus, the extract still began to synthesize protein again. Apparently the ribosomes from *E. coli* (and presumably from any organism) are not themselves specific. It is not to the ribosome that we must look for the genetic uniqueness of a species. The RNA (presumably including mRNA and sRNA) from other organisms is acceptable to the ribosomes of *E. coli* for the work of proton synthesis.

Finally, Nirenberg and Matthaei took RNA not from fresh *E. coli* or from yeasts or from a virus, but rather from a synthetic batch. They used a very simple polynucleotide, polyuridylic acid. In this RNA all the bases are uracils; it can be called poly-U. (Recall that uracil is found just in RNA, not in DNA. What they added, then, could serve only as RNA material.) When this poly-U was introduced as fresh RNA to the *E. coli* extract that had been treated to destroy its DNA but not its ribosomes or its enzymes, the extract proceeded to make polyphenylalanine. It

[4] Nobel prize, 1968. (See footnote 3, page 686.)

made a protein in which the only amino acids were phenylalanine units. Evidently, the code for phenylalanine was at least one uracil unit, or a sequence of such units.

Ochoa's group at the New York University School of Medicine, H. G. Khorana's group at the University of Wisconsin, and Crick's group at the Cavendish Laboratory, University of Cambridge, added their efforts to those of the Nirenberg-Matthaei (and P. Leder) group. The triplets of RNA bases that code for specific amino acids were quickly disclosed; Table 25.2 gives a summary. For most, more than one code word will do. Thus alanine can be recognized and transported by sRNA carrying any one of four coding triplets, GCC, GCU, GCA, or GGG. The code is therefore not a unique one; it is said to be *degenerate*. The significance of code degeneracy is not fully understood.

Table 25.2 RNA Codon Assignments*

Amino Acids	Codon Assignments					
Alanine	GCC	GCU	GCA	GGG		
Arginine	CGC	AGA	CGU	CGG	AGG	
Asparagine	AAC	AAU				
Aspartic acid	GAU	GAC				
Cysteine	UGU	UGA	AGU	UGC	UGG	AGC
Glutamic acid	GAA	GAG				
Glutamine	CAA	CAG				
Glycine	GGU	GGA	GGC	GGG		
Histidine	CAC	CAU				
Isoleucine	AUU	AUC				
Leucine	UUG	CUU	CUC	UUA	CUG	
Lysine	AAA	AAG				
Methionine	AUG	AUA				
Phenylalanine	UUU	UUC				
Proline	CCC	CCU	CCA	CCG		
Serine	UCU	UCC	UCG	AGU	UCA	AGC
Threonine	ACU	ACA	ACC	ACG		
Tryptophan	UGG	UGA				
Tyrosine	UAU	UAC				
Valine	GUU	GUC	GUA	GUG		

* These assignments are for the colon bacillus *Escherichia coli*, but a large number of them are known to be codon assignments for several other species. Data from W. A. Groves and E. S. Kempner, *Science*, Vol. 156, April 21, 1967, page 389.

Is the Genetic Code Universal? The Lipmann–von Ehrenstein Experiment. The codons in Table 25.2 were determined for *E. coli*. Many of them have been found to apply to other organisms, and evidence for a universal genetic code, although by no means conclusive, is mounting. If the code is universal, whenever the cells of whatever species engage in protein synthesis and the next amino acid to be required is, say, alanine, one of the four code "words" of Table 25.2, GCC, GCU, GCA or GGG, will be the codon. It will be the codon for alanine whatever the species. F. Lipmann and G. von Ehrenstein (1961) reported one piece of dramatic evidence supporting this. From *E. coli* they prepared a mixture of amino acyl$\sim$sRNAs charged with all the amino acids needed to make the globin of hemoglobin. One of the amino acyl groups, that of leucine, was made radioactive with carbon-14. In a

[5] Nobel prize, 1968. (See footnote 3, page 686.)

cell-free system Lipmann and von Ehrenstein added to this mixture from *E. coli* purified ribosomes and polysomes obtained from the immature red cells of a rabbit. These ribosomes make globin for hemoglobin. Guanosine triphosphate (high-energy phosphate) and appropriate enzymes for globin synthesis were also added. These enzymes do not make globin without a supply of amino acyl$\sim$sRNA. The question was whether the amino acyl$\sim$sRNA from the bacterium *E. coli* could serve in the synthesis of the globin for the rabbit? Globin was indeed made, and about half the carbon-14 was found in it. The experiment not only hinted at the possibility of a universal genetic code but also gave evidence that the polysome component, presumably the mRNA thereof, has the information needed to control protein synthesis.

HEREDITARY DEFECTS AND GENE-TO-ENZYME LINK

Several gene-to-enzyme-linked abnormalities in animal organisms illustrate how important are the accurate transcription and translation of the genetic code.

Phenylketonuria, PKU. Several of the pathways of phenylalanine and tyrosine metabolism are shown in Figure 25.21. Some babies in whom mental retardation develops are apparently born with a defective gene for making the enzyme needed to convert phenylalanine to tyrosine at (*A*) in Figure 25.21. Phenylalanine is therefore processed more frequently than normal by a pathway leading to phenylpyruvic acid. As this acid is formed in greater than normal concentrations, the kidneys remove some, and its appearance in urine is called *phenylketonuria* or PKU. Many people believe that the excess phenylpyruvic acid still circulating in the bloodstream causes the mental retardation often observed when PKU is detected. Unfortunately, the presence of phenylpyruvic acid in urine cannot be readily and reliably detected until an infant is six to eight weeks old. Efforts to reduce the dangers of PKU must probably be started earlier than this. Hence a simple blood test on a sample taken four or five days after birth was devised to detect excessive phenylalanine in circulation. About thirty-seven states have laws requiring that this test be made (as of 1968.)

The treatment of PKU that initially showed the most promise in preventing the onset of mental retardation was the exclusion of as much phenylalanine from the diet as possible. Finding proteins low in phenylalanine is very difficult, and the diet is not only austere but also potentially dangerous—no milk, eggs, bread, or meat. One slice of bread has all the phenylalanine a PKU baby is normally able to handle in a day. Special supplements are used, augmented by certain fruits, vegetables, and cereals.

Uncertainties in mass screenings for PKU among newborn infants, stemming in part from the unreliability of the blood test, together with the dangers of the PKU diet, have led the American Academy of Pediatrics to issue a formal statement of opposition to extension of PKU legislation.

Albinism. The pigments of hair, skin, and irises of eyes are polymeric substances called *melanins* which form from tyrosine. At point (*B*) in Figure 25.21, a gene is needed to make the enzyme for this reaction. If the gene is not properly constructed, the enzyme fails to be made normally, and the melanin pigments do not form as they should. People suffering from this seemingly minor chemical defect are albinos.

Sickle-Cell Anemia. Largely through the work of V. M. Ingram (Massachusetts Institute of Technology), Pauling and Itano (California Institute of Technology), and many others, we have a better understanding of how faulty DNA causes production of the abnormal hemoglobin of sickle-cell anemia. Hemophilia, agammaglobulinemia, and Wilson's disease are other hereditary diseases believed to be caused by recessive genes that fail to produce effective enzymes.

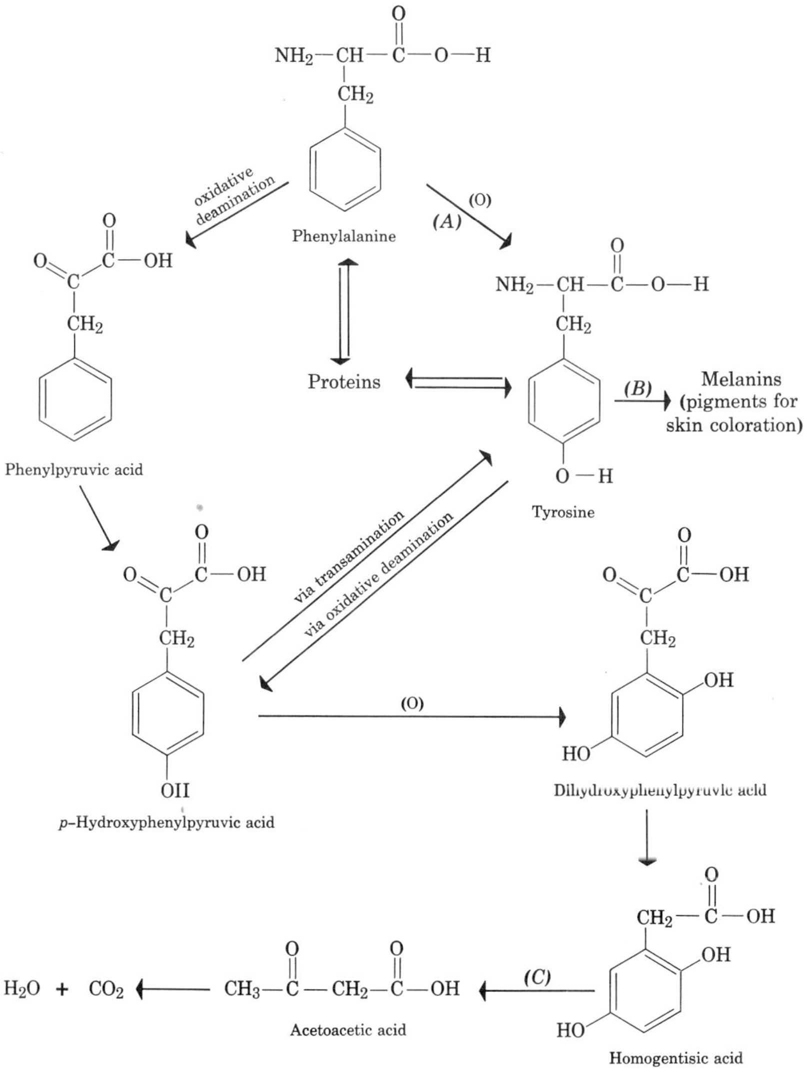

Phenylalanine
oxidative deamination
(O) (A)
Phenylpyruvic acid
Proteins
Tyrosine
(B) Melanins (pigments for skin coloration)
via transamination
via oxidative deamination
p-Hydroxyphenylpyruvic acid
(O)
Dihydroxyphenylpyruvic acid
Homogentisic acid
(C)
Acetoacetic acid
H_2O + CO_2

Figure 25.21 Pathways of phenylalanine and tyrosine metabolism. Letters in parentheses refer to textual discussion.

DNA AND MUTATIONS

We have learned that the genetic machinery of an organism must carry a code both faithful to the characteristics of the species and capable of accurate transmission from one generation to the next. At the same time it must also be capable of evolutionary change, which involves the appearance of mutants. Most mutants are too weak to survive; a few are improvements of the species. Whatever kind of mutation occurs must be transmitted to the next generation if it is to survive. Mutations, both those that survive and those that do not, are almost certainly related to changes in the structures of genetic molecules.

From what we have learned of the Watson-Crick theory, a change in the nucleotide sequence of a DNA molecule must necessarily affect the amino acid sequence of an enzyme. Of the important physical agencies for altering genetic material, atomic radiations and X-rays produce the most serious changes.

CHEMICAL BASIS FOR RADIATION DAMAGE

Ionizing radiations of any type are dangerous, but useful at the same time. This paradox has a chemical explanation. Before investigating it, we need to know more about the general symptoms of radiation sickness.

Radiation Sickness. Gross Symptoms. A drop in the level of the white cells in the blood; severe damage to the skin together with loss of hair and the appearance of ugly, nonhealing, ulcerating wounds; nausea, vomiting, diarrhea, and a feeling of weakness; internal bleeding—these are the principal symptoms of radiation illness. Even patients given deep X-ray or cobalt-ray treatment experience gastrointestinal disturbances and a falling white cell count.

These symptoms are so different in nature that it hardly seems possible one cause can have such diverse effects. Within the last three decades, however, scientists have done much to provide a chemical basis for understanding radiation sickness.

Radiation Sickness. A Chemical Explanation. When radiations penetrate living cells, they leave in their tracks a mixture of strange, unstable organic ions. As these seek stability, covalent bonds within them may break permanently. The ions may recombine either with themselves or with neighbors. New molecules, foreign to the cell, form.

Suppose that all these disruptions happen in a nucleic acid within a cell nucleus. These giant molecules simply cannot suffer structural changes and remain true to the genetic messages they bear. If a DNA molecule, a gene, is structurally altered, mutant daughter cells at the very least will form when the cell divides. At the worst, and this is common, the cell will not be able to divide at all and will be reproductively dead.

We could suppose that large enzymes also suffer similar structural damage. Although they sometimes do, the intensity of the radiation exposure needed to damage seriously an enzyme system is much greater than that needed to damage genes and chromosomes.

The best available evidence points to the conclusion that *the primary site of radiation damage is the genetic machinery in a cell's nucleus.* This damage occurs impartially in all cells exposed, but it is felt keenly only when they attempt to divide. If the exposure to radiation is intense, the changes in gene structure are likely to be so great that the cell is unable to divide when it comes time to do so. If the cells so damaged, as a result of deliberate exposure, are cancer cells, the cancer is halted. High doses of radiation aimed at cancerous cells during a reasonably short period stop the growth of cancer. On the other hand, if exposure is relatively mild and over a general area, damage to genes in cell nuclei is less likely to be profound. In fact, chromosomes and genes are capable of considerable self-repair. Minor changes that do remain are not likely to prevent cell division, but they may cause mutations. In future divisions these mutations may exhibit the wild, erratic behavior characteristic of cancer. Thus repeated low exposures to radiations are likely to induce cancerous growths in otherwise healthy tissue.

If the primary site of radiation damage is the genetic apparatus of a cell, the symptoms of overexposure will appear first among cells that most frequently undergo division. Cells in bone marrow are in this category. Since they are responsible for making white cells, it is not surprising that an early sign of radiation illness is a drop in the white cell count. Damage to other cells that divide less frequently becomes apparent as they reach their time for division. Whatever tissue they constitute will suffer. Of course, if the initial exposure is sufficiently severe, enough cells in enough parts of the body may be rendered reproductively dead to produce all effects of radiation illness quickly.

Damage to the contents of cell nuclei in *germ cells,* cells that produce sperm or eggs, can produce mutations in offspring. Whether or not these actually occur, and how important the changes may be, depends on the extent of the damage and on many other factors beyond the scope of this book. Any possibility of mutation is dangerous to the future of the species. So few mutations are in the direction of improving the species that claims for the desirability of an accelerated mutation rate must be considered unfounded. We should note, however, that man has always lived in the presence of radiations. Cosmic rays are ionizing radiations, and natural radioactivity has always been with us. The production of detrimental genes is unavoidable. Increases in the radiation level trouble all scientists and most informed people.

POSTSCRIPT

The excitement created in scientific circles by the Watson-Crick theory has been great. The enthusiasm of some workers and students, however, has sometimes led to exaggerated claims and statements. Barry Commoner, one scientist who has argued for balance, has voiced some of the following concerns.

Described as a "living molecule," as a "self-duplicating molecule," DNA is neither. It cannot be duplicated apart from suitable enzymes, primer, and triphosphate forms of the nucleotides. *In vitro* syntheses of DNA have required these as well as the watchful, thinking mind and hands of the scientist. Biological evidence has long supported the conclusion that the least complex entity capable of self-duplication is the living cell. None of the recent biochemical evidence refutes this. The DNA molecule is not self-sufficient. Biological specificity is only partly due to it. The DNA from one cell cannot be introduced willy-nilly into *any* other kind of cell to make the cell from that point on do what the DNA tells it to do. Carefully selected host cells can have their genetic apparatus taken over by "alien" DNA, but such selection is necessary. The cell does not otherwise exhibit the expected changes. For genetic specificity not just the DNA but also the appropriate enzymes and the nucleotides must be present. For eventual enzyme synthesis all the needed sRNAs, the amino acids, and the needed enzymes must be available. Physicists have long hoped to uncover the secrets of the atomic nucleus by examining the debris from atom-smashing experiments. Recently, however, some physicists have suggested that the unique properties of the atomic nucleus are too intimately associated with its very complexity to be explained by data obtained from the fragments. Commoner argues that the living cell is no less complex and that this very complexity means, as it does for the physicists, that the *function* of DNA in the cell cannot be completely extricated from the dynamic *organization* of the cell. Commoner objects to the statement "DNA is the secret of life." He prefers "Life is the secret of DNA."

The Kornberg-Goulian Experiment. Nucleic Acids and Viruses. The viruses that have been studied are chemical aggregates made of two types of materials, protein and nucleic acid. The protein serves as an overcoat for the nucleic acid; in some viruses the nucleic acid is RNA, in others DNA. Each type of virus has its own peculiar host cell. The tobacco mosaic virus, for example, infects only the leaves of tobacco plants and causes no apparent harm to people who plant, cultivate, and harvest tobacco.

When a virus encounters its host cell, its nucleic acid core penetrates the cell. Once inside the living cell, the viral nucleic acid directs the making of more of itself and of protein overcoats for the new cores. When enough new viral particles are made, the host cell bursts, and neighboring host cells are invaded by the cores of the newly made virus particles. Thus the viral infection spreads.

One virus that has been extensively studied is labeled ΦX174 and has E. coli as its host cell. The nucleic acid in ΦX174, which is DNA, has 5500 nucleotide units and exists as a single-stranded closed ring. This DNA core material can be isolated from the virus and yet retain the natural infectivity of the virus.

One question that has bothered molecular biologists is whether DNA is made with only the four nucleotides and their four amines, A, T, G, and C. Perhaps the analytical techniques missed another unit or missed a nonnucleotide building block. If there were only one or two such units among several thousand nucleotides in a DNA molecule, they might escape detection. In 1967 Kornberg and Goulian reported powerful evidence that only the four nucleotides are required to make DNA. They took natural DNA cores from the ΦX174 virus and grew new cores in an artificial medium in which *only* the four different nucleotides were provided. Two enzymes were required, a DNA polymerase to join the nucleotides together in a string and a DNA-joining enzyme to catalyze the closure of the string into the ring form. No other building block molecules for DNA were needed. The synthetic DNA viral cores, when allowed to be in contact with their host cells, *E. coli,* were infectious. They entered the host cells, and many new virus particles were produced.

The Kornberg-Goulian experiment provided information about DNA, about viruses, and about the enzymes necessary for making DNA. One of the important scientific advances of the twentieth century, it brought closer the day when synthetic modifications of existing, but (humanly speaking) imperfect, DNA may be made and introduced into defective tissue to bring about its repair. It hastened the day when synthetic but noninfectious DNA may be used to take other "hitchhiker" molecules inside cells of defective tissue to act in a chemotherapeutic way. The experiment did not, however, constitute the creation of life in a test tube. Growing potatoes is not the same thing as creating potatoes. Growing viruses is not the same thing as creating viruses or creating life. Kornberg of course made no such extravagant claim, but his experiment did provide one more dramatic example of the molecular basis of life.

REFERENCES AND ANNOTATED READING LIST

BOOKS

G. Beadle and M. Beadle. *The Language of Life.* Doubleday and Company, Garden City, N.Y., 1966. Written especially for the scientific layman, this book is for him probably the best single introduction to what modern genetics is all about.

J. D. Watson. *Molecular Biology of the Gene.* W. A. Benjamin, New York, 1965. This is an outstanding treatment of molecular biology by one of the originators of the Watson-Crick theory. (Paperback.)

J. D. Watson. *The Double Helix.* Atheneum, New York, 1968. One who made history gives a personal account of exciting times in the development of a biological theory.

V. M. Ingram. *The Biosynthesis of Macromolecules.* W. A. Benjamin, New York, 1966. Written as a supplement to a biochemistry course, this book is also an excellent reference. (Paperback.)

P. E. Hartman and S. R. Suskind. *Gene Action.* Prentice-Hall, Englewood Cliffs, N.J., 1965. The relation of molecular biology to classical genetics is highlighted. (Paperback.)

H. Stern and D. L. Nanney. *The Biology of Cells.* John Wiley and Sons, New York, 1965.

A. White, P. Handler, and E. L. Smith. *Principles of Biochemistry,* third edition. McGraw-Hill Book Company, New York, 1964.

B. Wallace and Th. Dobzhansky. *Radiation, Genes, and Man.* Holt, Rinehart, and Winston, New York, 1959. The book delineates the genetic aspects of radiation damage.

D. Grosch. *Biological Effects of Radiations.* Blaisdell Publishing Company, New York, 1965. After discussing the nature of radiations, their effects on molecules, cells, tissues, organs, organisms, and ecological communities are thoroughly described.

ARTICLES

J. D. Watson and F. H. C. Crick. "A Structure for Deoxyribose Nucleic Acid." *Nature,* Vol. 171, 1953, page 737. "Genetical Implications of the Structure of Deoxyribonucleic Acid." *Nature,* Vol. 171, 1953, page 964.

F. H. C. Crick. "The Genetic Code." *Scientific American,* October 1962, page 66.

M. W. Nirenberg. "The Genetic Code: II." *Scientific American,* March 1963, page 80.

F. H. C. Crick. "The Genetic Code: III." *Scientific American,* October 1966, page 55.

R. T. Hinegardner and J. Engelberg. "Rationale for a Universal Genetic Code." *Science,* Vol. 142, 1963, page 1083. Neither abrupt nor gradual changes in the genetic code seem to be rational, argue the authors.

J. Hurwitz and J. J. Furth. "Messenger RNA." *Scientific American,* February 1962, page 41. The discovery of mRNA is described.

A. Rich. "Polyribosomes." *Scientific American,* December 1963, page 44. The author discusses the emergence of the idea that not single ribosomes acting in isolation but collections of ribosomes working together are the "factories" of protein synthesis.

G. Zubay. "Molecular Model for Protein Synthesis." *Science,* Vol. 140, 1963, page 1092. The author links structural and biochemical information into one coherent pattern and offers a stereochemically sound model for the template mechanism in protein synthesis.

M. F. Singer. "In vitro Synthesis of DNA." *Science,* Vol. 158, December 22, 1967, page 1550. This is a report of the Kornberg-Goulian experiment.

A. Kornberg. "The Synthesis of DNA." *Scientific American,* October 1968, page 64. Kornberg describes the testtube synthesis of DNA that duplicates the viral activity of the DNA in the virus ΦX174.

R. B. Merrifield. "The Automatic Synthesis of Proteins." *Scientific American,* March 1968, page 56. The *in vitro* synthesis of small proteins is described. Peptide chains are assembled one amino acid unit at a time by a remarkable solid phase method in which the growing chains are anchored on small beads of polystyrene.

J. P. Changeux. "The Control of Biochemical Reactions." *Scientific American,* April 1965, page 36. Ways in which feedback systems regulate the biosynthesis of cell products, including the work of repressor genes, are discussed.

B. F. C. Clark and K. A. Marcker. "How Proteins Start." *Scientific American,* January 1968, page 36. The authors describe how protein synthesis in bacteria is initiated.

B. Commoner. "DNA and the Chemistry of Inheritance." *American Scientist,* Vol. 52, 1964, page 365. "Life is the secret of DNA" would be a better guide for biological investigations than "DNA is the secret of life"—this is the argument of Commoner. See also "The Elusive Code of Life: Is DNA Really the Master Key to Heredity?" *Saturday Review,* October 1, 1966, page 71.

R. W. Holley. "The Nucleotide Sequence of a Nucleic Acid." *Scientific American,* February 1966, page 30. The work of determining the nucleotide sequence in alanine transfer RNA is described.

E. H. Davidson. "Hormones and Genes." *Scientific American,* June 1965, page 36. Evidence that some hormones act by controlling the activities of genes is described and discussed.

A. G. Bearn and J. L. German III. "Chromosomes and Disease." *Scientific American,* November 1961, page 66. Mongolism is linked to abnormalities in human chromosomes and to a defect in the genetic machinery for passing heredity material from parents to offspring.

A. G. Bearn. "The Chemistry of Hereditary Disease." *Scientific American,* December 1956, page 127. Several relatively rare hereditary diseases are examined in the light of the idea that a single gene may control the synthesis of a single enzyme.

H. Fraenkel-Conrat. "The Genetic Code of a Virus." *Scientific American,* October 1964, page 47. The virus that infects tobacco leaves consists of a coiled strand of RNA surrounded by a coat of protein molecules.

S. E. Stewart. "The Polyoma Virus." *Scientific American,* November 1960, page 63. Viruses, which contain nucleic acid, appear to operate by deranging the genetic machinery of the host cell.

L. Gorini. "Antibiotics and the Genetic Code." *Scientific American,* April 1966, page 102. Streptomycin and related drugs can change the meaning of the code that directs protein synthesis.

T. T. Puck. "Radiation and the Human Cell." *Scientific American,* April 1960, page 142. Remarkable photomicrographs of radiation-induced chromosome damage illustrate this article.

PROBLEMS AND EXERCISES

1. Define each of the following terms.

(*a*) zygote	(*b*) mitosis	(*c*) germ cells
(*d*) somatic cells	(*e*) chromonemata	(*f*) chromosomes
(*g*) genes	(*h*) replication	(*i*) nucleotide
(*j*) nucleic acid	(*k*) DNA	(*l*) RNA
(*m*) base pairing	(*n*) template	(*o*) rRNA
(*p*) mRNA	(*q*) sRNA	(*r*) codon
(*s*) anticodon	(*t*) genetic code	(*u*) phenylketonuria
(*v*) melanins		

2. What structural features are common to all DNA molecules?
3. How do DNA molecules differ?
4. What structural features are common to all RNA molecules?
5. How do RNA molecules differ?
6. In terms of molecular structures, what specifically is meant by the genetic code?
7. By means of a "flow chart," discuss the one-gene, one-polypeptide theory.
8. If the sequences of heterocyclic bases on one section of a DNA molecule is A-T-C-G-G-T-T-A, what is the sequence in this region on the corresponding RNA molecule? (Remember that in RNA there is no thymine; uracil, U, takes its place.)
9. Discuss the particular contribution Watson and Crick made to our understanding of the relation between gene *structure* and gene *function.*
10. How does the Watson-Crick model correlate with Chargaff's data, $A/T = 1$ and $G/C = 1$, for the DNA of various species?
11. Discuss the relation of rRNA, mRNA, and sRNA to one another.
12. Discuss the relation of mRNA, ribosomes, and polysomes to one another.
13. What is believed to be the basis for mutations?
14. Liver cells are known to be particularly active in protein synthesis. They are also known to be particularly rich in RNA. How are these two facts correlated?
15. Explain in your own words how replication of DNA occurs.
16. How will small doses of radiations or X-rays over a long time possibly lead to cancer?
17. How does a well-focused "dose" of X-rays accomplish the arresting of cancer?
18. How is it that radiations can both cause cancer and cure it?

APPENDIX I

Exponentials

When numbers are either very large or very small, it becomes convenient to express them as powers of 10. For example:

Number	Exponential Form
1	1×10^0
10	1×10^1
100	1×10^2
1000	1×10^3
10,000	1×10^4
100,000	1×10^5
1,000,000	1×10^6
0.1	1×10^{-1}
0.01	1×10^{-2}
0.001	1×10^{-3}
0.0001	1×10^{-4}
0.00001	1×10^{-5}
0.000001	1×10^{-6}

The number placed up and to the right of the 10 is the *exponent.*[1]

If the exponent is positive, it tells us how many times the number before the 10 must be multiplied by 10 to give the full expression for the number involved. Thus

$$1 \times 10^4 = 1 \times \underbrace{10 \times 10 \times 10 \times 10}_{10^4} = 10{,}000$$

$$6 \times 10^3 = 6 \times 10 \times 10 \times 10 = 6{,}000$$

$$8.576 \times 10^2 = 8.576 \times 10 \times 10 = 857.6$$

[1] For a short programmed review of this topic see R. J. Flexer and A. S. Flexer, *Programmed Reviews of Mathematics,* Vol. 4, *Exponents and Square Roots,* Harper and Row, New York, 1967 (paperback).

If the exponent is negative, it tells us how many times the number before the 10 is to be divided by 10. Thus

$$1 \times 10^{-4} = 1 \div 10 \div 10 \div 10 \div 10$$
$$= \frac{1}{10 \times 10 \times 10 \times 10} = \frac{1}{10{,}000} = \frac{1}{10^4}$$
$$= 0.0001$$

$$6 \times 10^{-3} = 6 \div 10 \div 10 \div 10$$
$$= \frac{6}{10 \times 10 \times 10} = \frac{6}{1000}$$
$$= 0.006$$

$$8.576 \times 10^{-2} = \frac{8.576}{10 \times 10} = \frac{8.576}{100} = \frac{8.576}{10^2}$$
$$= 0.08576$$

To multiply numbers expressed as powers of 10

—Multiply the numbers *before* the tens.
—Add the exponents of the tens algebraically.

For example: $(2 \times 10^4) \times (3 \times 10^5) = 2 \times 3 \times 10^{4+5}$
$= 6 \times 10^9 = (2000)(300{,}000)$

For example:

$$6576 \times 2000 = 6.576 \times 10^3 \times 2 \times 10^3$$
$$= 13.152 \times 10^6$$
$$= 1.3152 \times 10^7$$

(Note that the number before the 10 is conventionally made into one having a value between 1 and 10 by appropriately adjusting the exponent. Thus $13 = 1.3 \times 10^1$ and $13 \times 10^6 = 1.3 \times 10^1 \times 10^6 = 1.3 \times 10^7$.

For example: $7{,}000 \times 0.02 = (7 \times 10^4) \times (2 \times 10^{-2})$
$= 7 \times 2 \times 10^{4+(-2)}$
$= 14 \times 10^2$
$= 1.4 \times 10^3$

Exercise I.1 For practice try the following products.

(*a*) $6{,}000{,}000 \times 0.0000002$
(*b*) $10^6 \times 10^{-7} \times 10^8 \times 10^{-7}$
(*c*) $0.003 \times 0.002 \times 0.000001$
(*d*) $1500 \times 3{,}000{,}000{,}000{,}000$
(*e*) $6 \times 10^{23} \times 2$

Ans. (*a*) 1.2, (*b*) 1, (*c*) 6×10^{-12}, (*d*) 4.5×10^{15}, (*e*) 1.2×10^{24}.

To divide numbers expressed as powers of 10

—Divide the numbers in front of the tens.
—Subtract the exponents of the tens.

For example: $8 \times 10^4 \div 2 \times 10^3 = (8 \div 2) \times 10^{4-3}$
$= 4 \times 10^1$

For example: $8 \times 10^4 \div 2 \times 10^{-3} = (8 \div 2) \times 10^{4-(-3)}$
$= 4 \times 10^7$

Exercise I.2 For practice try the following.

(*a*) $6{,}000{,}000 \div 1500$

(*b*) $7460 \div 0.0005$

(*c*) $\dfrac{3{,}000{,}000 \times 6{,}000{,}000{,}000}{20{,}000}$

(*d*) $\dfrac{0.016 \times 0.0006}{0.000008}$

(*e*) $\dfrac{400 \times 500 \times 0.002 \times 5000}{2{,}500{,}000}$

Ans. (*a*) 4×10^3, (*b*) 1.492×10^7, (*c*) 9×10^{11}, (*d*) 1.2, (*e*) 8×10^{-1}.

APPENDIX II

Common Logarithms

The logarithm of a number N to the base 10 is the exponent or the power to which 10 must be raised to equal the number N. If

$$N = 10^x$$

then the logarithm of N to the base 10 is simply x, or for short,

$$\log N = x$$

If	$N = 10$	$\log N = 1$	because $10 = 10^1$
	$= 100$	$\log N = 2$	because $100 = 10^2$
	$= 1000$	$\log N = 3$	because $1000 = 10^3$
	$= 0.1$	$\log N = -1$	because $0.1 = 10^{-1}$
	$= 0.00001$	$\log N = -5$	because $0.00001 = 10^{-5}$
If	$N = 10 \times 1000$	$\log N = 1 + 3$	because $10 \times 1000 = 10^1 \times 10^3$

Rule 1 In general, the logarithm of the product of two numbers is equal to the sum of the logarithms of the numbers:

$$\log (A \times B) = \log A + \log B$$

This is one of the major values of logarithms. By using them, we can carry out long and complicated multiplications (and/or divisions) by adding (or subtracting) numbers, the logarithms.

For a short programmed review of this topic see R. J. Flexer and A. S. Flexer, *Programmed Reviews of Mathematics,* Vol. 5, *Logarithms,* Harper and Row, New York, 1967 (paperback).

But what if N equals 4? What is the value of x in the equation $4 = 10^x$? Whatever its value, it is still called the logarithm of 4. Or what if N equals 40? What is the value of x in the equation $40 = 10^x$? As a start,

$$40 = 4 \times 10^1$$

Therefore

$$\log 40 = \log(4 \times 10^1)$$

But from Rule 1

$$\begin{aligned} \log 40 &= \log 4 + \log 10^1 \quad \text{(which we know to be 1)} \\ &= \log 4 + 1 \end{aligned}$$

Thus the problem of what is the logarithm of 40 resolves itself to the problem of what is the logarithm of 4.

In general:

$$\begin{aligned} \log(M \times 10^x) &= x + \log M \quad \text{where } M \text{ is a number between 1 and 10} \\ \log(M \times 10^{-x}) &= -x + \log M \end{aligned}$$

The value of x in these expressions is called the *characteristic* of the logarithm, and the value of log M is called the *mantissa*. To determine a mantissa, that is, to determine the value of the logarithm of a number between 1 and 10, we use a table of common logarithms. All our needs will be met by a simple two-place logarithm table. For more careful work tables of four- and five-place logarithms are available. Some examples will show how to use the table.

Examples What is the log of 4? *Ans.* 0.60

of 4.1 *Ans.* 0.61

of 4.8 *Ans.* 0.68

of 2.55 *Ans.* 0.41 (2.55 is halfway between 2.50 and 2.60. The mantissa for 2.50 is 0.40; for 2.60, 0.42. Halfway between is 0.41, the mantissa for 2.55.)

Find the log of 475.

Solution. $475 = 4.75 \times 10^2$

$$\begin{aligned} \log 475 &= \log 4.75 + \log 10^2 \\ &= \log 4.75 + 2 \quad \text{(the characteristic)} \\ &= 0.68 + 2 \\ &= 2.68 \end{aligned}$$

Find the log of 0.0475

Solution. $0.0475 = 4.75 \times 10^{-2}$

$$\begin{aligned} \log 0.0475 &= \log 4.75 + \log 10^{-2} \\ &= \log 4.75 + (-2) \\ &= 0.68 - 2 \\ &= -1.32 \end{aligned}$$

Having seen how to find the logarithm of a number, we must next learn how to find the number corresponding to a logarithm. The problem could arise, for example, from doing a multiplication. Suppose the logarithm of a number is 3; what is the number? *Ans.* 10^3. Remember that a logarithm is an exponent; it is the power to which 10 must be raised. But what if the logarithm is 1.60? The "1" comes from 10^1 in the number. What does the 0.60 refer to? From the table, working in a sense backward, the number whose logarithm is 0.60 is 4. Therefore the number whose logarithm is 1.60 must be 4×10^1 or simply 40.

Exercise II.1 What are the logs of the following numbers?

(*a*) 1.5 (*b*) 15 (*c*) 150
(*d*) 0.003 (*e*) 4.6 (*f*) 82
(*g*) 9130 (*h*) 53 (*i*) 13.5

Ans. (*a*) 0.18, (*b*) 1.18, (*c*) 2.18, (*d*) −2.53, (*e*) 0.66, (*f*) 1.91, (*g*) 3.96, (*h*) 1.72, (*i*) 1.13.

Exercise II.2 What are the numbers that have the following logs?

(*a*) 1.18 *Ans.* (*a*) 15
(*b*) −2.70 (*b*) 2×10^{-3} (note, −2.70 = −3 + .30)
(*c*) 5.23 (*c*) 1.7×10^5
(*d*) 0.46 (*d*) 2.9
(*e*) −7.56 (*e*) 2.75×10^{-8} (note, −7.56 = −8 + .44)

To multiply or divide using logarithms, we use substantially the same rules developed for exponentials.

$$\log (A \times B) = \log A + \log B$$
$$\log (A \div B) = \log A - \log B$$
$$\log A^p = p \log A$$

Examples What is 65 × 32?

$$\log 65 + \log 32 = 1.81 + 1.51 = 3.32$$

Ans. The number whose log is 3.32 is 2.1×10^3. (The precise answer is 2.08×10^3, which we would have obtained had we used more accurate logarithm tables.)

What is 800 ÷ 20?

$$\log 800 - \log 20 = 2.90 - 1.30 = 1.60$$

Ans. The number whose log is 1.60 is 40.

For drill make up a few of your own problems.

Two-Place Logarithm Table

N	0	1	2	3	4	5	6	7	8	9
1	00	04	08	11	15	18	20	23	26	28
2	30	32	34	36	38	40	42	43	45	46
3	47	49	51	52	53	54	56	57	59	59
4	60	61	62	63	64	65	66	67	68	69
5	70	71	72	72	73	73	74	75	76	77
6	78	79	79	80	80	81	81	82	83	84
7	84	85	86	86	87	88	88	89	89	90
8	90	91	91	92	92	93	93	94	94	95
9	95	96	96	97	97	98	98	99	99	99

INDEX

Page numbers in italics refer to tables.